Sustainable Construction Engineering and Management

Sustainable Construction Engineering and Management

Editors

Edmundas Kazimieras Zavadskas
Jurgita Antuchevičienė
Reza Hosseini
Igor Martek

MDPI • Basel • Beijing • Wuhan • Barcelona • Belgrade • Manchester • Tokyo • Cluj • Tianjin

MDPI

Editors
Edmundas Kazimieras Zavadskas
Vilnius Gediminas Technical University
Lithuania

Jurgita Antuchevičienė
Vilnius Gediminas Technical University
Lithuania

Reza Hosseini
Deakin University
Australia

Igor Martek
Deakin University Australia

Editorial Office
MDPI
St. Alban-Anlage 66
4052 Basel, Switzerland

This is a reprint of articles from the Special Issue published online in the open access journal *Sustainability* (ISSN 2071-1050) (available at: https://www.mdpi.com/journal/sustainability/special_issues/construction_engineering).

For citation purposes, cite each article independently as indicated on the article page online and as indicated below:

LastName, A.A.; LastName, B.B.; LastName, C.C. Article Title. *Journal Name* **Year**, *Volume Number*, Page Range.

ISBN 978-3-0365-2628-7 (Hbk)
ISBN 978-3-0365-2629-4 (PDF)

sustainability, better living conditions for individuals, and to pursue the highest possible economic value. There are many opportunities in the real estate market to increase the value of a property, e.g., by investing in the projects for the creation or renovation of infrastructure or building structures. The common goal is to renovate the building to improve living or working conditions and create a higher value of the object. Together with policy developments, the perceptions of real estate project stakeholders have an influence on project investor decisions in achieving ambitious goals. When compared to other industries, projects in the construction industry have been facing numerous risks, e.g., if they are not managed properly, they will fail in achieving main goals [40].

Construction projects differ in the budget, duration, variety of works, a number of implementers, and stakeholder [41]. The results of the projects also vary in following ways: some of them can be implemented successfully and others can be terminated with losses. The risks that are associated with the execution of construction projects influence each other. For example, the risks related to construction project delivery time can influence project costs and vice versa [42]. The success or failure of real estate investment decisions depends on the assessment and management of the inherent risk and uncertainty [43]. In some AHP method applications for a real estate investment problem, multi-criteria group decision-making method first uses analytic hierarchy process to construct decision preference matrix, and then it uses the hesitant probabilistic fuzzy linguistic set to model uncertainties.

The risks that are related to decision-making in construction projects were recently studied in many related works [42,44–47]. For example, Hatefi et al. [44] state that the source of the project risks is the presence of high uncertainty in construction projects. Because of the existence of factors that are associated with uncertainty, the appropriate models for decision-making are necessary. Hatefi & Tamošaitienė [42] applied an integrated fuzzy DEMATEL-fuzzy ANP model to assess the relative importance of risk factors and alternatives, as well as to prioritise construction projects. Ghasemi et al. [45] developed Bayesian network (BN) methodology for modelling and analysing the risks and then applied it to a project portfolio of a construction company. Asadi et al. [46] proposed a three-stage approach that is based on the fuzzy inference system for project risk evaluation. The approach combines different parameters (e.g., the time, cost, quality, contribution rate, resilience, and resistance) to assess the risk index. For sustainable risk analysis and decision-making in the construction sector [47], an alternative approach based on consistent fuzzy preference relation and ANP methodology provided. Multi-criteria decision-making methods are used by individuals and enterprises to achieve effective solutions for many of their problems, since they usually include subjective, intangible, and not easily quantifiable aspects [48]. The research [49] applies Fuzzy multi criteria decision-making methodology, called DEcision-Making Trial and Evaluation Laboratory (DEMATEL) and Analytic Network Process (ANP) method, to investigate the Critical Success Factors (CSFs) of project management. These factors were categorised into five criteria groups: (1) project, (2) project management, (3) organisation, (4) external environment, and (5) sustainability, from which the highest weights were assigned to the sub-criteria of top management and sponsors' support, stakeholders' expectations, and end users' imposed restrictions.

In addition to different measurable risks, the investor's strategy can be strongly affected by the possible cyclical recurring crises in the real estate market, like the housing crisis, which affected many economies in 2007–2008 [50]. In times of global financial crisis, the decision to invest at a certain point in time and the correct assessment of risks are key issues: investors need to know how to measure risks and identify the relationship between risks appearance and risks incentives demanded, according to their attitude towards risks [51]. To operate in unfavourable business conditions, it is appropriate to gather and analyse more information to make rational decisions. The use of information systems enables to collect necessary data and the coordinate solutions with business partners remotely. This can provide an advantage in carrying out construction project activities in a difficult situation. Because each construction project has specific complexities and uniqueness, it is necessary to take these factors into account while proposing an advanced decision-making tool for project managers [40].

One of the sophisticated tools for improving the decision-making process in construction projects is the decision tree method. The decision tree, as a construction projects risk management tool, can be useful in providing real estate market participants with the necessary information. The information advantage gives agent buyers a greater bargaining power when buying houses for their use at lower prices than other non-agent buyers (for example, agents bought houses at prices that are 2.54 lower than comparable houses that were bought by other buyers) [52].

With the diversification of Internet technology and cloud computing, an increasing number of Internet companies has started to afford users with a variety of remote services. Data analysis and prediction, including risk assessment, image recognition, and spam detection are among the most popular services. The functions that are mentioned above were achieved through machine learning classifiers that have recently attracted considerable attention [53]. Decision trees are among the most popular tools for learning and extracting classification rules from data [54]. When compared with other algorithms, decision trees require less effort for data preparation. The constructed decision tree model is intuitive and easy to explain to technical teams and stakeholders. At the same time, the result of the decision tree may be unstable because a little change in the data may lead to the creation of a completely different tree. A problem can be solved by using an ensemble decision tree [55]. Constructing a decision tree is usually a recursive procedure, where a function is repeatedly optimized and training data are partitioned into the root and internal nodes until a termination condition is met [56]. Usually, the termination condition is the logic disjunction of several stopping predicates that account for different kinds of imposed limitations, for example, on the branch length, on the possible information gain [57].

Analytic Hierarchy Process (AHP) is another tool for improving the decision-making process [58]. AHP is a broadly applied multi-criteria decision-making method for determining the weights of criteria and it priorities of alternatives in a structured manner that is based on pairwise comparison. Because subjective judgements during comparison might be imprecise, the modifications of AHP by combining fuzzy sets with AHP are proposed and referred to as fuzzy AHP or FAHP [59]. It is noteworthy that the AHP method is widely used for construction project assessment, for example, to develop a decision support system that embodies the relative preferences of the owner and architect among multiple key criteria [16], in order to assess the building performance in the field of anti-seismic behaviour [60], to evaluate building material suppliers while taking a large number of criteria that are often subjective and hard to measure into consideration [61].

Changes of decisions raise the uncertainties and can negatively affect the sustainability of the project results, e.g., the changes that were made in the middle of a project can significantly differ from its implementation from the initial plan, as it was pointed out in the above-mentioned articles. In order to analyse the possible shifts in the decisions, the analysis of model robustness must be performed, i.e., it is important to understand how sensitive decisions are to possible changes in project participants' behaviour. For example, the investor changes his/her priorities or the project can be transferred to/inherited by another investor. The resistance of the decisions during project implementation to the possible changes of the project conditions must be assessed. Thus, there is a lack of studies that provide clear solutions for the accurate assessment of the stability and robustness of the decision model in terms of its sensitivity to changes in conditions, including the change of the investor. Note that the current study is not necessily a source for the improvement over the mentioned works, it is hard to judge the usefulness of the requirement to perform the additional analysis mentioned above. Instead, we analyse the case where it is needed, and the lack of such analysis in other studies creates a gap, which this article intends to fill. It became especially topical in the context of the recent changes in the policy of EU regarding the sustainability of the investments.

Assessing investment against sustainability criteria has only just begun to be included in the debates, and new initiatives are emerging to assess the sustainability of investments, such as the EU classification system for green investments [62]. Sustainable investment decisions are commonly referred to as those that involve sustainability-related considerations [63]. However, until now, the assessment of investment solutions was strictly based on financial indicators, like NPV (Net Present Value), (IRR (Internal Rate of Return), PI (Profitability Index), and similar [64,65]. Lee et al. (2016) [66] used the rate of return, relative return, investment, and relative investment rate for the analysis of investment options. Some of the studies applied qualitative assessment, like SWOT (Strengths, Weaknesses, Opportunities, and Threats) analysis [67]. Seeking to ensure the attractiveness of investments in the future organization addressing the LEED (The Leadership in Energy & Environmental Design), BREEAM (Building Research Establishment Environmental Assessment Method), and similar certification methods for managing the construction projects [68]. However, these methods require a lot of time and effort to evaluate a single alternative and, due to this shortcoming, are rarely applied to the evaluation of multiple alternatives. The recently issued EU Corporate Sustainability Reporting Directive (COM (2021) 189) [69] requires companies to provide information on the sustainability of their business practices in a transparent and comparable way, which will also serve as a reliable guide for investment decisions. Market players should apply the requirements of this Directive and include sustainability criteria in their investment valuation models to contribute to the Green Deal objectives and get a chance to benefit from the financial support mechanisms that are offered by the EU. This study is dedicated to analyse how different conditions (innovations, restrictions) might affect the behaviour of market players (investors). Such an analysis could be applied as a part of methodology for developing the means for supporting the Green Deal objectives. The study of Duong et al. (2021) [70] revealed that there is a positive stock market reaction to incremental sustainability innovation; however, there is a lack of studies that focus on the analysis of the investor behaviour in the context of changes in the market.

To fill this gap, this study develops a new model that combines the decision-making process modelling with AHP method and includes the analysis of model stability in relation to stakeholders' behaviour. Accordingly, this study assesses possible patterns of project participants' behaviour, identifying the key aspects of sustainability in a project management context. It also helps to understand the importance from the project investor's perspective, at the same time taking into account other stakeholders' needs. The methodology was validated on a case study of eight roofing projects by assessing different options for their implementation. The sensitivity analysis was performed to determine possible scenarios and to investigate the robustness of decision-making processes, i.e., how resistant the decisions are to possible fluctuations of parameters.

3. The Methodology of the Research

In this section, we describe the main methods that were applied to the case study. The current research is based on the work [17], in which we have provided an example of how to apply the methods of a decision tree and AHP to select the best project. In this article, we further develop the topic of a decision tree and AHP methods application by addressing additional problems, such as the parametrisation of investor's behaviour, sensitivity analysis, and visualisation of results. The sensitivity analysis is especially important, since it permits assessing the possible changes of the investor's opinion and its impact on selection of the project. The methodology applied in the paper consists of following main steps:

1. Identification of the major stakeholders involved in the project execution process.
2. Development of a decision tree scheme to select the best project from the planned set of alternatives.
3. Creation of the algorithms for the decision tree solution.

4. Adaptation of AHP method to assess the investor's subjective opinion about the various projects, predicting whether the investor would like to sell the completed project or not.
5. The parametrisation of the investor's behaviour with a special mapping of the ratings scale to the monetary values and the sensitivity analysis.
6. Analysis of how the different mappings of the rating scale to the monetary values affect the assessment of the projects in the decision tree and visualisation of the obtained results.

The developed methodology was validated on the real case study to demonstrate how algorithms work in real conditions.

3.1. Identification of Major Stakeholders

The creation of the decision tree starts from the analysis of the participants' behaviour during the project execution. These participants are identified as stakeholders in the project outcomes.

Local governments, community-based organisations, foundations, neighbourhood and other advocacy groups (for example, Greenpeace organisation), construction companies, investors, commercial banks, tenants and their brokers, ecologists, media and unions—all of these are the city stakeholders and participants of the city development. They should be able to create feasible projects that generate benefits and reduce the risk that is involved in urban development [71].

It is important to consider the factors that are driven by stakeholders in order to manage construction projects. The impact of these factors can lead to the project success or failure. Depending on the directions of the research, various authors indicate the factors, components, and dimensions that need to be evaluated and controlled during project implementation.

In the research [72] that was performed by N. Srinivasan and S. Dhivya, the major factors that were concerned with stakeholder management in construction projects were identified, as follows: stakeholder participation, decision making powers, organisational structure, quality performance, customer related factors, and employee related factors.

S. Demirkesen and B. Ozorhon in article [73] identify integration components and dimensions that are important for successful construction project management. These components are presented, as follows: the development of a project charter, knowledge integration, process integration, staff integration, supply chain integration, and integration of changes. The dimensions of project management performance are presented, as follows: time, cost, quality, safety, and client satisfaction.

Various authors who study stakeholder theory, distinguish various groups of them. Table 1 presents the stakeholder groups that can influence roof installation projects.

Table 1. Groups of stakeholders that were researched in the sources related to construction projects and sustainability.

Sources	Researched Groups of Stakeholders	Content of the Research
Li et al. (2018) [74]	Buyers, sales personnel, financial institutions, developers, designers and drafting personnel, estimators, project managers/coordinators, regulators, superintendents, inspectors, trades/suppliers, home occupants and warranty staff	Stakeholder's studies and the social networks of NetZero energy homes
Zhao et al. (2012) [75]	Employees, customers, shareholders, creditors, suppliers and partners, environment and resources agencies, local communities, government, competitors and non-governmental organisations	A corporate social responsibility indicator system for construction enterprises

Table 1. *Cont.*

Sources	Researched Groups of Stakeholders	Content of the Research
Freudenreich et al. (2020) [76]	Social stakeholders, financial stakeholders, customers, business partners and employees	Value creation for sustainability
Chu et al. (2020) [77]	Government, developers and residents	Evolutionary game analysis on improving collaboration in sustainable urban regeneration
Vilutiene and Ignatavičius (2018) [78]	Finance institutions, local authorities, building owners, tenants, contractors, technology providers, material suppliers, consultants and facility managers	Key performance indicators for quality monitoring during sustainable renovation
Zheng et al. (2019) [79]	Governments, cost consultants, owners, building information modelling consultants, designers, general contractors and subcontractors	Quantifying and visualising value exchanges in building information modelling projects
Lin et al. (2019) [80]	Internal stakeholders (end users, developers and investors, main contractors, subcontractors, suppliers and employees) and external stakeholders (governments, non-governmental organizations, communities and the public)	Stakeholders' influence strategies on social responsibility implementation in construction projects
Maceika et al. (2020) [17]	Project owner, state organisations, building design company, interested community, construction company, suppliers, construction business partners, consultants and supervisors	The modelling of roof installation projects using decision trees and the AHP method

Based on the Table 1, especially with regard to the [17], the major stakeholders that are involved in the roofing project execution process were identified, as follows:

1. Investor in a roof installation project.
2. State organisations.
3. Building design company.
4. Interested community.
5. Construction company.
6. Suppliers.
7. Consultants.
8. Supervisors.

The interests of the major stakeholders and probability to impact were assessed using the expert method [81]. Figure 1 presents the obtained results of stakeholders' positioning. The investor was rated as the most interested and able to make the greatest impact on the project, as can be seen from Figure 1.

It is notable to mention that the building process is mainly implemented by a construction company, although with the involvement of building design company, suppliers, consultants, and supervisors. The investor is not directly involved in the building process; however, in the considered case, many key decisions must be made by the investor (more details will be provided later, see Section 4.2).

Therefore, investor's decisions were examined using both the AHP method and decision tree. However, the influence of other stakeholders was only assessed in the decision tree. Each of the stakeholders influenced the decision making. Each activity involving stakeholders generated revenue or expenses that were expressed in monetary units. The probable additional costs that might be incurred were also assessed with an appropriate probability. Such additional costs might arise if the project needing correction or the project could not be coordinated by interested local community, and state organisations, or construction and supply companies failed. The tree also calculated the probable loss if the project fails completely. It should be noted that the creation of the fragments of the decision tree is a non-trivial task, which is out of our research scope. As an example, the assessment to select supply organisation is needed and that information must be converted into a fragment in the decision tree. For this purpose, some advanced supplier selection techniques might be applied, as shown in [82].

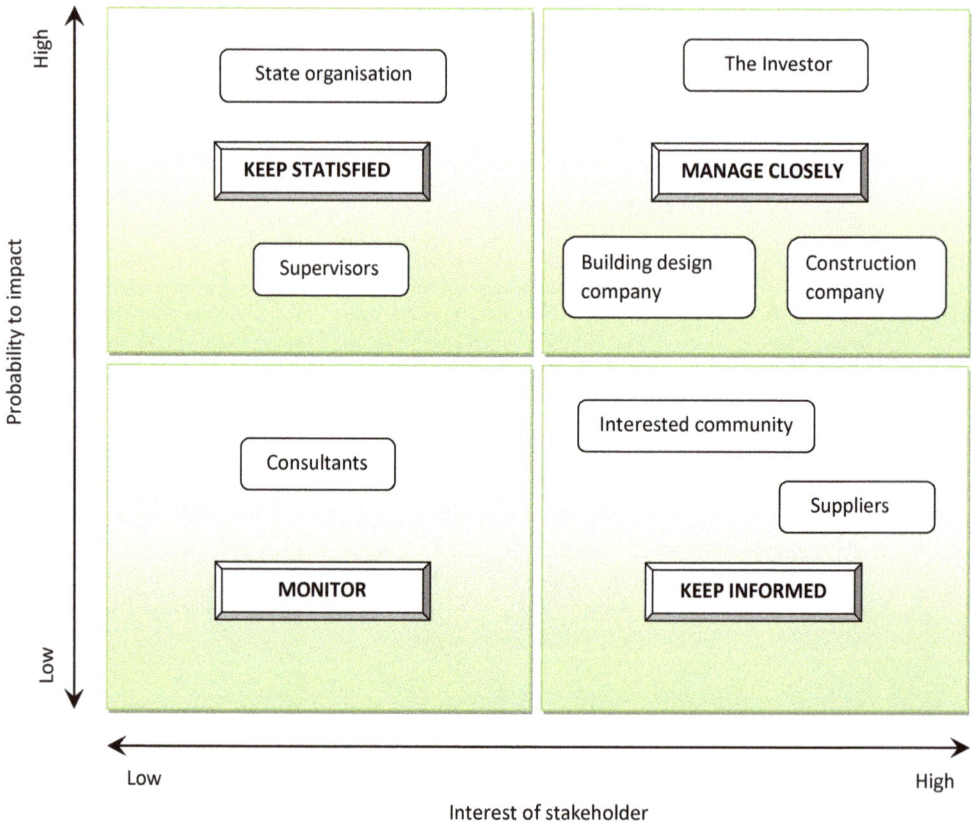

Figure 1. Positioning of major stakeholders involved in roof installation project according to interests and probability to impact.

3.2. The Decision Tree

Here, we briefly present the modelling of stochastic process that was introduced in the previous researches. The objective function is to maximize value (profit), which should be received by the investor (the model was developed for a private investor). The decisions were expressed in integer form by $X_{ijk} = 0$ or 1, $i = 1, 2, 3, \ldots, g$ (describes the decision tree level); $j = 1, 2, 3, \ldots, m$ (describes the branch group in the decision tree level); $k = 1, 2, 3, \ldots, n$ (describes the branch in the group); $X_{ijk} \geq 0$; $\sum_{k=1}^{n} X_{ijk} = 1$; if $X_{ijk} = 1$—alternative ijk is selected; if $X_{ijk} = 0$—alternative ijk is not selected; P_{ijk}—

describes probability of events, $P_{ijk} \geq 0$; $\sum_{k=1}^{n} P_{ijk} = 1$. We also measure the profit/losses $S_{(i-1)jk}$, which can be chosen as a better alternative:

$$S_{(i-1)jk} = \sum_{k=1}^{n} S_{ijk} X_{ijk}; \tag{1}$$

or it can be a probabilistic quantity:

$$S_{(i-1)jk} = \sum_{k=1}^{n} S_{ijk} P_{ijk}. \tag{2}$$

The mathematical model of the decision tree was constructed based on the algorithms that are presented in [4].

3.3. The Main Steps for the Sensitivity Analysis

To perform the modelling of the investor's behaviour, we propose applying a special mapping of the ratings scale to the monetary values, describing the influence of the wealthness on the decision of the individual. We show how to integrate that mapping into the model and evaluate the obtained results.

Initially, the model was developed for a single investor. Thus, possible changes in the assessment or parameters were taken into account, which would be the case if the number number of investors will increase. The parametrisation of investor's behaviour and sensitivity analysis of AHP method were performed in the following steps:

1. The investor's behavior was parameterized based on a subjective opinion regarding the monetary value of the project (more details will be provided later, see Section 4.4).
2. Subsequently, by changing the introduced monetary value for the economic criterion, a series of mappings with the rating scale were formed.
3. It was calculated how the parameters of the economic criterion would change if we apply different scale mapping multipliers.
4. It was analysed how the results (of the decision tree and AHP method application) were changed, depending on the changes in the scale.
5. It was analysed how the investor's behavior would change if we offer him various projects based on the market price.
6. The consequences of decisions regarding whether to sell project results or not were investigated. It was taken into account that not only the investor was interested in the results of the project, but there were also other stakeholders affected by the investor's changing opinion. It is well known that stakeholders were interested in how the starting market price was formed.
7. The visualisation of the results of the decision tree for various projects was performed.
8. It was described how the possible final result that was obtained in the decision tree might be changed, depending on possible changes in the investor's opinion.

Note that, in the steps mentioned above, only economic criteria can be seen explicitly. However, the assessment by the AHP method implicitly includes all of the criteria that were discussed in Section 4.3.

3.4. The AHP-Based Project Assessment

The data of the decision tree end nodes on the investor's subjective opinion when assessing possible options were obtained by applying the AHP method, as already shown in the article [17].

At this stage, we used the AHP method, because we wanted to evaluate the investor's subjective opinion regarding the value of the project. The AHP method was chosen because it was simple enough to use and it enabled to perform the sensitivity analysis based on [83,84].

The APH method was adapted according to the methodology that was proposed by [58]. Based on this methodology, a comparison matrix was initially created to determine the weights of the criteria of pair-wise C_{lm} elements:

$$\begin{bmatrix} C_{11} & C_{12} & ... & C_{1z} \\ C_{21} & C_{22} & ... & C_{2z} \\ ... & ... & ... & ... \\ C_{z1} & C_{z2} & ... & C_{zz} \end{bmatrix}; \tag{3}$$

all elements are positive ($C_{lm} > 0$) and reciprocal ($C_{lm} = 1/C_{lm}, \forall, m = 1, 2, \ldots, z$).

The relative value in comparison to the sum of the columns consisting of z criteria is then found:

$$Y_{lm} = \frac{C_{lm}}{\sum\limits_{l=1}^{z} C_{lm}}; \tag{4}$$

where z is the number of criteria.

A normalised matrix is formed for comparison:

$$\begin{bmatrix} Y_{11} & Y_{12} & ... & Y_{1z} \\ Y_{21} & Y_{22} & ... & Y_{2z} \\ ... & ... & ... & ... \\ Y_{z1} & Y_{z2} & ... & Y_{zz} \end{bmatrix}. \tag{5}$$

Dividing by the number of criteria gives a weight matrix:

$$W_l = \frac{\sum\limits_{m=1}^{z} Y_{lm}}{z}; \tag{6}$$

$$\begin{bmatrix} W_1 \\ W_2 \\ ... \\ W_z \end{bmatrix}, \tag{7}$$

The consistency vector is calculated according to the formulas:

$$Cv_1 = \frac{1}{W_1}[C_{11}W_1 + C_{12}W_2 + ... + C_{1z}W_z];$$

$$Cv_2 = \frac{1}{W_2}[C_{21}W_1 + C_{22}W_2 + ... + C_{2z}W_z];$$

$$...$$

$$Cv_z = \frac{1}{W_z}[C_{z1}W_1 + C_{z2}W_2 + ... + C_{zz}W_z]. \tag{8}$$

The Eigen value λ_{max} is found by:

$$\lambda_{max} = \frac{1}{z}\sum_{l=1}^{z} Cv_l \tag{9}$$

then a consistency index (*CI*) was calculated by formula:

$$CI = \frac{\lambda_{\max} - z}{z - 1}; \tag{10}$$

and consistency ratio (if *CR* of 0.1 or below, the results are acceptable):

$$CR = \frac{CI}{RI}; \tag{11}$$

where *RI* is the random index that is determined by [58], for example, in our case the number of criteria is 4, so *RI* = 0.9.

We determined the criterion of economic logic that we will use in the decision tree. Using the redesigned AHP method, when two possible alternatives—to sell the object or keep it for yourself—are the same (F = 0.5 weighted and normalised points):

$$F = \sum_{m=1}^{z} \frac{W_m V_m}{V_m + 1}, \tag{12}$$

where V_m is the rating value of m-th criteria before the normalisation. The rating value of the first criteria:

$$V_1 = \frac{a}{1-a}, a = \left(F - \sum_{m=2}^{z} \frac{W_m V_m}{V_m + 1}\right) / W_1, \tag{13}$$

Subsequently, we calculated the value (the rating is represented by a real value in intervals $[-9, -1]$ and $[1, 9]$):

$$\bar{V}_1(V_1) = \begin{cases} -1/V_1, & \text{when } V_1 < 1 \\ V_1, & \text{otherwise.} \end{cases} \tag{14}$$

If the rating value $\bar{V}_1$ exceeds the maximum permissible limit of 9 points or is less than -9 points, then this value is specified as equal to 9 or -9, depending on the situation. The implementation of this requirement must be done in formula (13) by limiting $a \leq 0.9$, because it is the value where the rating 9 is achieved, otherwise there is a possibility for a to obtain the negative values.

Regarding the rest of criteria, we assume that the rating values $\bar{V}_i, i = 2, \ldots, z$ are given, according to the fundamental rating scale that is used to evaluate the weights of criteria. Table 2 presents these criteria

Table 2. The fundamental rating scale for AHP elements assessment (based on [58]).

Rating	Definition	Explanation
9	Extreme importance of the first element	The evidence favouring the first element over the second is of highest possible order of affirmation
8	Very, very strong importance of the first element	Intermediate value between two adjacent judgments when a compromise exists
7	Very strong or demonstrated importance of the first element	The first element is strongly favoured and its dominance is demonstrated in practice
6	Strong plus importance of the first element	Intermediate value between two adjacent judgments when a compromise exists
5	Strong importance of the first element	Experience and judgement strongly favour the first element over the second
4	Moderate plus importance of the first element	Intermediate value between two adjacent judgments when a compromise exists
3	Moderate importance of the first element	Experience and judgement favour a little more the first element over the second
2	Weak importance of the first element	Intermediate value between two adjacent judgments when a compromise exists
1 and −1	Equal importance of the elements	Both elements contribute equally to the objective
−2	Weak importance of the second element	Intermediate value between two adjacent judgments when a compromise exists
−3	Moderate importance of the second element	Experience and judgement favour a little more the second element over the first

Table 2. *Cont.*

Rating	Definition	Explanation
−4	Moderate plus importance of the second element	Intermediate value between two adjacent judgments when a compromise exists
−5	Strong importance of the second element	Experience and judgement strongly favour the second element over the first
−6	Strong plus importance of the second element	Intermediate value between two adjacent judgments when a compromise exists
−7	Very strong or demonstrated importance of the second element	The second element is strongly favoured and its dominance is demonstrated in practice
−8	Very, very strong importance of the second element	Intermediate value between two adjacent judgments when a compromise exists
−9	Extreme importance of the second element	The evidence for favouring the second element over the first is of the highest possible order of affirmation

We used rating scale from 1 to 9 if the first comparable criterion was more important than the second, and from −9 to −1 if vice versa. A higher number on the rating scale means a higher degree of criterion importance. A rating scale from 1 to 9 was used in the standard AHP method. We extended that scale with the usage of negative values −1 to −9 for more convenient assessment of the criteria and interpretation of the obtained data.

Note that, in Formulas (12) and (13), the values $V_i, i = 2, \ldots, z$ are calculated from the inverse function of Formula (14) which is the same function:

$$V_m(\bar{V}_m) = \begin{cases} -1/\bar{V}_m, & \text{when } \bar{V}_m < 1 \\ \bar{V}_m, & \text{otherwise.} \end{cases} \tag{15}$$

Note that $\bar{V}_m < 1$ in Formula (15) means integer values from −9 to −1, however, in Formula (14) $V_1 < 1$ means real values from interval $[1/9, 1)$.

The AHP assessment in the context of this research can be summarised as follows:

1. Opinion data is written to Formula (3).
2. Weights are calculated by Formulas (4)–(7).
3. The consistency ratio (11) is checked to fulfill the AHP method requirements (otherwise the opinion data must be re-evaluated).
4. The value $\bar{V}_1$ representing the economic logic is calculated using Formula (14). Note that according to formula, this criterion expresses the remaining criteria.
5. Depending on the case, the value $\bar{V}_1$ is converted to a monetary value, e.g., using the scale and interpolation between scale values. See Section 4.4 for more details about the case which was studied in this research.

3.5. The Algorithms for a Solution

The basic element of our data structure (a tree) is a node with connections to parent and children nodes (Figure 2).

Algorithm 1 describes a data structure that includes a type of node, which can be equal to 1 if a positive decision was made, and 0 otherwise, a tariff that is used to calculate the costs of consulting, the price of the event and the additional price, the estimate of an event duration, and other data.

Algorithm 1: Data structure

```
struct {
    int type; //type of the node
    float p; //probability for this node to be selected by a parent
    float price; //price of the event
    float time; //time before the event starts
    float ap; //probability of the additional cost and duration
    float aprice; //additional price
    float atime; //additional time
    float tariff; //additional price per day
    node* parent; //the pointer to parent node
    vector< node* > children; //list of children
    float priceTotal; //accumulated price
    float timeTotal; //accumulated time
    float value; //value of the expected profit of the node
    float extime; //expected time of the node
    string project; //reference to the project data
} node;
```

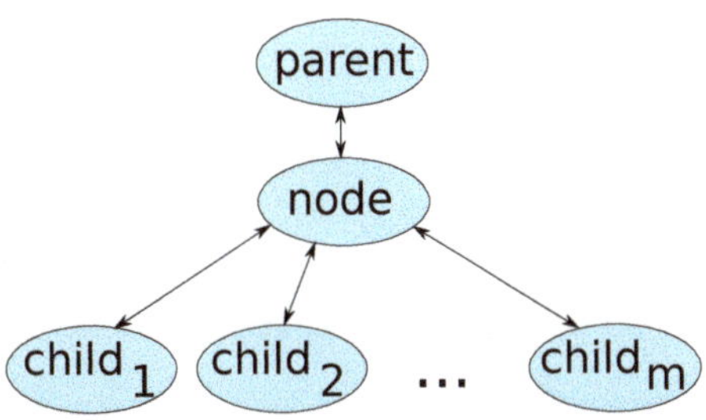

Figure 2. The relations between a node and its parent and children nodes (the picture was taken from [4]).

The details of the fields are provided in paper [17]. The field *project* describes a reference to the project data; in practical implementation, it can be a name of the file with the required data. We will also use algorithms from the mentioned work to define the solving procedure, more specifically function *CalcPars* from Algorithm 1, *CalcValues* from Algorithm 2. Algorithm 2 presents the general calculation procedure. Here, *ReadTree* defines the procedure for reading the tree data from the given input that is defined by the variable *data*. *GenerateScales* returns the list of the possible mappings of rating values to the monetary price values.

Algorithm 2: The general calculation procedure

```
input  :data
output:Array of functions F
tree = ReadTree(data);
mappings = GenerateScales(data);
for each scale p in mappings do
    ApplyAHP(tree, data, p);
    CalcPars(tree);
    CalcValues(tree);
    P = {};
    CollectProjectValues(tree, P);
    F.add(P);
end
```

The mapping p can be seen as a function describing the dependency of the monetary price values on rating values, since, later in Algorithm 3, the interpolation procedure is applied. The array P represents the evaluations of all projects in the tree with a given mapping p. Such evaluations are collected in the array functions F, where we assume that method *add* adds a single point for each function in the array.

Algorithms 3 and 4 both use typical preorder tree traversal to reach the leaves of the tree and apply the necessary operations to them—the AHP method in Algorithm 4 and collection of the project values as a final output in Algorithm 3.

Algorithm 3: Sensitivity analysis procedure

```
Function CollectProjectValues(node, P)
    if node.project! = NULL then
        P[node.name] = node.value;
    end
    if node.children! = NULL then
        for each node t in node.children do
            CollectProjectValues(t);
        end
    end
end
```

Algorithm 4: The algorithm for the application of AHP to the tree

```
Function ApplyAHP(node,data,p)
    if node.project! = NULL then
        evaluate v from AHP parameters from data according to formula (14);
        evaluate node.value by interpolating at the point v using mapping p;
    end
    if node.children! = NULL then
        for each node t in node.children do
            ApplyAHP(t);
        end
    end
end
```

4. Case Study: The Modelling of Investor's Behaviour

4.1. The Description of the Case

In this section, we briefly introduce the case and apply the previously described methodology to that case, more specifically, the general procedure from Algoritm 2 was applied. Algorithms are described by pseudo-code that can be easily converted using any general purpose programming language, like Python, Java, etc., depending on the needs of integrating this logic into existing systems or other reasons for the language preferences. In this research, the algorithms were implemented using C++ solely for the purposes of this research without any support of user interface or any other advanced features for the general user.

The case study was chosen as an empirical research method to demonstrate how algorithms work by applying them to a real-life example, i.e., the type the case study is illustrative in this research. The case study examines a real-life individual building project that was implemented in Vilnius, Lithuania. An individual residential house project with a usable area of 167.84 square meters was chosen as an example. It is very close to typical housing. For comparison: in Lithuania in 2019, the average usable living space in an individual house was 136.2 square meters, while in an apartment it was 57.3 square meters [85]. In the considered case, the old roof was removed and a new one was installed. Thus, this is a reconstruction project involving the installation of roof structures, which is the modelled process in this research. The reconstruction project in the presented case faced problems that are typical when the investor is a person, not a company. For example, it was necessary to adapt the project to meet the investor's requirements, paying special attention to the professional installation of the roof. The work performance was inspected by hired consultants and supervisors, who identified errors and helped to correct them. The possible alternatives for the roof implementation of this project were examined and presented in the decision tree as eight different projects, which were also evaluated using the AHP method. Groups of stakeholder were also identified, and their influence on project progress and decision-making were taken into account.

The roof installation projects were considered as a separate case of construction projects. The implementation of such projects covers the typical stages of construction projects [86]—feasibility, design, construction, and the beginning of the operation. Note that the selected illustrative case is a separate case of construction projects, which is a part of the real estate. Besides the mentioned case, the proposed model is suitable for a wide range of other projects within the real estate life cycle. Moreover, the methodology can be easily expanded to cover other types of projects, such as R&D projects, which support the return to the previous stages (for example, after testing stage) [87]—for that purpose, the cyclic expansion algorithm can be used [4].

Note that the considered case of roof installation projects does not represent all possible construction projects, and the projects must share these main properties:

1. The pool of possible construction solutions must be well-defined in order to define the leaves of the decision tree.
2. The investor must have the leading developers role, i.e., he must initialise the project and lead the process of the project implementation. The involvement of additional stakeholders might change the role of the investor making some possible decisions in the tree obsolete, the decision strategy might become trivial from the investor's point of view, and such modelling might become not useful.

There are no major limitations other than those listed above, for hypothetical possibility to apply the discussed approach to the projects modelling, as long as the process can be described as a Markov process, which is a very general way of process formalisation. Problems to apply the technique may arise because some of the cases might require a lot of information to be defined in order to fill the decision tree by possible events.

4.2. The Application of the Proposed Approach to the Case Study

During the project implementation, there are many decisions to be made by different project participants; however, in the context of the current research, the only the decisions made by investors were considered. Additionally, in this study, we analyse the small scale construction project, which implies the limited set of the decisions for the considered case. The peculiarity of other projects could greatly affect the decisions and their order, potentially creating a more complex graph of dependencies between decisions.

The investor must decide whether or not to carry out the roof reconstruction project, taking into account the potential losses if nothing is done. He also makes decisions regarding whether or not to hire consultants and supervisors. The investor can choose a building design company from options marked with the letters "A" and "B". The different roof installation projects to be selected are marked with the letters "A", "B", "C", "D", "E", "F", "G", and "H". The construction companies are connected to each of the projects according to the planned works, and they are marked "AA", "AB", "BA", "BB", "CA", "CB", "DA", "DB", "EA", "EB", "FA", "FB", "GA", "GB", "HA", and "HB". The supply companies are connected to the construction site of intended construction companies and they are marked "AAS", "ABS", "BAS", "BBS", "CAS", "CBS", "DAS", "DBS", "EAS", "EBS", "FAS", "FBS", "GAS", "GBS", "HAS", and "HBS" (Figure 3). We provide all decisions in the Table 3, where we sort them in the order in which they must be performed.

The decision tree scheme for selecting the best project (Figure 3) presents the chances of failure or success of a project execution, and the decisions that are related to the different stages of the project. Three types of nodes were used:

1. Decision nodes which are represented by rectangle shapes.
2. Chance nodes represented by circles.
3. End nodes represented by triangles.

Table 3. A description of the investor's decisions that form the basis of the decision tree.

The Order	Description of the Investor's Decisions
1	Decision whether to implement the project
2	Decision whether to hire the consultants and supervisor
3	Decisions whether to select the building design company "A" or "B" for the branches of the decision tree with or without consulting
4	Decisions whether to select the medium-cost project from a set of projects "A", "C", "E", "G" or expensive project from a set of projects "B", "D", "F", "H"
5	Decisions whether to select the construction company "AA" or "AB" for the project "A"; "BA" or "BB" for the project "B"; "CA" or "CB" for the project "C"; "DA" or "DB" for the project "D"; "EA" or "EB" for the project "E"; "FA" or "FB" for the project "F"; "GA" or "GB" for the project "G"; "HA" or "HB" for the project "H"
6	Decisions whether to select the supply company "AAS" or "ABS" for construction company "AA"; "AAS" or "ABS" for construction company "AB"; "BAS" or "BBS" for construction company "BA"; "BAS" or "BBS" for construction company "BB"; "CAS" or "CBS" for construction company "CA"; "CAS" or "CBS" for construction company "CB"; "DAS" or "DBS" for construction company "DA"; "DAS" or "DBS" for construction company "DB"; "EAS" or "EBS" for construction company "EA"; "EAS" or "EBS" for construction company "EB"; "FAS" or "FBS" for construction company "FA"; "FAS" or "FBS" for construction company "FB"; "GAS" or "GBS" for construction company "GA"; "GAS" or "GBS" for construction company "GB"; "HAS" or "HBS" for construction company "HA"; "HAS" or "HBS" for construction company "HB"
7	Decisions whether to sell the object of the project, based on the results of applying the AHP method

It is worth mentioning that the decisions are performed based on the values of the nodes in a tree; these values are computed as a probabilistic expectation by Formula (2). This means that the dynamic programming method allows for indirectly including the AHP evaluation into the whole tree, i.e., the criteria to decide are the ones that are used in AHP method.

The AHP output for the tree is represented by a single value; however, this value is not necessarily the actual price of the object—it is the result of the evaluation of all criteria that are included into AHP method. For example, the criteria that are connected to the environment might raise the value in the eyes of the investor, as a result—the values on the leaves change, followed by changes of the solution of the dynamic programming method, which could affect the decision to select the consultants. Next, the actual criteria that are included in the analysis will be discussed.

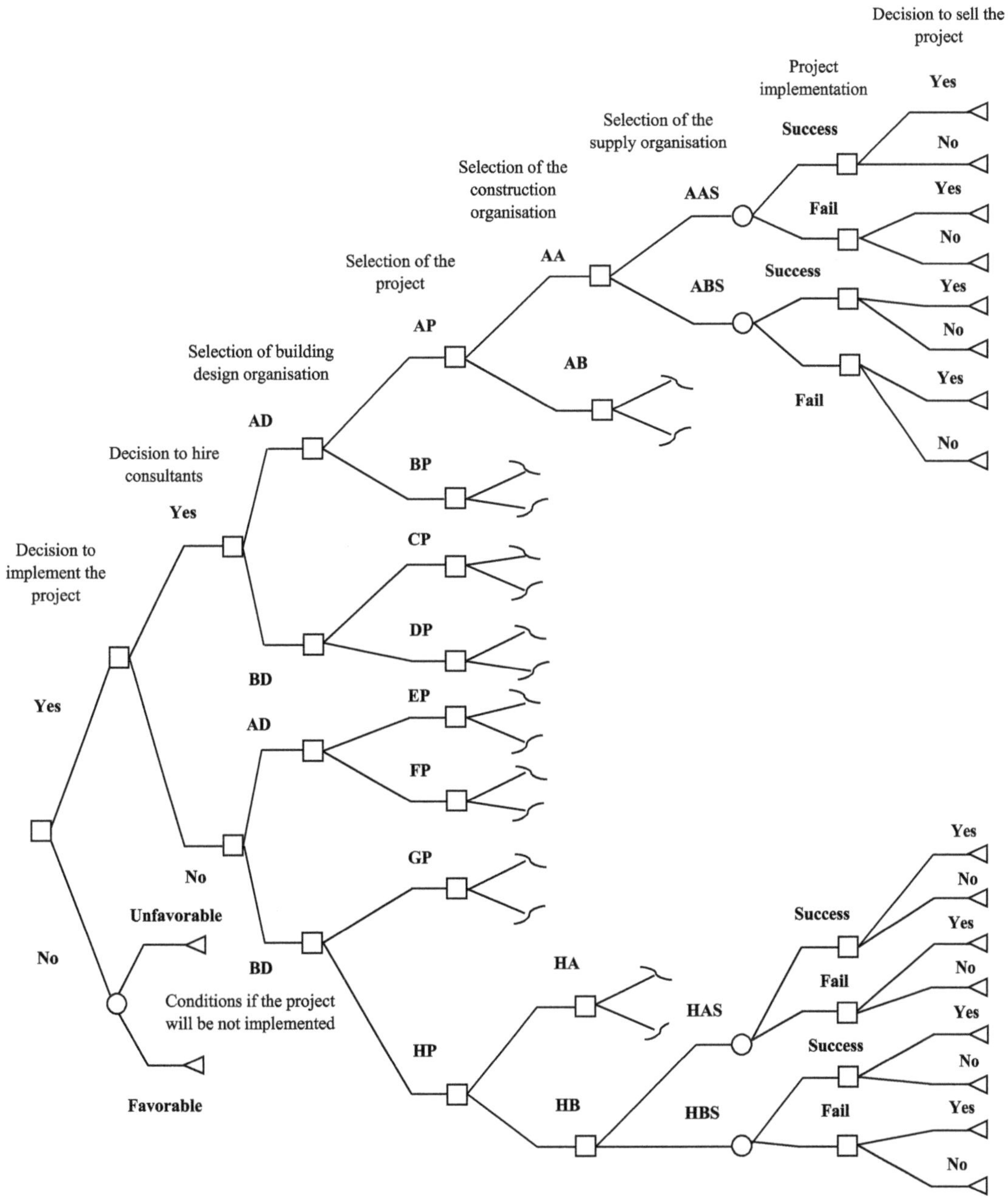

Figure 3. The decision tree scheme fragment for construction projects selection.

4.3. Criteria

The criteria for AHP method were selected based on the results of the article [17]. The main difference of current research is that one of these criteria (the criterion of compliance with the economic logic) is very special—it varies depending on the scale, and we consider a series of scales in order to perform the sensitivity analysis. Additionally, in this section, we provide deeper insight of the dependence of the criterion of compliance with the economic logic on the rest. These four criteria were selected:

1. The criterion of compliance with the psychological and social needs (situation of neighborhood, habitual place, status, lifestyle, appearance, romanticism, and history).
2. The criterion of compliance with the economic logic including an assessment of the object price.
3. The criterion of compliance with the strategic (political) objectives including an assessment of the investor's plans—whether the object will be rented, sold, or used for living.
4. The criterion of compliance with the best location option—best location option including an assessment of the site quality, accessibility, and amenities of public and private service.

It is notable that the selected criteria fit into the general sustainability categories (see Figure 4): Economic (compliance with the economic logic), Social (compliance with the psychological and social logic), and Environmental (compliance with the best location option) [88].

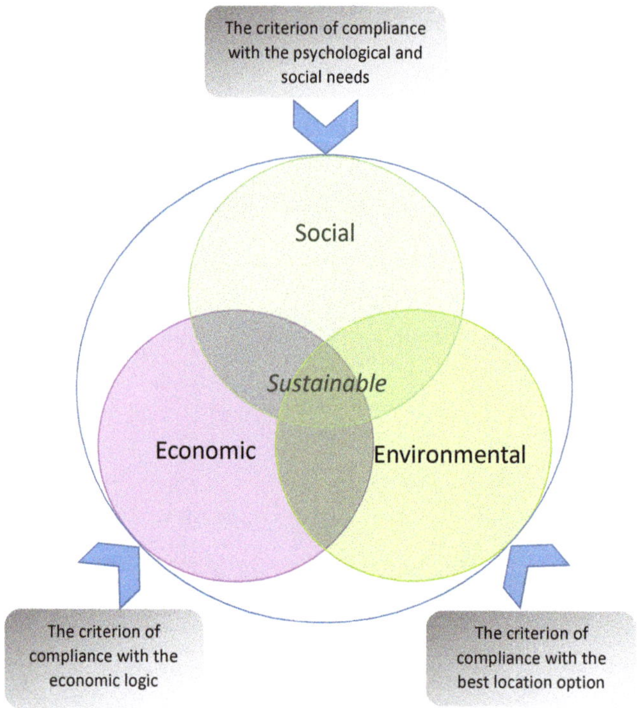

Figure 4. Sustainability categories for different criteria.

The selected criteria are weighted using a rating scale from 1 to 9 if the first comparable criterion is more important than the second, and from −9 to −1 if vice versa.

The investor compared the selected criteria according to their importance in choosing the best roof installation project, thus determining the weight of each criterion. Table 4 presents this assessment.

Table 4. Comparison table for weights of the criteria.

Criteria for Comparison	Importance and Intensity
Compliance with the psychological and social needs vs. Compliance with economic logic	−3
Compliance with the psychological and social needs vs. Compliance with strategic objectives	−3
Compliance with the psychological and social needs vs. Compliance with best location option	2
Compliance with economic logic vs. Compliance with strategic objectives	−1
Compliance with economic logic vs. Compliance with best location option	5
Compliance with strategic objectives vs. Compliance with best location option	5

The data from Table 4 are entered into the Formulas (3)–(7) according to AHP method. After applying these formulas, the weights of the criteria were compared and normalised; Table 5 presents the obtained results. Note that the consistency ratio (11) in that case is equal to 0.001541.

Table 5. Weights for the criteria.

Criteria	Compliance with the Psychological and Social Needs	Compliance with Economic Logic	Compliance with Strategic Objectives	Compliance with Best Location Option
Weights	0.1376	0.3935	0.3935	0.0754

The criteria for selecting projects were assessed using a rating value from 1 to 9 if the investor wanted to sell the project and from −9 to −1 if the sale was undesirable.

At this point, we assume that the evaluations of all ratings of criteria are known, with the exception of the compliance with the economic logic, i.e., we model a situation when this single parameter could differ, depending on various conditions, and the rest of values are fixed. We derive the single case where that value is such that the decision is on the edge, i.e., the weighted scores for both alternatives are equal, and they are equal to 0.5. Thus, the value of criterion of compliance with the economic logic is calculated from the rest of criteria, assuming that both alternatives whether to sell the object or not have the same weighted scores 0.5. In order to illustrate the dependency of the compliance with the economic logic (the main criteria) on the rest of criteria, we chose the same values for all three of them. Figure 5 presents the result function. The graph shows how the points of compliance with the economic logic criterion decrease as the rest of the criteria rise, i.e., for a better project, the calculated price (estimated according to the economic logic) is worse. It is also worth mentioning that, due to the AHP method restrictions, we have a maximum of assessment, which is nine points. It means that both of the alternatives are not necessarily equal anymore, and it leads to a natural restriction of the proposed approach.

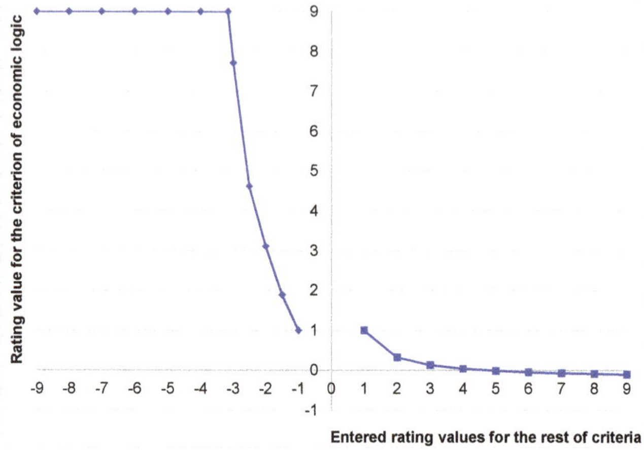

Figure 5. The dependency function of the compliance with the economic logic on the rest of criteria.

Table 6 presents the investor's assessment of the compliance of A-H projects criteria with psychological and social needs, strategic objectives and best location. The criterion of compliance with economic logic was calculated on the basis of the assessment of the above criteria using Formulas (12)–(14); it is shown at the bottom of Table 6. Here, it should be noted that, due to the restrictions of the AHP method when the investor evaluates the criteria as more appropriate for alternative do not to sell the project, and the scores are negative below the limits, the values of the economic logic criterion are nine points.

Table 6. THe assessment of the criteria by the investor (decision whether to sell the object).

Criteria	Values of Projects Criteria							
	A	B	C	D	E	F	G	H
Compliance with the psychological and social needs	−3	−9	−3	−9	−3	−9	−3	−9
Compliance with strategic objectives	−2	−5	−1	−4	−3	−6	−2	-5
Compliance with best location option	2	2	2	2	2	2	2	2
Compliance with economic logic, points from AHP when both alternatives are equal	2.6	9	1.25	9	4.14	9	2.6	9

4.4. The Sensitivity Analysis

This section presents the parametrisation of the investor's behaviour and sensitivity analysis of the AHP method. Sensitivity analysis helps to assess the robustness of the decision. We propose modelling the investor's behaviour using a special linking of the rating scale to the monetary values, describing the influence of the property on the individual's decision. We show how to integrate that mapping into the model and evaluate the obtained results.

The model was initially developed for a single investor. Thus, possible changes in the assessment or parameters that will occur if the number of the investors increases are taken into account.

For the sensitivity analysis, we introduce different mappings of the ratings scale to the monetary values of project prices, as presented in Table 7. For the sake of simplicity, we use a single parameter describing the differences between mappings—a multiplier which ranges from 0.2 to 2 with a step of 0.2; the values in the table are proportional to that multiplier. The investor has estimated project prices that correspond to the maximum and minimum ratings when the multiplier is equal to one. The rating scale is the result of the investor's survey. In our case, the investor agreed that the project intermediate prices will be determined by linear interpolation and assigned to the remaining ratings of the scale.

Table 7. The mappings of the scale to monetary values of the project for economic criteria.

Scale	Multiplier of the Scale Mapping and the Corresponding Monetary Values in Euros									
	0.2	**0.4**	**0.6**	**0.8**	**1**	**1.2**	**1.4**	**1.6**	**1.8**	**2**
9	8000	16,000	24,000	32,000	40,000	48,000	56,000	64,000	72,000	80,000
8	7625	15,250	22,875	30,500	38,125	45,750	53,375	61,000	68,625	76,250
7	7250	14,500	21,750	29,000	36,250	43,500	50,750	58,000	65,250	72,500
6	6875	13,750	20,625	27,500	34,375	41,250	48,125	55,000	61,875	68,750
5	6500	13,000	19,500	26,000	32,500	39,000	45,500	52,000	58,500	65,000
4	6125	12,250	18,375	24,500	30,625	36,750	42,875	49,000	55,125	61,250
3	5750	11,500	17,250	23,000	28,750	34,500	40,250	46,000	51,750	57,500
2	5375	10,750	16,125	21,500	26,875	32,250	37,625	43,000	48,375	53,750
1	5000	10,000	15,000	20,000	25,000	30,000	35,000	40,000	45,000	50,000
−1	5000	10,000	15,000	20,000	25,000	30,000	35,000	40,000	45,000	50,000
−2	4625	9250	13,875	18,500	23,125	27,750	32,375	37,000	41,625	46,250
−3	4250	8500	12,750	17,000	21,250	25,500	29,750	34,000	38,250	42,500
−4	3875	7750	11,625	15,500	19,375	23,250	27,125	31,000	34,875	38,750
−5	3500	7000	10,500	14,000	17,500	21,000	24,500	28,000	31,500	35,000
−6	3125	6250	9375	12,500	15,625	18,750	21,875	25,000	28,125	31,250
−7	2750	5500	8250	11,000	13,750	16,500	19,250	22,000	24,750	27,500
−8	2375	4750	7125	9500	11,875	14,250	16,625	19,000	21,375	23,750
−9	2000	4000	6000	8000	10,000	12,000	14,000	16,000	18,000	20,000

In Table 8, we present the market prices of projects A–H, which, in the decision tree, are compared with the prices that are determined by the AHP method based on the investor's subjective opinion. The market prices for the roof projects A–H were determined on the basis of an expert's assessment, while taking into account statistical indicators and the fact that the roof area of all projects was 63.2 square meters, the geographical location of the building was also identical, and the price varied due to the differences of roof and attic layout, materials, finishing, comfort, and design features. The detailed procedure of market pricing is outside the scope of our study and, therefore, is not presented in this paper.

Table 8. Market prices of the investigated roof installation projects to be used in the decision tree.

Criteria	Projects							
	A	B	C	D	E	F	G	H
Market price for sale (b)	27,313	42,341	28,468	43,501	16,326	80,231	27,486	41,401

Figure 6 shows the data that were obtained by solving the decision tree in the presence of various multipliers. Data on the values of projects for the investor were entered into the relevant end nodes of the decision tree. These values are determined by interpolating the monetary values from Table 7 at the points of the scores of the criterion of the economic logic value from Table 6. The market prices of the projects, which are presented in Table 8, were also entered into the relevant end nodes of the decision tree.

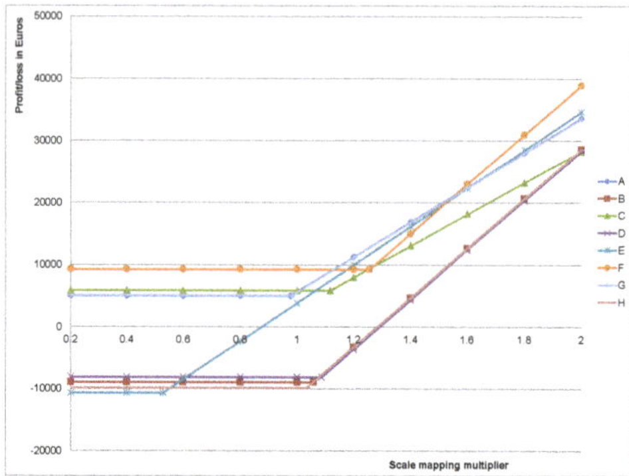

Figure 6. The best options of the projects A-H according to the investor.

As the scale changes, the results obtained in the decision tree for each project change differently, as can be seen from Figure 6. Initially, when the values of the multipliers are small, the main factors influencing the final result are the market price and project implementation costs; a little later, when the multipliers increase, the value for the investor exceeds the market price, and causes a breaking point on the graph. As the multiplier continues to increase, the final result is determined by the investor's subjective opinion and the cost of the project. Here, it is possible to find points of lines intersection where the monetary value of some different projects is the same. Analysis of the changes in the investor's subjective opinion is useful in the selection of parameters for scale mapping, in assessment how the final result depends on the chosen scale. The results of this analysis can be used to determine the price of the project and reconcile the stakeholders' interests. Depending on the investor's subjective opinion about the value of the project, brokers can make attractive offers to the investor that would lead to the sale of the project outcomes, if there is a corresponding need.

It is worth mentioning that the horizontal parts of the curves represent the cases when the decision to sell the project is made. With the mapping multiplier 1 and lower, the investor decides to select the project F and sell it, as can be seen from Figure 7. However, with higher values the decision not to sell the project is made, the results might differ, depending on value of the mapping multiplier: with values 1.2 and 1.4, A and G projects are selected, both having very similar values (A has approximately 0.7% higher values). This behaviour of changes in decision can be interpreted as:

1. Lower values of scale mapping mean a greater tendency to sell the project rather than not to.
2. When choosing the project with lower values for sale, the project F is the best option.
3. Higher mapping values mean the investor's bigger tolerance to the higher costs, thus he tends not to sell the project.
4. With values 1.2 and 1.4 the investor selects the medium-cost project, but does not sell it.
5. As the scale multiplier raise, the investor chooses to invest in expensive project and not to sell it.

4.5. An Example of Case Assessment Made by Independent Expert

Up to this point, decisions were made solely on the basis of the investor's opinion. In the context of the research, one investor is sufficient for performng the analysis, because it is necessary to analyse his behaviour. In other words, there is no need to aggregate the data of multiple experts' opinion, as it is useful to do for an estimation of some objects, where multiple opinions give a better and less subjective picture—in the case of the considered situation we want to assess the effect of investor's subjective opinion. However, in order to estimate the possible changes in the results in the case of the radical changes during the implementation of the project, such as the change of the investor, we had to consider all of the steps affected by the investor. What we assumed up to the point is that the mapping of the scale to the monetary values will change; however, it is obvious that the whole assessment could also change. Here, all possible parts of the model, affected by such changes, will be taken into account, and, for this purpose, to achieve, we will briefly follow all of the necessary steps.

A person with five years of experience in the field of construction marketing was selected as an independent expert. Further, we assume that the expert assesses the situation from the investor's point of view. Note that we do not want to enrich the results of the assessment of the considered case; instead, we provide an illustration of what needs to be changed if the investor changes. Further, we will provide the following steps that are needed for modification:

1. The independent expert compares the criteria for weights determination (analogically to Table 4). All of the criteria are rated as equally important, i.e., by 1 point.
2. Using AHP, the weights (Table 5) for criteria are calculated. The weights of the criteria in this case are the same and equal to 0.25.
3. Table 6 is re-evaluated by the expert, the results are provided in Table 9.
4. The table, analogous to Table 7, is filled according to the new opinion. It is important to note, that every field in it can be evaluated individually, however, it can be simplified as it was done already—the interpolation can be applied to form the base mapping, and the multiplier can produce the rest of mappings. We interpolated between two values to create the base mapping; the results are provided in Table 10.
5. Based on the achieved results, the new set of curves describing the dependency of profit/loss on the scale mappings are produced; the results are provided in Figure 7.

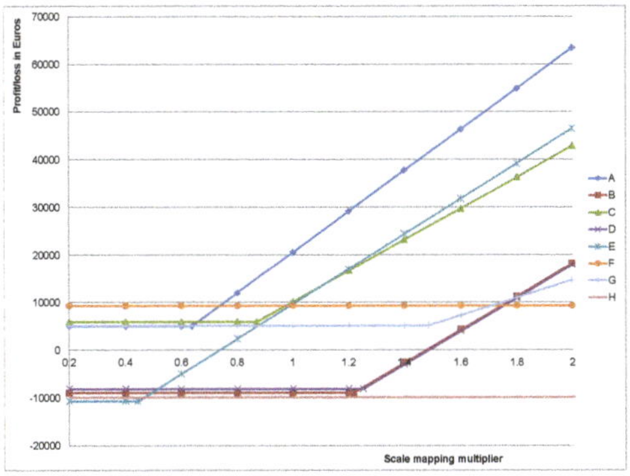

Figure 7. The best options of the projects A-H according to the expert.

With the mapping multiplier 0.8 and higher, the expert selects project A, as can be seen from Figure 7. The result shows that the expert's decision in favour of selecting project A and not selling the project is robust, and it does not change with higher multiplier values. However, with lower multiplier values, the expert decides to select the project F and sell it.

Table 9. Criteria assessment made by the expert (decision whether to sell the object).

Criteria	Values of the Projects Criteria							
	A	B	C	D	E	F	G	H
Compliance with psychological and social needs	−2	1	1	1	2	5	3	7
Compliance with strategic objectives	1	2	2	2	−2	5	2	3
Compliance with best location option	−2	−2	1	−2	−2	5	1	3
Compliance with economic logic, points from AHP when both alternatives are equal	5	1	−2	1	2	−9	−9	−9

Table 10. The mappings of the scale to monetary values of the project for economic criteria (based on expert's opinion).

Scale	Multiplier of the Scale Mapping and the Corresponding Monetary Values in Euros									
	0.2	0.4	0.6	0.8	1	1.2	1.4	1.6	1.8	2
9	10,200	20,400	30,600	40,800	51,000	61,200	71,400	81,600	91,800	102,000
8	9793.75	19,587.5	29,381.25	39,175	48,968.75	58,762.5	68,556.25	78,350	88,143.75	97,937.5
7	9387.5	18,775	28,162.5	37,550	46,937.5	56,325	65,712.5	75,100	84,487.5	93,875
6	8981.25	17,962.5	26,943.75	35,925	44,906.25	53,887.5	62,868.75	71,850	80,831.25	89,812.5
5	8575	17,150	25,725	34,300	42,875	51,450	60,025	68,600	77,175	85,750
4	8168.75	16,337.5	24,506.25	32,675	40,843.75	49,12.5	57,181.25	65,350	73,518.75	81,687.5
3	7762.5	15,525	23,287.5	31,050	38,812.5	46,575	54,337.5	62,100	69,862.5	77,625
2	7356.25	14,712.5	22,068.75	29,425	36,781.25	44,137.5	51,493.75	58,850	66,206.25	73,562.5
1	6950	13,900	20,850	27,800	34,750	41,700	48,650	55,600	62,550	69,500
−1	6950	13,900	20,850	27,800	34,750	41,700	48,650	55,600	62,550	69,500
−2	6543.75	13,087.5	19,631.25	26,175	32,718.75	39,262.5	45,806.25	52,350	58,893.75	65,437.5
−3	6137.5	12,275	18,412.5	24,550	30,687.5	36,825	42,962.5	49,100	55,237.5	61,375
−4	5731.25	11,462.5	17,193.75	22,925	28,656.25	34,387.5	40,118.75	45,850	51,581.25	57,312.5
−5	5325	10,650	15,975	21,300	26,625	31,950	37,275	42,600	47,925	53,250
−6	4918.75	9837.5	14,756.25	19,675	24,593.75	29,512.5	34,431.25	39,350	44,268.75	49,187.5
−7	4512.5	9025	13,537.5	18,050	22,562.5	27,075	31,587.5	36,100	40,612.5	45,125
−8	4106.25	8212.5	12,318.75	16,425	20,531.25	24,637.5	28,743.75	32,850	36,956.25	41,062.5
−9	3700	7400	11,100	14,800	18,500	22,200	25,900	29,600	33,300	37,000

4.6. The Application of the Model to Other Cases

Up to this point, it was focused on the selected case study that was used as a tool to demonstrate the proposed ideas. However, the modelling approach provides a powerful way to deal with a wide range of situations. Here are some key points that the applicator must take into account:

- The process being modelled must be described as a Markov process.
- All possible scenarios in different process states must be evaluated by the probabilities of their occurrence.
- The dependencies between states must form a tree, so, in some cases, some redundant information must be included (i.e., the structure of the tree branch is identical in both cases: either the consultant will be selected or not). However, there is an important feature—the model supports some process costs that occur as additional costs with some probabilities, it can be added directly without the duplication of some branches of a tree.

- The pool of projects must be formed with an appropriate assessment of each of them via AHP method.
- If one would like to an add additional parametrisation of the project, he should consider every set of parameters (e.g., some quality parameters) as a separate project entity and then add it as a leaf to the tree together with the needed branch.
- As was mentioned above, s wide variety of decisions can be added and the proposed algorithms are absolutely compatible with as many decision choices from the single state as needed, i.e., there is no requirement for the tree to be binary.
- The mappings of the scale to the monetary values are decisive for the outcome of the modelling. Thus, these mappings must be carefully thought of before the final sensitivity analysis.

5. Discussion

After studying a joint derivative of the Decision Tree and AHP methods, it was found that a combination of these methods can be used for project assessment. Although data collection and modelling is quite labour-intensive, the development of a model template opens up a wide range of possibilities for project assessment, exploiting the potential of information technology and remote working.

It was found that the rating values for the economic criterion, which were obtained using the AHP method, sometimes exceed the limits that are set by AHP method (i.e., the maximum score according to the considered scale is 9). This happens, for example, in the case when the rest of criteria have the same rating, and when the rating values for the remaining criteria reach -3.1588 points. This situation is typical for expensive projects, when the investor tends not to sell the results of the project, i.e., the economic criterion rating exceeds the maximum value in the rating scale—in such a case, the maximum value will be used.

When modelling the investor's behavior, it is necessary to examine his opinion in detail, as well as to know the real estimate of the project, which also poses certain difficulties. On the one hand, in order to solve this problem, the examples of known projects that are similar in content and value can be used; on the other hand, many typical projects can be created and their data can be applied to the model.

The methodology that was considered in the article, which combines AHP and decision tree methods, solves the problem of the selection of the most suitable project. As a part of the methodology, it was introduced a sensitivity analysis, which enables taking the investors' different opinions into account. Such sensitivity analysis shows how sensitive the possible decisions are, depending on changes of the situation (e.g., if the investor changes)—it is dedicated to the analysis of the robustness of decisions.

In this research, different options of medium-cost and expensive roofing projects were examined. The costs of all roof elements and project activities were provided and applied according to the methodology; these values were used in the decision tree. The AHP method was used to estimate the investor's subjective opinion.

It is notable to mention that the criteria for the AHP method were selected in accordance with the principles of sustainability, more specifically—the main categories of sustainability: social, environmental, and economic. The proposed approach was applied to estimate how one of the parameters (the economic one) reacts to changes of the rest parameters, provided that the assessment by the AHP method is on the edge between two decisions for two alternatives. In this research, we examined variations of mappings, which directly affected the economic criterion. The economic criterion is derived from the rest of criteria, thus the analysis can be extended by additional modification of the rest of criteria, which would greatly expand the analysis; however, from a methodological point of view, the current method does not need any modifications.

The considered case study has the primary typical attributes of building projects, thus we assume that it is sufficient to demonstrate the proposed approach. We believe that this approach can be applied to other cases—it is a topic for future research. Moreover, investor's and independent expert's opinions were used to produce a base mapping of the scale to the monetary values. The rest of the scales were derived using a simple multiplier representing different opinions, which was applied to a simple linear interpolation of the monetary values. Thus, the logical addition would be to use different scales without clear dependency between them; however, it is out of scope of this research, as it would not qualitatively improve the results—in this research, the illustrative type of the case study was performed. The set of the decisions and their order were selected from the point of view of the investor. Thus the set of the decisions to form the decisions strategy might look quite limited, because the decisions performed or affected by other stakeholders are included into probabilistic nodes and via additional costs and duration with probabilities.

It is worth mentioning that the AHP method was chosen, since it is widely used and the explicit formula (see Formula (13)) can be derived for the required assessment. Other methods of multi-criteria assessment can also be applied; however, it is not a straightforward task and it needs a separate investigation. The analysed case can be directly described as a Markov process that permits formulating the optimisation problem in order to maximise investor's profit—it was used to model the investor's behaviour. The solution of this problem was exact and it was achieved using the dynamic programming method. Thus, as long as the modelled process was the Markov process, and the computations were performed in the reasonable time, there was no reason to analyse any other methods to solve that optimisation problem. The AHP method gives the assessment of the economic criterion for the decision tree method; the usage of the alternative multi-criteria methods could be studied in future work.

In this research, the case covering most of the aspects that must be taken into account when filling the model with data was analysed. The investor is individual and the project is single. Additionally, it was assumed that the projects are well-defined, so the investor can choose one of them that absolutely defined the constructional solutions (unless some risks will trigger, for example, with mistakes, etc.). This means that all quality parameters or any project specific constructional details must be bundled with one of project and used in the model as a separate example. The considered case was a limited by a single project, the decisions were made using the probabilistic expectation values according to dynamic programming method that was defined by Formula (2), which can be altered to estimate the risks in a non-linear way. However, if the investor considers the pool of projects forming a possible portfolio, then the diversification of the investment can lead to a better tolerance to the possible risks, making the mentioned expectation values more suitable for the needs of the investor.

The main stakeholder in this research is the investor; thus, for the reader, it might create the suspicion that this research is dedicated to some sort of a tool for the investor. However, the mentioned sensitivity analysis considers the investor as a part of the model that might be prone to the changes, so the proposed techniques are potentially useful for any stakeholder who want to analyse the possible scenarios and changes. For example, the building design company might be interested to analyse what do investor choose and under which conditions, for example, if the investor tends to choose some specific type of projects, the company might want to prepare more variations for this type of projects.

The conducted research is not oriented to investor's needs; however, the investor might find the discussed techniques useful for:

1. including the possible changes of priorities, financial abilities, etc., into risk assessment routines; and,
2. getting recommendations for selection of the consultants and supervisors, companies of building design, construction, and supply, also different roof installation projects.

Summarising, the topic of such analysis of possible changes is important for hypothetical usage, for example, for these stakeholders:

- By investors to include the information into their risk assessment and obtain some recommendations.
- By building design or construction companies to improve their pool of potential projects.

- By government to model how possible restrictions might affect the priorities of investors.
- The representatives of interested community might be interested in predictable successful project implementation as a part of sustainable city development, so the lack of decision robustness might indicate an important information for these stakeholders.

Note that the exact procedure of how to apply the results in the mentioned cases is up for discussion, and it is out of scope of this research.

As it was already mentioned in introduction, the AHP application assumes that the criteria are evaluated by individual; thus, the current approach is limited to dealing with an opinion of a single evaluator. Thus, the reader should not try to apply it to some study where the collective evaluation is needed by the group of experts. This research considers the opposite—taking the bias of the individual investor into account in the case when the type of project implies the assessment by a single person. Certainly, it can be assumed that the evaluation made by that individual is performed with the help (consultations) of some sort experts; however, technically it is not important for the methods in the current research.

In the context of this research, the AHP assessment is dedicated to the decisions on whether to sell the project or not, which directly reflects the opinion of the investor about that project: if in the eyes of the investor the project is not worth the price, then he will tend to sell it. In the case when some of the criteria are very important for the investor, the value of the project will exceed the value of the price—here, the word "value" stands to represent the evaluation of the project or the money (this value might depend on the richness of the investor) that can be earned from it. The fact of the direct comparison between these two values might create the illusion that only the actual object price is included into decision directly; however, the value that comes from the AHP assessment has nothing to do with the actual price of the object—in fact, it is the opposite, it represents all of the remaining criteria and does not include any direct information about the market prices or the prices of materials, etc. In other words, the comparison in the nodes is a comparison between the actual price, which does not depend on the investor's opinion and the value of the rest criteria that is scaled, so it could be comparable to the value of the price. Thus, all of the criteria are directly included into the decision via AHP method. However, the evaluation of the rest of the decisions is derived from the last decision (to sell or not), so it is up for discussion as to whether the criteria should affect those decisions in the same manner as they do on the decision to sell the object or not. For example, the importance of hiring consultants might be driven by the importance of some criteria to a greater degree that it is in the case of the decision to sell the object. The proposed approach is unable to include a special role for different criteria for intermediate decisions directly, it needs a special analysis to identify the necessity for support of such decisions and the way to implement them—this could lead to the potential improvement of the proposed model; however, it is out of scope of the current research.

6. Conclusions

It was shown that different mappings influence the project assessment differently, leading to changes in the best project selection. It was found that, with a small mapping multiplier, more expensive projects were unprofitable, with the exception of E and F projects.

The effect of the investor's and the independent expert's subjective opinion on the selection of the best project was visualised. The obtained results are visually presented in the form of graphs (Figures 6 and 7). The graphs of the dependence of the projects assessment results on the special mapping of the ratings scale helped to understand the situation of the most appropriate project selection. It showed how close the potential projects were to each other in terms of the assessment, which was directly used for the selection of the best project. The visualisation of the obtained results can be useful in forecasting the investor's behaviour during project development and in negotiations for the sale of project outcomes.

After examining the results of the decisions tree, it was found that more expensive project F was the most attractive with a multiplier of 1. Project F, due to its design features, was rated significantly better from a market perspective than other more expensive projects. With multiplier values from 1.1 to 1.6, the profit value of this project was lower than the values of some other projects; in other cases, project F was more profitable. The medium-cost project E was rated badly by the market—it was not profitable to sell it, although the investor with some multiplier values rated it as the most suitable for him in comparison with other medium-cost projects. The independent expert's opinion was also evaluated. The expert's decisions differed from the investor's opinion—if multipliers were higher than 0.7, then the expert selected project A, but, if the multiplier was less than 0.7, both opinions almost coincided. A small increase in project implementation costs can affect the project selection—this means a possible lack of the robustness of the project decisions may appear.

The analysed methodology can be generalised for application to other project selection problems, not only for implementation of roofing projects. For the application, it is necessary to take the specifics of the projects, the costs of certain types of activities, the probabilities, the possible losses, the factors that affect the system, and the possible patterns of the process participants' behaviour into account. It is also worth considering the amount of time for project preparation and implementation, because, in some cases, this factor can be extremely important.

Author Contributions: Conceptualisation, A.M. and O.R.Š.; Data curation, A.M. and A.B.; Formal analysis, T.V.; Investigation, O.R.Š. and T.V.; Methodology, A.M., A.B. and O.R.Š.; Resources, A.M. and A.B.; Software, A.B.; Supervision, O.R.Š.; Validation, A.M. and A.B.; Visualisation, O.R.Š. and T.V.; Writing—original draft, A.M.; Writing—review & editing, A.B. and T.V. All authors have read and agreed to the published version of the manuscript.

Funding: This research was funded by Vilnius Gediminas Technical University.

Institutional Review Board Statement: Not applicable.

Informed Consent Statement: Not applicable.

Data Availability Statement: The data presented in this study are available on request from the corresponding author. The data are not publicly available due to privacy concerns.

Conflicts of Interest: The authors declare no conflict of interest.

References

1. Issa, U.H.; Mosaad, S.A.; Hassan, M.S. Evaluation and selection of construction projects based on risk analysis. In *Structures*; Elsevier: Amsterdam, The Netherlands, 2020; Volume 27, pp. 361–370.
2. Edjossan-Sossou, A.M.; Galvez, D.; Deck, O.; Al Heib, M.; Verdel, T.; Dupont, L.; Chery, O.; Camargo, M.; Morel, L. Sustainable risk management strategy selection using a fuzzy multi-criteria decision approach. *Int. J. Disaster Risk Reduct.* **2020**, *45*, 101474. [CrossRef]
3. Kalra, N.; Hallegatte, S.; Lempert, R.; Brown, C.; Fozzard, A.; Gill, S.; Shah, A. *Agreeing on Robust Decisions: New Processes for Decision Making under Deep Uncertainty*; The World Bank: Washington, DC, USA, 2014.
4. Bugajev, A.; Šostak, O. An algorithm for modelling the impact of the judicial conflict-resolution process on construction investment. *Sustainability* **2018**, *10*, 182. [CrossRef]
5. Khan, A.S.; Pruncu, C.I.; Khan, R.; Naeem, K.; Ghaffar, A.; Ashraf, P.; Room, S. A Trade-Off Analysis of Economic and Environmental Aspects of a Disruption Based Closed-Loop Supply Chain Network. *Sustainability* **2020**, *12*, 7056. [CrossRef]
6. Pedersen, S.; Gangås, K.E.; Chetri, M.; Andreassen, H.P. Economic Gain vs. Ecological Pain—Environmental Sustainability in Economies Based on Renewable Biological Resources. *Sustainability* **2020**, *12*, 3557. [CrossRef]
7. Kamali, F.P.; Borges, J.A.; Meuwissen, M.P.; de Boer, I.J.; Lansink, A.G.O. Sustainability assessment of agricultural systems: The validity of expert opinion and robustness of a multi-criteria analysis. *Agric. Syst.* **2017**, *157*, 118–128. [CrossRef]
8. Muñoz, B.; Romana, M.G.; Ordóñez, J. Sensitivity Analysis of Multicriteria Decision Making Methodology Developed for Selection of Typologies of Earth-retaining Walls in an Urban Highway. *Transp. Res. Procedia* **2016**, *18*, 135–139. [CrossRef]
9. Hughes, W.; Murdoch, J.R. *Roles in Construction Projects: Analysis and Terminology*; Construction Industry Publications: Birmingham, UK, 2001.
10. Chitkara, K. *Construction Project Management*; Tata McGraw-Hill Education: New York, NT, USA, 1998.
11. Banaitiene, N.; Banaitis, A. Risk management in construction projects. Risk Management–Current Issues and Challenges. In *Risk Management–Current Issues and Challenges*; Banaitiene, N., Ed.; InTech: Rijeka, Croatia, 2012; pp. 429–448.

12. Clough, R.H.; Sears, G.A.; Sears, S.K. *Construction Project Management*; John Wiley & Sons: Hoboken, NJ, USA, 2000.
13. Phoya, S.; Pietrzyk, K. Holistic view on multi-stakeholders' influence on health and safety risk management in construction projects in Tanzania. In *Risk Management in Construction Projects*; IntechOpen: London, UK, 2019.
14. Alothaimeen, I.; Arditi, D. Overview of multi-objective optimization approaches in construction project management. In *Multicriteria Optimization-Pareto-Optimality and Threshold-Optimality*; IntechOpen: London, UK, 2019.
15. Roberts, A. *Code of Practice for Project Management for Construction and Development*; John Wiley & Sons: Hoboken, NJ, USA, 2014.
16. Al-Saggaf, A.; Nasir, H.; Hegazy, T. An Analytical Hierarchy Process-based system to evaluate the life-cycle performance of buildings at early design stage. *J. Build. Eng.* **2020**, *31*, 101364. [CrossRef]
17. Maceika, A.; Bugajev, A.; Šostak, O.R. The Modelling of Roof Installation Projects Using Decision Trees and the AHP Method. *Sustainability* **2020**, *12*, 59. [CrossRef]
18. Che, L. Investor types and stock return volatility. *J. Empir. Financ.* **2018**, *47*, 139–161. [CrossRef]
19. da Silva, R.F.; Razzolini Filho, E. *Sustainability in the Decision Making Process: A Systematic Review of Literature*; Universities and Sustainable Communities: Meeting the Goals of the Agenda 2030; Springer: Berlin, Germany, 2020; pp. 291–305.
20. Cabot, J.; Easterbrook, S.; Horkoff, J.; Lessard, L.; Liaskos, S.; Mazón, J.N. Integrating sustainability in decision-making processes: A modelling strategy. In Proceedings of the 2009 31st International Conference on Software Engineering-Companion Volume, Vancouver, BC, Canada, 16–24 May 2009; pp. 207–210.
21. Cuadrado, J.; Zubizarreta, M.; Rojí, E.; García, H.; Larrauri, M. Sustainability-related decision making in industrial buildings: An AHP analysis. *Math. Probl. Eng.* **2015**, *2015*. [CrossRef]
22. MacDonald, A.; Clarke, A.; Huang, L. Multi-stakeholder partnerships for sustainability: Designing decision-making processes for partnership capacity. *J. Bus. Ethics* **2019**, *160*, 409–426. [CrossRef]
23. Beiragh, R.G.; Alizadeh, R.; Kaleibari, S.S.; Cavallaro, F.; Zolfani, S.H.; Bausys, R.; Mardani, A. An integrated multi-criteria decision making model for sustainability performance assessment for insurance companies. *Sustainability* **2020**, *12*, 789. [CrossRef]
24. European Commission.The European Green Deal. Communication from the Commission to the European Parliament, the European Council, the Council, the European Economic and Social Committee, the Committee of the Regions and the European Investment Bank. 2019. Available online: https://ec.europa.eu/info/sites/info/files/european-green-deal-communication_en.pdf (accessed on 13 May 2021).
25. Taha, F.F.; Hatem, W.A.; Jasim, N.A. Effectivity of BIM technology in using green energy strategies for construction projects. *Asian J. Civ. Eng.* **2020**, *21*, 995–1003. [CrossRef]
26. European Commission. The 11th Sustainable Development Goal. 2015. Available online: https://ec.europa.eu/international-partnerships/sdg/sustainable-cities-and-communities_en (accessed on 13 May 2021).
27. European Commission. *Roadmap to a Resource Efficient Europe*; European Commission: Brussels, Belgium, 2011.
28. GPM, G. Insights into Sustainable Project Management. 2019. Available online: https://greenprojectmanagement.org/2019-insights-into-sustainable-project-management (accessed on 1 December 2020).
29. Morioka, S.N.; Bolis, I.; Evans, S.; Carvalho, M.M. Transforming sustainability challenges into competitive advantage: Multiple case studies kaleidoscope converging into sustainable business models. *J. Clean. Prod.* **2017**, *167*, 723–738. [CrossRef]
30. Rosa, P.; Sassanelli, C.; Terzi, S. Towards Circular Business Models: A systematic literature review on classification frameworks and archetypes. *J. Clean. Prod.* **2019**, *236*, 117696. [CrossRef]
31. Bocken, N.M.; Short, S.W.; Rana, P.; Evans, S. A literature and practice review to develop sustainable business model archetypes. *J. Clean. Prod.* **2014**, *65*, 42–56. [CrossRef]
32. Kudratova, S.; Huang, X.; Kudratov, K.; Qudratov, S. Corporate sustainability and stakeholder value trade-offs in project selection through optimization modeling: Application of investment banking. *Corp. Soc. Responsib. Environ. Manag.* **2020**, *27*, 815–824. [CrossRef]
33. Martens, M.L.; Carvalho, M.M. Key factors of sustainability in project management context: A survey exploring the project managers' perspective. *Int. J. Proj. Manag.* **2017**, *35*, 1084–1102. [CrossRef]
34. Aarseth, W.; Ahola, T.; Aaltonen, K.; Økland, A.; Andersen, B. Project sustainability strategies: A systematic literature review. *Int. J. Proj. Manag.* **2017**, *35*, 1071–1083. [CrossRef]
35. Hatefi, S.M.; Tamošaitienė, J. Construction projects assessment based on the sustainable development criteria by an integrated fuzzy AHP and improved GRA model. *Sustainability* **2018**, *10*, 991. [CrossRef]
36. French, N. Decision theory and real estate investment: An analysis of the decision-making processes of real estate investment fund managers. *Manag. Decis. Econ.* **2001**, *22*, 399–410. [CrossRef]
37. Carbonara, S.; Stefano, D. An Operational Protocol for the Valorisation of Public Real Estate Assets in Italy. *Sustainability* **2020**, *12*, 732. [CrossRef]
38. Dobrovolskienė, N.; Tamošiūnienė, R.; Banaitis, A.; Ferreira, F.A.; Banaitienė, N.; Taujanskaitė, K.; Meidutė-Kavaliauskienė, I. Developing a composite sustainability index for real estate projects using multiple criteria decision making. *Oper. Res.* **2019**, *19*, 617–635. [CrossRef]
39. Mangialardo, A.; Micelli, E.; Saccani, F. Does Sustainability Affect Real Estate Market Values? Empirical Evidence from the Office Buildings Market in Milan (Italy). *Sustainability* **2019**, *11*, 12. [CrossRef]
40. Shojaei, P.; Haeri, S.A.S. Development of supply chain risk management approaches for construction projects: A grounded theory approach. *Comput. Ind. Eng.* **2019**, *128*, 837–850. [CrossRef]

41. Zavadskas, E.K.; Vilutienė, T.; Turskis, Z.; Šaparauskas, J. Multi-criteria analysis of Projects' performance in construction. *Arch. Civ. Mech. Eng.* **2014**, *14*, 114–121. [CrossRef]
42. Hatefi, S.M.; Tamošaitienė, J. An integrated fuzzy DEMATEL-fuzzy ANP model for evaluating construction projects by considering interrelationships among risk factors. *J. Civ. Eng. Manag.* **2019**, *25*, 114–131. [CrossRef]
43. Singh, A.; Beg, I.; Kumar, S. Analytic Hierarchy Process for Hesitant Probabilistic Fuzzy Linguistic Set with Applications to Multi-criteria Group Decision-Making Method. *Int. J. Fuzzy Syst.* **2020**, *22*, 1596–1606. [CrossRef]
44. Hatefi, S.M.; Basiri, M.E.; Tamošaitienė, J. An evidential model for environmental risk assessment in projects using dempster–shafer theory of evidence. *Sustainability* **2019**, *11*, 6329. [CrossRef]
45. Ghasemi, F.; Sari, M.H.M.; Yousefi, V.; Falsafi, R.; Tamošaitienė, J. Project portfolio risk identification and analysis, considering project risk interactions and using Bayesian networks. *Sustainability* **2018**, *10*, 1609. [CrossRef]
46. Asadi, P.; Zeidi, J.R.; Mojibi, T.; Yazdani-Chamzini, A.; Tamošaitienė, J. Project risk evaluation by using a new fuzzy model based on Elena guideline. *J. Civ. Eng. Manag.* **2018**, *24*, 284–300. [CrossRef]
47. Chatterjee, K.; Zavadskas, E.K.; Tamošaitienė, J.; Adhikary, K.; Kar, S. A hybrid MCDM technique for risk management in construction projects. *Symmetry* **2018**, *10*, 46. [CrossRef]
48. Benítez, J.; Carpitella, S.; Certa, A.; Izquierdo, J. Constrained consistency enforcement in AHP. *Appl. Math. Comput.* **2020**, *380*, 125273. [CrossRef]
49. Mavi, R.K.; Standing, C. Critical success factors of sustainable project management in construction: A fuzzy DEMATEL-ANP approach. *J. Clean. Prod.* **2018**, *194*, 751–765. [CrossRef]
50. Yunus, N. Dynamic Linkages Among US Real Estate Sectors Before and After the Housing Crisis. *J. Real Estate Financ. Econ.* **2019**, *58*, 264–289. [CrossRef]
51. D'alpaos, C.; Canesi, R. Risks assessment in real estate investments in times of global crisis. *WSEAS Trans. Bus. Econ.* **2014**, *11*, 369–379.
52. Agarwal, S.; He, J.; Sing, T.F.; Song, C. Do real estate agents have information advantages in housing markets? *J. Financ. Econ.* **2019**, *134*, 715–735. [CrossRef]
53. Wang, C.; Wang, A.; Xu, J.; Wang, Q.; Zhou, F. Outsourced privacy-preserving decision tree classification service over encrypted data. *J. Inf. Secur. Appl.* **2020**, *53*, 102517.
54. Mu, Y.; Liu, X.; Wang, L.; Zhou, J. A parallel fuzzy rule-base based decision tree in the framework of Map-Reduce. *Pattern Recognit.* **2020**, *103*, 107326. [CrossRef]
55. Tian, Z.; Xiao, J.; Feng, H.; Wei, Y. Credit Risk Assessment based on Gradient Boosting Decision Tree. *Procedia Comput. Sci.* **2020**, *174*, 150–160. [CrossRef]
56. Yan, J.; Zhang, Z.; Lin, K.; Yang, F.; Luo, X. A hybrid scheme-based one-vs-all decision trees for multi-class classification tasks. *Knowl.-Based Syst.* **2020**, *198*, 105922. [CrossRef]
57. Barsacchi, M.; Bechini, A.; Marcelloni, F. An analysis of boosted ensembles of binary fuzzy decision trees. *Expert Syst. Appl.* **2020**, *154*, 113436. [CrossRef]
58. Saaty, T.L. *Fundamentals of Decision Making and Priority Theory with the Analytic Hierarchy Process*; RWS Publications: Pittsburgh, PA, USA, 2000; Volume 6.
59. Liu, Y.; Eckert, C.M.; Earl, C. A review of fuzzy AHP methods for decision-making with subjective judgements. *Expert Syst. Appl.* **2020**, 113738. [CrossRef]
60. Sangiorgio, V.; Uva, G.; Fatiguso, F. Optimized AHP to overcome limits in weight calculation: Building performance application. *J. Constr. Eng. Manag.* **2018**, *144*, 04017101. [CrossRef]
61. Plebankiewicz, E.; Kubek, D. Multicriteria selection of the building material supplier using AHP and fuzzy AHP. *J. Constr. Eng. Manag.* **2016**, *142*, 04015057. [CrossRef]
62. European Commission. Communication from the Commission to the European Parliament, the Council, the European Economic and Social Committee and the Committee of the Regions. EU Taxonomy, Corporate Sustainability Reporting, Sustainability Preferences and Fiduciary Duties: Directing Finance towards the European Green Deal. Brussels, 21.4.2021. COM(2021) 188 Final. 2021. Available online: https://ec.europa.eu/finance/docs/law/210421-sustainable-finance-communication_en.pdf (accessed on 1 April 2021).
63. Migliorelli, M. What Do We Mean by Sustainable Finance? Assessing Existing Frameworks and Policy Risks. *Sustainability* **2021**, *13*, 975. [CrossRef]
64. Huang, X. Optimal project selection with random fuzzy parameters. *Int. J. Prod. Econ.* **2007**, *106*, 513–522. [CrossRef]
65. Puška, A.; Beganović, A.; Šadić, S. Model for investment decision making by applying the multi-criteria analysis method. *Serb. J. Manag.* **2018**, *13*, 7–28. [CrossRef]
66. Lee, S.; Ahn, S.; Park, C.; Park, Y.J. Development of a resource allocation model using competitive advantage. *Sustainability* **2016**, *8*, 217. [CrossRef]
67. Stojcetovic, B.; Nikolic, D.; Velinov, V.; Bogdanovic, D. Application of integrated strengths, weaknesses, opportunities, and threats and analytic hierarchy process methodology to renewable energy project selection in Serbia. *J. Renew. Sustain. Energy* **2016**, *8*, 035906. [CrossRef]
68. Wang, W. The concept of sustainable construction project management in international practice. *Environ. Dev. Sustain.* **2021**, 1–23. [CrossRef]

69. European Commission. Proposal for a Directive Amending Directive 2013/34/EU, Directive 2004/109/EC, Directive 2006/43/EC and Regulation (EU) No 537/2014, as Regards Sustainable Corporate Reporting by Certain Undertakings, COM (2021) 189. 2021. Available online: https://ec.europa.eu/finance/docs/law/210421-proposal-corporate-sustainability-reporting_en.pdf (accessed on 1 April 2021).
70. Duong, L.N.K.; Wang, J.X.; Wood, L.C.; Reiners, T.; Koushan, M. The value of incremental environmental sustainability innovation in the construction industry: An event study. *Constr. Manag. Econ.* **2021**, 1–21. [CrossRef]
71. Wojewnik-Filipkowska, A.; Węgrzyn, J. Understanding of Public–Private Partnership Stakeholders as a Condition of Sustainable Development. *Sustainability* **2019**, *11*, 1194. [CrossRef]
72. Srinivasan, N.; Dhivya, S. An empirical study on stakeholder management in construction projects. *Mater. Today Proc.* **2020**, *21*, 60–62. [CrossRef]
73. Demirkesen, S.; Ozorhon, B. Impact of integration management on construction project management performance. *Int. J. Proj. Manag.* **2017**, *35*, 1639–1654. [CrossRef]
74. Li, H.X.; Patel, D.; Al-Hussein, M.; Yu, H.; Gül, M. Stakeholder studies and the social networks of NetZero energy homes (NZEHs). *Sustain. Cities Soc.* **2018**, *38*, 9–17. [CrossRef]
75. Zhao, Z.Y.; Zhao, X.J.; Davidson, K.; Zuo, J. A corporate social responsibility indicator system for construction enterprises. *J. Clean. Prod.* **2012**, *29*, 277–289. [CrossRef]
76. Freudenreich, B.; Lüdeke-Freund, F.; Schaltegger, S. A stakeholder theory perspective on business models: Value creation for sustainability. *J. Bus. Ethics* **2020**, *166*, 3–18. [CrossRef]
77. Chu, X.; Shi, Z.; Yang, L.; Guo, S. Evolutionary Game Analysis on Improving Collaboration in Sustainable Urban Regeneration: A Multiple-Stakeholder Perspective. *J. Urban Plan. Dev.* **2020**, *146*, 04020046. [CrossRef]
78. Vilutiene, T.; Ignatavičius, Č. Towards sustainable renovation: Key performance indicators for quality monitoring. *Sustainability* **2018**, *10*, 1840. [CrossRef]
79. Zheng, X.; Lu, Y.; Li, Y.; Le, Y.; Xiao, J. Quantifying and visualizing value exchanges in building information modeling (BIM) projects. *Autom. Constr.* **2019**, *99*, 91–108. [CrossRef]
80. Lin, X.; McKenna, B.; Ho, C.M.; Shen, G.Q. Stakeholders' influence strategies on social responsibility implementation in construction projects. *J. Clean. Prod.* **2019**, *235*, 348–358. [CrossRef]
81. Stackpole, C.S. *A User's Manual to the PMBOK Guide*; John Wiley & Sons: Hoboken, NJ, USA, 2013.
82. Zhang, Z.; Liao, H.; Al-Barakati, A.; Zavadskas, E.K.; Antuchevičienė, J. Supplier selection for housing development by an integrated method with interval rough boundaries. *Int. J. Strateg. Prop. Manag.* **2020**, *24*, 269–284. [CrossRef]
83. Khalil, N.; Kamaruzzaman, S.N.; Baharum, M.R. Ranking the indicators of building performance and the users' risk via Analytical Hierarchy Process (AHP): Case of Malaysia. *Ecol. Indic.* **2016**, *71*, 567–576. [CrossRef]
84. Kamaruzzaman, S.N.; Lou, E.C.W.; Wong, P.F.; Wood, R.; Che-Ani, A.I. Developing weighting system for refurbishment building assessment scheme in Malaysia through analytic hierarchy process (AHP) approach. *Energy Policy* **2018**, *112*, 280–290. [CrossRef]
85. Portal, O.S. Building Construction. 2019. Available online: https://osp.stat.gov.lt/informaciniai-pranesimai?articleId=6908417 (accessed on 13 May 2021).
86. Antunes, R.; Gonzalez, V. A production model for construction: A theoretical framework. *Buildings* **2015**, *5*, 209–228. [CrossRef]
87. Farokhad, M.R.; Otegi-Olaso, J.R.; Pinilla, L.S.; Gandarias, N.T.; de Lacalle, L.N.L. Assessing the success of R&D projects and innovation projects through project management life cycle. In Proceedings of the 2019 10th IEEE International Conference on Intelligent Data Acquisition and Advanced Computing Systems: Technology and Applications (IDAACS), Metz, France, 18–21 September 2019; Volume 2, pp. 1104–1110.
88. Vilutiene, T.; Kumetaitis, G.; Kiaulakis, A.; Kalibatas, D. Assessing the Sustainability of Alternative Structural Solutions of a Building: A Case Study. *Buildings* **2020**, *10*, 36. [CrossRef]

sustainability

MDPI

Article

Digitalization as a Strategic Means of Achieving Sustainable Efficiencies in Construction Management: A Critical Review

Bahareh Nikmehr [1], M. Reza Hosseini [1,*], Igor Martek [1], Edmundas Kazimieras Zavadskas [2] and Jurgita Antucheviciene [3]

1 School of Architecture and Built Environment, Deakin University, Geelong, VIC 3220, Australia; bnikmehr@yahoo.com (B.N.); igor.martek@deakin.edu.au (I.M.)
2 Institute of Sustainable Construction, Vilnius Gediminas Technical University, LT-10223 Vilnius, Lithuania; edmundas.zavadskas@vilniustech.lt
3 Department of Construction Management and Real Estate, Vilnius Gediminas Technical University, LT-10223 Vilnius, Lithuania; jurgita.antucheviciene@vilniustech.lt
* Correspondence: reza.hosseini@deakin.edu.au

Abstract: Construction is a complex activity, characterized by high levels of capital investment, relatively long delivery durations, multitudinous risks and uncertainties, as well as requiring the integration of multiple skills delivering a huge volume of tasks and processes. All of these must be coordinated carefully if time, cost, and quality constraints are to be met. At the same time, construction is renowned for performing poorly regarding sustainability metrics. Construction activity generates high volumes of waste, requires vast amounts of resources and materials, while consuming a significant proportion of total energy generated. Digitalization of the construction workplace and construction activities has the potential of improving construction performance both in terms of business results as well as sustainability outcomes. This is because, to put it simply, reduced energy usage, for example, impacts economic and "green" performance, simultaneously. Firms tinkering with digitalization, however, do not always achieve the hoped-for outcomes. The challenge faced is that a digital transition of construction firms must be carried out at a strategic level—requiring a comprehensive change management protocol. What then does a digital strategy entail? This study puts forward an argument for the combined economic and sustainability dividends to be had from digitizing construction firm activities. It outlines the requirements for achieving digitalization. The elements of a comprehensive digitalization strategy are cataloged, while the various approaches to developing a digitalization strategy are discussed. This study offers practitioners a useful framework by which to consider their own firm-level efforts at digitalization transition.

Keywords: digital transformation; digital technology; sustainability; strategy; construction management; change management

Citation: Nikmehr, B.; Hosseini, M.R.; Martek, I.; Zavadskas, E.K.; Antucheviciene, J. Digitalization as a Strategic Means of Achieving Sustainable Efficiencies in Construction Management: A Critical Review. *Sustainability* **2021**, *13*, 5040. https://doi.org/10.3390/su13095040

Academic Editor: Sunkuk Kim

Received: 23 March 2021
Accepted: 27 April 2021
Published: 30 April 2021

1. Introduction

The construction industry is one of the largest sectors of the global economy [1]. On a global scale, construction-related spending accounts for 13%, and the total annual revenue of the sector is estimated to be around $10 trillion, predicted to be up to $14 trillion by 2025 [1]. The construction industry has also one of the greatest economic spillover effects, namely, it represents an additional economic benefit of $2.86 for every $1 of construction Gross Domestic Product (GDP) [2]. As a result, even a slight improvement in the sector will carry huge positive implications for the national economy [2,3].

Despite its significance, the construction industry is still struggling with a wide range of problems: high construction costs; unsatisfactory project performance [4]; poor site safety records; low construction productivity; a lack of creativity and innovation; and above all poor sustainability outcomes [3,5–7]. Several developments offer the potential of mitigating these pitfalls, of which, the most promising is industry reform through digitalization [5,8,9].

Digitalization benefits any industry in various ways: greater convenience, lower prices, variety of choice, better information, enhanced sustainability, and the profitability of existing business models and investments [10,11]. Better business models, cost reduction, improved quality of communications, enhanced customer satisfaction, and so forth are other advantages of digital technologies. Indeed, these are interrelated where improvements across any economic, social and environmental outcomes collectively contribute to a more sustainable construction industry [12–15]. Given these potentials, digitalization has emerged as instrumental in driving sector change [16]. Traditional resistance to efforts at making construction more sustainable has been that it costs too much. However, the evidence shows that the potential added value of digitalization could be around $25 billion annually in the years 2017–2027, in Australia alone, while at the same time improving sustainability outcomes [8].

Digitalization can be simply defined as the use of digital technologies to change business models to increase revenue and value-producing opportunities for companies and businesses; the term refers to the process of moving to a digital business [17,18]. Digitalization is largely seen as a powerful intervention into the core business of companies and is associated with organization-wide modernization efforts affecting all structures, systems, and processes within companies [18,19]. Similarly, sustainability is widely defined as meeting the needs of the present without compromising the ability of future generations to meet their future needs. In practical terms, this breaks down into three mutually inclusive "bottom-line" measures. Present economic outcomes must be improved in ways that do not undermine future economic growth; present use of resources must be undertaken in ways that do not degrade the environment of the future; and social justice must be pursued such that nobody is left excluded, disadvantaged, or otherwise denied from an equitable share of accrued societal gains.

This paper argues that in order to successfully absorb the economic and sustainability benefits of digitalization into construction companies, appropriate digitalization adoption strategies must be proactively adopted. Thus, the factors and steps required in order to implant a successful digitalization strategy are here outlined. To this end, this paper begins by identifying the driving forces compelling the digitalization of construction firms. Next, the role of digitalization strategies along with the various approaches to digitalization available to construction companies are discussed. The paper concludes by describing the required steps for the development of a construction company-specific digitalization strategy able to meet economic and sustainability performance improvement outcomes simultaneously.

2. The Digital Transformation of the Construction Industry

Innovations comprise a wide range of transformative systems, from lean concepts through to information and communications technologies [20]. These innovations are enablers that reduce industrial process emissions and energy consumption from manufacturing construction components, to construction, operations, and building decommissioning. Their underlying intelligent operating platforms and automated solutions have the potential to optimize task outcomes, and in so doing greatly limit the otherwise adverse effects of traditional systems and processes regarding greenhouse gas emissions, pollution, and indeed even on potentially exploitative or hazardous work practices [21].

Construction companies seek remedial solutions to these issues [22], as a result of which, the field observes an increasing shift towards the use of digital technologies in the construction industry [23]. Evidence shows that they can benefit from various technological innovations in delivering projects [23]. Technological innovations can overcome a wide range of challenges that affect the construction industry, including cost overruns, rework, low project performance, poor safety records, substandard quality and undesired productivity [16]. And digital technologies offer solutions. These include vast capabilities offered by the Internet of Things (IoT) [24,25]; unmanned aerial vehicles (UAVs) [26,27]; 3D printing [28,29]; augmented reality (AR) [30,31]; virtual reality (VR) [32–34]; mixed reality (MR) [35,36]; Building Information Modeling (BIM) [37–45]; Artificial Intelligence

(AI); and intelligent decision support systems (DSS) [46–49]. At the organization level, digital engineering (DE) is proposed to complement the limitations of BIM beyond the boundaries of projects. So too, digital platforms for material procurement; robots; digital marketing; and digital tools for administration purposes, have profound positive impacts on the business front of companies [16,50].

There are many real-life examples and success stories of using various digital system in construction projects. Standing at the forefront of digitalization, there is compelling evidence for the advantages and benefits of using BIM in dealing with the complexities of multidisciplinary teams, identifying clashes, and reducing rework in and large-sized projects from around the globe, from Australia to Norway [51,52]. A coalescence of BIM and blockchain is proven effective in enhancing the effectiveness of managing financial transactions, enabling modern procurement methods, increase profit and cost savings [53]. In view of the full range of benefits, the construction sector has also embraced the use of UAVs in laying out sites; conducting remote and hazardous observations and surveying; risk-free site inspections; and safety monitoring [23,26]. Recently, the construction industry is observing a shift to the adoption of AI, with many benefits documented in the literature [20]. With AI, sophisticated algorithms are trained to learn from big data, and apply the acquired knowledge in revolutionizing industry practice, and improve productivity [54].

Although digitalization improves a wide range of construction business aspects that also ultimately produce better sustainable outcomes [50], digitalization adoption is fraught with challenges. Chief among these is that construction companies may need to consider changing almost every aspect of their business to achieve "digital transformation." Digital transformation is defined as *"a process that aims to improve an entity by triggering significant changes to its properties through combinations of information, computing, communication, and connectivity technologies."* [55] Digital transformation can be achieved in construction companies by focusing on five key activities, as illustrated in [50].

In the construction context, implementation of a digital technology relies heavily on various types of information: enough knowledge about the company such as the firm's structure, type of work, and the characteristics of human resources [56]. As illustrated in Figure 1, the digitalization of a construction firm is synonymous with successful digital transformation. This, in turn, requires the completion of five key activities, as described below:

- Integration of digital technologies into existing systems to exchange information among all construction market contributors.
- Improve business procedures with the aim of smoother exchange of data and information, control of products, and managing methods.
- Modification of organizational structures and human resources with the aim of choosing skillful workforces according to the digital transformation needs.
- Ensure that digital transformation is supported by all the staff and business contributors.
- Digital transformation investments must be assessed according to both financial and economic activities, not just economic ones.

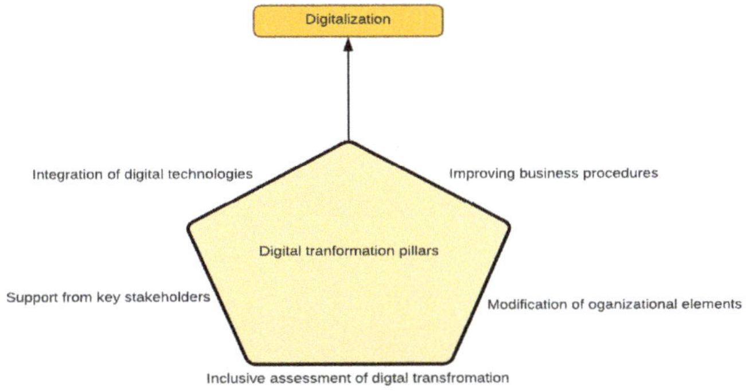

Figure 1. Digitalization and digital transformation activities.

Achieving the aims and objectives of each of these pillars requires the development of a robust strategy, as discussed next.

3. The Need for a Digital Transformation Strategy

The digitalization of an existing firm is much more difficult than the establishment of a new digital company [22]. Hence, there is a need to develop a suitable strategy as a solution for construction companies to identify the main objectives, roadmaps, relevant actions, and methods of assessment [19]. This strategy transforms several essential elements and dimensions of business, including customer experience, business procedures, operations, stakeholders, and networks [57]. A company without a digital transformation strategy completes some isolated and small-sized projects, with little effect, where much-needed resources are wasted [19].

Several definitions for a digital transformation strategy exist in the literature [58,59], yet no universal consensus has been reached to define it [60]. Simply, a digital transformation strategy is defined as the comprehensive vision of a firm in its move towards digitalization. To achieve this vision, a digital transformation strategy should include strategic measures; it should describe goals and tools for services, products, and value creation for a company/organization too [19]. Besides, adjustment of digital technologies' impacts and the nature of merging with internal firm's procedures and external firm's interfaces must be defined as essential elements of a digital transformation strategy.

Construction companies face many problems in developing a digital transformation strategy, mostly due to the novelty and complication of digitalization processes [61]. Familiarity with various dimensions of digital transformation and procedures of developing a successful digital transformation strategy are prerequisites, as discussed next.

4. Challenges in Developing Such a Strategy

The importance of digital transformation for companies, organizations, industries, and firms has resulted in the creation of a growing body of knowledge. However, recent studies suffer from a lack of a holistic approach to developing digital transformation strategies [62], where most existing studies represent a limited number of relevant factors and dimensions of digitalization. Bharadwaj, El Sawy, Pavlou, and Venkatraman [58], as will be discussed in great detail later on in the chapter, introduced four factors for providing a framework towards a digital transformation strategy for an organization: value creation, scope, scale, and speed of digital transformation. The speed of digital transformation is defined based on four factors: speed of product launching, speed of decision making, speed of supply chain orchestration, and speed of network formation and adaptation. Dimensions of a model for the development of a digital transformation strategy as provided by Matt, et al. [63] are the utilization of technology, structural changes, financial perspectives, and changes in value creation which can be defined as the effects of digital transformation on the companies' value chains due to the use of innovative technologies. These four items were believed to form a framework that informs companies in analyzing their existing capabilities, culminating in the development of a digital transformation strategy, as an ongoing procedure. Holotiuk and Beimborn [61] also developed a framework with eight dimensions: sales and customer experience; culture and leadership; abilities and human resources (HR) qualifications; forethought and vision; data and information technology (IT); functions; partners; and 40 critical success factors. Gimpel, et al. [64] proposed a framework of action fields following interviews with the chief digital officers from fifty organizations. This framework comprises six action fields: clients, value creation, functions, data, organization, and transformation management, in order to offer guidelines to engage in digital transformation.

Few studies have explored the development of a process for a digital transformation strategy. According to Schallmo, et al. [65], the integration of six steps results in the development of a digital transformation strategy. These are strategic principles, choices, forecasting, external and internal strategic analysis, and strategy formulization. Pflaum and Gölzer [66], also represent a four-stage process framework to facilitate the digital transformation of a firm, through the combination of top-down and bottom-up approaches. the steps are the business strategy step, knowledge creation, knowledge application, and the procedure of making decisions. However, they did not study the position of digital transformation strategy among the three levels of corporate, business, or functional strategies. As for finding the position of the digital transformation strategy, Lipsmeier, Kühn, Joppen, and Dumitrescu [19] argue that digital transformation strategy should be addressed at the corporate level. Hence, the digitalization of all business units should be aligned with the general strategic direction. The authors also conducted a process model for developing a digital transformation strategy as well as introducing the main factors of a digital transformation strategy. Albukhitan [67] also introduced a process for the development of a strategy by analyzing potential challenges that digital transformation attempts face, in the form of a process in six steps. Steps entailed identifying the vision, firms' digital transformation capability, customers and workforce experience, and analyzing and choosing alternative solutions, creating action plans, providing the required infrastructure and skilled human resources.

The construction industry is innately complex; it is project-based, unique in terms of high demand and supply variability [68]. Moreover, squeezed profit margins due to different forms of delays and accidents bring other challenges [69], where construction practitioners suffer from improper communication and issues with accountability [70]. In light of these challenges, providing a strategy to direct and inform the digitalization journey is of utmost importance, however, there are few studies that offer a strategy for digitalization for the construction industry. Among these, Stoyanova [71] provides suggestions for enhancing the likelihood of success in the digital transformation of construction firms. Koscheyev, Rapgof, and Vinogradova [50] introduce various dimensions of digital transformation. Ernstsen, et al. [72] represent three visions of efficient construction, user-data-driven built environment, and value-driven computational design for the digital transformation of construction firms in the UK.

5. Considerations in Formulating a Digital Transformation Strategy

Several studies have attempted to define the key dimensions of digital transformation, as the elements which need to be defined early in the process [58,63,73]. These are discussed next.

The scope of a digital business strategy includes references to products, business actions, and the functions of running a company. The scope of a digital business strategy defines the relationships between digital elements and companies, industries, IT infrastructures, and the external environment. Furthermore, it can help to facilitate the assessing of various impacts of digital technology on firms' business strategy [58].

The scale of a digital business strategy is used as a profitability driver. There are four ways that the scale of a digital business strategy can benefit a company: fast scale up or down according to dynamic market conditions, change of scale based on big data, rapid scale due to network effects, and better scale through alliances and partnership—sharing assets with other companies [58].

The speed of a digital business strategy has an important role in digital business management, recognized as an important item that can benefit firms in terms of strategic management. Speed should be considered through the speed of product launching, speed of making decisions, speed of supply chain arrangement, speed of network formation, and adaptation [58].

The utilization of technology refers to the attitude of the company to the innovative technology, and the capability of a firm to benefit from it [63].

Changes in value creation often occur with the application of innovative technologies. The digital business strategy brings increased value through information, multisided business models, coordinated business models in a network, and control of digital industry architecture [58,63].

As for structural changes, the application of digital activities results in improving products and services; however, they often need advanced technological skills and expose companies to various risks due to lack of experience in a new domain. Hence, utilization of various technologies and different forms of value creation require substantial structural changes to provide a sound basis for new activities. Structural changes consist of changes in a company's organizational arrangement, especially efforts to diffuse innovative digital function into all corporate elements [63].

Financial perspectives are needed to achieve all other dimensions. Companies that suffer from financial pressures may face difficulties in finding external financial ways to support digital transformation. Hence, companies need to consider digital transformation strategies alongside their resources and funding capacities [63].

6. Approaches to Defining a Digital Transformation Strategy

Two major approaches are suggested for digital transformation, comprising of top-down and bottom-up. The former, also referred to as strategy-driven, includes changes to the business model by using modernizers. This is a long-term approach that focuses on changing the existing value chain, value proposition, and revenue modeling. The bottom-up approach, or technology-driven approach, focuses on small or medium-sized changes in companies through using technology-driven tools and techniques. The objective of this approach is to drive slow improvements in productivity, employees' responsibility, and better customer experience and satisfaction [71]. The bottom-up approach is, however, inadequate for developing a successful digital transformation strategy, given the overarching impacts of digitalization on companies' fundamental components such as organizational structure, competencies, organizational procedures, and working culture. The bottom-up approach might have a detrimental impact on a company such as a productivity dip. The strategy-driven or top-down approach avoids such negative impacts and accelerates transformation. Nevertheless, developing a digital transformation strategy with a purely top-down approach is prone to some risks. These include increasing the likelihood of defining unrealistic objectives, scant attention to existing procedures, structures, initiatives, and a lack of buy-in from employees. The combination of these two approaches dominated by the bottom-up approach is suggested in the literature to tackle the problems facing digital transformation and speed up the process of digitalization [19,66].

Another important factor is the position of a digital transformation strategy among three levels of corporate, business, and functional levels, as illustrated in Figure 2.

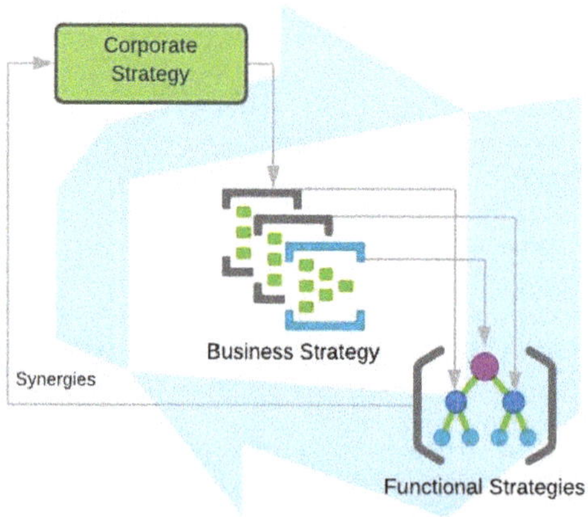

Figure 2. Strategy levels (adapter from Lipsmeier, Kühn, Joppen, and Dumitrescu [19]).

Corporate strategy corresponds to managing business units and the entire portfolio. The position of digital transformation strategy in relation to the corporate strategy can be considered in three different positions, including independent, being a part of it, or having the same weight as that of corporate strategy [65]. Most companies tend to develop their digital transformation strategy as a subsection of their corporate strategy [59]. With this, the strategy should be noted as an important consideration pertinent to the position of digital transformation. That is, digital transformation strategy should be defined at the corporate level and all other digitalization tools of business units should be aligned with the corporate strategy [19].

7. The Process of Developing a Digital Transformation Strategy

The development of a digital transformation strategy starts with defining a strategic direction according to digital guiding principles. This involves five components: digital vision, digital mission, digital policies, digital targets, and digital terms [19]. The first step is the definition of a strategic business vision for the proposed digital organization [66,67]. The strategic vision should consider long-term goals and short-term resources [67]. A digital vision can aim at the digital transformation of products and services or at value creation or both [19]. Vision should also involve business tools and digital use cases, based on business strategies and goals [66]. The vision for the digital transformation of construction firms has three dimensions, as follows [72]:

- Efficient construction focuses on expediting the construction procedure and enhancing efficiency. This can be achieved by concentrating on perspectives such as off-site construction, AI, BIM, lean construction, standardization, modularization, automation of design tasks, and alliancing business models.
- A user-data-driven built environment focuses on gathering real data, for instance, by IoT systems in the built environment. This vision requires the use of big data, IoT-based asset management, VR and 3D design, AR and maintenance, IoT-based energy utilization, sustainability, and health, and comfort of users.
- Value-driven computational design focuses on simulating various digital design alternatives and changing the design to satisfy various design criteria and clients' priorities within the construction procedure. This vision can be earned by concentrating on "digital fabrication on-site, gig economy, design simulations, blockchain, bespoke semi-automation, data-driven companies, distributed off-site production, and digital twin of the city."

The mission, pertinent to the digital transformation, however, seeks to find the reasons for the involvement of a company in digitalization; where digital policies propose regulations related to digital elements of a company such as digital management, digital initiative, data usage, information technology (IT), and safety, as well as the implementation of the digitalization process. Digital targets are initially described as qualitative values taken from digital vision, mission, and digital policies. In order to achieve the consistent realization of all these within a company, primary digital terms should be introduced [19].

After the identification of digital guiding principles, the next step is the assessment of the existing conditions of an organization in terms of digital transformation. To this end, systems, tools, and software applications should be assessed to evaluate their capabilities in fulfilling current and future requirements. The outcome of this step can facilitate decision-makers to understand which technologies, tools, and processes need to be improved [69]. Major tools for the assessment include market analysis and digital maturity assessment tools. Market analysis is of paramount importance. This tool provides the company with an up-to-date strategy [67]. A digital maturity tool should be provided to assess the framework of the IT infrastructure, organization, workforce, culture, partnering, technology, and functions, etc. in five levels of "unaware, conceptual, defined, integrated, and transformed" [66,67].

Subsequently, new systems and functions for facilitating employee jobs and clients' experiences should be provided. This can be achieved through proper use-cases of employees and new experiences for customers through digital technologies and platforms [67]. Use-cases should be prioritized and then implemented according to the allocated ranks [66].

A good tool for the structuring of required digital action plans—each functional area—is called "digital target" (Figure 3), which includes four factors. These are (1) digital vision which is a main objective of a functional area; (2) digital use-cases, which are digital tools for each working area; (3) strategic objectives, which should be determined in light of digitalization and aligns with digital guiding principles; (4) digital focus topics, extracted based on digital use cases, and work for communicating among functional areas and synchronizing primary actions in light of digitalization. Consequently, cross-functional topics are obtained according to the digital focus topics for each functional area. Cross-functional topics are extracted by the combination of digital focus topics, leading to determining the principal digitalization topics for the business unit. The qualitative goals are defined as cross-functional topics and then can be changed into quantitative values [19].

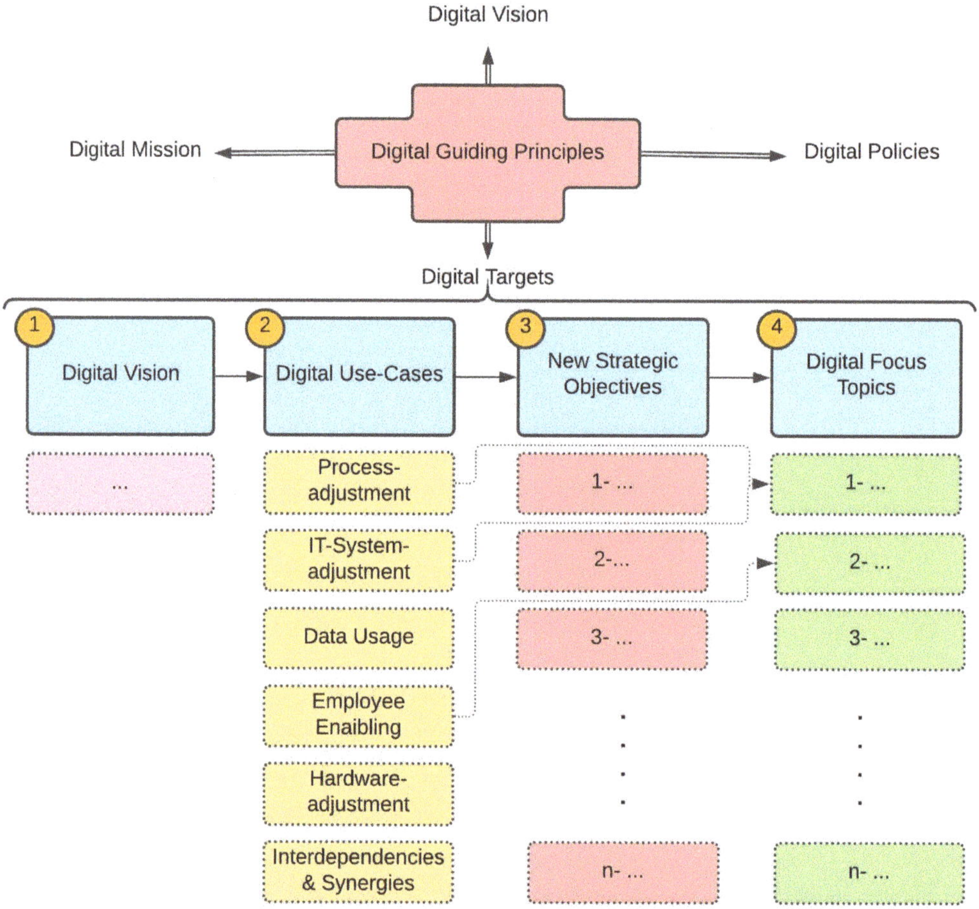

Figure 3. Digital target picture (adapted from Lipsmeier, Kühn, Joppen, and Dumitrescu [19]).

Knowledge creation is another step, which involves the modeling of all necessary data for possible problems related to use cases. The next step is knowledge application, in which, knowledge obtained from previous steps is analyzed by AI to forecast solutions for each use case [66]. Choosing the best solution that addresses digital objectives and people's experiences needs to be assessed according to their capabilities by a tool such as a comparison matrix of solutions [67].

The decision-making process is the last step, in which, the method of integration among knowledge-driven solutions and organizational decision processes is obtained (specification) [66]. Later, all of the digital objectives, solutions, and technologies are merged to form an action plan [67]. Vision, road map, and frameworks are also modified to be closer to the vision of a digital company [66].

It should be mentioned that firms need skilled staff members along the way towards digital transformation. These competent employees should work under the supervision of someone who possesses transformation leadership skills [63]. Therefore, preparation of the skillful human resources, with expertise in digital change management, is an important step for digital transformation [67].

8. Conclusions

Construction represents a significant proportion of the economy in both developing and developed nations. Yet, the profitability of construction firms, generally, remains below par when compared with other sectors within an economy. Moreover, construction activity is a major source of energy consumption, greenhouse gas emissions, and waste generation. Competitiveness and sustainability, therefore, are two key challenges the construction industry must face worldwide. Fortunately, there is a way forward and that is through the digitalization of the industry at the firm level. The other good news is that the efficiency gains to be had through the digitalization of construction firms can be expected to repair the poor sustainability record of the industry, as it currently operates. Specifically, cost reductions, efficiency gains, and better returns on assets that digitalization promises would be achieved through reduced energy consumption, better resource and material utilizations, and stronger control of waste, pollution, and carbon emissions.

However, meaningful digital transformation cannot be successfully realized without a strong commitment to such change. This means more is required than the frequently observed manner in which firms attempt to transition to a digitized business model; that is, employing IT experts to run introduced digitalization software on new hardware platforms, while also persisting with traditional forms of operations. Running parallel construction management practices within a firm—older analog systems and new digitized systems—has not been shown to be an effective means of evolving firms to full digitalization. What is needed is genuine change management. For that to be realized, a clear digitalization transition strategy must be formulated and implemented.

This study has documented the way forward in this regard. The need for change is argued. The quest for digitalization is shown to require firm-level digital transformation. The dimensions of a digital strategy are presented. Finally, the various approaches to embracing a digital transformation strategy are discussed. It is hoped that this paper will provide practitioners intent on making the transition to digitalization with a guiding framework for making the changes.

Author Contributions: Conceptualization, B.N. and M.R.H.; methodology, M.R.H. and I.M.; analysis of the literature, B.N.; writing—original draft preparation, B.N. and I.M.; writing—review and editing, M.R.H., J.A., and E.K.Z.; visualization, J.A.; supervision and quality control, E.K.Z. All authors have read and agreed to the published version of the manuscript.

Funding: This research received no external funding.

Institutional Review Board Statement: Not applicable.

Informed Consent Statement: Not applicable.

Data Availability Statement: Not applicable.

Conflicts of Interest: The authors declare no conflict of interest.

References

1. Barbosa, F.; Woetzel, J.; Mischke, J.; Ribeirinho, M.J.; Sridhar, M.; Parsons, M.; Bertram, N.; Brown, S. Reinventing Construction Through a Productivity Revolution. Available online: https://www.mckinsey.com/industries/capital-projects-and-infrastructure/our-insights/reinventing-construction-through-a-productivity-revolution (accessed on 12 September 2020).
2. Sategna, L.G.; Meinero, D.; Volontà, M. *Digitalising the Construction Sector*; Committee for European Construction Equipment: Brussels, Belgium, 2019.
3. Leviäkangas, P.; MokPaik, S.; Moon, S. Keeping up with the pace of digitization: The case of the Australian construction industry. *Technol. Soc.* **2017**, *50*, 33–43. [CrossRef]
4. Ghodoosi, F.; Bagchi, A.; Hosseini, M.R.; Vilutienė, T.; Zeynalian, M. Enhancement of bid decision-making in construction projects: A reliability analysis approach. *J. Civ. Eng. Manag.* **2021**, *27*, 149–161. [CrossRef]
5. Loosemore, M. Australia's Construction Industry Must Unite around a Cohesive Strategy. Available online: https://www.thefifthestate.com.au/innovation/building-construction/australias-construction-industry-must-unite-around-a-cohesive-strategy/ (accessed on 30 October 2020).
6. Fathalizadeh, A.; Hosseini, M.R.; Vaezzadeh, S.S.; Edwards, D.J.; Martek, I.; Shooshtarian, S. Barriers to sustainable construction project management: The case of Iran. *Smart Sustain. Built Environ.* **2021**. [CrossRef]
7. Hosseini, M.R.; Banihashemi, S.; Martek, I.; Golizadeh, H.; Ghodoosi, F. Sustainable Delivery of Mega projects in Iran: Integrated Model of Contextual Factors. *J. Manag. Eng.* **2018**, *34*, 05017011. [CrossRef]
8. Gruszka, A.; Jupp, J.R.; DeValence, G. Digital Foundations: How Technology Is Transforming Australia's Construction Sector. Available online: https://opus.lib.uts.edu.au/handle/10453/124861 (accessed on 30 October 2020).
9. Hampson, K.D.; Brandon, P. *Construction 2020—A Vision for Australia's Property and Construction Industry*; CRC Construction Innovation: Queensland University of Technology: Brisbane City, QLD, Australia, 2004.
10. The Productivity Commission Growing the Digital Economy in Australia and New Zealand: Maximising Opportunities for Small Medium Enterprises (SMEs). Available online: https://www.pc.gov.au/research/completed/growing-digital-economy (accessed on 30 October 2020).
11. Hilty, L.M.; Aebischer, B. Ict for sustainability: An emerging research field. In *ICT Innovations for Sustainability*; Springer: Berlin/Heidelberg, Germany, 2015; pp. 3–36.
12. Grubic, T.; Jennions, I. Remote monitoring technology and servitised strategies–factors characterizing the organizational application. *Int. J. Prod. Res.* **2018**, *56*, 2133–2149. [CrossRef]
13. Kaklauskas, A.; Zavadskas, E.K.; Binkyte-Veliene, A.; Kuzminske, A.; Cerkauskas, J.; Cerkauskiene, A.; Valaitiene, R. Multiple Criteria Evaluation of the EU Country Sustainable Construction Industry Lifecycles. *Appl. Sci.* **2020**, *10*, 3733. [CrossRef]
14. Nosratabadi, S.; Mosavi, A.; Shamshirband, S.; Zavadskas, E.K.; Rakotonirainy, A.; Chau, K.W. Sustainable business models: Areview. *Sustainability* **2019**, *11*, 1663. [CrossRef]
15. Stanujkic, D.; Popovic, G.; Zavadskas, E.K.; Karabasevic, D.; Binkyte-Veliene, A. Assessment of Progress towards Achieving Sustainable Development Goals of the "Agenda 2030" by Using the CoCoSo and the Shannon Entropy Methods: The Case of the EU Countries. *Sustainability* **2020**, *12*, 5717. [CrossRef]
16. Hosseini, M.R.; Martek, I.; Banihashemi, S.; Chan, A.P.; Darko, A.; Tahmasebi, M. Distinguishing characteristics of corruptionrisks in Iranian construction projects: A weighted correlation network analysis. *Sci. Eng. Ethics* **2020**, *26*, 205–231. [CrossRef] [PubMed]
17. Gartner Gartner Glossary: Digitalization. Available online: https://www.gartner.com/en/information-technology/glossary/digitalization (accessed on 5 January 2020).
18. Plekhanov, D.; Netland, T. Digitalisation stages in firms: Towards a framework. In Proceedings of the 26th EurOMA Conference, Helsinki, Finland, 17–19 June 2019.
19. Lipsmeier, A.; Kühn, A.; Joppen, R.; Dumitrescu, R. Process for the development of a digital strategy. *Procedia CIRP* **2020**, *88*, 173–178. [CrossRef]
20. Jafari, K.G.; Noorzai, E.; Hosseini, M.R. Assessing the capabilities of computing features in addressing the most common issues in the AEC industry. *Constr. Innov.* **2021**. [CrossRef]
21. Ghansah, F.A.; Owusu-Manu, D.-G.; Ayarkwa, J.; Edwards, D.J.; Hosseini, M.R. Exploration of latent barriers inhibiting project management processes in adopting smart building technologies (SBTs) in the developing countries. *Constr. Innov.* **2021**. [CrossRef]
22. Ianenko, M.; Ianenko, M.; Kirillova, T.; Amakhina, S.; Nikitina, N. Digital transformation strategies of trade enterprises: Key areas, development and implementation algorithms. In *IOP Conference Series: Materials Science and Engineering*; IOP Publishing: London, UK, 2020; Volume 940, p. 012051.
23. Elghaish, F.; Matarneh, S.; Talebi, S.; Kagioglou, M.; Hosseini, M.R.; Abrishami, S. Toward digitalization in the construction industry with immersive and drones technologies: A critical literature review. *Smart Sustain. Built Environ.* **2020**. [CrossRef]
24. Ghosh, A.; Hosseini, M.R.; Al-Ameri, R.; Kaklauskas, G.; Nikmehr, B. Internet of Things (IoT) for digital concrete quality control (DCQC): A conceptual framework. In *Proceedings of the 13th International Conference Modern Building Materials, Structures and Techniques*; VGTU Press: Vilnius, Lithuania, 2019.
25. Ghosh, A.; Edwards, D.J.; Hosseini, M.R. Patterns and trends in Internet of Things (IoT) research: Future applications in the construction industry. *Eng. Constr. Archit. Manag.* **2020**, *28*, 457–481. [CrossRef]
26. Golizadeh, H.; Hosseini, M.R.; Martek, I.; Edwards, D.; Gheisari, M.; Banihashemi, S.; Zhang, J. Scientometric analysis of research on "remotely piloted aircraft". *Eng. Constr. Archit. Manag.* **2019**, *27*, 634–657. [CrossRef]

27. York, D.D.; Al-Bayati, A.J.; Al-Shabbani, Z.Y. Potential Applications of UAV within the Construction Industry and the Challenges Limiting Implementation. In *Construction Research Congress 2020: Project Management and Controls, Materials, and Contracts, 2020*; American Society of Civil Engineers: Reston, VA, USA, 2020; pp. 31–39.
28. Romdhane, L.; El-Sayegh, S.M. 3D Printingin Construction: Benefits and Challenges. *Int. J. Struct. Civ. Eng. Res.* **2020**, *9*, 314–317. [CrossRef]
29. El-Sayegh, S.; Romdhane, L.; Manjikian, S. A critical review of 3D printing in construction: Benefits, challenges, and risks. *Arch. Civ. Mech. Eng.* **2020**, *20*, 34. [CrossRef]
30. Noghabaei, M.; Heydarian, A.; Balali, V.; Han, K. Trend Analysis on Adoption of Virtual and Augmented Reality in the Architecture, Engineering, and Construction Industry. *Data* **2020**, *5*, 26. [CrossRef]
31. Fenais, A.S.; Ariaratnam, S.T.; Ayer, S.; Smilovsky, N. A review of augmented reality applied to underground construction. *J. Inf. Technol. Constr.* **2020**, *25*, 308–324. [CrossRef]
32. Zhang, Y.; Liu, H.; Kang, S.-C.; Al-Hussein, M. Virtual reality applications for the built environment: Research trends and opportunities. *Autom. Constr.* **2020**, *118*, 103311. [CrossRef]
33. Lucas, J. Rapid development of Virtual Reality based construction sequence simulations: A case study. *ITcon* **2020**, *25*, 72–86. [CrossRef]
34. Ahmed, S. A review on using opportunities of augmented reality and virtual reality in construction project management. *Organ. Technol. Manag. Constr. Int. J.* **2018**, *10*, 1839–1852. [CrossRef]
35. Cheng, J.C.; Chen, K.; Chen, W. State-of-the-art review on mixed reality applications in the AECO industry. *J. Constr. Eng. Manag.* **2020**, *146*, 03119009. [CrossRef]
36. Rokhsaritalemi, S.; Sadeghi-Niaraki, A.; Choi, S.-M. A Review on Mixed Reality: Current Trends, Challenges and Prospects. *Appl. Sci.* **2020**, *10*, 636. [CrossRef]
37. Elghaish, F.; Abrishami, S.; Hosseini, M.R.; Abu-Samra, S. Revolutionising cost structure for integrated project delivery: A BIM-based solution. *Eng. Constr. Archit. Manag.* **2020**. [CrossRef]
38. Hamidavi, T.; Abrishami, S.; Hosseini, M.R. Towards intelligent structural design of buildings: A BIM-based solution. *J. Build. Eng.* **2020**, *32*, 101685. [CrossRef]
39. Khalesi, H.; Balali, A.; Valipour, A.; Antucheviciene, J.; Migilinskas, D.; Zigmund, V. Application of Hybrid SWARA–BIM in Reducing Reworks of Building Construction Projects from the Perspective of Time. *Sustainability* **2020**, *12*, 8927. [CrossRef]
40. Pavlovskis, M.; Migilinskas, D.; Antucheviciene, J.; Kutut, V. Ranking of heritage building conversion alternatives by applying BIM and MCDM: A case of Sapieha Palace in Vilnius. *Symmetry* **2019**, *11*, 973. [CrossRef]
41. Qin, X.; Shi, Y.; Lyu, K.; Mo, Y. Usinga TAM-TOE model to explore factors of Building Information Modelling (BIM) adoption in the construction industry. *J. Civ. Eng. Manag.* **2020**, *26*, 259–277. [CrossRef]
42. Wang, G.; Wang, P.; Cao, D.; Luo, X. Predicting behavioural resistance to BIM implementation in construction projects: Ane mpirical study integrating technology acceptance model and equity theory. *J. Civ. Eng. Manag.* **2020**, *26*, 651–665. [CrossRef]
43. Zhang, L.; Yuan, J.; Xia, N.; Ning, Y.; Ma, J.; Skibniewski, M.J. Measuring value-added-oriented BIM climate in construction projects: Dimensions and indicators. *J. Civ. Eng. Manag.* **2020**, *26*, 800–818. [CrossRef]
44. Vilutiene, T.; Kalibatiene, D.; Hosseini, M.R.; Pellicer, E.; Zavadskas, E.K. Building information modeling (BIM) for structural engineering: A bibliometric analysis of the literature. *Adv. Civ. Eng.* **2019**, *2019*, 5290690. [CrossRef]
45. Vilutiene, T.; Hosseini, M.R.; Pellicer, E.; Zavadskas, E.K. Advanced BIM applications in the construction industry. *Adv. Civ. Eng.* **2019**, *2019*, 6356107. [CrossRef]
46. Kaklauskas, A.; Abraham, A.; Dzemyda, G.; Raslanas, S.; Seniut, M.; Ubarte, I.; Kurasova, O.; Binkyte-Veliene, A.; Cerkauskas, J. Emotional, affective and biometrical states analytics of a built environment. *Eng. Appl. Artif. Intell.* **2020**, *91*, 103621. [CrossRef]
47. Stojčić, M.; Zavadskas, E.K.; Pamučar, D.; Stević, Ž.; Mardani, A. Application of MCDM methods in sustainability engineering: A literature review 2008–2018. *Symmetry* **2019**, *11*, 350. [CrossRef]
48. Zavadskas, E.K.; Antucheviciene, J.; Kar, S. Multi-objective and multi-attribute optimization for sustainable development decision aiding. *Sustainability.* **2019**, *11*, 3069. [CrossRef]
49. Fallahpour, A.; Wong, K.Y.; Rajoo, S.; Olugu, E.U.; Nilashi, M.; Turskis, Z. A fuzzy decision support system for sustainable construction project selection: An integrated FPP-FIS model. *J. Civ. Eng. Manag.* **2020**, *26*, 247–258. [CrossRef]
50. Koscheyev, V.; Rapgof, V.; Vinogradova, V. Digital transformation of construction organizations. In *IOP Conference Series: Materials Science and Engineering*; IOP Publishing: Saint-Petersburg, Russian, 2019; Volume 97, p. 012010.
51. Mignone, G.; Hosseini, M.R.; Chileshe, N.; Arashpour, M. Enhancing collaboration in BIM-based construction networks through organisational discontinuity theory: A case study of the new Royal Adelaide Hospital. *Archit. Eng. Des. Manag.* **2016**, *12*, 333–352. [CrossRef]
52. Merschbrock, C.; Hosseini, M.R.; Martek, I.; Arashpour, M.; Mignone, G. Collaborative role of sociotechnical components in BIM-based construction networks in two hospitals. *J. Manag. Eng.* **2018**, *34*, 05018006. [CrossRef]
53. Elghaish, F.; Abrishami, S.; Hosseini, M.R. Integrated project delivery with blockchain: An automated financial system. *Autom. Constr.* **2020**, *114*, 103182. [CrossRef]
54. Darko, A.; Chan, A.P.C.; Adabre, M.A.; Edwards, D.J.; Hosseini, M.R.; Ameyaw, E.E. Artificial intelligence in the AEC industry: Scientometric analysis and visualization of research activities. *Autom. Constr.* **2020**, *112*, 103081. [CrossRef]
55. Vial, G. Understanding digital transformation: A review and a research agenda. *J. Strateg. Inf. Syst.* **2019**, *28*, 118–144. [CrossRef]

56. Sackey, E.; Tuuli, M.; Dainty, A. Sociotechnical systems approach to BIM implementation in a multi-disciplinary construction context. *J. Manag. Eng.* **2015**, *31*, A4014005. [CrossRef]
57. Ismail, M.H.; Khater, M.; Zaki, M. Digital business transformation and strategy: What do we know so far. *Camb. Serv. Alliance* **2017**, *10*. [CrossRef]
58. Bharadwaj, A.; ElSawy, O.A.; Pavlou, P.A.; Venkatraman, N. Digital business strategy: Toward a next generation of insights. *MIS Q.* **2013**, *37*, 471–482. [CrossRef]
59. Schallmo, D.; Williams, C.A.; Lohse, J. Clarifying Digital Strategy–Detailed Literature Review of Existing Approaches. In *ISPIM Conference Proceedings, 2018*; The International Society for Professional Innovation Management (ISPIM): Stockholm, Sweden, 2018; pp. 1–21.
60. Dang, D.; Vartiainen, T. Digital strategy patterns in information systems research. In Proceedings of the PACIS 2019 Proceedings, Xi'an, China, 8–12 July 2019.
61. Holotiuk, F.; Beimborn, D. Critical success factors of digital business strategy. In Proceedings of the Track 9—Business Innovations and Business Models, St. Gallen, Switzerland, 12–15 February 2017.
62. Korachi, Z.; Bounabat, B. General Approach for Formulating a Digital Transformation Strategy. *J. Comput. Sci.* **2020**, *16*, 493–507.
63. Matt, C.; Hess, T.; Benlian, A. Digital transformation strategies. *Bus. Inf. Syst. Eng.* **2015**, *57*, 339–343. [CrossRef]
64. Gimpel, H.; Hosseini, S.; Huber, R.X.R.; Probst, L.; Röglinger, M.; Faisst, U. Structuring Digital Transformation: A Framework of Action Fields and its Application at ZEISS. *J. Inf. Technol. Theory Appl.* **2018**, *19*, 31–54.
65. Schallmo, D.; Williams, C.A.; Lohse, J. Digital Strategy—Integrated Approach and Generic Options. *Int. J. Innov. Manag.* **2019**, *23*, 1940005. [CrossRef]
66. Pflaum, A.A.; Gölzer, P. The IoT and digital transformation: Toward the data-driven enterprise. *IEEE Pervasive Comput.* **2018**, *17*, 87–91. [CrossRef]
67. Albukhitan, S. Developing Digital Transformation Strategy for Manufacturing. *Procedia Comput. Sci.* **2020**, *170*, 664–671. [CrossRef]
68. Morris, P.W. Project management in the construction industry. *Wiley Guide Manag. Proj.* **2004**, 1350–1367.
69. Behera, P.; Mohanty, R.P.; Prakash, A. An investigation of implementation issues, process phases and knowledge areas of project management in the performance of construction supply chains. *Int. J. Proj. Organ. Manag.* **2018**, *10*, 137–157. [CrossRef]
70. Patanakul, P.; Kwak, Y.H.; Zwikael, O.; Liu, M. What impacts the performance of large-scale government projects? *Int. J. Proj. Manag.* **2016**, *34*, 452–466. [CrossRef]
71. Stoyanova, M. Good Practices and Recommendations for Success in Construction Digitalization. *TEM J.* **2020**, *9*, 42–47.
72. Ernstsen, S.N.; Whyte, J.; Thuesen, C.; Maier, A. How Innovation Champions Frame the Future: Three Visions for Digital Transformation of Construction. *J. Constr. Eng. Manag.* **2020**, *147*, 05020022. [CrossRef]
73. Mitroulis, D.; Kitsios, F. Digital Transformation Strategy: A literature review. In Proceedings of the 6th National Student Conference of HELORS, Xanthi, Greece, 28 February–2 March 2019; pp. 59–61.

sustainability

MDPI

Article

Environmental Performance of Residential Buildings: A Life Cycle Assessment Study in Saudi Arabia

Hatem Alhazmi [1], Abdulilah K. Alduwais [1], Thamer Tabbakh [2], Saad Aljamlani [3], Bandar Alkahlan [3] and Abdulaziz Kurdi [3,*]

1 National Center for Environmental Technology, King Abdulaziz City for Science and Technology, P.O. Box 6086, Riyadh 11442, Saudi Arabia; halhazmi@kacst.edu.sa (H.A.); aaldowas@kacst.edu.sa (A.K.A.)
2 Material Science Institute, King Abdulaziz City for Science and Technology, P.O. Box 6086, Riyadh 11442, Saudi Arabia; ttabbakh@kacst.edu.sa
3 National Center for Building and Construction Technology, King Abdulaziz City for Science and Technology, P.O. Box 6086, Riyadh 11442, Saudi Arabia; saljamlani@kacst.edu.sa (S.A.); alkahlan@kacst.edu.sa (B.A.)
* Correspondence: akurdi@kacst.edu.sa

Abstract: The building and construction sector has a huge impact on the environment because of the enormous amounts of natural resources and energy consumed during the life cycle of construction projects. In this study, we evaluated the potential environmental impact of the construction of a villa, from cradle to grave, in the Saudi Arabian context. Centrum voor Milieukunde Leiden (CML) for Centre of Environmental Science of Leiden University-IA baseline v3.03 methods were used to obtain the environmental profile for the impact categories, and Cumulative Energy Demand v1.09 was used to measure the embodied energy of the villa life cycle. The analyzed midpoint impact categories include global warming (GWP100a), ozone layer depletion (ODP), acidification (AP), eutrophication (EP), photochemical oxidation (POCP), and indicator cumulative energy demand (CED). The operation use phase of the villa was found to have the highest global warming potential and acidification with 2.61×10^6 kg CO_2-eq and 1.75×10^4 kg SO_2-eq, respectively. Sensitivity analysis was performed on the Saudi Arabian plans to increase the share of renewable sources and reduce the amount of electricity generated from hydrocarbons, which currently represents 46% of the total installed power, by 2032. The results showed that compared with the current electricity environmental impact, the CO_2 emission from electricity will decrease by 53%, which represents a significant reduction in environmental impact. The findings will help with the life cycle assessment of structures during future planning and for energy conservation.

Keywords: sustainability; buildings; life cycle assessment; materials; greenhouse

Citation: Alhazmi, H.; Alduwais, A.K.; Tabbakh, T.; Aljamlani, S.; Alkahlan, B.; Kurdi, A. Environmental Performance of Residential Buildings: A Life Cycle Assessment Study in Saudi Arabia. *Sustainability* **2021**, *13*, 3542. https://doi.org/10.3390/su13063542

Academic Editors: Edmundas Kazimieras Zavadskas, Jurgita Antuchevičienė, M. Reza Hosseini and Igor Martek

Received: 22 February 2021
Accepted: 19 March 2021
Published: 23 March 2021

1. Introduction

The global focus on sustainability has increased in recent years given the dangers posed by climate change, global warming, and environmental degradation. Over 85% of the world's primary energy needs are still met using fossil fuels, making them the most significant contributors to greenhouse gas (GHG) emissions [1]. Overall, the efficient and effective use of energy and materials is needed across sectors. The United Nations Sustainability Development Goals for 2030 consolidate many of these challenges and highlight the need for inclusive development through building sustainability, resource conservation, and innovation in development [2]. The building sector is no exception to sustainable development. In developed economies, such as those of the United States and the European Union, buildings account for nearly 40% of all primary energy consumption [3]. An extensive study [4] across building types and climate conditions in the United States showed that interventions in the building sector can result in average energy savings of 29%, thus significantly reducing the overall emissions. In the European Union, the building construction industry annually consumes nearly half of all raw materials and one-third of the water used while generating 25–30% of the waste [5]. Better construction and other sustainable interventions could lead to a 42% reduction in final energy consumption and a reduction of over 35% in greenhouse gas emissions [6]. Similar potential savings can be achieved by countries around the globe. As a signatory to the Paris Agreement [7], Saudi Arabia is committed to reducing its greenhouse emissions through multiple interventions involving renewable energy, carbon capture, and energy efficiency management [8]. The building sector in Saudi Arabia consumes large quantities of materials and energy, with contracts estimated at $52.6 billion awarded in 2019 alone [9]. In Saudi Arabia, buildings consume nearly 80% of the overall electricity generated, with residential buildings accounting for 50% of the total [10]. Hence, there is room for significant energy and emission savings within this sector.

However, it is necessary to comprehend how buildings consume energy and resources throughout their lifetime to identify potential energy-saving interventions. Life cycle assessment (LCA) is an approach commonly employed in this context. This method allows the researcher to study the energy and resource consumption of a certain building starting from the stage of resource extraction up to the demolition of the building and waste management at the end of a building's life cycle [11]. Insights gained from the LCA can lead to optimized resource and energy use at all stages of a building's life cycle, leading to building LCA becoming a distinct area within the practice of life cycle assessment. Building LCA can help tackle specific characteristics that are unique to the construction industry. For example, the choice and sourcing of raw materials needed for construction can impact the energy and environmental footprint of a building. This includes the environmental degradation and energy footprint associated with the extraction, processing, packaging, and transportation of these materials [12]. This preconstruction phase is followed by construction, which generates significant waste and pollution. Because buildings have a long-life cycle, the next operational phase, in general, accounts for the most energy consumption of all phases. Studies estimate this value to be between 40% and 90% of the life cycle energy consumption depending on climatic conditions and usage habits [13,14]. This also includes any impact associated with the building maintenance operations. At the end of the building's life cycle, demolition activities consume energy and generate waste that can be recycled, reused, or sent to landfills. These impact the overall life cycle assessment of buildings [15]. All these stages are analyzed in a complete building LCA. A multitude of such studies can be found in the literature; a few pertaining to residential buildings is discussed below.

Life cycle assessments of residential buildings have been conducted for multiple climatic and economic conditions [16]. A study on various types of residential buildings (multifamily dwellings and single-family dwellings) in Brazil was conducted by Evangelista et al. [17]. The study found that single-family dwellings often have a higher potential environmental impact than multifamily dwellings of similar sizes and standards. In addition, for the same family size, high-standard dwellings have higher environmental costs. The study found that some aspects such as structures, foundations, and coatings have higher environmental costs than others, and that the operational phase is responsible for 80% of the energy demand. This is similar to other reported findings [18] on a three-bedroom house in Scotland, identifying concrete, timber, and tiles as the most energy-intensive materials used in its construction. They are extensively used for foundations, structures, and interior coatings. Evangelista et al., however, did not consider many options for the demolition phase and assumed that the entire building would end up in landfill. A similar study on a single-family home in Sweden showed that production stage and maintenance operations accounted for the largest footprint (67%), while the operational and end-of-life phases together accounted for less than 12% of GHG emissions from the building. This study, however, was extremely subjective as most of Sweden's electricity comes from renewables, and the house was a wooden construction [19]. A study on Canadian residential buildings [20] found a linear correlation between the operational energy footprint and the overall energy footprint of buildings regardless of their differences, much like in another study [18] where high-rise multifamily housing units performed better than single dwellings and low-rise apartments. The relatively high energy footprint associated with the operational phase was also highlighted, in agreement with the results obtained for Canadian houses [21]. The number of studies on buildings from the Middle East [22], Africa [23], and South Asia [24] is limited; much of the literature in this field is restricted to China [25], North America, and Europe [26]. The environmental loads associated with each phase of the entire life cycle of the residential building as defined by the European Committee for Standardization (EN 15804) [27,28]. In 2017, the Saudi Arabian government prepared a strategy called the 2030 Vision. The objective of the Saudi Vision 2030 (SV2030) is to set up renewable and sustainable energy (RnSE) projects to meet the electricity demand—which is expected to surpass 120 GW by 2032—by increasing the use of renewable resources, reducing dependency on fossil fuels, and reducing the country's CO_2 emissions. Concluding the existing research, no building life cycle assessment study, from cradle to grave, has been conducted in the context of Saudi Arabia. Therefore, to fill this gap, an LCA of a residential building in Saudi Arabia should be conducted.

The literature discussed so far shows that LCA can be a valuable tool for optimizing energy consumption in buildings. Given the size of Saudi Arabia's construction industry, significant energy savings can be achieved by better understanding this sector. However, there is a lack of examples in the literature of building LCAs (or a cradle-to-grave study of building energy consumption) pertaining specifically to Saudi Arabia; most studies are limited to Europe and North America. Because residential buildings consume 50% of the electricity generated in Saudi Arabia, in this study, LCA was used to better understand the energy footprint and environmental impact of a typical residential villa. Information obtained from this study will aid industry professionals and government agencies in incorporating environmental health and sustainability into planning and construction.

The aim of this study is to understand the potential environmental impact caused by the whole life cycle of a typical residential building (villa) in the Saudi Arabian context. The reference building taken into consideration is a Saudi Arabian villa built in the capital city, Riyadh, using the latest standards in construction techniques and conventional materials normally used in the local context. Thus, this study focuses on evaluating the potential environmental impact of a villa (a typical Saudi Arabian residential building) in five impact categories and one life cycle indicator: global warming (GWP100a), ozone layer depletion (ODP), acidification (AP), eutrophication (EP), photochemical oxidation (POCP), and indicator cumulative energy demand (CED). This study will help to analyze the performance of Villa buildings with reference to Life cycle assessment implementation.

2. Materials and Methods

The LCA methodology was used to evaluate the environmental impact of a typical residential building in Saudi Arabia considering the whole life cycle, from cradle to grave. The attributional LCA was conducted according to International Organization for Standardization (ISO) 14040 [29] and ISO 14044 [30]. SimaPro software version 9.1 was used to model the LCA [31]. The methods used to obtain the environmental profile of the average villa life cycle included CML-IA baseline v3.03 [32,33] for impact categories and Cumulative Energy Demand v1.09 to calculate the embodied energy in the life cycle of the villa. The CML methodology, developed by the Center of Environmental Science of Leiden University, is widely accepted; EN 15804 [27,28], the core standard for products categorized as construction products, takes the characterization factors from this method and allows for the comparison of results with those of other LCA studies. The midpoint impact categories that were analyzed with CML-IA baseline v3.03 methods were global warming (GWP100a), ozone layer depletion (ODP), acidification (AP), eutrophication (EP), photochemical oxidation (POCP), and indicator cumulative energy demand (CED) following the Cumulative Energy Demand v1.09 method. The villa was modeled using Revit software (Chetu, Plantation, FL, USA), the widely used building information modeling software. Data were obtained from local construction firms via questionnaires and interviews, ensuring the villa is representative of the current average residential building in Saudi Arabia. The Ecoinvent version 3.2 database [34] was used to model upstream processes and is globally recognized as one of the most consistent Life Cycle Inventory (LCI) databases available. The step wise methodology of this research can be expressed as:

Step.1: Selection of Case Study Area-Villa
Step.2: Description of Villa Characteristics
Step.3: Defining the System Boundaries
Step.4: Life Cycle Inventory and Assumptions
Step.5: Results and Assessment
Step.6: Decision Making.

2.1. Selection of Case Study Area-Villa

This study is designed to assess the potential environmental impact caused by the life cycle of a single-family house, called a villa, in the Saudi Arabian context. Because all stages are specific to the Saudi Arabian context, a comparison with similar buildings located in different regions was performed. Finally, because Saudi Arabia is committed to reducing its greenhouse emissions through multiple interventions—one of which is implementing renewable energy—a sensitivity analysis was conducted to assess how the reduction in the electricity impact would affect the building's life cycle. Similar to other studies [35–37] and the principles of Product Category Rule (PCR) 2014:02 for buildings [38], in this investigation, the functional unit (FU) is a villa with a total gross floor area (GFA) of 387 m^2 and a lifespan of 50 years.

2.2. Description of Villa Characteristics

The assessed villa is an average single-family building with a concrete-based structure. The GFA is 387 m^2. It is a two-floor villa with an open space on the second floor and five bedrooms and bathrooms. The building plans are provided in Figures 1–3.

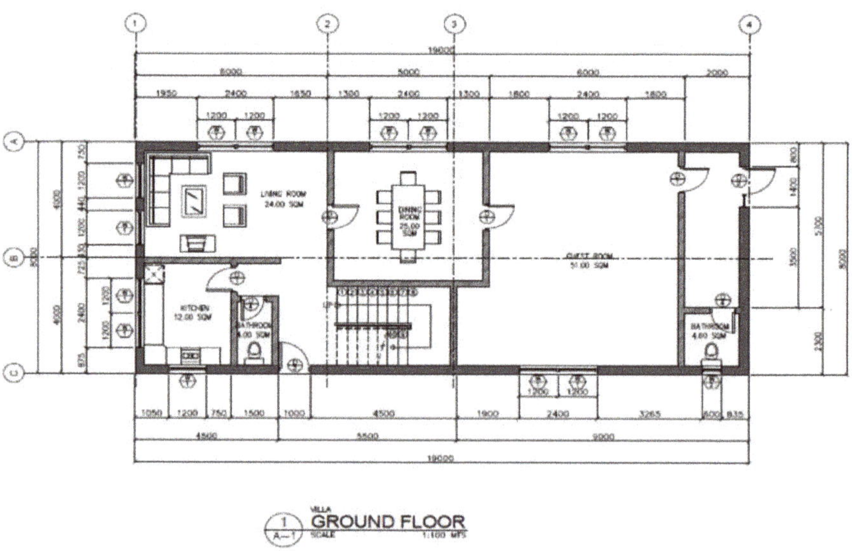

Figure 1. Villa ground floor plan.

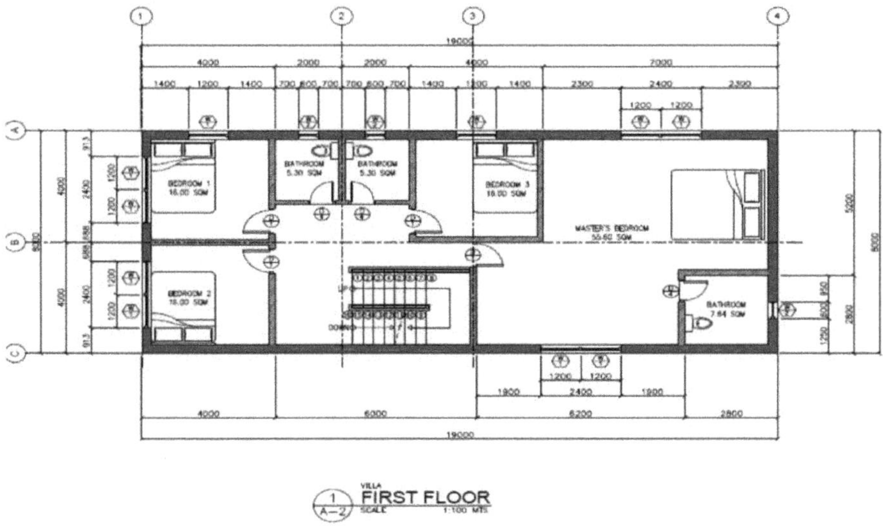

Figure 2. Villa first floor plan.

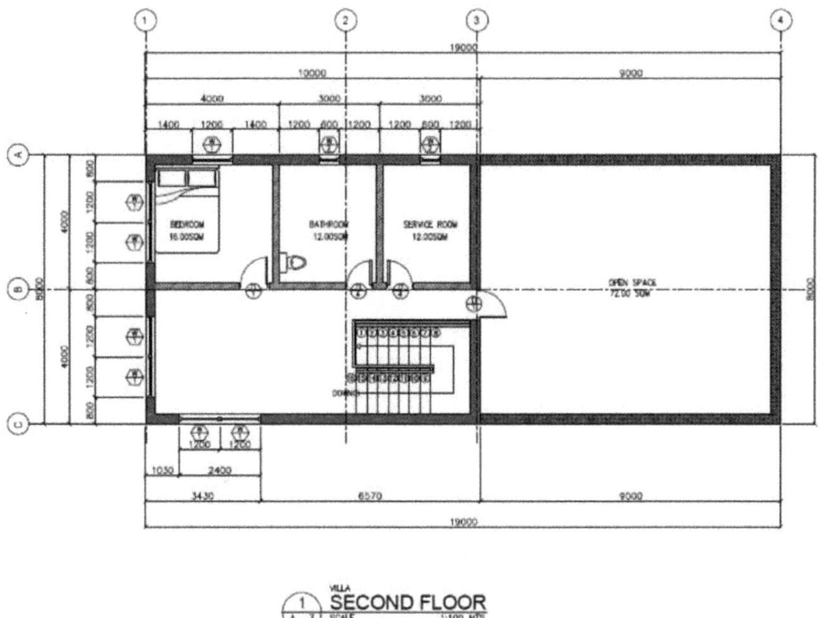

Figure 3. Villa second floor plan.

The villa was built mainly with concrete. In Table 1, the building components are described in terms of materials and quantities. Both the foundation and structure were made of reinforced concrete, whereas walls, both internal and external, were built with concrete blocks. The external concrete blocks contain extruded polystyrene (XPS), which provides thermal insulation to the building. Other materials used in villa construction were ceramic tiles, cement tiles, bitumen to provide waterproofing, gypsum plasterboards for ceilings, and paint, among others. The domestic appliances, such as washing machines, refrigerators, cooking appliances, and heating, ventilation, and air conditioning (HVAC), were excluded in line with EN 15804 as they represent less than 1% of the total mass input in the construction stage.

Table 1. Villa Components and Material Inventory.

Building Component	Component/Material	Units	Quantity
Foundation	Reinforced concrete slab on grade	m^3	20.75
Structure	Reinforced concrete	m^3	138.7
Roof and Open space	Layers: 20 mm cement tile + 40 mm mortar layer + 50 mm XPS + 10 mm bitumen sheet + 4 mm bitumen coating	m^2	138.60
Ceiling	12.5 mm gypsum board on metal furring + paint	m^2	322.81
Exterior walls and parapet	400 mm × 200 mm × 300 mm concrete block with insulation + adhesive mortar + 2 faced 20 mm cement plaster + ladder mesh	m^2	303.3
	200 mm parapet wall	m^2	23
	Paint	m^2	519.65
Internal walls	400 mm × 200 mm × 150 mm hollow concrete block + adhesive mortar + 2 faced 20 mm cement plaster + ladder mesh	m^2	231
	Paint	m^2	821.60
Floor and wall tiles	Dry areas: 10 mm ceramic tile + 40 mm mortar	m^2	302.2
	Wet areas: 10 mm ceramic tile + 40 mm mortar + 2 layered 5 mm bitumen sheet + 4 mm bitumen coating	m^2	30.78
	Wall tiles: 10 mm ceramic tile + 4 mm bitumen coating	m^2	81.258
Windows	Double glazed window with aluminum frame	windows	37
Doors	Steel door	doors	2
	Wood door	doors	16
Stairs	Welded tubular stainless steel	m	13.5
Electrical network	Cooper wire	m	360

2.3. Defining the System Boundaries

A cradle-to-grave evaluation was conducted for the whole life cycle of the villa within the system boundaries defined in Figure 4. Figure 4 shows the life cycle phases of the constructed building in conformance with UNE 15804 [25].

The pre-use phase consisted of the subphases of material production comprising raw material supply, transportation, and manufacturing (modules A1-A3 of EN 15804), and that of building construction, which consisted of the transport and assembly of components, energy consumption related to land soil preparation and excavation, and building material waste generation (modules A4–A5 of EN 15804). During the use stage, the operational use of energy and water are considered along with the repainting of the building and replacement of the floor (modules B6, B7, B2, and B4 of EN 15804). At the end of their life cycles, buildings are demolished, and building materials are transported and managed into landfill (modules C1, C2, and C4 of EN 15804) [39].

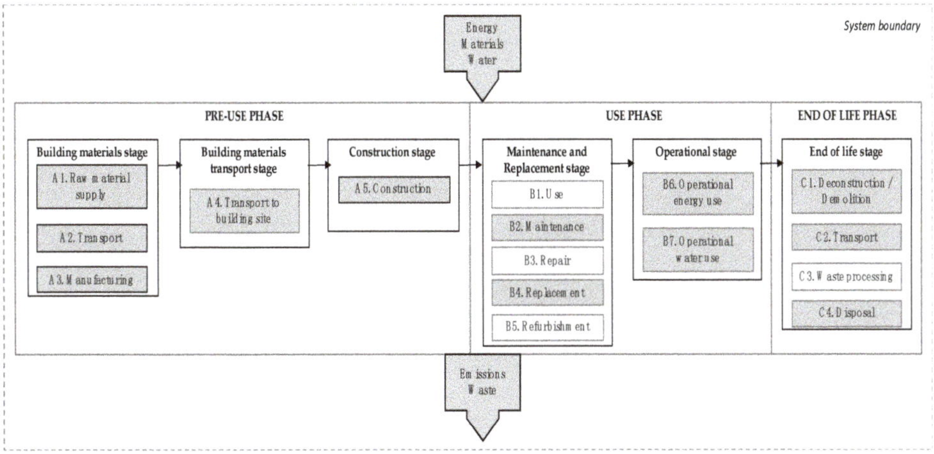

Figure 4. Life cycle assessment (LCA) system boundary based on EN 15804 modularity (included modules are shaded) and stages defined for the villa assessment.

The use phase includes any emissions to the environment (module B1); technical operations on the building: maintenance, repair, replacement, and refurbishment (respectively module B2 to B5); and operation of the building, divided into operational energy use (module B6) and operational water use (module B7). Only maintenance and replacement operations and operational energy and water use are considered relevant for the villa use phase.

At the end of its life, the entire building is deposited as waste in landfill, which means that the C3 module of waste processing is not relevant in this system.

Other life cycle processes were omitted because they account for less than 1% of the total environmental impact, and data availability was limited for infrastructure, construction, production equipment, and tools that are not directly consumed in the construction process; as well as for employee-related activities such as transport to and from work, packaging of construction products and packaging waste produced during the A5 module, communication installations, villa equipment, HVAC, and lamps.

2.4. Life Cycle Inventory and Assumptions

The inputs and outputs used to calculate the environmental impact of the average villa were compiled from the building's bill of materials. Specific data collected from local construction firms via questionnaires and interviews were used to model each life cycle stage and taken as representative of the Saudi Arabian construction process for this type of building. Generic data that were not based on measures or direct calculations for the specific processes or stages were obtained from the Ecoinvent version 3.2 database. The hypothesis of the Ecoinvent database was assumed, even though some processes were adapted to the Saudi Arabian context. Detailed process data were considered in this study during the life cycle for each material during manufacturing, transportation, and disposal [40,41].

2.4.1. Building Materials Stage

Quantities of materials specified in the bill of materials were used to model the building materials stage (Table 1). Because no specific information was available on the manufacturing of the construction products in Saudi Arabia and previous stages, the Ecoinvent database was used. The datasets were modified to include the Saudi Arabian electricity mix and water supply as recommended by other studies [40].

2.4.2. Building Materials Transport Stage

The scenario for the transportation stage was set from the local construction sector experience. The transportation distance was assumed to be 50 km from the manufacturer to the construction site for all materials because all materials would have been obtained from Riyadh's second industrial city. Materials transportation from manufacturers to the construction site was assumed to be carried by an average 16–32 ton truck.

2.4.3. Construction Stage

In the construction phase, waste material was added as an additional 10% of the overall quantity for the bill of quantities because during this stage, the wastage rate is assumed to be 10%. For this additional 10%, the same assumptions were used for the building materials and transportation stages. In Saudi Arabia, the common practice for building waste is the landfill process. Hence, the construction waste was assumed to be transported 50 km for disposal. Waste generated via transportation and management was assumed to be carried by an average 16–32 ton truck.

In the construction stage, the soil preparation and excavation were included based on the bill of materials information.

2.4.4. Maintenance and Replacement Stage

The proposed scenario for the use stage, which refers to technical operations, covers both maintenance and replacement. Based on the service life of components and materials and the building lifespan, the external walls will be repainted twice, while internal walls will be repainted three times. The replacement tasks cover the replacement of floor tiles and all layers that conform to the floor (twice during the building lifespan) and wall tiles (once) along with the required materials. The materials used for replacement and maintenance were assumed to be transported 50 km by an average of 3.5–7.5-ton truck. The replacement materials waste was assumed to be transported and disposed of. The same scenario defined in the construction stage was used in this stage.

2.4.5. Operational Stage

The energy consumption for the villa was collected by Energy Plus software conducting a one-year simulation based on the villa's characteristics. The data obtained and used for the scenario in the operational stage are shown in Table 2. It was assumed that the villa would be occupied by six people and the temperature inside the villa would be 21.3 °C for comfort. The most demanding uses are cooling, accounting for 62% of the electricity demand, and interior lighting, accounting for 14% of the consumption.

Table 2. End-use energy consumption (one-year simulation). GFA, gross floor area.

End Use	Electricity Consumption (kWh/year)
Cooling	29,221.30
Interior Lighting	6367.27
Exterior Lighting	2185.08
Interior Equipment	3719.76
Fans	1791.72
Pumps	0.57
Water Systems	3743.88
Total	47,029.58
Total/GFA (kWh/m^2)	121.52

Water is consumed during the operational stage. The tap water Ecoinvent dataset was modeled to consider part of the water supply coming from groundwater and the other part from seawater in Saudi Arabia.

2.4.6. End-of-Life Stage

In Saudi Arabia, the common practice for Construction and demolition waste (CDW) is the landfill process. Hence, the scenario that models the end-of-life stage states that the building is dismantled and all building materials are transported by truck and disposed of.

3. Results and Discussion

The life cycle of the villa was divided into the following stages as aligned with the European core rules for the product category of construction products EN 15804 and the construction sector: product (building materials), building materials transport, construction, operational, maintenance and replacement, and end of life. The midpoint impact categories that were analyzed with CML-IA baseline v3.03 methods were global warming (GWP100a), ozone layer depletion (ODP), acidification (AP), eutrophication (EP), photochemical oxidation (POCP), and indicator cumulative energy demand (CED) following the Cumulative Energy Demand v1.09 method.

3.1. Life Cycle Impact Assessment

3.1.1. Main Findings

Table 3 presents the results of the life cycle of the villa from a cradle-to-grave perspective for the functional unit and m^2 (GFA). Figure 5 depicts the contribution of each life cycle stage. The results show that for all impact categories, the operational use stage is the most important stage, with a contribution ranging from 91% (photochemical oxidation) to 96% (ozone depletion and acidification). A total of 7.26 tons CO_2-eq per m^2 (GFA) were potentially emitted during the villa life cycle, and 6.76 tons CO_2-eq were from the operational use stage, electricity, and water consumed over 50 years. The impact of the operational stage was obtained mainly from electricity consumption, which was modeled on an annual basis using Energy Plus. The obtained data show that 47,030 kWh was consumed annually, representing 122 kWh/m^2 (GFA). Of the total electricity consumed, 62% was used for cooling, whereas 14% was used for interior lighting. The villa was composed of 695 tons of materials, but the building materials stage, that is, the material supply and manufacturing, represented a maximum of 6% of the life cycle impact in the category of photochemical oxidation. This stage is analyzed in detail in the next section. The transport of building materials to the construction site did not have a significant impact considering the whole building life cycle, accounting for less than 1% of all impact categories. The transport distance was assumed as 50 km because factories are in Riyadh, the second industrial city, so all transport operations were optimized. During the construction stage, 10% of building materials were assumed to be transformed into waste materials. Thus, the extra number of materials is consumed as part of this stage as well as in waste transport and the management of landfills. Some other operations were included, such as excavation. Therefore, the contribution of this stage was highly dependent on the building material stage, but its contribution to the total life cycle of the villa was less than 1% for all impact categories and indicators. The maintenance and replacement stages considered the replacement of some building elements that have a shorter lifespan than that of the building. In this case, two relevant substitutions of floor tiles and one substitution of wall tiles and painting works of external and internal walls were considered. Even though the amount of material consumed during the use of the building was relevant—almost 12 tons of materials—the contribution of this stage was, at most, 1.5% of the impact of the photochemical oxidation impact category.

Table 3. Life cycle impact assessment (LCIA) of the life cycle of the villa considering a lifespan of 50 years (per FU) and per m^2 of GFA.

Impact Category or Indicator	Acronym	Units	Total per FU	Total per GFA (m^2)
Global warming	GWP 100 years	kg CO_2-eq	2,807,943.8	7255.668734
Ozone layer depletion	ODP	kg CFC-11-eq	0.35283721	0.000911724
Photochemical oxidation	POCP	kg C_2H_4	782.15594	2.02107478
Acidification	AP	kg SO_2-eq	18,279.348	47.23345736
Eutrophication	EP	kg PO_4^{3-}-eq	1577.4363	4.076062791
Cumulative energy demand	CED	MJ	43,015,866.13	111,152.1089

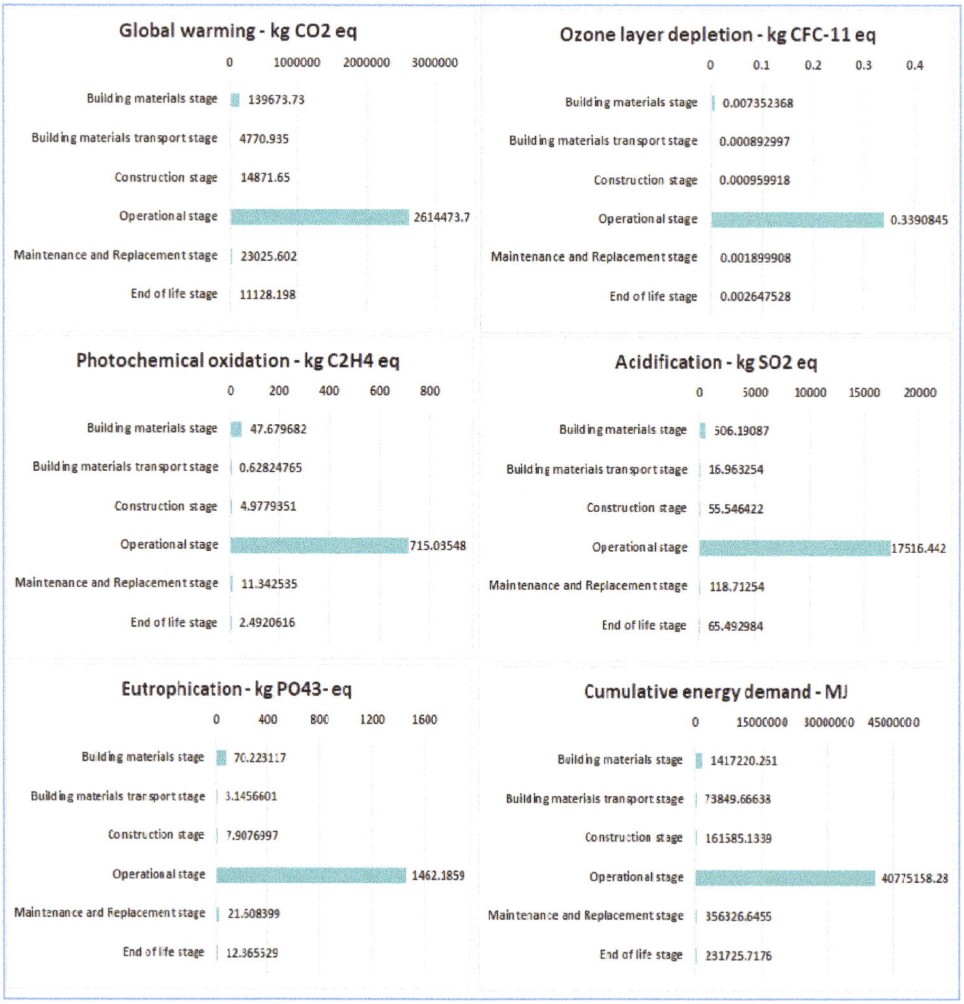

Figure 5. Contribution of each life cycle stage to the FU environmental impact.

At the end of the life cycle, it was assumed that all the building materials were landfilled because this the current protocol for handling construction and demolition waste in Saudi Arabia. However, the impact of this stage contributes to less than 1% of the total life cycle impact.

A graphical explanation of these factors is provided for analysis of the impact of indicators.

3.1.2. Building Materials Stage Impact Assessment

The LCA was developed from the villa bill of materials, so the specific contribution of the building elements is analyzed in Table 4 for each impact category. The average villa has a reinforced concrete structure, whereby the external and internal walls are made of concrete and, in this study, concrete blocks. Therefore, the most important material was concrete because it represented 83% of the building weight, 57% of the structure and foundation, 19% of the external walls, and 8% of the internal walls.

Table 4. Contribution of building elements to the LCIA of building materials stage.

Impact Category or Indicator	GWP 100 years	ODP	POCP	AP	EP	CED
Foundation	8%	7%	6%	6%	6%	6%
Structure	49%	48%	58%	47%	49%	46%
Roof and open space	0.9%	1.1%	0.9%	0.9%	0.6%	2.0%
Ceiling	1.7%	1.7%	2.6%	2.4%	1.8%	2.0%
Exterior walls and parapet	21%	20%	14%	18%	19%	20%
Internal walls	13%	14%	11%	13%	16%	13%
Floor and wall tiles	3%	4%	2%	4%	2%	4%
Windows	2%	3%	2%	4%	2%	3%
Doors	0.9%	1.1%	1.1%	1.6%	1.3%	2.9%
Stairs	0.9%	0.8%	1.0%	1.5%	0.9%	1.3%
Electrical network	0.04%	0.05%	0.4%	1.0%	2.0%	0.1%

In terms of contribution to the environmental impact of building materials, the same weight relation was followed. Table 4 shows that the contribution of the structure and foundation ranges from 52% of the impact of the material's cumulative energy demand to 64% for the photochemical oxidation impact category.

The structure and foundation are followed by exterior and parapet walls, with a contribution ranging from 14% (photochemical oxidation) to 21% (global warming), and internal walls, with an impact ranging from 11% (photochemical oxidation) to 16% (eutrophication).

In terms of the type of materials and their contribution to global warming (GWP), concrete accounted for 36% of the impact while contributing 83% of the building's weight. Steel, used mainly as reinforcement rebar, had a significant contribution of 34%, despite representing only 3% of the total consumed material. Approximately 13% of the impact came from cement mortar and plaster, which accounted for 11% of the materials' weight. Cement and mortar are used in several building elements, such as roof or floor, but the majority of the quantity is used in internal and external walls to plaster both faces of the concrete blocks and to paste them.

3.2. Sensitivity Analysis—2030 Vision

In Section 3.1, the results showed that the main life cycle stage contributing to the environmental impact was the operational use stage, where electricity and water were mainly consumed and wastewater was generated. For all categories, the impact came from electricity, contributing at least 70% (eutrophication). Currently, Saudi Arabia is highly dependent on fossil fuels to produce electricity. The electricity mix of the country is 1.072 kg CO_2-eq/kWh (obtained from the Ecoinvent v3.2 dataset and CML-IA baseline method).

In 2017, the Saudi Arabian government prepared a strategy called the 2030 Vision. The objective of the Saudi Vision 2030 (SV2030) is to set up renewable and sustainable energy (RnSE) projects to meet the electricity demand—which is expected to surpass 120 GW by

2032—by increasing the use of renewable resources, reducing dependency on fossil fuels, and reducing the country's CO_2 emissions.

In this sensitivity analysis, we modeled the impact of the electricity mix based on the share of energy sources proposed by the King Abdullah City for Atomic and Renewable Energy (K.A.CARE) to deliver clean energy by 2032, that is, 9 GW of wind, 41 GW of solar (25 GW of concentrated solar and 16 solar photovoltaic (PV) cells), 17.6. GW nuclear, 1 GW geothermal, and 3 GW WtE sources, with a total 60 GW hydrocarbon capacity to meet the expected future energy demand and supply [42,43]. Thus, power mix resources by 2032 are planned to be 41% from renewable sources (12% PV cells, 19% concentrated solar, 7% wind, 1% geothermal, and 2% from WtE), 13% nuclear, and 46% hydrocarbon. Based on the 2032 scenario, the carbon footprint of the electricity mix would be 0.501 kg CO_2-eq/kWh, which is 53% lower than the current impact.

Considering that a villa would consume 47,030 kWh annually based on the calculations with EnergyPlus, significant environmental savings could be achieved every year by improving the electricity carbon intensity.

Table 5 compares the current annual operational use stage and the 2030 vision operational use stage together with the reduction of the impact by implementing the strategy.

Table 5. Environmental impact of annual operational use stage with current electricity mix and 2030 Vision.

Impact Category	Units per Year	Current Operational Use Stage	2030 Vision Operational Use Stage	Variation 2030 Vision Compared to Current Situation
GWP 100 years	kg CO_2-eq	52,289.474	25,465.652	−51%
ODP	kg CFC-11-eq	0.00678169	0.003886419	−43%
POCP	kg C_2H_4	14.3007096	6.8689938	−52%
AP	kg SO_2-eq	350.32884	167.201548	−52%
EP	kg PO_4^{3-}-eq	29.243718	18.3043584	−37%
CED	MJ	815,503.1655	525,875.8671	−36%

The increase in the share of renewable sources in the Saudi Arabian electricity mix contributes to a reduction in the operational impact on a yearly basis, with a minimum of 36% for eutrophication and a maximum of 52% for photochemical oxidation and acidification.

Decarbonizing the electricity generated in Saudi Arabia that is used in the residential sector could positively impact the life cycle of the villa. It could produce up to half of the impact, which is the maximum reduction obtained, for the acidification impact category (Table 6 and Figure 6).

Table 6. Environmental impact of the villa's life cycle with current electricity mix and 2030 Vision.

Impact Category	Units per Year	Current Villa	Villa with 2030 Vision Operational Stage	Variation 2030 Vision Compared to Current Situation
GWP 100 years	kg CO_2-eq	2,807,943.8	1,466,752.7	−48%
ODP	kg CFC-11-eq	0.35283721	0.20807369	−41%
POCP	kg C_2H_4	782.15594	410.57015	−48%
AP	kg SO_2-eq	18,279.348	9122.9835	−50%
EP	kg PO_4^{3-}-eq	1577.4363	1030.4683	−35%
CED	MJ	43,015,866.13	28,534,501.27	−34%

Figure 6. GWP, ODP, POCP, AP, EP, and CED impacts by LCA stage, expressed per FU considering the current villa and a villa during the 2030 Vision operational stage.

Even though the benefits of improving the electricity mix per kWh per year and for 50 years are important, the operational stage under the improved mix would have remained the stage with the largest impact on the life cycle of the villa, with a contribution from 84% (photochemical oxidation) to 93% (ozone layer depletion), as shown in Figure 6.

3.3. Comparison with Previous Studies

From the obtained results, it can be seen that the impact of the Saudi Arabian villa is highly dependent on the region where it is located because the energy demand is mainly dedicated to maintaining thermal comfort (Table 2). As suggested in ISO 14044, data validation is an element of LCA methodology that could be performed by comparing the results with those of other published research studies. Because no LCA study has previously been conducted from cradle to grave in Saudi Arabia, we attempted to compare our results to those of other studies in different regions that considered similar materials and scope.

A cradle-to-grave comparison with a residential building in Malaysia [40] and Uruguay [44] was performed considering the impact of GWP, ODP, AP, and EP, as the three studies used the same LCA method, CM -baseline, and the scopes were similar, which makes the results relatively comparable.

The comparison of these four impact categories with those of other studies is presented in Figure 7.

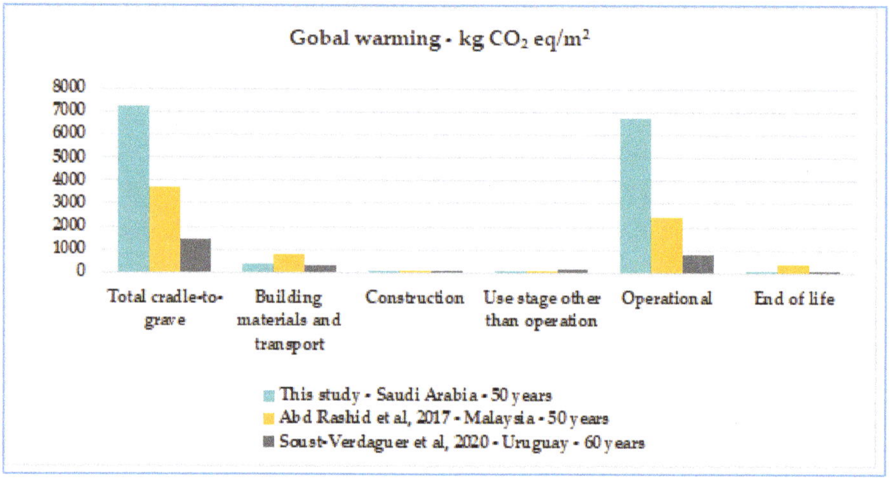

(a)

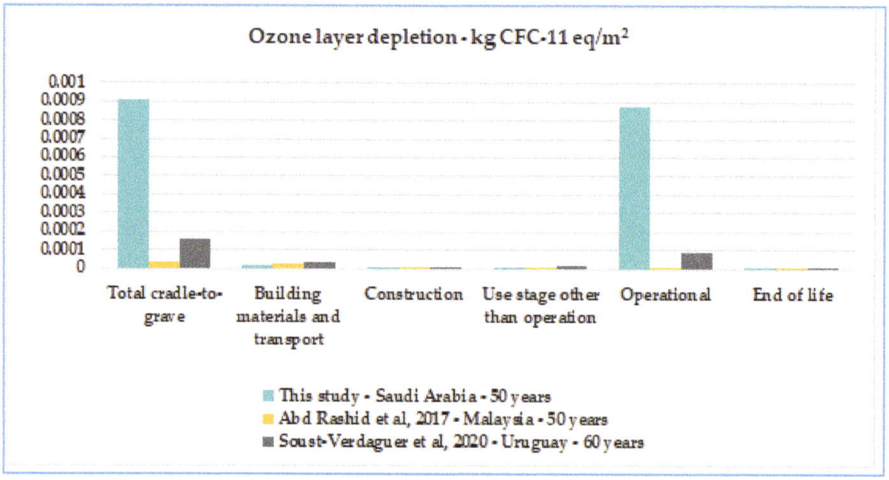

(b)

Figure 7. *Cont.*

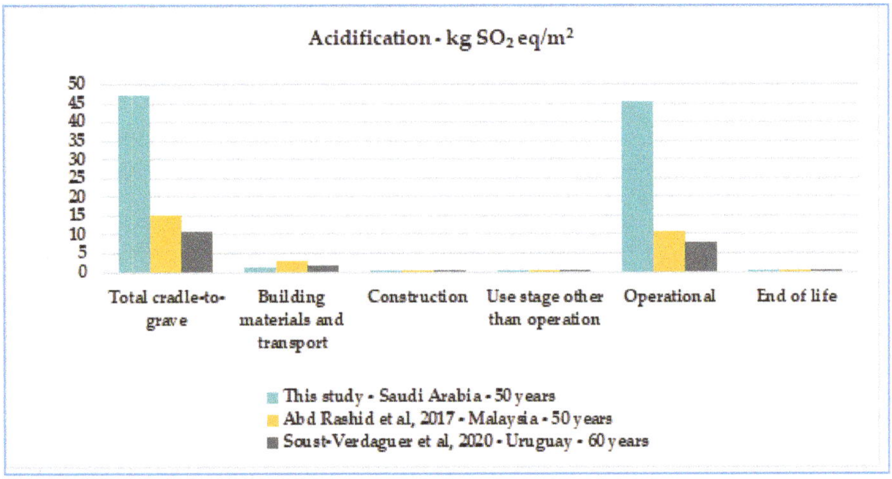

(c)

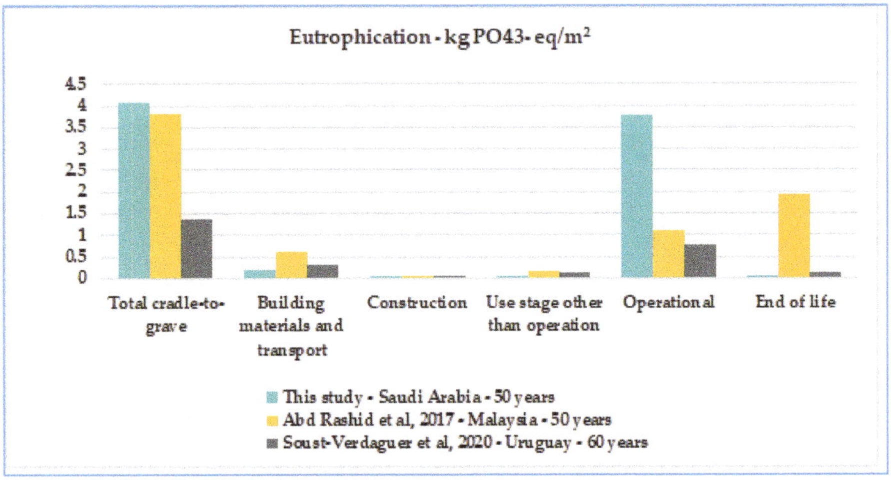

(d)

Figure 7. Comparison of (**a**) GWP, (**b**) ODP, (**c**) AP, and (**d**) EP impacts from cradle to grave of a Saudi Arabian villa with those of residential buildings from other studies in Malaysia and Uruguay that covered the entire building life cycle.

The three buildings are residential. The Malaysian case assessed a 246 m^2 GFA building, with a building frame structure of reinforced concrete and clay bricks as the building envelope. The LCA referred to the environmental performance of the building during a 50-year lifespan, including in the assessment of the pre-use, construction, maintenance and operation, and end of life phases.

The Uruguayan building included in this comparison was a COVISA house, a typical three-bedroom concrete masonry Uruguayan house with a 57 m^2 GFA. The environmental assessment considered the performance during its 60-year lifespan and included the entire life cycle except for the use (module B1 of EN 15804), refurbishment (B5), operational water use (B7), and waste processing (C3) modules.

For all analyzed impact categories, the comparison showed that the Saudi Arabian villa has the highest impact because of the impact of the operational stage, even though the

Malaysian and Uruguayan buildings are also highly dependent on the operational stage. The operational stage of the Uruguayan house contributed at least more than 50% to the total building life cycle, whereas building materials production and transport represent around 20% of the impact. The Malaysian case differed as the operational stage was the most significant for GWP and AP, but for ODP, it is the materials stage, and for EP, the end of life.

As observed in Figure 7, the operational stage in the Saudi Arabian villa represents more than 90% of the four impact categories used for the comparison. In this study, the total amount of electricity consumed per year was 121.52 kWh/m^2; this was 59 and 57 kWh/m^2 in Malaysia and Uruguay, respectively. In Malaysia, cooling energy demand represented 47% of the total electricity use, while in Saudi Arabia, it was 62%.

Regarding building materials, the Saudi Arabian residential building had the lowest impact for three of four impact categories. The villa has a larger GFA, which can mean less weight per m^2. Moreover, in the building foundation, a significant difference exists between the amount of concrete, a material that was widely used in the three buildings. For the Saudi Arabian villa, a slab on grade foundation was assumed, which is a method most commonly used in warmer climates where there is no seasonal freezing of the ground.

The comparison shows the results are comparable and confirm that the energy demand hotspot in the residential sector in Saudi Arabia is the operational stage.

3.4. Key Limitations

An LCA reflects the system analyzed and data used, so all limitations concerning data availability and system boundaries need to be considered. The bill of materials was exhaustively analyzed, and different partners were involved in the process, which influence all villa life cycle stages. The most significant stage, the operational stage, was modeled with EnergyPlus according to the characteristics of the villa defined using Revit software, so the limitations of the tools used were assumed when conducting the LCA. Even though secondary datasets were modeled to represent the Saudi Arabia context, the results are also sensitive to the datasets used in the assessment, particularly those describing materials manufacturing. Hence, all used datasets are from the same LCA database, Ecoinvent.

4. Conclusions

The LCA allowed us to analyze the whole life cycle of a typical single-family residential building in Saudi Arabia considering specific construction materials and scenarios of waste management and transport. The results showed that the operational stage has the most impact on energy consumption and the environment in the life cycle of the villa. These results align with previously published life cycle assessments of residential buildings. As indicated in Section 1 and as published in previous studies, the operational stage represents between 40% and 90% of the life cycle energy consumption depending on climatic conditions and usage habits. The operational stage of a typical Saudi Arabian villa represented 95% of the total energy demand, which is above this range. The significant contribution of this stage to the energy demand has implications in terms of environmental impact, which has two main causes. First, the amount of electricity consumed during the use stage is very high (122 kWh/m^2 (GFA) and year) mainly due to climatization requirements and, second, because electricity generation is highly dependent on fossil fuels.

In this study, a sensitivity analysis for the second cause was conducted. Saudi Arabia plans to increase the share of renewable energy sources and reduce the amount of electricity generated from hydrocarbons, which currently represent 46% of the total installed power, by 2032. Compared to the current electricity environmental impact, the CO_2 emission from electricity generation will decrease by 53%, which represents a significant reduction in impact. However, if we analyze how this reduction in environmental impact affects the residential building life cycle, we conclude that it is an impacting factor as a reduction in the impact is obtained, but the main factor is the amount of energy demand during the use stage. Therefore, the results from the sensitivity analysis showed that even with a reduction of 54% in hydrocarbons in electricity generation, the contribution of the operational stage to the total life cycle environmental impact of the villa remained unchanged. For the global warming impact category, its contribution decreases from 93%, with the current electricity mix, to 87% with the increase in renewable sources.

Further research needs to be conducted to identify the main cause of building life cycle environmental impact and energy demand, which is the amount of electricity required during the use stage. Energy efficiency strategies need to be developed and evaluated in the Saudi Arabian context to reduce the electricity demand of residential buildings, which now accounts for 50% of the country's electricity demand. The life cycle approach will allow us to evaluate how the focus on operational energy demand of these strategies affects all building life cycle stages, from building materials manufacturing to the end of life.

Author Contributions: Conceptualization, H.A., A.K. and B.A.; methodology, H.A., A.K.A.; formal analysis, A.K., S.A. and T.T.; writing—original draft writing, H.A., A.K.A., A.K. and T.T.; writing—review and editing, A.K.A. and B.A. All authors have read and agreed to the published version of the manuscript.

Funding: This research was funded by King Abdulaziz City for Science and Technology, Project number (20-0003) from the National Center for Building and Construction Technology.

Data Availability Statement: Data will be available on suitable demand.

Acknowledgments: The authors would like to express their deep and sincere gratitude to King Abdulaziz City for Sciences and Technology (KACST), for funding and supporting this research project. Project number (20-0003) from the National Center for Building and Construction Technology.

Conflicts of Interest: The authors declare no conflict of interests.

References

1. Dudley, B. *BP Statistical Review of World Energy*, 67th ed.; BP p.l.c.: London, UK, 2018.
2. United Nations. Sustainable Development Goals. Available online: https://www.un.org/sustainabledevelopment/sustainable-development-goals/ (accessed on 22 February 2021).
3. Cao, X.; Dai, X.; Liu, J. Building energy-consumption status worldwide and the state-of-the-art technologies for zero-energy buildings during the past decade. *Energy Build.* **2016**, *128*, 198–213. [CrossRef]
4. Fernandez, N.; Katipamula, S.; Wang, W.; Xie, Y.; Zhao, M. Energy savings potential from improved building controls for the US commercial building sector. *Energy Effic.* **2018**, *11*, 393–413. [CrossRef]
5. Vandecasteele, C.; Andres, A.; Coz, A. WASCON 2015: Resource efficiency in construction. *Waste Biomass Valoris.* **2017**, *8*, 1379–1380. [CrossRef]
6. Ruuska, A.; Häkkinen, T. Material efficiency of building construction. *Buildings* **2014**, *4*, 266–294. [CrossRef]
7. Climate Action Tracker. Saudi Arabia. Country Summary. Available online: https://climateactiontracker.org/countries/saudi-arabia/ (accessed on 22 February 2021).
8. Liu, H.; Tellez, B.G.; Atallah, T.; Barghouty, M. The role of CO_2 capture and storage in Saudi Arabia's energy future. *Int. J. Greenh. Gas Control* **2012**, *11*, 163–171. [CrossRef]
9. Oxford Business Group. Saudi Arabia's Construction Sector Bounces Back in 2019. Available online: https://oxfordbusinessgroup.com/overview/strong-foundations-sector-bounced-back-2019-infrastructure-works-housing-developments-and-large (accessed on 22 February 2021).
10. Felimban, A.; Prieto, A.; Knaack, U.; Klein, T.; Qaffas, Y. Assessment of current energy consumption in residential buildings in Jeddah, Saudi Arabia. *Buildings* **2019**, *9*, 163. [CrossRef]
11. Curran, M.A. Life cycle assessment: A review of the methodology and its application to sustainability. *Curr. Opin. Chem. Eng.* **2013**, *2*, 273–277. [CrossRef]
12. Halliday, S. *Sustainable Construction*; Butterworth-Heinemann: Sydney, Australia, 2008.
13. Hong, J.; Zhang, X.; Shen, Q.; Zhang, W.; Feng, Y. A multi-regional based hybrid method for assessing life cycle energy use of buildings: A case study. *J. Clean. Prod.* **2017**, *148*, 760–772. [CrossRef]
14. Guan, L.; Walmsely, M.; Chen, G. Life cycle energy analysis of eight residential houses in Brisbane, Australia. *Proc. Eng.* **2015**, *121*, 653–661. [CrossRef]
15. Ding, G.K. Life cycle assessment (LCA) of sustainable building materials: An overview. In *Eco-Efficient Construction and Building Materials*; Elsevier: Amsterdam, The Netherlands, 2014; pp. 38–62.
16. Babaizadeh, H.; Haghighi, N.; Asadi, S.; Broun, R.; Riley, D. Life cycle assessment of exterior window shadings in residential buildings in different climate zones. *Build. Environ.* **2015**, *90*, 168–177. [CrossRef]

17. Evangelista, P.P.; Kiperstok, A.; Torres, E.A.; Gonçalves, J.P. Environmental performance analysis of residential buildings in Brazil using life cycle assessment (LCA). *Constr. Build. Mater.* **2018**, *169*, 748–761. [CrossRef]
18. Asif, M.; Muneer, T.; Kelley, R. Life cycle assessment: A case study of a dwelling home in Scotland. *Build. Environ.* **2007**, *42*, 1391–1394. [CrossRef]
19. Petrovic, B.; Myhren, J.A.; Zhang, X.; Wallhagen, M.; Eriksson, O. Life cycle assessment of a wooden single-family house in Sweden. *Appl. Energy* **2019**, *251*, 113253. [CrossRef]
20. Kumar, V.; Hewage, K.; Sadiq, R. Life cycle assessment of residential buildings: A case study in Canada. *Int. J. Energy Environ.* **2015**, *9*, 1017–1025.
21. Zhang, W.; Tan, S.; Lei, Y.; Wang, S. Life cycle assessment of a single-family residential building in Canada: A case study. In *Building Simulation*; Springer International Publishing: Cham, Switzerland, 2014; Volume 7, pp. 429–438.
22. Asif, M.; Dehwah, A.H.A.; Ashraf, F.; Khan, H.S.; Shaukat, M.M.; Hassan, M.T. Life cycle assessment of a three-bedroom house in Saudi Arabia. *Environments* **2017**, *4*, 52. [CrossRef]
23. Berardi, U. Sustainability assessment in the construction sector: Rating systems and rated buildings. *Sustain. Dev.* **2012**, *20*, 411–424. [CrossRef]
24. Ramesh, T.; Prakash, R.; Shukla, K.K. Life cycle energy analysis of a residential building with different envelopes and climates in Indian context. *Appl. Energy* **2012**, *89*, 193–202. [CrossRef]
25. Yang, X.; Hu, M.; Wu, J.; Zhao, B. Building-information-modeling enabled life cycle assessment, a case study on carbon footprint accounting for a residential building in China. *J. Clean. Prod.* **2018**, *183*, 729–743. [CrossRef]
26. Islam, H.; Jollands, M.; Setunge, S. Life cycle assessment and life cycle cost implication of residential buildings—A review. *Renew. Sustain. Energy Rev.* **2015**, *42*, 129–140. [CrossRef]
27. CEN/TC 350. *Sustainability of Construction Works—Environmental Product Declarations—Core Rules for the Product Category of Construction Products EN 15804:2012+A1:2013/FprA2:2019*; European Committee for Standarization. s.l.: Brussels, Belgium, 2019.
28. EN-European Standard. *EN15978-Sustainability of Construction Works-Assessment of Environmental Performance of Buildings-Calculation Method*; No. EN 15978:2011; European Committee for Standarization. s.l.: Brussels, Belgium, 2011.
29. ISO-14040. *Environmental Management—Life Cycle Assessment—Principles and Framework*, 2nd ed.; International Organization for Standardization, ISO14040; ISO: Geneva, Switzerland, 2006.
30. ISO-14044. *Environmental Management—Life Cycle Assessment—Principles and Framework*, 1st ed.; International Organization for Standardization, ISO14044; ISO: Geneva, Switzerland, 2006.
31. PRé. SimaPro. Available online: http://www.pre-sustainability.com/simapro (accessed on 22 February 2021).
32. Van Oers, L. CML-IA Database, Characterisation and Normalisation Factors for Midpoint Impact Category Indicators. Version 4.5. April 2015. Available online: http://www.cml.leiden.edu/software/data-cmlia.html (accessed on 22 February 2021).
33. Guinée, J.B.; Jeroen, B.; Lindeijer, E. *Handbook on Life Cycle Assessment. Operational Guide to the ISO Standards*; Springer: Berlin/Heidelberg, Germany, 2002; Volume 7.
34. Wernet, G.; Bauer, C.; Steubing, B.; Reinhard, J.; Moreno-Ruiz, E.; Weidema, B. The Ecoinvent Database Version 3 (part I): Overview and Methodology. *Int. J. Life Cycle Assess.* **2016**, *21*, 1218–1230. [CrossRef]
35. Cuéllar-Franca, R.M.; Azapagic, A. Environmental impacts of the UK residential sector: Life cycle assessment of houses. *Build. Environ.* **2012**, *54*, 86–99. [CrossRef]
36. Adalberth, K.; Almgren, A.; Petersen, E.H. Life-cycle assessment of four multi-family buildings. *Low Energy Sustain. Build.* **2001**, *2*, 1–21.
37. Pierluca, V.; Umberto, A. An attributional life cycle assessment for an Italian residential multifamily building. *Environ. Technol.* **2018**, *39*, 3033–3045. [CrossRef]
38. EPD International AB. *Product Category Rules (PCR) 2014:2, Version 2.01 Buildings*; EPD International AB: Stockholm, Sweden, 2019.
39. Clift, R.; Doing, G.; Finnveden, G. The application of life cycle assessment to integrated solid waste management part1—methodology. *Process Saf. Environ. Prot.* **2000**, *78*, 279–287. [CrossRef]
40. Abd Rashid, A.F.; Idris, J.; Yusoff, S. Environmental Impact Analysis on Residential Building in Malaysia Using Life Cycle Assessment. *Sustainability* **2017**, *9*, 329. [CrossRef]
41. Frischknecht, R.; Jungbluth, N.; Althaus, H.J.; Doka, G.; Heck, T.; Hellweg, S.; Hischier, R.; Nemecek, T.; Rebitzer, G.; Spielmann, M. Overview and Methodology. Available online: http://www.ecoinvent.org/files/200712_frischknecht_jungbluth_overview_methodology_ecoinvent2.pdf (accessed on 22 February 2021).
42. Khan, M. Saudi Arabia's Vision 2030. *Def. J.* **2016**, *19*, 36–42.
43. Yamani, H.; Zell, E.; Gasim, S.; Wilcox, S.; Katamoura, S.; Stoffel, T.; Shibli, H.; Engel-Cox, J.; Subie, M.A. 2012. Energy Sustainability for Future Generatoins. KACARE. Assessment of solar radiation resources in Saudi Arabia. *Sol. Energy* **2015**. [CrossRef]
44. Soust-Verdaguer, B.; Llatas, C.; Moya, L. Comparative BIM-Based Life Cycle Assessment of Uruguayan Timber and Concrete-Masonry Single-Family Houses in Design Stage. *J. Clean. Prod.* **2020**, *277*, 121958. [CrossRef]

sustainability

Article

Determination of Optimal MR&R Strategy and Inspection Intervals to Support Infrastructure Maintenance Decision Making

Yingnan Yang [1] and Hongming Xie [2,*]

1 College of Civil Engineering and Architecture, Zhejiang University, Hangzhou 310058, China; yyn@zju.edu.cn
2 School of Management, Guangzhou University, Guangzhou 510006, China
* Correspondence: xiehm_gzhu@126.com

Abstract: In the commonly used approach to maintenance scheduling for infrastructure facilities, maintenance decisions are made under the assumptions that inspection frequency is periodical and fixed, and that the true state of a facility is revealed through inspections. This research addresses these limitations by proposing a decision-making approach for determining optimal maintenance, repair, and rehabilitation (MR&R) strategy and inspection intervals for infrastructure facilities that can explicitly take into account non-periodical inspections as well as previously considered periodical inspections. Four transition probabilities are proposed to represent four different MR&R strategies. Then, an optimization program is suggested to minimize MR&R and inspection costs of a bridge element network over a given time period, while keeping the condition states of the element network above a predetermined level. A case study was applied to illustrate the proposed approach. The results show that the proposal approach can support decision making in situations where non-periodical inspections and MR&R actions are incorporated into the model development. If employed properly, this may allow agencies to maintain their infrastructure more effectively, resulting in cost savings and reducing unnecessary waste of resources.

Keywords: maintenance, repair, and rehabilitation (MR&R); inspection; optimization; infrastructure; decision making

Citation: Yang, Y.; Xie, H. Determination of Optimal MR&R Strategy and Inspection Intervals to Support Infrastructure Maintenance Decision Making. *Sustainability* **2021**, *13*, 2664. https://doi.org/10.3390/su13052664

Academic Editor: Edmundas Kazimieras Zavadskas

Received: 8 January 2021
Accepted: 27 February 2021
Published: 2 March 2021

1. Introduction

Under continuous wear by traffic, environment and weather, infrastructure facilities inevitably deteriorate and require effective maintenance, although most agencies lack sufficient funding and effective decision-making approaches for allocating limited resources [1,2]. Given available resources, highway agencies are faced with a number of choices: (1) How often should the inspection be done? (2) After an inspection, what type of maintenance actions should be performed? (3) How to determine the optimal inspection interval and maintenance strategy to minimize costs [3]. With regard to infrastructure issues, these choices are based on the consequences of possible maintenance, repair, and rehabilitation (MR&R) actions on the future condition of the infrastructure facilities. Since information about the future condition of infrastructure facilities is not available, performance prediction models are used. This framework is common in the current maintenance decision making of infrastructure facilities, although the actual formulation of the performance prediction and optimization models may differ [4].

The main difficulty faced by the current maintenance decision making for infrastructure facilities is the lack of empirical data related to the infrastructure facilities' historical behaviour, which to a large extent relies on the experience of managers and technical personnel [5,6]. Second, the currently used approaches and models have some limitations, which affect the effectiveness of maintenance decision making. Frangopol, et al. [7] reviewed the research related to models and modelling approaches of maintenance decision making for infrastructure facilities. They concluded that no single model has yet proven to be generally applicable, and each model has its advantages and disadvantages. Therefore, several studies have suggested some advanced options to overcome these barriers. Third, infrastructure maintenance decisions are usually made by practitioners. Therefore, models in the context of infrastructure applications need to be interpreted in a more direct and simple manner, and should be easy to use in practice. However, the methods and models currently proposed for infrastructure maintenance decisions are usually quite theoretical in nature. It is evident that the difficulties faced by the current maintenance decision making arise from the data as well as the model. As mentioned by Mishalani and McCord [8], much more effort needs to be devoted to transferring the models and approaches into practice by better utilizing more pertinent data, and explicitly addressing limitations to practical implementation.

Therefore, the aim of this paper is to propose a decision-making approach for determining the optimal MR&R strategy and inspection intervals for infrastructure facilities that can explicitly take into account non-periodical inspections as well as previously considered periodical inspections, which mainly include two parts: (1) to identify the limitations of the current maintenance decision making used for infrastructure facilities based on a literature review; (2) to develop a decision-making approach that can address these identified limitations. The proposed approach is expected to extend current maintenance decision making by addressing non-periodical inspection issues, and hence enhance the capability and feasibility of the optimization module in current maintenance decision-making systems. The remainder of the paper is organized as follows. Section 2 reviews the current literature of maintenance decision making for infrastructure facilities, followed by a discussion of maintenance decision making for elements undergoing periodical inspection in Section 3. Section 4 develops an optimization model to support the maintenance decision making of elements undergoing non-periodical inspection. A general maintenance decision-making support architecture covering non-periodical as well as periodical inspections is proposed in Section 5. The last section concludes the paper by discussing the limitations and suggestions for future research.

2. Review of Maintenance Decision Making for Infrastructure Facilities

Based on the current state evaluation and future condition prediction of the infrastructure, the key of maintenance decision making for infrastructure facilities is to develop effective optimization models for programming maintenance and/or inspection schedules under a limited financial budget [9]. The current maintenance management of infrastructure facilities relies on information collected from periodical inspections, which can be used to assess the condition states and conduct maintenance optimization [10]. An important requirement when making maintenance decisions for infrastructure facilities is data availability [11]. However, resources to be invested in data collection and optimization analysis are usually limited in practice [12]. The Markov decision process (MDP) and Semi-Markov decision process (SMDP) are often used for maintenance decision analysis due to the following advantages: (1) they are state-based models suited to incorporate information from visual inspection [7]; (2) they are probabilistic models which can address both the uncertainties associated with deterioration patterns and the dynamic decisions of the model [5]; and (3) they can manipulate networks with a large number of facilities because of their computational efficiency and simplicity of use [13]. As a result, after an inspection, the decision maker can apply the MR&R activity specified by the optimal policy for that condition state of the facility [14,15].

In the MDP and SMDP, infrastructure maintenance decisions are made under the assumptions that inspections are performed at periodical intervals and that they reveal the true condition state of a facility, with no measurement error [14,16]. The assumptions raise several concerns which come from the simplification required to predict deterioration of the facilities and the uncertainty from the inspection [10,17]. The assumption that the performance prediction model depends on the state of a facility revealed through periodical inspections is an example of the former, while the uncertainty that exists in generating transition probabilities for the MDP and SMDP is an example of the latter [18,19]. The assumptions ignore the effects, the cost, and the complexity of MR&R actions and inspections, thus limiting the effectiveness of the approach in many situations. Examples include the implementation of certain types of infrastructure facilities that are susceptible to accidental damage and need frequent MR&R actions during the life cycle, but have not been studied extensively, e.g., expansion joints, pipeline segments, and parapets.

The assumption of error-free inspections has been demonstrated to be incorrect in several empirical studies [10,20]. Additionally, there is a large measurement uncertainty in visual inspection, and this uncertainty will affect maintenance decisions. This is because measurement errors will affect the performance prediction of the facilities and ultimately lead to the selection of the "wrong" activities. The assumption of periodical and fixed intervals forces decision makers to schedule inspections at the beginning of planning, regardless of the cost and effectiveness of the inspection. Indeed, parts of this assumption may be valid for certain bridge elements, e.g., beam, pier, and girder, those where the deterioration is primarily governed by mechanical processes. However, it is unrealistic for some bridge components, e.g., expansion joints, bearing, and parapets, which are susceptible to accidental damage and need to be inspected non-periodically [21,22]. Madanat and Ben-Akiva [23] identified that increasing the frequency of inspections increases inspection costs but enhances the quality of information available to the decision maker.

A latent Markov decision process (LMDP) was proposed for maintenance decisions that accounts for the presence of non-fixed time intervals and measurement uncertainty [10,20]. Although relaxing the assumptions, LMDP research still manifests some limitations in determining the optimal inspection and MR&R strategy. First, the effects of MR&R actions between two adjacent inspections are still not properly incorporated into the model development. This means that the state derived from LMDP may still not reveal the true condition state of the bridge elements due to inconsistency in the bridge deterioration profile. This is especially critical for some infrastructure facilities, e.g., expansion joints, pipeline segments, and parapets, which are susceptible to accidental damage and need to be inspected non-periodically and treated in time in order to provide safe and good service quality for users [21]. Second, the approaches to date assume a presumed and constrained inspection frequency in MDP [5,14], or only address the decision of whether to inspect in a given year or not in LMDP [10,20,24]. Nazari, Noruzoliaee, Zou and Mohammadian [16] used LMDP to seek the optimal facility-specific inspection intervals, and the MR&R policies, but focused only on inspection error associated with technology. As discussed above, due to the effects of non-periodical inspections and subsequent MR&R actions not being incorporated into the model development, a "wrong" maintenance decision will be made, resulting in unnecessary waste of resources and materials. In fact, as suggested by Hu and Samer [24], the effects of MR&R actions must be considered and conducted to ensure the performance of infrastructure at safe and satisfactory levels [24]. Third, the current LMDP models for infrastructure still seem quite theoretical, and hence appear difficult to use in practice. It can be seen that the above limitations affect the effectiveness of maintenance decision making and the use of the model. Therefore, it is necessary to conduct research to address the above limitations.

Furthermore, the proposed maintenance scheduling should take into account the multiple constraints, including technical and economical considerations, as well as the balanced development imperatives [25,26]. If a decision maker uses a single objective algorithm to optimize the inspection intervals or maintenance scheduling separately, the conflict may result in an unsatisfied demand.

3. Maintenance Decision Making for Elements Undergoing Periodical Inspection

MDPs and SMDPs can be used for maintenance-decision analysis of the elements undergoing periodical inspection with the following underlying assumptions [14]:

1. The deterioration is represented by transition probabilities;

2. Perfect inspection is carried out to identify the condition state of the elements;
3. The process has a finite state space-maintenance action, and it is assumed that at in every each period a set of maintenance actions is available;
4. The elements are inspected on a predetermined and fixed-time interval;
5. After the optimization analysis, the optimal periodical inspection interval and the optimal MR&R strategies can be determined.

MDPs have been extensively used for maintenance decision making for elements undergoing periodical inspections. The deterioration in this case is modelled as a Markov process. The inspection interval in most MDP-based MR&R optimization is pre-determined, usually 12 or 24 months. Some research has attempted to determine the optimal periodical inspection with MDP, for example, Kallen and Van Noortwijk [27] proposed a decision model to determine the optimal time between periodic inspections. Linear programming (LP) can be used for optimal MDP-based maintenance scheduling, which was first proposed by Golabi, et al. [28]. The decision variables of the optimization model are the fractions of the facilities in the network that are in various condition states, and to which different MR&R actions should be applied [29]. To improve computational efficiency, researchers have proposed evolutionary-based algorithms, e.g., the genetic algorithm (GA), for searching a near-optimum solution [30–32].

Compared with MDP, the generalized characteristics of SMDP make decision analysis more effective in choosing the optimal inspection interval. For example, Berenguer, et al. [33] used SMDP to derive a predictive maintenance policy which indicates, at each inspection and according to the observed value, whether a preventive maintenance is necessary and when the next inspection should be performed; Ge, Tomasevicz and Asgarpoor [3] used SMDP to determine the optimal inspection rate and maintenance policy by maximizing equipment availability and minimizing the cost.

4. Maintenance Decision Making for Elements Undergoing Non-Periodical Inspection

4.1. Problem Description

Referring to the rating system of PONTIS [5], a bridge network is not considered as a set of individual bridges but as a combination of bridge elements that interact with each other in various forms and quantities. Thus, each bridge can be defined as a combination of its constituent elements. Now, if one unit of an element is considered, it is possible to define the condition state of a bridge element at any time. This makes it possible to specify the MR&R actions that can be applied to the specific state of each element. In reality, since one unit of a bridge element can usually be in one of four or five condition states at any given time, there are only a few MR&R actions available to correspond. So, at any time, the possible discrete states and available MR&R actions may be associated with one unit of the bridge element.

Thus, it is possible to use different types of MR&R actions to discretize the condition state of a bridge element, i.e., States 1,2,3, and 4 are distinguished by different MR&R actions ("Do nothing", "Preventive maintenance", "Corrective maintenance", and "Rehabilitation or Replacement"). More information about this rating system is described by Yang, Pam and Kumaraswamy [22]. According to the rating system, the number of possible condition states is determined for one unit of a bridge element at any given time. More pertinent data from past inspections and maintenance records are added to the maintenance decision analysis. The states of the bridge element take into account the impact of non-periodical inspections and MR&R actions, which can be seen as an extension of the approach from PONTIS. In earlier MDP and SMDP, the state is an integer representing the condition of the infrastructure facility, assuming there are no measurement errors [27]. However, in the model presented in this study, the state is represented by the different types of MR&R actions. Data from actual MR&R actions were used to supplement information from non-periodical inspections, relaxing the assumption of no measurement errors. These aspects include the time since the last inspection, as well as the most recent MR&R actions and the impact of these actions.

Markov and semi-Markov models are used for deterioration modelling by defining discrete condition states and accumulating the transition probabilities from one condition state to another, over multiple discrete time intervals, in which the deterioration process is represented by transition probabilities [34]. The transition probabilities are integrated in a transition probability matrix (TPM). Based on the rating system defined above, the following four TPMs are derived, denoted as M_1, M_2, M_3, and M_4.These four TPMs are proposed to represent four different MR&R strategies, which are expected to simplify MR&R decision making, i.e., strategy I takes all the MR&R actions ("Do nothing", "Preventive maintenance", "Corrective maintenance", and "Rehabilitation or Replacement"); strategy II takes three MR&R actions ("Do nothing", "Corrective maintenance", and "Rehabilitation or Replacement") apart from preventive maintenance, strategy III takes three MR&R actions ("Do nothing", "Preventive maintenance", and "Rehabilitation or Replacement") apart from corrective maintenance, while strategy IV only takes "do nothing" and "rehabilitation/replacement" actions.

$$M_1 = \begin{bmatrix} \beta_{11} & \beta_{12} & \beta_{13} & \beta_{14} \\ \beta_{21} & \beta_{22} & \beta_{23} & \beta_{24} \\ \beta_{31} & \beta_{32} & \beta_{33} & \beta_{34} \\ 1 & 0 & 0 & 0 \end{bmatrix} \tag{1}$$

$$M_2 = \begin{bmatrix} \beta_{11} & \beta_{13} & \beta_{14} \\ \beta_{31} & \beta_{33} & \beta_{34} \\ 1 & 0 & 0 \end{bmatrix} \tag{2}$$

$$M_3 = \begin{bmatrix} \beta_{11} & \beta_{12} & \beta_{14} \\ \beta_{21} & \beta_{22} & \beta_{24} \\ 1 & 0 & 0 \end{bmatrix} \tag{3}$$

$$M_4 = \begin{bmatrix} \beta_{11} & \beta_{14} \\ 1 & 0 \end{bmatrix} \tag{4}$$

4.2. Objective Function

Discrete-time SMDP can be used for maintenance-decision analysis of the elements undergoing non-periodical inspection with the following assumptions:

1. The deterioration is represented by transition probabilities;
2. Perfect inspection is carried out to identify the state of the elements;
3. The process has a finite state space—MR&R actions—and it is assumed that at every period a set of MR&R actions is available;
4. The elements are inspected at non-periodical intervals;
5. After each analysis, two main issues are determined: (1) the optimal inspection interval and (2) the optimal MR&R strategies.

Based on the four TPMs derived above, an optimization program is described below.

$$Min \cdot \sum_{i=0}^{N} \alpha^i \left[S_i \cdot TPM_{(i,i+1)} \cdot C_M + C_I \right] \tag{5}$$

$$s \cdot t \cdot S_0 = \begin{bmatrix} 1 & 0 & 0 & 0 \end{bmatrix} \tag{6}$$

$$S_{k+1} = S_k \cdot TPM_{(k,k+1)}, \forall k = 0, 1, 2 \ldots i \tag{7}$$

$$S_i \leq S^{Thr} \tag{8}$$

where S_i = condition state vector (1 × 4) at the time point i; $TPM_{(i,i+1)}$ = TPM (4 × 4) from time point i to $i+1$; C_M = MR&R unit cost vector (4 × 1); C_I = inspection unit cost; α = discount rate; T = planning horizon; S_i^{Thr} = the threshold condition state vector (1 × 4).

5. Case Study

A bridge element (e.g., expansion joints) was used as an example to validate the approach, which was proposed to determine the optimal MR&R strategy and inspection intervals for a network of bridge element. The TPMs derived from Yang, et al. [35] were used as inputs to the optimization Equation (5). Searching for appropriate algorithms is the most important step factor in optimization. Network-level optimization problems of infrastructure are frequently formulated as MDPs, and the optimal MDP-based MR&R policy can be determined by short-term and/or long-term optimization. Long-term optimization is based on infinite planning horizons, while short-term optimization is based on a predetermined and finite planning horizon [14]. Considering the inspection frequency of the actual project, 6 months was used as a calculation step in this case. The planning period was set at 20 years. A common approach of solving an MDP-based model is by transforming it into a LP model. Therefore, a short-term LP optimization was considered in this case study, for which efficient algorithms exist (e.g., Golabi, Kulkarni and Way [28]; Smilowitz and Madanat [10,14]).

The objective function (5) minimizes the total expected cost of MR&R activities and inspection by selecting the optimal values of the decision variables over the planning horizon T. The required constraints for this minimization problem are listed in Equations (6)–(8). The first constraint (6) limits the initial variable of condition state vector. Constraint (7) shows the Chapman–Kolmogorov equation. Constraint (8) defines a predetermined threshold value to the condition states of the bridge element. More detailed calculation procedures can be seen in [14].

Referring to the real costs and the suggestions from the engineers, the inspection unit cost is assumed to be 10 (USD/m), and the ratio of MR&R action unit cost is assumed to be 0:10:100:1000 for Strategy I; 0: 100:1000 for Strategy II; and 0:10:1000 for Strategy III. The last value of the condition state vector is related to the structural safety and can be ensured by setting a critical value. According to Yang, Kumaraswamy, Pam and Xie [35], 1% can be used as a critical value for the last value of the condition state vector, i.e., if the last value of the condition state vector exceeds the critical value, a penalty in the form of an additional MR&R cost of USD 1000 per unit is applied. The above values can be adjusted according to the actual project.

As shown in Table 1, Strategy II is a good choice if the lowest total cost is considered, and Strategy I can be a good choice if good condition performance is considered. One possible explanation for the lowest total cost of Strategy II is that it does not use the preventive maintenance, reducing the frequency of inspections and associated costs that result from the preventive maintenance. In contrast, Strategy III has the worst conditional state and the highest total cost, making it an unsatisfactory choice.

Table 1. Output of analysis.

Strategy	Fractions of Condition State	Optimal Inspection Interval (Month)	Total Cost ($)
I	[0.060; 0.092; 0.765; 0.083]	12–24—30–36—42–78—120–240	3.657×10^3
II	[0.073; 0.837; 0.090]	6–24—78–240	2.026×10^3
III	[0.438; 0.386; 0.078]	12–114—120–126—210–240	4.328×10^3

The results of the case study also indicate that relaxing the frequency of inspections may contribute to cost savings, due to the increased costs of frequent inspections. In particular, for higher levels of measurement error, the increased costs may be higher. The results suggest that the instrumentation could be used to improve the accuracy of infrastructure inspections in the future and the importance of jointly optimizing inspection intervals and MR&R policies. The results of the case study have some similarities to those of previous research (e.g., [10]).

6. Discussion

A case study was performed to illustrate the proposed approach to programming and formulating an optimal inspection interval and MR&R strategy for a network of bridge elements. The optimization problem is formulated as a short-term LP problem. The optimization objective is to minimize the MR&R and inspection cost of a bridge element network over a given time period, while keeping the element network condition above a predefined threshold level.

For a given time period (e.g., 20 years), the results of the case study comprise the inspection intervals, the total cost, and fractions of condition state for a given time period. Decisions are thus made by balancing these items. This application illustrated the feasibility, efficiency, and capability of the proposed approach. A general maintenance decision-making support architecture covering non-periodical as well as periodical inspections was thus proposed (see Figure 1). Four modules are included in the architecture, which are (1) data input, (2) optimization process, (3) output, and (4) making decisions. Just like the optimization results generated by the case study, engineers can choose one of them and make maintenance decisions. If the optimization outputs are not satisfied, the model can be run again with different parameters adjusted and entered, e.g., adjusting MR&R and inspection costs, and planning horizon.

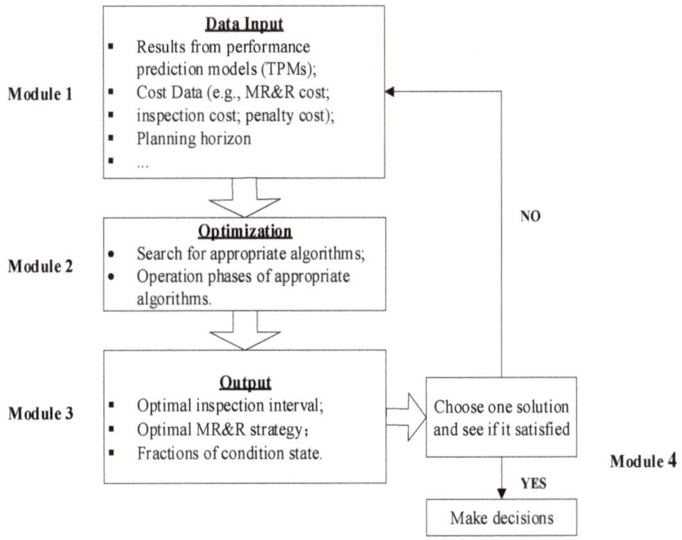

Figure 1. General maintenance decision-making support architecture.

7. Conclusions

Highway agencies are increasingly recognizing the need for an effective approach to allocate limited resources in a cost-effective and environmental-friendly way for infrastructure maintenance [2,36]. Much of the current research on sustainable infrastructure is focused on the design phase, encouraging more efficient use of natural resources and green design [37], and neglecting sustainable issues in the maintenance phase [38]. A literature review was used to identify the limitations of the current maintenance decision making used for infrastructure facilities. Based on this, a decision-making approach was proposed for determining optimal MR&R strategy and inspection intervals for infrastructure facilities that can explicitly take into account non-periodical inspections as well as previously considered periodical inspections. The optimization model enables highway agencies to determine the optimal maintenance policy for a given level of performance and minimum cost, while ensuring the security of critical infrastructures and reducing waste of resources.

After all, many materials and resources could be wasted, and infrastructure facilities could deteriorate more severely with improper MR&R policies. The outputs of the proposed decision-making support architecture comprise the inspection intervals, the total cost, and fractions of condition states for a given time period. Multi-criteria decisions can thus be made by optimizing the targeted balance between these items. This makes the proposed approach attractive, because it considers the engineering practices. It is impractical to make strong assumptions about performance prediction and maintenance optimization. For example, the existing approach to maintenance decision making for certain types of infrastructure elements, e.g., expansion joints, pipeline segments, and parapets, usually assumes that inspection is fixed, with no measurement error.

Through this approach, there is a great potential for highway agencies to make their infrastructure sustainable by saving costs and reducing unnecessary waste of resources. Compared with the optimization models presented in previous research (e.g., Memarzadeh and Pozzi [12]; Wu, Yuan, Kumfer and Liu [2]), the assumptions were addressed in this study; and the outputs include not only the optimal MR&R strategy but also the optimal inspection intervals.

The MR&R strategies vary greatly from one bridge element to another. The TPMs proposed in this study may not be applicable to another bridge element (e.g., bridge pier). Appropriate TPMs can be proposed for different types of bridge elements in future research. The inspection unit cost and the MR&R action unit cost are needed in the optimization model. Future studies will continue to collect cost data to ensure accurate and reasonable results. Additionally, the user cost (e.g., including traffic delay and resource use cost) and other social costs could be incorporated into the overall cost optimization. A comprehensive database could be developed to obtain such data. Assuming this is possible, it is also necessary to develop a cost model to incorporate all such pertinent cost data for overall optimization in future research. The proposed approach was only verified using the data from one important bridge element. More data from other bridge elements should be collected to verify the proposed approach in the future. Furthermore, superposing the additional variables and constraints will lead to an exponential increase in the number of variables, making it computationally expensive to reach an optimal solution. To improve computational efficiency, robust optimization techniques, e.g., the genetic algorithm (GA), could be tested in a future study.

Author Contributions: Y.Y. developed the concept based on the analysis and drafted the manuscript. H.X. provided constructability feedback and reviewed the manuscript. All of the authors read and approved the final manuscript. All authors have read and agreed to the published version of the manuscript.

Funding: This research was supported by the National Natural Science Foundation of China (Grant No. 71673240 and 71772163), Zhejiang Provincial Natural Science Foundation of China (Grant No. LY16G020009 and LY17G020024), and Guangzhou Philosophy and Social Science Foundation (Grant No.2020GZYB94).

Institutional Review Board Statement: Not applicable.

Informed Consent Statement: Not applicable.

Data Availability Statement: Not applicable.

Acknowledgments: The authors are grateful for the kind support and cooperation of the Highways Department of HKSAR in providing valuable data, relevant manuals, and other required information. The authors thank the anonymous reviews for their valuable suggestions.

Conflicts of Interest: The authors declare no conflict of interest.

References

1. Chu, J.C.; Huang, K.H. Mathematical programming framework for modeling and comparing network-level pavement maintenance strategies. *Transp. Res. Part B Methodol.* **2018**, *109*, 1–25. [CrossRef]
2. Wu, D.; Yuan, C.; Kumfer, W.; Liu, H. A life-cycle optimization model using semi-markov process for highway bridge maintenance. *Appl. Math. Model.* **2016**, *43*, 45–60. [CrossRef]
3. Ge, H.; Tomasevicz, C.L.; Asgarpoor, S. Optimum Maintenance Policy with Inspection by Semi-Markov Decision Processes. In Proceedings of the 39th North American Power Symposium, Las Cruces, NM, USA, 30 September–2 October 2007; pp. 541–546.
4. Robelin, C.-A.; Madanat, S.M. History-dependent bridge deck maintenance and replacement optimization with Markov decision processes. *J. Infrastruct. Syst.* **2007**, *13*, 195–201. [CrossRef]
5. Golabi, K.; Shepard, R. Pontis: A system for maintenance optimization and improvement of US bridge networks. *Interfaces* **1997**, *27*, 71–88. [CrossRef]
6. Madanat, S.; Smilowitz, K.; Lago, A. Optimal inspection and maintenance policies for infrastructure systems: Facility and network problems. *Transp. Res. Rec. J. Transp. Res. Board* **1999**, *1667*, 1–7. [CrossRef]
7. Frangopol, D.M.; Kallen, M.-J.; Van Noortwijk, J.M. Probabilistic Models for Life-Cycle Performance of Deteriorating Structures: Review and Future Directions. *Prog. Struct. Eng. Mater.* **2004**, *6*, 197–212. [CrossRef]
8. Mishalani, R.G.; McCord, M.R. Infrastructure condition assessment, deterioration modelling, and maintenance decision making: Methodological advances and practical considerations. *J. Infrastruct. Syst.* **2006**, *12*, 145–146. [CrossRef]
9. Fang, Y.; Sun, L. Developing A Semi-Markov Process Model for Bridge Deterioration Prediction in Shanghai. *Sustainability* **2019**, *11*, 5524. [CrossRef]
10. Smilowitz, K.; Madanat, S. Optimal inspection and maintenance policies for infrastructure networks. *Comput. Civ. Infrastruct. Eng.* **2000**, *15*, 5–13. [CrossRef]
11. Mandiartha, P.; Duffield, C.F.; Thompson, R.G.; Wigan, M.R. Measuring pavement maintenance effectiveness using markov chains analysis. *Struct. Infrastruct. Eng.* **2016**, *13*, 844–854. [CrossRef]
12. Memarzadeh, M.; Pozzi, M. Integrated inspection scheduling and maintenance planning for infrastructure systems. *Comput. Civ. Infrastruct. Eng.* **2016**, *31*, 403–415. [CrossRef]
13. Morcous, G. Performance prediction of bridge deck systems using markov chains. *J. Perform. Constr. Facil.* **2006**, *20*, 146–155. [CrossRef]
14. Madanat, S.M.; Park, S.; Kuhn, K. Adaptive Optimization and Systematic Probing of Infrastructure System Maintenance Policies under Model Uncertainty. *J. Infrastruct. Syst.* **2006**, *12*, 192–198. [CrossRef]
15. Durango-Cohen, P.L. Maintenance and repair decision making for infrastructure facilities without a deterioration model. *J. Infrastruct. Syst.* **2016**, *10*, 1–8. [CrossRef]
16. Nazari, F.; Noruzoliaee, M.; Zou, B.; Mohammadian, A. Optimal facility-specific inspection and maintenance decisions under measurement uncertainty: Unifying framework. *J. Infrastruct. Syst.* **2017**, *23*, 04017036. [CrossRef]
17. Kim, J.; Ahn, Y.; Yeo, H. A comparative study of time-based maintenance and condition-based maintenance for optimal choice of maintenance policy. *Struct. Infrastruct. Eng.* **2016**, *12*, 1525–1536. [CrossRef]
18. Hong, T.; Chae, M.J.; Kim, D.; Koo, C.; Lee, K.S.; Chin, K.H. Infrastructure asset management system for bridge projects in south korea. *KSCE J. Civ. Eng.* **2013**, *17*, 1551–1561. [CrossRef]
19. Moreira, A.V.; Fwa, T.F.; Oliveira, J.R.M.; Costa, L. Coordination of user and agency costs using two-level approach for pavement management optimization. *Transp. Res. Rec. J. Transp. Res. Board* **2017**, *2639*, 110–118. [CrossRef]
20. Durango-Cohen, P.L.; Madanat, S.M. Optimization of inspection and maintenance decisions for infrastructure facilities under performance model uncertainty: A quasi-bayes approach. *Transp. Res. Part A Policy Pr.* **2008**, *42*, 1074–1085. [CrossRef]
21. Mishalani, R.G.; Madanat, S.M. Computation of infrastructure transition probabilities using stochastic duration models. *J. Infrastruct. Syst.* **2002**, *8*, 139–148. [CrossRef]
22. Yang, Y.N.; Pam, H.J.; Kumaraswamy, M.M. Framework development of performance prediction models for concrete bridges. *J. Transp. Eng.* **2009**, *135*, 545–554. [CrossRef]
23. Madanat, S.M.; Ben-Akiva, M. Optimal inspection and repair policies for transportation facilities. *Transp. Sci.* **1994**, *28*, 55–62. [CrossRef]
24. Liang, W.Y.; Huang, C.C.; Lin, Y.C.; Chang, T.H.; Meng, H.S. The multi-objective label correcting algorithm for supply chain modeling. *Int. J. Prod. Econ.* **2013**, *142*, 172–178. [CrossRef]
25. Hu, X.; Samer, M. Determination of optimal mr&r policies for retaining life-cycle connectivity of bridge networks. *J. Infrastruct. Syst.* **2015**, *21*, 04014042.
26. Lee, S.Y.; Park, W.; Ok, S.Y.; Koh, H.M. Preference-based maintenance planning for deteriorating bridges under multi-objective optimisation framework. *Struct. Infrastruct. Eng.* **2011**, *7*, 633–644. [CrossRef]
27. Chen, L.; Henning, T.F.P.; Raith, A.; Shamseldin, A.Y. Multiobjective optimization for maintenance decision making in infrastructure asset management. *J. Manag. Eng.* **2015**, *31*, 04015015. [CrossRef]
28. Kallen, M.J.; Van Noortwijk, J.M. Optimal periodic inspection of a deterioration process with sequential condition states. *Int. J. Press. Vessel. Pip.* **2006**, *83*, 249–255. [CrossRef]
29. Golabi, K.; Kulkarni, R.B.; Way, G.B. A statewide pavement management system. *Interfaces* **1982**, *12*, 5–21. [CrossRef]

30. Yang, Y.N.; Kumaraswamy, M.M. Towards life-cycle focused infrastructure maintenance for concrete bridges. *Facilities* **2011**, *29*, 577–590. [CrossRef]
31. Morcous, G.; Lounis, Z. Maintenance optimization of infrastructure networks using genetic algorithms. *Autom. Constr.* **2005**, *14*, 129–142. [CrossRef]
32. Lee, C.K.; Kim, S.K. GA-based algorithm for selecting optimal repair and rehabilitation methods for reinforced concrete (RC) bridge decks. *Autom. Constr.* **2007**, *16*, 153–164. [CrossRef]
33. Cheng, T.M.; Yan, R.Z. Integrating messy genetic algorithms and simulation to optimize resource utilization. *Comput. Civ. Infrastruct. Eng.* **2009**, *24*, 401–415. [CrossRef]
34. Berenguer, C.; Chu, C.; Grall, A. Inspection and maintenance planning: An application of semi-Markov decision processes. *J. Intell. Manuf.* **1997**, *8*, 467–476. [CrossRef]
35. Thomas, O.; Sobanjo, J. Semi-markov models for the deterioration of bridge elements. *J. Infrastruct. Syst.* **2016**, *22*, 04016010. [CrossRef]
36. Yang, Y.N.; Kumaraswamy, M.M.; Pam, H.J.; Xie, H.M. Integrating semiparametric and parametric models in survival analysis of bridge element deterioration. *J. Infrastruct. Syst.* **2013**, *19*, 176–185. [CrossRef]
37. Benitez, P.; Rocha, E.; Talukdar, S.; Varum, H.; Rodrigues, F. Efficiency analysis of optimal inspection management for reinforced concrete structures under carbonation-induced corrosion risk. *Constr. Build. Mater.* **2019**, *211*, 1000–1012. [CrossRef]
38. Vicent, P.-P.; Tatiana, G.-S.; José, M.; Víctor, Y. An optimization-lca of a prestressed concrete precast bridge. *Sustainability* **2018**, *10*, 685.

sustainability

MDPI

Article

Fuzzy Synthetic Evaluation of the Critical Success Factors for the Sustainability of Public Private Partnership Projects in China

Binchao Deng [1,*], Dongjie Zhou [1], Jiachen Zhao [1], Yilin Yin [1,2] and Xiaoyu Li [1]

[1] School of Management, Tianjin University of Technology, Tianjin 300384, China; djzhou2015@sohu.com (D.Z.); zjchen_2018@sina.com (J.Z.); yinyilin@tjut.edu.cn (Y.Y.); xiaoyuli2020@sina.com (X.L.)

[2] School of Management, Tianjin University, Tianjin 300072, China

* Correspondence: dbchao1985@tju.edu.cn

Abstract: Public Private Partnership (PPP) projects have attracted wide attention from academia and industry over the past 20 years, however, they have been plagued by certain factors. This study identified, classified, and evaluated the success factors that may affect PPP projects for achieving sustainability. First, a list of 32 critical success factors were categorized into 3 groups, then a questionnaire survey was conducted, with 108 responses received from experts, researchers, and PPP project managers in China. Second, using a fuzzy synthetic evaluation (FSE) method, stakeholder relationships (A_1–A_{10}), external environmental (B_1–B_8), and project management of a special purpose vehicle (C_1–C_{14}) collected data at three different factor group locations in PPP projects were used in this evaluation. The results obtained nine top factors: private sector financing capacity, government credit, government commitment or guarantee, completeness of legal framework, available financial markets, the feasibility study report and implementation, effectiveness of risk management, project investment, and cost control and revenue distribution. It was demonstrated that fuzzy synthetic evaluation techniques are quite appropriate techniques for PPP projects. The research findings should impact on policy development towards PPP and Private Finance Initiative (PFI) project governance.

Keywords: public private partnership; critical success factors; fuzzy synthetic evaluation; sustainability; project governance

Citation: Deng, B.; Zhou, D.; Zhao, J.; Yin, Y.; Li, X. Fuzzy Synthetic Evaluation of the Critical Success Factors for the Sustainability of Public Private Partnership Projects in China. *Sustainability* **2021**, *13*, 2551. https://doi.org/10.3390/su13052551

Academic Editors: Edmundas Kazimieras Zavadskas and Fausto Cavallaro

Received: 6 January 2021
Accepted: 20 February 2021
Published: 26 February 2021

1. Introduction

Public private partnership (PPP) projects have been widely used to ease pressure on government finances in China since 2013. The core principles of a PPP project include win–win cooperation, risk allocation, sustainability, and revenue sharing [1–7]. However, PPP projects have been characterized as having a long implementation period [8], large investment scale [9], complex financing structure [10], financial and investment sustainability [11–13], and diverse participants [14]. The performance of PPP projects is closely related to the interests of the public and other stakeholders. A PPP project failure can cause a significant waste of social resources and affect the government's reputation. Based on a recent literature review [15–18], critical success factors over a long-term cooperation period were identified to help public and private stakeholders control PPP project performance risks.

PPP projects need a smoothly sustainable environment. However, it is not clear whether the reality matches the ideal with respect to the cooperation between the public and the private sectors, who, together, achieve value for money (VfM), project success, and sustainability. In particular, 348 PPP projects were forced to pull out of the project management library of China public private partnerships center (CPPPC) after "Implementation opinions on promoting standardized development of cooperation between the public and private sector" (No. 10, 2019 Ministry of Finance of China), because the public sector or private sector did not provide compliance documents or other non-conforming operations that are required by CPPPC. It is indicated that several risk factors impact PPP project success, including project demands, location, financing, legal and policy environment, taxation, design, construction and technology, operation and customer service interlinked risks, and other factors [19–22]. These risks significantly threaten PPP project success. According to the CPPPC report (http://www.cpppc.org:8086/pppcentral/map/toPPPMap.do, accessed on 30 December 2020), from 2013 to 2020, 99,930 PPP projects were finished in CPPPC, with a total capital expenditure of RMB 15,278.1 billion. This was mainly invested in more than 20 industries. Similarly, over the past 25 years, more than 6000 PPP projects have reached financial closure in developing countries [23].

Despite the great benefits of PPP, these projects have faced many problems (negative effects of risk management and risk sharing, technical capacity of the private sector, investment controls, lacking a complete legal framework, the lack of a feasible operation model, and lack of a government commitment or guarantee) and many of them failed or required renegotiation [24–28]. Many studies have investigated why PPP projects fail. These studies classified reasons for failure into the following areas: risk management and allocation [29,30], stakeholder management [31,32], feasibility of operation management [33–35], government commitment or guarantee [36–39], and completeness of legal and policy framework [36,40]. All of these areas have all been extensively explored by researchers worldwide.

The indicators above show the interest researchers have had in exploring the success factors involved in delivering PPP projects worldwide. In total, 18 Critical Success Factors (CSF) were examined using a factor analysis in the context of construction PPP and PFI projects in the United Kingdom [19,41] identified 29 reliable factors, and other studies introduced fuzzy synthetic evaluation to determine CSFs and assess the factors for particular critical risk groups [32,42,43]. Ng et al. indicated that addressing the tripartite expectations (public sector, private sector, and other stakeholders) has been indispensable in ensuring the feasibility and successfulness of PPP schemes in Hong Kong [44]. Zou et al. identified the CSFs associated with relationship management in PPP projects [45]. Another study examined stakeholder perceptions of CSFs in Nigeria [46]. Finally, Osei-Kyei et al. [15] reviewed studies on CSFs from 1990 to 2013; these indicated increased worldwide research interest in PPP projects. These research publications have provided practitioners and researchers with more insights into the critical success factors and sustainability of PPP projects. Therefore, inspired by the above literature and research, this study prioritized the factors significantly influencing PPP projects. This included applying a fuzzy synthetic evaluation analysis method to overcome the issues of interdependencies and feedback among different factor-ranking alternatives. This research also developed a checklist of CSFs for PPP, which could be adopted in the further empirical and sustainable research.

The remainder of this paper is structured as follows. Section 2 offers a brief background and identification of critical success factors for PPP projects, which uses a literature review and case study. Section 3 uses fuzzy comprehensive evaluation to analyze the data of success factors that were collected by a questionnaire survey. Then, the results are discussed in Section 4, the factors are ranked, and the top nine critical success factors are obtained. Lastly, Section 5 explains the implications, limitations, suggestions for future research, and conclusions of this paper.

2. The Identification of Critical Success Factors for PPP Projects

The main aim of this paper is to identify CSFs which influence the establishment of a sustainable PPP, and which will enable more efficient management of PPP processes in China. For the past few decades, a major area of PPP studies receiving significant attention from academic and managerial communities relates to critical success factors (CSF). Bing et al. [19] used a factor analysis to identify 18 potential factors most likely to affect PPP and PFI project success in the UK. They included: efficient procurement, the capacity of project implementation, government guarantees, favorable economic conditions, sustainable environment, and available financial markets. According to top tier academic journals from 1990 to 2013, Osei-Kyei and Chan [15] identified the following factors as being very significant for PPP project success: risk allocation and sharing, strong private consortia, political support, public support, and transparent procurement.

CSFs have also been categorized and assessed in studies in different countries, including: Iran [35], the UK [47], Ghana [34,48], Greece [49], Hong Kong [44,50,51], Nigeria [52], Australia [53], Vietnam [54], Malaysia [55,56], and China [43,57]. These studies, of success factors in those countries, found that different PPP projects are associated with different critical success factors. Therefore, this study identified CSFs from literature and case studies, and obtained 14 critical success factors using a comprehensive analysis, providing support for a fuzzy comprehensive evaluation.

2.1. Literature Review on Critical Success Factors of PPP Projects

To comprehensively research PPP projects, "critical success factor" and "PPP project" were utilized as search keywords to identify journal papers published from 2000 to 2019 in international journals using the China National Knowledge Infrastructure (CNKI) database in China, and the Web of Science database. We obtained 279 papers after the data-cleaning process, including 186 Chinese papers and 93 international journal papers.

From the above-selected literature, similarities of the success factors for PPPs are obvious, and priority is placed on nominating perceived CSFs based on perception of public and private sector participants. A large proportion of the reviewed studies arrived at their nominated CSFs based on their mean scores or experience analysis [58–84]. Therefore, it is imperative to establish and statistic the key principal success factors in life cycle of PPP projects, their interrelationships, management principles, and contribution to successful implementation of a candidate project. The researcher read these papers to ensure that no invalid records were included. Table 1 lists 30 critical success factors from the document analysis.

Table 1. Summary of Literature on Success Criteria from CNKI and Web of Science.

No.	Critical Success Factors	Authers	Sum
1	Effective risk management and risk sharing	Chan and Chan [58]; Chan et al. [59]; Yuan et al. [60]; Mladenovic et al. [25]; Liyanage and Villalba [49]; Dixon et al., 2005 [61]; Zhang(a) [62]; Lam and Javed [24]; Cheung et al. [63]; Meng et al. [64]; Zhang(b) [15]; Qiao et al. [65]; Zhen-Yu Zhao [66]; Robert et al. [15]; Wang et al. [56]; Hofmeister and Borchert [67]; Binquan and Tong [68]; Hongping and Sudong [69]; Jingfeng et al. [70]; Qian and Xinli [71]	20
2	Technical capacity of private sector	Chan and Chan [58]; Chan et al. [59]; Liu et al. [72]; Liyanage and Villalba [49]; Yuan et al. [60]; Dixon et al. [61]; Zhang(b) [15] Lam and Javed [24]; Cheung et al. [63]; Meng et al. [64]; Li et al. [73]; Qiao et al. [66]; Zhang [65]; Robert et al. [15]; Xueqing et al. [43]; Wang et al. [56]; Hofmeister and Borchert [67]; Binquan and Tong [68]; Hongping and Sudong [69]; Jingfeng et al. [70]; Qian and Xinli [71]; Chou and Pramudawardhani [17]; Osei-Kyei and Chan [16]; Keers and van Fenema [30]	24
3	Control of investment	Ahadzie et al. [74]; Chan and Chan [58]; Lim and Mohamed [75]; Bryde and Robinson [76]; Al-Tmeemy et al. [77]; Baccarini [78]; Cox et al. [79]; Chan et al. [59]; Yuan et al. [60]; Mladenovic et al. [25]; Liu et al. [72]; Liyanage and Villalba [49]; Dixon et al. [61]; Zhang(a) [62]; Lam and Javed [24]; Meng et al. [64]; Zhang(b) [15] Li et al. [73]; Qiao et al. [66]; Robert et al. [15]; Wang et al. [56]; Hofmeister and Borchert [67]; Jingfeng et al. [70]	23
4	Reasonable project cooperation period	Ahadzie et al. [74]; Chan and Chan [58]; Lim and Mohamed [75]; Al-Tmeemy et al. [77]; Baccarini [78]; Cox et al. [79]; Lai and Lam [80]; Chan et al. [59]; Yuan et al. [60]; Mladenovic et al. [25]; Liu et al. [72]; Liyanage and Villalba [49]; Bryde and Robinson [76]; Dixon et al. [61]; Zhang(a) [62]; Lam and Javed [24]; Cheung et al. [63]; Meng et al. [64]; Zhang(b) [15] Li et al. [73]; Robert et al. [15]; Jingfeng et al. [70]	22
5	Long-term market demand	Chan and Chan [58]; Chan et al. [59]; Yuan et al. [60]; Mladenovic et al. [25]; Liu et al. [72]; Zhang(a) [62]; Cheung et al. [63]; Meng et al. [64]; Zhang(b) [15] Li et al. [73]; Robert et al. [15]; Xueqing et al. [43]; Hongping and Sudong [69]; Jingfeng et al. [70]; Qian and Xinli [71]; Xia et al. [81]	16
6	Long-term relationship with cooperation between government and private sector	Chan and Chan [58]; Chan et al. [59]; Mladenovic et al. [25]; Liu et al. [72]; Liyanage and Villalba [49]; Dixon et al. [61]; Zhang(a) [62]; Lam and Javed [24]; Cheung et al. [63]; Meng et al. [64]; Zhang(b) [15] Li et al. [73]; Robert et al. [15]; Xueqing et al. [43]; Wang et al. [56]; Hofmeister and Borchert [67]; Qian and Xinli [71]	17
7	Financial resources for private sector	Liu et al. [72]; Qiao et al. [66]; Zhang [65]; Robert et al. [15]; Wang et al. [56]; Hofmeister and Borchert [67]; Hongping and Sudong [69]; Jingfeng et al. [70]; Qian and Xinli [71]	9
8	Reasonable income distribution	Chan and Chan [58]; Al-Tmeemy et al. [77]; Lai and Lam [80]; Chan et al. [59]; Yuan et al. [60]; Mladenovic et al. [25]; Liu et al. [72]; Liyanage and Villalba [49]; Dixon et al. [61]; Zhang(a) [62]; Lam and Javed [24]; Cheung et al. [63]; Meng et al. [64]; Zhang(b) [15] Li et al. [73]; Xia et al. [81]	15
9	Complete legal framework	Qiao et al. [66]; Zhang [65]; Robert et al. [15]; Wang et al. [56]; Hofmeister and Borchert [67]; Binquan and Tong [68]; Hongping and Sudong [69]; Qian and Xinli [71]; Xia et al. [81]	9

Table 1. *Cont.*

No.	Critical Success Factors	Authers	Sum
10	Reduced public and political protests	Chan and Chan [58]; Chan et al. [59]; Yuan et al. [60]; Mladenovic et al. [25]; Liu et al. [72]; Liyanage and Villalba [49]; Dixon et al. [61]; Zhang(a) [62]; Lam and Javed [24]; Cheung et al. [63]; Meng et al. [64]; Li et al. [73]; Robert et al. [15]	13
11	Feasible operating model	Chan and Chan [58]; Chan et al. [59]; Yuan et al. [60]; Mladenovic et al. [25]; Liu et al. [72]; Liyanage and Villalba [49]; Zhang(a) [62]; Lam and Javed [24]; Cheung et al. [63]; Li et al. [73]; Robert et al. [15]; Xueqing et al. [43] Osei-Kyei and Chan [15]; Ahmadabadi and Heravi [35]	14
12	Local economic development	Chan and Chan [58]; Chan et al. [59]; Yuan et al. [60]; Mladenovic et al. [25]; Liu et al. [72]; Liyanage and Villalba [49]; Dixon et al. [61]; Lam and Javed [24]; Cheung et al. [63]; Meng et al. [64]; Zhang(a) [62]; Li et al. [73]; Wang et al. [56]	14
13	Government commitment or guarantee	Qiao et al. [66]; Zhang [65]; Robert et al. [15]; Xueqing et al. [43]; Wang et al. [56]; Hofmeister and Borchert [67]; Binquan and Tong [68]; Qian and Xinli [71]; House [37]; Jiang [38]; Muhammad and Johar [82]; Ahmadabadi and Heravi [35]; Ameyaw and Chan [18]; Wang et al. [83]; Kwofie et al. [49]; Emmanuel [84]; Verhoest et al. [36]	17
14	Financing power for private sector	Qiao et al. [66]; Robert et al. [15]; Xueqing et al. [43]; Hongping and Sudong [69]; Jingfeng et al. [70]; Xia et al. [81]	6
15	Fair competition for procurement process	Robert et al. [15]; Hofmeister and Borchert [67]; Binquan and Tong [68]; Jingfeng et al. [70]	4
16	Purchasing procedure	Robert et al. [15]; Hofmeister and Borchert [67]; Binquan and Tong [68]; Qian and Xinli [71]	4
17	Reductions in litigation and arguments	Chan et al. [59]; Yuan et al. [60]; Mladenovic et al. [25]; Liu et al. [72]; Liyanage and Villalba [49]; Dixon et al. [61]; Lam and Javed [24]; Cheung et al. [63]; Meng et al. [64]; Zhang(b) [15] Li et al. [73]	10
18	Supervision mechanism	Wang et al. [56]; Hongping and Sudong [69]; Jingfeng et al. [70]; Qian and Xinli [71]	4
19	Government credit	Robert et al. [15,69]; Hofmeister and Borchert [67]; Qian and Xinli [71]; Xia et al. [81]	4
20	Project quality	Ahadzie et al. [74]; Chan and Chan [58]; Baccarini [78]; Cox et al. [79]; Lai and Lam [80]; Chan et al. [59]; Liyanage and Villalba [49]; Dixon et al. [61]; Jingfeng et al. [70]	9
21	Economic policy	Robert et al. [15]; Xueqing et al. [43]; Hofmeister and Borchert [67]; Binquan and Tong [68]; Jingfeng et al. [70]; Verhoest et al. [36]; Qian and Xinli [71]	7
22	Financial market	Robert et al. [15]; Xueqing et al. [43]; Hofmeister and Borchert [67]; Binquan and Tong [68]; Hongping and Sudong [69]; Qian and Xinli [71]	6
23	Feasibility study	Zhen-Yu Zhao [69]; Robert et al. [15]; Hofmeister and Borchert [67]; Binquan and Tong [68]; Jingfeng et al. [70]; Qian and Xinli [71]	6
24	Project performance assessment	Osei-Kyei and Chan [16]; Jingfeng et al. [70], Mladenovic et al. [25]; Liu et al. [72]; Liyanage and Villalba [49]; Dixon et al. [61]	6

Table 1. *Cont.*

No.	Critical Success Factors	Authers	Sum
25	Stability of project operation	Chan and Chan [58]; Lim and Mohamed [75]; Cox et al. [79]	3
26	Flexible pricing mechanism	Wang et al. [56]; Jingfeng et al. [70]; Qian and Xinli [71]; Xia et al. [81]	4
27	Feasible implemention scheme	Cheung et al. [63]; Binquan and Tong [69]	2
28	Public support	Jingfeng et al. [70]; Qian and Xinli [72]	2
29	Cost-benefit assessment	Hofmeister and Borchert [67]; Binquan and Tong [68]	2
30	Government approval process	Hongping and Sudong [69]; Xia et al. [81]	2

2.2. A Case Study Analysis of Critical Success Factors

A Delphi survey was conducted on PPP projects that implemented in 2013–2018, and analyze critical success factors and their processing modes for PPP projects in mainland China.

Cases study and telephone interviews were conducted out to collect data from 40 successful and failed PPP projects in China (Table 2). As a result, 17 critical success factors were identified based on the reasons for the success or failures of these cases. The analysis showed that these critical success factors were mainly related to political influence. These include the effectiveness of risk management and risk allocation, the technical capacity of the private sector, long-term market demand, a long-term cooperative relationship, financial resources for the private sector, reasonable revenue distribution, and a complete legal framework (Table 3).

Table 2. Typical case studies of PPP projects in China.

No.	Successful Case	No.	Failed Case
1	Beijing subway Line 4	1	National Sports Complex
2	Shenzhen subway Line 4	2	Taiwan North-South highway
3	Dali urban and rural garbage disposal integrated system project	3	Wuhan Tangshunhu Sewage Treatment Plant
4	Shanghai Xinzhuang CCHP project	4	Changchun Huijin Sewage Treatment Plant
5	Gu'an industrial park new urbanization project	5	Jinzhou Sewage Treatment Plant
6	Chengdu No. 6 waterworks	6	Beijing No. 10 waterworks
7	Hefei Wangxiaoying Sewage Treatment Plant	7	Qingdao Veolia Sewage Treatment Plant
8	Guangxi Laibin B Power Plant	8	Shenzhen Wutongshan Tunnel
9	Jiangxi Xiajiang water conservancy project	9	Guangdong Lianjiang Sino-French Water Plant
10	Guangzhou–Shenzhen Expressway	10	Shanghai Dachang waterworks
11	Jiuquan city district cogeneration central heating project	11	Jiangsu Wujiang waste incineration plant
12	Nanjing Yangtze river bridge	12	Shanghai Yan'an Road.(E) Tunnel
13	Shaanxi south gate water conservancy project	13	Yangpu Bridge
14	Chongqing Fuling-Fengdu expressway project	14	Fujian Quanzhou Citong Bridge
15	Shenzhen University games center project	15	Huangqiao power plant
16	Zhangjiajie Yangjiaxi Sewage Treatment Plant	16	Wuhan 3rd Yangtze River Bridge
17	Wuzhong-Jingmaiyuan waste-to-energy incineration project	17	Zunyi North Suburb water plant
18	Weinan natural gas utilization project	18	Hangzhou Bay Bridge
19	Transfer Project of Tianjin NorthWater Co. Ltd.	19	Nanjing 3rd Yangtze River Bridge
20	Shenzhen Shajiao B power plant	20	Beijing five ring highway

Successful cases were selected from the typical cases of PPP projects in the national development and reform commission website of China (https://www.ndrc.gov.cn/xwdt/ztzl/pppzl/dxal/pppdxal/, accessed on 30 December 2020). Failure cases were selected from the typical cases in the related literature in the CNKI Database. Next, the implementation effect of all the cases listed in Table 2 were analyzed, and the study sorted and determined which factors affect project success in the actual process, as shown in Table 3. The goal was to facilitate the success of more PPP projects in the total project life cycle.

Table 3. Critical success factors base on case analysis.

No.	Critical Success Factor	Successful Case	Failed Case
1	Effectiveness of risk management and risk allocation	1,16	
2	Technical capacity of the private sector	3,19	
3	Long-term market demand		18,19,20
4	Long-term cooperative relationship	7,9	
5	Financial resources for the private sector	13,10,17,20	
6	Reasonable revenue distribution	1,2	
7	Complete legal framework	3	
8	Commitment and trust between the public and private sector	7,9	6,7,12
9	Financing capacity of the private sector	3,9,13,15,19	
10	Fair competitive procurement procedures	7,9,13	
11	Transparent procurement procedures	7,9,13	
12	Effective monitoring mechanism	1,3,17	
13	Good government credit		3,12,16
14	Stable economic policy	1,11,18	15,17
15	Project Feasibility Study completed and implemented	7	1,2,5
16	Flexible pricing mechanism	1,4,6,8,10	8,10,14
17	Effective exit mechanism	10,12,20	4

2.3. *Key Success Factors for PPP Projects*

Many factors could impact on the success of PPP projects' success, and it is possible to rank and classify the relative importance of these factors. Identifying the list of critical success factors of PPP projects is done by reviewing existing literature research results and experience summaries for typical domestic PPP projects. The effect of the factors on project success can be represented as a pyramid relationship (see Figure 1), with connections between the public sector, people, and private sector. This triangular pyramid clearly shows two analytical perspectives: horizontal and vertical relationships. For the vertical perspective, the public sector, private sector, and people have a common goal: project success. This perspective mainly embodies three aspects: project governance ensured by the public sector; project management promoted by the private sector, and satisfactory feedback by people. For the horizontal perspective, the public sector and private sector work together under a project contract, and include providers who can offer high-quality public services.

Based on a literature review, case summaries, and the triangular pyramid relationship in PPP projects, this study divided project success factors into three dimensions: relationships between stakeholders, project management in a Special Purpose Vehicle (SPV), and the external environment of a PPP project. First, the relationship between stakeholders included each participant's behavior, and the partnership and contractual relationship, including the technical ability of the social sector, government credit, and other factors. Second, the project management of a SPV is composed of technology and management factors, impacting the project success in project life cycle management. Examples of this include risk allocation in risk management, investment control, and other factors. Third, the external environmental holds uncontrollable factors that affect the implementation effect of PPP projects, such as a sound legal framework and credible economic policies. Therefore, after collection, screening, and analysis processes, the literature research and case analysis yielded a final list of 32 CSFs (named A_i, B_i, or C_i) and grouped as A, B, and C on Table 4.

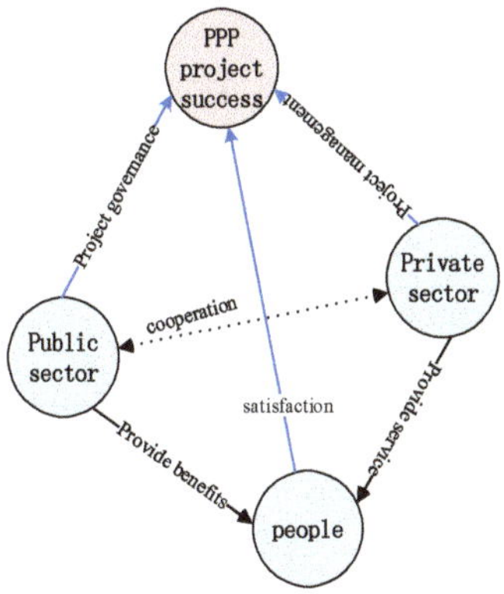

Figure 1. Triangular pyramid relationship in PPP projects.

Table 4. Critical Success Factors in a PPP project.

Factor Group	A: Stakeholder Relationships	B: External Environmental	C: Project Management of a Special Purpose Vehicle (SPV)
Factors	A_1: Technical capabilities of the social sector A_2: Government credit A_3: Examination and approval procedure A_4: Flexibility of pricing mechanism A_5: Financial resources of private sector A_6: Private sector financing capacity A_7: Management capabilities of the private sector A_8: Effective of the regulatory mechanism A_9: Government commitment or guarantee A_{10}: Long-term cooperative relationship	B_1: Completeness of legal framework B_2: Public opposition and political protest B_3: Economic policy change B_4: Local economic development level B_5: Available financial markets B_6: Favorable public support B_7: Long-term market demand B_8: Renegotiation and arbitration	C_1: Feasibility study and implementation plan C_2: Competitive bidding C_3: Transparency of bidding C_4: Effectiveness of risk management C_5: Project investment and cost control C_6: Project quality C_7: The feasibility of operation mode C_8: Terms of cooperation C_9: Revenue distribution C_{10}: Operational stability C_{11}: Project Feasibility Study Report C_{12}: Cost-benefit assessment C_{13}: Performance Evaluation C_{14}: Exit mechanism

3. Methodology

Fuzzy synthetic evaluation (FSE) is a branch of fuzzy set theory [85], it has been developed and extensively applied in different disciplines to quantify multi-evaluations and multi-attributes. These fields include including knowledge management [86], human resource management [87], and construction megaprojects [88], and risk management or risk assessment in PPP projects [16,18,89]. It is an analytical tool that objectifies the subjective judgment inherent in human decision-making. Therefore, this study applies this method to construct the project success index (PSI) equation to analyze the decision-making strategies between the public sector and the private sector.

3.1. Questionnaire Survey

A questionnaire survey was conducted to assess the significance of the identified project success factors; it was completed by scholars, experts and project managers for different types of infrastructure-focused PPP projects. This survey allows respondents to have time to carefully ponder over their responses without any interference from researchers.

The questionnaire survey was sent by email and conducted over 6 months, with a recovery rate of 72% (108 valid questionnaires from the 150 questionnaires distributed). Following those questionnaires, the critical success factors (CSF) influencing the establishment of a sustainable PPP were extracted.

Respondents represented the private sector, financial institutions, advisory institution universities, research institutions, and the public sector. Table 5 shows the sectors and experience levels in PPP projects. A total of 49.07% of respondents were from engineering advisory institutions; 24.07% came from universities or research institutions; 18.52% came from the private sector, 2.78% came from financial institutions, and 3.7% came from other types of organizations. The distribution of respondents was consistent with PPP project stakeholders, essentially representing all stakeholders across the PPP project life cycle.

Table 5. The Profile of companies, respondents, and projects.

Characteristics	Category	Number	Percentage
Sector of respondents	Private sector	20	18.52
	Financial institution	3	2.78
	Advisory institution	53	49.07
	Universities or research institutions	26	24.07
	Public sector	2	1.85
	Other	4	3.7
	Total	108	100
Years of working or research experience	2 years below	34	31.48
	2–5 years	54	50
	6 years and above	20	18.52
	Total	108	100

The data about the respondents' number of working years were as follows: 31.48% had less than 2 years of work experience; 50% had 2–5 years of experience; and the others had more than 5 years of experience. Among the 108 questionnaires managed by the respondents, 68.52% had more than 5 years of working years, with rich work experience. This screening information ensured quality, reduced the occurrence of potential risks, and improved the accuracy of the research conclusions.

Since respondents may be engaged in multiple types of PPP projects, in order to avoid the problem of overgeneralization, the author made multiple choice on the type of PPP projects the respondent has been engaged in questionnaire. The results showed that most of PPP projects engaged by respondents are distributed in the following Figure 2: such as 63 transportations, 46 water conservancy, 55 ecological construction and environmental protection, 65 municipal engineering, 42 government infrastructure construction, 29 comprehensive pipeline development, etc. This data conforms to the current development trend of PPP projects. Therefore, it is crucial to identify and analyze the key success factors of PPP projects.

The project success questionnaire included two parts: (1) the background information of respondents and their experience working on a PPP project, and (2) the Likert scale structured questions about the importance of the identified project success factors. For part one, the 108 questionnaires assessed experiences with different kinds of infrastructure types, including subways, waterworks, highways, energy, transportation, and water and waste treatment projects. This ensured the veracity and consistency of the research. For part two, respondents were requested to rate their degree of agreement against each of the identified CSFs, using a five-point Likert scale as follows: 1—Can be ignored or not important; 2—Maybe important; 3—Important; 4—Very important; 5—Most important.

Table 6 reports basic statistical parameters for the CSFs from the questionnaire, generated using SPSS24.0 software.

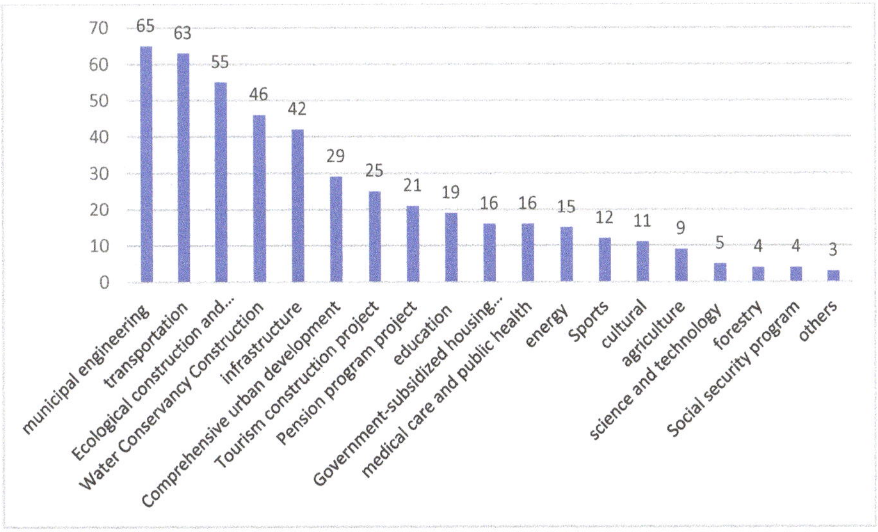

Figure 2. Types of PPP projects from respondents.

Table 6. Statistic of critical success factors.

Factor Group	Factor	Mean	Standard Deviation	Normalization	Rank	Weights
A:Stakeholders relationship	A_1: Technical capabilities of the social sector	4.2	0.733	0.5670	15	0.0989
	A_2: Government credit	4.58	0.657	0.9588	2	0.1078
	A_3: Examination and approval procedure	4.03	0.803	0.3918	25	0.0949
	A_4: Flexibility of pricing mechanism	3.97	0.703	0.3299	26	0.0935
	A_5: Financial resources of private sector	4.2	0.694	0.5670	16	0.0989
	A_6: Private sector financing capacity	4.62	0.575	1.0000	1	0.1088
	A_7: Management capability of private sector	4.25	0.672	0.6186	12	0.1
	A_8: Effectiveness of regulatory mechanism	4.17	0.755	0.5361	18	0.0982
	A_9: Government commitment or guarantee	4.37	0.705	0.7423	5	0.1029
	A_{10}: Long-term cooperative relationship	4.09	0.838	0.4536	22	0.0963

Table 6. *Cont.*

Factor Group	Factor	Mean	Standard Deviation	Normalization	Rank	Weights
B:External environmental	B_1: Completeness of legal framework	4.34	0.738	0.7113	6	0.1348
	B_2: Public opposition and political protest	3.9	0.853	0.2577	27	0.1211
	B_3: Economic policy change	4.07	0.732	0.4330	23	0.1264
	B_4: Local economic development level	4.04	0.76	0.4021	24	0.1255
	B_5: Available financial markets	4.31	0.703	0.6804	9	0.1339
	B_6: Favorable public support	3.65	0.868	0.0000	32	0.1134
	B_7: Long-term market demand	4.1	0.669	0.4639	21	0.1273
	B_8: Renegotiation and arbitration	3.79	0.724	0.1443	30	0.1177
C:Project management of Special Purpose Vehicle (SPV)	C_1: Feasibility study and implementation plan	4.42	0.643	0.7938	3	0.0881
	C_2: Competitive bidding	3.86	0.803	0.2165	29	0.0769
	C_3: Transparency of bidding	3.9	0.917	0.2577	28	0.0777
	C_4: Effectiveness of risk management	4.42	0.685	0.7938	4	0.0881
	C_5: Project investment and cost control	4.33	0.684	0.7010	7	0.0863
	C_6: Project quality	4.25	0.712	0.6186	13	0.0847
	C_7: The feasibility of operating mode	4.31	0.636	0.6804	10	0.0859
	C_8: Terms of cooperation	3.79	0.737	0.1443	31	0.0755
	C_9: Revenue distribution	4.32	0.609	0.6907	8	0.0861
	C_{10}: Operational stability	4.19	0.699	0.5567	17	0.0835
	C_{11}: Project Feasibility Study Report	4.12	0.758	0.4845	19	0.0821
	C_{12}: Cost-benefit assessment	4.28	0.681	0.6495	11	0.0853
	C_{13}: Performance Evaluation	4.21	0.749	0.5773	14	0.0881
	C_{14}: Exit mechanism	4.11	0.74	0.4742	20	0.0769

Table 6 shows that 25 critical success factors received a score at 4 or above; and 7 other factors scored between 3.65 and 4. This indicated there was some internal connection between 32 factors and project success in PPP projects. The top four scores included the financing capacity of the private sector, government credit, a feasibility study report and implementation plan, and the effectiveness of risk management, at 4.62, 4.58, 4.42, and 4.42, respectively. This indicates that respondents believe these factors have the greatest impact on PPP success projects. Therefore, PPP project participants should consider the above factors as a core concern, introducing the vitality of social capital, increasing market employment competition, improving infrastructure construction, and reducing financial pressure.

3.2. Data Analysis

The proposed fuzzy synthetic evaluation model is a multi-criteria evaluation model [16,43,90] for critical success factors, requiring six steps:

Step 1: Establish the set of basic critical success factors as $U = \{f_1, f_2, \cdots f_n\}$, where n is the number of critical success factors;

Step 2: Establish the grade alternatives as $L = \{L_1, L_2, \cdots L_5\}$, with the set of grade categories being the scale measurement. A 5-point Likert scale was used as the set of grade alternatives: L_1 is least important, L_2 is fairly important, L_3 is important, L_4 is very important, and L_5 is extremely important.

Step 3: Establish the set of basic critical success factors weight as $w = \{w_1, w_2, \cdots w_n\}$. The weighting ($w$) was determined from the survey using the following equation:

$$w_i = M_i / (\sum_{i=1}^{5} M_i),\ 0 \le w_i \le 1,\ 0 \le i \le 1,$$

In this expression, w_i is weighting and $\sum_{i=1}^{5} w_i = 1$, and M_i is mean score of a particular criterion or factor component.

In Step 3, the weights of each factors are calculated from the indexes obtained using SPSS. An example includes the technical capabilities for social sector (A_1)

$$W_{A_1} = \frac{4.2}{4.2 + 4.58 + 4.03 + 3.97 + 4.2 + 4.62 + 4.25 + 4.17 + 4.37 + 4.09} = 0.0989$$

Based on Step 3, we determine following weights of success factors (See Table 6).

Step 4: Generate a CSF evaluation matrix: $R_i = (r_{ij})_{m \times n}$, where r_{ij} denotes the degree to which the alternative L_j satisfies the criterion f_i. Let:

$$\mathrm{R_i} = \begin{pmatrix} MF_{u_{i1}} \\ MF_{u_{i2}} \\ \cdots \\ MF_{u_{in}} \end{pmatrix} \tag{1}$$

In this expression, $MF_{u_{i1}} = \left(\frac{N_{L_1}}{N}, \frac{N_{L_2}}{N}, \cdots, \frac{N_{L_5}}{N}\right)$; $N = 108$; MF is the membership function; and N_{L_i} is the number of critical success factors f_i from the questionnaires. For example, when examining the first critical success factor about technical capacity in the private sector, one person selected L_1 as the least important; no one selected L_2 as fairly important; 14 people selected L_3 as important; 54 people selected L_4 as very important; and 39 people selected L_5 as extremely important. This resulted in the following expression:

$$MF_{u_{11}} = \left(\frac{1}{108}, \frac{0}{108}, \frac{14}{108}, \frac{54}{108}, \frac{39}{108}\right) = (0.009, 0.000, 0.129, 0.500, 0.362)$$

Step 5: Calculate the data for the weights and evaluation results, shown in Table 7.

Step 6: Generate final fuzzy synthetic evaluation results for the evaluation by considering the weighting vector and the fuzzy evaluation matrix, using the following equation:

$$T = W \times R = (w_1, w_2, \cdots, w_n) \times \begin{pmatrix} r_{11} & r_{12} & \cdots & r_{1m} \\ r_{21} & r_{22} & \cdots & r_{1m} \\ \vdots & \vdots & \vdots & \vdots \\ r_{n1} & r_{n2} & \cdots & r_{nm} \end{pmatrix} = (t_1, t_2, \cdots, t_n) \tag{2}$$

Table 7. Fuzzy relational matrix data indicators for critical success factors.

	Stakeholders Relationship			Weight					Evaluation Result		
1	A_1: Technical capabilities of the social sector	0.009	0	0.13	0.5	0.36	0.005	0.008	0.13	0.44	0.42
2	A_2: Government credit	0.009	0	0.037	0.31	0.65	—	—	—	—	—
3	A_3: Examination and approval procedure	0.009	0.019	0.194	0.49	0.29	—	—	—	—	—
4	A_4: Flexibility of pricing mechanism	0.009	0.009	0.176	0.61	0.19	—	—	—	—	—
5	A_5: Financial resources of private sector	0	0.009	0.13	0.51	0.35	—	—	—	—	—
6	A_6: Private sector financing capacity	0	0	0.046	0.29	0.67	—	—	—	—	—
7	A_7: Management capability of the private sector	0	0	0.13	0.49	0.38	—	—	—	—	—
8	A_8: Effectiveness of regulatory mechanism	0	0.019	0.157	0.46	0.36	—	—	—	—	—
9	A_9: Government commitment or guarantee	0	0.009	0.102	0.40	0.49	—	—	—	—	—
10	$A1_0$: Long-term cooperative relationship	0.009	0.019	0.194	0.43	0.35	—	—	—	—	—
	External Environmental			Weight			0.003	0.021	0.21	0.47	0.30
11	B_1: Completeness of legal framework	0	0.009	0.13	0.37	0.49	—	—	—	—	—
12	B_2: Public opposition and political protest	0.009	0.037	0.25	0.45	0.25	—	—	—	—	—
13	B_3: Economic policy change	0	0.019	0.176	0.52	0.29	—	—	—	—	—
14	B_4: Local economic development level	0	0.019	0.213	0.48	0.29	—	—	—	—	—
15	B_5: Available financial markets	0	0.009	0.111	0.44	0.44	—	—	—	—	—
16	B_6: Favorable public support	0.019	0.046	0.361	0.42	0.16	—	—	—	—	—
17	B_7: Long-term market demand	0	0.009	0.148	0.57	0.27	—	—	—	—	—
18	B_8: Renegotiation and arbitration	0	0.028	0.306	0.52	0.15	—	—	—	—	—
	Project Management of Special Purpose Vehicle (SPV)			Weight			0.003	0.01	0.15	0.47	0.37
19	C_1: Feasibility study and implementation plan	0	0	0.083	0.417	0.5	—	—	—	—	—
20	C_2: Competitive bidding	0	0.019	0.343	0.4	0.24	—	—	—	—	—
21	C_3: Transparency of bidding	0.028	0.019	0.25	0.435	0.269	—	—	—	—	—
22	C_4: Effectiveness of risk management	0	0.009	0.083	0.389	0.519	—	—	—	—	—

Table 7. *Cont.*

Stakeholders Relationship		Weight					Evaluation Result				
23	C_5: Project investment and cost control	0	0.009	0.093	0.453	0.444	—	—	—	—	—
24	C_6: Project quality	0	0.009	0.13	0.463	0.398	—	—	—	—	—
25	C_7: The feasibility of operating mode	0	0	0.093	0.5	0.407	—	—	—	—	—
26	C_8: Terms of cooperation	0.009	0.019	0.29	0.546	0.139	—	—	—	—	—
27	C_9: Revenue distribution	0	0	0.074	0.528	0.398	—	—	—	—	—
28	C_{10}: Operational stability	0	0.009	0.139	0.509	0.43	—	—	—	—	—
29	C_{11}: Project Feasibility Study Report	0	0.019	0.176	0.472	0.333	—	—	—	—	—
30	C_{12}: Cost-benefit assessment	0	0.019	0.074	0.519	0.389	—	—	—	—	—
31	C_{13}: Performance Evaluation	0.009	0	0.139	0.472	0.38	—	—	—	—	—
32	C_{14}: Exit mechanism	0	0.019	0.167	0.5	0.315	—	—	—	—	—

In this expression. t_i is the fuzzy set of the membership, and "·" is the fuzzy operator. For example, we can calculate the membership of the external environment:

$$T_B = (0.135\ 0.121\ 0.127\ 0.125\ 0.134\ 0.113\ 0.127\ 0.118) \times \begin{bmatrix} 0 & 0.009 & 0.13 & 0.370 & 0.491 \\ 0.009 & 0.037 & 0.25 & 0.454 & 0.25 \\ 0 & 0.019 & 0.18 & 0.519 & 0.287 \\ 0 & 0.019 & 0.212 & 0.481 & 0.287 \\ 0 & 0.009 & 0.111 & 0.444 & 0.435 \\ 0.019 & 0.046 & 0.361 & 0.417 & 0.157 \\ 0 & 0.009 & 0.148 & 0.574 & 0.269 \\ 0 & 0.028 & 0.306 & 0.519 & 0.148 \end{bmatrix}$$
$$= (0.0032\ 0.02133\ 0.2073\ 0.4717\ 0.2965)$$

Step 7: Normalize the final FSE evaluation matrix and calculate a PSI for a particular factor component using the following equation:

$$PSI = \sum_{i=1}^{5} T \times L \tag{3}$$

From (3), we have

$$PSI_B = (0.0032\ 0.02133\ 0.2073\ 0.4717\ 0.2965) \times \begin{pmatrix} 1 \\ 2 \\ 3 \\ 4 \\ 5 \end{pmatrix} = 4.0368$$

Based on Step 6, we obtain the PSI of stakeholders' relationship and project management of Special Purpose Vehicle in Table 8.

Table 8. PSI index order.

No.	Success Factor Group	PSI Index	Coefficients	Rank
1	Stakeholder relationships	4.259	0.341	1
2	External environment	4.037	0.323	3
3	Project management of the Special Purpose Vehicle	4.188	0.336	2

The project success index for PPP projects in China is therefore expressed using the following equation:

$$\begin{aligned} PSI = \; & 0.341 \times stakeholders\, relationship \\ & +0.323 \times external\, environmental \\ & +0.336 \times project\, management\, of\, SPV \end{aligned} \quad (4)$$

4. Results

Equation (4) shows that stakeholder's relationships yielded the highest coefficient (0.341) in the evaluation model, followed by project management of a Special Purpose Vehicle (0.336) and external environment (0.323). The sum of all these coefficients is one, which fits within the unity threshold. This success index equation should significantly enable practitioners in China to evaluate the success level of their PPP projects in a practical and reliable manner. What is more, the evaluation model makes it possible for practitioners to compare the success levels of two or more projects at the same level. The application of this research should improve the implementation practices of PPP projects in China.

This section discusses the top nine critical success factors that has divide into three success groupings in formulating sustainable PPP. The top three factors concerning stakeholder relationships include private sector financing capacity, government credit, and government commitment or guarantee. The top two factors related to the external environment include: completeness of legal framework and available financial markets. The top four factors related to the project management of the social purpose vehicle included: the feasibility study report and implementation, effectiveness of risk management, project investment, and cost control and revenue distribution. The high overall confirmed that the PSI was necessary for PPP projects in China.

4.1. Stakeholder Relationships

The stakeholder relationship category had a PFI of 4.259 and a coefficient value of 0.341 in the critical success factors evaluation model. Previous studies have also noted the stakeholder relationship category as critical criteria for most traditional construction projects [91–94].

Among the 32 critical success factors, "private sector financing capacity" was ranked at the top, mainly attributed to the reduction in the financial burden on the government. The availability of flexible and attractive financial instruments is expected important to enable the private sector to finance PPP projects; these instruments include debt, equity, supplier and purchaser credit, and securities [64]. PPP projects are generally large infrastructure construction projects, and face a paradox due to uncertainty and the fact that available information is not aligned throughout the PPP projects' life cycle [95]. Additionally, PPP projects are funded by private financing; the public sector self-finances a certain proportion of the expenses. Self-financing for the public sector and private financing require significant synergies that can contribute to PPP project success.

Government credit was the second most important factor impacting PPP project success. A failure by public agencies to fulfill their obligations in the concession contract can directly or indirectly negatively affect the project. Government credit poses a critical risk to PPP projects in different sectors [96]. A perfect credit system could improve the efficiency of PPP implementation [97]. However, it has been reported that the probability of local public agencies breaching contracts has been relatively high in China [98,99]. In infrastructure PPP projects, good government credit is a critical factor impacting PPP project success [100]. There some PPP projects were not successful, such as failed cases 14, 17, and 19 in Table 2.

Government commitment or guarantee was ranked as the fifth most significant factor, and was attributed to improvements in the investment motivation of the private consortiums in PPP projects [101], and can influence the magnitude of political and regulatory risks [102]. Government guarantees include credit guarantees, material supply and price guarantees, minimum income guarantees, and guarantees related to exchange rates, interest rates, and inflation. PPP projects with government guarantees can maximize social-economic net present value and simultaneously optimize welfare [38]. However, a stable long-term plan for PPP projects requires enhanced certainty with respect to the government commitment or guarantee. For example, Treasury (2012) [103] launched PF2 (the latest version of PFI), which devoted a full chapter titled "Strengthening the Procurement Process." This chapter stipulates the government's commitment to 'ensuring that PF2 procurement is faster and cheaper than PFI procurement has been in the past, without sacrificing quality and competitiveness' (HM Treasury 2012 [103]). Meanwhile, government guarantees tend to stimulate an express expansion of PPP projects (MoF-China 2014 [104]); to this end, China's Ministry of Finance and the National Development and Reform Commission has promulgated a series of PPP policies since 2014.

4.2. External Environmental

Completeness of the legal framework was ranked sixth in importance, because of the immature legal systems in China [9]. The scholar and the practitioner have been aware of the urgent need for the Chinese government to establish a sound legal and institutional system to successfully apply PPP projects in China [105]. Meanwhile, an increasing number of renegotiations [22,51], contract variations, adjustments and arbitration [34], and early terminations [21,27,106] have already been reported by PPP project practices in China. Inadequate legal systems have been named as a critical factor restrict the development of PPP projects in China.

Available financial markets were ranked as the ninth most critical success factor for PPP projects. Many researchers [16,17,39,48,73] have found that project financing is a critical factor for private sector investment in PPP projects. The validity of an efficient and mature financial market, with the benefits of low financing costs and a diversified range of financial products would incentivize private sector pick-up of PPP projects. An unattractive financial market can create an obstacle to the implementation of PPP projects [15].

4.3. Project Management of Special Purpose Vehicle

Feasibility study completion and implementation planning was ranked as the third most important factor. The feasibility study provides project data and instruments that facilitate a profound analysis and that assist the PPP project's decision making process. Generally, the feasibility study is an appropriate means to illustrate the PPP project's practicability and operability. The implementation plan and data are extracted from the feasibility study for a PPP project [107]. In the life cycle of a PPP project, identifying an uncertain factor could be quite difficult, unless detailed feasibility studies have been done to assure the project's viability and enforceability [44], and it can easily lead to project failure.

Effective risk management was ranked as the fourth most important factor. PPP project risk management practices are highly variable, intuitive, subjective, and unsophisticated [108]; this is likely to lead to project failure. Many studies [109–112] have shown that risk management is a critical concern in PPP projects and the efficient allocation of risk remains problematic [113–115]. Furthermore, previous studies on PPP practices [44,89,116,117] have documented the prevalence of inefficiencies in risk allocation.

There is a clear understanding and recognition that the nature of PPP risk misallocation must be quantitatively represented and verified when investing in a life-cycle PPP project. In addition, it is necessary to balance and share the considerations needed for effective risk distribution between the public and private sectors. However, parties that facilitate project success are guided by the premise of the basic principles of sound risk management.

Project investment and cost control was ranked as the seventh most critical factor. PPP project investments depend heavily upon private capital markets for financing and depend on private firms for managerial expertise. Since 2013, when China's economic growth entered a transition, the risk of debt exposure emerged, and PPP projects became a main approach for infrastructure construction. Since China's promotion of PPP projects, PPP project investments have gotten out of control, leading to an increase in government expenditure responsibility. This directly affects earnings to public and private sector.

Revenue distribution was listed as the ninth most critical factor. When revenue is distributed, the two parties compete for interests and strive to minimize their own opportunity costs. Under market competition, public and private sector achieve a win–win situation through cooperation and competition [118]. Nonetheless, unreasonable revenue distribution can affect a project's normal operation. There is the need for a revenue distribution mechanism, where the government ensures extra revenues. Therefore, identifying revenue as an attribution mechanism is indispensable as a suitable mitigation strategy to mitigate traffic revenue risks in PPP transport infrastructure projects [112].

5. Conclusions

PPP projects have been implemented to support infrastructure development in both developed and developing countries with diverse results, and many researchers claimed that PPP can contribute to sustainability in China as it promotes long-term productive use of resources [119,120]. These provide a mechanism for investing in public infrastructure, while also effectively transferring the government function to the private sector. Meanwhile, this generates significant problems as an increasing number of project failures appear. In China, from 2013 to 2019 (years inclusive), CPPPC data (http://www.cpppc.org:8086/pppcentral/map/toPPPMap.do, accessed on 30 December 2020) show that the market capacity for PPP projects reached nearly 10,000 projects, with a total investment of more than 13.7 trillion yuan BRI data (http://www.bridata.com/, accessed on 30 December 2020) show that China's PPP projects occupy a market share of 15.4 trillion yuan, with the number of PPP projects reaching 10,226 projects. However, with the release of normative documents from central government in China, thousands of PPP projects have been withdrawn from the CPPPC library in the large PPP market. Those unreasonable exit phenomenon needs are more detailed identification of critical success factors for PPP projects. Then, this study defined and categorized the factors affecting project success and failure. From this classification and definition, we applied a fuzzy synthetic method to prioritize these factors and provide an evaluation criterion.

In fact, by using the fuzzy synthetic evaluation model for PPP projects, the most critical success factors for different types of PPP projects could be identified and both precautionary and remedial actions could be taken as soon as possible. Both the public sector and private sector can adopt this model to assess the risk level of their PPP projects. And the results can be used to compare the critical success factor levels with their counterparts for benchmarking purposes. Such an extension would provide a deeper understanding of managing different types of PPP projects. Since the critical success factor level may vary at various stages of a project life cycle, it is worthwhile to develop a PPP fuzzy synthetic evaluation model for measuring critical success factors across different stages of a project life cycle in future.

First, due to the wide range of success factors and categories amassed by researchers [121], this study reviewed recent literature and cases to define the success factors of PPP projects during the period 2000 to 2019, highlighting the research contributions by various countries with respect to their authors and institutions. Therefore, 32 success factors were sorted from recent literature and cases. These were then divided into three dimensions: stakeholder relationships, external environment, and the project management of a special purpose vehicle (SPV).

Second, based on 32 defined factors, a questionnaire was designed and distributed to experts, researchers, and PPP project managers. Survey data were then collected, and mean score values of the response data were used to rank the relative importance of 32 critical success factors in the China PPP environment. Then, 10 factors emerged as being most important in developing a successful China PPP: private sector financing capacity, government credit, feasibility study, effective risk management, government commitment, completeness of legal framework, project investment control, revenue distribution, available financial markets, and operational feasibility.

Finally, a fuzzy synthetic method was applied to prioritize the critical success factors. Despite the model's applications and the survey and case study results, this research did have some constraints. Extending the sample frame to other type of PPP projects could improve the validity of the research model. Examining a similar model with other projects and other countries and comparing them could yield practical results

Different success factors were identified using a questionnaire survey, case studies, literature review, fuzzy synthetic methods, and interviews and correspondence with worldwide PPP experts and practitioners. Furthermore, this assessment provides results in terms of the performance of dominant CSFs. This can be useful when prioritizing PPP project tasks. These approaches are valid, could be used globally for other PPP projects, and may also be evaluated with respect to CSFs in a PPP context.

Author Contributions: B.D., D.Z. and Y.Y., conceived the study; D.Z., J.Z. and X.L. conducted the literature review, developed the model, designed the experiments and performed the experiments; B.D., D.Z. and J.Z. analyzed the data and results; B.D., D.Z., J.Z. and X.L. edited the final version of the paper. All authors have read and agreed to the published version of the manuscript.

Funding: 1. the National Natural Science Foundation of China: 71602144; 2. the Tianjin education commission Project of Key Research Institute of Humanities and Social Sciences at Universities: 2017JWZD15; 3. the Program for Innovation Research Team in Universities of Tianjin: TD13-5019.

Institutional Review Board Statement: Not applicable.

Informed Consent Statement: Not applicable.

Data Availability Statement: Not applicable.

Acknowledgments: The authors thank the National Natural Science Foundation of China (Grant No. 71602144), the Tianjin education commission Project of Key Research Institute of Humanities and Social Sciences at Universities (Grant No. 2017JWZD15), and the Program for Innovation Research Team In Universities of Tianjin (TD13-5019).

Conflicts of Interest: The authors declare that they have no competing interests. Authors' contributions: all authors contributed equally and significantly in writing this paper. All authors read and approved the final manuscript.

References

1. Yuan, J.F.; Skibniewski, M.J.; Li, Q. The driving factors of china's public—Private partnership projects in Metropolitian transportation systems: Public sector's viewpoint. *J. Civ. Eng. Manag.* **2010**, *16*, 5–18. [CrossRef]
2. Jin, X.H.; Zhang, G. Modelling optimal risk allocation in PPP projects using artificial neural networks. *Int. J. Proj. Manag.* **2011**, *29*, 591–603. [CrossRef]
3. Zhu, L.; Zhao, X.; Chua, D.K.H. Agent-based debt terms' bargaining model to improve negotiation inefficiency in PPP projects. *J. Comput. Civ. Eng.* **2016**, *30*, 04016014. [CrossRef]
4. Yang, T.; Long, R.; Cui, X. Application of the public–private partnership model to urban sewage treatment. *J. Clean. Prod.* **2017**, *142*, 1065–1074. [CrossRef]
5. Lomoro, A.; Mossa, G.; Pellegrino, R. Optimizing Risk Allocation in Public-Private Partnership Projects by Project Finance Contracts. The Case of Put-or-Pay Contract for Stranded Posidonia Disposal in the Municipality of Bari. *Sustainability* **2020**, *12*, 806. [CrossRef]

6. Martiniello, L.; Morea, D.; Paolone, F. Energy Performance Contracting and Public-Private Partnership: How to Share Risks and Balance Benefits. *Energies* **2020**, *13*, 3625. [CrossRef]
7. Zhang, L.; Sun, X.; Xue, H. Identifying critical risks in Sponge City PPP projects using DEMATEL method: A case study of China. *J. Clean. Prod.* **2019**, *226*, 949–958. [CrossRef]
8. Almarri, K.; Blackwell, P. Improving risk sharing and investment appraisal for PPP procurement success in large green projects. *Procedia-Soc. Behav. Sci.* **2014**, *119*, 847–856. [CrossRef]
9. Zhang, S.; Chan, A.P.C.; Feng, Y. Critical review on PPP Research–A search from the Chinese and International Journals. *Int. J. Proj. Manag.* **2016**, *34*, 597–612. [CrossRef]
10. Grimsey, D.; Lewis, M.K. Evaluating the risks of public private partnerships for infrastructure projects. *Int. J. Proj. Manag.* **2002**, *20*, 107–118. [CrossRef]
11. Morea, D.; Balzarini, M. Bankability of a public private partnership in agricultural sector: A project in Sub Saharan Africa. *Agric. Econ. (Agricecon)* **2019**, *65*, 212–222. [CrossRef]
12. Morea, D.; Marino, B. Financial sustainability of a public-private partnership for an agricultural development project in Sub-Saharan Africa. *Agric. Econ. (Agricecon)* **2018**, *64*, 389–398.
13. Visconti, R.M.; Martiniello, L.; Morea, D. Can Public-Private Partnerships Foster Investment Sustainability in Smart Hospitals? *Sustainability* **2019**, *11*, 1704. [CrossRef]
14. Kurniawan, F.; Mudjanarko, S.W.; Ogunlana, S. Best practice for financial models of PPP projects. *Procedia Eng.* **2015**, *125*, 124–132. [CrossRef]
15. Osei-Kyei, R.; Chan, A.P.C. Review of studies on the Critical Success Factors for Public–Private Partnership (PPP) projects from 1990 to 2013. *Int. J. Proj. Manag.* **2015**, *33*, 1335–1346. [CrossRef]
16. Osei-Kyei, R.; Chan, A.P.C.; Ameyaw, E.E. A fuzzy synthetic evaluation analysis of operational management critical success factors for public-private partnership infrastructure projects. *Benchmarking: Int. J.* **2017**, *24*, 2092–2112. [CrossRef]
17. Chou, J.S.; Pramudawardhani, D. Cross-country comparisons of key drivers, critical success factors and risk allocation for public-private partnership projects. *Int. J. Proj. Manag.* **2015**, *33*, 1136–1150. [CrossRef]
18. Ameyaw, E.E.; Chan, A.P.C. Critical success factors for public-private partnership in water supply projects. *Facilities* **2016**, *34*, 124–160. [CrossRef]
19. Bing, L.; Akintoye, A.; Edwards, P.J.; Hardcast, C. The allocation of risk in PPP/PFI construction projects in the UK. *Int. J. Proj. Manag.* **2005**, *23*, 25–35. [CrossRef]
20. Pfnür, A.; Armonat, S. Modelling uncertain operational cash flows of real estate investments using simulations of stochastic processes. *J. Prop. Invest. Financ.* **2013**, *31*, 481–501. [CrossRef]
21. Song, J.; Zhang, H.; Dong, W. A review of emerging trends in global PPP research: Analysis and visualization. *Scientometrics* **2016**, *107*, 1111–1147. [CrossRef]
22. Xiong, W.; Zhao, X.; Yuan, J.F.; Luo, S. Ex post risk management in public private partnerships infrastructure projects. *Proj. Manag. J.* **2017**, *48*, 76–89. [CrossRef]
23. Guasch, J.L.; Benitez, D.; Portables, I.; Folr, L. *The Renegotiation of PPP Contracts: An overview of its recent evolution in Latin America*; International Transport Forum Discussion Papers; OECD Publishing: Paris, France, 2014.
24. Lam, P.T.I.; Javed, A.A. Comparative study on the use of output specifications for Australian and UK PPP/PFI projects. *J. Perform. Constr. Facil.* **2013**, *29*, 04014061. [CrossRef]
25. Mladenovic, G.; Vajdic, N.; Wündsch, B.; Temeljotov-Salaj, A. Use of key performance indicators for PPP transport projects to meet stakeholders' performance objectives. *Built Environ. Proj. Asset Manag.* **2013**, *3*, 228–249. [CrossRef]
26. Domingues, S.; Zlatkovic, D. Renegotiating PPP contracts: Reinforcing the 'p' in partnership. *Transp. Rev.* **2015**, *35*, 204–225. [CrossRef]
27. Xiong, W.; Zhang, X. The real option value of renegotiation in public–private partnerships. *J. Constr. Eng. Manag.* **2016**, *142*, 04016021. [CrossRef]
28. Soecipto, R.M.; Verhoest, K. Contract stability in European road infrastructure PPPs: How does governmental PPP support contribute to preventing contract renegotiation? *Public Manag. Rev.* **2018**, *20*, 1145–1164. [CrossRef]
29. Osei-Kyei, R.; Chan, A.P.C. Implementing public–private partnership (PPP) policy for public construction projects in Ghana: Critical success factors and policy implications. *Int. J. Constr. Manag.* **2017**, *17*, 113–123. [CrossRef]
30. Keers, B.B.M.; Fenema, P.C. Managing risks in public-private partnership formation projects. *Int. J. Proj. Manag.* **2018**, *36*, 861–875. [CrossRef]
31. Babatunde, S.O.; Perera, S.; Zhou, L. Methodology for developing capability maturity levels for PPP stakeholder organisations using critical success factors. *Constr. Innov.* **2016**, *16*, 81–110. [CrossRef]
32. Osei-Kyei, R.; Chan, A.P.C. Perceptions of stakeholders on the critical success factors for operational management of public-private partnership projects. *Facilities* **2017**, *35*, 21–38. [CrossRef]
33. Singh, A.; Shrivastava, P.; Kambekar, A.R. Financial Risk Assessment of Public Private Partnership Project. *Sustain. Infrastruct. Dev. Manag. (Sidm)* **2019**, *20109*, 10. [CrossRef]
34. Osei-Kyei, R.; Chan, A.P.C.; Yu, Y.; Chen, C.; Dansoh, A. Root causes of conflict and conflict resolution mechanisms in public-private partnerships: Comparative study between Ghana and China. *Cities* **2019**, *87*, 185–195. [CrossRef]

35. Ahmadabadi, A.A.; Heravi, G. The effect of critical success factors on project success in Public-Private Partnership projects: A case study of highway projects in Iran. *Transp. Policy* **2019**, *73*, 152–161. [CrossRef]
36. Verhoest, K.; Petersen, O.H.; Scherrer, W.; Soecipto, R.M. *Policy Commitment, Legal and Regulatory Framework, and Institutional Support. for PPP in International Comparison: Indexing Countries' Readiness for Taking Up PPP*; Working Papers in Economics and Finance; University of Salzburg: Salzburg, Austria, 2014.
37. House, S. Responsive regulation for water PPP: Balancing commitment and adaptability in the face of uncertainty. *Policy Soc.* **2016**, *35*, 179–191. [CrossRef]
38. Jiang, Y. Selection of PPP Projects in China Based on Government Guarantees and Fiscal Risk Control. *Int. J. Financ. Res.* **2017**, *8*, 99–111. [CrossRef]
39. Ahmadabadi, A.A.; Heravi, G. Risk assessment framework of PPP-megaprojects focusing on risk interaction and project success. *Transp. Res. Part. A Policy Pract.* **2019**, *124*, 169–188. [CrossRef]
40. Rostiyanti, S.; Coffey, V.; Pangeran, M.H.; Tamin, R. A Critical Perspective of the Indonesian Institutional Framework for PPP Toll Roads. *Univ. Cent. Lancs. (Uclan) Preston* **2013**, *2013*, 415.
41. Oyedele, L.O. Avoiding performance failure payment deductions in PFI/PPP projects: Model of critical success factors. *J. Perform. Constr. Facil.* **2012**, *27*, 283–294. [CrossRef]
42. Xu, Y.; Yeung, J.F.Y.; Chan, A.P.; Chan, D.W.; Wang, S.Q.; Ke, Y. Developing a risk assessment model for PPP projects in China-A fuzzy synthetic evaluation approach. *Autom. Constr.* **2010**, *19*, 929–943. [CrossRef]
43. Valipour, A.; Yahaya, N.; Md-Noor, N. A fuzzy analytic network process method for risk prioritization in freeway PPP projects: An Iranian case study. *J. Civ. Eng. Manag.* **2015**, *21*, 933–947. [CrossRef]
44. Ng, S.T.; Wong, Y.M.W.; Wong, J.M.W. Factors influencing the success of PPP at feasibility stage–a tripartite comparison study in Hong Kong. *Habitat Int.* **2012**, *36*, 423–432. [CrossRef]
45. Zou, W.; Kumaraswamy, M.; Chung, J.; Wong, J. Identifying the critical success factors for relationship management in PPP projects. *Int. J. Proj. Manag.* **2014**, *32*, 265–274. [CrossRef]
46. Babatunde, S.O.; Perera, S.; Zhou, L.; Udeaja, C. Stakeholder perceptions on critical success factors for public-private partnership projects in Nigeria. *Built Environ. Proj. Asset Manag.* **2016**, *6*, 74–91. [CrossRef]
47. Li, B.; Akintoye, A.; Edwards, P.J.; Hardcastle, C. Critical success factors for PPP/PFI projects in the UK construction industry. *Constr. Manag. Econ.* **2005**, *23*, 459–471. [CrossRef]
48. Kwofie, T.E.; Afram, S.; Botchway, E. A critical success model for PPP public housing delivery in Ghana. *Built Environ. Proj. Asset Manag.* **2016**, *6*, 58–73. [CrossRef]
49. Liyanage, C.; Villalba-Romero, F. Measuring success of PPP transport projects: A cross-case analysis of toll roads. *Transp. Rev.* **2015**, *35*, 140–161. [CrossRef]
50. Osei-Kyei, R.; Chan, A.P.C. Evaluating the project success index of public-private partnership projects in Hong Kong: The case of the Cross Harbour Tunnel. *Constr. Innov.* **2018**, *18*, 371–391. [CrossRef]
51. Liang, Y.; Jia, H. Key success indicators for PPP projects: Evidence from Hong Kong. *Adv. Civ. Eng.* **2018**, *2018*, 9576496. [CrossRef]
52. Olusola Babatunde, S.; Opawole, A.; Emmanuel Akinsiku, O. Critical success factors in public-private partnership (PPP) on infrastructure delivery in Nigeria. *J. Facil. Manag.* **2012**, *10*, 212–225. [CrossRef]
53. Jefferies, M. Critical success factors of public private sector partnerships: A case study of the Sydney Superdome. *Eng. Constr. Archit. Manag.* **2006**, *13*, 451–462. [CrossRef]
54. Duy-Nguyen, L.; Ogunlana, S.O.; Thi-Xuan-Lan, D. A study on project success factors in large construction projects in Vietnam. *Eng. Constr. Archit. Manag.* **2004**, *11*, 404–413. [CrossRef]
55. Abdul-Aziz, A.R.; Kassim, P.S.J. Objectives, success and failure factors of housing public–private partnerships in Malaysia. *Habitat Int.* **2011**, *35*, 150–157. [CrossRef]
56. Cheong-Yong, Y.; Emma-Mustaffa, N. Analysis of factors critical to construction project success in Malaysia. *Eng. Constr. Archit. Manag.* **2012**, *19*, 543–556. [CrossRef]
57. Chan, A.P.C.; Lam, P.T.I.; Chan, D.W.M.; Cheung, E.; Ke, Y. Critical success factors for PPPs in infrastructure developments: Chinese perspective. *J. Constr. Eng. Manag.* **2010**, *136*, 484–494. [CrossRef]
58. Chan, A.P.C.; Chan, A.P.L. Key performance indicators for measuring construction success. *Benchmarking Int. J.* **2004**, *11*, 203–221. [CrossRef]
59. Chan, A.P.C.; Scott, D.; Lam, E.W.M. Framework of success criteria for design/build projects. *J. Manag. Eng.* **2002**, *18*, 120–128. [CrossRef]
60. Yuan, J.; Zeng, A.Y.; Skibniewski, M.J.; Li, Q. Selection of performance objectives and key performance indicators in public–private partnership projects to achieve value for money. *Constr. Manag. Econ.* **2009**, *27*, 253–270. [CrossRef]
61. Dixon, T.; Pottinger, G.; Jordan, A. Lessons from the private finance initiative in the UK: Benefits, problems and critical success factors. *J. Prop. Invest. Financ.* **2005**, *23*, 412–423. [CrossRef]
62. Zhang, X. Factor analysis of public clients' best-value objective in public–privately partnered infrastructure projects. *J. Constr. Eng. Manag.* **2006**, *132*, 956–965. [CrossRef]
63. Cheung, S.O.; Tam, C.M.; Ndekugri, I.; Harris, F.C. Factors affecting clients' project dispute resolution satisfaction in Hong Kong. *Constr. Manag. Econ.* **2000**, *18*, 281–294. [CrossRef]

64. Meng, X.; Zhao, Q.; Shen, Q. Critical success factors for transfer-operate-transfer urban water supply projects in China. *J. Manag. Eng.* **2011**, *27*, 243–251. [CrossRef]
65. Zhang, X. Critical success factors for public–private partnerships in infrastructure development. *J. Constr. Eng. Manag.* **2005**, *131*, 3–14. [CrossRef]
66. Qiao, L.; Wang, S.Q.; Tiong, R.L.K. Framework for critical success factors of BOT projects in China. *J. Proj. Financ.* **2001**, *7*, 53–61. [CrossRef]
67. Hofmeister, A.; Borchert, H. Public–private partnerships in Switzerland: Crossing the bridge with the aid of a new governance approach. *Int. Rev. Adm. Sci.* **2004**, *70*, 217–232. [CrossRef]
68. Binquan, C.; Tong, P. Critical Success Factors of Public-private Partnerships in Transport Infrastructure Project. *Urban. Rapid Rall Transit.* **2010**, *23*, 17–22.
69. Hongping, Z.; Sudong, Y. Study on the Determinants and their Relationships of PPP Projects. *Sci. Technol. Manag. Res.* **2016**, *36*, 203–207.
70. Yuan, J.; Skibniewski, M.J.; Deng, X.; Ji, C.; Li, Q. The Identification of Key Performance Indicators in Public Private Partnership Projects based on Structural Equation Modeling. *J. Chongqing Univ. (Soc. Sci. Ed.)* **2012**, *18*, 56.
71. Qian, L.; Xinli, L. Identification of critical success factors of PPP projects based on process analysis. *J. Railw. Sci. Eng.* **2017**, *14*, 415–424.
72. Liu, J.; Li, Q.; Wang, Y. Risk analysis in ultra-deep scientific drilling project-A fuzzy synthetic evaluation approach. *Int. J. Proj. Manag.* **2013**, *31*, 449–458. [CrossRef]
73. Li, B. Risk Management of Construction Public Private Partnership Projects. Ph.D. Thesis, Glasgow Caledonian University, Glasgow, UK, 2003.
74. Ahadzie, D.K.; Proverbs, D.G.; Olomolaiye, P.O. Critical success criteria for mass house building projects in developing countries. *Int. J. Proj. Manag.* **2008**, *26*, 675–687. [CrossRef]
75. Lim, C.S.; Mohamed, M.Z. Criteria of project success: An exploratory re-examination. *Int. J. Proj. Manag.* **1999**, *17*, 243–248. [CrossRef]
76. Bryde, D.J.; Robinson, L. Client versus contractor perspectives on project success criteria. *Int. J. Proj. Manag.* **2005**, *23*, 622–629. [CrossRef]
77. Al-Tmeemy, S.M.H.M.; Abdul-Rahman, H.; Harun, Z. Future criteria for success of building projects in Malaysia. *Int. J. Proj. Manag.* **2011**, *29*, 337–348. [CrossRef]
78. Baccarini, D. The logical framework method for defining project success. *Proj. Manag. J.* **1999**, *30*, 25–32. [CrossRef]
79. Cox, R.F.; Issa, R.R.A.; Ahrens, D. Management's perception of key performance indicators for construction. *J. Constr. Eng. Manag.* **2003**, *129*, 142–151. [CrossRef]
80. Lai, I.K.W.; Lam, F.K.S. Perception of various performance criteria by stakeholders in the construction sector in Hong Kong. *Constr. Manag. Econ.* **2010**, *28*, 377–391. [CrossRef]
81. Xia, Y.; Yongjian, K.; Shouqing, W. Analysis on Critical Risk Factors Causing the Failures of China's PPP Project. *J. China Soft Sci.* **2009**, *5*, 107–113.
82. Muhammad, Z.; Johar, F. A Conceptual Framework for Evaluating the Success of Public-Private Partnership (PPP) Projects. *Adv. Sci. Lett.* **2017**, *23*, 9130–9134. [CrossRef]
83. Wang, K.; Ke, Y. Public-Private Partnerships in the Electric Vehicle Charging Infrastructure in China: An Illustrative Case Study. *Adv. Civ. Eng.* **2018**, *2018*, 9061647. [CrossRef]
84. Emmanuel, O.O. Critical success factors (CSF) determining the implementation of public-private partnership projects. *Covenant J. Res. Built Environ.* **2013**, *1*, 1–16.
85. Zadeh, L.A. Fuzzy sets. *Information and Control.* **1965**, *8*, 338–353. [CrossRef]
86. Lyu, H.; Zhou, Z.; Zhang, Z. Measuring knowledge management performance in organizations: An integrative framework of balanced scorecard and fuzzy evaluation. *Information* **2016**, *7*, 29. [CrossRef]
87. Chou, Y.C.; Sun, C.C.; Yen, H.Y. Evaluating the criteria for human resource for science and technology (HRST) based on an integrated fuzzy AHP and fuzzy DEMATEL approach. *Appl. Soft Comput.* **2012**, *12*, 64–71. [CrossRef]
88. Boateng, E.B.; Pillay, M.; Davis, P. Developing a Safety Culture Index for Construction Projects in Developing Countries: A Proposed Fuzzy Synthetic Evaluation Approach. In Proceedings of the International Conference on Applied Human Factors and Ergonomics, Washington, DC, USA, 24–28 July 2019; Springer: Cham, Switherlands, 2019; pp. 167–179.
89. Wu, Y.; Song, Z.; Li, L.; Xu, R. Risk management of public-private partnership charging infrastructure projects in China based on a three-dimension framework. *J. Energy* **2018**, *165*, 1089–1101. [CrossRef]
90. Mu, S.; Cheng, H.; Chohr, M.; Wei, P. Assessing risk management capability of contractors in subway projects in mainland China. *Int. J. Proj. Manag.* **2014**, *32*, 452–460. [CrossRef]
91. Mazur, A.; Pisarski, A.; Chang, A.; Ashkanasy, N.M. Rating defence major project success: The role of personal attributes and stakeholder relationships. *Int. J. Proj. Manag.* **2014**, *32*, 944–957. [CrossRef]
92. Davis, K. Different stakeholder groups and their perceptions of project success. *Int. J. Proj. Manag.* **2014**, *32*, 189–201. [CrossRef]
93. Nguyen, T.H.D.; Chileshe, N.; Rameezdeen, R.; Wood, A. External stakeholder strategic actions in projects: A multi-case study. *Int. J. Proj. Manag.* **2019**, *37*, 176–191. [CrossRef]

94. Oliveira, G.F.; Rabechini, J.R. Stakeholder management influence on trust in a project: A quantitative study. *Int. J. Proj. Manag.* **2019**, *37*, 131–144. [CrossRef]
95. Samset, K.; Volden, G.H. Front-end definition of projects: Ten paradoxes and some reflections regarding project management and project governance. *Int. J. Proj. Manag.* **2016**, *34*, 297–313. [CrossRef]
96. Cheung, E.; Chan, A.P.C. Risk factors of public-private partnership projects in China: Comparison between the water, power, and transportation sectors. *J. Urban. Plan. Dev.* **2011**, *137*, 409–415. [CrossRef]
97. Wang, Y. Evolution of public–private partnership models in American toll road development: Learning based on public institutions' risk management. *Int. J. Proj. Manag.* **2015**, *33*, 684–696. [CrossRef]
98. Li, S. *The Legal Environment and Risks for Foreign Investment in China*; Springer Science & Business Media: Berlin/Heidelberg, Germany, 2007.
99. Li, Y.; Wang, X. Risk assessment for public–private partnership projects: Using a fuzzy analytic hierarchical process method and expert opinion in China. *J. Risk Res.* **2018**, *21*, 952–973. [CrossRef]
100. Li, S.; Abraham, D.; Cai, H. Infrastructure financing with project bond and credit default swap under public-private partnerships. *Int. J. Proj. Manag.* **2017**, *35*, 406–419. [CrossRef]
101. Valéro, V. Government opportunism in public—Private partnerships. *J. Public Econ. Theory* **2015**, *17*, 111–135. [CrossRef]
102. Hwang, B.G.; Zhao, X.; Gay, M.J.S. Public private partnership projects in Singapore: Factors, critical risks and preferred risk allocation from the perspective of contractors. *Int. J. Proj. Manag.* **2013**, *31*, 424–433. [CrossRef]
103. HM Treasury. *A New Approach to Public Private Partnerships*; HMSO: London, UK, 2012.
104. Ministry of Finance (MoF) China. Experience from Asian Development Bank in applying PPP model and Suggestions. *China State Financ.* **2014**, *9*, 18–19. (In Chinese)
105. Peng, J.B. On Chin's current status of BOT legislation, existing problems and possible solutions. *J. Law Econ.* **2011**, *10*, 60–63. (In Chinese)
106. Liu, J.; Gao, R.; Cheah, C.Y.J. Pricing mechanism of early termination of PPP projects based on Real Option Theory. *J. Manag. Eng.* **2017**, *33*, 04017035. [CrossRef]
107. Xu, Y.; Sun, C.; Skibniewski, M.J.; Yeung, J.F.Y.; Hu, C. System Dynamics (SD)-based concession pricing model for PPP highway projects. *Int. J. Proj. Manag.* **2012**, *30*, 240–251. [CrossRef]
108. Akintoye, A.; Beck-Hardcastle, C.; Chinyio, E.; Assenova, D. *Framework for Risk Assessment and Management of Private Finance Initiative Projects*; Glasgow Caledonian University: Glasgow, UK, 2001.
109. Ke, Y.; Wang, S.Q.; Chan, A.P.C.; Lam, P.T.I. Preferred risk allocation in China's public–private partnership (PPP) projects. *Int. J. Proj. Manag.* **2010**, *28*, 482–492. [CrossRef]
110. Ke, Y.; Wang, S.Q.; Chan, A.P.C.; Cheung, E. Understanding the risks in China's PPP projects: Ranking of their probability and consequence. *Eng. Constr. Archit. Manag.* **2011**, *18*, 481–496. [CrossRef]
111. Roumboutsos, A.; Pantelias, A. Allocating revenue risk in transport infrastructure public private partnership projects: How it matters. *Transp. Rev.* **2015**, *35*, 183–203. [CrossRef]
112. Carbonara, N.; Management, M.E.; Costantino, N.; Gunnigan, L. Risk management in motorway PPP projects: Empirical-based guidelines. *Transp. Rev.* **2015**, *35*, 162–182. [CrossRef]
113. Sastoque, L.M.; Arboleda, C.A.; Ponz, J.L. A proposal for risk allocation in social infrastructure projects applying PPP in Colombia. *Procedia Eng.* **2016**, *145*, 1354–1361. [CrossRef]
114. Shrestha, A.; Chan, T.K.; Aibinu, A.A.; Chen, C.; Asce, A.M.; Martek, I. Risk allocation inefficiencies in Chinese PPP water projects. *J. Constr. Eng. Manag.* **2018**, *144*, 04018013. [CrossRef]
115. Xia, X.; Hui, A. *Government-Led Rural Infrastructure PPP Project Risk Allocation. C. The First International Symposium on Management and Social Sciences (ISMSS 2019)*; Atlantis Press: Paris, France, 2019; Volume 309, pp. 1–6.
116. Babatunde, S.O.; Perera, S.; Zhou, L.; Udeaja, C. Barriers to public private partnership projects in developing countries: A case of Nigeria. *Eng. Constr. Archit. Manag.* **2015**, *22*, 669–691. [CrossRef]
117. Chan, A.P.; Osei-Kyei, R.; Hu, Y.; Yun, L.E. A fuzzy model for assessing the risk exposure of procuring infrastructure mega-projects through public-private partnership: The case of Hong Kong-Zhuhai-Macao Bridge. *Front. Eng. Manag.* **2018**, *5*, 64–77. [CrossRef]
118. Du, Y.; Fang, J.; Hu, J. Research on the Equilibrium of a Revenue Sharing Contract in a Transfer-Operation-Transfer Project Based on the Theory of Share Tenancy. *Am. J. Ind. Bus. Manag.* **2019**, *9*, 1111–1135. [CrossRef]
119. Amovi, G.; Maksimovi, R.; Buni, S. Critical Success Factors for Sustainable Public-Private Partnership (PPP) in Transition Conditions: An Empirical Study in Bosnia and Herzegovina. *Sustainability* **2020**, *12*, 7121. [CrossRef]
120. Shiying, S.; Heap-Yih, C.; Lihong, L.; Ye, X. Examining the Interrelationship among Critical Success Factors of Public Private Partnership Infrastructure Projects. *Sustainability* **2016**, *8*, 1313.
121. Sepasgozar, S.M.E.; Bliemel, M.; Bemanian, M.R. Discussion of "Barriers of Implementing Modern Methods of Construction" by M. Motiar Rahman. *J. Manag. Eng.* **2016**, *32*, 7015001. [CrossRef]

sustainability

MDPI

Article

Cross-Organizational Learning Approach in the Sustainable Use of Fly Ash for Geopolymer in the Philippine Construction Industry

Jason Maximino C. Ongpeng [1,*], Ernesto J. Guades [2] and Michael Angelo B. Promentilla [3]

[1] Civil Engineering Department, De La Salle University, Manila 0922, Philippines
[2] Department of Civil Engineering, Technical University of Denmark, 2800 Kgs. Lyngby, Denmark; ergua@byg.dtu.dk
[3] Chemical Engineering Department, De La Salle University, Manila 0922, Philippines; michael.promentilla@dlsu.edu.ph
* Correspondence: jason.ongpeng@dlsu.edu.ph

Abstract: The construction industry faces a challenging situation in attaining sustainable development goals. The carbon footprint of the production and use of construction materials such as the use of ordinary Portland cement in concrete products is still on the rise despite of many alternatives and technologies. In this paper, the local cross-organizational learning approach (COLA) and a systematic review of academic and professional literatures were applied in analyzing the use of fly ash as a geopolymer in the Philippine construction industry. Three primary stakeholders were considered: academe, professional organizations, and industry. Documents from each stakeholder were collected, with keywords including sustainability, fly ash, and geopolymer. These documents included published materials, newsletters, department orders, codes, and policies. Text analytics throughout the documents were applied using the Latent Dirichlet Allocation model, which uses a hierarchal Bayesian-modelling process that groups set of items into topics to determine the maturity level of the organizational learning. An adoption framework is proposed aligning COLA with the awareness, interest, desire, and action (AIDA) funnel model. Results show that the organizational maturity until optimization of academe is sufficient towards interest and desire, while industry is highly encouraged to increase organizational maturity from managed to optimization towards desire and action. Factors such as organizational intelligence (OI) and organizational stupidity (OS) are to be considered in balancing critical thinking across organizations. Further studies are recommended by considering the use of COLA with ASEAN organizations in the development of sustainable construction materials.

Keywords: organizational learning; fly ash; geopolymer; environment; sustainable; construction

Citation: Ongpeng, J.M.C.; Guades, E.J.; Promentilla, M.A.B. Cross-Organizational Learning Approach in the Sustainable Use of Fly Ash for Geopolymer in the Philippine Construction Industry. *Sustainability* **2021**, *13*, 2454. https://doi.org/10.3390/su13052454

Academic Editors: Edmundas Kazimieras Zavadskas, Jurgita Antuchevičienė, Reza Hosseini and Igor Martek

Received: 28 December 2020
Accepted: 17 February 2021
Published: 24 February 2021

1. Introduction

The construction industry consumes half of the non-renewable resources of the globe [1] and undoubtedly plays a major role in addressing construction material sustainability. Sustainability can be broadly defined as using resources without depletion and can be described in three spheres: social, economic, and environmental. In order to monitor protection of the current and future state of our planet, adoption of 17 Sustainable Development Goals (SDGs) and 169 targets with 232 indicators can be considered [2]. Sustainable Development Goals are difficult to achieve because of several inherent constraints, including research and development resources on a national level. Insufficient or absence of effective techniques on data knowledge management towards organizational learning should be addressed. One of the solutions is through the use of digitalization to improve the untapped role of technological innovation and knowledge management [3].

Organizational learning (OL) is gaining interest in the construction industry [4] for its added benefits, as it can be integrated to develop effective techniques in developing sustainable construction materials. It is a theory of action that constitutes a conscious and repeatable entity-wide process of creating, acquiring, understanding, sharing, applying, improving, and managing social, tacit, and explicit knowledge in support of the organization's purpose, strategies, and goals [5] and is recognized as essential for an organization's enhanced performance [6]. In addition, it is defined as the continuous increase in performance level through systematic promotion of a learning culture in an organization that includes all stakeholders of all levels as an individual and as a collective group [7]. There are two major types of knowledge that can be used in OL: explicit, as codified knowledge usually recorded in project documents; and tacit, as non-codified knowledge from personal experience that may not be recorded in project documents.

In a controlled environment wherein OL is being practiced in the organization and employee turnover rate is minimal, both explicit and tacit knowledge can be used for practice knowledge management. Explicit knowledge includes documented reports in the organization and can easily be transmitted from one person to another, while tacit knowledge is difficult to write down and difficult to transmit within an organization. One of the common tools in organizational learning in construction in extracting both types of knowledge is through post project review, by passing on previous experience to enhance organizational practices applied to future practices [8]. However, post project review is difficult to achieve if there are no available data from the organizational process assets. Tools for conducting OL, particularly aiming for good project governance, have already been developed, including those of the cross-organizational learning approach (COLA) proposed by [9]. The construction industry is highly fragmented and focused on bringing value to the client; with COLA review process, stakeholders can benefit by realizing shared objectives through sustainable business partnership [9].

The construction industry is broad, and the application of OL can be general or specific to a particular scope. Some notable works with substantial numbers of citations are on quality [10], contractor performance [11], productivity [12], building information modeling [13], and safety. Sustainable construction, in a Scopus search from 1994 to 2018, showed a rapid increase after the year 2010. One of the popular topics in this review article is on the use of alternative materials as a solution to address sustainability in construction [14]. For instance, fly ash or geopolymer material has been considered as an alternative to cement-based construction material, since its production emits less CO_2. Compared to cement, it has excellent bond strength to the concrete substrate [15] and greater durability in severe environments [16]. Geopolymer has been widely used in building materials, nuclear waste disposal, and aerospace materials [17] and is now perceived as an alternative to conventional Portland cement. In the Philippines, the use of fly ash for geopolymer has not been fully documented and its contribution as construction material has not been clearly identified. To the knowledge of the authors, there is in particular a scarcity of information related to its sustainable use in the construction industry.

This paper presents an overview on the extent of fly ash used for geopolymer as sustainable material in the Philippine construction industry. The cross-organizational learning approach [9] was adopted as tool for analyzing the state of the art regarding various stakeholders to describe the sustainability of the considered construction materials.

Information obtained from the literature, project documents, reports, memos, and other data available served as primary inputs. In addition, bibliometric analysis was performed using MATLAB to obtain text analytics. This tool shows that topic modeling for sustainability using construction demolition and waste can raise awareness in a circular economy framework [18] and development of self-healing concrete [19]. It is expected that the present paper will lead to a coherent adoption framework for the development of the sustainable use of fly ash as a geopolymer in the Philippine construction industry.

In the Philippines, coal fly ash (CFA) is one of the main raw materials for geopolymer precursors. Millions of metric tons of CFA are generated annually from the 28 coal-fired power plants currently operating throughout the Philippines. Because of its pozzolanic properties, this industrial by-product is most commonly used as a component in blended ordinary Portland cement (OPC) and marketed as eco-friendly cement products. However, for an archipelago like the Philippines, the percentage utilization of fly ash is still low [20]. For example, power plants in the cities of Naga and Toledo and even a new plant in the Visayas region have projected remaining landfill capacities in the community of less than a decade given the huge amount of CFA generated. The annual coal ash generation in the country is projected to increase to 13 million metric tons by 2035 [21].

The huge amount of CFA can be transformed into a resource for building materials through geopolymer technology. The CFA in the Philippines can be classified as Class F or Class C depending on the relative composition of the major oxides. The presence of major oxides (SiO_2 plus Al_2O_3 plus Fe_2O_3) of more than 70% suggest the pozzolanic properties of the ash, which meets the chemical requirement of Class F fly ash typically derived from bituminous coal. The fly ash also can be considered as having a moderate calcium content, which is expected for coal ash obtained from sub-bituminous coal. Such fly ash is also expected to exhibit some degree of cementitious properties typical of class C fly ash if the presence of major oxides (SiO_2 plus Al_2O_3 plus Fe_2O_3) is more than 50%.

Based on the reported diffractogram of XRD analyses of these CFA, the major components are silicon (which appeared in quartz-SiO_2), aluminum (which appeared in tricalcium aluminate and melilite), calcium (which appeared in tricalcium aluminate, lime–CaO, and melilite), and iron (which appeared in magnetite and melilite and sodium from melilite). The diffractogram did not only provide the major crystalline phase but also suggests the presence of amorphous alumina and silica. These reactive alumino–silica components are important precursors for geopolymerization.

The built environment is a major contributor to loss of biodiversity, and it is thought that raising awareness, including engaging with stakeholders, is beneficial [22]. In order to arrive at effective sustainable project delivery, a framework integrating both organizational learning and sustainability [23], and a cross-organization learning approach [9] is pursued. The motivation of this paper is to measure the research university's OL in relation to the research and development of sustainable construction materials, specifically the use of fly ash for geopolymers.

2. Methodology

Included in the COLA are the primary stakeholders from professional organizations, academe, and industry, as seen in Figure 1. There are linkages between the stakeholders, such as seminar events, which happen at least once a year with a given theme. Seminars usually caters to large audiences and pull communication is practiced. Pull communication is a method wherein a receiver proactively retrieves the information wherein tacit knowledge is limited and explicit knowledge is varying depending on the motivation of the receiver to comprehend the references available.

For the academe, De la Salle University was chosen as the subject. There were two identified research groups focusing on the development of sustainable materials using CFA, namely the Geopolymers and Advance Materials Engineering Research and Sustainability (GAMERS) and Materials for Sustainable Construction and Recyclables Applied to Projects (M-SCRAP) from chemical and civil engineering departments, respectively. The focus of the research groups is the use and sustainable consumption of CFA as supplementary cementitious material (SCM) and development of specific materials such as geopolymers. The applications range from managing coal fly ash to developing products for bricks, mortars, and concrete. In addition, a seminar-workshop on waste utilization is practiced to share results and is published with the National Library of the Philippines. This seminar-workshop covers different fields and disciplines, from chemical and material sciences to the civil engineering field, and from academe, including Japanese universities, to industry.

For the professional organizations, two associations were considered as being at the forefront of the construction industry in the Philippines. These are the Philippine Institute of Civil Engineers (PICE), composed of local and international chapters, and the Philippine Constructors Association (PCA), with different chapters locally. Regular conferences and seminars are conducted by the associations with guest experts and speakers known in their field of specialization.

Figure 1. Cross-organizational learning approach shared across primary stakeholders.

Lastly, the Department of Public Works and Highways (DPWH) was chosen as the industry, and could also be treated as the policy maker represented in the COLA. The DPWH is mandated to undertake major infrastructure projects in the Philippines from planning to design, construction, and maintenance of national roads, bridges, flood control structures, water resources projects, and other public works. The stakeholder was chosen because of its function in providing material standards for the construction industry through one of its offices, the Bureau of Research and Standards (BRS). From the identified stakeholders, the cross-organizational learning approach was achieved to better develop an adoption framework for the research and development of the use of fly ash and geopolymers in the Philippine construction industry.

After identifying the stakeholders, data text analytics was applied using the Latent Dirichlet Allocation model that uses a hierarchical Bayesian modelling process, which groups sets of items into topics [24]. An archiving of available project documents from different stakeholders is displayed in Table 1. The documents are related to sustainable construction materials using fly ash, since this is the closest subject pertaining to geopolymers that can be gathered for both industry and professional organizations. Retrieved documents included building codes, laws/regulations/department orders, newsletters/circulars/reports, and a recent roadmap of the construction industry, 2020–2030. The documents used from academe were published documents in the form of abstracts, conferences, and journal papers with fly ash or geopolymer as the keyword. It is noted that the dates of the project documents were dispersed due to the limited documents available from each stakeholder that pertained to the keywords search used.

Additionally, a maturity level was adopted for each primary stakeholder. There are five levels of organizational maturity including initial, repeatable, defined, managed, and optimizing [25]. Three levels for measuring organizational maturity on the use of sustainable construction material with fly ash were used for simplicity in this paper. Organizational maturity was observed through text analytics, wherein frequent topics appeared in most documents. It was classified in this paper as Level 1, from initial to repeatable; Level 2, from initial to managed; and Level 3, from initial to optimizing. Progressive analysis was considered, and maturity level was established during this period.

Table 1. Primary stakeholders and documents.

Stakeholders	Description	Number of Documents	Type of Documents	Year
Academe	Utilization of Waste Seminar-Workshop	3	Book of abstracts	2013
				2018
				2019
	Engineering	45	Conference Papers Journal Papers	
Professional Organizations	Philippine Institute of Civil Engineers (PICE)	4	PD 1096	2004
			RA 544	2010
			Newsletter/Circulars	2020
			President's Report	2019
	Philippine Constructors Association (PCA)	1	Roadmap 2020–2030	2019
Industry	Department of Public Works and Highways (DPWH)	5	Research Symposium	2019
			DO 34 1991	1991
			SEM MO 2016	2016
			SEM MO 2014	2014
			Philippine Green Building Code	2015

Legend: RA—Republic Act; DO—Department Order; SEM—Social and Environmental Management Manual of Operation.

3. Results and Discussion

The documents were collected, processed, and analyzed through MATLAB text analytics. The outputs were word cloud, topic modelling, and topic mixtures. The word cloud shows the highest frequency of words in a large font. The topic modelling and mixtures show the word cloud with topics separated from a folder of documents, and the bar chart shows the probability of the topic for each document. These were used to determine the level of maturity of each stakeholder toward organizational learning using coal fly ash for geopolymer.

3.1. Academe

As seen in Table 1, the academe was separated into two parts: three seminar-workshops, and 45 published conference papers and journals. From the seminar-workshop with the theme "utilization of waste", a limited book of abstracts was available, and a word cloud was not essential in the analysis. Shown in Figure 2 is the topic modelling from the three seminar-workshops, where two topics were generated: material strength, and use of waste in concrete. It shows that the seminar-workshop covered utilization of waste in general with no specificity to the use of fly ash and or geopolymers. Shown in Figure 3 is the topic mixtures, showing that almost equal contributions from all abstracts considered both material strength and the use of waste in concrete. It is indicative from Figure 3 that waste was considered in the project documents but not much on the use of CFA towards geopolymers.

Figure 4 shows the word cloud of published materials from 45 engineering publications wherein fly ash was the main focus. Figure 5 shows the four topics generated from the documents. These were as follows: improving strength using fly ash, development of geopolymers, utilization of fly ash as additive or replacement to cement, and utilization and removal of waste. Results show that the direction of sustainable construction using fly ash is broad, including utilization of fly ash waste from industrial plants for cement replacement, development of geopolymers, and development of other materials such as mortars and bricks for a wide variety of applications.

Figure 2. Topic modelling for seminar-workshop from academe.

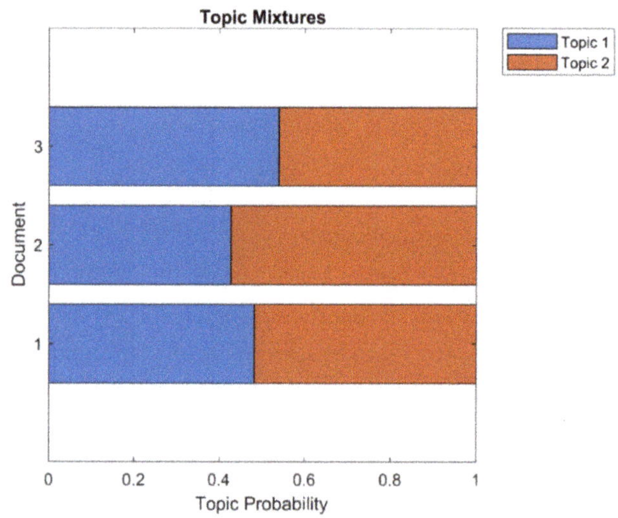

Figure 3. Topic mixtures for seminar-workshop from academe.

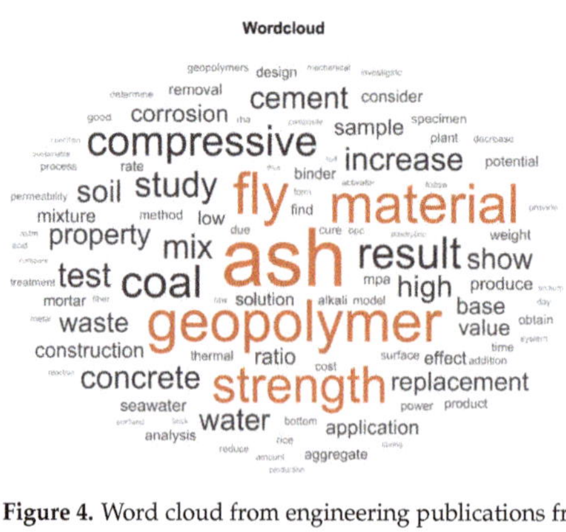

Figure 4. Word cloud from engineering publications from academe.

Figure 5. Topic modelling from engineering publications from academe.

Figure 6 shows the topic mixture of the 45 documents considered. For some documents, the development of geopolymers, which is topic 2 from the figure, was smaller compared to topic 3, which is using fly ash as an SCM in concrete products. It shows that the use of CFA towards geopolymers is evident. Detailed discussions on the 45 documents are described in Table 2.

The summary of work from the published materials is seen in Table 2. It shows that organizational maturity from academe was Level 3, from initial to optimizing from 2001 to present, which was the highest level. There were 16 out of 45 documents or 35.6% that studied utilization of fly ash alone or using it as an SCM in concrete products. The remaining documents, 29 out of 45 or 64.4%, explored the use of CFA for the development of geopolymers. The documents covered the use of CFA from an economic perspective; optimization of the use and design; CO_2 reduction; development of different materials including paste, mortar, bricks, soil, and concrete with ordinary Portland cement as binder; or fly ash with sodium hydroxide, sodium silicate, and other alternatives as binder. Tests differed from chemical and mechanical for the analysis of elements and compounds, compressive strength, flexural tests, permeability, and corrosion, among others.

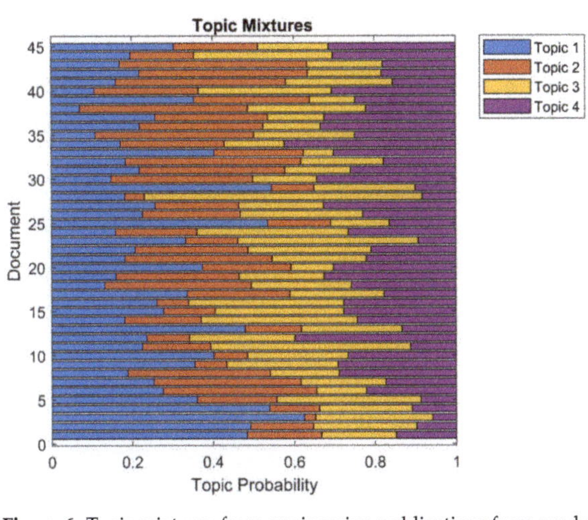

Figure 6. Topic mixtures from engineering publications from academe.

Table 2. Materials and tests from academic publications.

Reference	Materials						Test		Other Specific Tests	Remarks
	Used as		Used as Geopolymer				Chemical	Mechanical		
	CFA	SCM	M	S	B	C				
[26]	✓						✓		Leaching test	Phosphate removal
[27]		✓						✓	Corrosion test	Mortar with CFA
[28]	✓								Impact assessment	Coal ash
[29]	✓								Financial analysis	Fly and bottom ash
[30]		✓						✓		CFA and rubber crumbs
[31]		✓						✓	Corrosion test	Use of seawater
[32]			✓			✓				Optimization model
[33]						✓		✓		Fly ash and water glass solution
[34]			✓					✓		Varying sand-fly ash ratio
[35]					✓			✓	Shrinkage test	Optimization model
[20]			✓					✓		Varying sand-fly ash ratio
[36]						✓	✓	✓		Coal fly and bottom ash
[37]		✓					✓	✓		CFA, RHA, and sludge
[38]		✓						✓		CFA and ceramic waste
[39]		✓						✓		CFA/Mangima stone
[40]			✓				✓			Abaca fiber reinforced GP
[41]			✓					✓		Water treatment sludge
[42]						✓		✓		Different sizes of aggregates
[43]			✓			✓				Multi-objective optimization
[44]					✓			✓		Bottom and rice hush ash
[45]					✓		✓	✓		Gold mine tailings
[46]		✓					✓	✓		CFA and ceramic waste
[47]		✓						✓	Consolidated drain test	Road embankment
[48]				✓			✓	✓		Soil stabilizer
[49]				✓			✓	✓		Soil stabilizer
[50]				✓				✓	Leaching test	Soil and fly ash
[51]			✓				✓		Acid resistance test	Water treatment sludge

Table 2. *Cont.*

Reference	Materials						Test		Other Specific Tests	Remarks
	Used as		Used as Geopolymer				Chemical	Mechanical		
	CFA	SCM	M	S	B	C				
[52]				✓				✓	Unconfined compressive test	Silty sand embankment
[53]				✓				✓	Permeability test	Using soil mix
[54]				✓				✓	Permeability test	Using dredged soil
[55]				✓				✓		Using dredged soil
[56]				✓				✓		Using dredged soil
[57]		✓						✓	Permeability test	Pervious concrete wit CFA and sawdust
[58]	✓								CO_2 and cost evaluation	Transportation of CFA
[59]		✓					✓	✓		Autoclaved aerated
[60]			✓					✓		Self-healing using bacteri
[61]	✓				✓	✓		✓		Organic/ inorganic binders
[62]		✓						✓	Corrosion test	Use of seawater
[60]		✓						✓	Corrosion test	Use of seawater
[63]						✓			Corrosion test	GPC with reinforcement
[64]					✓			✓		Compressed earth blocks
[21]			✓				✓			Degradation of dye
[65]			✓				✓			Applied for acid treatment
[59]			✓				✓			Biomaterials for self-healing
[66]			✓				✓			Nickel-laterite as precursor

Legend: M—mortar; S—soil; B—brick/block; C—concrete; CFA—coal fly ash; SCM—supplementary cementitious material.

3.2. *Professional Organizations*

As seen in Table 1, a few documents were available from professional organizations. These documents were codes, newsletters, reports, and the roadmap 2020–2030. Organizational maturity was low at Level 1 from initial to repeatable. The stakeholder focused on sustainability in general and served as an ambassador to the members and the public; hence, there was not much specificity on the use of fly ash and development of geopolymers observed from text analytics.

3.3. *Industry*

From the documents taken from the DPWH, which is considered as the primary stakeholder for the industry, it showed that the word cloud in Figure 7 focused on environmental consideration in construction projects, which leads to the promotion of sustainability. As seen in Figure 8, the topic modelling from five documents were environmentally-friendly concrete, environmentally-friendly projects, generation of reports, and environmental impact of projects. Topic mixtures in Figure 9 were skewed differently for each topic. For example, documents 1 to 3 considered specifically the use of environmentally-friendly concrete as topic 1, while documents 4 to 5 considered in general environmentally-friendly projects with reports and impacts on the environment, which covered topic 2–4. The organizational maturity was Level 2, from initial to managed.

Figure 7. Word cloud from industry.

Figure 8. Topic modelling from industry.

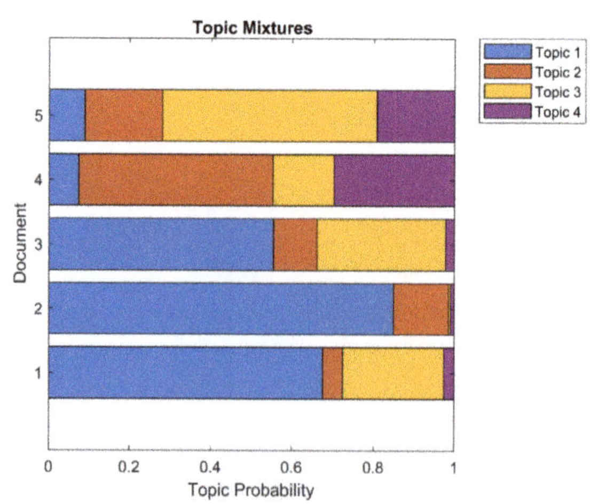

Figure 9. Topic mixtures from industry.

3.4. Cross-Organizational Learning Approach from Stakeholders

Organizational learning from academe, industry, and professional organizations was classified as Level 3, 2, and 1, respectively, based on the discussion in Sections 3.1–3.3. Academe was rated at Level 3, since published documents were managed well with room for optimization through continuous research and development. On the other hand, industry was Level 2, since the stakeholder deals with the material standards adopted by public works and highways nationwide, and professional organizations were Level 1, since while sustainability was emphasized, not much continuous development was defined. It is worth noting that the primary stakeholder from industry issued DPWH DO 23 in 1987, creating a coordinating committee for studying the application of fly ash as an admixture to concrete. After four years, another order was issued, DPWH DO 34 1991, which approved the use of fly ash that meets the requirements of ASTM C618 with 20% replacement of Portland cement in concrete mix. This shows that organizational learning towards sustainable construction materials was practiced in the industry.

Using the documents retrieved, topic modelling was generated and analyzed to align each stakeholder towards sustainability, the use of fly ash, and development of materials from the use of fly ash, such as geopolymers. Integration of COLA and an existing model, described in a later section, were considered in arriving at a proposed adoption framework.

Combining all 58 documents from Table 1 resulted in the word cloud shown in Figure 10 with fly ash in concrete materials and geopolymer development. Topic modelling is seen in Figure 11 and was categorized as follows: conducting environmentally friendly projects, utilization of fly ash in geopolymer concrete products, developing geopolymer using waste materials, and providing solutions for the construction industry. Figure 12 shows all the 58 document topic mixtures and shows varying skewness, with some documents showing prioritization on one topic over the other. For example, document 58 showed more of topic 1 (conducting environmentally friendly projects) over the use of fly ash as to whether it will be for SCM or geopolymer products. This shows that each stakeholder had different specific topics, and in general led to the development of sustainable construction materials.

After text analytics on each and consolidated knowledge from each stakeholder, a proposed adoption framework was considered. Many studies in each respective field use knowledge, competence, wisdom, talent, and learning, which are used to describe organization intelligence (OI) [67]. Opposite to OI is organizational stupidity (OS), which is considered an illnesses for organizations [68], wherein smart people pretend to be stupid [67,69] and is a concept contrary to critical thinking [70]. The OI and OS were not placed on the two extremes of a single spectrum; rather, these concepts moved hand in hand [71]. Generally, these are two counterparts, where increasing one of them leads to increases in the other.

Figure 10. Word cloud from all primary stakeholders.

Figure 11. Topic modelling from all primary stakeholders.

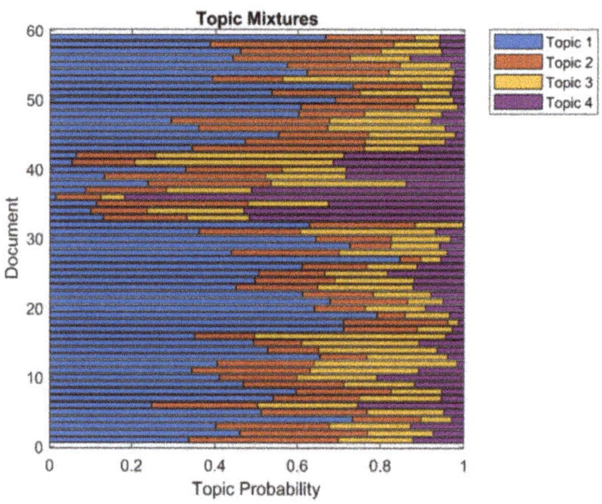

Figure 12. Topic mixtures from all primary stakeholders.

An existing model, called the awareness, interest, desire, and action (AIDA) purchase funnel model, was adopted in this paper. Integration of this model with COLA and factors such as OI and OS can lead to better understanding of an adoption framework in arriving at an OL towards sustainable construction materials. The AIDA purchase funnel model is a hierarchal diagram used for marketing wherein there is a wider opening at the top, which represents awareness, followed by narrowing down to interest, desire, and, lastly, action [72]. This model is used in marketing, where the customer journey is considered in purchasing goods or services. Figure 13 shows the COLA from the primary stakeholders and balancing its organizational maturity level, and the role that each primary stakeholder plays in the development of proper OL, where OI and OS exists in the organizations. Professional organizations with maturity from initial to repeatable is observed in the awareness and interest of using sustainable materials like fly ash and geopolymers. From the academe, the position on the funnel on awareness, interest, and desire is highly correlated with organizational maturity from initial to optimizing. On the other hand, development and motivation for industry is highly recommended, since the current maturity is from initial to managed with little or no action. This does not negate that there is no research and development on the field of CFA and other materials in the industry; in fact, from 2016 to present, numerous department orders from DPWH in relation to concrete were made, such as inclusion of macro synthetic fiber for Portland cement concrete pavement, standard specification of the use of roller-compacted concrete pavement, and use of one-day Portland cement concrete pavement, which shows that the cross-organizational learning approach is recommended to harmonize the learning process and attain sustainable development goals in the Philippine construction industry. It is recommended that interventions from new technology, market demand, legal requirements, and environmental considerations be used as motivators to move awareness to action for all stakeholders in the continuous learning process in the development of sustainable materials.

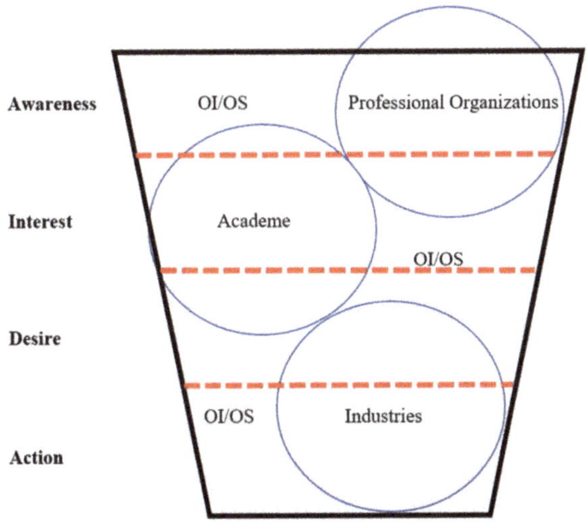

Figure 13. Primary stakeholders in the AIDA funnel model.

4. Conclusions

The construction industry has contributed substantially to CO_2 emissions compared to other sectors. The challenge of attaining sustainable development goals can be achieved with a good adoption framework, wherein different stakeholders can apply a cross-organizational learning approach (COLA). Results show that a local COLA and systematic review from academic and professional literatures using text analytics to understand explicit knowledge from documents can be achieved. Alignment of COLA's maturity level and the AIDA (awareness, interest, desire, and action) purchase funnel model can lead to an adoption framework.

It was found that the maturity of the sustainable use of fly ash for geopolymers according to the perspective of professional organizations is from initial to repeatable, as observed in the awareness and interest scale. Professional organizations can be the ambassadors for sustainable development goals wherein awareness and interest can be imparted to the members and the public. From the viewpoint of academe, the awareness of, interest in, and desire for the use of these alternative construction materials are highly correlated with organizational maturity, from initial to optimization. Academe can continue with research and development with interest and desire, considering organizational maturity until optimization. Lastly, industry, as one of the primary stakeholders, plays an important role in developing organizational maturity until optimization with desire and action in using fly ash and to develop more materials from it. It is recommended that interventions can be made, such as motivation coming from new technology, market demand, and environmental considerations. The movement from awareness to action for all stakeholders is the ideal setting in order to have a continuous learning process in the development of sustainable materials.

Further studies are recommended by considering COLA from wider organizations, locally and internationally, in developing sustainable construction materials. Benchmarking can bring about best practices on the adoption of new materials in the construction industry.

Author Contributions: Conceptualization, J.M.C.O. and E.J.G.; methodology, J.M.C.O. and E.J.G.; software, J.M.C.O.; validation, E.J.G. and M.A.B.P.; formal analysis, J.M.C.O.; investigation, J.M.C.O. and E.J.G.; resources, J.M.C.O., E.J.G., and M.A.B.P.; writing—original draft preparation, J.M.C.O.; writing—review and editing, E.J.G., and M.A.B.P.; visualization, J.M.C.O.; supervision, J.M.C.O.; project administration, J.M.C.O.; funding acquisition, J.M.C.O. All authors have read and agreed to the published version of the manuscript.

Funding: The is funded by the project Green Fiber-Reinforced Polymer (FRP) Composites as an Innovative Repair System for Earthquake-prone Historical Buildings with project account number: 40203.

Institutional Review Board Statement: Not applicable.

Informed Consent Statement: Not applicable.

Data Availability Statement: Not applicable.

Conflicts of Interest: The authors declare no conflict of interest.

References

1. Lundholm, J.T. Green roofs and facades: A habitat template approach. *Urban Habitats* **2006**, *4*, 87–101.
2. United Nations. Transforming Our World: The 2030 Agenda for Sustainable Development United Nations. 2015. Available online: https://sustainabledevelopment.un.org/content/documents/21252030%20Agenda%20for%20Sustainable%20Development%20web.pdf (accessed on 15 December 2020).
3. Del Río Castro, G.; González Fernández, M.C.; Uruburu Colsa, Á. Unleashing the convergence amid digitalization and sustainability towards pursuing the Sustainable Development Goals (SDGs): A holistic review. *J. Clean. Prod.* **2021**, *280*, 122204. [CrossRef]
4. Latham, M. Constructing the Team: Final Report of the Government/Industry. Review of Procurement and Contractual Arrangements in the UK Construction Industry. 1994. Available online: https://constructingexcellence.org.uk/wp-content/uploads/2014/10/Constructing-the-team-The-Latham-Report.pdf (accessed on 15 December 2020).
5. Strang, K. *Achieving Organizational Learning Across Projects*; Project Management Institute: Delaware, PA, USA, 2003.
6. Wong, P.S.P.; Cheung, S.O.; Yiu, R.L.Y.; Hardie, M. The unlearning dimension of organizational learning in construction projects. *Int. J. Proj. Manag.* **2012**, *30*, 94–104. [CrossRef]
7. Eken, G.; Bilgin, G.; Dikmen, I.; Birgonul, M.T. A lessons-learned Tool for Organizational Learning in Construction. *Autom. Constr.* **2020**, *110*, 102977. [CrossRef]
8. Cushman, M.; Venters, W.; Cornford, T.; Mitev, N. *Understanding Sustainability as Knowledge Practice*; British Academy of Management Conference: London, UK, 2002.
9. Cushman, M.; Cornford, T. Infrastructures for Construction collaboration: The Cross Organizational Learning Approach. *Int. J. Arch. Eng. Constr.* **2003**, *1*, 67–76.
10. Love, P.E.; Li, H.; Irani, Z.; Holt, G.D. Re-thinking TQM: Toward a framework for facilitating learning and change in construction organizations. *TQM Mag.* **2000**, *12*, 107–117. [CrossRef]
11. Kululanga, G.; Price, A.; Mccaffer, R. Empirical Investigation of Construction contractors' Organizational Learning. *J. Constr. Eng. Manag.* **2002**, *128*, 385–391. [CrossRef]
12. Love, P.E.; Josephson, P.E. Role of error-recovery process in projects. *J. Manag. Eng.* **2004**, *20*, 70–79. [CrossRef]
13. Chan, P.; Cooper, R.; Tzortzopoulos, P. Organizational learning: Conceptual challenges from a project perspective. *Constr. Manag. Econ.* **2005**, *23*, 747–756. [CrossRef]
14. Detudomsap, A.; Hallinger, P. A Bibliometric Review of Research on Sustainable construction, 1994–2018. *J. Clean. Prod.* **2020**, *254*, 120073. [CrossRef]
15. Hu, S.; Wang, H.; Zhang, G.; Ding, Q. Bonding and abrasion resistance of geopolymeric repair material made with steel slag. *Cem. Concr. Compos.* **2008**, *30*, 239–244. [CrossRef]
16. Ganesan, N.; Abraham, R.; Deepa Raj, S. Durability characteristics of steel fiber reinforced geopolymer concrete. *Constr. Build. Mater.* **2015**, *93*, 471–476. [CrossRef]
17. Guo, X.; Pan, X. Mechanical properties and mechanisms of fiber reinforced fly ash-steel slag based geopolymer mortar. *Constr. Build. Mater.* **2018**, *179*, 633–641. [CrossRef]
18. Ginga, C.; Ongpeng, J. Materials Circular Economy on Construction and Demolition waste: A Literature Review on Material Recovery and Production. *Materials* **2020**, *13*, 2970. [CrossRef] [PubMed]
19. Griño, A.A.; Daly, M.; Klarissa, M.; Ongpeng, J.M.C. Bio-Influenced Self-Healing Mechanism in Concrete and Its Testing: A Review. *Appl. Sci.* **2020**, *10*, 5161. [CrossRef]

20. Kalaw, M.E.; Culaba, A.; Hinode, H.; Kurniawan, W.; Gallardo, S.; Promentilla, M.A. Optimizing and characterizing geopolymers from ternary blend of Philippine coal fly ash, coal bottom ash and rice hull ash. *Materials* **2016**, *9*, 580. [CrossRef] [PubMed]
21. Jamora, J.B.; Gudia, S.E.L.; Go, A.W.; Giduquio, M.B.; Loretero, M.E. Potential CO2 reduction and cost evaluation in use and transport of coal ash as cement replacement: A case in the Philippines. *Waste Manag.* **2020**, *103*, 137–145. [CrossRef]
22. Opoku, A. Biodiversity and the built environment: Implications for the Sustainable Development Goals (SDGs). *Resour. Conserv. Recycl.* **2019**, *141*, 1–7. [CrossRef]
23. Opoku, A.; Fortune, C. Organisational Learning and Sustainability in the Construction Industry. *Built Hum. Environ. Rev.* **2011**, *4*, 98–107.
24. Roque, C.; Cardoso, J.; Connell, T.; Schermers, G.; Weber, R. Topic Analysis of Road Safety Inspections Using Latent Dirichlet allocation: A Case Study of Roadside Safety in Irish Main Roads. *Accid. Anal. Prev.* **2019**, *131*, 336–349. [CrossRef] [PubMed]
25. Von Zedtwitz, M. Organizational learning through post–project reviews in R&D. *RD Manag.* **2002**, *32*, 255–268.
26. Yamada, K.; Haraguchi, K.; Gacho, C.C.; Salinas, L.S.; Silverio, C.M.; Wongsiri, B.P. Removal of PhosPhate from aqueous solutions by crystallization using coal fly ash. In *2001 International Ash Utilization Symposium, Center for Applied Energy Research*; Paper 7; University of Kentucky: Lexington, KY, USA, 2001.
27. Madlangbayan, M.; Otsuki, N.; Nishida, T.; Baccay, M. Corrosion behavior of steel bar in chloride contaminated mortars with fly ash. *Philipp. Eng. J.* **2005**, *26*, 13–24.
28. Gallardo, S.; Dungca, J.; Gallardo, R.; Kalaw, M. *Sustainability Issues Due to Coal Ash from Coal Fired Power Plants Phase 1: Impact Assessment of Coal Ash Dumping in a Typical Power Generating Facility*; Interdisciplinary Research Final Report; University Research Coordination Office (URCO), De la Salle University: Manila, Philippines, 2012.
29. Clark, E. Economic Implications of the Partial Substitution of Coal Fly Ash and Bottom Ash for Cement and Aggregate Construction Materials. In *Sustainable Issue due to Coal Ash from Coal Fired Power Plants in the Philippines: Phase 2–Sustainable Solutions for an Industry Partner*; University Research Coordination Office (URCO), De la Salle University: Manila, Philippines, 2013; pp. 87–146.
30. Bustamante, A.; Dablo, G.M.; Sia, R.; Arazo, R. Physical and mechanical properties of composite brick from cement mortar, fly ash and rubber crumbs. *Int. Res. Eng. Technol.* **2015**, *4*, 1–5.
31. Lim, E.D.; Roxas, C.L.; Gallardo, R.; Nishida, T.; Otsuki, N. Strength and corrosion behavior of mortar mixed and/or cured with seawater with various fly ash replacement ratios. *Asian J. Civ. Eng.* **2015**, *16*, 835–849.
32. Sumabat, A.K.R.; Mañalac, A.J.; Nguyen, N.T.; Kalaw, M.E.; Tan, R.R.; Promentilla, M.A.B. Optimizing geopolymer-based material for industrial application with analytic hierarchy process and multi-response surface analysis. *Chem. Eng. Trans.* **2015**, *45*, 1147–1152.
33. Kalaw, M.E.L.; Culaba, A.B.; Nguyen, H.T.; Nguyen, K.; Hinode, H.; Kurniawan, W.; Gallardo, S.M.; Promentilla, M.A.B. Mechanical and thermal properties of geopolymers from mixtures of coal ash and rice hull ash using water glass solution as activator. *ASEAN J. Chem. Eng.* **2015**, *15*, 51–61. [CrossRef]
34. Guades, E.J. Experimental investigation of the compressive and tensile strengths of geopolymer mortar: The effect of sand/fly ash (S/FA) ratio. *Constr. Build. Mater.* **2016**, *127*, 484–493. [CrossRef]
35. Promentilla, M.A.B.; Thang, N.H.; Kien, P.T.; Hinode, H.; Bacani, F.T.; Gallardo, S.M. Optimizing ternary-blended geopolymers with multi-response surface analysis. *Waste Biomass Valorization* **2016**, *7*, 929–939. [CrossRef]
36. Nguyen, T.H.; Promentilla, M.A.B. Producing Geopolymer-based Concrete from Coal Fly Ash and Bottom Ash. In Proceedings of the 23rd Regional Symposium on Chemical Engineering, AUN-SEED/NET, Vung Tau, Vietnam, 27–28 October 2016.
37. Opiso, E.M.; Sato, T.; Otake, T. Microstructural properties of hardened cement paste blended with coal fly ash, sugar mill lime sludge and rice hull ash. *Adv. Concr. Constr.* **2017**, *5*, 289.
38. Gallardo, R.S.; Elevado, K.J.T. Cost-Benefit Analysis of Concrete Mixed with Waste Ceramic Tiles and Fly Ash. In Proceedings of the 5th Seminar on Utilization of Waste Materials, De la Salle University, Manila, Philippines, 7–8 September 2017.
39. Cabahug, R.R.; Ape, J.L.; Curiba, A.M.G.; Egama, V.J.D.; Mabano, M.C.; Uy, R.M.A. Coal Fly Ash as Cement Replacement on Mortar Mixed with Mangima Stone and Conventional Fine Aggregates. *Mindanao J. Sci. Technol.* **2017**, *15*, 137–150.
40. Malenab, R.A.J.; Ngo, J.P.S.; Promentilla, M.A.B. Chemical treatment of waste abaca for natural fiber-reinforced geopolymer composite. *Materials* **2017**, *10*, 579. [CrossRef]
41. Promentilla, M.A.B.; Kalaw, M.E.; Nguyen, H.T.; Aviso, K.B.; Tan, R.R. A fuzzy programming approach to multi-objective optimization for geopolymer product design. *Comput. Aided Chem. Eng.* **2017**, *40*, 1015–1020.
42. Nguyen, H.T.; Pham, T.K.; Promentilla, M.A.B. Development of Geopolymer-Based Materials from Coal Bottom Ash and Rice Husk Ash with Sodium Silicate Solutions. In Proceedings of the 4th Congrès International de Géotechnique–Ouvrages–Structures, Singapore, 26–27 October 2017; pp. 402–410.
43. Aseniero, J.P.J.; Opiso, E.M.; Banda, M.H.T.; Tabelin, C.B. Potential utilization of artisanal gold-mine tailings as geopolymeric source material: Preliminary investigation. *SN Appl. Sci.* **2018**, *1*, 35. [CrossRef]
44. Ho, V.; Orbecido, A.; Promentilla, M.A.B. Investigation on mixture design of one-part geopolymer from fly ash and water treatment sludge. In *MATEC Web of Conferences*; EDP Sciences: Ulis, France, 2018; Volume 156, p. 05009. [CrossRef]
45. Elevado, K.; Galupino, J.G.; Gallardo, R.S. Compressive strength modelling of concrete mixed with fly ash and waste ceramics using K-nearest neighbor algorithm. *Int. J. Geomate* **2018**, *15*, 169–174. [CrossRef]
46. Uy, E.E.S.; Dungca, J.R. Hyperbolic Model Parameters of Philippine Coal Ash. *Int. J.* **2018**, *15*, 95–102.

47. Ngo, J.P.; Promentilla, M.A.B. Development of abaca fiber-reinforced foamed fly ash geopolymer. In Proceedings of the 24th Regional Symposium on Chemical Engineering, Semarang, Indonesia, 14 March 2018; pp. 15–16.
48. Tigue, A.A.S.; Dungca, J.R.; Hinode, H.; Kurniawan, W.; Promentilla, M.A.B. Synthesis of a one-part geopolymer system for soil stabilizer using fly ash and volcanic ash. In *MATEC Web of Conferences*; EDP Sciences: Ulis, France, 2018; Volume 156, p. 05017.
49. Tigue, A.A.S.; Malenab, R.A.J.; Dungca, J.R.; Yu, D.E.C.; Promentilla, M.A.B. Chemical stability and leaching behavior of one-part geopolymer from soil and coal fly ash mixtures. *Minerals* **2018**, *8*, 411. [CrossRef]
50. Orbecido, A.; Ho, V.; Hinode, H.; Kurniawan, W.; Nguyen, L.; Promentilla, M.A.B. Characterization and acid resistance test of one-part geopolymer from fly ash and water treatment sludge. In *MATEC Web of Conferences*; EDP Sciences: Ulis, France, 2018; Volume 156, p. 05003.
51. Dungca, J.R.; Codilla, E.E.T., II. Fly-ash-based geopolymer as stabilizer for silty sand embankment materials. *Int. J. Geomate* **2018**, *14*, 143–149. [CrossRef]
52. Dungca, J.R.; Ang, K.D.; Isaac, A.M.L.; Joven, J.J.R.; Sollano, M.B.T. USE OF DRY MIXING METHOD IN FLY ASH BASED GEOPOLYMER AS A STABILIZER FOR DREDGED SOIL. *Int. J.* **2019**, *16*, 9–14.
53. Guades, E.J. Effect of coarse aggregate size on the compressive behaviour of geopolymer concrete. *Eur. J. Environ. Civ. Eng.* **2019**, *23*, 693–709. [CrossRef]
54. Dungca, J.R.; Lao, W.D.T.; Lim, M.; Lu, M.D.; Redelicia, J.C.P. Vertical permeability of dredged soil stabilized with fly-ash based geopolymer for road embankment. *Int. J. Geomate* **2019**, *17*, 8–14. [CrossRef]
55. Dungca, J.R.; Lao, W.D.T.; Lim, M.; Lu, W.D.; Redelicia, J.C.P. Radial flow permeameter: A proposed apparatus to measure horizontal hydraulic gradient of fly-ash based geopolymer-soil mix. *Int. J. Geomate* **2019**, *16*, 218–223. [CrossRef]
56. Lira, B.C.S.; Dellosa, S.B.A.; Toh, C.I.L.; Al Patrick, A.Q.; Nidoy, A.L.S.; Cerna, K.D.; Yu, D.E.C.; Janairo, J.I.B.; Promentilla, M.A.B. Coal Fly Ash-based Geopolymer Spheres Coated with Amoxicillin and Nanosilver for Potential Antibacterial Applications. *ASEAN J. Chem. Eng.* **2019**, *19*, 25–37. [CrossRef]
57. Opiso, E.M.; Supremo, R.P.; Perodes, J.R. Effects of coal fly ash and fine sawdust on the performance of pervious concrete. *Heliyon* **2019**, *5*, 02783. [CrossRef] [PubMed]
58. Balangao, J.K.B.; Ramos, M.S.K. Production of Autoclaved Aerated Concretes with Fly Ash from a Coal Power Plant in Misamis Oriental, Philippines. *Indian J. Sci. Technol.* **2019**, *12*, 14. [CrossRef]
59. Doctolero, J.Z.S.; Benedicto, M.Y.G.R.; Samson, M.L.C.; Ongpeng, J.M.C.; Promentilla, M.A.B. Self-Healing Potential of Fly Ash-Based Geopolymers with Directly Added Bacterial Agents. In Proceedings of the 4th International Symposium on Concrete and Structures for Next Generation, Kanazawa, Japan, 17–19 June 2019.
60. Ongpeng, J. Next Generation Materials and Testing in Concrete. In Proceedings of the 4th International Symposium on Concrete and Structures for Next Generation, Kanazawa, Japan, 17–19 June 2019.
61. Roxas, C.; Lejano, B.; Ongpeng, J. Corrosion performance of seawater concrete with fly ash under impressed current. In Proceedings of the Fifth International Conference on Sustainable Construction Materials and Technologies, Kingston upon Thames, UK, 15–17 July 2019.
62. Temew, K.C.; Roxas, C.L.C.; Lejano, B.; Ongpeng, J. Evaluation of Macrocell Corrosion of Steel Segments in Fly Ash Concrete Mixed with Seawater. In Proceedings of the 4th International Symposium on Concrete and Structures for Next Generation, Kanazawa, Japan, 17–19 June 2019.
63. Ongpeng, J.; Gapuz, E.; Andres, J.J.S.; Prudencio, D.; Cuadlisan, J.; Tadina, M.; Zacarias, A.; Benauro, D.; Pabustan, A. Alkali-activated binder as stabilizer in compressed earth blocks. In *IOP Conference Series: Materials Science and Engineering*; IOP Publishing: Bristol, UK, 2020; Volume 849, p. 012042.
64. Shimizu, E.; Promentilla, M.A.B.; Yu, D.E. Utilization of Coal Fly Ash and Rice Hull Ash as Geopolymer Matrix-cum-Metal Dopant Applied to Visible-Light-Active Nanotitania Photocatalyst System for Degradation of Dye in Wastewater. *Catalysts* **2020**, *10*, 240. [CrossRef]
65. Tigue, A.A.S.; Malenab, R.A.J.; Promentilla, M.A.B. A Systematic Mapping and Scoping Review on Geopolymer and Permeable Reactive Barrier for Acid Mine Drainage Treatment Research. *Process. Integr. Optim. Sustain.* **2020**, *4*, 15–35. [CrossRef]
66. Longos, A.; Tigue, A.A.; Malenab, R.A.; Dollente, I.J.; Promentilla, M.A.B. Mechanical and thermal activation of nickel-laterite mine waste as a precursor for geopolymer synthesis. *Results Eng.* **2020**, *7*, 100148. [CrossRef]
67. Albrecht, K. *The Power of Minds at Work: Organizational Intelligence in Action*; Amacom Books: New York, NY, USA, 2003.
68. Forester, J. On fieldwork in a Habermasian way: Critical ethnography and the extra-ordinary character of ordinary professional work. In *Studying Management Critically*; Sage Publications Ltd.: London, UK, 2003; pp. 46–65.
69. Ten Bos, R. The vitality of stupidity. *Soc. Epistemol.* **2007**, *21*, 139–150. [CrossRef]
70. Paulsen, R. Slipping into functional stupidity: The bifocality of organizational compliance. *Hum. Relat.* **2017**, *70*, 185–210. [CrossRef]
71. Karimi-Ghartemani, S.; Khani, N.; Esfahani, A.N. Stupidity for organizational developing a framework. *J. High. Technol. Manag. Res.* **2020**, *31*, 100392. [CrossRef]
72. St. Elmo Lewis, E. *Financial Advertising*; Levey Bros. & Company: Indianapolis, IN, USA, 1908.

sustainability

MDPI

Article

Sustainability Indicator Selection by a Novel Triangular Intuitionistic Fuzzy Decision-Making Approach in Highway Construction Projects

Hassan Hashemi [1], Parviz Ghoddousi [1,*] and Farnad Nasirzadeh [2]

[1] School of Civil Engineering, Iran University of Science and Technology, P.O. Box 16765-163, Narmak, Tehran 1684613114, Iran; hashemi.h@live.com

[2] School of Architecture and Built Environment, Deakin University, Geelong, VIC 3220, Australia; farnad.nasirzadeh@deakin.edu.au

* Correspondence: ghoddousi@iust.ac.ir; Tel.: +98-21-77240398

Abstract: The construction industry has been criticized as being a non-sustainable industry that requires effective tools to monitor and improve its sustainability performance. The multiplicity of indicators of the three pillars of sustainability—economic, social, and environmental—complicates construction sustainability assessments for project managers. Therefore, prioritizing and selecting appropriate sustainability indicators (SIs) is essential prior to conducting a construction sustainability assessment. The main purpose of this research is to select the most appropriate set of SIs to address all three pillars of highway sustainability by a new group decision-making approach. The proposed approach accounts for risk attitudes of experts and entropy measures under a triangular intuitionistic fuzzy (TIF) environment, to handle the inherent uncertainty and vagueness that is present throughout the evaluation process. Furthermore, new separation measures and ranking scores are introduced to distinguish the preference order of SIs. Eventually, the approach is implemented in a case study of highway construction projects and the applicability of the approach is examined. To investigate the stability and validity of computational results, a sensitivity analysis is carried out and a comparison is made between the obtained ranking outcomes and the traditional decision-making methods.

Keywords: sustainable highway construction; sustainability indicators; triangular intuitionistic fuzzy; multi-criteria decision-making; entropy measure; risk attitudes

Citation: Hashemi, H.; Ghoddousi, P.; Nasirzadeh, F. Sustainability Indicator Selection by a Novel Triangular Intuitionistic Fuzzy Decision-Making Approach in Highway Construction Projects. *Sustainability* **2020**, *13*, 1477. https://dx.doi.org/10.3390/su13031477

Received: 1 October 2020
Accepted: 13 November 2020
Published: 1 February 2021

1. Introduction

The preliminary concept of sustainable development was introduced in the 1980s [1]. According to the World Commission on Environment and Development (WCED) report, sustainable development refers to development that can be useful for nature, not harmful and aids in meeting the needs of present generations without compromising the needs of future generations. Sustainable development is generally balanced among three aspects or pillars: economic, environmental and social sustainability and aims to meet all these needs/objectives simultaneously [2].

In recent decades, sustainable construction—as a fundamental contributor towards sustainable development—has been the focus of a great deal of research. Recent research efforts have largely concentrated on the performance measurement process and sustainability assessment in building and construction projects, based on analytical and computational evaluation approaches, sustainable construction tools, standards and rating systems, or a combination of these. Indeed, these studies have attempted—by various techniques—to aid the construction industry in reaching sustainable development ideals and goals. As illustrated below, some of these research studies have been reviewed. Yu et al. [3] provided the project management team with planning strategies using a sustainability-assessing system. The proposed system was developed to monitor and evaluate the sustainability of whole activities throughout the construction projects' life cycle. Goubran and Cucuzzella [4] presented two analytical mapping tools for design teams of building projects to utilize Sustainable Development Goals (SDGs) as a sustainability analyzing framework in the design process. The first tool was constructed based on distinguishing between the architectural, engineering and operational concerns, while the second tool was designed based on the characteristics of the design approach (either product or human-focused) and its inspiration (history vs. future driven). Karaca et al. [5] developed a rapid sustainability assessment method using indicators and their relative weights attained from stakeholders, and an assessment approach based on the responses of buildings' occupants to measure the sustainability performance of residential buildings in Nur-Sultan, Kazakhstan. Li et al. [6] provided a comprehensive analysis of various stakeholder groups associated with sustainable construction in China. In addition, the level of stakeholder influence in decision/evaluations was measured using semi-structured interviews and the Delphi technique. Omer and Noguchi [7] developed a conceptual framework for the selection of appropriate building materials considering the implementation of the 2030 Agenda for Sustainable Development. Indeed, they presented a knowledge-based decision support system to assist policymakers, designers and construction stakeholders in making appropriate decisions towards the achievement of SDGs. Xu et al. [8] evaluated the sustainability of the construction industry by an assessment model based on the entropy method in China. The level of sustainability in construction projects was determined by two indices named the social, economic, and environmental benefits index and the ecological costs index. Illankoon et al. [9] suggested a scoring model regarding the inter-links between the Leadership in Energy and Environmental Design (LEED) credits and SDGs to evaluate buildings constructed in Australia. Their proposed model identified a Comprehensive Contribution to Development Index (CCDI) to policymakers as a guideline for evaluating building projects in order to achieve the United Nations (UN)'s SDGs. Olawumi et al. [10] introduced a grading system of buildings in Nigeria, named the Building Sustainability Assessment Method (BSAM) scheme. The scheme involves the identification of key sustainability assessment criteria and assigns weighted-scores to the various criteria by the multi-expert consultation method. Mansell et al. [11] used empirical evidence to identify a golden thread between sustainability reporting frameworks at the project level and the organizational level, and impacts of the UN's SDGs. The frameworks benefit from the Ceequal reporting methodology at the project level and the Global Reporting Initiative (GRI) methodology at the organizational level. Accordingly, a database of indicators was extracted that aligned with the specific SDG targets. Additionally, a robust investment appraisal was provided for the design stage of infrastructure projects.

Since highway projects are one of the most important aspects for the development of transportation infrastructure—necessary due to higher population concentrations and greater transportation demands in urban areas [12,13]—they have been considered as one of the most crucial components of sustainable development. Moreover, highway construction projects use a vast quantity of energy and natural materials, generate waste, and produce greenhouse gases that can greatly affect the sustainability of the construction industry.

The sustainability indicators (SIs) are significant factors in the sustainability assessment of highway projects. Various studies and tools have introduced numerous SIs for the assessment of the sustainability of construction projects that has led to the complication of the assessment process. Therefore, the prioritization of indicators and the adoption of an optimal number of SIs are major issues. In addition, the evaluation of SIs is complex for decision-makers, owing to inadequate evidence and uncertainty surrounding highway construction projects [13–15]. Hence, Multi-Criteria Decision-Making (MCDM) techniques under uncertainty are useful tools to cope with these problems [16–20]. While the construction industry plays an important role in global sustainable development, numerous research efforts have studied different subjects regarding the sustainability of the construction industry and the use of multi-criteria decision making to evaluate sustainability in this industry.

Huang and Yeh [21] developed a framework to analyze the green highway projects applying the max-min fuzzy Delphi method to recognize the main classifications and related items. Chen et al. [22] proposed a model called the construction method selection model aimed at lending support to assess the prefabrication feasibility at the initial level utilizing the simple multi-attribute rating technique and subsequently to adopt the best strategy to employ prefabrication at the following level. Reza et al. [23] proposed a thorough assessment technique using Triple Bottom Line (TBL) criteria to assess flooring systems with respect to the combination of Analytic Hierarchy Process (AHP) and Life Cycle Analysis (LCA) techniques. Waris et al. [24] established criteria for selecting sustainable construction equipment based on qualitative and quantitative feedbacks of construction industry experts and finally selected the top five criteria. Li et al. [25] proposed a comprehensive methodology using entropy, which is suitable for calculating weights, and the Technique for Order of Preference by Similarity to Ideal Solution (TOPSIS) methods at the same time to evaluate the development of highway transportation. Kucukvar et al. [26] presented a fuzzy MCDM method for prioritizing pavements and selecting the best one based on the respective sustainability performance using the TOPSIS method. Medineckiene et al. [27] proposed a novel MCDM technique to adopt criteria from which their sets and weights are determined in accordance with the Swedish certification system Miljöbyggnad and used AHP for building sustainability assessment. Kamali and Hewage [28] identified sustainability performance indicators to evaluate life cycle sustainability. Subsequently, an organized framework was developed based on designing and conducting a survey to choose the most suitable sustainability performance indicators for modular and conventional construction methods in North America. Pan et al. [29] developed a sustainability indicator framework to reliably assess the performance of construction automation and robotics in the building industry context. Indeed, the study proposed guidelines for sustainable automated and robotic options for advanced construction technology. Zolfani et al. [30] presented a hybrid MCDM methodology applying Step-wise Weight Assessment Ratio Analysis (SWARA) and Complex Proportional Assessment (COPRAS) for criteria weights and prioritizing alternatives, respectively.

Liu and Qian [31] developed an integrated sustainability assessment methodology in accordance with the life cycle sustainability assessment framework. In addition, a combination of AHP and ELimination Et Choice Translating REality (ELECTRE) was applied to derive criteria weights and prioritize alternatives. Reddy et al. [32] introduced a decision-making method to adopt a sustainable material without Life Cycle Inventory (LCI) information requirements. In this method, criteria that highly influence material sustainability were investigated and consequently applied to analyze the performance of materials in different aspects of the material life cycle to develop a sustainable material performance index utilizing AHP. Chen [33] used a new multi-criteria assessment approach integrating the Grey Relational Analysis (GRA) and TOPSIS techniques, which operate according to the intuitionistic fuzzy entropy method, for selection of the appropriate sustainable supplier of construction materials. Roy et al. [34] developed a combinative distance-based evaluation method utilizing Interval-Valued Intuitionistic Fuzzy Numbers (IVIFNs) to decide comprehensively and logically to deal with the problem of material adoption under uncertainty. Tseng et al. [35] introduced various features and measures to build a model and assess the construction projects in Ecuador, employing fuzzy decision-making trials with Decision Making Trial and Evaluation Laboratory (DEMATEL) in addition to an Analytic Network Process (ANP) to evaluate interdependence between the features of a sustainable product–service system. Hendiani and Bagherpour [14] presented a novel social sustainability performance assessment in construction projects using fuzzy numbers to evaluate the present social sustainability position associated with construction. Furthermore, the barriers that reduce the value of the social sustainability index were recognized and addressed. Alawneh et al. [36] proposed a novel framework that identifies and weighs SIs for sustainable non-residential buildings and contributes to achieving the SDGs in Jordan. The framework applies the Delphi technique to identify and categorize SIs and then integrates AHP and Relative Importance Index (RII) methods to weigh SIs. In addition, a management tool (Gantt chart) integrates SIs into the project phases towards sustainable construction management. Dabous et al. [37] proposed a multi-criteria decision-support approach to handle decision-making in sustainable pavement adoption. The main sustainable decision factors were recognized through a hierarchy structure in their approach. In addition, the AHP technique in combination with multi-attribute utility theory was used to rank the networks of pavement sections. For sustainable landfill site selection, Rahimi et al. [38] introduced a Geographical Information System (GIS)-MCDM methodology considering the group fuzzy Best-Worst Method (BWM), fuzzy MULTIMOORA method and GIS-based suitability maps. The methodology was employed in Mahallat city, Iran, and it could provide suitable guidance for the waste management department of municipalities. Navarro et al. [39] developed an assessment methodology to measure the sustainability performance of the concrete bridge deck based on a neutrosophic group AHP approach. In addition, the TOPSIS technique was utilized to aggregate the sustainability criteria.

The aforementioned studies demonstrate that the previous research did not pay much attention to the sustainability performance of highway construction projects. Furthermore, there are no studies focusing on prioritizing and selecting the indicators of the three sustainability pillars. For the recognized gaps, a new multi-criteria weighting and ranking approach, according to group decision-making, is presented in the present study to analyze and adopt SIs in highway construction projects. Initially, SIs and criteria are collected and listed concerning experts' views and the literature review. Thus, triangular intuitionistic fuzzy (TIF) decision matrices are constructed based on experts' views in terms of linguistic variables. Subsequently, the weights of experts are gained according to the concept of entropy. Afterward, primary weight vectors of criteria are specified by entropy measures and experts' views. Finally, SIs are ranked based on the positive and negative ideal separation matrices via presenting a new ranking score. Moreover, a case study in highway construction projects is addressed to demonstrate the efficiency of the presented approach.

The rest of this paper is organized as follows. In Section 2, a novel multi-criteria group decision-making approach is proposed and applied in a case study of a highway construction project. Section 3 presents the results of the approach implementation in detail. The obtained results are compared with the prevalent decision-making methods and other mentioned SIs in the cited literature, and the sensitivity analyses are conducted in Section 4. To conclude the paper, Section 5 depicts the concluding remarks.

2. Materials and Methods

2.1. TIF Group Decision-Making Approach

The proposed approach aims to assist project managers in selecting the most significant SIs for sustainability assessment of highway construction projects based on a novel TIF group decision-making approach. Figure 1 presents the proposed approach for the selection of SIs.

The phases of the presented approach are as follows:

Step 1. Constitute a group of experts $(E_e; e = 1, 2, \ldots, t)$, whose views and judgments will be employed to build and assess the problem.

Step 2. Gather a list of indicators that are possible to be applied for the sustainability evaluation of highway construction projects $(I_i; i = 1, 2, \ldots, m)$.

Step 3. Recognize a set of criteria for analyzing SIs through consensus of experts' views $(C_j; j = 1, 2, \ldots, n)$.

Step 4. Assign the risk attitude to each expert and incorporate it into the related triangular intuitionistic fuzzy numbers (TIFNs) (Definition A1).

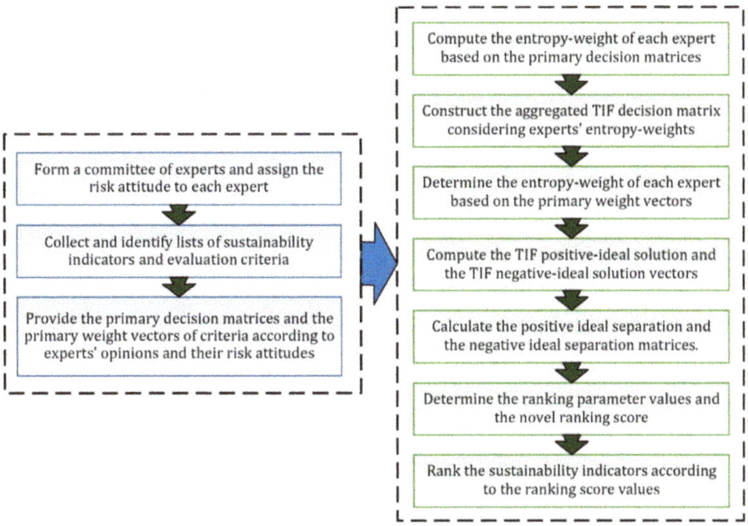

Figure 1. The proposed approach for the selection of sustainability indicators (SIs).

Each expert is assigned a risk attitude according to his or her character. The risk attitudes are able to be specified by a higher management level and expressed by linguistic variables, like absolutely optimistic (AO), optimistic (O), neutral (N), pessimistic (P), and absolutely pessimistic (AP) [40,41], for 5-scale TIFNs (Table 1).

Table 1. Linguistic variables of the risk attitudes assigned to each expert for 5-scale triangular intuitionistic fuzzy numbers (TIFNs).

Linguistic Variables	TIFN Derived from $\langle (a,b,c);\mu,v\rangle$ for Benefit Criteria	TIFN Derived from $\langle (a,b,c);\mu,v\rangle$ for Cost Criteria
Absolutely optimistic (AO)	$\langle (a,\ c,\ c);\mu+\pi,\ v\rangle$	$\langle (a,\ a,\ c);\mu,\ v+\pi\rangle$
Optimistic (O)	$\langle (a,\ (b+c)/2,\ c);\mu+\pi/2, v\rangle$	$\langle (a,\ (a+b)/2,\ c);\mu, v+\pi/2\rangle$
Neutral (N)	$\langle (a,b,\ c);\mu,\ v\rangle$	$\langle (a,b,\ c);\mu,\ v\rangle$
Pessimistic (P)	$\langle (a,\ (a+b)/2,\ c);\mu, v+\pi/2\rangle$	$\langle (a,\ (b+c)/2,\ c);\mu+\pi/2, v\rangle$
Absolutely pessimistic (AP)	$\langle (a,\ a,\ c);\mu,\ v+\pi\rangle$	$\langle (a,\ c,\ c);\mu+\pi,\ v\rangle$

In Table 1, π indicates the hesitation degree of TIFN $\langle (a,b,c);\mu,\ v\rangle$ and is equal to $1-\mu-v$.

Step 5. Construct the primary decision matrices based on the experts' views.

The primary decision matrices are constructed from the performance rating of each indicator versus each criterion based on the experts' view in terms of linguistic terms (Table 2) and converted into TIFNs.

$$\widetilde{X}^e = \left[\widetilde{x}^e_{ij}\right]_{m\times n} = \left[\left\langle \left(a_{\widetilde{x}^e_{ij}},\ b_{\widetilde{x}^e_{ij}},\ c_{\widetilde{x}^e_{ij}}\right);\mu_{\widetilde{x}^e_{ij}},\ v_{\widetilde{x}^e_{ij}}\right\rangle\right]_{m\times n} \quad (1)$$

Table 2. Linguistic variables applied for the rating of SIs.

Linguistic Variables	Triangular Intuitionistic Fuzzy Numbers
Extremely high (EH)	$\langle(0.95,\ 1.00,\ 1.00);0.95,\ 0.05\rangle$
Very very high (VVH)	$\langle(0.90,\ 1.00,\ 1.00);0.90,\ 0.10\rangle$
Very high (VH)	$\langle(0.80,\ 0.90,\ 1.00);0.80,\ 0.10\rangle$
High (H)	$\langle(0.70,\ 0.80,\ 0.90);0.70,\ 0.20\rangle$
Medium high (MH)	$\langle(0.50,\ 0.60,\ 0.70);0.60,\ 0.30\rangle$
Medium (M)	$\langle(0.30,\ 0.50,\ 0.70);0.50,\ 0.40\rangle$
Medium low (ML)	$\langle(0.30,\ 0.40,\ 0.50);0.40,\ 0.50\rangle$
Low (L)	$\langle(0.10,\ 0.20,\ 0.30);0.25,\ 0.60\rangle$
Very low (VL)	$\langle(0.00,\ 0.10,\ 0.20);0.10,\ 0.75\rangle$
Very very low (VVL)	$\langle(0.00,\ 0.00,\ 0.10);0.10,\ 0.90\rangle$

Step 6. Convert the primary decision matrices to the individual decision matrices based on each expert's risk attitude.

The primary decision matrices are converted to decision matrices taking into account each expert's risk attitude according to Table 1.

$$\widetilde{R}^e = \left[\widetilde{r}^e_{ij}\right]_{m\times n} = \left[\left\langle\left(a_{\widetilde{r}^e_{ij}},\ b_{\widetilde{r}^e_{ij}},\ c_{\widetilde{r}^e_{ij}}\right);\mu_{\widetilde{r}^e_{ij}},\ \nu_{\widetilde{r}^e_{ij}}\right\rangle\right]_{m\times n} \tag{2}$$

Step 7. Compute each expert's entropy-weight according to the individual decision matrices.

The entropy measure F^e_{ij} is calculated by [42]:

$$F^e_{ij} = -\frac{f^e_{ij} ln\left(f^e_{ij}\right)}{ln(t)}, \tag{3}$$

where

$$f^e_{ij} = \frac{\left(a_{\widetilde{r}^e_{ij}} + b_{\widetilde{r}^e_{ij}} + c_{\widetilde{r}^e_{ij}}\right)\times\left(1+\mu_{\widetilde{r}^e_{ij}} - \nu_{\widetilde{r}^e_{ij}}\right)}{\sum_{e=1}^{t}\left[\left(a_{\widetilde{r}^e_{ij}} + b_{\widetilde{r}^e_{ij}} + c_{\widetilde{r}^e_{ij}}\right)\times\left(1+\mu_{\widetilde{r}^e_{ij}} - \nu_{\widetilde{r}^e_{ij}}\right)\right]}. \tag{4}$$

Thus, each expert's entropy-weight is determined as follows [42,43]:

$$\alpha^e_{ij} = \frac{\sum_{e=1}^{t} F^e_{ij} + 1 - 2\times F^e_{ij}}{\sum_{e=1}^{t}\left(\sum_{e=1}^{t} F^e_{ij} + 1 - 2\times F^e_{ij}\right)}, \tag{5}$$

where $0 \le \alpha^e_{ij} \le 1$, and $\sum_{e=1}^{t}\alpha^e_{ij} = 1$.

Step 8. Build the aggregated TIF decision matrix taking into account the entropy-weights of experts.

According to the TIF weighted geometric aggregation (TIFWGA) operator [44], the aggregated TIF decision matrix concerning the entropy-weights of experts is gained as follows:

$$\widetilde{R} = \left[\widetilde{r}_{ij}\right]_{m\times n} = \left[\left\langle\left(a_{\widetilde{r}_{ij}},\ b_{\widetilde{r}_{ij}},\ c_{\widetilde{r}_{ij}}\right);\mu_{\widetilde{r}_{ij}},\ \nu_{\widetilde{r}_{ij}}\right\rangle\right]_{m\times n} \tag{6}$$

where $\widetilde{r}_{ij} = \left(\widetilde{r}^1_{ij}\right)^{\alpha^1_{ij}} \otimes \left(\widetilde{r}^2_{ij}\right)^{\alpha^2_{ij}} \otimes \cdots \otimes \left(\widetilde{r}^t_{ij}\right)^{\alpha^t_{ij}}$.

Step 9. Construct the primary weight vectors of criteria based on experts' views.

The significance of criteria is provided based on the experts' views in terms of linguistic terms (Table 3).

$$\widetilde{\omega}^e = \left\{\widetilde{\omega}^e_j\right\} = \left\{\left\langle\left(a_{\widetilde{\omega}^e_j},\ b_{\widetilde{\omega}^e_j},\ c_{\widetilde{\omega}^e_j}\right);\mu_{\widetilde{\omega}^e_j},\ \nu_{\widetilde{\omega}^e_j}\right\rangle\right\}, \tag{7}$$

Table 3. Linguistic variables applied for rating the significance of criteria.

Linguistic Variables	Triangular Intuitionistic Fuzzy Numbers
Very important (VI)	$\langle(0.80,\ 0.90,\ 1.00);0.90,\ 0.10\rangle$
Important (I)	$\langle(0.60,\ 0.70,\ 0.80);0.75, 0.20\rangle$
Medium (M)	$\langle(0.40,\ 0.50,\ 0.60);0.50,\ 0.45\rangle$
Unimportant (UI)	$\langle(0.20,\ 0.30,\ 0.40);0.35,\ 0.60\rangle$
Very unimportant (VUI)	$\langle(0.00,\ 0.10,\ 0.20);0.10,\ 0.90\rangle$

Step 10. Convert the primary weight vectors to the individual weight vectors based on each expert's risk attitude.

The primary weight vectors are converted to the individual weight vectors taking into account each expert's risk attitude according to Table 1.

$$\widetilde{w}^e = \left\{\widetilde{w}_j^e\right\} = \left\{\left\langle\left(a_{\widetilde{w}_j^e},\ b_{\widetilde{w}_j^e},\ c_{\widetilde{w}_j^e}\right);\mu_{\widetilde{w}_j^e},\ \nu_{\widetilde{w}_j^e}\right\rangle\right\}, \tag{8}$$

Step 11. Compute each expert's entropy-weight according to the weight vectors.

The entropy measure G_j^e is calculated by [42]:

$$G_j^e = -\frac{g_j^e ln\left(g_j^e\right)}{ln(t)}, \tag{9}$$

where

$$g_j^e = \frac{\left(a_{\widetilde{w}_j^e} + b_{\widetilde{w}_j^e} + c_{\widetilde{w}_j^e}\right) \times \left(1 + \mu_{\widetilde{w}_j^e} - \nu_{\widetilde{w}_j^e}\right)}{\sum_{e=1}^{t}\left[\left(a_{\widetilde{w}_j^e} + b_{\widetilde{w}_j^e} + c_{\widetilde{w}_j^e}\right) \times \left(1 + \mu_{\widetilde{w}_j^e} - \nu_{\widetilde{w}_j^e}\right)\right]}. \tag{10}$$

Thus, each expert's entropy-weight according to the expert-based weight vector is determined as follows [42,43]:

$$\beta_j^e = \frac{\sum_{e=1}^{t} G_j^e + 1 - 2 \times G_j^e}{\sum_{e=1}^{t}\left(\sum_{e=1}^{t} G_j^e + 1 - 2 \times G_j^e\right)}, \tag{11}$$

where $0 \leq \beta_j^e \leq 1$, and $\sum_{e=1}^{t} \beta_j^e = 1$.

Step 12. Provide the TIF weight vector of the criteria.

The TIF weight vector $\widetilde{W}$ is built according to experts' entropy-weight by using Definition A2 as follows:

$$\widetilde{W} = \left\{\widetilde{W}_j\right\} = \left\{\widetilde{W}_1, \widetilde{W}_2, \cdots, \widetilde{W}_n\right\}, \tag{12}$$

where

$$\widetilde{W}_j = \left\langle\left(a_{\widetilde{W}_j},\ b_{\widetilde{W}_j},\ c_{\widetilde{W}_j}\right);\mu_{\widetilde{W}_j},\ \nu_{\widetilde{W}_j}\right\rangle = \left(\widetilde{w}_j^1\right)^{\beta_j^1} \otimes \left(\widetilde{w}_j^2\right)^{\beta_j^2} \otimes \cdots \otimes \left(\widetilde{w}_j^t\right)^{\beta_j^t}. \tag{13}$$

Step 13. Compute the TIF positive-ideal solution (PIS) and the TIF negative-ideal solution (NIS) vectors.

The TIF PIS $\widetilde{r}_j^*$ and the TIF NIS $\widetilde{r}_j^-$ are, respectively, defined as follows:

$$\widetilde{r}_j^* = \left\{\left\langle\left(a_{\widetilde{r}_j^*},\ b_{\widetilde{r}_j^*},\ c_{\widetilde{r}_j^*}\right);\mu_{\widetilde{r}_j^*},\ \nu_{\widetilde{r}_j^*}\right\rangle\right\} = \begin{cases} \left\{\left\langle\left(\max_i\left(c_{\widetilde{r}_{ij}}\right),\ \max_i\left(c_{\widetilde{r}_{ij}}\right),\ \max_i\left(c_{\widetilde{r}_{ij}}\right)\right);\max_i\left(\mu_{\widetilde{r}_{ij}}\right), \min_i\left(\nu_{\widetilde{r}_{ij}}\right)\right\rangle\right\} & for\ j \in J_1 \\ \left\{\left\langle\left(\min_i\left(a_{\widetilde{r}_{ij}}\right),\ \min_i\left(a_{\widetilde{r}_{ij}}\right),\ \min_i\left(a_{\widetilde{r}_{ij}}\right)\right);\min_i\left(\mu_{\widetilde{r}_{ij}}\right), \max_i\left(\nu_{\widetilde{r}_{ij}}\right)\right\rangle\right\} & for\ j \in J_2, \end{cases} \tag{14}$$

and

$$\widetilde{r_j^-} = \left\{\left\langle \left(a_{\widetilde{r_j^-}}, b_{\widetilde{r_j^-}}, c_{\widetilde{r_j^-}}\right); \mu_{\widetilde{r_j^-}}, \nu_{\widetilde{r_j^-}}\right\rangle\right\} = \begin{cases} \left\{\left\langle \left(\min_i\left(c_{\widetilde{r}_{ij}}\right), \min_i\left(c_{\widetilde{r}_{ij}}\right), \min_i\left(c_{\widetilde{r}_{ij}}\right)\right); \min_i\left(\mu_{\widetilde{r}_{ij}}\right), \max_i\left(\nu_{\widetilde{r}_{ij}}\right)\right\rangle\right\} & for\ j \in J_1 \\ \left\{\left\langle \left(\max_i\left(a_{\widetilde{r}_{ij}}\right), \max_i\left(a_{\widetilde{r}_{ij}}\right), \max_i\left(a_{\widetilde{r}_{ij}}\right)\right); \max_i\left(\mu_{\widetilde{r}_{ij}}\right), \min_i\left(\nu_{\widetilde{r}_{ij}}\right)\right\rangle\right\} & for\ j \in J_2, \end{cases} \quad (15)$$

where J_1 and J_2 are the benefit criteria and cost criteria, respectively.

Step 14. Determine the positive-ideal separation (*PISE*) and the negative-ideal separation (*NISE*) matrices.

The *PISE* matrix (Δ^*) and the *NISE* matrix (Δ^-) are defined based on hamming distance [45].

$$\Delta^* = \left[\Delta_{ij}^*\right]_{m\times n} = \begin{bmatrix} \Delta\left(\widetilde{r}_1^*, \widetilde{r}_{11}\right) & \Delta\left(\widetilde{r}_2^*, \widetilde{r}_{12}\right) & \cdots & \Delta\left(\widetilde{r}_n^*, \widetilde{r}_{1n}\right) \\ \Delta\left(\widetilde{r}_1^*, \widetilde{r}_{21}\right) & \Delta\left(\widetilde{r}_2^*, \widetilde{r}_{22}\right) & \cdots & \Delta\left(\widetilde{r}_n^*, \widetilde{r}_{2n}\right) \\ \vdots & \vdots & \ddots & \\ \Delta\left(\widetilde{r}_1^*, \widetilde{r}_{m1}\right) & \Delta\left(\widetilde{r}_2^*, \widetilde{r}_{m2}\right) & \cdots & \Delta\left(\widetilde{r}_n^*, \widetilde{r}_{mn}\right) \end{bmatrix}, \quad (16)$$

and

$$\Delta^- = \left[\Delta_{ij}^-\right]_{m\times n} = \begin{bmatrix} \Delta\left(\widetilde{r}_1^-, \widetilde{r}_{11}\right) & \Delta\left(\widetilde{r}_2^-, \widetilde{r}_{12}\right) & \cdots & \Delta\left(\widetilde{r}_n^-, \widetilde{r}_{1n}\right) \\ \Delta\left(\widetilde{r}_1^-, \widetilde{r}_{21}\right) & \Delta\left(\widetilde{r}_2^-, \widetilde{r}_{22}\right) & \cdots & \Delta\left(\widetilde{r}_n^-, \widetilde{r}_{2n}\right) \\ \vdots & \vdots & \ddots & \\ \Delta\left(\widetilde{r}_1^-, \widetilde{r}_{m1}\right) & \Delta\left(\widetilde{r}_2^-, \widetilde{r}_{m2}\right) & \cdots & \Delta\left(\widetilde{r}_n^-, \widetilde{r}_{mn}\right) \end{bmatrix}. \quad (17)$$

Step 15. Compute the $\mathfrak{A}_i$, $\mathfrak{B}_i$, $\mathfrak{A}_i'$, and $\mathfrak{B}_i'$ values.

The $\mathfrak{A}_i$, $\mathfrak{B}_i$, $\mathfrak{A}_i'$, and $\mathfrak{B}_i'$ values are computed according to the score function [46] as follows:

$$\begin{gathered} \mathfrak{A}_i = \sum_{j=1}^n \Delta_{ij}^* \cdot \widetilde{W}_j \\ = \tfrac{1}{4}\left(\sum_{j=1}^n \Delta_{ij}^* \cdot a_{\widetilde{W}_j} + 2\Delta_{ij}^* \cdot b_{\widetilde{W}_j} + \Delta_{ij}^* \cdot c_{\widetilde{W}_j}\right)\left(1 - \prod_{j=1}^n \left(1-\mu_{\widetilde{W}_j}\right)^{\Delta_{ij}^*} - \prod_{j=1}^n \nu_{\widetilde{W}_j}^{\Delta_{ij}^*}\right) \end{gathered} \quad (18)$$

$$\begin{gathered} \mathfrak{B}_i = \max_j\left(d_{ij}^* \cdot \widetilde{W}_j\right) \\ = \max_j\left(\tfrac{1}{4}\left(\Delta_{ij}^* \cdot a_{\widetilde{W}_j} + 2\Delta_{ij}^* \cdot b_{\widetilde{W}_j} + \Delta_{ij}^* \cdot c_{\widetilde{W}_j}\right)\left(1 - \left(1-\mu_{\widetilde{W}_j}\right)^{\Delta_{ij}^*} - \nu_{\widetilde{W}_j}^{\Delta_{ij}^*}\right)\right) \end{gathered} \quad (19)$$

$$\begin{gathered} \mathfrak{A}_i' = \sum_{j=1}^n \Delta_{ij}^- \cdot \widetilde{W}_j \\ = \tfrac{1}{4}\left(\sum_{j=1}^n \Delta_{ij}^- \cdot a_{\widetilde{W}_j} + 2\Delta_{ij}^- \cdot b_{\widetilde{W}_j} + \Delta_{ij}^- \cdot c_{\widetilde{W}_j}\right)\left(1 - \prod_{j=1}^n \left(1-\mu_{\widetilde{W}_j}\right)^{\Delta_{ij}^-} - \prod_{j=1}^n \nu_{\widetilde{W}_j}^{\Delta_{ij}^-}\right) \end{gathered} \quad (20)$$

$$\begin{gathered} \mathfrak{B}_i' = \max_j\left(\Delta_{ij}^- \cdot \widetilde{W}_j\right) \\ = \max_j\left(\tfrac{1}{4}\left(\Delta_{ij}^- \cdot a_{\widetilde{W}_j} + 2\Delta_{ij}^- \cdot b_{\widetilde{W}_j} + \Delta_{ij}^- \cdot c_{\widetilde{W}_j}\right)\left(1 - \left(1-\mu_{\widetilde{W}_j}\right)^{\Delta_{ij}^-} - \nu_{\widetilde{W}_j}^{\Delta_{ij}^-}\right)\right) \end{gathered} \quad (21)$$

Step 16. Calculate the κ_i and ϑ_i values.

The values of indices κ_i and ϑ_i are calculated as follows:

$$\kappa_i = \chi\left(\frac{\mathfrak{A}_i - \mathfrak{A}^*}{\mathfrak{A}^- - \mathfrak{A}^*}\right) + (1-\chi)\left(\frac{\mathfrak{B}_i - \mathfrak{B}^*}{\mathfrak{B}^- - \mathfrak{B}^*}\right), \quad (22)$$

and

$$\vartheta_i = \psi\left(\frac{\mathfrak{A}_i' - \mathfrak{A}'^-}{\mathfrak{A}'^* - \mathfrak{A}'^-}\right) + (1-\psi)\left(\frac{\mathfrak{B}_i' - \mathfrak{B}'^-}{\mathfrak{B}'^* - \mathfrak{B}'^-}\right), \quad (23)$$

where $\begin{cases} \mathfrak{A}^* = \min\limits_i \mathfrak{A}_i \\ \mathfrak{A}^- = \max\limits_i \mathfrak{A}_i \end{cases}$, $\begin{cases} \mathfrak{B}^* = \min\limits_i \mathfrak{B}_i \\ \mathfrak{B}^- = \max\limits_i \mathfrak{B}_i \end{cases}$, $\begin{cases} \mathfrak{A}'^* = \max\limits_i \mathfrak{A}'_i \\ \mathfrak{A}'^- = \min\limits_i \mathfrak{A}'_i \end{cases}$, $\begin{cases} \mathfrak{B}'^* = \max\limits_i \mathfrak{B}'_i \\ \mathfrak{B}'^- = \min\limits_i \mathfrak{B}'_i \end{cases}$, χ and ψ are regarded as the relative importance for the strategy of the majority attributes, whereas $1-\chi$ and $1-\psi$ are the relative importance of the individual regret.

Step 17. Compute the novel ranking score.

The ranking scores $\mathfrak{C}_i$ are defined as follows:

$$\mathfrak{c}_i = \eta\left(\frac{\kappa_i - \kappa^*}{\kappa^- - \kappa^*} + \frac{\vartheta^* - \vartheta_i}{\vartheta^* - \vartheta^-}\right) + (1-\eta)\left(\frac{\kappa_i - \kappa^*}{\kappa^- - \kappa^*} \times \frac{\vartheta^* - \vartheta_i}{\vartheta^* - \vartheta^-}\right), \tag{24}$$

$$\mathfrak{C}_i = \frac{\mathfrak{c}_i - \gamma}{\lambda - \gamma} \tag{25}$$

where $\begin{cases} \kappa^* = \min\limits_i \kappa_i \\ \kappa^- = \max\limits_i \kappa_i \end{cases}$, $\begin{cases} \vartheta^* = \max\limits_i \vartheta_i \\ \vartheta^- = \min\limits_i \vartheta_i \end{cases}$, $\gamma = \min\limits_i \mathfrak{c}_i$, $\lambda = \max\limits_i \mathfrak{c}_i$ and $0 \leq \eta \leq 1$.

Step 18. Rank the SIs according to the ranking score ($\mathfrak{C}_i$ values).

The SIs are sorted by the $\mathfrak{C}_i$ values in decreasing order. The maximum value of the $\mathfrak{C}_i$ indicates the higher importance.

2.2. Case Study

The efficiency of the presented approach was examined through a case study of a highway construction project. To that end, an Iranian construction firm was involved in transportation infrastructures. The firm has numerous highway construction projects being built in various areas of the country. To evaluate the projects according to sustainable construction principles, the firm managers intended to recognize and prioritize SIs to adopt the key evaluation indicators from a pool of numerous SIs in these projects.

According to step 1, five experts working on highway projects were adopted from employees of the firm. The participants comprised construction project managers and sustainable construction experts. They had enough experience and knowledge of nearly all the sustainable aspects of construction projects. As such, a group of five experts (E_1, E_2, … , E_5) was considered for analyzing potential SIs. After forming the committee, the experts picked out a set of potential SIs in addition to a set of relevant criteria for SIs assessment (Steps 2 and 3). To that end, a brainstorming session was held with the experts and thirty sustainability indicators (SoI_1, SoI_2, … , SoI_9, EcI_1, EcI_2, … , EcI_9, EnI_1, EnI_2, … , EnI_{12}), as well as seven criteria (C_1, C_2, … ,C_7) were obtained from the various investigations in the literature (e.g., [22,47–64]) and the consensus opinion of the group members (Tables 4 and 5). Furthermore, the project manager utilizing Table 1 specifies the experts' risk attitude according to his or her recognition of them. The outcomes are represented in Table 6 (Step 4).

Table 4. List of obtained sustainability indicators.

		Sustainability Indicators	Description
Sustainability aspects	Social	SoI_1: Health	Highlighting on-site sanitation, and the provision of health care
		SoI_2: Education	Number and time of training course to different levels of employees
		SoI_3: Culture and heritage	Measure of negative impacts from construction operations on any cultural heritage
		SoI_4: Safety	Number of accidents, the supply rate of on-site supervision and training course to employees to provide a safe and reliable workplace
		SoI_5: Stakeholder satisfaction	Measure of stakeholder satisfaction by using stakeholder management models
		SoI_6: Job opportunities	Providing direct and indirect jobs
		SoI_7: Tourism	Impacts on tourism development
		SoI_8: Traffic	Vehicle traffic congestion
		SoI_9: Access to public transportation	Extension of public transportation services and proximity to it
	Economic	EcI_1: Net present value (NPV)	$NPV = \sum_{t=1}^{n} \frac{R_t}{(1+i)^t}$ where R_t is the net cash inflow-outflows during a single period t, i is the discount rate of return that could be earned in alternative investments and t is the number of time periods
		EcI_2: Payback period	Initial Investment/Net Cash Flow per Period
		EcI_3: Investment planning	Compliance with the investment plan
		EcI_4: Benefit–cost ratio	Relationship between the relative costs and benefits of a proposed project expressed in monetary or qualitative terms
		EcI_5: Debt–asset ratio	(Short-term Debt + Long-term Debt)/Total Assets
		EcI_6: Project budget	Compliance with budget
		EcI_7: Internal rate of return (IRR)	$NPV = \sum_{t=1}^{T} \frac{C_t}{(1+IRR)^t} - C_0 = 0$ where C_t is the net cash inflow during the period t, C_0 is the total initial investment cost and t is the number of time periods
		EcI_8: Financial risk	Possibility of losing money on the investment
		EcI_9: Life-cycle cost	Total cost for a construction project over its life
	Environmental	EnI_1: Material consumption	Efficiency rate of using materials and resources
		EnI_2: Air pollution	Measure of mixture of solid particles and gases in the air
		EnI_3: landscape respect	Protection of landscape features during construction
		EnI_4: Noise emissions	Rate of noise pollution during the construction phase in the environment of the project
		EnI_5: Erosion	Rate of soil erosion during the construction phase in the environment of the project
		EnI_6: Ecological impacts	Measure of negative impacts from project to flora, fauna, and ecosystems
		EnI_7: Habitat loss and damage	Destructive effects on the living environment for both human being and animals
		EnI_8: Soil contamination	Measure of alteration in the physical, chemical and biological characteristics of the soil environment
		EnI_9: Aesthetical and visual impacts	Aesthetic quality of the project during the construction phase
		EnI_{10}: Water pollution	Measure of alteration in the physical, chemical and biological characteristics of water environment
		EnI_{11}: Water saving	Rate of reduction water consumption during the construction phase
		EnI_{12}: Hazardous waste	Production rate of hazardous waste

Table 5. List of obtained criteria.

Criteria	Criteria Type		Description
	Benefit	Cost	
C_1: Measurability	✓		Measurability in qualitative or quantitative terms
C_2: Applicability	✓		Practicality and straightforward use of sustainability indicator (SI) for evaluation
C_3: Data availability	✓		Relative simplicity to gather the necessary data for evaluation of SI
C_4: Acceptant	✓		Acceptance of SI by major stakeholders
C_5: Complexity		✓	Relative difficulty in meaningful interpretation of SI
C_6: Time consuming		✓	Required time for the evaluation of SI
C_7: Uncertainty		✓	Ambiguity in assigning the value to SI during evaluation

Table 6. Experts' risk attitudes.

Experts	E_1	E_2	E_3	E_4	E_5
Risk attitudes	Neutral	Absolutely optimistic	Pessimistic	Optimistic	Neutral

3. Results

The primary decision matrices are constructed by experts employing linguistic terms in Table 2. The matrices are shown in Table 7 (Step 5). Afterward, the primary decision matrices are converted to decision matrices taking into account each expert's risk attitude according to Table 1 (Table 8) (Step 6). Owing to space limitations, only the outcomes associated with three SIs (Sol_6, EcI_4 and EnI_{10}) are shown as a sample of each dimension of sustainability in some of the following tables.

Then, each expert's entropy-weight and aggregated TIF decision matrix is calculated. The outcomes of these steps are represented in Tables 9 and 10, respectively (Steps 7 and 8). In addition, the criteria weight vector is achieved according to experts' preferences and is illustrated in Table 11 (Step 9). According to steps 10 to 12, based on criteria weight vector, expert's risk attitude and expert's entropy-weight, the TIF criteria weight vectors are built. The results are presented in Table 12. Next, the TIF PIS and NIS vectors are specified as given in Table 13 (Step 13). Then, as shown in Table 14, the PISE matrix (D^*) and NISE matrix (D^-) are built (Step 14).

Table 7. Performance rating of each indicator versus each criterion based on experts' views in terms of linguistic terms.

SIs	Experts	Criteria						
		C_1	C_2	C_3	C_4	C_5	C_6	C_7
Sol_1	E_1	H	VH	ML	VH	L	VL	M
	E_2	H	VH	M	VVH	VL	VL	MH
	E_3	VH	H	M	VVH	VL	VVL	M
	E_4	H	VVH	MH	VH	VVL	L	ML
	E_5	VH	VVH	M	VVH	VVL	VL	ML
Sol_2	E_1	H	ML	MH	H	L	L	M
	E_2	M	M	M	MH	VL	ML	M
	E_3	MH	M	H	H	L	ML	ML
	E_4	M	ML	M	MH	ML	M	M
	E_5	MH	L	MH	MH	L	ML	M

Table 7. *Cont.*

SIs	Experts	Criteria						
		C_1	C_2	C_3	C_4	C_5	C_6	C_7
SoI_3	E_1	VL	ML	MH	MH	H	MH	VH
	E_2	L	L	H	H	MH	H	H
	E_3	VVL	L	H	VH	VH	H	VH
	E_4	VL	VL	VH	H	H	MH	VH
	E_5	VL	L	VH	VH	MH	MH	H
SoI_4	E_1	H	MH	H	H	ML	L	ML
	E_2	VH	MH	VH	VH	M	VVL	L
	E_3	H	M	VVH	VH	ML	VL	ML
	E_4	H	M	VH	H	L	VVL	L
	E_5	VH	MH	VH	VVH	ML	VL	L
SoI_5	E_1	L	H	ML	VH	VH	H	H
	E_2	VL	VH	L	H	H	MH	VVH
	E_3	ML	VH	VL	VVH	H	VH	VH
	E_4	ML	VVH	VL	VH	VH	H	VH
	E_5	L	VVH	ML	VH	VVH	VH	H
SoI_6	E_1	VH	VH	H	MH	L	L	L
	E_2	MH	H	H	H	VL	ML	L
	E_3	H	H	MH	MH	VL	L	M
	E_4	H	VH	MH	M	L	ML	M
	E_5	VH	VVH	H	H	L	L	L
SoI_7	E_1	VH	MH	M	H	L	MH	H
	E_2	VH	H	M	MH	VL	MH	VH
	E_3	VVH	H	ML	VH	VL	H	H
	E_4	VH	MH	ML	H	VVL	H	MH
	E_5	VVH	MH	ML	H	VL	VH	MH
SoI_8	E_1	H	MH	M	MH	L	H	VH
	E_2	MH	H	MH	MH	VL	VH	VVH
	E_3	H	H	MH	M	VL	H	VH
	E_4	H	VH	MH	M	VVL	VH	VVH
	E_5	VH	H	M	H	L	VH	VH
SoI_9	E_1	VH	MH	L	ML	VL	M	H
	E_2	H	H	ML	MH	VL	ML	M
	E_3	VH	MH	ML	ML	VL	L	M
	E_4	H	VH	ML	M	VL	L	H
	E_5	H	H	M	M	VVL	ML	MH
EcI_1	E_1	EH	EH	H	ML	VL	VVL	H
	E_2	EH	VVH	VVH	M	VVL	VVL	MH
	E_3	EH	EH	VH	MH	VL	VVL	MH
	E_4	EH	VH	VH	MH	VVL	VVL	M
	E_5	EH	EH	VVH	ML	VVL	VVL	H
EcI_2	E_1	VVH	H	H	M	L	VL	MH
	E_2	VH	VH	H	ML	VL	L	H
	E_3	VH	H	H	L	L	L	VH
	E_4	VH	H	H	L	L	VL	H
	E_5	VVH	H	VH	ML	VL	VL	VH
EcI_3	E_1	H	MH	MH	L	M	ML	H
	E_2	MH	MH	M	VL	M	ML	MH
	E_3	H	H	M	L	MH	M	VH
	E_4	VH	H	ML	VL	MH	M	MH
	E_5	VH	H	M	ML	MH	MH	H

Table 7. *Cont.*

SIs	Experts	Criteria						
		C_1	C_2	C_3	C_4	C_5	C_6	C_7
EcI_4	E_1	EH	EH	VH	M	L	ML	ML
	E_2	EH	VVH	VVH	MH	ML	L	M
	E_3	EH	VVH	VVH	M	L	L	ML
	E_4	EH	EH	VVH	MH	L	ML	ML
	E_5	EH	EH	EH	MH	VL	L	L
EcI_5	E_1	VH	H	H	ML	L	ML	H
	E_2	H	H	MH	L	L	ML	MH
	E_3	VVH	MH	H	L	ML	M	MH
	E_4	H	MH	MH	L	L	M	H
	E_5	H	H	MH	ML	L	M	MH
EcI_6	E_1	H	VH	H	ML	L	M	MH
	E_2	VH	VH	H	M	VL	MH	H
	E_3	H	H	MH	ML	VVL	M	H
	E_4	VH	VH	VH	ML	VL	MH	MH
	E_5	VH	VVH	VH	M	VVL	ML	H
EcI_7	E_1	VVH	EH	VH	ML	VL	VL	MH
	E_2	EH	EH	VH	ML	VVL	VL	H
	E_3	EH	EH	VH	M	VVL	VL	MH
	E_4	EH	EH	H	M	VL	VVL	MH
	E_5	EH	EH	VVH	MH	VVL	VVL	ML
EcI_8	E_1	ML	VVH	M	ML	H	MH	EH
	E_2	M	H	ML	ML	MH	MH	VVH
	E_3	M	H	ML	L	MH	H	VH
	E_4	ML	MH	M	L	M	H	VVH
	E_5	M	VVH	M	ML	M	MH	VH
EcI_9	E_1	MH	H	M	MH	M	VH	H
	E_2	MH	VH	ML	H	ML	H	VH
	E_3	H	VH	ML	H	MH	VH	VH
	E_4	H	H	L	VH	ML	VVH	H
	E_5	H	VVH	ML	H	MH	MH	MH
EnI_1	E_1	H	MH	H	MH	L	L	H
	E_2	H	MH	VH	ML	L	VL	MH
	E_3	H	MH	H	ML	VL	VL	M
	E_4	H	H	VH	MH	L	L	H
	E_5	VH	H	VVH	M	VL	VVL	M
EnI_2	E_1	VH	VH	ML	MH	M	ML	VH
	E_2	H	H	MH	MH	MH	M	H
	E_3	H	H	MH	H	M	ML	VH
	E_4	VH	VH	ML	H	MH	ML	H
	E_5	H	VVH	MH	H	M	L	H
EnI_3	E_1	ML	L	ML	M	MH	H	VH
	E_2	L	L	ML	MH	H	H	H
	E_3	L	ML	ML	M	MH	VH	MH
	E_4	VL	L	ML	M	H	H	MH
	E_5	ML	ML	M	MH	MH	H	VH
EnI_4	E_1	M	M	L	M	ML	MH	H
	E_2	MH	ML	VL	MH	M	M	H
	E_3	ML	ML	L	M	ML	ML	MH
	E_4	ML	M	L	ML	ML	M	VH
	E_5	MH	M	ML	MH	L	ML	MH

Table 7. *Cont.*

SIs	Experts	Criteria						
		C_1	C_2	C_3	C_4	C_5	C_6	C_7
EnI_5	E_1	MH	ML	L	ML	H	H	VH
	E_2	M	L	VL	L	MH	VH	H
	E_3	ML	VL	VL	L	H	VH	VH
	E_4	ML	L	L	VL	VH	H	H
	E_5	MH	L	ML	ML	MH	H	MH
EnI_6	E_1	MH	H	M	M	H	H	H
	E_2	M	MH	ML	MH	VH	MH	VH
	E_3	M	H	M	M	H	MH	H
	E_4	MH	H	M	ML	VH	H	MH
	E_5	H	VH	MH	MH	VVH	VH	MH
EnI_7	E_1	ML	MH	ML	M	ML	MH	H
	E_2	M	ML	L	ML	ML	M	MH
	E_3	MH	MH	ML	L	M	MH	MH
	E_4	M	MH	M	L	M	M	H
	E_5	MH	M	M	M	ML	M	H
EnI_8	E_1	H	VH	H	MH	ML	MH	H
	E_2	MH	H	MH	MH	ML	M	MH
	E_3	H	MH	M	H	VL	M	H
	E_4	MH	MH	MH	H	ML	MH	VH
	E_5	VH	VH	H	H	L	ML	MH
EnI_9	E_1	M	MH	ML	MH	MH	M	VH
	E_2	ML	H	L	H	H	M	H
	E_3	ML	MH	VL	H	H	MH	H
	E_4	ML	MH	L	VH	MH	H	MH
	E_5	M	H	ML	VH	MH	M	MH
EnI_{10}	E_1	H	H	MH	H	M	M	M
	E_2	VH	H	H	H	M	MH	MH
	E_3	H	MH	M	VH	M	M	MH
	E_4	VH	H	M	H	ML	MH	M
	E_5	VVH	VH	MH	VH	ML	M	ML
EnI_{11}	E_1	H	H	MH	H	VL	M	H
	E_2	H	MH	H	MH	VL	ML	VH
	E_3	H	MH	MH	H	VL	MH	H
	E_4	H	H	M	MH	L	ML	H
	E_5	VH	VH	MH	VH	VL	ML	MH
EnI_{12}	E_1	VH	L	MH	MH	VL	ML	VH
	E_2	H	ML	M	H	L	L	H
	E_3	MH	L	H	M	VL	ML	VH
	E_4	MH	ML	MH	MH	VL	ML	H
	E_5	H	L	H	H	VVL	L	MH

Table 8. Experts' view concerning the rating of sample indicators with respect to the criteria by taking into account each expert's risk attitude.

Criteria	Experts	SIs		
		SoI_6	EcI_4	EnI_{10}
C_1	E_1	⟨(0.800, 0.900, 1.000);0.800, 0.100⟩	⟨(0.950, 1.000, 1.000);0.950, 0.050⟩	⟨(0.700, 0.800, 0.900);0.700, 0.200⟩
	E_2	⟨(0.500, 0.700, 0.700);0.700, 0.300⟩	⟨(0.950, 1.000, 1.000);0.950, 0.050⟩	⟨(0.800, 1.000, 1.000);0.900, 0.100⟩
	E_3	⟨(0.700, 0.750, 0.900);0.700, 0.250⟩	⟨(0.950, 0.975, 1.000);0.950, 0.050⟩	⟨(0.700, 0.750, 0.900);0.700, 0.250⟩
	E_4	⟨(0.700, 0.850, 0.900);0.750, 0.200⟩	⟨(0.950, 1.000, 1.000);0.950, 0.050⟩	⟨(0.800, 0.950, 1.000);0.850, 0.100⟩
	E_5	⟨(0.800, 0.900, 1.000);0.800, 0.100⟩	⟨(0.950, 1.000, 1.000);0.950, 0.050⟩	⟨(0.900, 1.000, 1.000);0.900, 0.100⟩
C_2	E_1	⟨(0.800, 0.900, 1.000);0.800, 0.100⟩	⟨(0.950, 1.000, 1.000);0.950, 0.050⟩	⟨(0.700, 0.800, 0.900);0.700, 0.200⟩
	E_2	⟨(0.700, 0.900, 0.900);0.800, 0.200⟩	⟨(0.900, 1.000, 1.000);0.900, 0.100⟩	⟨(0.700, 0.900, 0.900);0.800, 0.200⟩
	E_3	⟨(0.700, 0.750, 0.900);0.700, 0.250⟩	⟨(0.900, 0.950, 1.000);0.900, 0.100⟩	⟨(0.500, 0.550, 0.700);0.600, 0.350⟩
	E_4	⟨(0.800, 0.950, 1.000);0.850, 0.100⟩	⟨(0.950, 1.000, 1.000);0.950, 0.050⟩	⟨(0.700, 0.850, 0.900);0.750, 0.200⟩
	E_5	⟨(0.900, 1.000, 1.000);0.900, 0.100⟩	⟨(0.950, 1.000, 1.000);0.950, 0.050⟩	⟨(0.800, 0.900, 1.000);0.800, 0.100⟩
C_3	E_1	⟨(0.700, 0.800, 0.900);0.700, 0.200⟩	⟨(0.800, 0.900, 1.000);0.800, 0.100⟩	⟨(0.500, 0.600, 0.700);0.600, 0.300⟩
	E_2	⟨(0.700, 0.900, 0.900);0.800, 0.200⟩	⟨(0.900, 1.000, 1.000);0.900, 0.100⟩	⟨(0.700, 0.900, 0.900);0.800, 0.200⟩
	E_3	⟨(0.500, 0.550, 0.700);0.600, 0.350⟩	⟨(0.900, 0.950, 1.000);0.900, 0.100⟩	⟨(0.300, 0.400, 0.700);0.500, 0.450⟩
	E_4	⟨(0.500, 0.650, 0.700);0.650, 0.300⟩	⟨(0.900, 1.000, 1.000);0.900, 0.100⟩	⟨(0.300, 0.600, 0.700);0.550, 0.400⟩
	E_5	⟨(0.700, 0.800, 0.900);0.700, 0.200⟩	⟨(0.950, 1.000, 1.000);0.950, 0.050⟩	⟨(0.500, 0.600, 0.700);0.600, 0.300⟩
C_4	E_1	⟨(0.500, 0.600, 0.700);0.600, 0.300⟩	⟨(0.300, 0.500, 0.700);0.500, 0.400⟩	⟨(0.700, 0.800, 0.900);0.700, 0.200⟩
	E_2	⟨(0.700, 0.900, 0.900);0.800, 0.200⟩	⟨(0.500, 0.700, 0.700);0.700, 0.300⟩	⟨(0.700, 0.900, 0.900);0.800, 0.200⟩
	E_3	⟨(0.500, 0.550, 0.700);0.600, 0.350⟩	⟨(0.300, 0.400, 0.700);0.500, 0.450⟩	⟨(0.800, 0.850, 1.000);0.800, 0.150⟩
	E_4	⟨(0.300, 0.600, 0.700);0.550, 0.400⟩	⟨(0.500, 0.650, 0.700);0.650, 0.300⟩	⟨(0.700, 0.850, 0.900);0.750, 0.200⟩
	E_5	⟨(0.700, 0.800, 0.900);0.700, 0.200⟩	⟨(0.500, 0.600, 0.700);0.600, 0.300⟩	⟨(0.800, 0.900, 1.000);0.800, 0.100⟩
C_5	E_1	⟨(0.100, 0.200, 0.300);0.250, 0.600⟩	⟨(0.100, 0.200, 0.300);0.250, 0.600⟩	⟨(0.300, 0.500, 0.700);0.500, 0.400⟩
	E_2	⟨(0.000, 0.000, 0.200);0.100, 0.900⟩	⟨(0.300, 0.300, 0.500);0.400, 0.600⟩	⟨(0.300, 0.300, 0.700);0.500, 0.500⟩
	E_3	⟨(0.000, 0.100, 0.200);0.175, 0.750⟩	⟨(0.100, 0.167, 0.300);0.325, 0.600⟩	⟨(0.300, 0.400, 0.700);0.550, 0.400⟩
	E_4	⟨(0.100, 0.150, 0.300);0.250, 0.675⟩	⟨(0.100, 0.150, 0.300);0.250, 0.675⟩	⟨(0.300, 0.350, 0.500);0.400, 0.550⟩
	E_5	⟨(0.100, 0.200, 0.300);0.250, 0.600⟩	⟨(0.000, 0.100, 0.200);0.100, 0.750⟩	⟨(0.300, 0.400, 0.500);0.400, 0.500⟩
C_6	E_1	⟨(0.100, 0.200, 0.300);0.250, 0.600⟩	⟨(0.300, 0.400, 0.500);0.400, 0.500⟩	⟨(0.300, 0.500, 0.700);0.500, 0.400⟩
	E_2	⟨(0.300, 0.300, 0.500);0.400, 0.600⟩	⟨(0.100, 0.100, 0.300);0.250, 0.750⟩	⟨(0.500, 0.500, 0.700);0.600, 0.400⟩
	E_3	⟨(0.100, 0.167, 0.300);0.325, 0.600⟩	⟨(0.100, 0.167, 0.300);0.325, 0.600⟩	⟨(0.300, 0.400, 0.700);0.550, 0.400⟩
	E_4	⟨(0.300, 0.350, 0.500);0.400, 0.550⟩	⟨(0.300, 0.350, 0.500);0.400, 0.550⟩	⟨(0.500, 0.550, 0.700);0.600, 0.350⟩
	E_5	⟨(0.100, 0.200, 0.300);0.250, 0.600⟩	⟨(0.100, 0.200, 0.300);0.250, 0.600⟩	⟨(0.300, 0.500, 0.700);0.500, 0.400⟩
C_7	E_1	⟨(0.100, 0.200, 0.300);0.250, 0.600⟩	⟨(0.300, 0.400, 0.500);0.400, 0.500⟩	⟨(0.300, 0.500, 0.700);0.500, 0.400⟩
	E_2	⟨(0.100, 0.100, 0.300);0.250, 0.750⟩	⟨(0.300, 0.300, 0.700);0.500, 0.500⟩	⟨(0.500, 0.500, 0.700);0.600, 0.400⟩
	E_3	⟨(0.300, 0.400, 0.700);0.550, 0.400⟩	⟨(0.300, 0.300, 0.500);0.450, 0.500⟩	⟨(0.500, 0.433, 0.700);0.650, 0.300⟩
	E_4	⟨(0.300, 0.400, 0.700);0.500, 0.450⟩	⟨(0.300, 0.350, 0.500);0.400, 0.550⟩	⟨(0.300, 0.400, 0.700);0.500, 0.450⟩
	E_5	⟨(0.100, 0.200, 0.300);0.250, 0.600⟩	⟨(0.100, 0.200, 0.300);0.250, 0.600⟩	⟨(0.300, 0.400, 0.500);0.400, 0.500⟩

Table 9. Entropy-weight assigned to each expert.

SIs	Experts	Criteria						
		C_1	C_2	C_3	C_4	C_5	C_6	C_7
SoI_6	E_1	0.196	0.199	0.197	0.202	0.187	0.206	0.208
	E_2	0.207	0.202	0.195	0.193	0.228	0.191	0.217
	E_3	0.202	0.205	0.207	0.204	0.208	0.205	0.184
	E_4	0.199	0.198	0.204	0.207	0.190	0.190	0.184
	E_5	0.196	0.197	0.197	0.194	0.187	0.206	0.208
EcI_4	E_1	0.200	0.199	0.202	0.204	0.196	0.188	0.196
	E_2	0.200	0.201	0.200	0.195	0.187	0.214	0.193
	E_3	0.200	0.201	0.200	0.207	0.195	0.204	0.197
	E_4	0.200	0.199	0.200	0.196	0.200	0.189	0.198
	E_5	0.200	0.199	0.198	0.197	0.222	0.205	0.217
EnI_{10}	E_1	0.204	0.199	0.199	0.202	0.196	0.202	0.200
	E_2	0.197	0.197	0.191	0.200	0.200	0.198	0.196
	E_3	0.205	0.210	0.208	0.199	0.196	0.202	0.195
	E_4	0.198	0.198	0.203	0.201	0.205	0.197	0.202
	E_5	0.197	0.195	0.199	0.198	0.203	0.202	0.208

Table 10. Aggregated triangular intuitionistic fuzzy (TIF) decision matrix ($\widetilde{R}$).

Criteria	SIs		
	SoI_6	EcI_4	EnI_{10}
C_1	$\langle(0.688, 0.814, 0.890); 0.748, 0.196\rangle$	$\langle(0.950, 0.995, 1.000); 0.950, 0.050\rangle$	$\langle(0.775, 0.892, 0.958); 0.803, 0.154\rangle$
C_2	$\langle(0.776, 0.895, 0.958); 0.806, 0.153\rangle$	$\langle(0.930, 0.990, 1.000); 0.930, 0.070\rangle$	$\langle(0.670, 0.784, 0.872); 0.724, 0.216\rangle$
C_3	$\langle(0.610, 0.726, 0.812); 0.685, 0.254\rangle$	$\langle(0.888, 0.969, 1.000); 0.888, 0.090\rangle$	$\langle(0.432, 0.596, 0.734); 0.600, 0.338\rangle$
C_4	$\langle(0.512, 0.674, 0.772); 0.642, 0.297\rangle$	$\langle(0.405, 0.557, 0.700); 0.583, 0.355\rangle$	$\langle(0.738, 0.859, 0.938); 0.769, 0.171\rangle$
C_5	$\langle(0.000, 0.000, 0.251); 0.188, 0.746\rangle$	$\langle(0.000, 0.169, 0.302); 0.234, 0.654\rangle$	$\langle(0.300, 0.384, 0.610); 0.465, 0.475\rangle$
C_6	$\langle(0.152, 0.232, 0.365); 0.316, 0.591\rangle$	$\langle(0.151, 0.210, 0.364); 0.315, 0.614\rangle$	$\langle(0.367, 0.487, 0.700); 0.548, 0.390\rangle$
C_7	$\langle(0.150, 0.222, 0.410); 0.328, 0.587\rangle$	$\langle(0.236, 0.300, 0.478); 0.386, 0.533\rangle$	$\langle(0.366, 0.444, 0.653); 0.521, 0.415\rangle$

Table 11. Criteria weight vector based on experts' preferences.

Experts	Criteria						
	C_1	C_2	C_3	C_4	C_5	C_6	C_7
E_1	I	I	I	I	M	M	I
E_2	M	VI	I	VI	I	UI	I
E_3	I	I	I	I	I	M	I
E_4	M	I	VI	I	M	M	VI
E_5	I	VI	I	I	I	M	VI

Table 12. TIF and crisp criteria weight vectors.

Criteria	Weight Vectors
	$\tilde{W}$
C_1	$\langle(0.509,\ 0.637,\ 0.712);0.655,\ 0.317\rangle$
C_2	$\langle(0.672,\ 0.789,\ 0.874);0.811,\ 0.167\rangle$
C_3	$\langle(0.635,\ 0.752,\ 0.836);0.788,\ 0.186\rangle$
C_4	$\langle(0.635,\ 0.750,\ 0.836);0.783,\ 0.186\rangle$
C_5	$\langle(0.509,\ 0.587,\ 0.712);0.640,\ 0.328\rangle$
C_6	$\langle(0.346,\ 0.413,\ 0.552);0.469,\ 0.504\rangle$
C_7	$\langle(0.672,\ 0.751,\ 0.874);0.811,\ 0.173\rangle$

Table 13. TIF positive-ideal solution (PIS) and TIF negative-ideal solution (NIS) vectors.

Criteria	Ideal Solutions	
	TIF PIS	TIF NIS
C_1	$\langle(0.950,\ 0.995,\ 1.000);0.950,\ 0.050\rangle$	$\langle(0.000,\ 0.248,\ 0.332);0.302,\ 0.625\rangle$
C_2	$\langle(0.950,\ 0.995,\ 1.000);0.950,\ 0.050\rangle$	$\langle(0.151,\ 0.263,\ 0.363);0.318,\ 0.584\rangle$
C_3	$\langle(0.888,\ 0.969,\ 1.000);0.888,\ 0.090\rangle$	$\langle(0.000,\ 0.190,\ 0.301);0.243,\ 0.676\rangle$
C_4	$\langle(0.858,\ 0.959,\ 1.000);0.869,\ 0.100\rangle$	$\langle(0.000,\ 0.202,\ 0.278);0.253,\ 0.672\rangle$
C_5	$\langle(0.123,\ 0.176,\ 0.330);0.279,\ 0.637\rangle$	$\langle(0.775,\ 0.788,\ 0.958);0.786,\ 0.171\rangle$
C_6	$\langle(0.151,\ 0.210,\ 0.364);0.315,\ 0.614\rangle$	$\langle(0.758,\ 0.773,\ 0.958);0.769,\ 0.171\rangle$
C_7	$\langle(0.151,\ 0.199,\ 0.363);0.305,\ 0.623\rangle$	$\langle(0.867,\ 0.865,\ 1.000);0.878,\ 0.090\rangle$

Table 14. Positive-ideal separation (PISE) and negative-ideal separation (NISE) matrices.

Ideal Separation	SIs	Criteria						
		C_1	C_2	C_3	C_4	C_5	C_6	C_7
PISE	Sol_6	0.313	0.208	0.344	0.392	0.049	0.006	0.016
	Ecl_4	0.000	0.028	0.000	0.490	0.022	0.000	0.063
	Enl_{10}	0.211	0.348	0.486	0.156	0.146	0.215	0.188
NISE	Sol_6	0.554	0.629	0.466	0.392	0.660	0.572	0.718
	Ecl_4	0.867	0.810	0.810	0.294	0.633	0.578	0.670
	Enl_{10}	0.656	0.490	0.324	0.628	0.465	0.363	0.545

Ultimately, the $\mathfrak{A}_i$, $\mathfrak{B}_i$, $\mathfrak{A}'_i$, $\mathfrak{B}'_i$, κ_i and ϑ_i values are calculated (χ and ψ are considered 0.5). Then, the novel ranking score is calculated (η considered 0.5), and SIs are prioritized according to ranking score ($\mathfrak{C}_i$ values). The gathered results are presented in Table 15 (Step 15 to 18).

Table 15. Computational results of the proposed approach and final ranking of SIs.

SIs	$\mathfrak{A}_i$	$\mathfrak{B}_i$	$\mathfrak{A}_i^{'}$	$\mathfrak{B}_i^{'}$	κ_i	ϑ_i	$\mathfrak{C}_i$	Final Ranking
Sol_1	0.627	0.093	2.766	0.250	0.208	0.814	0.786	5
Sol_2	1.620	0.273	1.799	0.116	0.601	0.475	0.363	16
Sol_3	2.415	0.384	1.041	0.084	0.873	0.306	0.154	28
Sol_4	0.553	0.128	2.752	0.240	0.239	0.798	0.756	6
Sol_5	2.304	0.276	1.026	0.200	0.717	0.458	0.295	21
Sol_6	0.684	−0.002	2.639	0.227	0.095	0.761	0.822	4
Sol_7	1.502	0.134	1.937	0.066	0.405	0.432	0.438	14
Sol_8	1.960	0.213	1.466	0.055	0.580	0.339	0.302	20
Sol_9	1.595	0.181	1.848	0.023	0.479	0.360	0.361	17
EcI_1	0.548	0.075	2.850	0.308	0.171	0.903	0.870	3
EcI_2	1.276	0.183	2.199	0.104	0.429	0.526	0.480	13
EcI_3	2.003	0.250	1.336	0.011	0.635	0.259	0.238	23
EcI_4	0.106	0.032	3.121	0.320	0.044	0.965	1.000	1
EcI_5	1.527	0.205	1.837	0.009	0.498	0.340	0.341	18
EcI_6	1.171	0.113	2.244	0.189	0.322	0.645	0.610	9
EcI_7	0.480	0.088	2.902	0.347	0.176	0.963	0.907	2
EcI_8	2.489	0.240	0.777	0.100	0.702	0.283	0.220	24
EcI_9	1.742	0.177	1.621	0.160	0.499	0.504	0.430	15
EnI_1	1.180	0.077	2.276	0.164	0.278	0.618	0.618	8
EnI_2	1.444	0.090	1.917	0.160	0.338	0.553	0.544	10
EnI_3	2.818	0.338	0.396	0.000	0.881	0.088	0.062	29
EnI_4	2.346	0.267	0.933	−0.008	0.713	0.167	0.161	27
EnI_5	3.135	0.390	0.119	−0.004	1.000	0.036	0.000	30
EnI_6	2.265	0.092	1.061	0.058	0.476	0.275	0.315	19
EnI_7	2.318	0.174	0.972	−0.031	0.590	0.142	0.197	26
EnI_8	1.364	0.007	2.008	0.040	0.220	0.408	0.515	11
EnI_9	2.152	0.266	1.182	0.082	0.680	0.326	0.251	22
EnI_{10}	1.004	0.031	2.309	0.125	0.191	0.572	0.637	7
EnI_{11}	1.354	0.037	2.073	0.035	0.256	0.413	0.500	12
EnI_{12}	1.857	0.347	1.572	0.008	0.733	0.294	0.211	25

4. Discussion

The results achieved from the presented approach are examined comprehensively, and the accuracy, precision, and sensitivity of the answers are investigated in this section. A thorough sensitivity analysis is performed on ten SIs with a higher priority for various values of approach variables. Furthermore, the comparisons are made between the outcomes of the presented approach and other cited literature.

4.1. Sensitivity Analysis

A comprehensive sensitivity analysis is conducted in this subsection. First, the sensitivity of ranking score ($\mathfrak{C}$ values) and ranking orders are investigated for values of χ, ψ and η ranging from 0 to 1 (Figures 2 and 3).

In Figure 2a,b, the $\mathfrak{C}$ values and ranking orders are represented for various values of χ and ψ ranging from 0 to 1, respectively. As represented in Figure 2a, the graph of the $\mathfrak{C}$ values for various SIs versus χ and ψ values ranging from 0 to 1 has three states of almost constant, ascending or descending. However, in most cases, the graphs of indicators are parallel, and only a few intersections are represented for χ and ψ values above 0.7.

As can be seen in Figure 2b, changing the graph for χ and ψ values above 0.7 leads to few changes in the ranking order of SIs. In addition, for variations of χ and ψ between 0.2 and 0.7, the rank of all SIs remains unchanged, and also a set of the top ten SIs remains in a range from 1 to 10. Hence, the conclusion can be drawn that the top ten SIs have the lowest sensitivity to the values of χ and ψ between 0.2 and 0.7, and the assumed value of 0.5 for this variable in the case study is suitable.

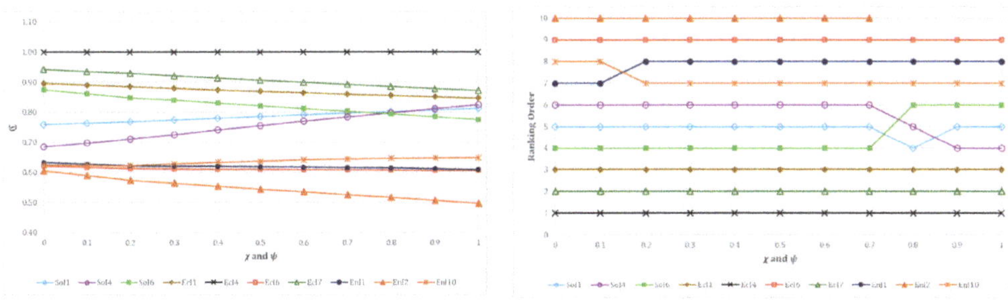

(**a**) (**b**)

Figure 2. (**a**) Sensitivity analysis on the $\mathfrak{C}$ values and (**b**) preference ranking order of top ten SIs related to majority attributes (χ and ψ).

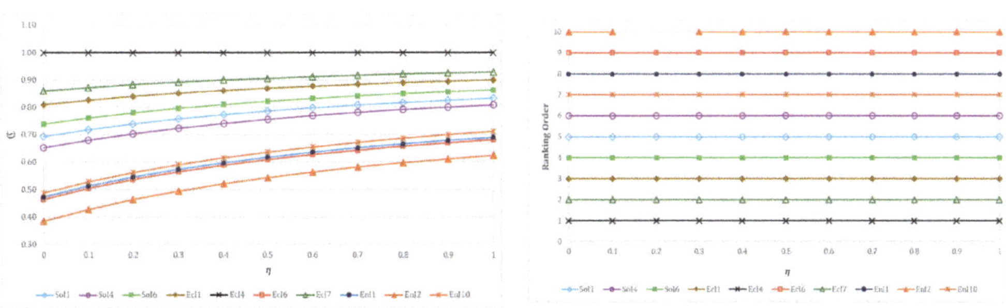

(**a**) (**b**)

Figure 3. (**a**) Sensitivity analysis on the $\mathfrak{C}$ values and (**b**) preference ranking order of top ten SIs related to η coefficient.

In Figure 3a,b, $\mathfrak{C}$ values and ranking orders versus η values ranging from 0 to 1 are represented, respectively. Figure 3a represents that the graph of $\mathfrak{C}$ values for all SIs is ascending by increasing the η value from 0 to 1. However, according to the figure, the gap between the values of $\mathfrak{C}$ increases by decreasing η. Thus, for smaller values of η, the gap between the $\mathfrak{C}$ values of different SIs is larger, allowing an accurate distinction for decision-makers. In addition, according to Figure 3b, it can be concluded that the ranking order of all top SIs is constant for η values other than EnI_2 for the value of $\eta = 0.2$.

With these in mind, this conclusion can be drawn that $\mathfrak{C}$ values and ranking orders have no sensitivity to η. Hence, choosing 0.5 as a median number of the interval for η in the case study is a suitable choice.

4.2. Comparison between the Proposed Approach and Other Cited Literature

To validate the presented approach outcomes, a comparison is made between the achieved results by the proposed approach and the traditional fuzzy MCDM methods. Table 16 shows the outcomes of this comparison. The comparison results for the top ten SIs are also represented in Figure 4.

Table 16. Comparative outcomes of the presented approach and other traditional fuzzy multi-criteria decision-making (MCDM) methods.

SIs	Proposed Approach		Fuzzy MCDM Methods					
			Fuzzy VIKOR [65]		Fuzzy SAW [66]		Fuzzy TOPSIS [67]	
	Ranking Score	Preference Order Ranking	Ranking Score	Preference Order Ranking	Ranking Score	Preference Order Ranking	Ranking Score	Preference Order Ranking
Sol_1	0.786	5	0.178	5	0.866	5	0.648	5
Sol_2	0.363	16	0.486	15	0.756	16	0.483	16
Sol_3	0.154	28	0.721	25	0.678	24	0.391	23
Sol_4	0.756	6	0.160	4	0.876	3	0.680	2
Sol_5	0.295	21	0.755	26	0.685	23	0.401	21
Sol_6	0.822	4	0.241	6	0.851	6	0.628	6
Sol_7	0.438	14	0.490	16	0.757	15	0.490	15
Sol_8	0.302	20	0.707	23	0.701	21	0.389	24
Sol_9	0.361	17	0.491	17	0.755	17	0.465	17
EcI_1	0.870	3	0.150	3	0.875	4	0.667	4
EcI_2	0.480	13	0.380	10	0.796	10	0.537	10
EcI_3	0.238	23	0.607	20	0.712	20	0.405	20
EcI_4	1.000	1	0.000	1	0.937	1	0.735	1
EcI_5	0.341	18	0.459	14	0.766	14	0.492	14
EcI_6	0.610	9	0.367	9	0.798	9	0.560	9
EcI_7	0.907	2	0.125	2	0.885	2	0.674	3
EcI_8	0.220	24	0.839	28	0.646	28	0.306	28
EcI_9	0.430	15	0.586	19	0.731	18	0.457	18
EnI_1	0.618	8	0.353	8	0.802	8	0.564	8
EnI_2	0.544	10	0.450	13	0.770	13	0.503	13
EnI_3	0.062	29	0.916	29	0.612	29	0.250	29
EnI_4	0.161	27	0.704	22	0.677	25	0.370	25
EnI_5	0.000	30	1.000	30	0.588	30	0.195	30
EnI_6	0.315	19	0.802	27	0.662	27	0.310	27
EnI_7	0.197	26	0.712	24	0.674	26	0.364	26
EnI_8	0.515	11	0.426	12	0.775	12	0.529	11
EnI_9	0.251	22	0.661	21	0.692	22	0.396	22
EnI_{10}	0.637	7	0.304	7	0.820	7	0.603	7
EnI_{11}	0.500	12	0.420	11	0.778	11	0.527	12
EnI_{12}	0.211	25	0.557	18	0.730	19	0.441	19

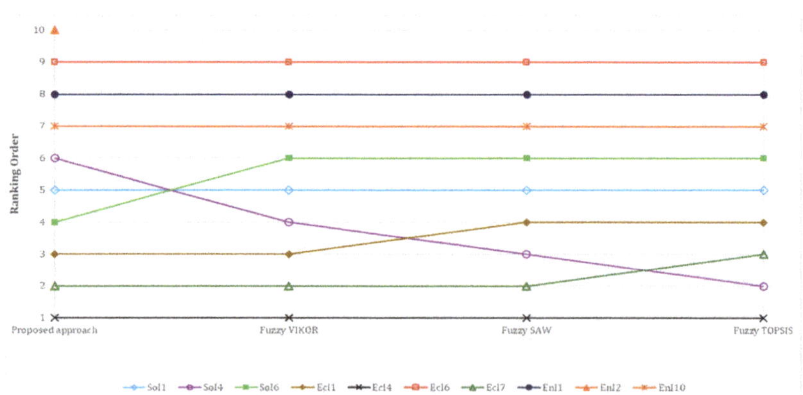

Figure 4. Preference order ranking of top ten SIs prioritized by the proposed approach and other fuzzy MCDM methods.

As presented in Table 16 and Figure 4, the priority of SIs derived from the presented approach is not very different from other methods in most cases (in most cases, the number of ranks is changed by up to three or four rank shifts in SIs priority). Besides, EcI_4 has the first priority in all methods, and EnI_5 has the last priority in the presented approach and all methods. For the ten first priorities that are key indicators, despite some changes of ranks (at most four ranks) in some methods, the priorities of indicators remain within the ten first priorities except EnI_2. Furthermore, as can be observed in the figure, most key indicators have the same ranks in the proposed approach and all traditional methods. However, SoI_4 had relatively more changes, which can be considered as the most sensitive key indicator.

From the above, the conclusion can be drawn that the proposed approach is reliable and its results benefit from the merits of taking into account risk attitudes of experts, concepts of entropy in determining weights of experts, and a new TIFS-ranking approach concurrently.

As another aspect, the results of the approach for ten SIs with a higher priority are compared with the indicators provided by seven cited literature studies and tools. The comparison results are indicated in Table 17.

Table 17. Comparison of ten SIs with higher priority and other literature studies and tools.

Related Literature		Social			Economic				Environmental		
		SoI_1	SoI_4	SoI_6	EcI_1	EcI_4	EcI_6	EcI_7	EnI_1	EnI_2	EnI_{10}
Awasthi et al. [47]	S **	✓*	✓	–*	–	✓	–	–	–	✓	–
Shen et al. [56]	S	✓	✓	✓	✓	–	✓	✓	–	✓	✓
Shen et al. [57]	S	✓	✓	✓	–	–	✓	✓	–	✓	✓
Yao et al. [59]	S	✓	✓	✓	✓	✓	✓	✓	–	✓	✓
CEEQUAL [62]	T **	✓	✓	✓	–	✓	–	–	✓	✓	✓
Invest [63]	T	✓	✓	–	–	✓	–	–	✓	✓	✓
Envision [64]	T	✓	✓	✓	–	✓	✓	–	✓	✓	✓

* Note: The symbol ✓indicates that the study/tool includes the SI, whereas—indicates that it does not. ** S: Study; T: Tool.

As presented in Table 17, most of the ten first priorities have been utilized as SIs' assessment in the cited literature. Much higher adaptation is related to Yao et al. [59] and Envision [64] and less adaptation is related to Awasthi et al. [47]. Three SIs of ten key indicators, SoI_1, SoI_4 and EnI_2, exist in all cited literature. In addition, EnI_{10} is introduced in all the literature except Awasthi et al. [47]. In addition, it can be observed that the social and environmental indicators have been incorporated in all cited tools (except SoI_6 in Invest) but economic indicators have not been considered in the cited tools (except EcI_4). These comparisons demonstrate that the outcomes of the approach are reliable and can be employed in a sustainable assessment of highway construction projects.

5. Concluding Remarks

Analyzing sustainability indicators (SIs) in construction projects between different potential indicators and considering various assessment criteria concurrently can be considered as a complicated group decision problem. A new triangular intuitionistic fuzzy set (TIFS) group decision approach for the multi-criteria evaluation is presented in this study to deal with this problem under uncertainty. A novel multi-criteria group decision-making approach considers experts' risk attitudes and views and entropy concepts were developed in the TIFS environment. Furthermore, new ranking scores were proposed through similarity to ideal solutions by the concept of closeness coefficient to prioritize and choose the sustainable indicators. A case study regarding highway construction projects was presented to analyze the sustainable indicators under uncertainty. The considered case study was solved using the introduced group-decision approach.

The primary aim of this paper is to present a sound approach for the assessment and adoption of SIs in highway construction projects. The principal novelties of this study are as follows:

- To cope with uncertainty in highway construction projects, triangular intuitionistic fuzzy sets (TIFSs) are used. The TIFSs make the process of decision-making more flexible regarding degrees of agreement, disagreement, and hesitancy utilizing a triangular function.
- Risk attitudes of experts are considered within the assessment and process of group decision-making because they can have various perspectives, such as optimistic or pessimistic, in their views owing to their various backgrounds and characteristics.
- A novel methodology is proposed to specify experts' weights within the process of group decision-making based on the concepts of entropy.
- A new compromise ranking score is proposed to evaluate and choose sustainability indicators in highway construction projects.

Ultimately, some sensitivity analyses were performed on the preference order ranking of the top ten SIs in a case study according to the change of approach coefficients and different risk attitudes of experts. The drawn conclusion of the sensitivity analyses was that approach coefficients selected in the case study were suitable choices. Moreover, the presented approach was compared with the traditional fuzzy MCDM techniques, including fuzzy SAW and fuzzy VIKOR. The computational results represented that there was no major difference between the proposed approach and other fuzzy MCDM techniques regarding the priority of SIs in most cases. In addition, both the first and last priorities derived from this approach were the same in all the aforementioned methods.

The introduced comprehensive approach has proposed an efficient decision-making method for highway construction regarding sustainable development principles. In fact, it presented a dependable model in which the results benefited from the merits of taking into account the risk attitudes of experts and the new TIFS-ranking method. Furthermore, the applied fundamental concepts were intelligible to the committee of experts and project managers, and the required calculations were straightforward. Hence, by introducing evaluated sustainable indicators, this paper helps project managers improve highway projects' sustainability and make the most sustainable decisions. As future research, a holistic framework can be developed that utilizes the mentioned criteria and considers environmental and social impacts as criteria in the evaluation of sustainability indicators. In addition, the ranked SIs with higher priority can be used as key indicators in the sustainability assessment of highway construction projects.

Author Contributions: H.H. presented the research methodology, performed the development and experiments of this study; P.G. and F.N. provided extensive advice throughout the study. All authors have read and agreed to the published version of the manuscript.

Funding: This research received no external funding.

Conflicts of Interest: The authors declare no conflict of interest.

Appendix A

Definition A1. [68] *The membership function of TIFN* $\widetilde{A} = \langle (a,b,c);\mu,\nu \rangle$ *is defined as follows:*

$$\mu_{\widetilde{A}}(x) = \begin{cases} \frac{x-a}{b-a}\mu & if\ a \le x < b \\ \mu & if\ x = b \\ \frac{c-x}{c-b}\mu & if\ b < x \le c \\ 0 & otherwise \end{cases} \quad \text{(A1)}$$

and non-membership function is defined as follows:

$$\nu_{\tilde{A}}(x) = \begin{cases} \frac{b-x+(x-a)\nu}{b-a} & if\ a \leq x < b \\ \nu & if\ x = b \\ \frac{x-b+(c-x)\nu}{c-b} & if\ b < x \leq c \\ 0 & otherwise \end{cases} \quad \text{(A2)}$$

where a, b and c are real numbers, $0 \leq \mu \leq 1, 0 \leq \nu \leq 1$ *and* $0 \leq \mu + \nu \leq 1$.

Definition A2. [44] *Let* $\widetilde{A} = \langle (a,b,c);\mu,\nu \rangle$ *and* $\widetilde{B} = \langle (a',b',c');\mu',\nu' \rangle$ *be two TIFNs, then the arithmetic operations are defined as follows:*

$$\widetilde{A} \oplus \widetilde{B} = \langle (a+a', b+b', c+c');\mu+\mu'-\mu.\mu', \nu.\nu' \rangle, \quad \text{(A3)}$$

$$\widetilde{A} \otimes \widetilde{B} = \langle (a.a', b.b', c.c');\mu.\mu', \nu+\nu'-\nu.\nu' \rangle, \quad \text{(A4)}$$

$$\lambda\widetilde{A} = \langle (\lambda a, \lambda b, \lambda c);1-(1-\mu)^{\lambda}, \nu^{\lambda} \rangle \ (\lambda \geq 0), \quad \text{(A5)}$$

$$\widetilde{A}^{\lambda} = \left\langle \left(a^{\lambda}, b^{\lambda}, c^{\lambda}\right);\mu^{\lambda}, 1-(1-\nu)^{\lambda} \right\rangle \ (\lambda \geq 0). \quad \text{(A6)}$$

References

1. WCED. *Our Common Future-Brundtland Report*; Oxford University Press: Oxford, UK, 1987.
2. Shen, L.Y.; Li Hao, J.; Tam, V.W.Y.; Yao, H. A checklist for assessing sustainability performance of construction projects. *J. Civ. Eng. Manag.* **2007**, *13*, 273–281. [CrossRef]
3. Yu, W.D.; Cheng, S.T.; Ho, W.C.; Chang, Y.H. Measuring the sustainability of construction projects throughout their lifecycle: A Taiwan lesson. *Sustainability* **2018**, *10*, 1523. [CrossRef]
4. Goubran, S.; Cucuzzella, C. Integrating the sustainable development goals in building projects. *J. Sustain. Res.* **2019**, *1*, e190010. [CrossRef]
5. Karaca, F.; Guney, M.; Kumisbek, A.; Kaskina, D.; Tokbolat, S. A new stakeholder opinion-based rapid sustainability assessment method (RSAM) for existing residential buildings. *Sustain. Cities Soc.* **2020**, *60*, 102155. [CrossRef]
6. Li, H.; Zhang, X.; Ng, S.T.; Skitmore, M. Quantifying stakeholder influence in decision/evaluations relating to sustainable construction in China—A Delphi approach. *J. Clean. Prod.* **2018**, *173*, 160–170. [CrossRef]
7. Omer, M.A.; Noguchi, T. A conceptual framework for understanding the contribution of building materials in the achievement of Sustainable Development Goals (SDGs). *Sustain. Cities Soc.* **2020**, *52*, 101869. [CrossRef]
8. Xu, X.; Wang, Y.; Tao, L. Comprehensive evaluation of sustainable development of regional construction industry in China. *J. Clean. Prod.* **2019**, *211*, 1078–1087. [CrossRef]
9. Illankoon, I.C.S.; Tam, V.W.; Le, K.N. United Nation's sustainable development goals: Establishing baseline for Australian building sector. *Intell. Build. Int.* **2020**. [CrossRef]
10. Olawumi, T.O.; Chan, D.W.; Chan, A.P.; Wong, J.K. Development of a building sustainability assessment method (BSAM) for developing countries in sub-Saharan Africa. *J. Clean. Prod.* **2020**, *263*, 121514. [CrossRef]
11. Mansell, P.; Philbin, S.P.; Broyd, T.; Nicholson, I. Assessing the impact of infrastructure projects on global sustainable development goals. *Proc. Inst. Civ. Eng. Eng. Sustain.* **2020**, *173*, 196–212. [CrossRef]
12. Mousavi, S.M.; Tavakkoli-Moghaddam, R.; Azaron, A.; Mojtahedi, S.M.H.; Hashemi, H. Risk assessment for highway projects using jackknife technique. *Expert Syst. Appl.* **2011**, *38*, 5514–5524. [CrossRef]
13. Ghoddousi, P.; Nasirzadeh, F.; Hashemi, H. Evaluating Highway Construction Projects' Sustainability Using a Multicriteria Group Decision-Making Model Based on Bootstrap Simulation. *J. Constr. Eng. Manag.* **2018**, *144*, 04018092. [CrossRef]
14. Hendiani, S.; Bagherpour, M. Developing an integrated index to assess social sustainability in construction industry using fuzzy logic. *J. Clean. Prod.* **2019**, *230*, 647–662. [CrossRef]
15. Rostamnezhad, M.; Nasirzadeh, F.; Khanzadi, M.; Jarban, M.J.; Ghayoumian, M. Modeling social sustainability in construction projects by integrating system dynamics and fuzzy-DEMATEL method: A case study of highway project. *Eng. Constr. Archit. Manag.* **2020**, *27*, 1595–1618. [CrossRef]
16. Hashemi, H.; Mousavi, S.M.; Zavadskas, E.K.; Chalekaee, A.; Turskis, Z. A new group decision model based on grey-intuitionistic fuzzy-ELECTRE and VIKOR for contractor assessment problem. *Sustainability* **2018**, *10*, 1635. [CrossRef]
17. Mousavi, S.M.; Antuchevičienė, J.; Zavadskas, E.K.; Vahdani, B.; Hashemi, H. A new decision model for cross-docking center location in logistics networks under interval-valued intuitionistic fuzzy uncertainty. *Transport* **2019**, *34*, 30–40. [CrossRef]

18. Stojčić, M.; Zavadskas, E.K.; Pamučar, D.; Stević, Ž.; Mardani, A. Application of MCDM methods in sustainability engineering: A literature review 2008–2018. *Symmetry* **2019**, *11*, 350. [CrossRef]
19. Zavadskas, E.K.; Antucheviciene, J.; Kar, S. Multi-Objective and Multi-Attribute Optimization for Sustainable Development Decision Aiding. *Sustainability* **2019**, *11*, 3069. [CrossRef]
20. Hashemi, H.; Bazargan, J.; Mousavi, S.M.; Vahdani, B. An extended compromise ratio model with an application to reservoir flood control operation under an interval-valued intuitionistic fuzzy environment. *Appl. Math. Model.* **2014**, *38*, 3495–3511. [CrossRef]
21. Huang, R.Y.; Yeh, C.H. Development of an assessment framework for green highway construction. *J. Chin. Inst. Eng.* **2008**, *31*, 573–585. [CrossRef]
22. Chen, Y.; Okudan, G.E.; Riley, D.R. Sustainable performance criteria for construction method selection in concrete buildings. *Autom. Constr.* **2010**, *19*, 235–244. [CrossRef]
23. Reza, B.; Sadiq, R.; Hewage, K. Sustainability assessment of flooring systems in the city of Tehran: An AHP-based life cycle analysis. *Constr. Build. Mater.* **2011**, *25*, 2053–2066. [CrossRef]
24. Waris, M.; Liew, M.S.; Khamidi, M.F.; Idrus, A. Criteria for the selection of sustainable onsite construction equipment. *Int. J. Sustain. Built Environ.* **2014**, *3*, 96–110. [CrossRef]
25. Li, Y.; Zhao, L.; Suo, J. Comprehensive assessment on sustainable development of highway transportation capacity based on entropy weight and TOPSIS. *Sustainability* **2014**, *6*, 4685–4693. [CrossRef]
26. Kucukvar, M.; Gumus, S.; Egilmez, G.; Tatari, O. Ranking the sustainability performance of pavements: An intuitionistic fuzzy decision making method. *Autom. Constr.* **2014**, *40*, 33–43. [CrossRef]
27. Medineckiene, M.; Zavadskas, E.K.; Björk, F.; Turskis, Z. Multi-criteria decision-making system for sustainable building assessment/certification. *Arch. Civ. Mech. Eng.* **2015**, *15*, 11–18. [CrossRef]
28. Kamali, M.; Hewage, K. Development of performance criteria for sustainability evaluation of modular versus conventional construction methods. *J. Clean. Prod.* **2017**, *142*, 3592–3606. [CrossRef]
29. Pan, M.; Linner, T.; Pan, W.; Cheng, H.; Bock, T. A framework of indicators for assessing construction automation and robotics in the sustainability context. *J. Clean. Prod.* **2018**, *182*, 82–95. [CrossRef]
30. Zolfani, S.H.; Pourhossein, M.; Yazdani, M.; Zavadskas, E.K. Evaluating construction projects of hotels based on environmental sustainability with MCDM framework. *Alex. Eng. J.* **2018**, *57*, 357–365. [CrossRef]
31. Liu, S.; Qian, S. Towards sustainability-oriented decision making: Model development and its validation via a comparative case study on building construction methods. *Sustain. Dev.* **2019**, *27*, 860–872. [CrossRef]
32. Reddy, A.S.; Kumar, P.R.; Raj, P.A. Preference based multi-criteria framework for developing a Sustainable Material Performance Index (SMPI). *Int. J. Sustain. Eng.* **2019**, *12*, 390–403. [CrossRef]
33. Chen, C.H. A new multi-criteria assessment model combining GRA techniques with intuitionistic fuzzy entropy-based TOPSIS method for sustainable building materials supplier selection. *Sustainability* **2019**, *11*, 2265. [CrossRef]
34. Roy, J.; Das, S.; Kar, S.; Pamučar, D.; Roy, J.; Das, S.; Kar, S.; Pamučar, D. An extension of the CODAS approach using interval-valued intuitionistic fuzzy set for sustainable material selection in construction projects with incomplete weight information. *Symmetry* **2019**, *11*, 393. [CrossRef]
35. Tseng, M.L.; Lin, S.; Chen, C.C.; Sarmiento, L.S.C.; Tan, C.L. A causal sustainable product-service system using hierarchical structure with linguistic preferences in the Ecuadorian construction industry. *J. Clean. Prod.* **2019**, *230*, 477–487. [CrossRef]
36. Alawneh, R.; Ghazali, F.; Ali, H.; Sadullah, A.F. A novel framework for integrating United Nations Sustainable Development Goals into sustainable non-residential building assessment and management in Jordan. *Sustain. Cities Soc.* **2019**, *49*, 101612. [CrossRef]
37. Dabous, S.A.; Zeiada, W.; Zayed, T.; Al-Ruzouq, R. Sustainability-informed multi-criteria decision support framework for ranking and prioritization of pavement sections. *J. Clean. Prod.* **2020**, *244*, 118755. [CrossRef]
38. Rahimi, S.; Hafezalkotob, A.; Monavari, S.M.; Hafezalkotob, A.; Rahimi, R. Sustainable landfill site selection for municipal solid waste based on a hybrid decision-making approach: Fuzzy group BWM-MULTIMOORA-GIS. *J. Clean. Prod.* **2020**, *248*, 119186. [CrossRef]
39. Navarro, I.J.; Yepes, V.; Martí, J.V. Sustainability assessment of concrete bridge deck designs in coastal environments using neutrosophic criteria weights. *Struct. Infrastruct. Eng.* **2020**, *16*, 949–967. [CrossRef]
40. Ebrahimnejad, S.; Mousavi, S.M.; Tavakkoli-Moghaddam, R.; Hashemi, H.; Vahdani, B. A novel two-phase group decision making approach for construction project selection in a fuzzy environment. *Appl. Math. Model.* **2012**, *36*, 4197–4217. [CrossRef]
41. Fenton, N.; Wang, W. Risk and confidence analysis for fuzzy multicriteria decision making. *Knowl. Based Syst.* **2006**, *19*, 430–437. [CrossRef]
42. Zhang, X.; Jin, F.; Liu, P. A grey relational projection method for multi-attribute decision making based on intuitionistic trapezoidal fuzzy number. *Appl. Math. Model.* **2013**, *37*, 3467–3477. [CrossRef]
43. Zhou, H.C.; Zhang, G.H.; Wang, G.L. Multi-objective decision making approach based on entropy weights for reservoir flood control operation. *J. Hydraul. Eng.* **2007**, *38*, 100–106.
44. Chen, Y.; Li, B. Dynamic multi-attribute decision making model based on triangular intuitionistic fuzzy numbers. *Sci. Iran.* **2011**, *18*, 268–274. [CrossRef]
45. Wan, S.P.; Wang, F.; Lin, L.L.; Dong, J.Y. Some new generalized aggregation operators for triangular intuitionistic fuzzy numbers and application to multi-attribute group decision making. *Comput. Ind. Eng.* **2016**, *93*, 286–301. [CrossRef]

46. Wang, J.Q.; Nie, R.; Zhang, H.Y.; Chen, X.H. New operators on triangular intuitionistic fuzzy numbers and their applications in system fault analysis. *Inf. Sci.* **2013**, *251*, 79–95. [CrossRef]
47. Awasthi, A.; Chauhan, S.S.; Omrani, H. Application of fuzzy TOPSIS in evaluating sustainable transportation systems. *Expert Syst. Appl.* **2011**, *38*, 12270–12280. [CrossRef]
48. Diabat, A.; Kannan, D.; Mathiyazhagan, K. Analysis of enablers for implementation of sustainable supply chain management—A textile case. *J. Clean. Prod.* **2014**, *83*, 391–403. [CrossRef]
49. Govindan, K.; Shankar, K.M.; Kannan, D. Sustainable material selection for construction industry—A hybrid multi criteria decision making approach. *Renew. Sustain. Energy Rev.* **2016**, *55*, 1274–1288. [CrossRef]
50. Kandziora, M.; Burkhard, B.; Müller, F. Interactions of ecosystem properties, ecosystem integrity and ecosystem service indicators—A theoretical matrix exercise. *Ecol. Indic.* **2013**, *28*, 54–78. [CrossRef]
51. Michael, F.L.; Noor, Z.Z.; Figueroa, M.J. Review of urban sustainability indicators assessment—Case study between Asian countries. *Habitat Int.* **2014**, *44*, 491–500. [CrossRef]
52. Nasirzadeh, F.; Ghayoumian, M.; Khanzadi, M.; Rostamnezhad Cherati, M. Modelling the social dimension of sustainable development using fuzzy cognitive maps. *Int. J. Constr. Manag.* **2020**, *20*, 223–236. [CrossRef]
53. Niemeijer, D.; de Groot, R.S. A conceptual framework for selecting environmental indicator sets. *Ecol. Indic.* **2008**, *8*, 14–25. [CrossRef]
54. Oltean-Dumbrava, C.; Watts, G.; Miah, A. Transport infrastructure: Making more sustainable decisions for noise reduction. *J. Clean. Prod.* **2013**, *42*, 58–68. [CrossRef]
55. Roy, R.; Chan, N.W. An assessment of agricultural sustainability indicators in Bangladesh: Review and synthesis. *Environmentalist* **2012**, *32*, 99–110. [CrossRef]
56. Shen, L.Y.; Tam, V.W.; Tam, L.; Ji, Y.B. Project feasibility study: The key to successful implementation of sustainable and socially responsible construction management practice. *J. Clean. Prod.* **2010**, *18*, 254–259. [CrossRef]
57. Shen, L.; Wu, Y.; Zhang, X. Key assessment indicators for the sustainability of infrastructure projects. *J. Constr. Eng. Manag.* **2010**, *137*, 441–451. [CrossRef]
58. Ugwu, O.O.; Haupt, T.C. Key performance indicators and assessment methods for infrastructure sustainability—A South African construction industry perspective. *Build. Environ.* **2007**, *42*, 665–680. [CrossRef]
59. Yao, H.; Shen, L.; Tan, Y.; Hao, J. Simulating the impacts of policy scenarios on the sustainability performance of infrastructure projects. *Autom. Constr.* **2011**, *20*, 1060–1069. [CrossRef]
60. Yuan, H. Key indicators for assessing the effectiveness of waste management in construction projects. *Ecol. Indic.* **2013**, *24*, 476–484. [CrossRef]
61. Zhao, Z.Y.; Zhao, X.J.; Davidson, K.; Zuo, J. A corporate social responsibility indicator system for construction enterprises. *J. Clean. Prod.* **2012**, *29*, 277–289. [CrossRef]
62. CEEQUAL Version 6: Technical Manual—International Projects. Watford, UK. Available online: https://www.ceequal.com/version-6/ (accessed on 30 September 2019).
63. Reid, L.; Bevan, T.; Davis, A.; Neuman, T.; Penney, K.; Seskin, S.; VanZerr, M.; Anderson, J.; Muench, S.; Weiland, C.; et al. Invest v1.3: Sustainable Highways Self-Evaluation Tool. Available online: https://www.sustainablehighways.org/files/4735.pdf (accessed on 12 February 2019).
64. Envision v3: Sustainable Infrastructure Framework Manual. Available online: https://sustainableinfrastructure.org/envision-version-3/ (accessed on 18 February 2019).
65. Opricovic, S. Fuzzy VIKOR with an application to water resources planning. *Expert Syst. Appl.* **2011**, *38*, 12983–12990. [CrossRef]
66. Chou, S.Y.; Chang, Y.H.; Shen, C.Y. A fuzzy simple additive weighting system under group decision-making for facility location selection with objective/subjective attributes. *Eur. J. Oper. Res.* **2008**, *189*, 132–145. [CrossRef]
67. Chu, T.C. Facility location selection using fuzzy TOPSIS under group decisions. *Int. J. Uncertain. Fuzziness Knowl. Based* **2002**, *10*, 687–701. [CrossRef]
68. Li, D.F. A ratio ranking method of triangular intuitionistic fuzzy numbers and its application to MADM problems. *Comput. Math. Appl.* **2010**, *60*, 1557–1570. [CrossRef]

sustainability MDPI

Article

A Systematic Literature Review of Multi-Criteria Decision-Making Methods for Sustainable Selection of Insulation Materials in Buildings

Indre Siksnelyte-Butkiene, Dalia Streimikiene *, Tomas Balezentis and Virgilijus Skulskis

Lithuanian Centre for Social Sciences, Institute of Economics and Rural Development, A. Vivulskio g. 4A-13, LT-03220 Vilnius, Lithuania; indre.siksnelyte@knf.vu.lt (I.S.-B.); tomas.balezentis@laei.lt (T.B.); virgilijus.skulskis@laei.lt (V.S.)
* Correspondence: dalia@mail.lei.lt

Abstract: The European Commission has recently adopted the Renovation Wave Strategy, aiming at the improvement of the energy performance of buildings. The strategy aims to at least double renovation rates in the next ten years and make sure that renovations lead to higher energy and resource efficiency. The choice of appropriate thermal insulation materials is one of the simplest and, at the same time, the most popular strategies that effectively reduce the energy demand of buildings. Today, the spectrum of insulation materials is quite wide, and each material has its own specific characteristics. It is recognized that the selection of materials is one of the most challenging and difficult steps of a building project. This paper aims to give an in-depth view of existing multi-criteria decision-making (MCDM) applications for the selection of insulation materials and to provide major insights in order to simplify the process of methods and criteria selection for future research. A systematic literature review is performed based on the Search, Appraisal, Synthesis and Analysis (SALSA) framework and the Preferred Reporting Items for Systematic Reviews and Meta-Analyses (PRISMA) statement. In order to determine which MCDM method is the most appropriate for different questions, the main advantages and disadvantages of different methods are provided.

Keywords: thermal insulation; multi criteria analysis; MCDM; SALSA; buildings; Renovation Wave

Citation: Siksnelyte-Butkiene, I.; Streimikiene, D.; Balezentis, T.; Skulskis, V. A Systematic Literature Review of Multi-Criteria Decision-Making Methods for Sustainable Selection of Insulation Materials in Buildings. *Sustainability* **2021**, *13*, 737. https://doi.org/10.3390/su13020737

Received: 18 December 2020
Accepted: 12 January 2021
Published: 14 January 2021

1. Introduction

The issue of sustainable energy development is one of the most important in various political documents. The construction sector, which consumes about 40% of the total primary energy [1,2] and emits 10% of CO_2 emissions [3], plays a significant role in addressing these issues. Renovation of buildings is a priority of the EU Renovation Wave Strategy adopted in 2020 [4]. The Renovation Wave Strategy aims to at least double renovation rates in the next ten years and ensure that energy renovations of buildings will provide higher energy efficiency and significant GHG emission reduction. Therefore, optimization of energy needs in buildings is an important aspect in the fight against climate change [5]. Most of the energy in buildings is used to meet the needs of heating, ventilation, and air conditioning [6]. Significant energy savings in buildings can be achieved by choosing appropriate building design solutions. Heat consumption is effectively reduced by improving the insulation properties of buildings; therefore, increasing the energy efficiency of buildings has become an important aspect of national energy strategies in many countries [7]. A lot of initiatives focus on the construction sector and there are many objectives aimed at promoting technological innovation, improving energy efficiency [8], reducing environmental impact [9], and improving life quality criteria [10]. Although extensive attention in the construction of new buildings has been paid to energy efficiency issues, new buildings account for only about 1% of the housing market annually [3]. Therefore, in order to reduce energy consumption, old buildings must be renovated with a strong focus on energy efficiency issues. In the European Union, the new Energy Performance of Buildings Directive (EPBD) 2018/844 highlights the issue of energy efficiency in buildings and sets out certain requirements and objectives to be pursued [11]. The aim is that both new and renovated buildings become zero-energy buildings, which have high energy efficiency, and in which renewable energy sources meet the greatest energy demand.

Building insulation materials play a particularly significant role in achieving the goals of energy efficiency in buildings. The choice of appropriate thermal insulation materials is one of the simplest and at the same time the most popular strategies that effectively reduce the energy demand of buildings [12,13]. The choice of insulation materials depends not only on the thermal efficiency of the building. The choice of materials can also determine the aspects related to the quality of life and the impact on the environment [14]. Today, the spectrum of insulation materials is quite wide, and each material has its own specific characteristics. Some materials are environmentally friendly, while others are more economically acceptable, and the rest have better thermal insulation properties [14–18]. The choice of materials in the case of a particular project and individual country depends on different factors, such as price, material availability factors, transportation costs, construction rules in the country, climatic conditions, and type of heating of the building. For example, in Europe, more than 60% of the consumed thermal insulation materials are glass wool, stone wool, and inorganic fibrous materials, while the use of polystyrene, organic foamy materials, expanded and extruded polystyrene constitutes less than 30% of the total [12].

It is recognized that the selection of materials is one of the most challenging and difficult steps of a building project [19]. At both the practical and scientific level, studies can be found in the literature focusing on finding the materials which are most suitable for a particular project. The Sustainable Development Goals have been pursued in different areas of economic activity; therefore, when choosing materials for the construction of buildings, not only are their physical and technical characteristics as well as economic factors taken into account, but also their social and environmental impacts [20]. A multi-criteria evaluation has become one of the most important tools in energy development studies in the last decade, allowing the comparison of different alternatives [21]. In this type of evaluation, the choice of methodology and its logical justification play a very important role. A correct choice of the evaluation method and the criteria on which the evaluation will be based can solve complex issues relating to the chosen alternatives.

This paper aims to give an in-depth view of existing multi-criteria decision-making (MCDM) applications for the selection of insulation materials and to provide major insights in order to simplify the process of method selection for future research. A systematic literature review is performed based on the Search, Appraisal, Synthesis and Analysis (SALSA) framework and the Preferred Reporting Items for Systematic Reviews and Meta-Analyses (PRISMA) statement [22]. In order to determine which MCDM method is the most appropriate for different insulation problems, the main advantages and disadvantages of different methods are provided. In order to achieve this purpose, Section 2 provides the methodology. Section 3 presents an analysis of the selected articles for review: the techniques used in the studies in order to select criteria for evaluation and determining their weights are provided; the criteria used are overviewed and arranged around four dimensions. Section 4 focuses on the advantages and disadvantages of different MCDM methods.

2. Methodology

A systematic literature search and analysis was carried out in accordance with the SALSA framework. The methodology of SALSA allows one to minimize the possible factor of subjectivity and is indicated as one of the most suitable tools for identifying, evaluating, and systematizing literature [23], and guarantees the methodological precision and completeness [24]. The accuracy and completeness of the research are also ensured by the PRISMA statement [22]. The framework for the systematic literature search and review in this research is provided in Table 1.

Table 1. The framework for systematic literature search and review.

Stage	Description
Search	Key actions: keywords identification; search database. Research scope: MCDM methods for solving questions of sustainable insulation.
Appraisal	Key actions: papers selection through the PRISMA statement.
Synthesis	Key actions: data extraction and categorization.
Analysis	Key actions: analysis of the data, result comparison and conclusions.

Before starting the search through databases, it is important to define the scope of the research and to identify the appropriate keywords that will be used during the search process. The literature search was carried out in the Web of Science (WoS) database based on a combination of topics: "insulation" + "multi criteria". In order to carry out the widest analysis of the literature as possible and to include as many as possible research papers corresponding to the topic in the search, the search for papers was carried out in all WoS database categories.

The papers obtained during the search were evaluated and the PRISMA statement recommendations for selection of papers were followed. The inclusion criteria of the articles are as follows: keywords are in the title, the keywords section or the abstract of the paper, and the paper is published in a scientific peer-reviewed journal. Accordingly, exclusion criteria are as follows: review articles, conference proceedings; editorial letters; non English papers, and papers which were not primary research. These papers were excluded from the further analysis. Thus, 34 conference proceedings papers and 3 non-English papers were excluded from the content analysis. One hundred and nineteen articles were found by the search combination "insulation" and "multi criteria", 82 of which met the inclusion criteria. Articles that were included in the content analysis were mostly published in *Energy and Buildings* (10), *Building and Environment* (6) and *Sustainability* (5).

Content analysis was performed for the 82 articles found in the search. A snowballing method was also applied. Therefore, content analysis was performed for other articles that were not found during the search. Seven additional papers were found. A total of 18 relevant scientific studies were found where different MCDM methods for insulation materials were applied. A flow of information is provided in Figure 1.

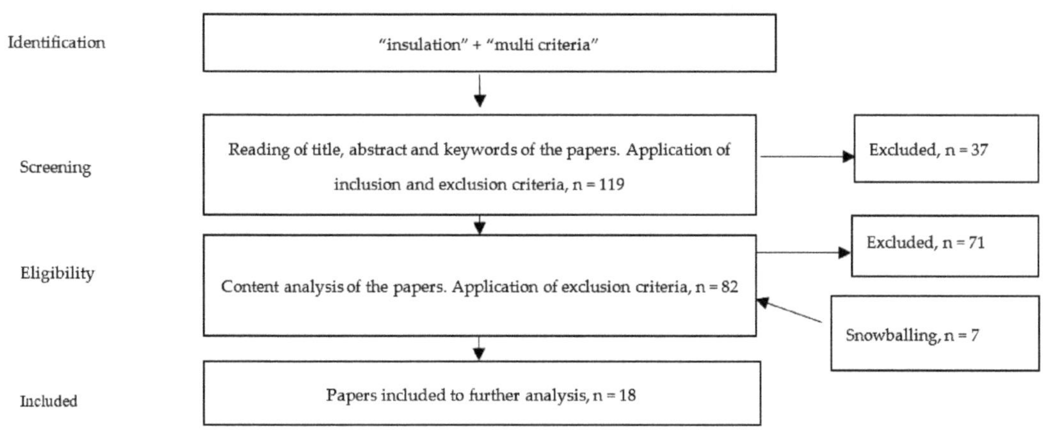

Figure 1. Flow of information (according to PRISMA).

The data of the selected articles were extracted and categorized according to the categories. Overall details of the reviewed studies are presented in Table 2. The next section provides detailed data on the analyzed articles.

Table 2. Overall data on the reviewed studies.

Application Areas	Methods Used	Groups of Indicators	Locations	Years of Publications
• Sustainability assessment • Guidelines for professionals • Suitability assessment	• The Technique for Order of Preference by Similarity to the Ideal Solution (TOPSIS) [25] • Analytical Hierarchy Process (AHP) [26] • Weighted Sum Method (WSM) [27] • VIKOR (an acronym in Serbian for multi-criteria optimization and compromise solution) [28] • Preference Ranking Organization Method for Enriching Evaluation V (PROMETHEE V) [29] • TODIM (an acronym in Portuguese for Interactive and Multi-criteria decision-making) [30] • Multi-Objective Optimisation by Ratio analysis (MOORA) [31,32] • Full Multiplicative Form of Multi-Objective Optimization by Ratio analysis (MULTIMOORA) [33] • Elimination and Choice Transcribing Reality (ELECTRE) [34,35] • Step-Wise Weight Assessment Ratio Analysis (SWARA) [36] • Simple Additive Weighting (SAW) [37] • Complex Proportional Assessment (COPRAS) [38]	• Economic • Social • Technological • Environmental • Performance • Energetic • Architectural • Not specified	• Vilnius, Lithuania • Montreal, Canada • Poznan, Poland • Turkey • Sarajevo, Serbia • Central Italy • Spain • Oran, Algeria • Riga, Latvia	• 2008 (2) • 2012 (1) • 2013 (1) • 2014 (3) • 2016 (1) • 2017 (1) • 2018 (2) • 2019 (2) • 2020 (5)

3. Literature Review

In order to carry out detailed literature analysis and systematically provide insights about the methods, evaluation criteria, and evaluation procedures used in practice, the publications discussed below are first categorized by application area. The following sub-section provides detailed analysis of the criteria and characteristics of the evaluation process (involvement of experts, motives for the selection of the criteria, methods for determining weights).

3.1. Assessment of Insulation Materials

According to the aim of the research, the papers could be grouped in three categories: sustainability assessment, suitability assessment and methods selection. Although sustainability assessment articles account only for 20% of all selected articles, the studies in this group are new, and this therefore shows the relevance of the topic. Sustainability assessment articles are summarized in Table 3.

Table 3. Sustainability assessment category.

Source	Aim of the Study	MCDM Method	Evaluation Level	Case Study Location	Materials Assessed	Main Contribution of the Study:
[39]	To evaluate the impact of sustainable insulations on the environment and their economic suitability.	ELECTRE TRI-rC	Local	A farmhouse in central Italy	Hard fiberboard, mineralized wood, polystyrene foam slab, cork slab, rock wool, glass wool, kenaf fibers, hemp fibers, expanded perlite, polyurethane, expanded vermiculite, cellulose	An original framework for the assessment is presented. The overall sustainability of insulating materials was evaluated, applying energy and comfort optimization, life cycle assessment (LCA) and life cycle costing (LCC) analysis for criteria selection and multi-criteria approach for ranking the alternatives. The most desirable materials are polystyrene foam slabs, kenaf fibers, hemp fibers, and cellulose.
[40]	To assess sustainability of flat roof types according of indicators aligned to the Sustainable Development Goals of the United Nations.	TOPSIS, AHP	National	Three weather scenarios in Spain (Mediterranean, Oceanic, Continental)	Four representative flat roof types (self-protected roof, gravel finishing roof, floating flooring roof and green roof).	The sustainability of four flat roof types was evaluated, based on indicators reflecting the Sustainable Development Goals of the United Nations. A green roof is the most sustainable alternative for all the scenarios evaluated.
[41]	To identify the factors that influence the selection of building roof system and to evaluate traditional and green roof systems.	AHP	Global	-	Traditional roof, green roof	The most significant criteria are related to performance criteria group. According to criteria outline by experts, a green roof is selected as a better option than a traditional roof.
[42]	To introduce a framework for the evaluation of sustainability of buildings insulation materials and to assess organic and inorganic building insulation materials in the context of sustainability.	interval TOPSIS	Global	-	Rock wool, expanded polystyrene, extruded polystyrene, kenaf, sheep wool, recycled cotton, recycled glass, recycled PET, recycled textile	A framework is presented and sustainable insulations are evaluated. Recycled glass and sheep wool are the best options for building insulation materials in the context of sustainability.

An original framework for the assessment of sustainability of insulating materials was presented by Rocchi et al. [39]. The case study of a farmhouse in central Italy considers the sustainability of twelve solutions for roof insulation according to seven criteria. The criteria for the assessment included combining energy and thermal comfort optimization with the environmental and economic LLA and LCC analysis. The ELECTRE TRI-rC method is used for ranking the selected organic and inorganic building insulation alternatives. The results show that the most favorable materials are polystyrene foam slabs, kenaf fibers, hemp fibers, and cellulose.

Guzman-Sanchez et al. [40] prepared a set of seventeen indicators for the assessment of the sustainability of different flat roof types, based on indicators reflecting the Sustainable Development Goals of the United Nations. The authors combined two MCDM techniques—the Analytical Hierarchy Process (AHP) and the Technique for Order of Preference by Similarity to the Ideal Solution (TOPSIS). In order to determine the relative importance of indicators, the AHP method was used for weighting. The TOPSIS technique was used for ranking the alternatives. The assessment was carried out under different weather scenarios. The results show that green roofs are the most sustainable choice for all the scenarios analyzed, by virtue of their insulation, possibility to recycle, life cycle cost, embodied energy, water purification and ecosystem-related aspects.

Rosasco and Perini [41] identified factors influencing the selection of building roof systems and applied the AHP technique to evaluate traditional and green roof systems. Experts identified the criteria and their weights for the assessment, and the most significant criteria are related to the performance criteria group. According to the criteria selected, evaluation demonstrates that a green roof is a better option than a traditional roof.

Streimikiene et al. [42] applied the interval TOPSIS method for sustainability evaluation and ranking of organic and inorganic building insulation materials. The authors carried out the sensitivity analysis by applying four different scenarios (equal, balanced, technological and environmental) with different weights for the selected criteria. The assessment shows that the best alternative according to the three scenarios (equal, balanced and technological) is recycled glass. According to the assessment, sheep wool is the best option in the environmental scenario.

Suitability assessment articles account for 40% of all selected articles and are summarized in Table 4.

Civic and Vucijak [43] applied the VIKOR technique for the evaluation of eight insulation materials. The authors selected seven criteria, which represent technical and environmental aspects. In this study, both the selection of criteria and their weighting are based on the selection of authors. The results show that the most preferred option is styrofoam, second place was taken by glass wool, and the third best is wood wool.

Zagorskas et al. [44] applied the TOPSIS Grey method for ranking five modern insulation materials (eco wool, flax/hemp fiber, thermo wool, aerogel, and a vacuum panel) for retrofitting historical buildings. Eco-wool was ranked as the best insulation solution. However, the results of the other alternatives are quite similar.

Ruzgys et al. [45] analyzed design solutions of modernized buildings in Lithuania. The authors ranked six external wall insulation alternatives for building modernization (polystyrene foam and thin plaster; mineral wool and fiber cement panels), applying the integrated SWARA-TODIM method. It was found that the best alternative for residential building modernization is a ventilated system with 130 mm thickness mineral wool insulation and fibrocement panels.

Table 4. Suitability assessment category.

Source	Aim of the Study	MCDM Method	Evaluation Level	Case Study Location	Materials Assessed	Main Contribution of the Study:
[43]	To emphasize the importance of energy management in buildings and to evaluate selected insulation materials on criteria selected.	VIKOR	National	Sarajevo, Serbia	Styrofoam, mineral wool (stone wool and glass wool), pluto panels, polyester, polyurethane, perlite, wood wool	The best alternative, according the criteria selected, is styrofoam, second place is taken by glass wool and third place is occupied by wood wool.
[44]	To rank five modern insulation materials for retrofitting historical buildings.	TOPSIS	National	Riga, Latvia	Eco wool, flax/hemp fiber, thermo wool, aerogel, vacuum panel	Eco-wool was ranked as the best insulation solution for retrofitting the historical buildings.
[45]	To analyze design solutions of modernized buildings.	TODIM	Project	Vilnius, Lithuania	Polystyrene foam and thin plaster, mineral wool and fiber cement panels	The best alternative for residential building modernization is a ventilated system with 130 mm thickness mineral wool insulation and fibrocement panels.
[46]	To assess double-skin façade systems reflecting the experience of experts who have applied them.	AHP	National	Turkey	The four types of double-skin façades (multistorey, corridor, shaft-box, box window)	The box window took first place, second place was taken by the corridor, the multi-storey double-skin façade was third, and the shaft-box took last place in the assessment.
[47]	To introduce and characterize new polymer-based composite materials.	AHP	-	-	Different formulations of rice husk and cork granules	New polymer-based composite materials were presented and characterized according to thermal conductivity and stability, vapour resistance, heat capacity, and acoustic characteristics.
[48]	To analyze building thermo-modernization solutions.	WSM	Local	Poznan, Poland	External polystyrene, mineral wool, extruded polystyrene	The best ranked solution is the variant of additional thermal insulation of extruded polystyrene with an additional thickness of 30 cm and wood windows.
[49]	To assess double-skin façade systems reflecting the experience of experts who have applied them and to compare the results with a previous study.	Fuzzy AHP	National	Turkey	The four types of double-skin façades (multistorey, corridor, shaft-box, box window)	The box window took first place, second place was taken by the corridor, the multi-storey double-skin façade was third, and the shaft-box took last place in the assessment.

Marques et al. [47] introduced innovative composite materials that incorporate rice husk and cork granules. The materials presented comprise a sustainable building solution. The AHP method was applied for different formulations with different ratios of materials. The results of the experiment show that a higher portion of rice husk in the composite formulations can provide better acoustic performance. Expanded cork granules reduce the thermal conductivity.

The four types of double-skin façade (multistorey, corridor, shaft-box, box window) were evaluated by Bostancioglu [49] The alternatives were ranked according to fuzzy AHP. The box window the first place, second place was taken by the corridor, the multi-storey double-skin façade was third, and the shaft-box took last place in the assessment.. It was found that a box window is the best alternative according to three criteria (noise and thermal insulation, fire protection). The results of the study were compared with previous research, where double-skin façades were evaluated with the AHP method [46]. The ranking of alternatives was unchanged.

Basinska et al. [48] analyzed building thermo-modernization solutions. The authors used the WSM method to find the best solution in regard to economic, energy-related, and environmental criteria. A total more than 400 possible solutions were analyzed. It was determined that the best solution is the variant of additional thermal insulation of extruded polystyrene with additional thickness of 30 cm and wood windows. The results show that the use of insulation with a thickness above 36 cm does not provide a significant energy or economic effect.

Methods selection articles account for 40% of all selected articles and are summarized in Table 5.

Table 5. Methods selection category.

Source	Aim of the Study	MCDM Method	Evaluation Level	Case Study Location	Materials Assessed	Main Contribution of the Study
[50]	To present an approach for the assessment of wall insulation alternatives and to find the best wall insulation solution.	SAW, TOPSIS, VIKOR, COPRAS	Project	Vilnius, Lithuania	Wall insulation (not specified)	The method for ranking alternatives was proposed and applied.
[51]	To present a methodology that allows one to rank different design solutions of a building's external walls, evaluating qualitative and quantitative attributes.	COPRAS, COPRAS-G (COPRAS with Grey relations)	-	-	Four external wall alternatives with insulation of rock wool or expanded polystyrene	The method for ranking alternatives was proposed and applied.
[52]	To find an optimal alternative for building renovation.	MOORA and MULTIMOORA	National	Vilnius, Lithuania	Walls, roofs, ceilings, windows (not specified)	The method for ranking alternatives was proposed and applied.
[53]	To present an approach for the assessment and ranking of technologies in the construction sector.	ELECTRE IV, MULTIMOORA, TOPSIS, ELECTRE III, VIKOR	National	Vilnius, Lithuania	Six alternatives (mineral wool, polystyrene foam)	The method for ranking alternatives was proposed and applied.
[54]	To introduce a tool for ranking different renovation solutions and exemplify it by evaluating a real-life case building.	PROMETHEE V	Local	Oran, Algeria	Exterior insulation of the facade or roof with expanded polystyrene, cellular concrete, wood fiber, lime hemp plaster; double glazing window; double windows; secondary glazing	The tool for ranking renovation solutions is presented and fifteen different insulation alternatives are evaluated. The results of the assessment show that the best solution is the exterior insulation of the roof with expanded polystyrene.
[55]	To create an assessment tool for the residential house construction materials selection.	MULTIMOORA-SVNS (Multiobjective Optimisation by Ratio Analysis Plus Full Multiplicative Form—Single-Valued Neutrosophic Set)	National	Lithuania	Houses with different thermal insulation alternatives (walls, roofs, ceilings, windows)	The method was proposed and applied.
[56]	To provide guidelines in achieving a high-performance facade system.	AHP	Local	Montreal, Canada	Four facade alternatives (combinations of 2 wall and 2 window systems)	The guidelines for each design phase are provided. An approach for decision making relating to the design of building facades is introduced.

Zavadskas et al. [51] presented a methodology that allows one to rank different design solutions of a building's external walls. The methodology involves qualitative and quantitative attributes and is based on the COPRAS technique. Ginevicius et al. [50] applied several MCDM methods (SAW, TOPSIS, VIKOR, COPRAS) for ranking five external wall insulation solutions and to select the most economically effective alternative for the renovation of a building. The study evaluates offers from subcontractors. Zavadskas et al. [53] presented an approach for the assessment and ranking of technologies in the construction sector. The authors evaluated six alternatives to mineral wool and polystyrene foam for thermal insulation of external walls. The assessment was based on ELECTRE IV, MULTIMOORA and hybrid SWARA-TOPSIS, SWARA-ELECTRE III and SWARA-VIKOR approaches. Another study by Zavadskas et al. [55] introduced a tool for the residential house construction materials selection based on MULTIMOORA and Neutrosophic sets. The proposed new extension of MULTIMOORA was named MULTIMOORA-SVNS. The study by Brauers et al. [52] evaluated twenty alternatives for external walls, roofs, ceilings, and windows in order to find the best alternative for the renovation of masonry buildings in Lithuania. The multi-criteria evaluation was carried out based on MOORA and MULTIMOORA.

Seddiki et al. [54] introduced a tool for ranking different renovation solutions. The tool is based on the MCDM PROMETHEE technique and combines Delphi method for criteria selection and Swing method for the determination of the weights of the criteria selected. A case study of a building in Algeria is provided and fifteen insulation alternatives are evaluated. It was determined that the best solution is the exterior insulation of the roof with expanded polystyrene.

Moghtadernejad et al. [56] presented an approach for the decision making of the design of a building façade. The approach integrated the MCDM tool AHP and Choquet integrals. The guidelines for each design phase are presented in the paper. The assessment also includes the assessment of building insulation materials as one of the components of the building façade. The criteria for evaluation are selected according to the objectives of the project and are not necessarily focused on the goals of sustainability.

3.2. *Criteria for Assessment in MCDM Models*

The majority of studies (67%) relied on experts (from 3 to 50) for evaluation. Most often, experts from the construction sector are involved. Some authors also relied on scientists and employees of state authorities who work in the field of construction or cultural heritage. Expert assistance can be used both in the selection of criteria and in determining the weights of the selected criteria. All studies that involved experts in the evaluation process used expert assistance in determining the weights of the criteria, but not all used experts to select the criteria. For the determination of the weight of criteria, an expert survey is usually used, in which the importance of the criteria is measured by pairwise comparison (scale 1–9, from 1 as "equally important" up to 9 as "extremely more important") (33% of studies), or by ranking from the most important to the least important (22% of studies). Some authors used their own estimation and expert surveys to determine weights [45,50], while others used Simon Roy Figueira's procedure [39], or the Swing method [54]. Evaluations which were made without the help of experts were based on the choice of the authors of the study by assigning weights to the criteria. In some studies (22%), experts participated in the selection of criteria [41,50,53,54]. Surveys, the Delphi method and cross-group discussion (brainstorming technique) were used for this purpose.

Articles in the methods selection category also use the concordance coefficient by Kendal calculation [50,51] and the determination of criteria weights by the SWARA method [45,53,55] to reasonably and logically determine criteria weights. The techniques used in the studies in order to select criteria for evaluation and to determine their weights are given in Table 6.

Table 6. Selection of criteria and determination of their weights in assessing insulation materials.

Source	MCDM Method	Supporting Methods	Way of Weighting	Experts	Type of Stakeholders	Number of Experts	Criteria Selection Process	Criteria
[50]	SAW, TOPSIS, VIKOR, COPRAS	Concordance coefficient by Kendal	Own estimation and expert survey (rating from the most important to the least important)	Yes	Experts in construction (from the Certification Centre of Construction Products, construction and reconstruction enterprises, researchers)	16	Experts survey	Not specified criteria
[51]	COPRAS, COPRAS-G	Concordance coefficient by Kendal	Experts survey (rating from the most important to the least important)	Yes	Experts (not specified)	39	Own selection	Not specified criteria
[52]	MOORA, MULTIMOORA	-	N/A	No	-	-	Own selection	Not specified criteria
[53]	ELECTRE IV, MULTI-MOORA, TOPSIS, ELECTRE III, VIKOR	Determination of criteria weights by SWARA method	Experts survey (rating from the most important to the least important)	Yes	Experts in civil engineers and in heating, ventilation, and air conditioning	25	Experts (Delphi method)	Not specified criteria
[44]	TOPSIS	-	Experts (pairwise comparison)	Yes	Experts in the cultural heritage, climate change and energy sectors	5	Own selection	Not specified criteria
[45]	TODIM	Determination of criteria weights by SWARA method	Own estimation and expert survey (rating from the most important to the least important)	Yes	N/A	25	Own selection	Not specified criteria
[43]	VIKOR	-	Own estimation	No	-		Own selection	Not specified criteria
[54]	PROMETHEE V	Sensitivity analysis	Swing method	Yes	Experts in the building and energy sector	4	Experts (Delphi method)	Economic, energetic and architectural criteria
[55]	MULTIMOORA-SVNS	Determination of criteria weights by SWARA method; sensitivity analysis; Neutrosophic sets	Experts (pairwise comparison)	Yes	Experts in house design (architects, engineers, and designers)	10	Own selection	Not specified criteria

Table 6. *Cont.*

Source	MCDM Method	Supporting Methods	Way of Weighting	Experts	Type of Stakeholders	Number of Experts	Criteria Selection Process	Criteria
[39]	ELECTRE TRI-rC	Energy and comfort optimization, LCA, LCC analysis, sensitivity analysis	Experts (Simon Roy Figueira procedure)	Yes	Experts (not specified)	3	Derived from the hybrid method developed (LCC analysis and LCA)	Economic and environmental criteria
[40]	TOPSIS, AHP	Sensitivity analysis (different weighting scenarios)	Experts (Questionnaire)	Yes	Experts in the building sector	50	Literature—the United Nations Sustainable Development Goals	Not specified criteria
[41]	AHP	-	Experts (pairwise comparison)	Yes	Experts in the building sector (architects, engineers, and researchers)	30	Experts (cross-group discussion—brainstorming technique); Literature	Economic, social, environmental and performance criteria
[46]	AHP	-	Experts (pairwise comparison)	Yes	Experts in the building sector	21	Literature	Not specified criteria
[49]	Fuzzy AHP	-	Experts (pairwise comparison)	Yes	Experts in the building sector	21	Literature	Not specified criteria
[42]	interval TOPSIS	Sensitivity analysis (different weighting scenarios)	Own estimation (different weighting scenarios)	No	-	-	Own selection; Literature	Technological and Environmental criteria
[56]	AHP	The Choquet integral	Experts (pairwise comparison)	No	-	-	Own selection; Literature	Not specified criteria
[47]	AHP	Different weighting scenarios	Own estimation (different weighting scenarios)	No	-	-	Own selection	Not specified criteria
[48]	WSM	LCA, different weighting scenarios	The method presented by Mroz [57]	No	-	-	Own selection; Literature	Not specified criteria

It should be noted that the criteria selected for evaluation are not categorized in most studies (almost 80%). Only four researchers divided the criteria into groups representing different evaluation dimensions. Seddiki et al. [54] divided the criteria into economic, energetic and architectural criteria to assess different alternatives for the renovation of the facade of the building. Rocchi et al. [39] singled out economic and environmental criteria groups to evaluate the impact of sustainable insulations on the environment and economic suitability. Rosasco and Perini [41] identified economic, social, environmental and performance criteria to identify factors that have the greatest influence when choosing building roof systems. Streimikiene et al. [42], in assessing the sustainability of organic and inorganic building insulation materials, identified the groups of technological and environmental criteria.

As previously mentioned, sustainability issues became particularly relevant in the construction sector. Although authors did not divide the criteria into groups in their assessments, this can be done in order to determine the popularity of the applied criteria and representation for different sustainability dimensions. Table 7 provides information on the criteria used in the evaluations, which are divided into four categories representing the essence of sustainable development. The popularity of the applied criteria is also estimated.

Table 7. Overview of criteria (arranged around four dimensions).

Dimension	Criteria	Popularity, %	Source
Economic	Investment cost, price	72	[41,43–46,49–56]
	Energy losses, heat losses, energy consumption decrease, energy saving	28	[39,41,45,52,54]
	Payback period	17	[45,52,53]
	Maintenance and disposal cost, operations and maintenance costs, decommissioning costs;	11	[41,56]
	Annual energy consumption, primary energy index	11	[48,56]
	Total amount saved per year	6	[52]
	Life cycle cost	6	[40]
	Comfort performance	6	[39]
	Net present value	6	[39]
	Tax incentives	6	[41]
	Real estate benefit	6	[41]
	Global cost	6	[48]
Social	Aesthetic	39	[40,41,46,49,54–56]
	Health, respiratory inorganics	17	[39,41,42]
	Air quality and heat island reduction	6	[41]
Technological	Thermal transmittance, thermal resistance, thermal conductivity, heat transfer, thermal insulation, heat capacity, insulation properties	78	[40–44,46,47,49–53,55,56]
	Water absorption coefficient, water vapour diffusion, Moisture properties	44	[42–45,47,50,53,56]
	Duration of works, construction process, complexity of the installation	44	[44–46,49–51,53,56]
	Durability, risk of the fabric	33	[41,50,51,54–56]
	Fire protection, fire classification	33	[42,46,49,56]
	Acoustic noise reduction, noise control, noise insulation, sound transmission class	33	[40,41,46,47,49,56]
	Weight, dead load	33	[40,41,50,51,55,56]
	Loss of space, total thickness	11	[44,56]
	Density	11	[42,43]
	Specific heat	11	[42,43]
	Wind pressure resistance	11	[46,49]
	Daylight	11	[46,49]
	Adhesive joint strength	6	[50]
	Extraction force of a pin fixing thermal insulating board to solid materials	6	[50]
	Warranty period	6	[50]

Table 7. *Cont.*

Dimension	Criteria	Popularity, %	Source
	Wall load-bearing capacity	6	[55]
	Protection	6	[40]
Environmental	CO_2 emissions	22	[41–43,48]
	Environmental friendliness of materials, resource sustainability, recycled materials	22	[40,41,55,56]
	Solar power, window solar performance	11	[40,56]
	Biodiversity	11	[40,41]
	Non-renewable energy	6	[39]
	Ozone layer depletion	6	[39]
	Global warming	6	[39]
	Albedo coefficient	6	[40]
	Carbon sequestration	6	[40]
	Embodied carbon	6	[40]
	Embodied energy	6	[40]
	Runoff attenuation	6	[40]
	Water purification	6	[40]
	Reduction in runoff temperature	6	[40]
	Agricultural productivity	6	[40]

All studies used indicators of insulation materials reflecting technological aspects. Overall, 78% of studies included thermal insulation characteristics in the evaluation. The use of the water absorption coefficient (44%) and duration of works (44%) took second place in terms of popularity. In addition, one third of studies included durability (33%), fire classification (33%), noise insulation (33%) and weight (33%). Economic indicators were included in 89% of the studies. The economic dimension is most often reflected by the investment cost or price criteria used by different authors. Overall, 72% of studies included this criterion in the assessment of insulation materials. The second criterion in terms of popularity is energy losses or energy saving (28%), while the third is payback period (17%). The criteria for social dimension were evaluated in 45% of studies. The following two criteria were also used: aesthetic (39%) and health (17%). Indicators representing the environmental dimension were also included in 45% of studies. The most commonly applied indicators were CO_2 emissions (22%) and environmental friendliness of insulation materials (22%).

4. Comparison of MCDM Models

The literature review revealed twelve different MCDM methods that were used in order to choose the most suitable insulation materials for buildings based on different criteria. These methods have different characteristics and different possibilities to include data in the estimations. Table 8 provides pros and cons of the MCDM techniques that were used for assessment of insulation materials.

The most popular AHP technique, developed by Saaty [26], helps to solve multi-criteria tasks using a pairwise comparison scale. The calculation technique of this method is quite simple and calculation results are obtained relatively quickly compared to other methods; the method is easily applied in various fields (tasks of construction, energy and other sectors) [58], and is logical and based on a hierarchical structure, and therefore focuses on all selected criteria. However, it should be noted that experience data of decision-makers plays a very important role here to determine the weights of the criteria. This can complicate the evaluation process if there is more than one decision-maker. In addition, additional analysis is required to verify the results of the evaluation [59–62].

Table 8. Comparative evaluation of MCDM methods.

MCDM Models		AHP	TOPSIS	MULTIMOORA	VIKOR	ELECTRE	COPRAS	MOORA	PROMETHEE	WSM	SWARA	SAW	TODIM
Popularity for Selection of Insulation Materials in Buildings, %		**33**	**28**	**17**	**11**	**11**	**11**	**6**	**6**	**6**	**6**	**6**	**6**
Pros	Easy to calculate	x	x	x	x			x		x		x	
	Non-compensatory			x		x		x	x				
	Comprehensible logic of calculations		x	x	x			x					
	Robust to outliers				x				x				
Cons	For verification additional analysis is required	x				x				x		x	
	Requires subjective assumptions	x				x			x		x	x	

The TOPSIS method is the second most popular method used when choosing insulation materials. The technique presented by Hwang and Yoon [25] is based on measuring the distance to the ideal solution [63]. As seen in the previously discussed technique, the TOPSIS is distinguished for fairly simple calculations and quickly obtains evaluation results, and the logic of calculation is rational and understandable, expressed in a fairly simple mathematical form. Therefore, it is easy for the decision-maker to interpret the results obtained and to understand the significance of the evaluation criteria for the final result. However, the TOPSIS is based on the Euclidean distance; therefore, positive and negative values of criteria are not reflected in the calculations. It is important to mention the fact that a significant deviation from the ideal solution in one evaluation criterion has a significant impact on the final results of the evaluation [64,65], and therefore the method is not suitable for evaluation when the indicators differ significantly among themselves.

MOORA was presented by Brauers in 2004 [31] and is identified as an objective tool to select alternatives. This approach is based on the ratio system and the reference point techniques. The method uses desirable and undesirable criteria simultaneously for ranking. Due to its objectivity, comprehensible logic of calculations, and simplicity, the method is widely used and is more robust than other MCDM techniques. The full multiplicative form was added to the MOORA by Brauers and Zavadskas [33], and the new method was named MULTIMOORA. Consequently, MULTIMOORA consists of three approaches: the ratio system and the reference point techniques, and the full multiplicative form [66]. Like its basis, MOORA, the method developed on its basis is widely used to solve problems in different areas.

The multi-criteria assessment technique VIKOR was presented by Opricovic [28] in 1998; this method is widely used in various fields of decision making. In addition, it is popular to integrate VIKOR with other MCDM techniques [67]. The method is based on seeking to determine the positive and the negative ideal solution (closeness to the ideal). Unlike the TOPSIS method, the VIKOR technique takes into account the relative importance of the distances from the positive and the negative ideal solution [68]. It is recognized that the VIKOR technique is understandable and the computation process is quite simple, compared with other methods. Despite that, the results could be affected by the normalization procedure and weight strategy.

The ELECTRE method was introduced by Roy in 1968 [34]. ELECTRE requires the determination of the concordance and discordance indices, which involves lengthy computations. The method needs to be subjected to human intervention, because the decision maker has to select threshold values for the calculation of concordance and discordance indices [69]. It is also recognized that for verification of the results, additional analysis is required.

COPRAS was introduced by Zavadskas et al. in 1994 [38]. It is one of the compromise methods, because COPRAS determines the ratio to the best ideal solution and the ratio to the worst ideal solution. The MCDM technique uses a stepwise ranking and evaluation procedure in terms of significance and utility degree. In addition, it is worth mentioning that qualitative and quantitative information can be used in calculations.

The methods of the PROMETHEE group are recognized as one of the most accurate methods. Currently, several versions of it are being developed. The first version was created in 1986. It was proposed by Brans et al. [70]. Calculations allow the use of qualitative and quantitative information as well as the use of uncertain information. In addition, alternatives that are highly interchangeable can be compared [71–73]. It is recognized that it is an accurate and effective multi-criteria evaluation technique; however, it has complex mathematical expressions [62,74], requires specific abilities, and results are not obtained as quickly as, for example, in the case of the TOPSIS or AHP. In principle, the method is intended for professionals engaged in this type of calculation.

The WSM method introduced by Zadeh [27] became popular due to its simple form and easy calculation [75]. This method is quite primitive and is designed to solve single-dimensional issues [76,77]. The WSM can be used as a separate method or as a component of other methods [78]. However, the issue of insulation material does not cover a single dimension that should be evaluated; therefore, it is basically more suitable for use as a component of other methods.

The SWARA is a relatively new method introduced by Kersuliene et al. [36]. The method is based on the logical calculation of weights and relative importance of the criteria selected. The greatest attention in the calculations is focused on the involvement of experts and the justification for participation in determining the weights of the evaluation criteria [79]. It can be said that experts have a key role in decision making. Although the method is new, it is widely used when solving different multi-criteria tasks [74]. The method is useful for collecting and coordinating information from experts [80].

One of the oldest, simplest, most commonly used and widely known MCDM technique is SAW [37]. This method is based on the weighted average, where the overall score of an alternative is determined by the weighted sum of selected criteria values. The calculation algorithm is very easy and do not requires specific knowledge. One of the advantages of this method is the proportional linear transformation of the raw data. Despite this, the result of the assessment may not be logical, when the values of one or several criteria differ from others. Additional analysis is required for verification of the results.

The TODIM technique was presented by Gomes and Lima in 1991 [30] and is based on a pairwise comparison. Although the method was introduced 30 years ago, it is not very popular in solving multi-criteria problems. The extended technique has the possibility to incorporate uncertain information [81–83]. TODIM is also distinguished by a long and complex calculation process [84] and less experience in the field of decision-making.

Depending on the available data, the experience of the decision-maker, the accuracy of the desired result and of the possible cost of time, the highlighted characteristics of the MCDM methods provide alternatives that allow faster evaluation process in future research.

5. Conclusions

A content analysis of articles has revealed that one third of studies used the AHP method for evaluation. The AHP method is used in half of all evaluations in the categories of sustainability assessment and suitability assessment. Meanwhile, articles in the method selection category offer more diverse, complex methods, requiring specific knowledge and skills. The second most popular MCDM method is TOPSIS, which is applied in 28% of all studies. Both methods are quite simple and easy to apply in practice. They do not require complex calculations, high costs in terms of time, or specific knowledge of the person seeking the solution. Although articles of the method selection category offer more complex calculation algorithms, they are much more methodologically accurate and logical when there is a need to select criteria for evaluation and determining criteria weights.

The majority of studies relied on experts for evaluation. All studies that involved experts in the evaluation process used expert assistance in determining the weights of the criteria, but not all used experts in the criteria selection process. For the determination of the weight of criteria, an expert survey is usually used, in which the importance of the criteria is measured by pairwise comparison or by ranking from the most important to the least important. For criteria selection, surveys, the Delphi method, and cross-group discussion (brainstorming technique) were used. Involvement of experts in the evaluation process reduces the subjectivity of the research and allows one to look at the problem being solved from different perspectives. The use of experts is recommended not only for the determination of weights, but also for criteria selection. In order to justify the involvement of experts in the evaluation process, scientific methods both for calculating the coincidence of expert opinion and for conducting the survey of experts should be used.

It should be noted that the criteria selected for evaluation are not categorized in most studies. All studies used indicators of insulation materials reflecting technological aspects, where thermal insulation characteristics were the most popular criteria. The economic dimension was evaluated in 89% of studies and mostly was reflected by the investment cost or price. The criteria for social and environmental dimensions were evaluated in 45% of studies. In order to carry out a comprehensive assessment of insulation materials, criteria representing different dimensions of sustainability should be used. The review of the evaluation criteria and their grouping by representing different dimensions makes it easier to select criteria for this type of assessment and ensures conformity of the evaluation with the current sustainability issues, which include the achievement of economic goals, energy efficiency, technological characteristics, and the impact on the environment and human health.

The conducted study provides an important input in guiding future studies on decision making for sustainable selection of insulation materials in buildings, which is the major issue in the Renovation Wave Strategy, aiming to improve the energy performance of buildings and at least doubling the renovation rates in the next ten years. As this strategy seeks to enhance the quality of life for people living in and using the buildings, the sustainability of materials needs to be properly addressed.

Author Contributions: The contribution of all authors is equal. I.S.-B. made formal analysis and prepared original draft, D.S. designed and supervised research, T.B. reviewed the paper, and V.S. reviewed the paper. All authors have read and agreed to the published version of the manuscript.

Funding: This research received no external funding.

Institutional Review Board Statement: Not applicable.

Informed Consent Statement: Not applicable.

Data Availability Statement: Not applicable.

Acknowledgments: Authors are thankful for the reviewers' comments and valuable suggestions.

Conflicts of Interest: The authors declare no conflict of interest.

References

1. Moran, P.; Goggins, J.; Hajdukiewicz, M. Super-insulate or use renewable technology? Life cycle cost, energy and global warming potential analysis of nearly zero energy buildings (NZEB) in a temperate oceanic climate. *Energy Build.* **2017**, *139*, 590–607. [CrossRef]
2. Grygierek, K.; Ferdyn-Grygierek, J. Multi-Objective Optimization of the Envelope of Building with Natural Ventilation. *Energies* **2018**, *11*, 1383. [CrossRef]
3. Serghides, D.K.; Dimitriou, S.; Katafygiotou, M.C.; Michaelidou, M. Energy efficient refurbishment towards nearly zero energy houses, for the Mediterranean region. *Energy Procedia* **2015**, *83*, 533–543. [CrossRef]
4. European Commission. *A Renovation Wave for Europe—Greening our Buildings, Creating Jobs, Improving Lives*; COM (2020) 662 Final; Communication from the Commission to the European Parliament, the Council, the European Economic and Social Committee and the Committee of the Region: Brussels, Belgium, 2020.
5. Santamouris, M. Innovating to zero the building sector in Europe: Minimising the energy consumption, eradication of the energy poverty and mitigating the local climate change. *Sol. Energy* **2016**, *128*, 61–94. [CrossRef]
6. Manzano-Agugliaro, F.; Montoya, F.G.; Sabio-Ortega, A.; García-Cruz, A. Review of bioclimatic architecture strategies for achieving thermal comfort. *Renew. Sustain. Energy Rev.* **2015**, *49*, 736–755. [CrossRef]
7. Cao, X.; Dai, X.; Liu, J. Building energy-consumption status worldwide and the state-of-the-art technologies for zero-energy buildings during the past decade. *Energy Build.* **2016**, *128*, 198–213. [CrossRef]
8. Noailly, J. Improving the energy efficiency of buildings: The impact of environmental policy on technological innovation. *Energy Econ.* **2012**, *34*, 795–806. [CrossRef]
9. Goulden, S.; Erell, E.; Garb, Y.; Pearlmutter, D. Green building standards as socio-technical actors in municipal environmental policy. *Build. Res. Inf.* **2017**, *45*, 414–425. [CrossRef]
10. Bonamente, E.; Brunelli, C.; Castellani, F.; Garinei, A.; Biondi, L.; Marconi, M.; Piccioni, E. A life-cycle approach for multi-objective optimisation in building design: Methodology and application to a case study. *Civ. Eng. Environ. Syst.* **2018**, *35*, 158–179. [CrossRef]
11. EU Commission and Parliament. Directive 2018/844 of the European Parliament and the Council of the 30 May 2018 Amending Directive 2010/31/EU on the Energy Performance of Buildings and Directive 2012/27/EU on Energy Efficiency. *Off. J. Eur. Union* **2018**, *156*, 75–91.
12. Amani, N.; Kiaee, E. Developing a two-criteria framework to rank thermal insulation materials in nearly zero energy buildings using multi-objective optimization approach. *J. Clean. Prod.* **2020**, *276*, 122592. [CrossRef]

13. Bisegna, F.; Mattoni, B.; Gori, P.; Asdrubali, F.; Guattari, C.; Evangelisti, L.; Sambuco, S.; Bianchi, F. Influence of Insulating Materials on Green Building Rating System Results. *Energies* **2016**, *9*, 712. [CrossRef]
14. Aditya, L.; Mahlia, T.M.I.; Rismanchi, B.; Ng, H.M.; Hasan, M.H.; Metselaar, H.S.C.; Muraza, O.; Aditiya, H.B. A review on insulation materials for energy conservation in buildings. *Renew. Sustain. Energy Rev.* **2017**, *73*, 1352–1365. [CrossRef]
15. Al-Homoud, M.S. Performance characteristics and practical applications of common building thermal insulation materials. *Build. Environ.* **2005**, *40*, 353–366. [CrossRef]
16. Patnaik, A.; Mvubu, M.; Muniyasamy, S.; Botha, A.; Anandjiwala, R.D. Thermal and sound insulation materials from waste wool and recycled polyester fibers and their biodegradation studies. *Energy Build.* **2015**, *92*, 161–169. [CrossRef]
17. Asdrubali, F.; D'Alessandro, F.; Schiavoni, S. A review of unconventional sustainable building insulation materials. *Sustain. Mater. Technol.* **2015**, *4*, 1–17. [CrossRef]
18. Gullbrekken, L.; Grynning, S.; Gaarder, J.E. Thermal Performance of Insulated Constructions—Experimental Studies. *Buildings* **2019**, *9*, 49. [CrossRef]
19. Saghafi, M.D.; Teshnizi, Z.S.H. Recycling value of building materials in building assessment systems. *Energy Build.* **2011**, *43*, 3181–3188. [CrossRef]
20. Samani, P.; Mendes, A.; Leal, V.; Guedes, J.M.; Correia, N. A sustainability assessment of advanced materials for novel housing solutions. *Build. Environ.* **2015**, *92*, 182–191. [CrossRef]
21. Siksnelyte, I.; Zavadskas, E.K.; Streimikiene, D.; Sharma, D. An Overview of Multi-Criteria Decision-Making Methods in Dealing with Sustainable Energy Development Issues. *Energies* **2018**, *11*, 2754. [CrossRef]
22. Moher, D.; Liberati, A.; Tetzlaff, J.; Altman, D.G. Preferred reporting items for systematic reviews and meta-analyses: The PRISMA statement. *Int. J. Surg.* **2010**, *8*, 336–341. [CrossRef] [PubMed]
23. Amo, I.F.; Erkoyuncu, J.A.; Roy, R.; Palmarini, R.; Onoufriou, D. A systematic review of Augmented Reality content-related techniques for knowledge transfer in maintenance applications. *Comput. Ind.* **2018**, *103*, 47–71.
24. Grant, M.J.; Booth, A. A typology of reviews: An analysis of 14 review types and associated methodologies. *Health Inf. Libr. J.* **2009**, *26*, 91–108. [CrossRef]
25. Hwang, C.L.; Yoon, K. *Multiple Attributes Decision Making Methods and Applications*; Springer: Berlin/Heidelberg, Germany, 1981; pp. 22–51.
26. Saaty, T.L. *The Analytic Hierarchy Process*; McGraw-Hill: New York, NY, USA, 1980; pp. 11–29.
27. Zadeh, L.A. Optimality and non-scalar-valued performance criteria. *IEEE Trans. Autom. Control* **1963**, *8*, 59–60. [CrossRef]
28. Opricovic, S. Multicriteria Optimization of Civil Engineering Systems. Ph.D. Thesis, Faculty of Civil Engineering, Belgrade, Serbia, 1998; 302p.
29. Mareschal, B.; Brans, J.P. *PROMETHEE V: MCDM Problems with Segmentation Constrains*; Universite Libre de Brusells: Brussels, Belgium, 1992; pp. 13–30.
30. Gomes, L.F.A.M.; Lima, M.M.P.P. TODIM: Basics and application to multicriteria ranking of projects with environmental impacts. *Found. Comput. Decis. Sci.* **1991**, *16*, 113–127.
31. Brauers, W.K. Optimization Methods for a Stakeholder Society. In *A Revolution in Economic Thinking by Multiobjective Optimization*; Kluwer Academic Publishers: Boston, MA, USA, 2004; 352p.
32. Brauers, W.K.M.; Zavadskas, E.K. The MOORA method and its application to privatization in transition economy. *Control Cybern.* **2006**, *35*, 443–468.
33. Brauers, W.K.M.; Zavadskas, E.K. Project Management by MULTIMOORA as an Instrument for Transition Economies. *Technol. Econ. Dev. Econ.* **2010**, *16*, 5–24. [CrossRef]
34. Roy, B. La methode ELECTRE. *Rev. D'Inform. Et. De Rech. Oper. (Riro)* **1968**, *8*, 57–75.
35. Vallée, D.; Zielniewicz, P. *ELECTRE III-IV, Version 3.x, Aspects Méthodologiques (Tome 1), Guide D'utilisation (Tome 2)*; Document du LAMSADE 85 et 85 bis; Université Paris Dauphine: Paris, France, 1994.
36. Kersuliene, V.; Zavadskas, E.K.; Turskis, Z. Selection of rational dispute resolution method by applying new step-wise weight assessment ratio analysis (Swara). *J. Bus. Econ. Manag.* **2010**, *11*, 243–258. [CrossRef]
37. MacCrimon, K.R. *Decision Marking among Multiple-Attribute Alternatives: A Survey and Consolidated Approach*; RAND Memorandum, RM-4823-ARPA; The Rand Corporation: Santa Monica, CA, USA, 1968; 63p.
38. Zavadskas, E.K.; Kaklauskas, A.; Sarka, V. The new method of multicriteria complex proportional assessment of projects. *Technol. Econ. Dev. Econ.* **1994**, *1*, 131–139.
39. Rocchi, L.; Kadzinski, M.; Menconi, M.E.; Grohmann, D.; Miebs, G.; Paolotti, L.; Boggia, A. Sustainability evaluation of retrofitting solutions for rural buildings through life cycle approach and multi-criteria analysis. *Energy Build.* **2018**, *173*, 281–290. [CrossRef]
40. Guzman-Sanchez, S.; Jato-Espino, D.; Lombillo, I.; Diaz-Sarachaga, J.M. Assessment of the contributions of different flat roof types to achieving sustainable development. *Build. Environ.* **2018**, *141*, 182–192. [CrossRef]
41. Rosasco, P.; Perini, K. Selection of (Green) Roof Systems: A Sustainability-Based Multi-Criteria Analysis. *Buildings* **2019**, *9*, 134. [CrossRef]
42. Streimikiene, D.; Skulskis, V.; Balezentis, T.; Agnusdei, G.P. Uncertain multi-criteria sustainability assessment of green building insulation materials. *Energy Build.* **2020**, *219*, 110021. [CrossRef]
43. Civic, A.; Vucijak, B. Multi-criteria Optimization of Insulation Options for Warmth of Buildings to Increase Energy Efficiency. *Procedia Eng.* **2014**, *69*, 911–920. [CrossRef]

44. Zagorskas, J.; Zavadskas, E.K.; Turskis, Z.; Burinskiene, M.; Blumberga, A.; Blumberga, D. Thermal insulation alternatives of historic brick buildingsin Baltic Sea Region. *Energy Build.* **2014**, *78*, 35–42. [CrossRef]
45. Ruzgys, A.; Volvaciovas, R.; Ignatavicius, C.; Turskis, Z. Integrated evaluation of external wall insulation in residential buildings using SWARA-TODIM MCDM method. *J. Civ. Eng. Manag.* **2014**, *20*, 103–110. [CrossRef]
46. Bostancioglu, E.; Onder, N.P. Applying analytic hierarchy process to the evaluation of double skin façades. *Archit. Eng. Des. Manag.* **2019**, *15*, 66–82. [CrossRef]
47. Marques, B.; Tadeu, A.; Antonio, J.; Almeida, J.; de Brito, J. Mechanical, thermal and acoustic behaviour of polymer-based composite materials produced with rice husk and expanded cork by-products. *Constr. Build. Mater.* **2020**, *239*, 117851. [CrossRef]
48. Basinska, M.; Kaczorek, D.; Koczyk, H. Building Thermo-Modernisation Solution Based on the Multi-Objective Optimisation Method. *Energies* **2020**, *13*, 1433. [CrossRef]
49. Bostancioglu, E. Double skin facade assessment by fuzzy AHP and comparison with AHP. *Archit. Eng. Des. Manag.* **2020**. [CrossRef]
50. Ginevicius, R.; Podvezko, V.; Raslanas, S. Evaluating the Alternative Solutions of Wall Insulation by Multicriteria Methods. *J. Civ. Eng. Manag.* **2008**, *14*, 217–226. [CrossRef]
51. Zavadskas, E.K.; Kaklauskas, A.; Turskis, Z.; Tamosaitiene, J. Selection of the effective dwelling house walls by applying attributes values determined at intervals. *J. Civ. Eng. Manag.* **2008**, *14*, 85–93. [CrossRef]
52. Brauers, W.K.M.; Kracka, M.; Zavadskas, E.K. Lithuanian Case Study of Masonry Buildings from the Soviet Period. *J. Civ. Eng. Manag.* **2012**, *18*, 444–456. [CrossRef]
53. Zavadskas, E.K.; Turskis, Z.; Volvaciovas, R.; Kildiene, S. Multi-criteria Assessment Model of Technologies. *Stud. Inform. Control* **2013**, *22*, 249–258. [CrossRef]
54. Seddiki, M.; Anouche, K.; Bennadji, A.; Boateng, P. A multi-criteria group decision-making method for the thermal renovation of masonry buildings: The case of Algeria. *Energy Build.* **2016**, *129*, 471–483. [CrossRef]
55. Zavadskas, E.K.; Bausys, R.; Juodagalviene, B.; Garnyte-Sapranaviciene, I. Model for residential house element and material selection by neutrosophic MULTIMOORA method. *Eng. Appl. Artif. Intell.* **2017**, *64*, 315–324. [CrossRef]
56. Moghtadernejad, S.; Chouinard, L.E.; Mirza, M.S. Design strategies using multi-criteria decision-making tools to enhance the performance of building facades. *J. Build. Eng.* **2020**, *30*, 101274. [CrossRef]
57. Mroz, T.M. *Energy Management in Built Environment: Tools and Evaluation Procedures*; Poznan University of Technology: Poznan, Poland, 2013; p. 138, ISBN 8377752387.
58. Kaya, I.; Çolak, M.; Terzi, F. Use of MCDM techniques for energy policy and decision-making problems: A review. *Int. J. Energy Res.* **2018**, *42*, 2344–2372. [CrossRef]
59. Saaty, T.L. Decision making-the analytic hierarchy and network processes (AHP/ANP). *J. Syst. Sci. Syst. Eng.* **2004**, *13*, 1–35. [CrossRef]
60. Ishizaka, A.; Labib, A. Analytic hierarchy process and expert choice: Benefits and limitations. *Or Insight* **2009**, *22*, 201–220. [CrossRef]
61. Shahroodi, K.; Keramatpanah, A.; Amini, S.; Sayyad Haghighi, K. Application of analytical hierarchy process (AHP) technique to evaluate and selecting suppliers in an effective supply chain. *Kuwait Chapter Arab. J. Bus. Manag. Rev.* **2012**, *1*, 119–132.
62. Kumar, A.; Sah, B.; Singh, A.R.; Deng, Y.; He, X.; Kumar, P.; Bansal, R.C. A review of multi criteria decision making (MCDM) towards sustainable renewable energy development. *Renew. Sustain. Energy Rev.* **2017**, *69*, 596–609. [CrossRef]
63. Jato-Espino, D.; Castillo-Lopez, E.; Rodriguez-Hernandez, J.; Canteras-Jordana, J.C. A review of application of multi-criteria decision making methods in construction. *Autom. Constr.* **2014**, *45*, 151–162. [CrossRef]
64. Shih, H.S.; Shyur, H.J.; Lee, E.S. An extension of TOPSIS for group decision making. *Math. Comput. Model.* **2007**, *45*, 801–813. [CrossRef]
65. Boran, F.E.; Genc, S.; Kurt, M.; Akay, D. A multi-criteria intuitionistic fuzzy group decision making for supplier selection with TOPSIS method. *Expert Syst. Appl.* **2009**, *36*, 11363–11368. [CrossRef]
66. Zavadskas, E.K.; Antucheviciene, J.; Hajiagha, S.H.R.; Hashemi, S.S. The Interval-Valued Intuitionistic Fuzzy MULTIMOORA Method for Group Decision Making in Engineering. *Math. Probl. Eng.* **2015**, *2015*, 560690. [CrossRef]
67. Mardani, A.; Zavadskas, E.K.; Govindan, K.; Senin, A.A.; Jusoh, A. VIKOR Technique: A Systematic Review of the State of the Art Literature on Methodologies and Applications. *Sustainability* **2016**, *8*, 37. [CrossRef]
68. Opricovic, S.; Tzeng, G.H. Extended VIKOR method in comparison with outranking methods. *Eur. J. Oper. Res.* **2007**, *178*, 514–529. [CrossRef]
69. Karande, P.; Chakraborty, S. Application of multi-objective optimization on the basis of ratio analysis (MOORA) method for materials selection. *Mater. Des.* **2012**, *37*, 317–324. [CrossRef]
70. Brans, J.P.; Vincke, P.; Mareshal, B. How to select and how to rank projects: The Promethee method. *Eur. J. Oper. Res.* **1986**, *24*, 228–238. [CrossRef]
71. Wang, M.; Lin, S.J.; Lo, Y.C. The comparison between MAUT and PROMETHEE. In Proceedings of the International Conference on Industrial Engineering and Engineering Management (IEEM), Macao, China, 7–10 December 2010; pp. 753–757.
72. Amaral, T.M.; Costa, A.P. Improving decision-making and management of hospital resources: An application of the PROMETHEE II method in an Emergency Department. *Oper. Res. Health Care* **2014**, *3*, 1–6. [CrossRef]

73. Brans, J.P.; De Smet, Y. PROMETHEE Methods. In *Multiple Criteria Decision Analysis*; Greco, S., Ehrgott, M., Figueira, J., Eds.; Springer: New York, NY, USA, 2016; Volume 233, pp. 187–219.
74. Alinezhad, A.; Khalili, J. *New Methods and Applications in Multiple Attribute Decision Making (MADM)*; International Series in Operations Research & Management Science 227; Springer: Cham, Switzerland, 2019; 203p.
75. Marler, R.T.; Arora, J.S. The weighted sum method for multi-objective optimization: New insights. *Struct. Multidiscip. Optim.* **2010**, *41*, 853–862. [CrossRef]
76. Misra, S.K.; Ray, A. Comparative study on different multi-criteria decision making tools in software project selection scenario. *Int. J. Adv. Res. Comput. Sci.* **2012**, *3*, 172–178.
77. Wimmler, C.; Hejazi, G.; de Oliveira Fernandes, E.; Moreira, C.; Connors, S. Multi-Criteria decision support methods for renewable energy systems on Islands. *J. Clean Energy Technol.* **2015**, *3*, 185–195. [CrossRef]
78. Wang, R.; Zhou, Z.; Ishibuchi, H. Localized Weighted Sum Method for Many-Objective Optimization. *IEEE Trans. Evol. Comput.* **2018**, *22*, 3–18. [CrossRef]
79. Mishra, A.R.; Rani, P.; Pandey, K.; Mardani, A.; Streimikis, J.; Streimikiene, D.; Alrasheedi, M. Novel Multi-Criteria Intuitionistic Fuzzy SWARA-COPRAS Approach for Sustainability Evaluation of the Bioenergy Production Process. *Sustainability* **2020**, *12*, 4155. [CrossRef]
80. Zolfani, S.H.; Saparauskas, J. New Application of SWARA Method in Prioritizing Sustainability Assessment Indicators of Energy System. *Inz. Ekon. Eng. Econ.* **2013**, *24*, 408–414. [CrossRef]
81. Zhang, X.; Xu, Z. The TODIM analysis approach based on novel measured functions under hesitant fuzzy environment. *Knowl. Based Syst.* **2014**, *61*, 48–58. [CrossRef]
82. Qin, J.; Liu, X.; Pedrycz, W. An extended TODIM multi-criteria group decision making method for green supplier selection in interval type-2 fuzzy environment. *Eur. J. Oper. Res.* **2017**, *258*, 626–638. [CrossRef]
83. Yu, S.M.; Wang, J.; Wang, J.Q. An extended TODIM approach with intuitionistic linguistic numbers. *Int. Trans. Oper. Res.* **2018**, *25*, 781805. [CrossRef]
84. Llamazares, B. An analysis of the generalized TODIM method. *Eur. J. Oper. Res.* **2018**, *269*, 1041–1049. [CrossRef]

sustainability

MDPI

Article

ACS: Construction Data Auto-Correction System—Taiwan Public Construction Data Example

Meng-Lin Yu and Meng-Han Tsai *

Department of Civil and Construction Engineering, National Taiwan University of Science and Technology, Taipei 106, Taiwan; m10705510@ntust.edu.tw
* Correspondence: menghan@mail.ntust.edu.tw; Tel.: +8862-2737-6356

Abstract: This study aims to develop an automatic data correction system for correcting the public construction data. The unstructured nature of the construction data presents challenges for its management. The different user habits, time-consuming system operation, and long pretraining time all make the data management system full of data in an inconsistent format or even incorrect data. Processing the construction data into a machine-readable format is not only time-consuming but also labor-intensive. Therefore, this study used Taiwan's public construction data as an example case to develop a natural language processing (NLP) and machine learning-based text classification system, coined as automatic correction system (ACS). The developed system is designed to automatically correct the public construction data, meanwhile improving the efficiency of manual data correction. The ACS has two main features: data correction that converts unstructured data into structured data; a recommendation function that provides users with a recommendation list for manual data correction. For implementation, the developed system was used to correct the data in the public construction cost estimation system (PCCES) in Taiwan. We expect that the ACS can improve the accuracy of the data in the public construction database to increase the efficiency of the practitioners in executing projects. The results show that the system can correct 18,511 data points with an accuracy of 76%. Additionally, the system was also validated to reduce the system operation time by 51.69%.

Keywords: natural language processing; construction data management; machine learning

Citation: Yu, M.-L.; Tsai, M.-H. ACS: Construction Data Auto-Correction System—Taiwan Public Construction Data Example. *Sustainability* **2021**, *13*, 362. https://doi.org/10.3390/su13010362

Received: 21 November 2020
Accepted: 18 December 2020
Published: 3 January 2021

Publisher's Note: MDPI stays neutral with regard to jurisdictional claims in published maps and institutional affiliations.

1. Introduction

1.1. Construction Data Management

Data management has been considered as one of the most vital tasks in the construction industry [1]. The unstructured nature (e.g., plain text) of the construction data presents challenges for its management. Processing the construction data into a machine-readable format is time-consuming and labor-intensive, as it requires lots of paperwork to structuralize data from various sources. Nowadays, many popular tools and standards have been developed by experts to help people manage construction data. Management tools, such as Microsoft Project and Primavera, are prevalent. The MasterFormat and UniFormat standards are common in the U.S. and Canada. However, engineers still need to follow the rules and form formats provided by standards and tools to manually transform unstructured data into structured data.

Even though many tools and standards are ready to follow, the quality of processed data is not sufficient due to the various backgrounds of the related personnel. Engineers from different fields may have their own interpretations of the standards, thus causing inconsistency of the data format or even incorrect data. For instance, Taiwan government provides a tool to its employees and contractors to manage public construction projects using a coding system similar to MasterFormat. However, due to the different interpretations of the coding standard, the average data accuracy of 7592 public construction projects is only 48%.

With the advancement of information technology, the improvement of computer performance, as well as large data storage, technologies such as artificial intelligence and deep learning have flourished for solving complex data management problems. Such improvement of these technologies can automatically transform unstructured data into structured data.

1.2. Public Construction Cost Estimation System in Taiwan

In Taiwan, it is required by law that for public projects with bids over NTD 10 million, the project documents need to be managed using the public construction cost estimation system (PCCES). Additionally, as the PCCES is a coding system, the law also requires that the accuracy rate of data encoded in the project documents should be at least 40%. However, according to the statistics of the Public Construction Commission, Executive Yuan (PCC), Taiwan, from 1 December 2016 to 30 June 2020, there were 422 public construction works projects that needed to use the PCCES. Among the 422 cases, 215 projects had statistics on the correct rate of coding, and the average correct rate was 37.47% [2].

The Taiwan government has been planning to make public construction work projects more transparent since 1995. The solution to this issue is making its own coding standard for the construction field. The Taiwan government imitated the MasterFormat coding standard formulated by the Construction Specifications Institute (CSI) and formulated its own coding standard [3]. According to the coding structure of MasterFormat, the PCCES coding standard is divided into chapters 00 to 16, for a total of 17 chapters. Its organization is arranged according to engineering practice and the experience of engineers. According to the work breakdown structure (WBS) [4], it is also divided into five code and four layer structures. The coding architecture is shown in Figure 1. In the first layer, the first and the second codes are the numbers of the chapter. In the second layer, the third code is a major category. In the third layer, the fourth code is the detailed classification code under each major category. For the fourth layer, the fifth code is for the work item, which is related to the third layer. Users can customize the code, but it should be recognized by the engineering committee's public engineering technology database control management.

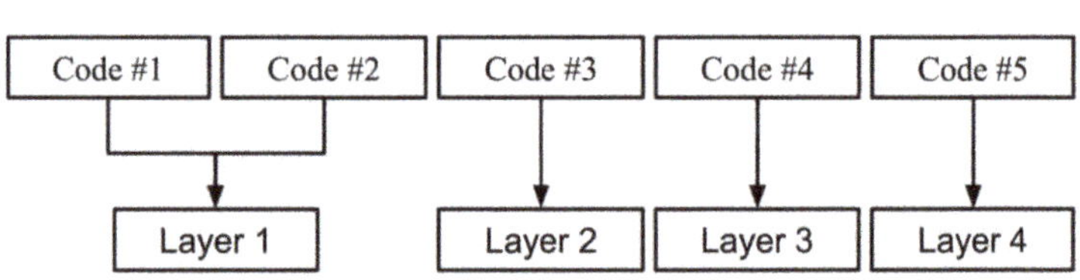

Figure 1. The coding architecture of the public construction cost estimation system (PCCES) [3].

The PCCES code considers work items and resource items. Resource items include materials, human resources, and machines. Work items are included in the composition of materials, machinery, labor, and miscellaneous. The coding of work items and resource items can be used differently. As there are many projects in public works, the scale of these projects varies, ranging from airports, dams, and tunnels to planting sidewalk trees, paving bricks on sidewalks, and dredging trenches, all within the scope of public works. Each project requires the participation of many upstream and downstream manufacturers, office staff, and government officials.

Due to different roles and divisions of labor, the perspectives of project management vary. Projects of various sizes, workers from different sources, and different roles in the project will cause these people to have a different understanding of the PCCES encoding system when using it. These differences in understanding have resulted in non-compliant materials in the project. However, these non-compliant materials have accumulated over time, presenting challenges for the database. Due to these reasons, with the accumulation of days and months, the database has become full of non-compliant data.

1.3. Benefits and Challenges of the PCCES

The PCCES occupies a very important position in the field of public construction works in Taiwan. Public construction projects require the use of the PCCES in accordance with the law, so these projects have uniform specifications that can be followed. The PCCES offers some benefits after collecting a large number of projects and integrating data. For example, according to the PCC announcement [3], the PCCES can improve the efficiency and credibility of fund review and fund comparison, avoid repeating the establishment of system software and hardware by various agencies, save national public funds, and reduce the opportunities for restricting competition and restricting bidding during project bidding.

Although the PCCES does bring these benefits and helps users a lot, some researches have pointed out problems with the system. For instance, users are not used to the standardized work items and resource codes in the PCCES [5]. Users need longer education and training to be familiar with and operate the PCCES [6]. Many practical applications are still not included in the PCCES [7,8]. The operation of the PCCES is time-consuming, requires a lot of manpower, and has a high error rate [9]. Moreover, in the production of PCCES documents there are general contradictions between architects and construction companies in measurement and calculation methods, and some measurement and calculation methods do not conform to the PCCES coding standard [10].

1.4. Research Objectives

This study focuses on the development of a machine-learning-based system to auto-correct a public database of the construction field that contains the PCCES data in Taiwan. People cannot use the data in the public database as the database is filled with messy data. In actuality, while people use many tools for managing construction data, the data are still chaotic. The quality of these data depends on how familiar a user is with the tools. There is no classification method that can consider the meaning of construction specifications in the construction classification system's database and then automatically correct the wrong data. Therefore, this study aims to develop a machine-learning-based system to improve the performance of people who work on the construction files and need to use construction classification systems. The developed method should achieve the following goals:

(1) Develop an auto-correct function to automatically correct the data from public databases related to the construction field and the unstructured construction project data. Users could use this function to correct data that they obtained from the open data database made by the government.
(2) Develop a recommender function, which can help users to perform their job efficiently without having experience in construction classification systems.

2. Literature Review

In order to develop a system that can automatically structuralize the construction data, this study first reviews the characteristics and challenges of construction data management (Section 2.1). Then, the novel methods for unistructural data processing proposed by other studies is discussed in Section 2.2. Lastly, the review of state-of-the-art machine-learning-based approaches to construction data processing is conducted in Section 2.3.

2.1. Challenges in Construction Data Management

Processing unstructured data is considered one of the most critical challenges in construction data management. Project owners, architects, contractors, and suppliers communicate and coordinate through various documents. It is common that documents have different naming conventions for the same object [11]. The different naming conventions increase the time and cost of processing the data as the personnel need to organize the raw materials for further communications. Attempts are being made by some to use a construction classification system (CCS), such as the MasterFormat [12], UniFormat [13], and OmniClass [12], to lower the extra cost of data processing. CCS usually uses layers of coding to categorize various materials and adds narrative descriptions. This system allows all team members to use the storage efficiently and use retrieval mechanism codes in the system to reference specific parts of any document, reducing the cost [14]. However, even users who are working on the same project will have different CCS systems according to their various roles in the project. It is not feasible to make all users familiar with the CCS systems used by each other: the learning cost is too high. Not all of the team members are familiar with these specifications, which may cause the inputted data to remain unstructured.

In the field of construction engineering, there is also the problem of dealing with the unstructured data [15–17]. Unstructured data mean the data set is not stored in the structured format in the database; an alternative definition may be that the data do not follow a predefined data model composition. This makes the data irregular, ambiguous, and difficult to understand using traditional computer programs [18]. Many documents are generated in a construction project, including images and text. The text-based unstructured materials include contract templates, construction specifications, quality documents, and material management documents [17]. Among them, only 20% of the available data are structured and stored in a relational database, while approximately 80% are unstructured text and stored in various forms of documents [19]. Traditionally, unstructured data are expected to be converted to structured data by manual work. However, acquiring knowledge from unstructured data is usually painful and expensive [20]. Therefore, several studies have been conducted to determine a simpler and cheaper means of retrieving useful information from unstructured data.

2.2. Unstructured Data Processing

Unstructured data have been considered a critical challenge in data management for decades. Several studies have been conducted on transferring the unstructured data into useful information in many fields. For instance, Kharrazi et al. used natural language processing (NLP) to solve the problem of unstructured electronic health records [21]. Luo et al. used the knowledge from data specialists and computer data modules to extract structured data from unstructured medical records [22]. In the business field, Farhadloo et al. attempted to discover the relative importance of each service or unique product using the Bayesian method for a customer review system [23]. These investigators used an online analytical processing (OLAP) system to analyze unstructured data from multiple perspectives, including text mining (TM), information retrieval (IR), and information extraction (IE), in an attempt to extract business intelligence from unstructured data.

In the field of construction engineering, researchers have also tried to solve the problem of unstructured data. In the study of [16], a view-based method was used, with metadata models to convert documents to structured data. Alsubaey et al. presented a Naïve Bayes text mining approach to identify early warnings of failure from meeting records [24]. Kim and Chi developed a system based on natural language processing (NLP) to extract hidden knowledge from construction accident cases [25]. Even though studies have been conducted that address unstructured data, none of these results can solve the problem that the CCS faces.

The issue for the CCS is in the material codes and in challenges in providing accurate description of materials. For materials in the CCS, there is a coding system for specifications, and the CCS uses specific terms to describe the specifications. However, in actuality, not everyone can master the coding system and become familiar with these terms. People use unprofessional terms in construction files at work, and it works fine as these terms are readable. Even though the codes and descriptions in construction files are invalid in the CCS, these construction files contain data with coding errors yet proper descriptions of unstructured data.

Machine learning is a popular solution to convert unstructured data into structured data. Machine learning is flourishing due to the improvement of hardware computing power, the reduction of data storage costs, and the innovation of various algorithms.

With the advancement of machine learning, various machine learning algorithms enable the model to learn from data, which makes computers able to handle more and more tasks. After training the model with a large number of examples, classifying the data or predicting the model training, the model can extract information from samples to learn and can complete specific tasks or prediction. These abilities allow a computer to complete specific tasks or make predictions for a variety of applications. Machine learning can gradually replace human resources in specific tasks, such as in autonomous vehicles [26], voice recognition [27], weather prediction [28], face recognition [29], lie detection [7], image processing [30], etc.

2.3. Machine-Learning-Based Methods For Construction Data Processing

A typical construction project may have thousands of outstanding issues. An artificial intelligence program that can help humans systematize these problems and the data accompanying them would greatly improve the efficiency of work. There are many things that artificial intelligence can accomplish. For images, in the unstructured data processing of images, some algorithms have performed classification of pictures directly [31], and some studies have algorithms that give these images a text caption, then these images became text data [32]. For text, Wu et al. used natural language analysis to automatically extract keyword lists from pathological examination reports [33]. Nandhakumar et al. used the characteristics of words or sentences and conditional random field (CRF) models to extract important parts of medical reports [34]. The methods described above are attempts made to structure data.

3. ACS Methodology

3.1. System Overview

This study proposes an automatic correction system (ACS) that can correct the data automatically or provide a recommended list to users for manual correction. The proposed system includes three primary modules: a data processing module, a search processing module, and a mapping processing module. Figure 2 shows the system overview of the ACS. The data processing module processes the raw data collected from different resources and then stores them in the result database. The main job of the data processing module is to train a word embedding model and establish a result data set. The search processing module processes the target data input by the user and then puts them into the word embedding model to find data with higher similarity. The data processing module and the search processing module both use the same text processor [35]. The following sections will describe each module sequentially.

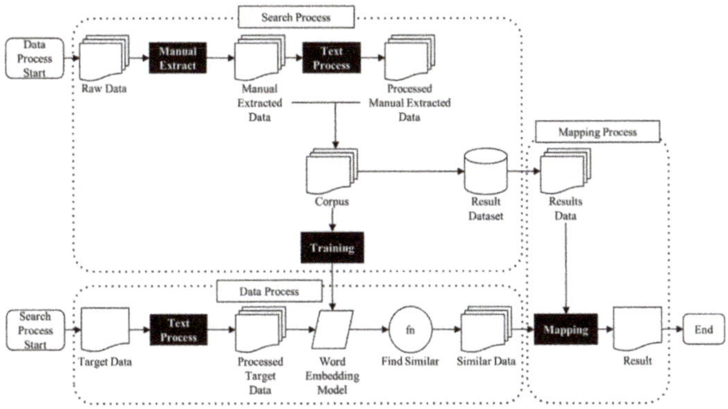

Figure 2. The designed system structures for the automatic correction system (ACS).

3.2. *Data Processing Module*

The data processing module processes the raw data then uses the processed raw data to train a word embedding model. This study used the corrected blank valuations and the CCS system manual as the raw data. The data processing module uses the text processor to normalize the data that were manually extracted from the raw data. These processed data will then be inserted into a corpus. Subsequently, correct code, correct name, correct name's segmentation, and original name's segmentation are extracted from the corpus as the result data and stored in the result database. At the end of this module, the corpus is taken as the input for training the word embedding model. Figure 3 shows the system overview of the ACS.

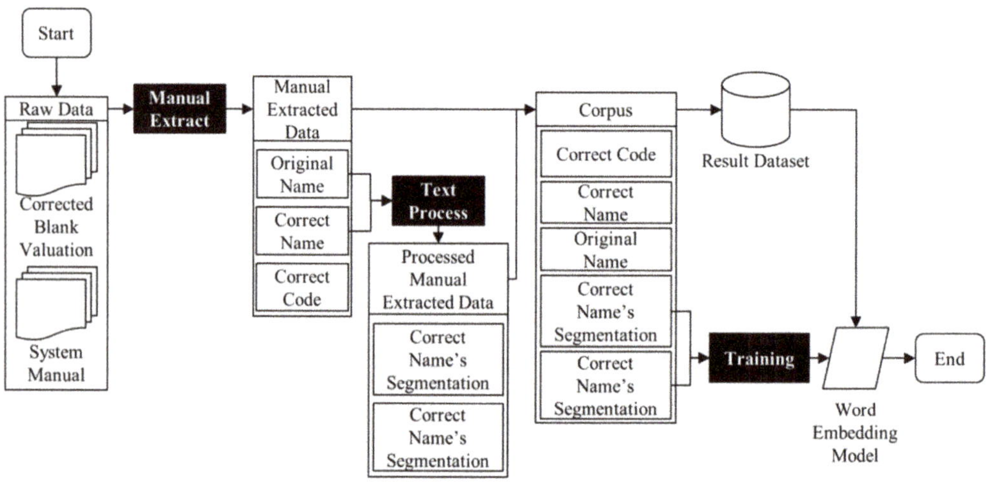

Figure 3. The workflow of the ACS's data processing module.

3.2.1. Raw Data Collection

For the raw data collection, the system manuals and blank valuations were collected that related to a specific CCS. The ACS has a particular use in correcting a public database associated with one specific CCS, and the performance of the ACS depends on the quality of the collected data. System descriptive documents were collected from the institutions that manage CCS, and blank valuations that use CCS were received from a private company's real cases. For the descriptive documents of the ACS, they at least contain specifications, descriptions, and coding systems. These correct codes and descriptions were extracted as a data column to two data fields in the "Manual Extracted Data," termed as "Correct Code" and "Correct Name." The inaccurate description was also extracted as a data column, termed as "Original Name".

3.2.2. Text Processor

The text preprocessing is advantageous for the subsequent classification results and can reduce the complexity of the calculation. Figure 4 shows workflow of the ACS's text processor. The text processor handles the text preprocessing and contains the following tasks: stop word removal, lowercase conversion, normalization, and tokenization. After users enter data, such as a sentence, the text processor removes stop words [36], converts uppercase and lowercase letters to be consistent, normalizes the preprocessing data obtained thus far, and finally uses tokenization to segment the data. An alias dictionary is utilized for the tokenizing. The dictionary includes synonyms, which can increase the quality of statements [35].

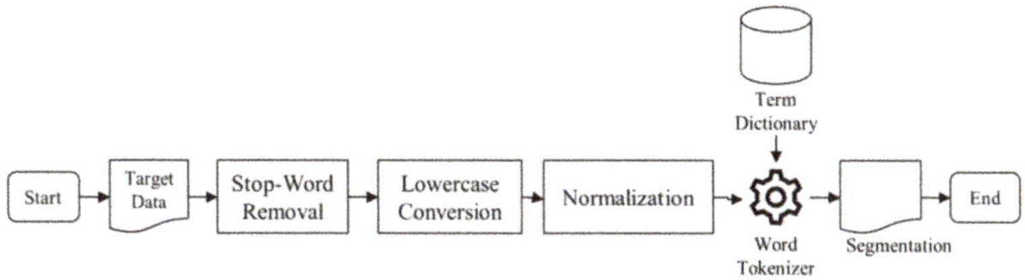

Figure 4. The workflow of the text processor.

3.3. Search Processing Module

For the search processing module, the input is the target data that the user keys in, and the output data are data with higher similarity obtained from the word embedding model. After the data processing module trains the word embedding model, the search processing module uses the text processor to handle the input target data and convert the target data into segmentations. Subsequently, the segmentation is taken as the input to the word embedding model to find similar data. The output of this module is the similar data that the model found. Figure 5 shows the system overview of the ACS.

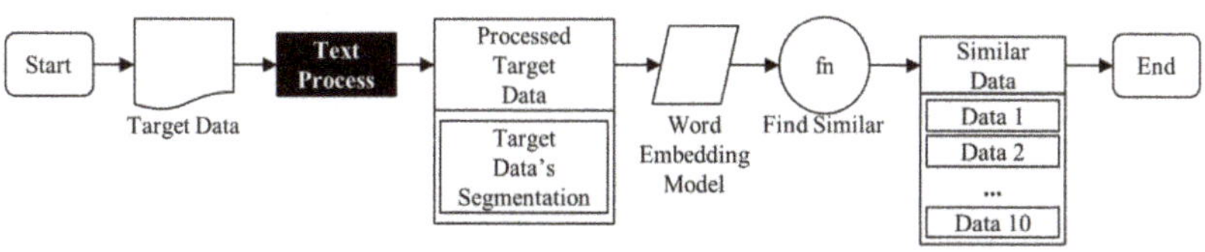

Figure 5. The workflow of the search process module.

3.3.1. Word Embedding

Word embedding is one of the most popular representations in documentation vocabulary [37]. It can capture the context, semantic, and syntactic similarity of words in documents, and the relationships with other words. Roughly speaking, it is vector representations of specific words. The program cannot directly use the text contained in an electronic file, and thus the text needs to be converted into a format that the computer can handle, with word vector representation being one of these conversion methods. A series of processing steps will be performed on the text until the text becomes a sentence or word. These sentences or words are given an independent code, and the code is a vector.

There are many ways to do word embedding, such as hashing vectorizer [38], count vectorizer [39], and Term Frequency – Inverse Document Frequency (TF-IDF) vectorizer [40]. The method we used in this system was a shallow neural network, where by learning a large amount of text, words are transformed into vectors in vector space. Then, a distribution of a large number of vectors in vector space is used to calculate similarity and find words with similar meanings. Words with the same meanings have identical representations. This representation is considered a fundamental breakthrough in machine learning for natural language processing problems [41]. Here, the characteristics of word embedding technologies are taken as the core application of the system.

Word embedding methods include dimensionality reduction of word co-occurrence matrices [42–44], probability models [45], and explicit representation of the context in which words are located [46]. In the ACS, we used the continuous bag of words (CBOW) [47] as the language model to obtain the vector matrix. Figure 6 shows the system overview of the ACS.

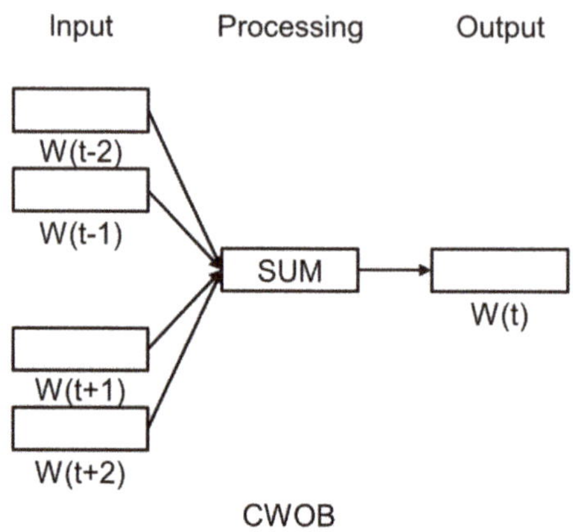

Figure 6. The architecture of the continuous bag of words (CBOW) model.

In the CBOW model, the context of surrounding words is used to predict the word in the middle. The input layer is one-hot encoding [48] and the size is equal to N. Each element in the input layer corresponds to the words in a vocabulary. Zero means no input, and a 1 means a word is input. The hidden layer is the input layer multiplied by the weight matrix ($1 \times N * N \times V = 1 \times V$). Since the input is a hot-coded vector, the hidden layer is the result of superimposing multiple rows from the weight matrix. The row index of these rows is equal to the index, with the input element being 1. Therefore, the input layer is like a lookup table for the weight matrix row search.

3.3.2. Similarity Calculation

The ACS uses text similarity to correct the data. For example, there are data that have been classified as wrong data, since the description is not sufficiently accurate even if it is semantically close. The system can replace the incorrect data with the most similar data if they can be found from the model.

The ACS deals with the text that users entered as a result and then inserts the results into the word embedding model. The result is the segmentation of the text. The model will give a vector to the result, which can represent the result in the model's vector space. Then, this study uses the cosine similarity [49–51] to calculate the similarity of vectors. The ACS takes the results in the result database, inputs them into the model to obtain vectors, and then uses the vectors from the results and the vectors from user model entries. After the vectors are input to the model, the ACS calculates the normalized dot-product from the cosine angle. Given two data a and b represented as two vectors V_a and V_b, the cosine similarity can be calculated as Equation (1).

$$cosine(V_a,\ V_b) = \frac{\sum_{i=1}^{N} V_a \cdot V_b}{\sqrt{\sum_{i=1}^{N} {V_{ai}}^2}\sqrt{\sum_{i=1}^{N} {V_{bi}}^2}}. \quad (1)$$

V_a is the frequency of each word in the user statement after being disassembled. V_b is the frequency of each word in the corpus statement after being disassembled. The $cosine(V_a,\ V_b)$ can represent the similarity of these two vectors. The closer the angle is to zero degrees, the more similar the two vectors are.

3.4. Mapping Process Module

The mapping process module finds the correct code from the result data set depending on the similar data from the search processing module. The system obtains a data set and data after finishing the above two modules. The result database contains the data processed by the text processor in the data processing module. The data in the database are similar to the target data processed by the search processing module. The mapping process module will pick the data from the similar data one by one as keywords to query the result database. If it finds the data that contain the segmentation that 100% matches the keyword, it takes the correct code from the data as the output. Figure 7 shows the system overview of the ACS.

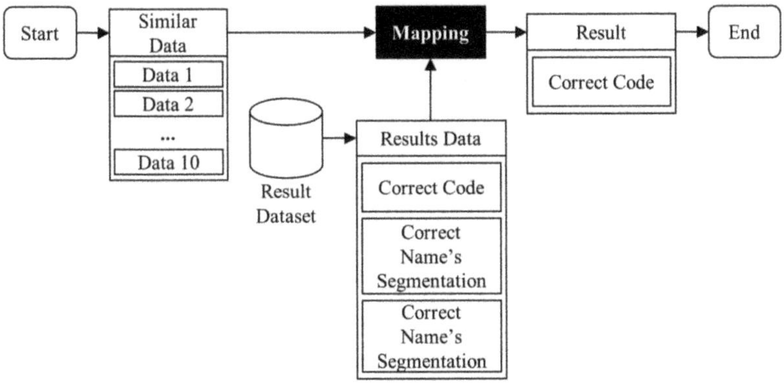

Figure 7. The workflow of the mapping process module.

4. Implementation

This research used the PCCES data as the training data to implement the developed system. The following subsections will describe the training data and the implementation of the ACS's three modules.

4.1. Training Data

For data training, in this study two types of data were collected for training the word embedding model: the PCCES manual and manually corrected actual project data. The latter was used, as these data were more in line with real work scenarios.

4.1.1. PCCES System Manuals

In the PCCES manuals, there are many of the specification codes and instructions. Anyone who wants to use the PCCES to estimate the project budget needs to know how to use the PCCES manuals. People need to be trained to know which objects they want to evaluate and where to place the objects, such as human resources, machines, materials, methods, and the environment. Then, they need to select the correct manual chapter. The specification code and description contain different layers, and thus a code and specification are needed for picking objects layer by layer. Finally, after combining these picked codes and specifications, the data processing is completed with 100% accuracy.

One-hundred percent correct information was obtained from the manual. There are 18 chapters in PCCES manuals, which cover the tender documents and contract items such as general requirements and fieldwork. This study only selected concrete chapters for implementations.

In a chapter there are multiple sections, and each section has a code designed for the name, specification, and unit of the material. Permutations were used to generate all possible data. The grid shown in Table 1 was used to choose one code from each column and combine them as a specification code with a description. For example, the code "03330/4/2/0/0/2" means "Building Concrete/Ready-Mixed Underwater Concrete/140 kgf/cm^2/M^2." The exhaustive method in the table was used to list all specification codes, with these specification codes being one of the training data for the word embedding model.

Table 1. The example of the PCCES manuals.

Chapter No.	Chapter Name	Class (6th)	Compressive Strength (7th)	Cement Type (8th)	Chemical Admixture (9th)	Valuation Unit (10th)
03330	Building Concrete	(0)	(0)	(0)	(0)	M (1)
		Machine-mixed (1)	80 kgf/cm^2 (1)	-	-	M2 (2)
		Ready-mixed (2)	140 kgf/cm^2 (2)	-	-	3 M (3)
		Machine-mixed underwater (3)	175 kgf/cm^2 (3)	-	-	Lump (4)
		Ready-mixed underwater (4)	280 kgf/cm^2 (4)	-	-	T (5)
		Ready-mixed, under 10F (5)	315 kgf/cm^2 (5)	-	-	Piece (6)
		Ready-mixed, under 20F (6)	350 kgf/cm^2 (6)	-	-	Each (7)
		Ready-mixed, under 30F (7)	400 kgf/cm^2 (7)	-	-	Set (8)
		-	-	-	-	KG (9)

4.1.2. Blank Valuations

Real project documents were obtained from two companies, Knowledge Analysis Space Exploration, Inc., Taipei, Taiwan and United Geotech, Inc., Taipei, Taiwan, which were undertaking public works, and these documents were one of the word embedding model training data sets. Each construction project contained a large number of documents and data files. The desired files in these documents and files were the price valuation files. The price valuation file contained names, terms, and aliases of objects or materials in the construction files. In actuality, the price valuation file may contain vague information that people can understand while the PCCES cannot. The system is unable to process the information and needs humans to correct it for the system to understand. Those corrected files were the desired files for collection as they contained two types of information: vague information and manually structured data. These two types of information can point to each other. Table 2 lists the example of the valuation we collected.

Table 2. The example of corrected blank valuation.

Wrong Data		Correct Data	
Work Item/Material	Code	Work Item/Material	Code
Technician	L00000520A5	Senior Worker	L000006200001
Unskilled Laborer	L0000061005	Junior Worker	L000006100001
Plastic Road Marking	M02898A003	Product, Road Marking, Glass Ball	M0289801009
Reflective Glass Ball	M02898B003	Road Marking, Glass Ball	02898B0009
Adhesive	M02900C000F	Product, Road Marking Adhesive	M0289800019
Equipment Fee	E0512450001	Not Classified Machinery	E000001000004
Tool Wear	W0127120004	Tool Wear	W0127120004

4.2. System Implementation

The ACS can be divided into four parts: the text processor, the database, the model training, and the searching and mapping. In this section, the implementation of these four parts is introduced.

4.2.1. Text Processing

The text processor used in the ACS was developed and used in our previous work [35]. In the previous work, a proper term dictionary was already constructed to improve the quality of data segmentation. This dictionary was used to develop a text processor, which could be used to handle any data imported into the system. Since the PCCES is written in Chinese, we implemented specific Chinese text processors, especially for Taiwan's construction field.

We constructed a dictionary for terms in Taiwan's construction field to solve these issues. This dictionary was used in the text processor to normalize the data. The text processor unified inconsistent units in Chinese, removed and replaced Chinese symbols and stop words, and replaced unit symbols with the real letter or number. The results are shown in Table 3. For example, if the input text was "Building Concrete and Ready-Mixed Underwater Concrete/140 kgf/cm^2/m^2," the output would be "Building Concrete/Ready-Mixed Underwater Concrete/140 kgf/cm^2/M^2."

Table 3. The characteristics that this study processed in the text processor.

	Before	Action
Stop words	"." (period)	Remove
Symbols	"m^2" (square meter)	Replace with "m^2"
Unnecessary prepositions	"and", "or", "included"	Remove
Full width character	"(" (left parenthesis), " 。" (period)	Remove
Unify units	"3000psi"	Replace with "210 kgf/cm^2"

4.2.2. Database

Specific information was extracted from the raw data and stored in a database for use in training the word embedding model. For the training data, there were already the PCCES manuals and blank valuations. However, these data were raw data that could not be used to train the word embedding model directly, and thus there was a need to first preprocess these data. One-hundred percent correct data were already generated from PCCES manuals. Furthermore, there were the original and corrected blank valuations. The 100% accurate data had two data types: specification code and specification description. For the fixed blank valuations, there were four data types: correct specification code, accurate specification description, invalid specification code, and invalid specification description. These four different data types were combined as four data table fields. The correct specification description and incorrect specification description were used to extend the other data fields, termed as "correct description segmentation" and "original description segmentation." This data field was used to store the result of the text preprocessing module after processing the specification text.

For example, consider three data strings. The first was from 100% correct data and was "0331043003, Building Concrete and Ready-Mixed Underwater Concrete 140 kgf/cm^2 m^3." The others were from the corrected blank valuations and were "0331000003, 140 kgf/cm^2 premix concrete" and "0331023003, Building Concrete and Ready-Mixed Concrete 140 kgf/cm^2 m^3." After segmenting the data in the text processor, the data and data segmentation were placed in the data table, as shown in Tables 4 and 5.

Table 4. The example of 100% correct data that this study made from the PCCES manuals.

Field Name	Value
Correct_Code	0331043003
Correct_Desc	Building Concrete and Ready-Mixed Underwater Concrete 140 kgf/cm^{2}m^2
Original_Code	N/A
Original_Desc	N/A
Correct_Desc_Seg	"building concrete," "ready-mixed," "underwater," "concrete," "140 kgf/cm^3,""m^3"
Original_Desc_Seg	N/A

Table 5. The example of the corrected blank valuation that this study obtained.

Field Name	Value
Correct_Code	0331023003
Correct_Desc	Building Concrete and Ready-Mixed Concrete 140 kgf/cm^{2}m^2
Original_Code	331000003
Original_Desc	140 kgf/cm^2 premix concrete
Correct_Desc_Seg	"building concrete," "ready-mixed," "concrete," "140 kgf/cm^3,""m^3"
Original_Desc_Seg	"140 kgf/cm^2," "ready-mixed"

4.2.3. Model Training

In this study, data segmentation and the CBOW were used to train the model. The segmentation was stored in the previously constructed database. The language model used was CBOW, which was developed by Tomas Mikolov [47]. Because the scope of application of this research was for a small field, the use of CBOW was enough for the goal of solving this research, and compared with other models such as BERT [52] and GPT-2 [53], the cost of CBOW was low, due to reduced hardware requirements, reduced data volume, and more constant training time, so this study chose CBOW. The CBOW language model has been implemented in the gensim package provided in Python. This study used the gensim package to train the CBOW model.

In the database, the correct description segmentation and original description segmentation were already present. The segmentation was combined as a huge list that contained 18,513 data rows. This list was the input of the CBOW, and the output was the word embedding model. The listing method is shown in Table 6.

Table 6. The list of segmentation that this study used to train the word embedding model.

Index	Segmentation List
1	0331023003
2	"building concrete," "ready-mixed," "underwater," "concrete," "140 kgf/cm^3,""m^3"
...	...
n	"140 kgf/cm^2," "ready-mixed"

4.2.4. Searching

The goal of the search function was to find the 10 most similar data and return the data to the user. In the search function, after the user entered a term, sentence, or messy text, the system used a text processor to process the text entered by the user. After processing the input text using a text processor, the text processor generated a word segmentation, with the quality of the word segmentation equal to that of the training data. With these word segmentations, the training data with the correct data were extracted from the database and segmented. Then, in the word embedding model, one by one, the user input text segmentation and training data segmentation performed calculations to find the 10 most similar data.

4.2.5. Mapping

The comparison function found the correct coding and narrative from the database according to the 10 most similar words. In the search function, the system completed the search and provided the 10 most similar words to the comparison function. The type of data was word segmentation, and it could not be used directly. At this time, the comparison function was needed to restore the word segmentation to the original text narrative, which was why the segmentation was stored in the database. SQL statements were used to compose database query commands, to query 10 segmentations of similar data separately, to find the correct code and correct description, and then to display it on the user interface for the user.

5. Validation

To validate that the ACS could structuralize and correct the existing data in the PCCES, this research conducted a system evaluation to evaluate the performance of the ACS. Additionally, a user test was also conducted to test if the recommendation function of the ACS could help related personnel to correct the data with higher efficiency. Two tests were designed to verify the usability of the ACS, one for the system evaluation and one for the user test. Whether the auto-correction function was complete and feasible was first tested, followed by testing of the recommendation function.

5.1. System Evaluation

A system evaluation test was conducted to validate the accuracy of the developed auto-correction function of the ACS. This study used the PCCES as an example and used real data from two companies in Taiwan as the data sources. The following subsections will describe the data source, classification, and the results of the test.

5.1.1. Data Source

The real data was obtained from two companies in Taiwan: Knowledge Analysis Space Exploration, Inc. and United Geotech, Inc. Furthermore, we generated 10,906 pieces of 100% correct data from the PCCES manuals. Totals of 5847 and 1382 raw data points were obtained from Knowledge Analysis Space Exploration, Inc. and United Geotech, Inc., respectively. The reason that these data were desirable was that they were from actual work projects; regardless of whether the code or the description of these data were correct, the project is still working fine, which means that the semantics in the data is accurate or close to the object.

For the 100% correct data, as per the codes and specifications listed in Table 1, six layers were included in the table. The PCCES manuals were used by picking one code and one description from each layer, depending on whether the description met the target material, and then combining these components. Finally, the data that perfectly fit the PCCES manuals were obtained. The user needed to insert this material in the project documentation if there was new material used in the construction project. For example, for a premixed concrete that had no additives and a strength of 4000 psi, then Table 1 could be used to generate the code and description for this material. The code of this material would be "0333024003," and the description would be "Building Concrete, Ready-Mixed Concrete, 280 kgf/cm^2, M3."

5.1.2. Classification

The corrected data were obtained after the raw data were processed by the developed system. These corrected data were used to calculate the accuracy of the automatic construction function to validate the performance and accuracy of the ACS. The obtained data included the raw data and correct data to calculate the accuracy of the corrected data.

Furthermore, the raw data and accurate data could correspond to each other. With the correct answers, we could know whether the system modified the data correctly or not.

Four rules were used to determine whether the automatically corrected data was correct to evaluate the accuracy of the automatic correction function: (1) if the raw data, manual correction data, and system correction data were all the same, it was correct; (2) if the manual calibration data and the machine calibration data were the same, it was accurate; (3) if the original data and the machine calibration data were the same, it was correct; and (4) after finishing the comparison with the first three rules, the information that was not included in the above three rules would be manually checked.

5.1.3. Results and Discussion for System Evaluation

In this section, the test result of the auto-correction function is presented. For the auto-correction function, the above four rules were used to judge whether the corrected data were correct. After completing the comparisons, 7392 and 1532 data points were obtained based on the first and second rules, respectively. Additionally, 4268 data points were obtained based on the third rule. Then, 1025 data points were confirmed manually. A total of 14,217 correct data points were obtained after the system processed the 18,551 input data points for a 76.64% accuracy. The results are shown in Table 7.

Table 7. The results of the auto-correction function evaluation.

Total (A)	Rule 1	Rule 2	Rule 3	Rule 4	Subtotal (B)	Correct Rate (B/Ax100)
18,551	7392	1532	4268	1025	14,217	76.64%

In the automatic correction function, the system took out the most similar data from the 10 most similar data and used these data to correct the wrong data. But in the failure cases, we observed that the correct answers to some of the failure cases were actually included in the 10 most similar data, but they were not the number one answers. In addition, there were some failure cases because they did not exist in the training data, such as tremie pipe and sprayed concrete, etc., so these cases were not applicable to this system.

5.2. User Test

Besides the auto-correction function, the ACS also provided a recommendation function to help the user correct the data manually. In order to test the usability of the recommendation function, this study designed a test and invited eight actual practitioners and students from the Department of Civil and Construction Engineering to use the ACS. The subjects were asked to conduct nine tasks by using both the PCCES and ACS. The operation times of each task were recorded for further analysis.

5.2.1. Background of the Subjects

For the user test, we invited eight users who had many years of industry experience and could use the PCCES proficiently at work, as well as students who had no work experience and no experience in using the PCCES. These users tested whether people with the same background in a specific field were familiar with the PCCES differently and whether there were differences in test results. The subjects included five civil engineering students without any experience in using the PCCES, and three civil engineers with 1, 4, and 8 years' experience. The details of these users are shown in Table 8.

Table 8. The background information of the invited subjects.

Subject	Background	Experience in Using PCCES (Years)
A	Undergrad student from Civil Engineering Department	0
B	Graduate students from Civil Engineering Department	0
C	Graduate students from Civil Engineering Department	0
D	Graduate students from Civil Engineering Department	0
E	Graduate students from Civil Engineering Department	0
F	Civil Engineer	1
G	Civil Engineer	4
H	Civil Engineer	8

5.2.2. Testing Scenario

In actuality, users often use email and Microsoft Excel when they are working. The resulting electronic documents contain various kinds of information. Users need to retrieve useful information from the electronic documents and then manually extract this information from the manual, finding the correct code and specification description in the PCCES to meet the work specification.

In this test scenario, we simulated the user's process of using the PCCES to create data in accordance with the description of materials, manpower, and equipment in the Excel file. Ten raw materials were extracted from some actual work project data to simulate the real work situation (Table 9). The eight testers were asked to use the PCCES and ACS to find their correct codes and specifications. At the same time, the time when they found the code was recorded, giving a usage time, and these statistics were further used to evaluate the benefits of the ACS for its users.

Table 9. List of the 10 tasks that were used in the user test.

Task	Raw Data
T1	Structure concrete, ready-mixed, 210 kgf/cm^2
T2	Structure concrete, including placing and compacting
T3	210 kg/cm^2 ready-mixed concrete
T4	Concrete placing and compacting
T5	Structure concrete, ready-mixed, 210 kgf/cm^2, nighttime construction
T6	Structure concrete, ready-mixed, 140 kgf/cm^2, nighttime construction
T7	140 kgf/cm^2 ready-mixed concrete
T8	Structure concrete, ready-mixed, 140 kgf/cm^2, daytime construction
T9	175 kg/cm^2 ready-mixed concrete
T10	280 kg/cm^2 ready-mixed concrete

5.2.3. Results and Discussion for the User Test

In this section, the results of asking users to check the recommendation function are presented. For the recommendation function, we mainly compared whether the ACS could help users work more efficiently than the PCCES, so the operating time was examined.

For the recommendation function, the times taken by users to process 10 raw data points using the two methods of the PCCES and ACS were recorded, and the processing times of the users according to each topic were averaged, as shown in Tables 10 and 11. Table 10 shows that after repeated operation of the PCCES, all the users minimized their operating times, and this minimum value could not be lowered. Additionally, as a result, as shown in Table 11, no matter how good a user was at operating the PCCES, users had really low operating times in the ACS. The average operating times of each question in the two systems are shown in Figure 8. Comparing these two systems, even if the users were allowed to repeat the operation 10 times in the PCCES, the average working time of the PCCES was much higher than that of the ACS.

Table 10. Results of the user test with the PCCES.

User	Operation Time (Sec.)											
	T1	**T2**	**T3**	**T4**	**T5**	**T6**	**T7**	**T8**	**T9**	**T10**	**Total**	**Avg.**
Q1	62.80	59.40	72.60	61.60	65.20	37.70	31.30	113.11	35.90	52.50	592.11	59.21
Q2	74.77	64.20	52.54	43.08	51.73	33.18	29.81	37.28	29.64	27.45	443.68	44.37
Q3	102.64	50.81	68.09	33.33	59.56	44.06	31.66	33.97	32.54	36.55	493.20	49.32
Q4	62.75	85.04	72.15	45.26	53.25	45.78	41.16	40.51	41.15	32.60	519.65	51.97
Q5	69.04	87.15	51.87	42.07	53.88	49.01	37.02	42.18	35.76	41.19	509.17	50.92
Q6	83.82	71.67	49.50	56.35	51.79	47.19	45.06	39.62	33.76	29.30	518.06	51.81
Q7	43.23	37.09	30.50	31.44	33.35	29.57	35.95	28.88	28.82	28.92	327.75	32.78
Q8	86.27	97.47	99.77	63.55	67.59	46.86	49.89	52.99	52.99	39.44	646.57	64.66
Total	585.32	552.83	497.01	376.68	446.35	333.35	301.85	388.54	388.54	287.95	-	-
Avg.	73.17	69.10	62.13	47.09	55.79	41.67	37.73	48.57	48.57	35.99	-	-

Table 11. Results of the user test with the ACS.

User	Operation Time (Sec.)											
	T1	**T2**	**T3**	**T4**	**T5**	**T6**	**T7**	**T8**	**T9**	**T10**	**Total**	**Avg.**
Q1	12.15	17.90	22.11	22.72	30.81	28.11	27.25	21.60	23.02	22.30	227.97	22.80
Q2	11.14	22.83	20.38	19.20	24.52	21.78	21.29	21.96	20.44	21.37	204.91	20.49
Q3	18.30	15.12	29.43	30.68	28.01	24.92	22.82	27.51	25.14	21.39	243.32	24.33
Q4	16.76	13.09	24.14	22.76	28.61	27.13	37.92	23.62	18.42	22.33	234.78	23.48
Q5	15.61	20.36	31.42	25.92	27.95	28.04	29.01	26.26	25.32	27.97	257.86	25.79
Q6	21.34	16.82	38.53	38.23	32.84	31.97	28.09	28.72	26.69	25.01	288.24	28.82
Q7	12.01	19.42	17.52	16.43	23.14	17.75	17.57	22.18	15.68	16.66	178.36	17.84
Q8	25.80	16.63	23.10	28.34	37.57	20.23	20.33	25.06	23.69	19.50	240.25	24.03
Total	133.11	142.17	206.63	204.28	233.45	199.93	204.28	196.91	178.40	176.53	-	-
Avg.	16.64	17.77	25.83	25.54	29.18	24.99	25.54	24.61	22.30	22.07	-	-

In this testing, we found that for a continuously operating system, as the familiarity increases, the user's operating time will decrease to a point at which, finally, there is a bottleneck that prevents further reductions. The decline rate of the operating time of the PCCES was significant; however, the rate of decrease for the ACS was less obvious, and the user's operating time was almost the same. It could be seen that the recommendation function of the ACS greatly eliminated this proficiency factor. Compared with the PCCES, using the recommendation function of the ACS helped users save 54% of the operation time.

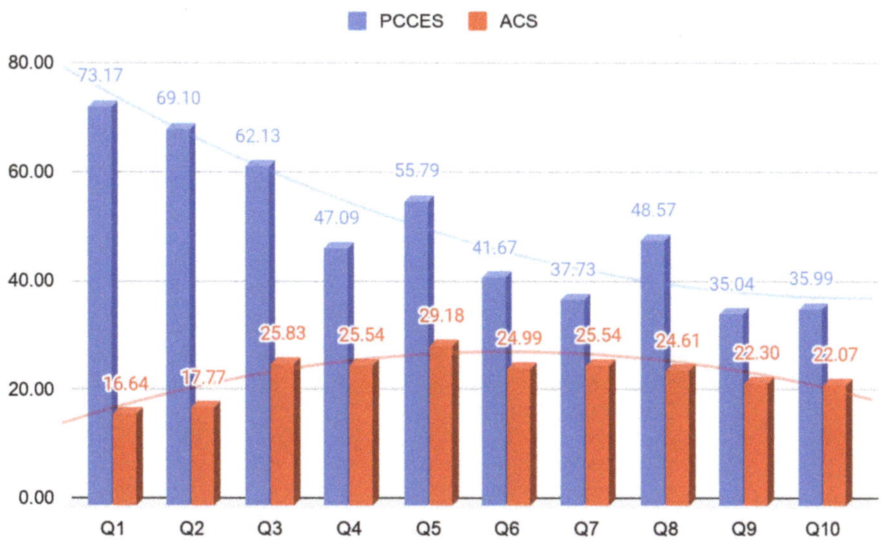

Figure 8. The average operating time for each task by eight users in two systems.

6. Discussions

The main contribution of this research is the development of a system that can automatically correct public databases in the field of construction engineering in Taiwan for a database that is full of chaotic data. Two main functions are implemented in the ACS: the auto-correction function and the recommendation function. No manual intervention is required during the processing of the auto-correction function, which can reduce the user's workload. The recommendation function allows users to input different keywords or similar words. The function will process the input text data and automatically check to find the most similar and standard data to provide to the user. The language model was trained based on the collected 18,551 data points, used to calculate the similarity between the user input text and the correct data, and then used the accurate data in the automatic correction function or provided a list of user recommendations. The auto-correction function was used to summarize 18,551 data points at an accuracy of 76%. Ten data out of the 18,551 were taken for user operation. Compared with the PCCES, using the recommendation function can help users save 54% in operation time.

6.1. Contributions

In the existing public database in Taiwan, there is a large amount of messy data, so that the Taiwan government needs to stipulate the accuracy rate of the data by law. We attempted to use the auto-correction function of the ACS to solve the problem of this messy data.

(1) The ACS can automatically correct and structuralize the untrusted construction data. Many studies have been conducted to structuralize the semi-structured data. For instance, Woo et al. used the text clustering method to clean the large-scaled medical report data [54]. Soto et al. proposed the ViTA-SSD system that allows the user to explore the insightful patterns in the semi-structured data by providing a visualized analysis method [55]. However, these methods still require lots of human interactions on results checking, correcting, and exploring. Instead, the ACS used a machine-learning-based method that can automatically correct and structuralize the construction data. It allows the ACS to improve quality and output efficiency and does not require human intervention in the processing process, which can reduce the workload of users.

Furthermore, the automation of operations to achieve the maximum benefit is a current industrial target.

(2) The ACS provides a recommendation function that can improve the related personnel's working efficiency and accuracy. As not everyone can be proficient in understanding the PCCES coding specifications, the produced data are often non-standard. Even for users who are skilled in using the PCCES, due to the operation steps of the PCCES, the operation time will encounter a bottleneck and cannot continue to be reduced. The recommendation function of the ACS is an attempt to solve the above two problems. By providing a more user-friendly experience and removing the need for familiarity of coding standards, the user's operating time and data accuracy are improved.

(3) The ACS successfully reduces the threshold of operating the data management system. It can be seen from the test of the recommendation function that, regardless of whether users have used the PCCES or not, when operating the PCCES, as the number of operations increases, the proficiency will increase, and the operation time will not continue to reduce past a certain level. This is because the PCCES code is attached and the operation interface has a lower operating time, even if the user is already skilled. The recommendation function implemented by the ACS relies on the user entering divergent keywords or similar words, such as material names, aliases, or specifications, and other obscure data. The system then provides the results that meet the PCCES specifications for the user to select. The benefit of the recommendation function is that it removes the variable proficiency levels. Regardless of whether the user has a professional background or is familiar with coding standards, he or she can rely on keywords to find the correct result.

(4) The system we developed is not limited to the PCCES, it can be applied to different CCS systems. Since the system is based on the CBOW language model, the data we collect will determine the application direction of this system. In this study, we use the data related to the PCCES system to obtain the model after training and use this model to classify and correct the wrong PCCES data. In other words, if the training data are replaced by another CCS system, this system can also be utilized on other CCS systems.

6.2. Limitations

(1) In this study, the data collected limited the scope of the ACS in the PCCES. In machine learning, data are the foundation of everything, where only by having more data can the application scope and accuracy be expanded. In this study, we only implemented one section of the PCCES coding specification. Most of the collected data belong to this category, as it is one of the most commonly used codes in this section. Furthermore, the specification description contained in this section has a description of the strength of the concrete. If this section can be improved, the coding accuracy rate will be beneficial for the PCCES as a whole.

(2) The accuracy of the ACS is not sufficient. In this study, we used 18,551 PCCES data to train the CBOW model, applied it to a small range of the PCCES, and got a 76.64% correct rate. The application range of the ACS in the PCCES depends on the data we collected, and although the accuracy of 76.64% was much higher than the 37.47% announced by PCC, it is not good enough to meet expectations. The application range of the ACS in the PCCES is not that wide. The average accuracy rate will decrease when the application range of the ACS is expanded and extended to a range that is difficult to automatically correct if the higher accuracy rate is not achieved in the application range for now. For the system that uses machine learning technology, the amount of data collected will affect the trained model and then affect its accuracy. If we can obtain more information, we may be able to improve the accuracy rate.

7. Conclusions

This research proposed a data correction system, the ACS, for automatically correcting the public construction data. The system we developed provides data auto-correction and recommendation features to improve human working performance and reduce the threshold of operating the data management system. A text classification system, the ACS, was developed using language models based on natural language processing and machine learning to correct public databases in the field of construction in Taiwan; at the same time, this system is also proposed to help users produce correct data more efficiently. By using a machine-learning-based language model to analyze text semantics and provide higher accuracy and efficiency information, the ACS can improve the efficiency of actual users and the accuracy of data in construction projects. In the automatic calibration test, a 76% accuracy rate was obtained after correcting 18,551 data points. A user test was also conducted on the recommendation function. A question was provided containing 10 real data points as well as a questionnaire to perform a user test on eight participants to observe them solving the problem under the two systems. After the trial, for the average processing time for each data point, 51.95% of the time was saved. From the test results, it was found that users using the ACS were more efficient than when using the original system and could accurately produce materials that meet the specifications. The results show that the ACS can effectively save operation time of the CCS and thus reduce the threshold of operating the data management system by providing a recommendation function. The proposed method can not only be used in the PCCES but can also be deployed to different CCS systems.

Author Contributions: Conceptualization, M.-H.T.; data curation, M.-H.T. and M.-L.Y.; methodology, M.-H.T. and M.-L.Y.; software, M.-L.Y.; validation, M.-L.Y.; writing—original draft, M.-L.Y.; writing—review and editing, M.-H.T. All authors have read and agreed to the published version of the manuscript.

Funding: This research was funded by Taiwan's Ministry of Science and Technology (MOST) under contract 109-2124-M-002-005.

Acknowledgments: We would like to thank Ming-Dar Tsai of Knowledge Analysis Space Exploration and Liang-Yuan Li of National Taiwan University of Science and Technology for providing their feedback and for their assistance in the development of the system.

Conflicts of Interest: The authors declare no conflict of interest.

References

1. Aziz, R.F.; Hafez, S.M. Applying lean thinking in construction and performance improvement. *Alex. Eng. J.* **2013**, *52*, 679–695. [CrossRef]
2. Public Construction Commission. Turnkey Cases That Already Been Awarded between 1 December 2017 and 30 June 2020. Available online: https://pcces2.pcc.gov.tw/PCC_MRP/Announcement/AnnDetail/9dec7761-dc2c-4dad-8cd4-be61fa44e5a7 (accessed on 18 November 2020). (In Chinese)
3. Public Construction Commission 2020. Public Construction Cost Estimation System. Available online: https://pcces.pcc.gov.tw/CSInew/Default.aspx?FunID=Fun_12&SearchType=E (accessed on 18 November 2020).
4. Norman, E.S.; Brotherton, S.A.; Fried, R.T. *Breakdown Structures: The Foundation for Project Management Excellence*; John Wiley and Sons, Inc.: Hoboken, NJ, USA, 2008. [CrossRef]
5. Liao, M. Study of Current State of Software for Drafting and Estimating in Construction. National Kaohsiung University. 2011. Available online: https://hdl.handle.net/11296/t7u695 (accessed on 18 November 2020).
6. Yu, S.H. Evaluation of the Public Works Procedure Efficiency of E-Procurement of Government of Kaohsiung. National Sun Yat-sen University. 2004. Available online: https://hdl.handle.net/11296/w4h5xd (accessed on 18 November 2020).
7. Davatzikos, C.; Ruparel, K.; Fan, Y.; Shen, D.G.; Acharyya, M.; Loughead, J.W.; Gur, R.C.; Langlebenb, D.D. Classifying Spatial Patterns of Brain Activity with Machine Learning Methods: Application to Lie Detection. *NeuroImage* **2005**, *28*, 663–668. [CrossRef] [PubMed]
8. Lin, C.C. Study on Budget Rationality and Supervision Practice of Construction Safety and Health. 2014. Available online: https://www.grb.gov.tw/search/planDetail?id=8390698 (accessed on 18 November 2020).

9. Chen, Y.C. Combination of Public Construction Coding System and Building Information Modeling for Budget Estimate. Master's Thesis, National Taiwan University, Taipei, Taiwan. 19 June 2013. Available online: https://hdl.handle.net/11296/khec2y (accessed on 18 November 2020).
10. Huang, C.H.; Yang, I.T.; Wang, C.Y.; Wu, M.J. A Study of Introducing Omniclass on BIM-based Building Design Checking, Architecture and Building Research Institute, Ministry of the Interior, ROC (Taiwan): Taipei, Taiwan. 2017. Available online: https://www.grb.gov.tw/search/planDetail?id=12066805 (accessed on 18 November 2020).
11. Caldas, C.H.; Soibelman, L. Automating hierarchical document classification for construction management information systems. *Autom. Constr.* **2003**, *12*, 395–406. [CrossRef]
12. Construction Specifications Institute. CSI MasterFormat. 2008. Available online: https://www.csiresources.org/home (accessed on 18 November 2020).
13. Charette, R.P.; Marshall, H.E. *UNIFORMAT II Elemental Classification for Building Specifications, Cost Estimating and Cost Analysis*; U.S. Department of Commerce: Washington, DC, USA, 1999; NIST Interagency or Internal Reports (NISTIR) 6389.
14. Ioannou, P.G.; Liu, L.Y. Advanced Construction Technology System—ACTS. *J. Constr. Eng. Manag.* **1993**, *119*, 288–306. [CrossRef]
15. Russell, A.D.; Chiu, C.Y.; Korde, T. Visual Representation of Construction Management Data. *Autom. Constr.* **2009**, *18*, 1045–1062. [CrossRef]
16. Mao, W.; Zhu, Y.; Ahmad, I. Applying Metadata Models to Unstructured Content of Construction Documents: A View-Based Approach. *Autom. Constr.* **2007**, *16*, 242–252. [CrossRef]
17. Soibelman, L.; Wu, J.; Caldas, C.; Brilakis, I.; Lin, K.Y. Management and Analysis of Unstructured Construction Data Types. *Adv. Eng. Inform.* **2008**, *22*, 15–27. [CrossRef]
18. Sint, R.; Schaffert, S.; Stroka, S.; Ferstl, R. Combining unstructured, fully structured and semi-structured information in Semantic Wikis. In *CEUR Workshop Proceedings*; Heraklion: Crete, Greece, 2009; pp. 73–87.
19. Chen, H.M.; Schütz, R.; Kazman, R.; Matthes, F. Amazon in the Air: Innovating with Big Data at Lufthansa. In Proceedings of the 2016 49th Hawaii International Conference on System Sciences (HICSS), Koloa, HI, USA, 5–8 January 2016; pp. 5096–5105. [CrossRef]
20. Rusu, O.; Halcu, I.; Grigoriu, O.; Neculoiu, G.; Sandulescu, V.; Marinescu, M.; Marinescu, V. Converting Unstructured and Semi-Structured Data into Knowledge. In Proceedings of the 2013 11th RoEduNet International Conference, Sinaia, Romania, 17–19 January 2013; pp. 1–4. [CrossRef]
21. Kharrazi, H.; Anzaldi, L.J.; Hernandez, L.; Davison, A.; Boyd, C.M.; Leff, B.; Kimura, J.; Weiner, J.P. The Value of Unstructured Electronic Health Record Data in Geriatric Syndrome Case Identification. *J. Am. Geriatr. Soc.* **2018**, *66*, 1499–1507. [CrossRef]
22. Luo, L.; Li, L.; Hu, J.; Wang, X.; Hou, B.; Zhang, T.; Zhao, L.P. A Hybrid Solution for Extracting Structured Medical Information from Unstructured Data in Medical Records via a Double-Reading/Entry System. *BMC Med. Inform. Decis. Mak.* **2016**, *16*, 114. [CrossRef]
23. Farhadloo, M.; Patterson, R.A.; Rolland, E. Modeling Customer Satisfaction from Unstructured Data Using a Bayesian Approach. *Decis. Support Syst.* **2016**, *90*, 1–11. [CrossRef]
24. Alsubaey, M.; Asadi, A.; Makatsoris, H. A Naïve Bayes Approach for EWS Detection by Text Mining of Unstructured Data: A Construction Project Case. In Proceedings of the IntelliSys 2015—Proceedings of 2015 SAI Intelligent Systems Conference, London, UK, 10–11 November 2015; pp. 164–168. [CrossRef]
25. Kim, T.; Chi, S. Accident Case Retrieval and analyses: Using natural language processing in the construction industry. *J. Constr. Eng. Manag.* **2019**, *145*, 04019004. [CrossRef]
26. Navarro, P.J.; Fernandez, C.; Borraz, R.; Alonso, D. A Machine Learning Approach to Pedestrian Detection for Autonomous Vehicles Using High-Definition 3D Range Data. *Sensors* **2016**, *17*, 18. [CrossRef]
27. Sainath, T.N.; Weiss, R.J.; Senior, A.; Wilson, K.W.; Vinyals, O. Learning the Speech Front-End with Raw Waveform CLDNNs. In Proceedings of the Annual Conference of the International Speech Communication Association, INTERSPEECH, Dresden, Germany, 6–10 September 2015; pp. 1–5. Available online: https://www.isca-speech.org/archive/interspeech_2015/i15_0001.html (accessed on 18 November 2020).
28. Krasnopolsky, V.M.; Fox-Rabinovitz, M.S. Complex Hybrid Models Combining Deterministic and Machine Learning Components for Numerical Climate Modeling and Weather Prediction. *Neural Netw.* **2006**, *19*, 122–134. [CrossRef]
29. Sharif, M.; Bhagavatula, S.; Bauer, L.; Reiter, M.K. Accessorize to a crime: Real and stealthy attacks on state-of-the-art face recognition. In *Proceedings of the 2016 ACM SIGSAC Conference on Computer and Communications Security*; Association for Computing Machinery: New York, NY, USA, 2016; pp. 1528–1540. [CrossRef]
30. Krizhevsky, A.; Sutskever, I.; Hinton, G.E. ImageNet Classification with Deep Convolutional Neural Networks. *Commun. ACM* **2017**, *60*, 84–90. [CrossRef]
31. Lu, D.; Weng, Q. A Survey of Image Classification Methods and Techniques for Improving Classification Performance. *Int. J. Remote Sens.* **2007**, *28*, 823–870. [CrossRef]
32. Lu, J.; Yang, J.; Batra, D.; Parikh, D. Neural Baby Talk. In Proceedings of the 2018 IEEE/CVF Conference on Computer Vision and Pattern Recognition, Salt Lake City, UT, USA, 18–23 June 2018; pp. 7219–7228. [CrossRef]
33. Wu, P.H.; Yu, A.; Tsai, C.W.; Koh, J.L.; Kuo, C.C.; Chen, A.L.P. Keyword Extraction and Structuralization of Medical Reports. *Health Inf. Sci. Syst.* **2020**, *8*, 18. [CrossRef]

34. Nandhakumar, N.; Sherkat, E.; Milios, E.E.; Gu, H.; Butler, M. Clinically Significant Information Extraction from Radiology Reports. In Proceedings of the 2017 ACM Symposium on Document Engineering—DocEng '17, Valletta, Malta, 4–7 September 2017; pp. 153–162. [CrossRef]
35. Yu, M.L.; Chan, H.Y.; Tsai, M.H. NLP-Based Method for Auto-Correcting Public Constructions Data. In Proceedings of the 2019 4th International Conference on Civil and Building Engineering Informatics, Sendai, Miyagi, Japan, 6–9 November 2019.
36. Hao, L.; Hao, L. Automatic Identification of Stop Words in Chinese Text Classification. In Proceedings of the 2008 International Conference on Computer Science and Software Engineering, IEEE, Wuhan, China, 12–14 December 2008; pp. 718–722. [CrossRef]
37. Yin, Z.; Shen, Y. On the Dimensionality of Word Embedding. In Proceedings of the Advances in Neural Information Processing Systems, Montreal, Canada, 2–8 December 2018; pp. 887–898. Available online: http://papers.nips.cc/paper/7368-on-the-dimensionality-of-word-embedd (accessed on 21 December 2020).
38. Hasan, M.; Islam, I.; Hasan, K.M.A. Sentiment Analysis Using Out of Core Learning. In Proceedings of the 2019 International Conference on Electrical, Computer and Communication Engineering (ECCE), IEEE, Cox's Bazar, Bangladesh, 7–9 February 2019; pp. 1–6. [CrossRef]
39. Kulkarni, A.; Shivananda, A. Converting Text to Features. In *Natural Language Processing Recipes*; Apress: Berkeley, CA, USA, 2019; pp. 67–96. [CrossRef]
40. IEEE Photonics Technology Letters Information for Authors. *IEEE Photonics Technol. Lett.* **2009**, *21*, C3. [CrossRef]
41. Li, W.; Zhu, L.; Guo, K.; Shi, Y.; Zheng, Y. Build a Tourism-Specific Sentiment Lexicon via Word2vec. *Ann. Data Sci.* **2018**, *5*, 1–7. [CrossRef]
42. Lebret, R.; Collobert, R. Word Emdeddings through Hellinger PCA. In Proceedings of the 14th Conference of the European Chapter of the Association for Computational Linguistics Gothenburg, Gothenburg, Sweden, 26–30 April 2014; Association for Computational Linguistics: Stroudsburg, PA, USA; pp. 482–490. [CrossRef]
43. Levy, O.; Goldberg, Y. Neural Word Embedding as Implicit Matrix Factorization. In *Advances in Neural Information Processing Systems*; Mit Press: Montreal, QC, Canada, 2014; pp. 2177–2185. Available online: http://papers.nips.cc/paper/5477-neural-word-embedding-as (accessed on 21 December 2020).
44. Li, Y.; Xu, L.; Tian, F.; Jiang, L.; Zhong, X.; Chen, E. Word Embedding Revisited: A New Representation Learning and Explicit Matrix Factorization Perspective. In Proceedings of the IJCAI International Joint Conference on Artificial Intelligence, Buenos Aires, Argentina, 25–31 July 2015; pp. 3650–3656. Available online: https://www.ijcai.org/Proceedings/15/Papers/513.pdf (accessed on 21 December 2020).
45. Globerson, A.; Chechik, G.; Pereira, F.; Tishby, N. Euclidean Embedding of Co-Occurrence Data. *J. Mach. Learn. Res.* **2007**, *8*, 2265–2295. Available online: https://www.jmlr.org/papers/v8/globerson07a.html (accessed on 21 December 2020).
46. Qureshi, M.A.; Greene, D. EVE: Explainable Vector Based Embedding Technique Using Wikipedia. *J. Intell. Inf. Syst.* **2019**, *53*, 137–165. [CrossRef]
47. Mikolov, T.; Chen, K.; Corrado, G.; Dean, J. Efficient Estimation of Word Representations in Vector Space. In Proceedings of the 1st International Conference on Learning Representations, ICLR 2013—Workshop Track Proceedings, Scottsdale, AZ, USA, 2–4 May 2013; Available online: http://arxiv.org/abs/1301.3781 (accessed on 21 December 2020).
48. Hatamian, M.; Hung, D.; Kurmas, Z.; Frenzel, J.; Pinter-Lucke, J.; and Zhao, P. In Praise of Digital Design and Computer Architecture. In *Digital Design and Computer Architecture*; Morgan Kaufmann: Burlington, MA, USA, 2016; pp. i–ii. [CrossRef]
49. Sidorov, G.; Gelbukh, A.; Gómez-Adorno, H.; Pinto, D. Soft Similarity and Soft Cosine Measure: Similarity of Features in Vector Space Model. *Comput. Sist.* **2014**, *18*, 491–504. [CrossRef]
50. Basu, T.; Murthy, C.D. Effective Text Classification by a Supervised Feature Selection Approach. In Proceedings of the 2012 IEEE 12th International Conference on Data Mining Workshops, Brussels, Belgium, 10 December 2012; pp. 918–925. [CrossRef]
51. Kim, K.; Chung, B.S.; Choi, Y.; Lee, S.; Jung, J.Y.; Park, J. Language Independent Semantic Kernels for Short-Text Classification. *Expert Syst. Appl.* **2014**, *41*, 735–743. [CrossRef]
52. Devlin, J.; Chang, M.W.; Lee, K.; Toutanova, K. BERT: Pre-Training of Deep Bidirectional Transformers for Language Understanding. In Proceedings of the NAACL HLT 2019—2019 Conference of the North American Chapter of the Association for Computational Linguistics: Human Language Technologies—Proceedings of the Conference 1 (October), Minneapolis, MN, USA, 2–7 June 2019; Available online: http://arxiv.org/abs/1810.04805 (accessed on 21 December 2020).
53. Radford, A.; Wu, J.; Child, R.; Luan, D.; Amodei, D.; Sutskever, I. Language Models Are Unsupervised Multitask Learners. *OpenAI Blog* **2020**, *1*, 9.
54. Woo, H.; Kim, K.; Cha, K.; Lee, J.Y.; Mun, H.; Cho, S.J.; Chung, J.I.; Pyo, J.H.; Lee, K.C.; Kang, M. Application of Efficient Data Cleaning Using Text Clustering for Semistructured Medical Reports to Large-Scale Stool Examination Reports: Methodology Study. *J. Med. Internet Res.* **2019**, *21*, e10013. [CrossRef]
55. Soto, A.J.; Kiros, R.; Kešelj, V.; Milios, E. Exploratory visual analysis and interactive pattern extraction from semi-structured data. *ACM Trans. Interact. Intell. Syst.* **2015**, *5*, 1–36. [CrossRef]

sustainability

MDPI

Article

Influence Mechanism of Organizational Flexibility on Enterprise Competitiveness: The Mediating Role of Organizational Innovation

Guodong Ni [1,2,*], Heng Xu [1], Qingbin Cui [3], Yaning Qiao [1,2], Ziyao Zhang [1], Huaikun Li [1] and Paul J. Hickey [3]

1 School of Mechanics and Civil Engineering, China University of Mining and Technology, Xuzhou 221116, China; TS18030227A31@cumt.edu.cn (H.X.); yaning.qiao@cumt.edu.cn (Y.Q.); zhangziyao@cumt.edu.cn (Z.Z.); lihuaikun@cumt.edu.cn (H.L.)
2 Research Center for Digitalized Construction and Knowledge Engineering, School of Mechanics and Civil Engineering, China University of Mining and Technology, Xuzhou 221116, China
3 Department of Civil and Environmental Engineering, A. James Clark School of Engineering, University of Maryland, College Park, MD 20742, USA; cui@umd.edu (Q.C.); pauljhickey@verizon.net (P.J.H.)
* Correspondence: niguodong@cumt.edu.cn

Abstract: Organizational success heavily relies on the competitiveness of products and services under rapidly changing market conditions. This enterprise competitiveness becomes more critical for project-based enterprises as modernization of the Chinese construction industry creates greater challenges and uncertainty in construction operations, which determines the sustainable advantages of enterprises to a certain degree. Traditional wisdom focuses on cost efficiency, asset differentiation, and service performance to gain competitive advantages. This paper explores the influence of organizational flexibility and organizational innovation on enterprise competitiveness for Chinese construction organizations. A designed structured questionnaire was developed and conducted targeting the project-based enterprises in China's construction industry and is accompanied by a structural equation modeling analysis. Results indicate a positive impact of organizational flexibility on enterprise competitiveness along with a mediation role of organizational innovation. The study concludes that new organizational strategies are required for Chinese project-based enterprises to maintain enterprise competitiveness in order to realize the sustainable development of enterprises.

Keywords: enterprise competitiveness; organizational flexibility; organizational innovation; modernization of construction industry; structural equation modeling

Citation: Ni, G.; Xu, H.; Cui, Q.; Qiao, Y.; Zhang, Z.; Li, H.; Hickey, P.J. Influence Mechanism of Organizational Flexibility on Enterprise Competitiveness: The Mediating Role of Organizational Innovation. *Sustainability* **2021**, *13*, 176. https://dx.doi.org/10.3390/su13010176

Received: 8 November 2020
Accepted: 23 December 2020
Published: 27 December 2020

1. Introduction

In today's competitive business environment, how to enhance enterprise competitiveness has been a hot topic [1]. Existing studies show that strengthening leadership [2], establishing good relationships with partners [3], technological renovations [4], and capital investment [5] are all effective for the promotion of enterprise competitiveness. However, due to fast technological advancement, increased competition, market fluctuations [6], and even the impact of infectious diseases such as COVID-19, traditional measures are struggling to maintain competitiveness effectively for enterprises in turbulent and complex business environments. Flexibility plays an important role in keeping firms alive and prospering in changing market environments [7]. Thanks to flexibility, firms can react to environmental changes and uncertainties more quickly: They can absorb change, integrate, develop, and restructure resources and capabilities in the organization in a short time [8]. Besides, improving flexibility is considered a crucial tool that can predict organizational performance [9]. Therefore, enhancing flexibility serves as a feasible solution for promotion of enterprise competitiveness in a turbulent environment. Although there are sporadic clues to indicate that strengthening flexibility inside an organization is conducive to enterprise survival and development, to our knowledge, few empirical studies connect organizational flexibility with competitiveness, which turns out to be a barrier for further understanding of the competitiveness promotion mechanism under the flexibility perspective.

Generally, it takes a relatively long time to complete a construction project. The longer the project duration, the greater the risk and uncertainties, such as increases in the price of labor and material, payments delays, safety, quality, and poor estimating [10]. Long project duration brings more uncertainties and dynamic risks and makes it even more difficult to manage project results and profitability for Project-Based Enterprises (PBEs) in the construction industry, which generally includes construction, design, engineering management, real estate development, and consulting firms [11,12]. Thus, on the one hand, it is more difficult to remain flexible for construction industry PBEs compared with the manufacturing company. On the other hand, keeping flexibility would make more sense for construction industry PBEs. Once construction industry PBEs are more flexible, they can prevent risks and loss or seize the opportunity caused by uncertainty and changes in advance via rapid and active response, which can not only improve their profitability but also enable them to gain a stronger market competitive advantage.

For Chinese construction industry PBEs, the advantage of flexibility in facing uncertainties and promoting enterprise competitiveness is more prominent. Under the trend of deep integration into the global economy, with China's proposal of the "Belt and Road" initiative in 2013, the construction industry, as a supporting industry of the national economy in China [13], has accelerated the pace of undertaking projects overseas. Obviously, faced with the uncertainties and risks in overseas markets, continuous competitive improvement holds the key to survival and development of PBEs in China's construction industry [14,15]. The sustainable development of China's construction industry concerns not only the interests of China, but also the healthy development of the whole world economy. Meanwhile, the rapid development of construction industry standards demands PBEs constantly dealing with a large number of changes in the operating environment [16]. Contributing factors include fluctuating construction demands [16], changing procurement trends [17], clients' higher performance standards for building services [10], and higher technical requirements [16]. The combined effect of these changes forces firms to improve their viability and competitiveness in the business environment [7]. Similar situations are particularly prominent in China's construction industry. Since the strategy of Modernization of Construction Industry (MCI) was first proposed by Chinese authorities in 2013, governments have focused on deepening the construction industry reform. MCI refers to the transformation and upgrade of the construction industry, taking technological innovation as a guide, modern management as support, informatization as the means, and new construction industrialization as the core, the strategy aims to renew, transform, and upgrad the whole industrial chain of construction, thus realizing the transformation from the traditional production mode to the modern industrial production mode and improving the quality, efficiency, and benefits comprehensively. MCI is the key to realizing sustainable development and upgrading the construction industry. MCI involves many aspects, such as prefabricated building [18], construction industry informatization and industrialization [19], design and construction integration, green development concept, and sustainable strategy [15,20,21]. Current development trends in construction can fundamentally change the traditional production modes of the Chinese industry but also create many uncertainties and challenges to PBEs. Therefore, maintaining the advantage in competition and avoid elimination in continuously changing and uncertain market environments emerges as an urgent issue for PBEs leaders and managers in China's construction industry.

As is well-known, innovation might be one of a few lasting sources of enterprise competitive advantage [22,23]. Promoting organizational innovation (OI) actively within the enterprise aids in coping with environmental changes and uncertainties. Organizational innovation means that an organization adopts a new idea or behavior [24], which generally includes technological environment advancements and management modernization of the organization [25]. Implementing organizational innovation activities is conducive to increase productivity and profitability as well as to expand existing market shares and exploiting new markets [26]. Organizational innovation capacity is of vital importance in shaping PBEs competitiveness in the construction industry [15]. Meanwhile, there is evidence indicating that innovation partially mediates the relationship between organizational flexibility and project portfolio performance [9]. PBEs with flexibility are likely to use various organizational resources to fully realize the benefits of technological innovation and management innovation. Therefore, organizational innovation, together with organizational flexibility, will play a critical role in enhancing the PBEs' competitiveness in the current complex situation.

However, it remains unclear whether enterprise competitiveness can be promoted effectively by organizational flexibility in China's construction industry and whether organizational innovation can play a mediating role between organizational flexibility and enterprise competitiveness. Therefore, this paper tries to figure out these two questions and aims to explore the mechanism for improving enterprise competitiveness in China's construction industry with a specific focus on organizational flexibility and organizational innovation. On the basis of theoretical analysis, this study will set up a theoretical model of the influence mechanism of organizational flexibility on the enterprise competitiveness of PBEs through putting forward the research hypotheses. Then, based on China's construction industry under the background of MCI, it will test the theoretical model and research hypotheses using the standard paradigm of empirical research, clarify the influence mechanism of organizational flexibility on enterprise competitiveness, and further verify the mediating role of organizational innovation.

2. Theoretical Background

2.1. Enterprise Competitiveness

Competitiveness can be regarded as an independent, dependent, or intermediary variable, depending on the perspective one takes in dealing with the problem. In the construction industry, the analysis of competitiveness is based on four levels: country, industry, firm, and project [2]. Among them, the firm's competitiveness provides the greatest interest for practitioners and attracts the most attention of researchers [27]. Many researchers emphasize the importance of competitiveness at the firm level [21,28]. Enterprise competitiveness for a coal firm, for example, refers to the enterprise's advantages compared with competitors in design, production, sale of products, personnel, technology, and management, considering price and non-price factors [29]. But for contractors, the firm's competitive advantage comes from its competitive strategy when facing strengths, weaknesses, opportunities, and threats, as well as its unique, irreplaceable, and inimitable resources [30]. The understandings of enterprise competitiveness are not the same in view of the differences between industry and research perspectives. It is necessary to clarify the connotation of the PBEs' competitiveness in the context of MCI.

Prior researchers tended to focus on how to measure and improve competitiveness [14,31]. According to [27], assets and processes within an organization (the measurement of competitive advantage) can be either tangible or intangible, including brand, reputation, culture, human resources, technology, market share, and profitability. In addition, Ghobadian et al. [32] proposed that identifying business opportunities such as increased market share, increased profitability, or reduced cost structure could lead to enterprise competitive advantage. In the construction sector, underlying contractor contributors can be divided into three parts: core competence, company strategy, and project performance [15]. According to [3], measurement indexes of enterprise competitiveness include firm image, financial ability, marketing ability, technical ability, management skill, and human resources advantage. Deng et al. [14] developed potential factors of the Chinese construction industry: domestic stable market, supply chain management, corporate management practices, qualified professionals, sound business climate, and migrant workers. These results indicate that the advantages in human resource, finance, market share, and management can form sources of enterprise competitiveness. In order to make the construction industry more competitive, Chinese PBEs need to not only learn from international practices but also to adjust and complete them according to the Chinese market environment [33]. Under the background of MCI, new technologies and construction requirements bring great uncertainty to project investment and construction. Strengthening the project management ability of construction engineering enterprises is crucial to ensuring the output of high-quality products and services and the stable operation and decision-making efficiency of project teams. Building on the literature review and the PBEs characteristic of the construction industry, this study establishes the measurement structure of enterprise competitiveness about PBEs from the following five competitive advantages: Talent Competition Advantage (TCA), Financial Operation Advantage (FOA), Market Share Advantage (MSA), Enterprise Management Advantage (EMA), and Project Management Advantage (PMA), which will be tested by empirical research later. In this study, enterprise competitiveness is defined as providing products or services more effectively than other competitors, with the help of the advantages in talents, financial operation, market, organizational management, and project management, so as to gain profits and sustainable development.

2.2. Organizational Flexibility

Usually, flexibility is considered as an independent variable to predict the effectiveness of an organization [34] and refers to the degree to organizational capability to make positive changes and adapt to environmental disturbances [35]. What is similar to flexibility is the concept of agility, which appears later than flexibility [36]. Some researchers regard flexibility as the center of agility [37], other studies see agility as an extension of flexibility [38]. To eliminate this confusion, Abdelilah et al. [36] analyzed the development of flexibility and agility and their relationship. The study pointed out that flexibility is just a part of agility, and other types of flexibility can also be viewed as agility sub-capabilities or as agility enablers. Flexibility is considered an operational ability, while agility is a strategic ability that enables a firm to build a long-term strategic vision [36]. However, the above findings are only discussed in manufacturing and the supply chain. It is still urgent to figure out the connotation of flexibility in the construction industry.

Firms' potential primarily rests with intrinsic flexibility of its resources and its ability to coordinate the use of those resources to achieve strategic goals [39], and more importantly, firms can generate sustainable competitive advantages by effectively controlling and using their unique, irreplaceable, and inimitable resource [40]. Theoretically, situation awareness, management of key vulnerabilities, and adaptive ability are the main three attributes of organizational flexibility and associated performance improvement processes [41]. Flexibility management is vital for firms to survive in turbulent market environments [42]. Organizational attributes such as human resources [43], organizational learning [44], organizational structure and management style [45], technologies capabilities [46,47], and supply chain capabilities [48] may affect the firm's organizational flexibility. In the field of engineering construction, this topic attracted the attention of scholars, although research works are relatively few compared with other fields [6,7,49–51]. Organizational flexibility in construction can be defined as an ability for organization to make use of resources effectively to respond and adapt to environmental changes via continuous learning in a timely and reversible manner [7].

Prior research offers a multitude of approaches on the structural dimensions of organizational flexibility. Ozer [43] pointed out some critical elements of flexibility: human resources, operations, market, finance, technology, and management. According to Maghool [35], organizational flexibility can be divided into four dimensions: operational flexibility, financial flexibility, structural flexibility, and technological flexibility. Lim et al. [7] deemed that the definition of organizational flexibility for construction firms includes various dimensions. It can be interpreted as "operational", "tactical", or "strategic" flexibility. Other studies involved product development flexibility [52], supply chain flexibility [53], human resource flexibility [54], process flexibility [47,55], leadership flexibility [56,57], team flexibility [58], cross-cultural flexibility [59], and contractual flexibility [60].

However, specific structural dimensions of organizational flexibility need to be adjusted to match specific enterprise or organization type.

For PBEs in the construction industry, organizational structure, resource, leadership, and technological flexibility can be intuitively considered as necessary elements. Furthermore, corporate culture is a strategic asset [59], and ideal organizational culture builds a learning organization that encourages and fosters organizational growth [44]. Study results show cultural change relates closely to incremental and radical innovation [61]. Hence, cultural flexibility encourages individuals and organizations to learn and innovate, generating an open atmosphere in PBEs. Besides, the ability to innovate is a component of flexibility capabilities [62]. Considering the significance of organizational climate in construction innovation [63], the authors conclude that innovation flexibility is one of the most important dimensions for PBEs' organizational flexibility under the changeable construction market environment, which is verified in this study. In summary, organizational flexibility measurement metrics for construction industry PBEs include six dimensions in this study: Structural Flexibility (SF), Resource Flexibility (RF), Leadership Flexibility (LF), Cultural Flexibility (CF), Technological Flexibility (TF), and Innovative Flexibility (IF).

- Structural Flexibility—Ability of the organization to restructure [35].
- Resource Flexibility—Ability to transform resources into other beneficial uses, providing a buffer for organizations to adapt to changes in uncertain environments [64].
- Leadership Flexibility—Leadership's capability to play several different roles, sometimes even opposite roles, to meet the demand of rapid pace and diversity of activities in various contexts [57], and their ability to adapt to by adjusting goals with their own knowledge and ability.
- Cultural Flexibility—Ability to adjust corporate culture to form a mental model, sense of worth, and learning atmosphere in order to adapt to environmental changes and uncertainties effectively.
- Technological Flexibility—Ability to change technical capacity in line with the competitive requirements [35].
- Innovative Flexibility—Ability to develop new products or services to quickly adapt to market demands at low cost [65].

Measurement of organizational flexibility developed in this study will also be tested by future follow-up empirical research.

2.3. *Organizational Innovation*

For a long time, innovation has been considered as one of the critical factors to improve national economic growth, firms' competitiveness and living advantages [26,66–69]. The process of innovation generally includes three basic steps: starting with a preliminary idea, deciding to adopt the concept, and finally, implementing the innovation [24]. In the previous study, the development of new products and new technologies is regarded as an innovation, so innovation was initially considered as a technical term, a synonym for invention. However, the concept has expanded to cover many domains and features, especially in non-technological areas [70], among which management innovation deserves a place. Evan [25] theorized that both executive administrators and working level employees initiate improvements, depending upon the type of proposed change. Organizations can adopt inspiration originating from both ends of the organizational hierarchy: Management Innovation (MI) originates near the top of the hierarchy and trickles down while Technological Innovation (TI) emerges near the bottom of the hierarchy and trickles up [25]. Technological innovation is the act of turning an idea for a new product, process or service into a reality [25]; management innovation, on the other hand, is usually recognized as the adoption of new management practices, processes, rules, methods, and structures with the aim of achieving organizational goals [71,72] and benefitting long-term performance promotion [70]. The interest in management innovation has been growing recently [73–75].

In the construction industry, innovation is shaped by project demands and forced by various environmental factors and is regarded as a means of performance improvement [66,76].

Adopting new methods presents challenges because of the disjointed and project-based characteristics of construction projects [77]. Therefore, various practices are applied to promote innovation within an organization. Liu and Chan [78] verified the crucial role of learning transfer climate in promoting innovation in construction. Lijauco et al. [79] tested the potential relationships between cultural factors and tendencies for adopting new concepts in small to medium enterprises of the construction industry. The results show that corporate relationships, market orientation, and workforce capacity are the main factors. Ozorhon et al. [77] conducted a case study and reported that denying changes, insufficient experience, and lack of innovative products can restrict adoption. Based on the characteristics of construction organizations, components of organizational innovation capacity include entrepreneurship, culture, learning organization, human resources, and information management [15]. Using Thailand's Bang Na Expressway case study, Brockmann et al. [80] summed up the categories of innovation as product or construction technology within technical, management, or legal organizations. Meng and Brown [81] argued that innovation strategies in construction firms fall into four categories: technology, management, resource, and marketing. MCI brings the motive power for PBEs to search for new innovative sources in China, and these firms should carefully consider organizational innovation. In this paper, organizational innovation refers to integration and utilization of new products and services in technology or management, change of existing production methods, renewal of new management policies, or the implementation of new management systems [82]. Considering the equally important role of technology and management in the construction industry, this research constructed measurements of organizational innovation structures as two dimensions according to Evan [25]: technological innovation and organizational innovation, which will also be tested in follow-on investigations.

3. Research Hypotheses and Theoretical Model

3.1. Organizational Flexibility and Enterprise Competitiveness

Through continuous learning, a flexible firm can maintain its competitiveness [83] because organizational flexibility serves as the most important source of competitive advantage for firms in today's dynamic environment [7]. Flexibility has usually been recognized as an ability for individuals, communities, or organizations to deal with, adapt to, and recover from a disaster event [84]. An organization with heightened flexibility weathers both daily business problems and successfully navigates crisis situations, gaining advantages in the fierce market competition [41]. Organizational flexibility can improve a firm's maneuvering capacity, and it is also beneficial to adapt existing systems and processes to environmental changes [7]. Project management practices that focus on flexibility based on collaboration, exploratory learning, and adaptation improve time performance in complex infrastructure projects [51]. Changes resulting from MCI in China induced a complex process of responses. Demands for effective organizational structure, rapid technological advancement, recruitment of new talents, and management mode innovation place significant challenges on industry PBEs. Organizational flexibility should also be required to improve enterprise competitiveness and response to these challenges for PBEs, so as to remain viable in the business environment. Therefore, this study submits the following hypothesis:

Hypothesis 1 (H1). *Organizational flexibility has a significant positive impact on enterprise competitiveness in the construction industry.*

3.2. Organizational Flexibility and Organizational Innovation

Organizational nimbleness generates creative thought according to Vickery et al. [53], allowing the firm to develop new products or services and adapt to market demands quickly at low cost. Flexibility offers a solution to environmental uncertainty as members can adjust the project to possible consequences brought by uncertain circumstances [50]. The six dimensions of the flexibility construct established in this paper are closely connected with organizational innovation. Effective leadership encourages creativity in individual project teams across the construction sector [63,77], and leadership fosters ingenuity throughout organizations [85]. The strongest driving forces of change are technical problems in construction projects, client demands, and top management [86]. Establishing an open, productive culture improves relationships between company leadership and outside firms [63]. Cultural factors drive propensity for positive change within enterprises in the construction sector [79]. The contribution of tactical business strategies to innovation performance has been confirmed by firms in the Australian construction industry [87]; furthermore, the positive relationship between external human resource flexibility and innovation has also been tested [54]. Within the context of MCI in China, organizational flexibility will be conducive to taking advantage of all kinds of PBEs organizational resources to fully realize the benefits of technological innovation and management innovation. Thus, the hypotheses are put forward:

Hypothesis 2 (H2). *Organizational flexibility has a significant positive impact on technological innovation in the construction industry.*

Hypothesis 3 (H3). *Organizational flexibility has a significant positive impact on management innovation in the construction industry.*

3.3. Organizational Innovation and Enterprise Competitiveness

Innovation serves as an organizational key performance indicator; resourcefulness enhances productivity and financial outcomes, and competitiveness has an essential role in the construction sector in enhancing work effectiveness, efficiency, and business performance [63]. According to Ozorhon et al. [76], innovation can bring many benefits, such as increased productivity and client satisfaction at the project level, improved corporate image, enhanced technical and management capability, and experience acquired at the firm-level. The relationship between innovation and competition is the focus of many academic studies concerned with economic growth and development, and companies need innovative skills in creating, producing, marketing, and managing to gain competitive advantage in global markets [26]. With accelerating fluctuations in market economies, firms must pursue revolutionary changes to gain and maintain competitive advantage [22]. Development and sustainability of competitiveness are based on speed and efficiency of the implementation of innovations to some extent [4]. Whether based on passive or active responses, organizational innovation evolves into an important factor impacting enterprise competitiveness and determines organizational viability [88], which can be harvested through building firms' capacity [15]. Both management innovation and technological innovation contribute to organization performance positively [89]. Economists have widely recognized technological innovation as a source of economic growth [90]. Existing research shows the significant positive effect of green technological innovation ability on enterprise competitiveness [91], and technological innovation can increase the economic benefits of construction firms [92]. The significant role of management innovation in boosting enterprise performance and competitive advantage has been verified [74]. Thus, it can be concluded that technological innovation and management innovation are helpful in promoting the formation of enterprise competitiveness in the construction industry. Therefore, the following hypotheses are proposed:

Hypothesis 4 (H4). *Technological innovation has a significant positive impact on enterprise competitiveness in the construction industry.*

Hypothesis 5 (H5). *Management innovation has a significant positive impact on enterprise competitiveness in the construction industry.*

3.4. Technological Innovation and Management Innovation

Researchers approved and adopted the viewpoint of the double cores model in regards to organizational innovation (e.g., technological innovation and management innovation) [24]. The impact of technological innovation and management innovation on organizational performance has been tested [89], which indicates that technological innovation and management innovation have a combined impact on performance and that a close relationship develops between technological innovation and management innovation [93]. The literature review for this study identified limited existing studies exploring the direct relationship between technological innovation and management innovation. Against the background of MCI in China, PBEs should master many evolving technologies highlighted by prefabricated construction technology, Building Information Model (BIM) technology, Radio Frequency Identification (RFID), green construction technology, and sponge city construction technology. Technological revolution requires the support of management reform, which means that technological innovation is conducive to the development of management innovation and management innovation needs to be consistent with technological innovation. Therefore, the following hypothesis is proposed:

Hypothesis 6 (H6). *Technological innovation has a significant positive impact on management innovation in the construction industry.*

3.5. Mediating Role of Technological Innovation and Management Innovation

For business operations and industrial development, the institutional environment is an important external environment and can influence the performance of the enterprise. The existing study shows that institutional environment changes may have an effect on enterprise technology innovation motivation [94]. Flexibility offers a solution to environmental uncertainty as members can adjust the project to possible consequences brought by uncertain circumstances [50]. It means that enhancing flexibility is helpful for keeping technological innovation motivation. As proposed in this paper, technology flexibility, the main component of organizational flexibility, refers to firms' ability to change the technical capacity in line with the competitive requirements [35]. Technological adjustments according to market demands change are likely to lead to innovation, and technology innovation is usually viewed as an effective assessment indicator of competitiveness [91]. Therefore, surmising the mediator role of technological innovation is reasonable.

Although there is no direct evidence indicating the role of management innovation between organizational flexibility and enterprise competitiveness, previous studies provide useful information to sort out the underlying logic. Inspiration of management innovation is considered as originating from the top of the hierarchy [25] when it comes to non-technological areas, such as management practices, processes, rules, methods, and structures [71,72]. Leaders possessing flexibility would be more likely to play several different roles to meet the demand for rapid pace and diversity of activities in various contexts [57]. Besides, the adaption of organizational culture—such as common values, beliefs and attitudes, and work practices at the organizational and national levels—to management innovations has be viewed as important [95]. Companies with flexibility in culture can align with management innovation practices quickly, and management innovation is found to have a positive relationship with firms' overall performance and financial performance [96]. It is worth noting that a mediating role of management innovation between the effects of manufacturing flexibility on organizational performance has been tested by an empirical study of 159 Spanish firms [97].

Within the context of MCI in China, organizational flexibility offers a solution to environmental uncertainty. PBEs with flexibility can respond to market demands promptly by making active adjustments in culture, resources, structure, and technology [35], which provides driving force for innovation. The dynamic environment can effectively improve the learning ability of the organization and increase the probability of implementing innovative activities. Technology innovation has been always recognized as an important approach to gain competitive advantage [90]. However, the adoption of new technology poses a higher challenge to the management ability of enterprises as the financial benefit of technology will be reduced without the support of management regulation, process, and method. Sometimes, technology innovation is an accelerator of the management revolution. For example, blockchain technology applied in prefabricated buildings can effectively stimulate innovation in quality management [98].

Based on the analysis above, three hypotheses are proposed:

Hypothesis 7a (H7a). *Technological innovation mediates the relation between organizational flexibility and enterprise competitiveness in the construction industry.*

Hypothesis 7b (H7b). *Management innovation mediates the relation between organizational flexibility and enterprise competitiveness in the construction industry.*

Hypothesis 7c (H7c). *Technological innovation and management innovation serially mediate the relation between organizational flexibility and enterprise competitiveness in the construction industry.*

Based on the above analysis and hypotheses, the theoretical model is established which is shown in Figure 1.

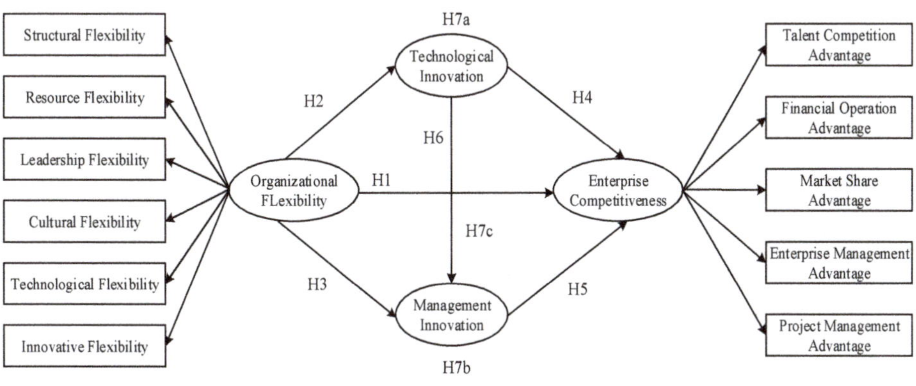

Figure 1. Theoretical model.

4. Methodology

4.1. Measures and Instruments

The scales in this study were developed based on reviewed literature to ensure the credibility of the research tool, and the localization of scale was fully considered according to advices of experts. Measurement methodology for each factor implemented a five-item scale adopted from SF—Young-Ybarra et al. [99] and Maghool [35]; RF—Mathews [64], and Martínez-Sánchez et al. [54]; LF—Phillips and Wright [100] and Baron et al. [57]; CF—Phillips and Wright [100]; TF—Ozer [43] and Maghool [35]; IF—Vickery et al. [53] and Zhu and Cheung [15]; TI—Daft [24] and Ozorhon et al. [76]; MI—Hamel [101], Crossan and Apaydin [82] and Zhu and Cheung [15]; EC(TCA/FOA/MSA/EMA/PMA)—Shen et al. [102], Tan et al. [103], Lu et al. [30], Orozco et al. [77], and Zhu and Cheung [15]. Based on the measurement scale mentioned above, this study developed a questionnaire with a total of 13 scales and 65 items. The items in the questionnaire are listed in Appendix A.

4.2. Sample and Data Collection

The hypotheses were tested by structured questionnaire surveys. As OF and EC of PBEs are interactive, various employees in China's construction industry comprised the sample respondent pool. Survey samples mainly involve senior managers, middle managers, junior managers, and common employees from various PBEs, specifically, construction, engineering management, real estate development, design, engineering consulting, and other firms in China's construction industry.

The questionnaire was originally done in English. The authors used a back-translation approach [104], which translated the original tool into Chinese according to the Chinese context. After modification and confirmation by two independent professors, the validity of the measurement project was tested. Then, the authors conducted an initial pilot test by selecting some employees in the construction industry through random selection in order to check whether respondents could accurately understand the wording and meaning of the questions. The survey questions were revised according to the pilot questionnaire analysis data. Finally, after the preliminary questionnaire was revised, the final Chinese questionnaire was developed for data collection. The Likert-type scale was used for the questionnaire, which ranged from 1 (strongly disagree) to 5 (strongly agree). The survey lasted from January to May 2019, during which 1000 questionnaires were distributed in 29 provinces and cities, including Beijing, Shanghai, Shandong, Chongqing, and Jiangsu. Among the 1000 questionnaires issued, 644 were responded to (a response rate of 64.4%). After eliminating invalid questionnaires, 463 were found to have been answered effectively (an effective rate of 71.9%). The demographic information of the interviewees is shown in Table 1.

Table 1. Demographic information of respondents (N = 463).

Variable	Category	Number of Cases	Frequency (%)
Gender	Male	343	74.1
	Female	120	25.9
Age	≤25	108	23.3
	26–30	137	29.6
	31–35	92	19.9
	36–40	63	13.6
	>40	63	13.6
Education	PhD	13	2.8
	Postgraduate	155	33.5
	Bachelor's degree	255	55.1
	Junior college student	34	7.3
	Technical secondary school or below	6	1.3
Years of experience	≤5 year	294	63.5
	5–10years	62	13.4
	>10 years	107	23.1
Position	Senior manager	22	4.8
	Middle manager	92	19.9
	Junior manager	188	40.6
	Common employees	146	31.5
	Others	15	3.2
Firm type	Construction firms	169	36.5
	Engineering management firms	45	9.7
	Real estate development firms	107	23.1
	Design firms	60	13.0
	Engineering consulting firms	23	5.0
	Other firms	59	12.7

4.3. Data Analysis Methods

Data analysis adopted Confirmatory Factor Analysis (CFA) and Structural Equation Modeling (SEM), using the SPSS 25.0 and AMOS 21.0 software. In statistics, CFA is the most commonly used special form of factor analysis in social research. The purpose of CFA test is to confirm that the data collected by the study conform to a hypothetical measurement model [105]. In this study, CFA was implemented to test the convergence and discriminant validity of the measured structural model. SEM provides support for these hypotheses and can perform path analysis.

5. Research Results

5.1. CFA

CFA modeling is used to verify the effectiveness of the overall measurement model. Both convergent and discriminant validity are considered in the study. The indicators for testing the convergent validity include factor loading, which should exceed 0.6 and be significant at the level of 0.01, and Composite Reliability (CR), which should be greater than 0.8. Average Variance Extracted (AVE) of all constructs should be larger than 0.5 [106]. In addition, Cronbach's Alpha coefficient was used to evaluate construct reliability in this study, which should be greater than 0.70 [105]. According to CFA, factor loading of SF1 and SF2 are lower than 0.6, so these two items are excluded from SF scale.

Table 2 presented the final results for the remining constructs, including Cronbach's alpha coefficients, factor loading, CR, and AVE. It can be seen that factor loading was always greater than 0.6 at the 0.001 significant level. The Cronbach's alpha coefficients, CR, and AVE of all items exceeded 0.7, 0.8, and 0.5 respectively. To conclude, the results show that the measurement model fully meets the reliability standard and convergent validity.

Table 2. Construct Validity and Reliability.

Construct	Item	Cronbach's Alpha	Factor Loading	CR	AVE
SF	SF3	0.822	0.736	0.825	0.611
	SF4		0.826		
	SF5		0.780		
RF	RF1	0.910	0.795	0.912	0.675
	RF2		0.846		
	RF3		0.836		
	RF4		0.847		
	RF5		0.780		
LF	LF1	0.907	0.766	0.898	0.637
	LF2		0.770		
	LF3		0.854		
	LF4		0.820		
	LF5		0.778		
CF	CF1	0.911	0.792	0.914	0.680
	CF2		0.859		
	CF3		0.864		
	CF4		0.829		
	CF5		0.777		
TF	TF1	0.922	0.776	0.923	0.707
	TF2		0.855		
	TF3		0.893		
	TF4		0.844		
	TF5		0.833		

Table 2. *Cont.*

Construct	Item	Cronbach's Alpha	Factor Loading	CR	AVE
IF	IF1 IF2 IF3 IF4 IF5	0.916	0.816 0.872 0.867 0.804 0.760	0.914	0.680
TI	TI1 TI2 TI3 TI4 TI5	0.936	0.830 0.804 0.882 0.863 0.894	0.931	0.731
MI	MI1 MI2 MI3 MI4 MI5	0.948	0.814 0.854 0.916 0.916 0.914	0.947	0.781
TCA	TCA1 TCA2 TCA3 TCA4 TCA5	0.925	0.808 0.855 0.846 0.867 0.849	0.926	0.715
FOA	FOA1 FOA2 FOA3 FOA4 FOA5	0.924	0.855 0.858 0.829 0.801 0.871	0.925	0.711
MSA	MSA1 MSA2 MSA3 MSA4 MSA5	0.919	0.857 0.835 0.844 0.837 0.800	0.920	0.697
EMA	EMA1 EMA2 EMA3 EMA4 EMA5	0.944	0.859 0.894 0.884 0.912 0.848	0.945	0.774
PMA	PMA1 PMA2 PMA3 PMA4 PMA5	0.941	0.844 0.891 0.879 0.868 0.878	0.941	0.761

Note: SF = Structural Flexibility; RF = Resource Flexibility; LF = Leadership Flexibility; CF = Cultural Flexibility; TF = Technological Flexibility; IF = Innovative Flexibility; TI = Technological Innovation; MI = Management Innovation; TCA = Talent Competition Advantage; FOA = Financial Operation Advantage; MSA = Market Share Advantage; EMA = Enterprise Management Advantage; PMA = Project Management Advantage.

Two criteria for evaluating discriminant validity were identified [106,107]. This study adopted the one proposed by [107], which suggested that the AVE of each latent variable is greater than the correlation coefficients between the same construct and any other construct. Table 3 presents the results of the descriptive statistics and correlation analysis. It can be inferred that the measurement model achieves appropriate discriminant validity as the diagonal elements are found to exceed respective off-diagonal elements. Besides, all correlation coefficients reveal significant positive correlations between all variables. Furthermore, this study assessed the collinearity between all measure variables by adopting the collinearity diagnostics. The results show that the maximal Variance Inflation Factor

(VIF) value is 5.587, well below the recommended cut-off of 10 [108], which indicates that there is no evident multicollinearity problem between measure variables.

Table 3. Descriptive Statistics and Correlation Analysis (N = 463).

Variable	Mean	SD	SF	RF	LF	CF	TF	IF	TI	MI	TCA	FOA	MSA	EMA	PMA
SF	3.846	0.733	*0.782*												
RF	3.715	0.764	0.668 **	*0.822*											
LF	3.767	0.740	0.739 **	0.768 **	0.798										
CF	3.854	0.732	0.738 **	0.666 **	0.845 **	*0.825*									
TF	3.747	0.740	0.633 **	0.824 **	0.798 **	0.713 **	*0.841*								
IF	3.691	0.752	0.666 **	0.722 **	0.748 **	0.734 **	0.824 **	*0.825*							
TI	3.588	0.784	0.572 **	0.741 **	0.699 **	0.629 **	0.791 **	0.791 **	*0.855*						
MI	3.625	0.773	0.609 **	0.708 **	0.702 **	0.623 **	0.739 **	0.757 **	0.818 **	*0.884*					
TCA	3.646	0.758	0.625 **	0.650 **	0.713 **	0.686 **	0.719 **	0.754 **	0.722 **	0.744 **	*0.846*				
FOA	3.651	0.778	0.554 **	0.555 **	0.623 **	0.614 **	0.633 **	0.625 **	0.585 **	0.597 **	0.742 **	*0.843*			
MSA	3.682	0.722	0.593 **	0.644 **	0.684 **	0.638 **	0.711 **	0.681 **	0.713 **	0.699 **	0.740 **	0.785 **	*0.835*		
EMA	3.638	0.760	0.615 **	0.593 **	0.678 **	0.653 **	0.666 **	0.695 **	0.666 **	0.682 **	0.752 **	0.759 **	0.805 **	*0.880*	
PMA	3.695	0.724	0.616 **	0.603 **	0.685 **	0.675 **	0.677 **	0.690 **	0.652 **	0.658 **	0.757 **	0.742 **	0.785 **	0.869 **	*0.872*

Note: SD = Standard Deviations. The diagonal italic values are the square roots of AVE. ** = Equals significant at the 0.01 level (two-tailed).

Adopting Hair et al.'s [109] approach, the measurement model fit is evaluated using indicators/parameters as follows:

- Absolute fit measures: Chi-square degree of freedom ratio (χ^2/df), Root Mean Square (RMS), Residual (RMR), Goodness-Of-Fit index (GFI), and RMS Error of Approximation (RMSEA);
- Incremental fit measures: Incremental Fit Index (IFI), Tucker-Lewis Index (TLI), and Adjusted Goodness-Of-Fit index (AGFI) and Comparative Fit Index (CFI);
- Parsimonious fit measures: Parsimony Goodness-Of-Fit index (PGFI), and Parsimony Comparative Fit Index (PCFI).

According to the data in Table 4, the fit indexes of the measurement models all reach the standard [44,105], which means the measurement model fit the survey data well. Therefore, the measurement models proposed in this paper are completely applicable to test the research hypothesis.

Table 4. Overall Fit Indices of the Scales.

Fit Index		χ^2/df	RMR	RMSEA	GFI	AGFI	TLI	IFI	CFI	PGFI	PCFI
scale	OF	3.193	0.033	0.069	0.848	0.815	0.929	0.938	0.938	0.694	0.824
	OI	2.764	0.017	0.062	0.964	0.936	0.983	0.989	0.989	0.543	0.681
	EC	2.562	0.019	0.058	0.894	0.870	0.960	0.965	0.965	0.729	0.852
Recommended cutoff value		$\leq 3^{a}$; $\leq 5^{b}$	≤ 0.05	$<0.08^{a}$; $<0.1^{b}$	$\geq 0.9^{a}$; $\geq 0.8^{b}$	$\geq 0.9^{a}$; $\geq 0.8^{b}$	≥ 0.9	≥ 0.9	≥ 0.9	≥ 0.5	≥ 0.5
Fit?		YES	YES	YES	YES	YES	YES	YES	YES	YES	YES

Note: OF = Organizational Flexibility; OI = Organizational Innovation; EC = Enterprise Competitiveness; a = Equals acceptable; b = Equals marginal.

5.2. SEM Analysis and Hypotheses Testing

SEM analysis (again using AMOS 21.0) was performed to verify the hypotheses. The critical ratio and the *p* value are two significant indicators for testing the hypothesis. Critical ratio should be higher than 1.96 [44]. The results of hypotheses testing are presented in Table 5 and Figure 2. Except for H4, all the critical ratio values exceed 1.96. OF has significant effect on EC, TI, and MI ($\beta = 0.633$, $p < 0.001$; $\beta = 0.868$, $p < 0.001$; $\beta = 0.351$, $p < 0.001$), which support H1, H2, and H3, respectively. However, influence of TI on EC is not significant ($\beta = 0.041$, $p > 0.05$); therefore, H4 is rejected. MI has a significant impact on EC ($\beta = 0.234$, $p < 0.001$); therefore, H5 is supported. H6 is also supported as the relationship of TI on MI is also found to be significant ($\beta = 0.556$, $p < 0.001$). In conclusion, except for H4, all direct effect hypotheses proposed in this paper have been confirmed.

Table 5. Hypothesis Testing Results.

Hypothesis	Path	Path Coefficient β	Critical Ratio	*p*	Remarks
H1	OF–EC	0.633	7.872	***	Supported
H2	OF–TI	0.868	14.137	***	Supported
H3	OF–MI	0.351	5.338	***	Supported
H4	TI–EC	0.041	0.540	0.589	Rejected
H5	MI–EC	0.234	3.584	***	Supported
H6	TI–MI	0.556	8.503	***	Supported

Note: *** = Equals significant at the 0.001 level (two tailed).

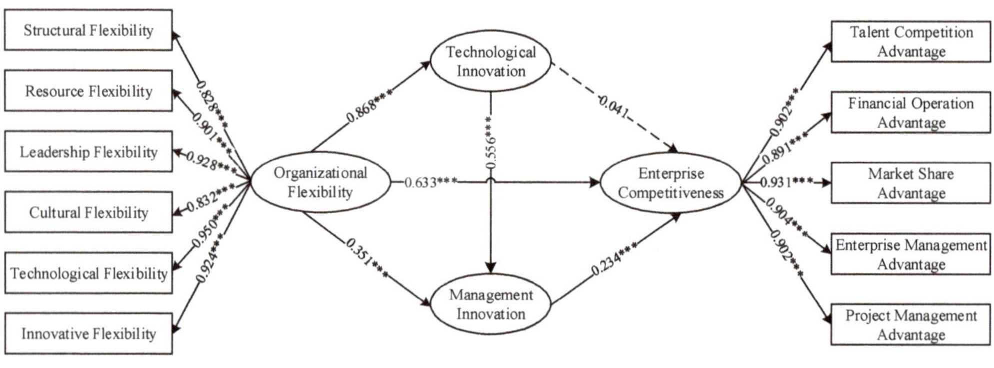

Figure 2. Research model and results of hypotheses test.

5.3. Bootstrapping

The indirect effect was tested using bootstrapping analysis with a sample of 5000. When the 95% confidence interval (CI) does not include zero, it means that the indirect mediation effect is significant at a 5% level [110]. In this study, three specific mediation effects are tested. First, the CI for the effect on the pathway from OF via TI to EC included zero (95% CI = [−0.099, 0.172]). The result indicates that the indirect mediation effect of TI is not significant at a level of 5%, thus rejecting Hypothesis 7a. Second, MI significantly mediates the effect of OF and EC, with 95% CI [0.030, 0.144]. Thus, Hypothesis 7b is supported. Third, the CI for the effect on the pathway from OF via TI and MI to EC excluded zero (95% CI = [0.047, 0.198]). It means the serial mediation effect of TI and MI between OF and EC is significant at a level of 5%, thus supporting Hypothesis 7c.

6. Discussion

6.1. Summary of Findings

This study is aimed at exploring the mechanism for EC promotion of PBEs in Chinese construction industry against the background of MCI with a specific focus on OF and OI when OI is considered including TI and MI. Five major findings are presented as follows after carrying out empirical analysis.

- OF has a significant and positive effect on EC with the influence effect of 0.633.
- OF is positively related with TI and MI, with the influence effect of 0.868 and 0.351, respectively.
- The influence effect of MI on EC is 0.234, while the direct influence between TI and EC is not supported.
- TI has a direct influence on MI with the influence effect of 0.556.

- MI significantly mediates the relation between OF and EC. The indirect mediation effect of TI is not significant. TI and MI play a serial mediation role in the relationship between OF and EC.

The results show that OF is an effective predictor of EC, which is in line with previous evidence [7]. Organizations with flexibility can react to environmental changes and uncertainties (e.g., velocity of technological advancement, increased competition, market fluctuations [6]) more quickly by absorbing change, integrating, and developing and restructuring resources and capabilities in a short time [8], which is beneficial to successfully navigating crisis situations and gaining advantages in the fierce market competition [41]. The uncertainty of an environment could be an opportunity or challenge depending on whether an organization can adjust according to new market demands effectively. In generally, firms with higher flexibility tend to exist and prosper [7].

Besides, the findings of this study show that OF is conducive to promoting TI and MI. OF generates creativity [53], allowing the firm to develop new products or services and make adjustments in management regulation, process, and method to adapt to market demands quickly [9]. Furthermore, previous studies demonstrated that the dimensions of organizational flexibility that are established in this paper (SF/RF/LF/CF/TF/IF) can positively promote OI. For example, effective leadership encourages creativity and foster ingenuity in individual project teams across the construction sector [15,77]. External human resource flexibility is also positively related with innovation [54].

Moreover, MI is found to have a direct influence on EC, which is aligned with a previous study [74]. The indirect effect testing results indicate that MI serves as a significant mediating function in the relationship between OF and EC. As analyzed in this study, firms with flexibility in some aspects (e.g., leadership [57], culture [95]) are more likely to respond to environmental changes actively by making adjustments in management practices to satisfy market acquirements and gain competitive advantages.

The empirical results show that TI has no direct positive influence on EC, which is the opposite to the conclusion drawn by Li et al. (2019) [91], and that the indirect mediation effect of TI is not significant. Interestingly, the indirect mediation effect of MI between OF and EC is significant, both in the serial mediation of TI and MI. The authors attempt to make the following explanations. In theory, TI in other literatures may be defined in an overly broad fashion, involving innovation in management, which is not consistent with this study. Moreover, the concept of competitiveness proposed in this paper is aimed at PBEs in the construction industry (see Section 2.1), which is not the same as existing definition [31]. It is universally acknowledged that the industry background and market development trend need to be considered fully when establishing competitiveness assessment indicators [29]. Therefore, the direct contribution of TI on EC may be limited to a certain industry or a special situation, which is not an eternal truth. Besides, it has been found that the influence of technological innovation output on the competitiveness of enterprises has a threshold effect and lag. Only when the number of patent applications of enterprise reaches a certain level can the innovation output promote enterprise competitiveness effectively [111]. As a practical matter, in a complex business environment, TI is a double-edged sword. On the one hand, it provides opportunities for firms to profit from innovation and improve their capabilities. On the other hand, innovation may change traditional supply chains, destroy some immature business processes, and bring expected losses to enterprises [112]. Furthermore, cycle of TI has been greatly shortened currently, and the technology penetration rate is increasing [113]. In the context of MCI, the competition in technology is intensified, and the possibility of maintaining the growth and competitiveness of enterprises through technical means is decreasing. Innovation in non-technical fields (management, etc.) is gradually considered a new breakthrough to enhance firm performance [75,114]. In addition, the application of new technologies poses a higher challenge to the management ability of enterprises. The introduction of innovative technologies is a common tool to promote the development of firms. However, the effectiveness of this work mainly depends on the absorption ability of technology receivers. The application of technology needs the support of the management practices within the organization. In other words, maximizing the value of TI requires the support of MI. Therefore, TI can effectively promote MI to some extent, which is also partly supported by research of Zhang et al. [98]. In a word, the above analysis supports the results of hypotheses H6 and H7c, that is, TI can promote MI, and TI and MI play an important serial mediation between EC and OF.

6.2. Theoretical Implications

Existing literature presented effective research of EC promotion and obtained significant research outcomes in the construction industry. Although researchers in the construction industry agree on the importance of gaining OF in a constantly changing business environment [6,7,47,49,50], specific effects of OF on EC require additional consideration and testing. The mediating role of OI, including TI and MI, also needs further study. This paper presents the first attempt to establish the theoretical connection between OF, OI, and EC in the construction industry in a single research framework. This study clarifies the direct influence effect of OF on EC, while verifying important mediating role of MI and serial mediating functions of TI and MI between them. These are the outstanding theoretical contribution to PBEs based on the context of MCI in China. However, the mediating role of TI between OF and EC is not significant, which is an interesting result. The author has given some explanations in theory and practice, which is helpful to re-examine the role of TI in enhancing EC.

In addition, previous research explored multiple types of flexibility from different perspectives and their importance. Key elements include product development [52], supply chain [53], process [47,55], operational [115,116], strategic [117], leadership [56,57], project team [58], financial, structural, technological [35], cross-cultural [59], contractual [60], and human resource flexibilities [54]. However, existing literature exploring construction industry PBEs OF is scarce. According to existing research, OF measurement structures are not uniform. Applying MCI to Chinese construction industry, OF measurement structure is grouped into six dimensions: SF, RF, LF, CF, TF, and IF. New and beneficial attempts in the construction industry have been well verified and contribute to literature on OF theory, especially dimensions of CF and IF proposed by this paper in particular. Similarly, considering the difference between PBEs and other types of organizations, this manuscript establishes the measurement structure about EC from five competitive perspectives: TCA, FOA, MSA, EMA, and PMA. All of them return good reliability and validity results and can pass the empirical test. Therefore, the proposed measurement scales (of OF and EC) can provide a reference for scholars to perform similar research.

6.3. Practical Implications

In the process of MCI in China, PBEs construction leaders and managers should consider implementing EC in order to survive and prosper in turbulent market environments. Based on empirical evidence, the findings of this study provide guidelines for PBEs' senior management, as well as practitioners, to make policies and strategies for gaining sustainable development and competitiveness. Researching and discussing EC improvement mechanisms from the comprehensive perspective of OF and OI can inspire the construction industry in China. Faced with the uncertainty in the context of MCI, OF provides an effective solution. By strengthening OF in terms of structure, resources, leadership, culture, technology, and innovation the environmental adaptability of firms can be significantly enhanced, which is conducive to seizing opportunities or avoiding underlying risks through dynamic learning so as to gain competitiveness. In addition, the serial mediating function of TI and MI in improving EC has been tested. Thus, OI should be valued by embracing TI as a power source, which further contributes in promoting MI. Coupling these components amplifies the benefits within the organization. Therefore, it is recommend that PBEs emphasize TI and MI to improve sustainability and performance, rather than focusing only on TI.

6.4. Limitations and Future Research

First of all, this paper adopted the questionnaire survey to test the theoretical model of China's construction industry with cross-sectional data. The data merely reflect the relationship, as well as regular rules, in the short term with a Chinese context. Future research is needed to conduct research in a long termed and longitudinal perspective. The research results should be examined via repeated research subjects, so as to verify the validity of the results. Although the results of this study could be an important and valued reference for Chinese construction PBEs leaders and managers who want to keep enterprise competitive advantage when responding to uncertainties and challenges brought by MCI. However, how to apply the theoretical results better to practice for enterprises is also a problem to be explored and solved.

Furthermore, this paper studied the common influence laws of OF and OI on EC in PBEs because the number of research samples in this study is insufficient to distinguish different types of PBEs for comparative analysis. However, it can be speculated that different types of enterprises may require different types of organizational flexibility. Therefore, if possible, more data can be collected for a specific focus on one type of enterprise (e.g., construction, design, engineering management, etc.).

Finally, this paper exclusively investigated the influence mechanism of OF and OI on EC in PBEs, but offers an interesting topic to analyze and study the factors affecting OF for PBEs. The results show that OF has a significant positive effect on EC, which provides a reference for improving EC from the perspective of flexibility. However, this paper only defines the connotation and develops the measurement dimension of OF, but the formation mechanism of OF and related influencing factors are not covered in this paper. Therefore, the authors suggest that more attention should be paid to these two aspects and how to improve the PBEs' OF should be further analyzed. In addition, the optimal level of the OF for an enterprise is also an important and interesting issue that need to be explored.

7. Conclusions

The MCI strategy influences improvement and development of China's construction industry, causing numerous uncertainties and complex external environment changes for various PBEs. Improving the enterprise competitiveness constantly and surviving and developing in fierce market competition are the focus of the leaders and managers of PBEs in China at present. Therefore, greater attention should be paid to improving PBEs' enterprise competitiveness. This paper explored improvement mechanism of enterprise competitiveness, focusing on organizational flexibility and organizational innovation. The following conclusions are drawn:

(1) Organizational flexibility can positively affect enterprise competitiveness significantly in China's construction industry. This indicates the need to improve PBEs' organizational flexibility so organizations can cope with the challenges and opportunities generated by MCI.

(2) Technological innovation and management innovation can play an important serial mediating role between organizational flexibility and enterprise competitiveness in the construction industry, and strengthening technological innovation and management innovation improves the enterprise competitiveness of PBEs. Furthermore, technological innovation positively facilitates management innovation, and resultant development strategy, organizational structure, management system, management process, and management method innovations should be compatible with technological innovation in PBEs.

(3) The measurement structure of organizational flexibility including structural flexibility, resource flexibility, leadership flexibility, cultural flexibility, technological flexibility, and innovative flexibility has been well verified, and measurement scales offer proven reliability and validity. Strengthening the structural flexibility, resource flexibility, leadership flexibility, cultural flexibility, technological flexibility, and innovative flexibility of the organization enhances organizational flexibility capabilities.

(4) The measurement structure of enterprise competitiveness involving talent competition advantage, financial operation advantage, market share advantage, enterprise management advantage, and project management advantage has also been well tested, and measurement scales were found to have ideal reliability and validity. Therefore, through strengthening the advantages associated with talent competition, financial operation, market share, enterprise management, and project management enterprise competitiveness of PBEs can be reinforced further.

Author Contributions: Conceptualization, G.N.; methodology, G.N. and H.X.; software, H.X.; validation, G.N., Z.Z. and H.L.; formal analysis, G.N. and Q.C.; investigation, G.N., Z.Z. and H.L.; resources, G.N.; data curation, Y.Q. and H.X.; writing—original draft preparation, G.N., H.X., Z.Z. and H.L.; writing—review and editing, Q.C., Y.Q. and P.J.H.; supervision, G.N. and Q.C.; project administration, G.N.; funding acquisition, G.N. All authors have read and agreed to the published version of the manuscript.

Funding: This research was funded by the National Natural Science Foundation of China, grant number 72071201 and the Fundamental Research Funds for the Central Universities in China, grant number 2020ZDPYMS30.

Institutional Review Board Statement: Not applicable.

Informed Consent Statement: Not applicable.

Data Availability Statement: The data presented in this study are available on request from the corresponding author.

Conflicts of Interest: The authors declare no conflict of interest.

Appendix A

Table A1. Questionnaire items in measurement scales.

Construct	Code	Measurement Item
SF	SF1	Your firm carries out organizational structure reform in response to the MCI.
	SF2	Your firm is able to actively prepare or perfect specialized organizations to cope with the MCI.
	SF3	Your firm advocates inter-departmental cooperation and communication within the organization.
	SF4	Your firm can fully arise the initiative of functional departments in response to MCI.
	SF5	Your firm establishes strategic cooperation with other firms to adapt to the MCI.
RF	RF1	Leaders and employees in your firm are good at learning and applying relevant knowledge of the MCI.
	RF2	Your firm attaches great importance to the training and education of employees related to the MCI.
	RF3	Your firm can be equipped with all kinds of equipment required for the MCI timely.
	RF4	Your firm is good at capturing and utilizing information related to the MCI.
	RF5	Your firm can raise funds to meet the needs of the MCI timely.
LF	LF1	Leaders of your firm pay close attention to the relevant policies and development status of the MCI.
	LF2	Leaders of your firm often adopt advices on facilitating the MCI.
	LF3	Leaders of your firm are good at motivating employees and display their subjective initiative.
	LF4	Leaders of your firm can respect, trust, understand and care for employees.
	LF5	Leaders of your firm can treat the success and failure of his subordinates objectively and fairly.
CF	CF1	Your firm is open and willing to develop the MCI.
	CF2	Your firm encourages free communication between superiors and subordinates.
	CF3	Your firm advocates teamwork and win-win cooperation.
	CF4	Your firm encourages employees to learn and communicate across functional areas.
	CF5	Your firm attaches great importance to humanistic care for employees.
TF	TF1	Your firm attaches great importance to the impact of technological revolution on economic benefits.
	TF2	Your firm is good at introducing relevant technologies to develop the MCI.
	TF3	Your firm can apply the relevant technologies of the MCI timely.
	TF4	Your firm is good at predicting the development trend of technology in the construction industry.
	TF5	Your firm values investment in technology research and development.

Table A1. *Cont.*

Construct	Code	Measurement Item
IF	IF1	Your firm attaches great importance to the recruitment and introduction of innovative talents.
	IF2	Your firm pays attention to the cultivation of employees' innovative consciousness and ability.
	IF3	Your firm advocates continuous innovation in the reform of the construction industry.
	IF4	Your firm can provide financial support for employees to do innovative work.
	IF5	Your firm focuses on rewarding employees for their innovations.
TI	TI1	Your firm has implemented technological innovation related to the MCI, independently or cooperatively.
	TI2	Your firm has obtained a number of patents or unique technologies related to the MCI.
	TI3	Your firm has improved traditional construction or management techniques in response to the MCI.
	TI4	Your firm has adapted traditional production or management tools in response to the MCI.
	TI5	Your firm can launch new products or services to cope with the market changes timely.
MI	MI1	Your firm has adjusted development strategy in response to the MCI.
	MI2	Your firm has innovated organizational structure in response to the MCI.
	MI3	Your firm has innovated management regulation in response to the MCI.
	MI4	Your firm has innovated management procedure in response to the MCI.
	MI5	Your firm has innovated management methods in response to the MCI.
TCA	TCA1	Your firm has more talents to meet the needs of MCI.
	TCA2	Employees have higher comprehensive quality in your firm.
	TCA3	Employees have strong learning ability in your firm.
	TCA4	Your firm has a reserve of talents
	TCA5	Your firm receives more attention or favor from job seekers.
FOA	FOA1	You firm has more solid financial support.
	FOA2	You firm has stronger financing capacity.
	FOA3	Your firm is in good financial condition.
	FOA4	You firm has higher cost control ability.
	FOA5	Your firm is more capable of capital appreciation.
MSA	MSA1	Your firm is able to respond to market needs or opportunities timely.
	MSA2	Business scope is more suitable for the MCI in your firm.
	MSA3	Your firm has a higher winning rate in the market.
	MSA4	Your firm has better marketing and public relations ability.
	MSA5	Your firm has a better corporate reputation and image.
EMA	EMA1	Modern enterprise system is more sound in your firm.
	EMA2	Internal processes are more efficient in your firm.
	EMA3	Internal communication mechanism is better in your firm.
	EMA4	Functional departments and project teams in your firm can cooperate effectively.
	EMA5	There are closer relationships between your firm and partners.
PMA	PMA1	Your firm gives project team more powers and responsibilities.
	PMA2	Project team has a stronger comprehensive strength in your firm.
	PMA3	Project team' cohesion is stronger in your firm.
	PMA4	Operation mechanism of project team is more reasonable in your firm.
	PMA5	Project team can provide a higher quality product or service in your firm.

References

1. Drobyazko, S.; Barwińska-Małajowicz, A.; Ślusarczyk, B.; Zavidna, L.; Danylovych-Kropyvnytska, M. Innovative entrepreneurship models in the management system of enterprise competitiveness. *J. Entrep. Educ.* **2019**, *22*, 1–6.
2. Orozco, F.A.; Serpell, A.F.; Molenaar, K.R.; Forcael, E. Modeling competitiveness factors and indexes for construction companies: Findings of Chile. *J. Constr. Eng. Manag.* **2014**, *140*, B4013002. [CrossRef]
3. Tan, Y.; Xue, B.; Cheung, Y.T. Relationships between main contractors and subcontractors and their impacts on main contractor competitiveness: An empirical study in Hong Kong. *J. Constr. Eng. Manag.* **2017**, *143*, 05017007. [CrossRef]
4. Stoyanova, T.; Angelova, M. Impact of the internal factors on the competitiveness of business organizations. In Proceedings of the 2018 International Conference on High Technology for Sustainable Development (HiTech) IEEE, Sofia, Bulgaria, 11–14 June 2018; pp. 1–3.
5. Dubey, R.; Gunasekaran, A.; Childe, S.J.; Fosso Wamba, S.; Roubaud, D.; Foropon, C. Empirical investigation of data analytics capability and organizational flexibility as complements to supply chain resilience. *Int. J. Prod. Res.* **2019**, 1–19. [CrossRef]

6. Lim, B.T.; Ling, F.Y.; Ibbs, C.W.; Raphael, B.; Ofori, G. Mathematical models for predicting organizational flexibility of construction firms in Singapore. *J. Constr. Eng. Manag.* **2012**, *138*, 361–375. [CrossRef]
7. Lim, B.T.; Ling, F.Y.; Ibbs, C.W.; Raphael, B.; Ofori, G. Empirical analysis of the determinants of organizational flexibility in the construction business. *J. Constr. Eng. Manag.* **2011**, *137*, 225–237. [CrossRef]
8. Gonçalves, J.M.; Ferreira, F.A.F.; Ferreira, J.J.M.; Farinha, L.M.C. A multiple criteria group decision-making approach for the assessment of small and medium-sized enterprise competitiveness. *Manag. Decis.* **2019**, *57*, 480–500. [CrossRef]
9. Saeed, M.A.; Jiao, Y.; Zahid, M.M.; Tabassum, H.; Nauman, S. Organizational flexibility and project portfolio performance: The roles of innovation, absorptive capacity and environmental dynamism. *Int. J. Manag. Proj. Bus.* **2020**. [CrossRef]
10. Liu, J.; Li, B.; Lin, B.; Nguyen, V. Key issues and challenges of risk management and insurance in China's construction industry: An empirical study. *Ind. Manag. Data Syst.* **2007**, *107*, 382–396. [CrossRef]
11. Gann, D.M.; Salter, A.J. Innovation in project-based, service-enhanced firms: The construction of complex products and systems. *Res. Policy.* **2000**, *29*, 955–972. [CrossRef]
12. Javernick-Will, A. Local embeddedness and knowledge management strategies for project-based multi-national firms. *Eng. Manag. J.* **2013**, *25*, 16–26. [CrossRef]
13. Luo, M.; Fan, H.; Liu, G. Measuring regional differences of construction productive efficiency in China: A distance friction minimization approach. *Eng. Constr. Archit. Manag.* **2019**, *27*, 952–974. [CrossRef]
14. Deng, F.; Liu, G.; Jin, Z. Factors Formulating the Competitiveness of the Chinese Construction Industry: Empirical Investigation. *J. Manag. Eng.* **2013**, *29*, 435–445. [CrossRef]
15. Zhu, L.; Cheung, S.O. Harvesting competitiveness through building organizational innovation capacity. *J. Manag. Eng.* **2017**, *33*. [CrossRef]
16. Gruneberg, S. Construction markets in a changing world economy. In *Economics for the Modern Built Environment*; Taylor & Francis Group: Oxfordshire, UK, 2008; pp. 153–167.
17. Cartlidge, D. *Procurement of Built Assets*; Butterworth Heinemann: Oxford, UK, 2004.
18. Luo, T.; Xue, X.; Wang, Y.; Xue, W.; Tan, Y. A systematic overview of prefabricated construction policies in China. *J. Clean Prod.* **2020**, *280*. [CrossRef]
19. Huang, J. Research on information industry innovation model under the background of integration of industrialization and informatization. In Proceedings of the 7th International Conference on Innovation & Management, Wuhan, China, 4–5 December 2010; pp. 588–592.
20. Zhai, X.; Reed, R.; Mills, A. Addressing sustainable challenges in China: The contribution of off-site industrialization. *Smart Sustain. Built Environ.* **2014**, *3*, 261–274. [CrossRef]
21. Cao, D.; Li, H.; Wang, G.; Luo, X.; Tan, D. Relationship network structure and organizational competitiveness: Evidence from BIM implementation practices in the construction industry. *J. Manag. Eng.* **2018**, *34*, 04018005. [CrossRef]
22. Dess, G.G.; Picken, J.C. Changing roles: Leadership in the 21st century. *Organ. Dyn.* **2000**, *28*, 18–34. [CrossRef]
23. Ungerman, O.; Dedkova, J.; Gurinova, K. The impact of marketing innovation on the competitiveness of enterprises in the context of industry 4.0. *J. Compet.* **2018**, *10*, 132–148. [CrossRef]
24. Daft, R.L. A dual-core model of organizational innovation. *Acad. Manag. J.* **1978**, *21*, 193–210.
25. Evan, W.M. Organizational lag. *Hum. Organ.* **1966**, *25*, 51–53. [CrossRef]
26. Dereli, D.D. Innovation management in global competition and competitive advantage. *Procedia Soc. Behav. Sci.* **2015**, *195*, 1365–1370. [CrossRef]
27. Ajitabh, A.; Momaya, K. Competitiveness of firms: Review of theory, frameworks and models. *Singap. Manag. Rev.* **2004**, *26*, 45–61.
28. Falciola, J.; Jansen, M.; Rollo, V. Defining firm competitiveness: A multidimensional framework. *World Dev.* **2020**, *129*, 104857. [CrossRef]
29. Afanasieva, N.V.; Rodionov, D.G.; Vasilev, Y.N. System of indicators of coal enterprise competitiveness assessment. *Espacios* **2018**, *39*, 10.
30. Lu, W.; Shen, L.; Yam, M.C. Critical success factors for competitiveness of contractors: China study. *J. Constr. Eng. Manag.* **2008**, *134*, 972–982. [CrossRef]
31. Yamova, O.; Maramygin, M.; Sharova, I.; Nesterenko, J.; Sobina, N. Integral valuation of an enterprise's competitiveness in the industrial economy. *Eur. Res. Stud.* **2018**, *21*, 777–786.
32. Ghobadian, A.; Viney, H.; Liu, J.; James, P. Extending linear approaches to mapping corporate environmental behaviour. *Bus. Strategy Environ.* **1998**, *7*, 13–23. [CrossRef]
33. Xu, T.; Tiong, R.L.; Chew, D.A.; Smith, N.J. Development model for competitive construction industry in the People's Republic of China. *J. Constr. Eng. Manag.* **2005**, *131*, 844–853. [CrossRef]
34. Dikmen, I.; Birgonul, M.T.; Kiziltas, S. Prediction of organizational effectiveness in construction companies. *J. Constr. Eng. Manag.* **2005**, *131*, 252–261. [CrossRef]
35. Maghool, A. Investigating the effect of flexibility (operational, financial, structural and technological) required for Banking Industry on the correlation between the strategic planning and the organization efficiency. *Am. J. Sci.* **2013**, *9*, 102–109.
36. Abdelilah, B.; El Korchi, A.; Balambo, M.A. Flexibility and agility: Evolution and relationship. *J. Manuf. Technol. Mana.* **2018**, *29*, 1138–1162. [CrossRef]

37. Prater, E.; Biehl, M.; Smith, M.A. International supply chain agility-Tradeoffs between flexibility and uncertainty. *Int. J. Oper. Prod. Man.* **2001**, *21*, 823–839. [CrossRef]
38. Backhouse, C.J.; Burns, N.D. Agile value chains for manufacturing-implications for performance measures. *Int. J. Agil. Manag. Systems* **1999**, *1*, 76–82. [CrossRef]
39. Koev, S.R.; Pavliuk, S.; Derhaliuk, M.; Sokolova, L.; Portna, O. Resource Strategy for Enterprise Management as a Tool to Ensure Its Competitiveness. *Acad. Strateg. Manag. J.* **2020**, *19*, 1–8.
40. Lin, H.; Zeng, S.X.; Ma, H.Y.; Qi, G.Y.; Tam, V.W. Can political capital drive corporate green innovation? Lessons from China. *J. Clean. Prod.* **2014**, *64*, 63–72. [CrossRef]
41. McManus, S.; Seville, E.; Vargo, J.; Brunsdon, D. Facilitated process for improving organizational resilience. *Nat. Hazards Rev.* **2008**, *9*, 81–90. [CrossRef]
42. Úbeda-García, M.; Claver-Cortés, E.; Marco-Lajara, B.; Zaragoza-Sáez, P.; García-Lillo, F. High performance work system and performance: Opening the black box through the organizational ambidexterity and human resource flexibility. *J. Bus. Res.* **2018**, *88*, 397–406. [CrossRef]
43. Ozer, M. The role of flexibility in online business. *Bus. Horiz.* **2002**, *45*, 61–69. [CrossRef]
44. Ni, G.; Cui, Q.; Sang, L.; Wang, W.; Xia, D. Knowledge-sharing culture, project-team interaction, and knowledge-sharing performance among project members. *J. Manag. Eng.* **2018**, *34*, 04017065. [CrossRef]
45. Lansley, P.R. Corporate strategy and survival in the UK construction industry. *Constr. Manag. Econ.* **1987**, *5*, 141–155. [CrossRef]
46. Ho, S.P.; Liu, L.Y. How to evaluate and invest in emerging A/E/C technologies under uncertainty. *J. Constr. Eng. Manag.* **2003**, *129*, 16–24. [CrossRef]
47. Gil, N.; Tommelein, I.D.; Stout, A.; Garrett, T. Embodying product and process flexibility to cope with challenging project deliveries. *J. Constr. Eng. Manag.* **2005**, *131*, 439–448. [CrossRef]
48. Langford, D.A.; Male, S.P. *Strategic Management in Construction*, 2nd ed.; Blackwell Science: London, UK, 2001.
49. Lansley, P.R. A practical approach to auditing organizational flexibility. *Constr. Manag. Econ.* **2006**, *1*, 145–156. [CrossRef]
50. Olsson, N.O. Management of flexibility in projects. *Int. J. Proj. Manag.* **2006**, *24*, 66–74. [CrossRef]
51. Eriksson, P.E.; Larsson, J.; Pesamaa, O. Managing complex projects in the infrastructure sector-a structural equation model for flexibility-focused project management. *Int. J. Proj. Manag.* **2017**, *35*, 1512–1523. [CrossRef]
52. Thomke, S.; Reinertsen, D. Agile product development: Managing development flexibility in uncertain environments. *Calif. Manag. Rev.* **1998**, *41*, 8–30. [CrossRef]
53. Vickery, S.N.; Calantone, R.; Dröge, C. Supply chain flexibility: An empirical study. *J. Supply Chain Manag.* **1999**, *35*, 16–24. [CrossRef]
54. Martínez-Sánchez, Á.; Vela-Jimenez, M.; Abella-Garces, S.; Gorgemans, S. Flexibility and innovation: Moderator effects of cooperation and dynamism. *Pers. Rev.* **2019**, *48*, 1548–1564. [CrossRef]
55. Chou, M.C.; Teo, C.P.; Zheng, H. Process flexibility: Design, evaluation, and applications. *Flex. Serv. Manuf.* **2008**, *20*, 59–94. [CrossRef]
56. Singh, A.; Jampel, G. Leadership flexibility space. *J. Manag. Eng.* **2010**, *26*, 176–188. [CrossRef]
57. Baron, L.; Rouleau, V.; Grégoire, S.; Baron, C. Mindfulness and leadership flexibility. *J. Manag. Dev.* **2018**, *37*, 165–177. [CrossRef]
58. Zhang, L.; He, J.; Zhou, S. Sharing tacit knowledge for integrated project team flexibility: Case study of integrated project delivery. *J. Constr. Eng. Manag.* **2013**, *139*, 795–804. [CrossRef]
59. Cray, D.; McKay, R.; Mittelman, R. Cultural intelligence and mindfulness: Teaching MBAs in Iran. *J. Int. Educ. Bus.* **2018**, *11*, 220–240. [CrossRef]
60. Song, H.; Zhu, F.; Klakegg, O.J.; Wang, P. Relationship between contractual flexibility and contractor's cooperative behavior. *Int. J. Manag. Proj. Bus.* **2018**, *11*, 382–405. [CrossRef]
61. Moreno-Luzon, M.D.; Gil-Marques, M.; Valls-Pasola, J. TQM, innovation and the role of cultural change. *Ind. Manag. Data Syst.* **2013**, *113*, 1149–1168. [CrossRef]
62. Dai, Y.; Goodale, J.C.; Byun, G.; Ding, F. Strategic flexibility in new high-technology ventures. *J. Manag. Stud.* **2018**, *55*, 265–294. [CrossRef]
63. Liu, A.M.M.; Chan, I.Y.S. Understanding the interplay of organizational climate and leadership in construction innovation. *J. Manag. Eng.* **2017**, *33*, 04017021. [CrossRef]
64. Mathews, J.A. Competitive advantages of the latecomer firm: A resource-based account of industrial catch-up strategies. *Asia Pac. J. Manag.* **2002**, *19*, 467–488. [CrossRef]
65. Lund, R. *Organizational and Innovative Flexibility Mechanisms and Their Impact upon Organizational Effectiveness*; DRUID Working Paper 1998 (23); Aalborg University: Aalborg, Denmark, 1998.
66. Ozorhon, B. Analysis of construction innovation process at project level. *J. Manag. Eng.* **2013**, *29*, 455–463. [CrossRef]
67. Yu, M.C.; Lu, Y.J.; Li, C.; Lin, H.; Shapira, P. More is less? The curvilinear effects of political ties on corporate innovation performance. *Technol. Econ. Dev. Econ.* **2019**, *25*, 1309–1335. [CrossRef]
68. Chen, J.; Yin, X.; Li, J. Firm innovation system: Paths for enhancing corporate indigenous innovation capability. *Front. Eng. Manag.* **2020**, *7*, 404–412. [CrossRef]
69. Lin, H.; Zeng, S.; Liu, H.; Li, C. Bridging the gaps or fecklessness? A moderated mediating examination of intermediaries' effects on corporate innovation. *Technovation* **2020**, *94*, 102018. [CrossRef]

70. Mothe, C.; Thi, T.U.N. The link between non-technological innovations and technological innovation. *Eur. J. Innov. Manag.* **2010**, *13*, 313–332. [CrossRef]
71. Birkinshaw, J.M.; Mol, M.J. How management innovation happens. *MIT Sloan Manag. Rev.* **2006**, *47*, 81–88.
72. Birkinshaw, J.; Hamel, G.; Mol, M.J. Management innovation. *Acad. Manag. Rev.* **2008**, *33*, 825–845. [CrossRef]
73. Su, S.; Baird, K. The role of leaders in generating management innovation. *Int. J. Hum. Resour. Manag.* **2017**, *29*, 2758–2779. [CrossRef]
74. Kraśnicka, T.; Głód, W.; Wronka-Pośpiech, M. Management innovation, pro-innovation organisational culture and enterprise performance: Testing the mediation effect. *Rev. Manag. Sci.* **2018**, *12*, 737–769. [CrossRef]
75. Allahar, H. A management innovation approach to project planning. *Technol. Innov. Manag. Rev.* **2019**, *9*, 4–13. [CrossRef]
76. Ozorhon, B.; Oral, K.; Demirkesen, S. Investigating the components of innovation in construction projects. *J. Manag. Eng.* **2016**, *32*, 04015052. [CrossRef]
77. Ozorhon, B.; Abbott, C.; Aouad, G. Integration and leadership as enablers of innovation in construction: Case study. *J. Manag. Eng.* **2014**, *30*, 256–263. [CrossRef]
78. Liu, A.M.M.; Chan, I.Y.S. Critical role of the learning transfer climate in fostering innovation in construction. *J. Manag. Eng.* **2017**, *33*, 04016050. [CrossRef]
79. Lijauco, F.; Gajendran, T.; Brewer, G.; Rasoolimanesh, S.M. Impacts of culture on innovation propensity in small to medium enterprises in construction. *J. Constr. Eng. Manag.* **2020**, *146*, 04019116. [CrossRef]
80. Brockmann, C.; Brezinski, H.; Erbe, A. Innovation in construction megaprojects. *J. Constr. Eng. Manage.* **2016**, *142*, 04016059. [CrossRef]
81. Meng, X.; Brown, A. Innovation in construction firms of different sizes: Drivers and strategies. *Eng. Constr. Archit. Manag.* **2018**, *25*, 1210–1225. [CrossRef]
82. Crossan, M.M.; Apaydin, M. A multi-dimensional framework of organizational innovation: A systematic review of the literature. *J. Manag. Stud.* **2010**, *47*, 1154–1191. [CrossRef]
83. Oke, A. A framework for analysing manufacturing flexibility. *Int. J. Oper. Prod. Manag.* **2005**, *25*, 973–996. [CrossRef]
84. Riolli, L.; Savicki, V. Information system organizational resilience. *Omega* **2003**, *31*, 227–233. [CrossRef]
85. Chan, I.Y.S.; Liu, A.M.M.; Fellows, R. Role of leadership in fostering an innovation climate in construction firms. *J. Manag. Eng.* **2014**, *30*, 06014003. [CrossRef]
86. Pellicer, E.; Yepes, V.; Correa, C.L.; Alarcón, L.F. Model for systematic innovation in construction companies. *J. Constr. Eng. Manag.* **2014**, *140*, B4014001. [CrossRef]
87. Manley, K.; McFallan, S.; Kajewski, S. Relationship between construction firm strategies and innovation outcomes. *J. Constr. Eng. Manag.* **2009**, *135*, 764–771. [CrossRef]
88. Subramaniam, M.; Youndt, M.A. The influence of intellectual capital on the types of innovative capabilities. *Acad. Manag. J.* **2005**, *48*, 450–463. [CrossRef]
89. Zhang, Y.; Khan, U.; Lee, S.; Salik, M. The influence of management innovation and technological innovation on organization performance: A mediating role of sustainability. *Sustainability* **2019**, *11*, 495. [CrossRef]
90. Zhang, J.; Chang, Y.; Zhang, L.; Li, D. Do technological innovations promote urban green development?—A spatial econometric analysis of 105 cities in China. *J. Clean. Prod.* **2018**, *182*, 395–403. [CrossRef]
91. Li, G.; Wang, X.; Su, S.; Su, Y. How green technological innovation ability influences enterprise competitiveness. *Technol. Soc.* **2019**, *59*, 101136. [CrossRef]
92. Wen, Q.; Chen, Y.; Hong, J.; Chen, Y.; Ni, D.; Shen, Q. Spillover effect of technological innovation on CO_2 emissions in China's construction industry. *Build. Environ.* **2020**, *171*, 106653. [CrossRef]
93. Damanpour, F.; Szabat, K.A.; Evan, W.M. The relationship between types of innovation and organizational performance. *J. Manag. Stud.* **1989**, *26*, 587–602. [CrossRef]
94. Du, Z. A Literature Review on Institutional Environment and Technological Innovation. *Am. J. Ind. Bus. Manag.* **2018**, *8*, 1941–1950. [CrossRef]
95. Alofan, F.; Chen, S.; Tan, H. National cultural distance, organizational culture, and adaptation of management innovations in foreign subsidiaries: A fuzzy set analysis of TQM implementation in Saudi Arabia. *J. Bus. Res.* **2020**, *109*, 184–199. [CrossRef]
96. Khosravi, P.; Newton, C.; Rezvani, A. Management innovation: A systematic review and meta-analysis of past decades of research. *Eur. Manag. J.* **2019**, *37*, 694–707. [CrossRef]
97. Camison, C.; López, A.V. An examination of the relationship between manufacturing flexibility and firm performance. *Int. J. Oper. Prod. Manag.* **2010**, *30*, 853–878. [CrossRef]
98. Zhang, Z.; Yuan, Z.; Ni, G.; Lin, H.; Lu, Y. The quality traceability system for prefabricated buildings using blockchain: An integrated framework. *Front. Eng. Manag.* **2020**, *7*, 528–546. [CrossRef]
99. Young-Ybarra, C.; Wiersema, M. Strategic flexibility in information technology alliances: The influence of transaction cost economics and social exchange theory. *Organ. Sci.* **1999**, *10*, 439–459. [CrossRef]
100. Phillips, P.A.; Wright, C. E-business's impact on organizational flexibility. *J. Bus. Res.* **2009**, *62*, 1071–1080. [CrossRef]
101. Hamel, G. The why, what, and how of management innovation. *Harv. Bus. Rev.* **2006**, *84*, 72–84. [PubMed]
102. Shen, L.Y.; Lu, W.; Shen, Q.; Li, H. A computer-aided decision support system for assessing a contractor's competitiveness. *Autom. Constr.* **2003**, *12*, 577–587. [CrossRef]

103. Tan, Y.T.; Shen, L.Y.; Yam, M.C.; Lo, A.A. Contractor key competitiveness indicators (KCIs): A Hong Kong study. *Build. Environ.* **2007**, *18*, 17–32.
104. Yang, J. The impact of knowledge sharing on organizational learning and effectiveness. *J. Knowl. Manage.* **2007**, *11*, 83–90. [CrossRef]
105. Fang, D.; Wu, C.; Wu, H. Impact of the supervisor on worker safety behavior in construction projects. *J. Manag. Eng.* **2015**, *31*, 04015001. [CrossRef]
106. Fornell, C.; Larcker, D.F. Evaluating structural equation models with unobservable variables and measurement error. *J. Mark. Res.* **1981**, *18*, 39–50. [CrossRef]
107. Chin, W.W.; Marcolin, B.L.; Newsted, P.R. A partial least squares latent variable modeling approach for measuring interaction effects: Results from a Monte Carlo simulation study and an electronic-mail emotion adoption study. *Inf. Syst. Res.* **2003**, *14*, 189–217. [CrossRef]
108. Wei, Z.; Song, X.; Wang, D. Manufacturing flexibility, business model design, and firm performance. *Int. J. Prod. Econ.* **2017**, *193*, 87–97. [CrossRef]
109. Hair, J.F.; Anderson, R.E.; Tatham, R.L.; Black, W.C. *Multivariate Data Analysis*, 5th ed.; Prentice Hall: Upper SaddleRiver, NJ, USA, 1998.
110. Kim, B.J. Unstable jobs cannot cultivate good organizational citizens: The sequential mediating role of organizational trust and identification. *Int. J. Environ. Res. Public Health* **2019**, *16*, 1102. [CrossRef] [PubMed]
111. Wang, L.; Wang, S.; Tao, P. R&D Investment, Innovation Output and Enterprise Competitiveness: A Perspective of Employee Education. *J. Phys. Conf. Ser.* **2020**, *1616*. [CrossRef]
112. Yang, Y.; Guo, L.; Zhong, Z.; Zhang, M. Selection of Technological Innovation for Service-Orientated Enterprises. *Sustainability* **2018**, *10*, 3906. [CrossRef]
113. Kim, D.G.; Choi, S.O. Impact of Construction IT Technology convergence innovation on business performance. *Sustainability* **2018**, *10*, 3972. [CrossRef]
114. Giuliani, P.; Robert, M.; Roy, F.L. Reinvention of management innovation for successful implementation. *Int. J. Entrep. Small Bus.* **2018**, *34*, 343–361. [CrossRef]
115. Calvo, R.; Domingo, R.; Sebastián, M.A. Operational flexibility quantification in a make-to-order assembly system. *Int. J. Flex. Manuf. Sys.* **2007**, *19*, 247–263. [CrossRef]
116. Fisch, J.H.; Zschoche, M. The role of operational flexibility in the expansion of international production networks. *Strateg. Manage. J.* **2012**, *33*, 1540–1556. [CrossRef]
117. Javalgi, R.; Whipple, T.W.; Ghosh, A.K.; Young, R. Market orientation, strategic flexibility, and performance: Implications for services providers. *J. Serv. Mark.* **2005**, *19*, 212–221. [CrossRef]

sustainability

MDPI

Article

Multi-Criteria Ranking of Green Materials According to the Goals of Sustainable Development

Amirhossein Balali [1], Alireza Valipour [1,*], Edmundas Kazimieras Zavadskas [2,*] and Zenonas Turskis [2]

1 Department of Civil Engineering, Shiraz Branch, Islamic Azad University, Shiraz 5-71993, Iran; a.balali@iaushiraz.ac.ir

2 Institute of Sustainable Construction, Vilnius Gediminas Technical University, LT–10223 Vilnius, Lithuania; zenonas.turskis@vgtu.lt

* Correspondence: vali@iaushiraz.ac.ir (A.V.); edmundas.zavadskas@vgtu.lt (E.K.Z.)

Received: 19 October 2020; Accepted: 12 November 2020; Published: 14 November 2020

Abstract: Modern, well-educated and experienced policy-makers support and promote the use of environmentally friendly materials and resources. The use of green resources is an exceptional and inevitable strategy to meet the needs of a rapidly growing Earth population. The growing population raises the need for new housing construction and urban infrastructure development. Such substances in construction refer to green building materials (GBMs). The environmental impact is lower if GBMs replace non-GBMs. Here, ranking among GBMs can facilitate and support the selection process. This study aimed to contribute to the body of knowledge to introduce a method for identifying and prioritizing GBMs in the construction industry to use in green building. The required data were collected using existing literature, interviews and questionnaires. Relevant Sustainable Development Goals (SDGs) are the first criteria for assessing GBM selection criteria. Critical weighted GBM selection criteria are the second criteria for prioritizing GBMs. The results show that "Natural, Plentiful and Renewable", "Affordability from cradle to gate" and "Affordability during operation" are the top three GBM selection criteria. The real case study helped select "Stramit Strawboard", "Aluminium Composite Panels (ACPs)" and "Solar Roof Tiles" as the most suitable GBMs for use in the context of the study. The model and results presented in this study will help actors of the construction industry to select and use GBMs more quickly and thus achieve a better level of construction sustainability, as well as environmental friendliness, than before.

Keywords: green building materials (GBMs); building industry; Sustainable Development Goals (SDGs); construction industry; MCDM; SWARA method; COPRAS method; real case study

1. Introduction

The vast majority of human activities in the modern world affect the environment. In most cases, this is a negative factor. It is essential to find the best solutions that cause the least possible conflict between people's wellbeing, their activities and the environment instead of looking for answers to such disputes [1]. Buildings are an integral part of all societies as they provide housing for people. Unfortunately, the building sector is known as one of the biggest energy-consuming sectors, and exploiting energy contributes to global climatic change and other environmental issues [2,3]. Previous studies illustrated that the building sector is responsible for consuming over 40% of the total final energy, using approximately 30% of the total resources, producing 45–65% of the waste disposed to landfills and emitting more than 30% of the greenhouse gases in developed countries [4–9].

Although constructing buildings results in environmental issues, there are some ways to decrease its negative impacts. One of the ways to achieve this goal is to consider sustainability in various parts of a building project. According to the definition of the United Nations' World Commission on Environment and Development (WCED) in 1987, sustainability is "development that meets the needs of the present without comprising the ability of the future generation to meet their own needs". The given definition can be thus linked to the low cost of operation and maintenance (O & M), long service life and high energy efficiency as pillars of a sustainable and green building [10].

The consideration of sustainability in the construction industry has delivered some valuable benefits by reducing the extensive impact on the environment through the use of renewable energy, analysing the consequences of design choices over the entire building life cycle, revised energy codes and low environmental impact materials [11]. Low environmental impact materials, also known as green building materials (GBMs), are widely used in the construction industry in order to alleviate the negative impacts of constructing buildings [12].

These materials are usually considered environmentally friendly and environmentally responsible [13,14]. GBMs not only promote health but also help in meeting sustainability goals [15]. Generally, various definitions of greenness in building materials can be summarised into possessing two main concepts, including "being sustainable during whole life-cycle" and "not being hazardous for human health". The former concept can be quantified by the life-cycle assessment (LCA) methodology, in a "cradle to grave" perspective [16]. With regards to the latter concept, GBMs must not lead to indoor types of pollutions constituting radon emissions, biological pollutions, uncomfortable indoor climate conditions and hazardous fibre dispersion [17,18]. Exploiting GBMs in the construction industry results in achieving Sustainable Development Goals (SDGs) in both direct and indirect ways. SDGs are discussed in the next paragraphs.

Sustainable Development Goals (SDGs) consist of 17 primary goals and 169 targets in various parts of sustainability. The mentioned goals are illustrated in Figure 1 [19–22].

Figure 1. Sustainable Development Goals (SDGs) [21].

Some of the SDGs are directly or indirectly related to the construction industry. For instance, Goal 11 is about sustainable cities and communities. Figure 2 illustrates the targets of this goal, which show a sensible relationship [21,23]. The construction industry consumes a large amount of natural resources, water, energy and materials; it can have a dramatic impact on achieving SDGs and some of the global challenges such as climate change, health and wellbeing [24,25]. The World Green Building Council (WGBC) illustrated how using GBMs can help the construction industry to attain SDGs. The mentioned contribution is illustrated in Figure 3 [26].

11.1 By 2030, ensure access for all to adequate, safe and affordable housing and basic services and upgrade slums

11.2 By 2030, provide access to safe, affordable, accessible and sustainable transport systems for all, improving road safety, notably by expanding public transport, with special attention to the needs of those in vulnerable situations, women, children, persons with disabilities and older persons

11.3 By 2030, enhance inclusive and sustainable urbanization and capacity for participatory, integrated and sustainable human settlement planning and management in all countries

11.4 Strengthen efforts to protect and safeguard the world's cultural and natural heritage

11.5 By 2030, significantly reduce the number of deaths and the number of people affected and substantially decrease the direct economic losses relative to global gross domestic product caused by disasters, including water-related disasters, with a focus on protecting the poor and people in vulnerable situations

11.6 By 2030, reduce the adverse per capita environmental impact of cities, including by paying special attention to air quality and municipal and other waste management

11.7 By 2030, provide universal access to safe, inclusive and accessible, green and public spaces, in particular for women and children, older persons and persons with disabilities

11.A Support positive economic, social and environmental links between urban, peri-urban and rural areas by strengthening national and regional development planning

11.B By 2020, substantially increase the number of cities and human settlements adopting and implementing integrated policies and plans towards inclusion, resource efficiency, mitigation and adaptation to climate change, resilience to disasters, and develop and implement, in line with the Sendai Framework for Disaster Risk Reduction 2015-2030, holistic disaster risk management at all levels

11.C Support least developed countries, including through financial and technical assistance, in building sustainable and resilient buildings utilizing local materials

Figure 2. Targets of SDG 11 [21].

SGDs are less considered in all the fields and especially in the construction industry of Iran. An extensive study was conducted through the existing literature in order to find papers that are relevant to SDGs, but only a few were found, and no relevant studies on GBMs within Iran could be identified.

To be able to select the optimum GBM for use in the construction industry, GBM selection criteria are required. Various studies have been conducted to introduce such criteria since 2009 [14,27]. Khoshnava et al. identified GBM's selection criteria using three pillars of sustainability in 2018. According to their findings, the mentioned criteria can be divided into five critical categorisations by considering their characteristics. These categorisations were AF (Affordability), WC (Water Efficiency), EE (Energy Efficiency), IAQ (Indoor Air Quality) and RE (Resource Efficiency), which stand for Affordability, Water Conservation, Energy Efficiency, Indoor Air Quality and Resource Efficiency, respectively [12]. Figure 4 illustrates this categorisation.

Mokal et al. studied the advantages, disadvantages, durability and economic impacts of various GBMs consisting of lime, sand-lime bricks, eco-friendly tiles, coloured lime plaster and reflectasol glass, concluding that GBMs reduce the adverse effects on construction projects [28]. Chauhan and Kamboj (2016) identified different means and needs to go green in the world's construction industry and found that exploiting green materials in the mentioned industry plays an important role in bringing benefits to both humans and the environment [29]. Another study was conducted in order to assess the relative fungal resistance of four pairs of GBMs. It was illustrated that the presence of organic matter in GBMs plays a significant role in their environmental impacts [30].

The general lack of related studies in the context of Iran and the absence of available GBMs makes it hard for regional comparison conceptually and the local industry practices to embrace sustainable practices in the material selection. Therefore, this paper aims to identify GBMs as well as rank them in the construction industry of Shiraz, Iran. The novelty of this study is that two groups of selection criteria were being used together to conduct this ranking. The first group of selection criteria was SDGs. Both relevant SDGs and the existing GBM selection criteria, which exist in the literature, were used to weight selection criteria. Then, these weighted criteria were exploited to rank GBMs. The SWARA and the COPRAS methods were used as the data analysing tools due to their success in solving complex decision-making problems. The findings of this study can be widely used by designers, engineers, managers and contractors in the construction industry in both existing and new buildings.

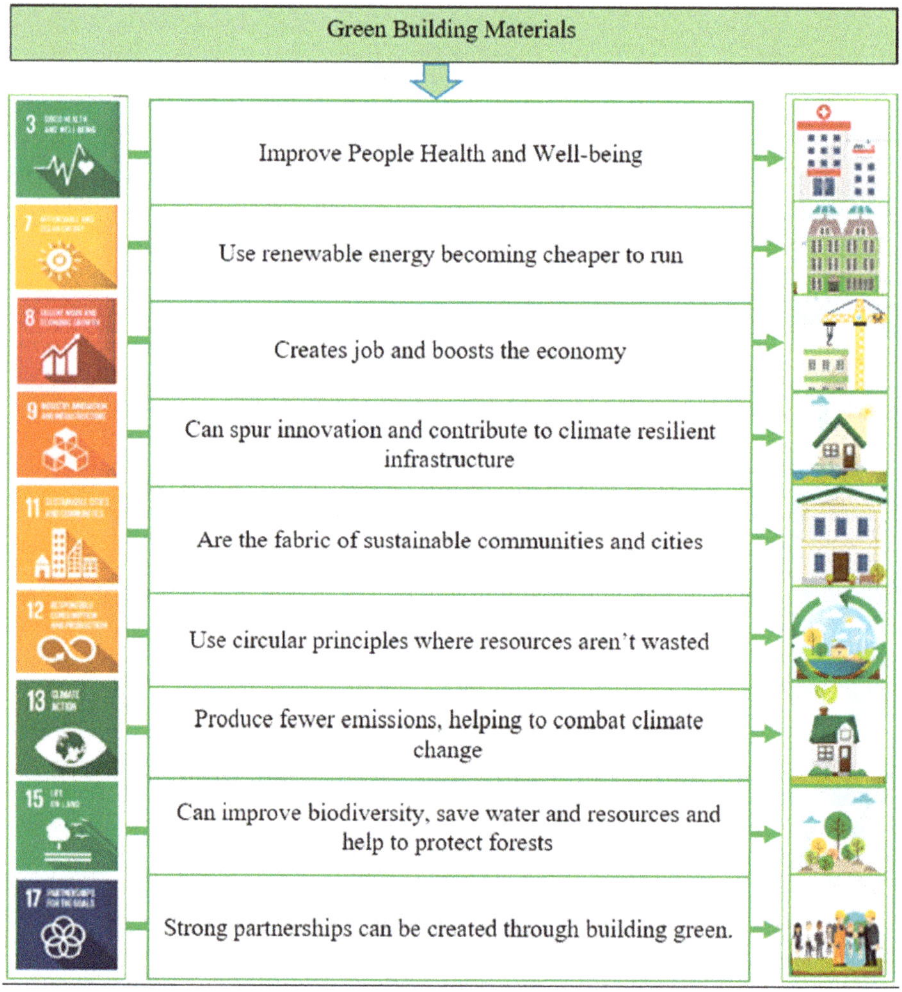

Figure 3. Relationship between green building materials (GBMs) and SDGs [26].

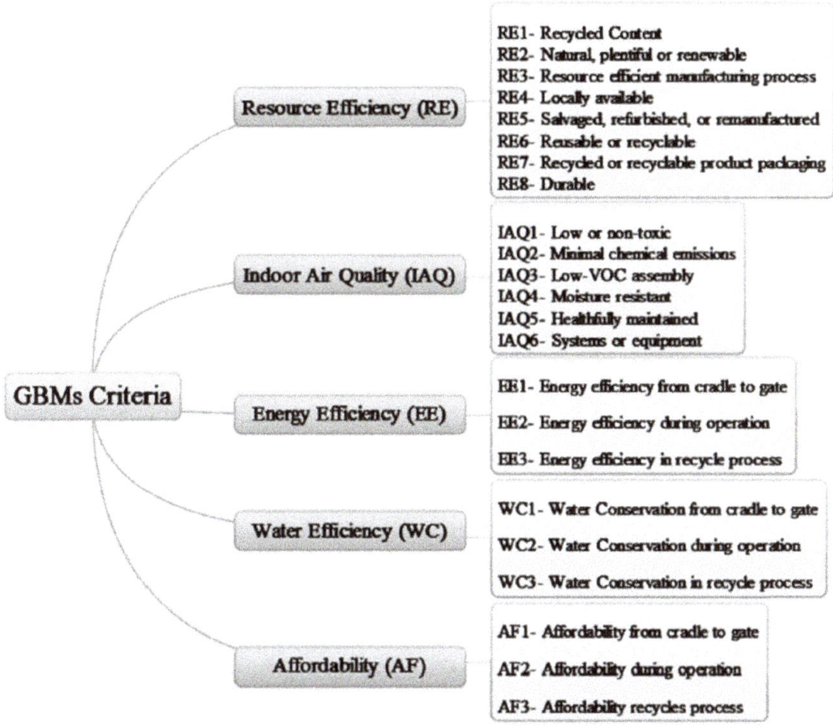

Figure 4. Categorization of GBM's selection criteria [12].

2. Research Methodology

The current paper's research methodology can be divided into four main steps. Firstly, green building materials were identified through conducting an extensive study on the existing literature, including journal papers, books, interviews with experts and online resources [14,16,28]. Then, in the second step, Sustainable Development Goals (SDGs) were studied precisely to identify relevant goals for the construction industry. To do so, several experts were interviewed. Then, using a questionnaire distributed among experts, the relevant goals were weighted using the SWARA method. Data of the questionnaires were gathered, analysed and put in the SWARA process for weighting. The third step focused on identifying and weighting the identified GBM selection criteria according to the specified SDGs using the previous studies and considering experts' opinions. The second questionnaire was used, and the COPRAS method was exploited as the analysis method. Concerning the usage of the COPRAS method, the weighted SDGs and GBM selection criteria were considered simultaneously and put in the process of the COPRAS method. In the final step, the weighted GBM selection criteria were exploited to prioritize GBMs through the second questionnaire using the COPRAS method. The COPRAS method assumes direct and proportional dependence of the significance and utility degree of considered versions based on a system of criteria proportionally explaining the alternatives and on weights and values of the criteria; this is the superiority of the COPRAS method compared to the other MCDM methods. Figure 5 illustrates the research methodology of this study.

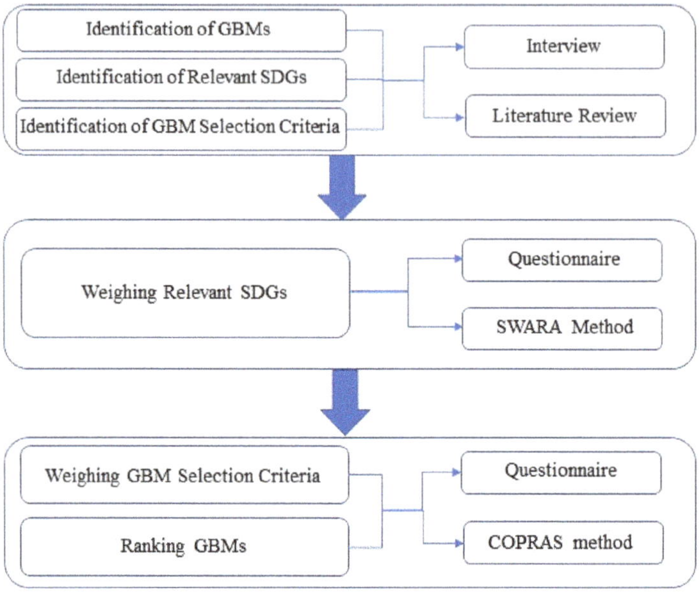

Figure 5. Research methodology.

2.1. *Questionnaire*

Questionnaires are widely used in different research as information collecting tools. In fact, questionnaires provide raw data, which will be analysed later. In this study, three types of questionnaires were exploited. Respondents' general information, including sex, years of experience, educational level and working background, can be seen in Section 3.2. Questionnaire A was used to weight the identified relevant SDGs. In Questionnaire B, the aim was to weight GBM selection criteria according to the weighted SDGs. Finally, questionnaire C was exploited to prioritise GBMs. In all the designed questionnaires, experts were supposed to give scores from 1 to 5, in which 1 and 5 stand for "very inappropriate" and "very appropriate", respectively.

Questionnaires must be reliable. Otherwise, the results are not viable. One of the ways to attain this goal is to compute Cronbach's alpha. Questionnaires that are more reliable possess a higher value of Cronbach's alpha. The value of 0.7 is regarded as an acceptable value [31–33]. In this study, the mentioned coefficient was calculated by SPSS software. Table 1 shows the computed values.

Table 1. Cronbach's alpha values of questionnaires.

Questionnaire	Purpose	Value
A	Obtaining weight of relevant SDGs	0.912
B	Obtaining weight of GBM selection criteria	0.871
C	Ranking GBMs	0.934

2.2. *SWARA (Step-Wise Weight Assessment Ratio Analysis) Method*

Keršuliene et al. introduced the SWARA method in 2010. The technique is exploited to weight criteria. This method has been exploited by numerous researchers [34,35]. Balali et al. used the SWARA method as part of their study to weight passive energy consumption strategies in Iran [36]. Akhanova et al. assessed the building's sustainability by SWARA in Kazakhstan [37]. Prajapati et al. prioritised the solutions of reverse logistics implementation to mitigate its barriers in India using the SWARA method [35]. Valipour et al. assessed risks of deep foundation excavation projects in Malaysia by using this method [38]. Maghsoodi et al. used SWARA to select dam materials in Iran [39]. Jaber assessed construction projects' risks in Iraq by using the SWARA method [40]. Readers are referred to the following papers [41–47] to observe more usages of the SWARA method.

In this study, the identified Sustainable Development Goals were weighted using SWARA. These goals were first ranked by experts from 1 to 5, where 1 and 5 stand for the most and least important goals, respectively. Average values of the returned questionnaires were used for analysis. The procedure of the SWARA method is illustrated below [35,48–53]:

1. Selection criteria are identified.
2. Identified criteria are sorted in terms of relative importance in descending order according to the respondents' points of view.
3. Comparative average value (s_j) is calculated. To do so, the second important ($j-1$) criterion is compared to the first criterion (j), and its relative importance is expressed. The same trend is continued for all the criteria.
4. Coefficient k_j, which stands for comparative importance, is computed according to the following formula:

$$k_j = \begin{cases} 1 & j=1 \\ s_j+1 & j>1 \end{cases} \quad (1)$$

5. Recalculated weights (q_j) are determined:

$$q_j = \begin{cases} 1 & j=1 \\ \frac{q_{j-1}}{k_j} & j>1 \end{cases} \quad (2)$$

6. Relative weights of the selection criteria (w_j) are computed as follows:

$$w_j = \frac{q_j}{\sum_{m=1}^{n} q_m} \quad (3)$$

where n stands for the number of selection criteria.

2.3. The COPRAS Method

Zavadskas and Kaklauskas introduced the complex proportional assessment (COPRAS) in 1994, which is a powerful and useful multi-criteria decision-making (MCDM) tool [54]. This method is usually used as the second tool due to its need for weighted selection criteria. Criteria are supposed to be weighted by other methods like the analytic hierarchy process (AHP), the analytic network process (ANP), the step-wise weight assessment ratio analysis (SWARA) or any other method [55,56] and software [57]. The COPRAS method has been used by many researchers. For instance, Ghose and Pradhan analysed renewable energy sources in India using the fuzzy COPRAS [58]. Tolga and Durak used the fuzzy COPRAS to evaluate innovation projects [59]. Amoozad Mahdiraji et al. exploited the COPRAS to identify and prioritise sustainable architecture factors in Iran [60]. Kundakcı and Işık used the COPRAS as part of their study for selecting a textile company's air compressor [61].

In this study, the COPRAS method was exploited to weight GBM selection criteria, as well as ranking GBMs. The procedure of using the method is presented as follows [54,62–64]:

1. Weighted selection criteria (q_i) were calculated before. Here, alternatives are determined.

2. Matrix X is constructed as below:

$$X = \begin{bmatrix} x_{11} & x_{12} & \cdots & x_{1m} \\ x_{21} & x_{22} & \cdots & x_{2m} \\ \vdots & \vdots & \ddots & \vdots \\ x_{n1} & x_{n2} & \cdots & x_{nm} \end{bmatrix}; i = \overline{1,n} \text{ and } j = \overline{1,m}$$

where i, j, m and n stand for an alternative, its corresponding criteria, number of alternatives and number of criteria, respectively.

3. Decision matrix $\overline{X}$ is normalised as follows:

$$\overline{X}_{ij} = \frac{x_{ij}}{\sum_{j=1}^{n} x_{ij}};\ i = \overline{1,n} \quad (4)$$

4. Weighted-normalised decision matrix $(\hat{X})$ is calculated in which the values are computed according to the formula below:

$$\hat{x}_{ij} = \overline{x}_{ij}.q_j;\ i = \overline{1,n} \text{ and } j = \overline{1,m} \quad (5)$$

where the importance of the ith criterion is shown with q_i.

5. Beneficial and non-beneficial (positive and negative) attributes are calculated using the following formulas:

$$P_{i\,+} = \sum_{j=1}^{k} \hat{x}_{ij} \quad (6)$$

$$P_{i-} = \sum_{j=k\,+\,1}^{m} x_{ij} \quad (7)$$

6. Minimum value of P_{i-} is calculated as follows:

$$P_{min-} = \min P_{i-};\ i = \overline{1,n} \quad (8)$$

7. The importance degree of each alternative is calculated and illustrated by Q_i:

$$P_{min-} = \min P_{i-};\ i = \overline{1,n} \quad (9)$$

8. Optimality criterion (K) is determined as below:

$$K = maxQ_i;\ i = \overline{1,n} \quad (10)$$

9. Alternatives' order ranking is determined according to Q_i.
10. Finally, the utility degree of each alternative is computed:

$$N_i = \frac{Q_i}{Q_{max}} \times 100\% \quad (11)$$

3. Application of the Model

The current study's purpose was identifying and ranking various green building materials for Shiraz, Iran. The generated results of this paper can be used in the buildings that are located in Shiraz and other cities that possess similar conditions. Two hundred building specialists were identified and contributed to the survey to attain this goal. By using the SWARA and the COPRAS methods, relevant SDGs to this topic, GBM selection criteria and GBMs themselves were weighted and ranked.

3.1. Case Study

Shiraz is one of the most populous cities in southwestern Iran [65]. The Shiraz climate is moderate [66,67]. Humidity and temperature difference between days and nights in Shiraz are vital factors influencing building materials and construction projects [68]. The municipality of Shiraz has reported that, according to the latest statistics, more than 1,500,000 people live in Shiraz [38]. Due to the growing demand for housing, many buildings are growing. It seems necessary to try to use GBMs in such projects. In this way, less damage would be done to the environment, leading to greater sustainability than before.

3.2. Sample Size

Building specialists who take part in building construction projects of Shiraz, Iran, were considered as the sample size of this study. The formula used for achieving the number of required specialists is shown as follows [69]:

$$SS = \frac{z^2 p(1-p)}{c^2} \tag{12}$$

where SS stands for the calculated sample size, z stands for the confidence level value, p stands for percentage picking a choice, and c stands for a confidence interval. Then, the corrected sample size was computed as follows:

$$Corrected\ SS = \frac{SS}{1 + \left(\frac{SS-1}{pop}\right)} \tag{13}$$

where rr stands for response rate.

Two hundred professional building experts were considered as the sample size. In this study, to get an acceptable result, the values of the variables were as follows. Percentage picking a choice (p) was considered as 0.5. The confidence level value (z) was taken as 95%. The confidence interval was also considered 10%. According to the conducted calculations, this survey required at least 116 questionnaires to be filled out by experts, which was considered. Table 2 describes general information about specialists.

Table 2. Cronbach's alpha values of questionnaires.

Category	Classification	Number
Occupation	Academia	33
	Manager	48
	Contractor	21
	Technician	22
Sex	Male	74
	Female	50
Experience (years)	<5	22
	5–10	14
	10–15	30
	>15	58

4. Results and Discussion

4.1. Identification and Allocation of Weights to the Relevant Sustainable Development Goals (SDGs)

Sustainable Development Goals include 17 primary goals, as well as 169 corresponding targets in vast areas [21]. Some of these goals and targets are related to the construction industry. Due to the profound impact of the construction industry on the environment and society, it is important to find relevant SDGs and consider them in decision-making problems. Various studies have taken place that showed the impact of the construction industry on achieving SDGs [24,25].

Thus, identifying relevant SDGs was the first stage of this research. A large number of building specialists, constituting both academic and building industry experts, were identified and interviewed. Finally, five relevant SDGs were identified according to three pillars of sustainability goals such as economy, environment and society. In the interviews' questions, the mentioned pillars were considered. The identified SDGs were G7, G9, G11, G12 and G17. These goals are illustrated in Table 3.

Table 3. Relevant Identified SDGs.

Sign	SDG	Nature
G7	Affordable and Clean Energy	Benefit
G9	Industry, Innovation and Infrastructure	Benefit
G11	Sustainable Cities and Communities	Benefit
G12	Responsible Consumption and Production	Benefit
G17	Partnerships for the Goals	Benefit

These goals were then analysed by the SWARA method to obtain their weights [34]. To do that, Questionnaire A was distributed among specialists to prioritise the goals. Table 4 shows the outcome of this part of the study.

Table 4. Weight of each selection criterion.

Criteria	S_j	$K_j = s_j+1$	q_j	w_j	Rank
G9	—	1	1	0.43	1
G7	0.63	1.63	0.61	0.26	2
G11	0.63	1.63	0.38	0.16	3
G12	0.78	1.78	0.21	0.09	4
G17	0.74	1.73	0.12	0.05	5

The results show that "Industry, Innovation and Infrastructure" (G9), "Affordable and Clean Energy" (G7), "Sustainable Cities and Communities" (G11), "Responsible Consumption and Production" (G12) and "Partnerships for the Goals" (G17) were ranked first to fifth important SDGs, respectively. Weights of the SDGs are shown in Table 4.

4.2. Identification and Weighting GBM Selection Criteria

GBM selection criteria must be identified in decision-making problems regarding GBMs. Various studies have been conducted to do so [14,27]. For instance, in one study in 2018, GBM selection criteria were identified and categorised into five main groups including affordability (AF), water conservation (WC), energy efficiency (EE), indoor air quality (IAQ) and resource efficiency (RE) [12].

In the current study, after obtaining the weights of the identified relevant SDGs, the next stage was to identify and weight GBM selection criteria. To identify GBM selection criteria, relevant identified SDGs from the last step were considered. GBM selection criteria were put into five categorisations using the findings of a study on the existing literature and conducting interviews with specialists, as well as considering the aims of the relevant SDGs. The descriptions of the relevant SDGs are illustrated in Figure 6 [21]. Finally, 19 criteria were identified [12]. These criteria and their categorisation are illustrated in Table 5.

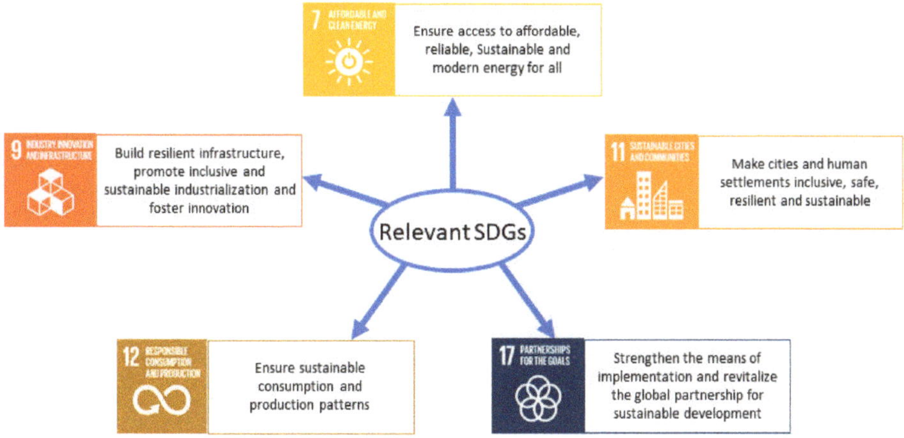

Figure 6. Descriptions of the relevant SDGs [21].

Table 5. Identified GBM selection criteria.

Main GBM Selection Criteria	Sign	GBM Selection Criteria
Resource Efficiency (RE)	RE1	Recycled content
	RE2	Natural, plentiful or renewable
	RE3	Resource-efficient manufacturing process
	RE4	Locally available
	RE5	Salvaged, refurbished or remanufactured
	RE6	Reusable or recyclable
	RE7	Recycled or recyclable product packaging
	RE8	Durable
Indoor Air Quality (IAQ)	IAQ1	Low or non-toxic
	IAQ2	Minimal chemical emissions
	IAQ3	Moisture resistant
	IAQ4	Healthfully maintained
	IAQ5	Systems or equipment
Energy Efficiency (EE)	EE1	Energy efficiency from cradle to gate
	EE2	Energy efficiency during operation
	EE3	Energy efficiency in the recycling process
Affordability (AF)	AF1	Affordability from cradle to gate
	AF2	Affordability during operation
	AF3	Affordability recycles process

Questionnaire B was given to specialists to weight GBM selection criteria, and the analysis took place through the COPRAS method. By considering the calculated criteria weights, the mentioned criteria were ranked. As it was mentioned before, due to the beneficial nature of SDGs, sums of normalised values for non-beneficial criteria do not exist in this study. Therefore, the sums of the weighted normalised values for beneficial criteria (pi +) are calculated and presented in Table 6. Priority values, as well as the quantitative utility of GBM selection criteria, are also calculated and presented in Table 7.

Table 6. The COPRAS results for GBM selection criteria.

GBM Selection Criteria	$P_{i+} = \sum_{j=1}^{k} x_{ij}$
RE1	0.055
RE2	0.070
RE3	0.053
RE4	0.036
RE5	0.043
RE6	0.054
RE7	0.044
RE8	0.041
IAQ1	0.056
IAQ2	0.051
IAQ3	0.060
IAQ4	0.039
IAQ5	0.047
EE1	0.062
EE2	0.045
EE3	0.055
AF1	0.067
AF2	0.064
AF3	0.056

Table 7. GBM selection criteria weights and ranking.

GBM Selection Criteria	$Q_i = P_i + \frac{\sum_{i=1}^{n} M_i}{R_i \sum_{i=1}^{n} \frac{1}{R_i}}$	$N_i = \frac{Q_i}{Q_{max}} 100\%$	Rank
RE1	0.055	78.07	9
RE2	0.070	100	1
RE3	0.053	75.27	11
RE4	0.036	51.69	19
RE5	0.043	61.85	16
RE6	0.054	76.96	10
RE7	0.044	62.69	15
RE8	0.041	58.58	17
IAQ1	0.056	79.92	7
IAQ2	0.051	71.98	12
IAQ3	0.060	85.65	5
IAQ4	0.039	55.12	18
IAQ5	0.047	67.06	13
EE1	0.062	88.46	4
EE2	0.045	63.66	14
EE3	0.055	78.41	8
AF1	0.069	95.05	2
AF2	0.064	91.81	3
AF3	0.056	80.34	6

According to the results, it is illustrated that the top three GBM selection criteria were "Natural, plentiful and renewable" (RE2), "Affordability from cradle to gate" (AF1) and "Affordability during operation" (AF2), respectively.

4.3. Identification and Prioritisation of GBMs

The final stage of this study was identifying and ranking GBMs. Many researchers have identified and investigated GBMs [14]. For instance, according to a survey conducted in India, five GBMs were investigated in terms of economic effects, durability, pros and cons. The mentioned GBMs were reflecting sol glass, coloured lime plaster, eco-friendly tiles, sand-lime bricks and lime [28]. However, it is worth noting that any new building material can be regarded as a green material if it possesses the required properties. Thus, no study can claim that it has identified and investigated all the GBMs. In this study, by considering previous research, as well as specialists' opinions, nine green building materials were finally identified for Shiraz and are shown in Table 8.

Table 8. Identified GBMs.

Sign	GBM Selection Criteria
M1	Cement Plast Artificial Stone
M2	Sand-Lime Bricks
M3	Fibre-Reinforced Concrete
M4	Solar Roof Tiles
M5	Thermochromic Windows
M6	Grasscrete
M7	Stramit Strawboard
M8	Aluminium Composite Panels (ACPs)
M9	Fly Ash Concrete

The purpose of the last questionnaire, questionnaire C, was ranking GBMs. Like in the previous part of the study, experts were asked to use their knowledge, expertise and experience to complete the questionnaire. Ranking of GBMs was conducted using GBM selection criteria weights and exploiting the COPRAS method. Weighted normalised values for beneficial criteria (pi +) were computed and are shown in Table 7.

To weight GBM selection criteria, Questionnaire B was exploited. Each specialist was asked to complete the questionnaire according to their own knowledge and experience, and the scores were analysed by the COPRAS method. As mentioned before, due to the beneficial nature of SDGs, sums of normalised values for non-beneficial criteria did not exist in this study. Therefore, the sums of the weighted normalised values for beneficial criteria (pi +) were computed and are illustrated in Table 9. Priority values, as well as the quantitative utility of GBMs, were also calculated and are presented in Table 10.

Table 9. The COPRAS results for GBM ranking.

GBMs	$P_{i+}=\sum_{j=1}^{k} x_{ij}$
M1	0.107
M2	0.113
M3	0.104
M4	0.122
M5	0.086
M6	0.104
M7	0.130
M8	0.124
M9	0.111

According to the results, it is shown that the top three GBMs were "Stramit Strawboard" (M7), "Aluminium Composite Panels (ACPs)" (M8) and "Solar Roof Tiles" (M4), respectively. As it was mentioned in the previous parts of the study, there has not been a ranking for selecting GBMs. Although GBMs have been discussed separately in other papers, and their benefits and advantages illustrate that the ranking of this study seems sensible and accurate. This ranking can be used by members of the construction industry to assess GBMs in other regions.

Table 10. The COPRAS results for GBM ranking.

GBMs	$Q_i = P_i + \frac{\sum_{i=1}^{n} M_i}{R_i \sum_{i=1}^{n} \frac{1}{R_i}}$	$N_i = \frac{Q_i}{Q_{max}} 100\%$	Rank
M1	0.107	82.87	6
M2	0.113	87.68	4
M3	0.104	80.92	8
M4	0.122	94.97	3
M5	0.086	66.49	9
M6	0.104	81.26	7
M7	0.129	100	1
M8	0.124	96.53	2
M9	0.111	86.36	5

5. Conclusions

Due to a growing number of people and therefore constructing a large number of buildings, it is necessary to attain sustainability using green building materials, which are environmentally friendly. To do so, the existence of a ranking of green building materials can help decision-makers to select suitable GBMs for their projects more manageable. This paper identified several GBMs and prioritised those using the SWARA and the COPRAS methods as tools to analyse options. To move through sustainability, relevant Sustainable Development Goals (SGDs) to the research topic were identified and used as the first selection criteria. These goals were "Affordable and Clean Energy" (G7), "Industry, Innovation and Infrastructure" (G9), "Sustainable Cities and Communities" (G11), "Responsible Consumption and Production" (G12), and "Partnerships for the Goals" (G14). These goals were weighted using the SWARA method, and it was shown that G9, G7 and G11 were the top three goals with weights of 0.431, 0.264 and 0.162, respectively. The next part of the research was associated with identifying and weighting GBM selection criteria according to the weights of the identified SDGs using the COPRAS method. Results show that "Natural, plentiful and renewable" (RE2), "Affordability from cradle to gate" (AF1) and "Affordability during operation" were the top three GBM selection criteria with weights of 0.070, 0.067 and 0.064, respectively. The last part of the study focused on the identification and prioritisation of GBMs. "Stramit Strawboard" (M7), "Aluminium Composite Panels (ACPs)" (M8) and "Solar Roof Tiles" (M4) were the top three GBMs, respectively.

The method used in this study is an appropriate one that can be exploited in other construction industry problems. Members of the construction industry in Shiraz, as well as all the cities possessing similar climatic and economic situations, can use this paper's results. The GBMs identified by this study are highly suggested to be used to move towards sustainability more than before. One of the limitations of this study was considering residential buildings. Therefore, it is suggested that prospective researchers conduct similar studies about other types of buildings such as commercial and industrial buildings. The authors also suggest using other MCDM tools and comparing their obtained results with this study.

Author Contributions: A.B.; Conceptualization, methodology, software, investigation, writing—original draft, A.V.; methodology, visualization, validation, investigation, writing—reviewing and editing, supervision, Z.T.; writing—reviewing and editing, investigation, writing—reviewing and editing, supervision, E.K.Z. All authors have read and agreed to the published version of the manuscript.

Funding: This research received no external funding.

Conflicts of Interest: The authors declare no conflict of interest. The funders had no role in the design of the study; in the collection, analyses, or interpretation of data; in the writing of the manuscript, or in the decision to publish the results.

References

1. Zolfani, S.H.; Zavadskas, E.K.; Turskis, Z. Design of Products with Both International and Local Perspectives based on Yin-Yang Balance Theory and Swara Method. *Econ. Res.-Ekon. Istraživanja* **2013**, *26*, 153–166. [CrossRef]
2. Wang, W.; Zmeureanu, R.; Rivard, H. Applying multi-objective genetic algorithms in green building design optimization. *Build. Environ.* **2005**, *40*, 1512–1525. [CrossRef]
3. Turskis, Z.; Lazauskas, M.; Zavadskas, E.K. Fuzzy multiple criteria assessment of construction site alternatives for non-hazardous waste incineration plant in Vilnius city, applying ARAS-F and AHP methods. *J. Environ. Eng. Landsc. Manag.* **2012**, *20*, 110–120. [CrossRef]
4. Ruzgys, A.; Volvačiovas, R.; Ignatavičius, Č.; Turskis, Z. Integrated evaluation of external wall insulation in residential buildings using SWARA-TODIM MCDM method. *J. Civ. Eng. Manag.* **2014**, *20*, 103–110. [CrossRef]
5. Zagorskas, J.; Zavadskas, E.K.; Turskis, Z.; Burinskienė, M.; Blumberga, A.; Blumberga, D. Thermal insulation alternatives of historic brick buildings in Baltic Sea Region. *Energy Build.* **2014**, *78*, 35–42. [CrossRef]
6. Katsinde, S.M.; Srinivas, S.C. Breast feeding and The Sustainable Development agenda. *Indian J. Pharm. Pr.* **2016**, *9*, 144–146. [CrossRef]
7. Castro-Lacouture, D.; Sefair, J.; Florez, L.; Medaglia, A.L. Optimization model for the selection of materials using a LEED-based green building rating system in Colombia. *Build. Environ.* **2009**, *44*, 1162–1170. [CrossRef]
8. Medineckiene, M.; Zavadskas, E.; Turskis, Z. Dwelling selection by applying fuzzy game theory. *Arch. Civ. Mech. Eng.* **2011**, *11*, 681–697. [CrossRef]
9. Medineckienė, M.; Turskis, Z.; Zavadskas, E.K. Sustainable construction taking into account the building impact on the environment. *J. Environ. Eng. Landsc. Manag.* **2010**, *18*, 118–127. [CrossRef]
10. Medineckiene, M.; Zavadskas, E.; Björk, F.; Turskis, Z. Multi-criteria decision-making system for sustainable building assessment/certification. *Arch. Civ. Mech. Eng.* **2015**, *15*, 11–18. [CrossRef]
11. Bagočius, V.; Zavadskas, E.K.; Turskis, Z. Multi-person selection of the best wind turbine based on the multi-criteria integrated additive-multiplicative utility function. *J. Civ. Eng. Manag.* **2014**, *20*, 590–599. [CrossRef]
12. Khoshnava, S.M.; Rostami, R.; Valipour, A.; Ismail, M.; Rahmat, A.R. Rank of green building material criteria based on the three pillars of sustainability using the hybrid multi criteria decision making method. *J. Clean. Prod.* **2018**, *173*, 82–99. [CrossRef]
13. Dahooei, J.H.; Zavadskas, E.K.; Abolhasani, M.; Vanaki, A.; Turskis, Z. A Novel Approach for Evaluation of Projects Using an Interval–Valued Fuzzy Additive Ratio Assessment (ARAS) Method: A Case Study of Oil and Gas Well Drilling Projects. *Symmetry* **2018**, *10*, 45. [CrossRef]
14. Spiegel, R.; Meadows, D. *Green Building Materials: A Guide to Product Selection and Specification*; John Wiley & Sons: Hoboken, NJ, USA, 2010.
15. Akhtar, N.; Patel, S. *Agro-Industrial Discards and Invasive Weed-Based Lignocelluloses as Green Building Materials: A Pertinent Review*; Springer Science and Business Media LLC: Berlin/Heidelberg, Germany, 2018; pp. 121–130.
16. Franzoni, E. Materials Selection for Green Buildings: Which Tools for Engineers and Architects? *Procedia Eng.* **2011**, *21*, 883–890. [CrossRef]
17. Directive, H.A.T. Council Directive 89/106/EEC of 21 December 1988 on the approximation of laws, regulations and administrative provisions of the Member States relating to construction products. *Off. J. L* **1989**, *40*, 0012–0026.
18. Monti, C. Il Progetto Ecosostenibile: Metodi e Soluzioni per la Casa e la città, Il Progetto Ecosostenibile. 2008. Available online: https://ec.europa.eu/environment/archives/life/publications/otherpub/documents/life_successo_web.pdf (accessed on 12 June 2020).
19. Ede, A.N.; Bamigboye, G.; Olofinnade, O.M.; Omole, D.O.; Adeyemi, G.A.; Ngene, B.U. Impact of reliable built structures in driving the sustainable development goals: A look at Nigerian building structures. In *Proceedings of the 3rd International Conference on African Development Issues (CU-ICADI) Ota-Nigeria, 11–13 May 2016*; Covenant University Press: Ota, Nigeria, 2016; pp. 350–353.
20. McArthur, J.W.; Rasmussen, K. Change of pace: Accelerations and advances during the Millennium Development Goal era. *World Dev.* **2018**, *105*, 132–143. [CrossRef]

21. Sayamov, Y.; Lomonosov Moscow State University, M.V.; Teplov, I. UNESCO and Sustainable Development Goals. *Her. Belgorod Univ. Coop. Econ. Law* **2020**, *4*. [CrossRef]
22. Wofford, D.; Macdonald, S.; Rodehau, C. A call to action on women's health: Putting corporate CSR standards for workplace health on the global health agenda. *Glob. Heal.* **2016**, *12*, 1–12. [CrossRef]
23. United Nations. Make Cities Inclusive, Safe, Resilient and Sustainable. Available online: https://www.un.org/sustainabledevelopment/cities/ (accessed on 16 July 2020).
24. Alawneh, R.; Ghazali, F.; Ali, H.; Asif, M. Assessing the contribution of water and energy efficiency in green buildings to achieve United Nations Sustainable Development Goals in Jordan. *Build. Environ.* **2018**, *146*, 119–132. [CrossRef]
25. Hurlimann, A.; Warren-Myers, G.; Browne, G.R. Is the Australian construction industry prepared for climate change? *Build. Environ.* **2019**, *153*, 128–137. [CrossRef]
26. Omer, M.A.; Noguchi, T. A conceptual framework for understanding the contribution of building materials in the achievement of Sustainable Development Goals (SDGs). *Sustain. Cities Soc.* **2020**, *52*, 101869. [CrossRef]
27. Zhou, C.-C.; Yin, G.-F.; Hu, X.-B. Multi-objective optimization of material selection for sustainable products: Artificial neural networks and genetic algorithm approach. *Mater. Des.* **2009**, *30*, 1209–1215. [CrossRef]
28. Mokal, A.B.; Shaikh, A.I.; Raundal, S.S.; Prajapati, S.J.; Phatak, U.J. Green building materials-a way towards sustainable construction. *Int. J. Appl. Innovat. Eng. Manag.* **2015**, *4*, 244–249.
29. Chauhan, S.; Kamboj, J. A way to Go Sustainable: Identifying Different Means & Need to Go Green in the Sector of Construction World. *Int. J. Civ. Eng. Technol.* **2016**, *7*, 22–32.
30. Hoang, C.P.; Kinney, K.A.; Corsi, R.L.; Szaniszlo, P.J. Resistance of green building materials to fungal growth. *Int. Biodeterior. Biodegradation* **2010**, *64*, 104–113. [CrossRef]
31. Santos, J.R.A. Cronbach's alpha: A tool for assessing the reliability of scales. *J. Ext.* **1999**, *37*, 1–5.
32. Nunnally, J.C. Psychometric theory—25 years ago and now. *Educ. Res.* **1975**, *4*, 7–21.
33. Gliem, J.A.; Gliem, R.R. Calculating, interpreting, and reporting Cronbach's alpha reliability coefficient for Likert-type scales. In *Proceedings of 29th Annual Midwest Research-to-Practice Conference in Adult, Continuing, and Community, Michigan State University, 26–28 September 2010*; Glowacki-Dudka, M., Ed.; Michigan State University: East Lansing, MI, USA, 2010.
34. Keršulienė, V.; Zavadskas, E.K.; Turskis, Z. Selection of rational dispute resolution method by applying new step-wise weight assessment ratio analysis (SWARA). *J. Bus. Econ. Manag.* **2010**, *11*, 243–258. [CrossRef]
35. Prajapati, H.; Kant, R.; Shankar, R. Prioritizing the solutions of reverse logistics implementation to mitigate its barriers: A hybrid modified SWARA and WASPAS approach. *J. Clean. Prod.* **2019**, *240*, 118219. [CrossRef]
36. Balali, A.; Hakimelahi, A.; Valipour, A. Identification and prioritization of passive energy consumption optimization measures in the building industry: An Iranian case study. *J. Build. Eng.* **2020**, *30*, 101239. [CrossRef]
37. Akhanova, G.; Nadeem, A.; Kim, J.R.; Azhar, S. A multi-criteria decision-making framework for building sustainability assessment in Kazakhstan. *Sustain. Cities Soc.* **2020**, *52*, 101842. [CrossRef]
38. Valipour, A.; Yahaya, N.; Noor, N.M.; Antuchevičienė, J.; Tamošaitienė, J. Hybrid SWARA-COPRAS method for risk assessment in deep foundation excavation project: An iranian case study. *J. Civ. Eng. Manag.* **2017**, *23*, 524–532. [CrossRef]
39. Maghsoodi, A.I.; Maghsoodi, A.I.; Poursoltan, P.; Antucheviciene, J.; Turskis, Z. Dam construction material selection by implementing the integrated SWARA–CODAS approach with target-based attributes. *Arch. Civ. Mech. Eng.* **2019**, *19*, 1194–1210. [CrossRef]
40. Jaber, A.Z. Assessment Risk in Construction Projects in Iraq Using COPRAS-SWARA Combined Method. *J. Southwest Jiaotong Univ.* **2019**, *54*. [CrossRef]
41. Karabasevic, D.; Stanujkic, D.; Urosevic, S.; Maksimovic, M. Selection of candidates in the mining industry based on the application of the SWARA and the MULTIMOORA methods. *Acta Montan. Slovaca* **2015**, *20*, 116–124.
42. Alimardani, M.; Zolfani, S.H.; Aghdaie, M.H.; Tamošaitienė, J. A novel hybrid SWARA and VIKOR methodology for supplier selection in an agile environment. *Technol. Econ. Dev. Econ.* **2013**, *19*, 533–548. [CrossRef]
43. Perçin, S. An integrated fuzzy SWARA and fuzzy AD approach for outsourcing provider selection. *J. Manuf. Technol. Manag.* **2019**, *30*, 531–552. [CrossRef]

44. Zarbakhshnia, N.; Soleimani, H.; Ghaderi, H. Sustainable third-party reverse logistics provider evaluation and selection using fuzzy SWARA and developed fuzzy COPRAS in the presence of risk criteria. *Appl. Soft Comput.* **2018**, *65*, 307–319. [CrossRef]
45. Zolfani, S.H.; Pourhossein, M.; Yazdani, M.; Zavadskas, E.K. Evaluating construction projects of hotels based on environmental sustainability with MCDM framework. *Alex. Eng. J.* **2018**, *57*, 357–365. [CrossRef]
46. Ghorabaee, M.K.; Amiri, M.; Zavadskas, E.K.; Antucheviciene, J. A new hybrid fuzzy MCDM approach for evaluation of construction equipment with sustainability considerations. *Arch. Civ. Mech. Eng.* **2018**, *18*, 32–49. [CrossRef]
47. Ighravwe, D.E.; Oke, S.A. A multi-criteria decision-making framework for selecting a suitable maintenance strategy for public buildings using sustainability criteria. *J. Build. Eng.* **2019**, *24*, 100753. [CrossRef]
48. Stanujkic, D.; Karabasevic, D.; Zavadskas, E.K. A framework for the Selection of a packaging design based on the SWARA method. *Eng. Econ.* **2015**, *26*, 181–187. [CrossRef]
49. Vafaeipour, M.; Zolfani, S.H.; Varzandeh, M.H.M.; Derakhti, A.; Eshkalag, M.K. Assessment of regions priority for implementation of solar projects in Iran: New application of a hybrid multi-criteria decision making approach. *Energy Convers. Manag.* **2014**, *86*, 653–663. [CrossRef]
50. Keršulienė, V.; Turskis, Z. Integrated fuzzy multiple criteria decision making model for architect selection. *Technol. Econ. Dev. Econ.* **2012**, *17*, 645–666. [CrossRef]
51. Alinezhad, A.; Khalili, J. SWARA Method. In *Fundamentals of Traffic Simulation*; Springer Science and Business Media LLC: Berlin/Heidelberg, Germany, 2019; pp. 99–102.
52. Zavadskas, E.K.; Čereška, A.; Matijošius, J.; Rimkus, A.; Bausys, R. Internal Combustion Engine Analysis of Energy Ecological Parameters by Neutrosophic MULTIMOORA and SWARA Methods. *Energies* **2019**, *12*, 1415. [CrossRef]
53. Ghenai, C.; AlBawab, M.; Bettayeb, M. Sustainability indicators for renewable energy systems using multi-criteria decision-making model and extended SWARA/ARAS hybrid method. *Renew. Energy* **2020**, *146*, 580–597. [CrossRef]
54. Zavadskas, E.K.; Kaklauskas, A. Determination of an efficient contractor by using the new method of multicriteria assessment. In *International Symposium for "The Organization and Management of Construction". Shaping Theory and Practice*; Langford, D.A., Retik, A., Eds.; St. Edmudsbury Press: Bury St. Edmunds, Suffolk, UK, 1996; Volume 1, pp. 94–104.
55. Turskis, Z.; Dzitac, S.; Stankiuviene, A.; Šukys, R. A Fuzzy Group Decision-making Model for Determining the Most Influential Persons in the Sustainable Prevention of Accidents in the Construction SMEs. *Int. J. Comput. Commun. Control.* **2019**, *14*, 90–106. [CrossRef]
56. Zemlickienė, V.; Turskis, Z. Evaluation of the expediency of technology commercialization: A case of information technology and biotechnology. *Technol. Econ. Dev. Econ.* **2020**, *26*, 271–289. [CrossRef]
57. Erdogan, S.A.; Šaparauskas, J.; Turskis, Z. Decision Making in Construction Management: AHP and Expert Choice Approach. *Procedia Eng.* **2017**, *172*, 270–276. [CrossRef]
58. Ghose, D.; Pradhan, S.; Shabbiruddin. A Fuzzy-COPRAS Model for Analysis of Renewable Energy Sources in West Bengal, India. In *Proceedings of the 2019 IEEE 1st International Conference on Energy, Systems and Information Processing (ICESIP)*; Institute of Electrical and Electronics Engineers (IEEE): New York, NY, USA, 2019; pp. 1–6.
59. Tolga, A.C.; Durak, G. Evaluating Innovation Projects in Air Cargo Sector with Fuzzy COPRAS. In *Proceedings of the Advances in Intelligent Systems and Computing*; Springer Science and Business Media LLC: Berlin/Heidelberg, Germany, 2019; pp. 702–710.
60. Mahdiraji, H.A.; Arzaghi, S.; Stauskis, G.; Zavadskas, E.K. A Hybrid Fuzzy BWM-COPRAS Method for Analyzing Key Factors of Sustainable Architecture. *Sustainability* **2018**, *10*, 1626. [CrossRef]
61. Kundakci, N.; Işık, A.T. Integration of MACBETH and COPRAS methods to select air compressor for a textile company. *Decis. Sci. Lett.* **2016**, *5*, 381–394. [CrossRef]
62. Goswami, S.S.; Mitra, S. Selecting the best mobile model by applying AHP-COPRAS and AHP-ARAS decision making methodology. *Int. J. Data Netw. Sci.* **2020**, *4*, 27–42. [CrossRef]
63. Alinezhad, A.; Khalili, J. COPRAS Method. In *Fundamentals of Traffic Simulation*; Barcelo, K., Ed.; (International Series in Operations Research and Management Science); Springer Science and Business Media LLC: New York, NY, USA, 2019; pp. 87–91.

64. Şahin, R. COPRAS method with neutrosophic sets. In *Fuzzy Multi-criteria Decision-Making Using Neutrosophic Sets*; Kahraman, C., Oray, I., Eds.; (Series, Studies in Fuzziness and Soft Computin); Springer: Cham, Switzerland, 2019; pp. 487–524.
65. Shiraz University of Medical Sciences, Discovering Shiraz. 2015. Available online: http://www.educationiran.com/en/Shiraz/sums/page/18924/Discovering-Shiraz (accessed on 20 July 2020).
66. Rahimi, F.; Goli, A.; Rezaee, R. Hospital location-allocation in Shiraz using Geographical Information System (GIS). *Shiraz E-Med. J.* **2017**, *18*, e57572. [CrossRef]
67. Aref, F. Residents' Attitudes Towards Tourism Impacts: A Case Study of Shiraz, Iran. *Tour. Anal.* **2010**, *15*, 253–261. [CrossRef]
68. Masoudi, M.; Ordibeheshti, F.; Rajaipoor, N. Status and prediction of nitrogen oxides in the air of Shiraz city, Iran. *JJEES* **2019**, *10*, 85–91.
69. Al-Tmeemy, S.M.H.; Rahman, H.A.; Harun, Z. Contractors' perception of the use of costs of quality system in Malaysian building construction projects. *Int. J. Proj. Manag.* **2012**, *30*, 827–838. [CrossRef]

Publisher's Note: MDPI stays neutral with regard to jurisdictional claims in published maps and institutional affiliations.

sustainability

MDPI

Article

Identifying and Assessing Sustainable Value Management Implementation Activities in Developing Countries: The Case of Egypt

Ahmed Farouk Kineber [1,*], Idris Othman [1], Ayodeji Emmanuel Oke [2], Nicholas Chileshe [3] and Mohanad Kamil Buniya [1]

1 Department of Civil & Environmental Engineering, Universiti Technologi PETRONAS, Seri Iskandar 32610, Malaysia; idris_othman@utp.edu.my (I.O.); mohanad_18000491@utp.edu.my (M.K.B.)
2 Department of Quantity Surveying, Federal University of Technology Akure, Akure 340271, Nigeria; emayok@gmail.com
3 UniSA STEM, Scarce Resources and Circular Economy (ScaRCE), University of South Australia, Adelaide 5001, Australia; Nicholas.Chileshe@unisa.edu.au
* Correspondence: Ahmed_17008588@utp.edu.my or A.farouk.kineber@gmail.com

Received: 8 October 2020; Accepted: 28 October 2020; Published: 3 November 2020

Abstract: Value management (VM) has become a useful tool for achieving sustainability in many countries. This paper aims to assess VM implementation and its activities towards achieving sustainable building projects in Egypt. Data were obtained from the literature, followed by a qualitative approach through a semi-structured interview and a quantitative approach via a questionnaire survey. In Cairo and Giza, data were collected from a sample of 200 building professionals using a questionnaire, while exploration of the country's VM activities practice was completed using exploratory factor analysis (EFA), including descriptive statistics such as "frequency analysis" and "measures of central tendencies". The results show that VM awareness was demonstrated by 64% of the sample, which means that they know about VM. On the other hand, most respondents, 85.3%, did not adopt VM and did not receive any VM training. The results also show, through EFA, that the correlations between these activities show five main components: an information phase, a function phase, a creativity phase, an evaluation phase, and a development/presentation phase. This study will help building professionals to eliminate unwanted costs and enhance project sustainability by adopting VM in building projects in developing countries. Finally, the results of this study will enhance building management through the implementation of VM elements, with a view to ensuring value for money and meeting sustainability goals.

Keywords: building projects; sustainability; sustainable construction; value management; exploratory factor analysis; construction management

---

1. Introduction

Building projects significantly influence culture, the environment and the economy over their whole life cycle. Buildings utilize more than 40% of universal power and energy and represent 30% of the total greenhouse gas (GHG) emissions in developed and developing nations [1]. Over 40% of Europe's and the USA's absolute power and energy is consumed via buildings [2,3]. In developing countries, the building field's project sustainability is subservient. These countries have experienced rapid development; therefore, there is no doubt that the building industry provides a crucial role in providing basic living [4]. This field's success indicators are measured by quality, cost, and time of construction projects [5]. The construction industry has experienced significant changes in meeting the developing countries' commercial requirements and goals [6]. It has been reported that the majority of the financial procedures of developing countries are still in the process of up-gradation [7]. Building projects in these cultures are frequently faced with various challenges, including failure to meet the required targets, schedule delays, budget over-runs, and insufficient sustainability [8,9]. Additionally, efforts to study the social budget of building projects in urban residential fields are still inadequate [10]. Taken altogether, the construction industry does not fulfill the aspirations of their respective government, clients, and sustainability targets in developing nations unlike other sectors like banking and manufacturing [11].

Egypt is a developing country with one of the most extreme examples of hazard marketplaces due to joblessness and the low salary scale of the employed [12]. These problems are a result of currency fluctuations, an absence of occupational selections and restrictions in financing standards [13]. It is also one of the most populated countries in North Africa, with rapid population increases occurring from 1950 to 2020 [14]. It is reported that the projected population by 2020 will be more than five times that of the 1950 population, according to average estimates. Consequently, building project shortages and meeting sustainability requirements represent the most significant and rising challenges facing policymakers in Egypt [15]. This problem puts pressure on the government to meet the residents' demands for adequate housing, with sufficient sustainability [16]. The building market is expected to grow due to population growth and urbanization, and between 2008 and 2013, the Egyptian population increased by 9% [15]. Moreover, rural urbanization between 2001 and 2012 has increased by 0.9% [15]. This has stressed the importance of and directed the improvement of "sustainable buildings" that are environmentally friendly and resource-efficient over their development procedures.

Brundtland et al. [17] defined sustainability as "achieving the needs of the current population without compromising future generations' ability to satisfy their own needs". Kibert [18] expressed sustainable construction as building a healthy ecosystem using ecological concepts, and with efficient use of resources. Sustainable construction has commonly been explained as a procedure that initiates before execution and continues after the practical completion of the construction project [19]. Wolstenholme et al. [20] agreed that the building industry should be modernized through implementing effective, integrated, novel, and sustainable building practices. Enhancing sustainability knowledge and understanding at the beginning of a building project is hugely recommended to control its course. Value management (VM) can integrate a sustainable approach at the initial and design stages of project procedures [21]. Society of American Value Engineers (SAVE) [22] recommended that VM is a mechanism that has been confirmed to improve the sustainable value of a project.

VM is a structured, function-oriented, systemic team approach to evaluate the functions, objectives, and costs of a process or facility to maximize its efficiency, by performing the necessary functions defined by customers at the least possible overall expense, coherent with performance criteria [23]. VM studies are frequently organized in the initial stages of a project to accomplish the greatest benefits and resource savings [24]. While costs are the essential objective of VM studies, VM is recognized as a cost-cutting method and a value-improvement technique [25]. The project's budget should not be decreased at the expense of its sustainability and functionality purposes, which would then reduce its value [26]. VM targets to accomplish the anticipated value with lower prices, without compromising the building's quality and performance [27]. VM adoption allows an in-depth evaluation of the objectives and expectations of projects from the customer or investor's perception [28].

Although previous studies have discussed VM benefits, activities, and technological efficiency in several countries, no effort has been made to determine the effects of VM implementation in developing countries. VM methods have not received comparable coverage in the majority of developing countries, including Egypt [29]. It is vital to remember that there is a research gap in this area. There is no formal VM study for the purpose of VM implementation and awareness in Egypt [30]. Subsequently, the implementation of VM standards is vital, as the country experiences low environmentally sustainable projects [31]. Moreover, Egypt had planned to be a country with a prosperous, balanced, and competitive economy. By 2030, the Government of Egypt aims to make Egypt one of the best thirty nations in the world, through the application of various management approaches like VM [32]. As a result, there is a need for VM implementation in Egyptian construction projects [33]. It is was hypothesized that there is a concordance on the importance of VM activities in building projects. Therefore, this paper aims to evaluate the application of VM and explore its activities in building projects in developing countries. For this to happen, mixed-method research is proposed for the first time, to explore VM activities in the Egyptian construction industry. The results from this study will be useful to assist decision-makers in succeeding in their building projects by eliminating unwanted costs and enhancing sustainability by implementing VM. As such, the results could be a game-changer in building projects—not only in Egypt but also in developed nations where buildings projects are implemented through a similar style and procedure [34].

2. VM and the Sustainable Built Environment Industries in Previous Studies

The subject of sustainability has been stressed by existing studies [35]. Transforming strategic sustainability targets and strategies procedures for projects is a complicated procedure [36] and a balance is essential among the social, economic and environmental aspects of sustainability [35,37]. The emergence of sustainability in the building industry has led to a search for practical ways to infuse this concept into existing working environments [21]. The need for sustainable improvement and the innovative corporate social responsibility ethic implemented through companies are drivers that could also encourage the adoption of VM at the primary strategic phases of building projects [38]. VM is an organized and analytical procedure designed to improve value for money by delivering the required functions of projects at the least cost while also paying attention to sustainability requirements [39]. Contemporary viewpoints, however, indicate a more significant role for VM in the identification, explanation, and confirmation of customer expectations and priorities early in the procurement stage [40]. This view aligns VM typically to the project briefing phase [41], but it is not clear how these are understood by professionals in the built environment.

The VM practice focuses on organized workshops from 4-hours (half a day) to 5 days [42]. The modifications in VM workshop days can be affected by factors such as the VM scope (i.e., size and system of the project, the extent to be objective for VM), and the VM stages/phases to be adopted. The methodological approach differs, for example, the US value engineers society (SAVE), highlights a three-step methodology "pre-study, value study, and post-study". The value study aims to implement six phases of the workshop, that is, information phase; function analysis phase; creative phase; evaluation phase; development phase; and presentation phase [22]. However, of these six phases, information, development, and presentation phases could be performed outside the workshop more efficiently and virtually [40]. While sustainability might not be regularly used in VM studies, sustainability dimensions such as a healthy indoor environment, minimization of waste, energy efficiency, aesthetic influences, user comfort, air and quality of water, and low life-cycle costs are regularly considered [43]. Sustainability thoughts can differ from one VM study to another because of the owners' distinctive objectives, the obligation to construction and execution viewpoints, awareness of the VM team members, and time available for the study [43]. For instance, sustainability principles have been adopted in many UK developments as part of a value management study, such as a sustainable building project in Crianlarich, Startfilan [44]; sustainable residential buildings and services in Stirlingshire; and the Katrine Water Project at Loch Katrine, Scotland [45]. Hayles [46] highlighted that, by encouraging major clients to implement VM concepts, a sustainable procedure to building decision making could be achieved. Al-Yousefi [47] revealed many advantages of handling VM as a framework to encourage and launch sustainability ideologies, such as increasing the effective use of tools and resources, enhancing applications and operational maintenance. Kelly et al. [48] concluded that commitment of multi-disciplinary stakeholder members, organized and formal VM study, sustainable concepts adoption as project objectives and focus on project cost delivery encourages the combination of VM with sustainability. Therefore, adopting sustainability via VM is feasible and advisable [49].

In the past three decades, substantial work has been carried out in VM adoption in the construction industry. However, there is a lack of studies comparing VM's current practice and application by the built environment stakeholders; especially in developing countries like Egypt. This study, therefore, explores this gap. The aim of this study therefore is to examine VM activities' understanding and current practice in the built environment with a view to achieving sustainable delivery of building projects.

3. Research Methods

The aim of this research is to enhance the sustainable delivery of building projects in the Egyptian construction industry through the adoption of VM. However, this paper is part of bigger research aimed at examining the impact of VM on the overall success of construction projects. The paper explores necessary VM activities require for sustainable delivery of construction projects through EFA analysis. The study adopted exploratory sequential mixed-method research. A mixed-method study was adopted with a view to enhancing the findings of the study for appropriate discussion. [50]. It is confirmed that using mixed methods to explore the same issue will help discover many recurrent patterns or consistent correlations between variables. Figure 1, adopted from Buniya et al. [51], shows the research process for the study. The study commenced with a critical review of previous studies and then qualitative research was performed by interviewing fifteen specialists from the Egyptian construction industry to refine selected activities obtained through previous studies. Validity and reliability tests were thereafter carried out; findings were also discussed while various conclusions and recommendations were discussed. These items are explained in detail in the subsequent sections of the paper. For instance, Section 2 discusses relevant literature while Section 4.1 explains the qualitative aspect of the study where the questionnaire was adopted. The sections for each of the research process items are indicated in Figure 1.

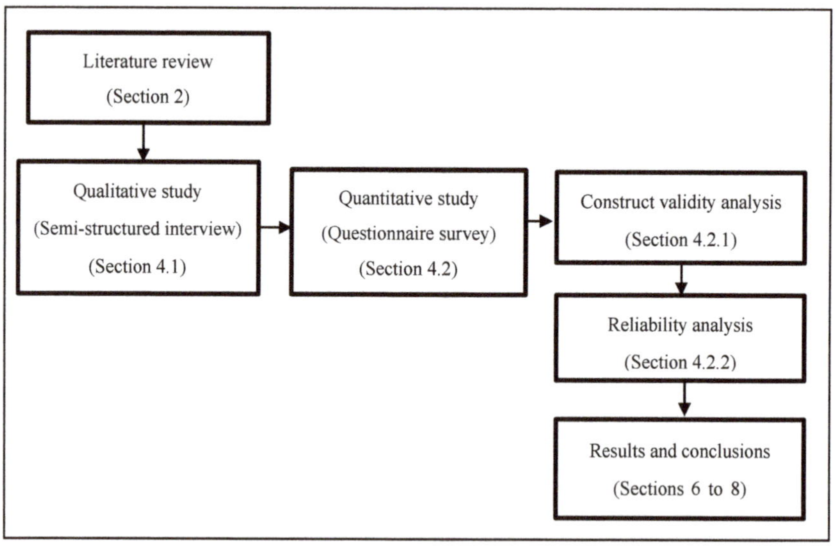

Figure 1. Research flowchart.

4. Research Process

4.1. *Qualitative Study (Semi-Structured Interview)*

VM measurement instrument adopted for the study was drawn from several studies in the area of VM adoption and its activities in the construction industry. A qualitative approach through 15 semi-structured interviews was undertaken with industry specialists to understand the activities that may influence VM implementation in building projects through a purposive sampling approach. This approach allows researchers to achieve the research objectives and control the variation levels between interviewees [52]. Additionally, while 15 interviews can appear to be moderate samples, Mason [53] reports that the sample size is irrelevant in qualitative research because its value is based on data quality. Many studies agreed that from ten to sixteen interviews were considered adequate [33,54].

In light of the different positions played in building projects, the interviewees' profiles indicated that they possess the required skills from the required organizations to contribute adequately to the study. Furthermore, as shown in Table 1, there is a fair representation of various individuals eligible to be members of the construction value management team. This implies that the interviewees possess the required knowledge and experience to provide meaningful and adequate information for the study [55]. This study adopted the abductive approach, a new approach currently becoming popular among modern researchers [56]. This technique uses previous studies to establish a theoretical basis from which a research methodology and analysis can be developed [57]. In other words, if issues can be described as a set of literature proposals, the definition of a framework is similar to the hypothetical deductive concept [56].

Table 1. Semi-structured interviewees' profile.

No.	Position	Education	Experience	Subdivision	Organization
1	Director	Ph.D.	30	Private	Contractor
2	Project manager/Professor	Ph.D.	28	Government	Owner
3	Senior Quantity surveyor	M.Sc.	20	Government	Contractor
4	Senior director	B.Sc.	24	Private	Owner
5	Principle Consultant/Professor	Ph.D.	40	Independent Consultant	Consultant
6	Senior project manager/Associate Professor	Ph.D.	30	Independent Consultant	Consultant
7	President/Professor	Ph.D.	35	Independent Consultant	Consultant
8	Architect	M.Sc.	15	Government	Owner
9	Civil engineer/Associate Professor	Ph.D.	28	Independent Consultant	Consultant
10	Civil engineer	M.Sc.	12	Independent Consultant	Consultant
11	Quantity surveyor	B.Sc.	9	Private	Owner
12	Associate Principle	M.Sc.	25	Independent Consultant	Consultant
13	Architect/Associate Professor	Ph.D.	22	Private sector	Contractor
14	Cost manager	M.Sc.	15	Private	Contractor
15	Civil engineer	B.Sc.	10	Government	Owner

Data analysis is to test the framework, enrich it and possibly extend it [57]. In the current study, existing literature was used to develop theoretical frameworks (VM activities) to design new ideas and take advantage of the necessary operating principles to evaluate the existing concepts. The interviews that follow enrich the framework and extend it. More so, the abduction approach was followed so that the evidence and current theories could be re-examined and analyzed in a local context. Consequently, VM activities were modified and categorized.

4.2. Quantitative Study (Questionnaire Survey)

To validate VM's categorizations and its activities obtained from the semi-structured interview, a pilot study was subsequently adopted before the main study, poised to examine the phases of VM and its activities using a questionnaire survey. This questionnaire survey helped to evaluate the following aspects: (1) Behaviors, opinions, and organizational norms, as well as (2) The link among various variables, primarily cause-and-effect relationships [58]. The questionnaire was subjected to a pilot test as suggested by Fellows and Liu [59] to check the questionnaire's intelligibility, ease of response, and clearness as well as to determine the required time for the survey. The pilot study focused on the perceptions and involvement of developers, consultants, and clients in the delivery of construction projects.

4.2.1. Construct Validity Analysis

The Confirmatory Factor Analysis (CFA) and Exploratory Factor Analysis (EFA) techniques are normally used for factor analysis. In this study, CFA was used for evaluating the structure underlying the adopted variables with a view to properly test the proposed hypotheses. On the other hand, EFA was used to gather information about the relationship among variables and reduce the variables into a few underlying structures. It is one of the functions built into the Statistical Package for the Social Sciences (SPSS) [60].

In the current study, the main multivariate analysis method, that is, EFA, was employed to explore the primary constructs of VM phases after categorizing the same from the interviews. It was used to assess the constructs' validity by evaluating the adequacy of the measurement items of individual constructs (i.e., measurement models) regarding their un-dimensionality, reliability, and validity. It is important to note that Principal Component Analysis (PCA) was selected over Principal Axis Factoring (PAF), image factoring, maximum probability, and alpha factoring since PCA is more accurate and less conceptually complex [61]. PCA is advocated when there is no prior theory or model and when preliminary solutions are found in EFA [60]. Thompson [62] reported that, in many statistical programs, PCA is the default form and is thus most widely used in EFA. The varimax rotation method was preferred over the direct oblimin or Promax because varimax's rotation attempts to optimize load dispersion between variables. Varimax is also suitable for simple factor analysis and is an excellent general approach that simplifies the clarification of factors [63]. More so, the number of participants can be used as a representative sample within acceptable ranges [64]. The 21 variables as well as 200 participants used in the current study, are considered suitable for factor analysis. It is vital to highlight that the sample size and methodology adopted for this research is similar to the study by Kim et al. [65] with 100participants; Luvara and Mwemezi [66] with 231 participants; and Shen and Liu [67] with 200 participants.

4.2.2. Reliability Analysis

A research method's validity and reliability are essential and must be taken into account to cover and ensure accurate results are obtained [68]. However, face and content validity are two traditional validity measures used to assess the extent to which the research instrument elements are significant and are reflective of a targeted research structure [69]. Twelve research experts from both the academic and the building sector were randomly chosen to validate the research instrument. According to Sushil and Verma [70], face validity is evaluated by having experts check the contents of the test to ensure that the items and questions are adequate. Furthermore, Cronbach's α test was used to check the research instrument's reliability.

5. Data Collection

5.1. Semi-Structured Interviews

The contributors were selected according to their years of practice, experience, education, and type of organization, as outlined in Table 1. Information in Table 1 also indicates that the respondents possess extensive knowledge of the construction industry.

Each interview lasted approximately 40–90 min and was recorded with the permission of the interviewee. The research followed a semi-structured interview method for the interviewees to stay concentrated and focused on VM activities [71]. The confirmation and checking procedures were undertaken between iterative interviews and data analysis [72]. This combined process is like the participation checks and validation as noted and adopted in the previous studies [71].

The interview method consisted of a variety of open questions in accordance with the study's objectives. The instrument of the interview was divided into four sections, that is, background information about the experts, knowledge, application and implementation of VM in the organizations, VM activities in the construction and building projects, benefits of implementing VM in the construction industry and factors affecting the adoption of VM for project sustainability.

To analyze the interview transcripts, a content analysis technique was used. Transcription is considered the first step toward the descriptive process of the results in which the interview contents are recalled and transformed into text [73]. As a result, interviewed experts argued that a more formal system should guide the implementation of the VM in projects and categorized VM phases activities under five phases, as shown in Table 2. Moreover, several activities were modified, and three activities were added to the list, as indicated in Table 2. The updated and added activities were used to develop the questionnaire.

5.2. Questionnaire Survey

Following the interview with experts, a pilot study via EFA analysis was conducted to check the new constructs as obtained from the interview sessions, with a view to possibly modify the content and design of the main questionnaire.

For this research, a stratified sampling method of the three major building boundaries in Giza and Cairo was adopted. Sharma [74] approved that the significant benefits of stratified sampling are as follows: (1) decreasing the biases in the choosing of the sample cases, which indicates that the sample will have a highly representative description of the population under study; (2) allow simplification of samples to the population (i.e., statistical inferences), since the selected cases are chosen based on a probabilistic approach, and this is a significant benefit because the generalization appears externally valid; and (3) to ensure that enough sampling points are used to support the independent study of all strata. As a result, a pre-qualification analysis with the various organizations was performed through telephone calls. Over 280 companies were generated in the screening study, but only 215 companies agreed to participate. This test also reviewed positions, addresses, and assurances that the companies selected have reached their five-year survival level. The selected companies are with 9 to 250 staff, self-employed, non-foreign corporations and they were established between 1994 and 2010. This is to overcome any impact from the parent group's international policy [75].

For the number of samples required for EFA analysis, there is no consensus among previous studies of a larger survey has been recommended [64,76]. Furthermore, the factor analysis is acceptable for 20–50 elements as factors are properly analyzed when the number of elements is within this number [77]. However, much literature has found that fewer variables may be studied if the sample size is large enough [29,78]. Consequently, in this study, 200 questionnaires were distributed but only 150 were returned filled and fit for analysis, this represents about 75%, which is well within acceptable level [64]. Some of the observations from the pilot study include improper use of dot lines, inaccurate counting of variables, spelling errors, and orthographic errors. Such views, observations, recommendations, and corrections were reported and included in the final draft of the final survey instruments.

The first section of the data collection was to collect demographic information of respondents and their projects. In the second section, VM adoption activities were rating using a 5-point Likert scale with one = very low, two = low, three = average four = high, and, five = very high as adopted in similar studies regarding the implementation of VM [79,80]. The activities identified from the literature review and the expert interviews are listed in Table 2. Finally, for the Egyptian building projects, 21 possible VM adoption activities were identified; hence, the questionnaire was validated on a face-to-face basis. The Cronbach α analysis was also used to evaluate the research tool's reliability, following EFA's generation.

Table 2. Factors and associated literature.

Group Names	Code	Activity Name	References
Information phase	VM.IP1	Carry out a site visit	[81]
	VM.IP2	Collect related historical information on the proposed project	[82]
	VM.IP3	Establish the time period and scope of the project	[81]
	VM.IP4	Involve stakeholders in the initial stage of the project;	[83]
	VM.IP5	Involve and allocate duties to construction specialists at the initial stage of the project;	[81]
	VM.IP6	Clarify relevant details and limitations of the project	[82]
	VM.IP7	Share project knowledge between professionals	[83]
	VM.IP8	Identify the project's high-cost areas	[84]
Function phase	VM.FP1	Make client express the scale and predictions of the project explicitly	[84]
	VM.FP2	Presentation by stakeholders of project restrictions	[83]
	VM.FP3	Express and understand the goals and roles of the project	[81]
	VM.FP4	Create and identify functions with their related costs into essential and secondary objects	[82]
Evaluation phase	VM.EP1	Estimate the cost of each alternate life cycle	Interview
	VM.EP2	Assess brainstormed alternatives to fulfill the desired functions	[84]
	VM.EP3	Investigate the alternative assessment criterion	Interview
Creativity phase	VM.CP1	Brainstorm on solutions and concepts to accomplish the desired functions and costs.	[82]
	VM.CP2	Categorize brainstormed session alternatives and suggestions into realistically appropriate to be adopted	[82]
	VM.CP3	Defining the project procurement and contract strategy approach	[84]
Development/presentation phase	VM.DP1	Establish a short-term alternative action plan	[84]
	VM.DP2	Meet and request a review of the action plan	[82]
	VM.DP3	Track a VM output action plan	Interview

6. Results and Findings

6.1. Respondents' Characteristics and Demographic Profiles

The authors classified the participants in this research based on their years of work experience, professionalism, present positions, level of education, and organizational function, as shown in Table 3. Regarding the profession of respondents, civil engineers accounted for the highest number (30.7%), followed by Architect (26.7%) while the least are Quantity Surveyors. Findings of the present position indicated that the Site Engineer had the maximum frequency (36.0%) followed by the Manager (30.0%), while the least frequency was observed for the director (6.0%). For the organization, 38% are from clients' organizations followed by contractors and consultants. For respondent's qualifications, 10.7%, 47.3%, and 23.3% are Bachelor, Master, and Ph.D. degree holders respectively. Table 3 also shows that around 18.7% of respondents had worked from one year to less than five years. Respondents with work experience ranging from 5 to 10 years, 11 to 15 years, and more than 25 years were approximately 16.0%, 27.3% and 15.3% respectively. This indicates that the participants in this study are experienced to provide the required information.

Table 3. Demographic characteristic frequency distribution.

Variable	Characteristics	Number of Respondents	(%)
Work experience (Years)	Less than five	28	18.7
	5–10	24	16.0
	11–15	41	27.3
	16–25	34	22.7
	More than 25	23	15.3
Professional field	Architect	40	26.7
	Civil Engineer	46	30.7
	Electrical Engineer	28	18.7
	Mechanical Engineer	24	16.0
	Quantity surveying	12	8.0
Current position	Director	9	6.0
	Senior Manager	14	9.3
	Manager	45	30.0
	Design Engineer	28	18.7
	Site Engineer	54	36.0
Educational level	Diploma	9	6.0
	Bachelor's degree	16	10.7
	M.Sc.	71	47.3
	Ph.D.	35	23.3
	Others	28	18.7
Organization function	Client	57	38.0
	Consultant	43	28.7
	Contractor	50	33.3
VM workshop adopting and attending	Yes	22	14.7
	No	128	85.3
Formal training on VM	Yes	12	8.0
	No	138	92.0
Awareness	Totally Familiar	2	1.3
	Familiar	94	62.7
	Moderately familiar	28	18.7
	Not familiar	10	6.7
Perception	Technique	16	10.7
	A concept	68	45.3
	A profession	66	44.0

6.2. Level of Awareness and Implementation of VM in the Egyptian Construction Industry

This research examined participants' understanding and awareness of the VM method implementation, as showed in Table 3. Observation of the result illustrates that 45.3% of respondents considered the perception of value management or value engineering as a concept while 44.0% of the respondents considered it as a profession. This finding is consistent with the finding of awareness of VM or value engineering where 62.7% of the respondents were familiar and 1.3% were totally familiar with the practice of VM. In general, the respondents had moderate VM awareness, with a knowledge level of 64.7%, slightly higher than the 50% average, which indicates an adequate awareness level between stakeholders. On the other hand, most of the respondents (about 85.3%) did not attend a VM workshop or received a formal VM workshop, 92.0% did not receive any VM training in respect to adopting and participating in VM workshops. This implies that these corporations did not adopt the VM. Most companies investigated reported that they do not use VM because of a variety of reasons like lack of cost and low level of awareness.

6.3. EFA for VM Implementation Activities

This study's primary aim is to explore the VM activities in the building projects, and this was achieved through EFA analysis. However, the test of data normality is vital before launching the EFA study. In the current study, the normality of data was measured as an elementary assumption, and the results of the normality test for VM activities are shown in Table 4. Byrne [85] concluded that if the kurtosis result is between −7 to +7 and skewness results are between −2 to +2, the data are regarded as normal. As shown in Table 4, the skewness ranged from −1.51 to −0.68, and the kurtosis ranged from 0.05 to 1.66, which indicates that all variables are normally distributed.

Table 4. Result of Normality Test.

Variable	Skewness	Std. Error	Kurtosis	Std. Error
Information phase	−1.51	0.17	1.66	0.33
Function phase	−1.15	0.17	0.76	0.33
Evaluation phase	−0.86	0.17	0.36	0.33
Creativity phase	−0.90	0.17	0.19	0.33
Development and presentation phase	−0.68	0.17	0.05	0.33

EFA was used to evaluate the structure of the factor across twenty-one VM implementation activities. Many well-known constraints for the factorability of a connection were used. KMO was adopted to assess factor homogeneity and is popularly adopted to evaluate whether the variables' partial correlations are minimum [86]. Table 5 illustrates that the sampling adequacy measure of KMO was 0.755, which is exceeding the suggested value of 0.6, and Bartlett's Test of Sphericity was significant (x^2 (210) = 1204.837, $p < 0.05$) [64,87,88].

Table 5. Kaiser–Meyer–Olkin (KMO) and Bartlett's test result related to value management (VM) activities.

KMO and Bartlett's Test		
Kaiser–Meyer–Olkin Measure of Sampling Adequacy		**0.755**
Bartlett's Test of Sphericity	Approx. Chi-Square	1254.261
	Df	210
	Sig.	0.000

The anti-image correlation matrix diagonals were just above 0.5 indicating that all the elements can be included in the factor analysis. Initial communalities were meant to assess the variance in every activity reflected by all components, and the slight values (<0.3) show variables that do not appropriately fit with the factor solution. Consequently, the results here show that every part of the initial communities exceeded the threshold, and all loadings were more than 0.5 as indicated in Table 6.

Table 6. Communalities of VM activities.

Item	Communalities	Item	Communalities
VM.IP1	0.626	VM.FP4	0.731
VM.IP2	0.700	VM.EP1	0.728
VM.IP3	0.416	VM.EP2	0.698
VM.IP4	0.669	VM.EP3	0.758
VM.IP5	0.790	VM.CP1	0.694
VM.IP6	0.630	VM.CP2	0.714
VM.IP7	0.647	VM.CP3	0.763
VM.IP8	0.483	VM.DP1	0.698
VM.FP1	0.670	VM.DP2	0.660
VM.FP2	0.768	VM.DP3	0.715
VM.FP3	0.723		

The findings of the EFA for VM activities resulted in only six factors with values greater than 1 for eigenvalues with a total variance of 68.03%. Findings from Varimax rotation demonstrated that 17.66% of the variance was explained by the first factor (information phase) and 13.18% of the variance by the second factor (function phase). The third component contained another subscale called the evaluation phase, which explained 11.04% of the variance, followed by component four (creativity phase), which explained 10.25% of the total variance. The fifth component, which relates to the development and presentation phase, was able to explain 9.75% of the total variance. It is important to note that just one item (VM.IP5) was eliminated. It was included as the last component and initially part of the information phase. Therefore, for this study, only five extraction components were found adequate. Pallant [87], therefore, proposed that the screen plot and matrix should be analyzed objectively to assess the components (factors) that are extracted and determined. A shift (or Elbow) in plot shape is detected when examining the screen plot, and only sections above this level are kept. Figure 2 further reveals that the six aspects are modified.

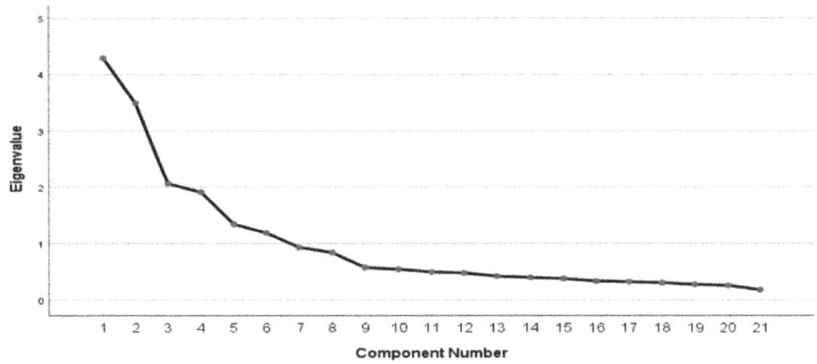

Figure 2. Scree plot result for VM activities components.

6.4. Reliability Analysis

For the factors extracted through EFA, the reliability statistics were determined. Table 7 indicates the results. The value of the Cronbach alpha becomes more acceptable as it tends towards 1.0 [89] Consequently as shown in Table 7, all VM activities have appropriate reliability since alpha Cronbach levels are greater than 0.6 [61].

Table 7. Reliability statistics for extracted factors (Cronbach's alpha).

Factor (VM Phases)	No. of Variables	Cronbach's Alpha
Information Phase	7	0.827
Function Phase	4	0.835
Creativity Phase	3	0.771
Evaluation Phase	3	0.799
Development and Presentation Phases	3	0.620

7. Discussion

Building projects impact the economy, society, and the environment over their whole life cycles [90], hence the link with sustainability. To accomplish the building project sustainability, there is a need for techniques that provide a clear vision of the projects' situation and realize other project objectives [10]. VM has been recognized as having the possibility to incorporate sustainability for building projects because it utilizes various knowledge resources, significant procedures, facilitated environment, strategic timing, and stakeholders and professional disciplines [49]. However, studies on the implementation of VM have focused on factors influencing the effective implementation of VM. Nevertheless, no study has measured or assessed the activities and elements that encourage the adoption of VM for sustainable building projects [91], particularly in Egypt.

7.1. Identify the Level of VM Awareness

Awareness of VM and its processing in construction projects significantly influence top management's decision to implement VM. Implementing VM is a complex activity requiring numerous details from different parties, but knowledge of the process can help overcome implementation challenges. From this study, it is evident that 64% of the respondents have adequate knowledge of VM, which means that they have a moderate amount of knowledge regarding VM. This research is contrary to the observation of Khodeir and El Ghandour [15] that about 51.6% of their respondents have light-sized knowledge of VM. The reasons can be due to the difference between the findings of both studies. Khodeir and El Ghandour [15] only covered thirty-five participants, the current study examines the topic more widely by assessing building experts' knowledge and collecting 150 questionnaires in the country's two largest regions (Cairo and Giza). Therefore, the higher the number of people covered, the greater the chances of more professionals that are aware of the subject.

7.2. Factor Analysis Results

Five phases of VM in Egyptian building projects were extracted through EFA. The activities executed are extracted under information, function, creativity, evaluation, and development/presentation phases. This finding did not match with the study of Tanko et al. [92] which was carried out in the Nigerian construction industry. Their findings show that the VM is extracted under three phases, that is, information/function phase, creativity/evaluation and development/presentation phase. These findings necessitate a need to enhance the application of VM in developing countries since the professionals in those countries did not consider VM phases according to VM standard generated by SAVE [22]. The extracted phases in this research are hereby discussed:

7.2.1. Information Phase

The first principal component is the information phase, which contains seven items, and accounts for 17.66% of the variance explained. This includes activities such as clarify project background information and constraints (0.769), involving clients and stakeholders at the early phase of the project (0.764), site visitation (0.732), share project information between stakeholders and professionals, identify high-cost areas of the project (0.699), involving and assigning duties to construction professionals at the early phase of the project (0.672), defining the time frame and scope of the project (0.612), collecting related background information on the suggested project (0.600). It is crucial for the exercise of VM that the information phase supports and embraces the adoption of VM. Every VM analysis consists of three types of contributors: decision-makers (owners or owners' representatives), VM team leader (facilitators), and team members [93]. At this stage of the workshop, information sharing is vital, and clients and end-users should obviously and explicitly state the scope, aims, requirements, and anticipations of their planned projects. The responsibility played by every member can provide positively towards the improvement of sustainability [94]. Methods for procurement, duration of projects, environment, and performance are also set out here by these participants. Furthermore, numerous green alternatives were combined into the early design phase in the information phase [49], and clients are expected to inform their objectives. Leung and Liu [95] reported that project objectives influence the VM participants' behavior and the outcome. Additionally, the VM team leader holds a strategic plan to enhance sustainability knowledge and awareness [96]. Communication among clients and VM facilitators increases the chance to motivate owners to commit to sustainability [94].

Consequently, details and information on the context, design, projected costs, and limitations of the project are provided [97]. An inquiry by Mohamad Ramly and Shen [84] revealed that, at this point, construction workers, such as the facilitator's team, provided essential information on their fields. However, different participants may also have identified limitations to their project [98].

7.2.2. Function Phase

The second principal component, which is the function phase comprised four items and accounted for 13.178% of the variance explained. It includes activities such as presenting project restrictions and limitations to stakeholders (0.849), creating and classifying functions/items as basic and secondary items with their associated costs (0.812), expressing and understanding the aims and functions of the project (0.802), making the client express clearly the scope and anticipations of the project (0.710). It is important to note that the most critical part of this phase is a logical approach that defines and fulfills the criteria, needs, aims, and anticipated goals of the undertaking. Moreover, four variables for the function phase, namely, mission, space, elements, and shape, will be investigated [99]. Regarding the task, the phase is concentrated on how the sponsor, owner, client, and other main parties to the project primarily view or distinguish a particular task [100]. This step aims to create, identify, and categorize primary and secondary functions [84]. Team members are encouraged during this phase to write down the buildings' functions, which can still be applicable for the next 50 years [90]. The basic and secondary functions with their associated costs are defined and categorized in this way in order to comprehend the sustainable aspects of the project. In the hierarchical function structure known as the functional tree diagram, project functions are defined, analyzed, and the main objective of the process is to recognize the project through the project functions [82]. Furthermore, through this diagram, the critical building objectives and functions established by the owner have been involved in this stage, and the sustainability of the building might itself be an essential aim and function [90]. Sustainability dimensions would be an essential element of the project's aims and functions [49]. Additionally, the study demonstrated that the dimensions of sustainability, such as user comfort, environmental influence and impact, accessibility, society, and life cycle costs, are integrated in the VM decision matrix [42].

7.2.3. Evaluation Phase

The third principal component is the phase of evaluation, which consists of three items, and accounts for 10.253% of the variance explained. It consists of activities such as estimating the total life-cycle cost of each alternative (0.815), investigating the criteria for assessing alternatives (0.778), and evaluating brainstormed alternatives to achieve the desired functions/elements (0.765). Therefore, the VM team must be interdisciplinary in assessing all alternatives during this VM workshop phase to achieve the project's sustainable objective. At this stage, the proposals made are analyzed and evaluated for each of the suggestions and ideas identified in the previous stage (creative stage). A review, assessment, and short-listing are carried out at this stage and to examine less promising ideas, it is necessary to assess each idea against functional requirements [101]. This is supported by the conclusion of Mohamad Ramly and Shen [84] that further testing of proposals and suggestions is best performed to determine how to achieve the project's desired goals and sustainable objectives.

7.2.4. Creativity Phase

The fourth principal component is the creativity phase. It consists of three items and accounts for 11.041% of the total variance. It includes: creating brainstorm alternatives and ideas to achieve the desired functions/elements and related costs (0.801); description of the procurement and contract strategy of the project (0.811) and categorizing the alternatives from the brainstormed session as feasible to be adopted (0.797). It could be noted that ideas are produced and created during the creative stage to fulfill the necessary and preferred functions of a practical tree structure. The members of the VM Workshop team investigate, explore, generates, and tests alternative methods and ways in which tasks are performed throughout these processes [102]. Innovative approaches and techniques, including brainstorming, Synectic, and side thinking, are being applied at this level. The VM facilitator must create a positive environment by ensuring that criticism and repression among VM team members are not possible. Through the creativity phase, the team member can have an excellent opportunity to achieve their objective by enhancing innovation ideas. Innovation is one of the highest drivers that could be sustained by the practical approach to improve resource [103]. Moreover, economic and investment sustainability must not be disconnected from community and environmental sustainability in recommending alternatives, solutions, and ideas [90].

7.2.5. Development and Presentation Phase

The final primary component, which comprises three items is the development and presentation phase. It is not surprising that the major activity of the phase as obtained from the EFA analysis is that Egypt's professionals did not comply with the VM standards. This combined phase represents 9.748% of the variance explained. It includes activities such as the generate an action plan on short-listed alternatives (0.811), the follow-up of an action plan on VM output (0.786), and the holding and presentation of an action plan review meeting (0.783), The members of the VM team are responsible for generating short-listed proposals and suggestions. The team preparations, manufacturers, and producers are sketches, descriptions/materials, drawings, specifications, and details as structured VM proposals. Each brief concept is regarded as a feasible, practical, and realistic solution. This step deals with all the drawbacks, benefits, and potential of the concepts and ideas in the guidelines, estimates, and cost of life-cycles measurements. Moreover, at this point, the strategic plan involves each part or function of the development phase. The finding is supported by Oke and Aigbavboa [100] where the development phase was noted to be a crucial step of the VM study and that consideration should be given to ensure that the development and presentation phases are carefully investigated and considered. On the other hand, some authors recommended presentation as the last step of VM. However, this phase's purpose is to send these suggestions to the body that authorized the study as a way of minimalist-feedback [100].

7.3. *VM Activities Implementation for Building Sustainable Success*

Recent building developments have brought more effective and sustainable methods, specialized techniques, and materials [104]. Building direction in the construction industry requires substantial and sustainable development [105]. However, measuring sustainability and performance during usage is becoming increasingly relevant [106]. The strategies to incorporate the concept of sustainable development must be established [107]. Moreover, VM can perform an effective method for achieving building sustainability. This study evaluated VM activities through EFA analysis to suggest VM implementation activities and stages, which can lead to sustainable delivery of building projects. VM study allowed sustainability principles to be included in the conceptual and initial design phases [38]. Through VM study, alternatives are anticipated to promote a healthy and safe ecosystem for the residents [108]. Additionally, VM offers multidisciplinary professionals' chances to focus on issues and matters regarding community, society and the environment [33,90] and effectively enhance sustainability concepts during the life cycle of building projects [90]. The conventional way of considering project success is the so-called iron-triangle of time, cost, and quality, which are frequently reviled in the late decades [109]. Thus, by implementing VM, building companies can balance time, expense, and quality as VM reduces costs, but not at the sacrifice of benefits [110]. Kelly and Male [111] suggested that time measured within the sustainable value system is part of the project's success, and VM can achieve the optimum project time. For instance, in the roadway field, Atabay and Galipogullari [112] used VM as a sustainable tool for saving around 12 months of project time. These results figures amount to 6% of the total budget and 17% of the schedule. Not surprising that, in the marine construction field, Tang and Bittner [113] confirmed that VM uses as a useful technique for improving the quality. Moreover, cost and financial sustainability can be targeted in building projects by adopting VM. Lin et al. [114] said that VM is commonly used as a supportive method to address challenges, such as limited financial resources and strict planning in the building industry, as financial analysis represents the primary method for evaluating a company's sustainable improvement and value creation [115]. VM provides sustainability to be included in a construction project [21,116]. In the sustainability sense of construction firms, it was found that they face several obstacles as they mature and have a positive social, environmental and economic impact [117]. These firms should also search for swift and new means of sustaining customer confidence [107]. The proposed VM activities give these companies practitioners the opportunity to take a good decision for achieving their company's success. These decisions are able to have a long-term effect on corporate practices [118]. Moreover, VM activities could be applied to the extent of traditional capital, operating costs, and the project's duration to expose customer value systems. These companies can only pay attention to sustainability aspects if the value created exceeds the damage caused [119]. Achievements of success by enterprises and their competence to enhance company value over a long time depend on having an effective business model [120], and this model can be achieved through the adoption of VM activities.

8. Conclusions

Construction projects, especially in developing countries like Egypt, are characterized by low quality and are not delivered in accordance with sustainable principles. This research showed that VM is a vital option to address this threat. It was hypothesized that the VM activities implementation is crucial for building professionals to achieve their sustainable goals. To examine this hypothesis, mixed methods research was conducted through a semi-structured interview of fifteen experts and a questionnaire survey. The study collectively deduced that the level of implementation of VM in building projects is relatively poor. The study also showed that building participants in Egypt have reasonable awareness (67.7 percent), and the majority of participants viewed VM as a concept. This indicates that the challenge of VM in the Egyptian construction industry is not of awareness but that of adoption. From the questionnaire analysis through EFA, the study shows that information; function, creativity, evaluation, and development/presentation phases are classified as five phases of VM adoption in Egyptian building projects The study investigated the new activities generated by the experts through the adoption of a semi-structured interview, which was not mentioned in the previous studies. These include estimating total life-cycle costs for each alternative, research evaluation criteria for alternatives and follow up an action plan for VM output. The EFA analysis also leads to the removal of one activity, that is, involve and assign responsibilities to construction professionals at the initial stage of the project. To this end, VM activities are essential for the sustainable delivery of building projects. The study results can guide policymakers and top managers in building projects to systematically understand the relative importance of VM activities and efficient subsequent implementation. Furthermore, the study can support developing countries to consider VM implementation for achieving their overall sustainable success.

Author Contributions: A.F.K. initiated the research project, carried out the mixed methods empirical research, and drafted the manuscript. I.O. supervised the project and provided feedback on the manuscript drafts. A.E.O. supervised the project and provided feedback on the manuscript drafts. N.C. provided feedback on the manuscript drafts. M.K.B. provided feedback on the manuscript drafts. All authors have read and agreed to the published version of the manuscript.

Funding: This research received no external funding.

Acknowledgments: The authors would like to express their utmost gratitude to FRGS grant Ministry of Higher Education Malaysia for supporting this research and Universiti Teknologi PETRONAS, 32610 Seri Iskandar, Perak, Malaysia.

Conflicts of Interest: The author declares no conflict of interest.

Abbreviations

The following abbreviations are used in this manuscript:

MDPI	Multidisciplinary Digital Publishing Institute
VM	Value Management
EFA	Exploratory Factor Analysis
GHG	Green House Gas emissions
CFA	Confirmatory Factor Analysis
SPSS	Statistical Package for the Social Sciences
PCA	Principal Component Analysis
PAF	Principal Axis Factoring

References

1. Sbci, U. *Buildings and Climate Change: Summary for Decision-Makers;* United Nations Environmental Programme, Sustainable Buildings and Climate Initiative: Paris, France, 2009; pp. 1–62.
2. DoE, U. *Energy Efficiency Trends in Residential and Commercial Buildings;* US Department of Energy: Washington, DC, USA, 2008. Available online: http://apps1.eere.energy.gov/buildings/publications/pdfs/corporate/bt_stateindustry.pdf (accessed on 28 October 2020).
3. Atanasiu, B.; Attia, S. Principles for nearly zero-energy buildings: Paving the way for effective implementation of policy requirements. In *Principles for Nearly Zero-Energy Buildings: Paving the Way for Effective Implementation of Policy Requirements;* Buildings Performance Institute Europe (BPIE): Brussels, Belgium, 2011; p. 124.
4. Durdyev, S.; Ismail, S.; Ihtiyar, A.; Bakar, N.F.S.A.; Darko, A. A partial least squares structural equation modeling (PLS-SEM) of barriers to sustainable construction in Malaysia. *J. Clean. Prod.* **2018**, *204*, 564–572. [CrossRef]
5. Yaseen, Z.M.; Ali, Z.H.; Salih, S.Q.; Al-Ansari, N. Prediction of risk delay in construction projects using a hybrid artificial intelligence model. *Sustainability* **2020**, *12*, 1514. [CrossRef]
6. Mousa, A. A Business approach for transformation to sustainable construction: An implementation on a developing country. *Resour. Conserv. Recycl.* **2015**, *101*, 9–19. [CrossRef]
7. Fang, Z.; Gao, X.; Sun, C. Do financial development, urbanization and trade affect environmental quality? Evidence from China. *J. Clean. Prod.* **2020**, 120892. [CrossRef]
8. Kissi, E.; Boateng, E.; Adjei-Kumi, T. Strategies for implementing value management in the construction industry of Ghana. In Proceedings of the DII-2015 Conference on Infrastructure Development and Investment Strategies for Africa, Livingstone, Zambia, 16–18 September 2015; pp. 255–267.
9. Adeyemi, L.A.; Idoko, M. Developing Local Capacity For Project Management–Key To Social And Business Transformation In Developing Countries. In Proceedings of the PMI® Global Congress 2008—EMEA, St. Julian's, Malta, 19–21 May 2008; Project Management Institute: Newtown Square, PA, USA, 2008.
10. Maceika, A.; Bugajev, A.; Šostak, O.R. The Modelling of Roof Installation Projects Using Decision Trees and the AHP Method. *Sustainability* **2020**, *12*, 59. [CrossRef]
11. Jekale, W. *Performance for Public Construction Projects in Developing Countries: Federal Road and Educational Building Projects in Ethiopia;* Norwegian University of Science & Technology: Trondheim, Norway, 2004.
12. Barakat, M.S.; Naayem, J.H.; Baba, S.S.; Kanso, F.A.; Borgi, S.F.; Arabian, G.H.; Nahlawi, F.N. Egypt Economic Report: Between the Recovery of the Domestic Economy and the Burden of External Sector Challenges. Available online: http://www.bankaudigroup.com (accessed on 28 October 2020).
13. Soliman, M.M.A.I. Risk Management in International Construction Joint Ventures in Egypt. Ph.D. Dissertation, University of Leeds, Leeds, UK, 2014.
14. Luo, P.; Sun, Y.; Wang, S.; Wang, S.; Lyu, J.; Zhou, M.; Nakagami, K.; Takara, K.; Nover, D. Historical assessment and future sustainability challenges of Egyptian water resources management. *J. Clean. Prod.* **2020**, *263*, 121154. [CrossRef]
15. Khodeir, L.M.; El Ghandour, A. Examining the role of value management in controlling cost overrun [application on residential construction projects in Egypt]. *Ain Shams Eng. J.* **2019**, *10*, 471–479. [CrossRef]
16. Ministry of Housing. *Urban Facilities and Facilities;* Ministry of Housing, Utilities & Urban Communities: Cairo, Egypt, 1999.

17. Brundtland, G.H.; Khalid, M.; Agnelli, S.; Al-Athel, S.; Chidzero, B. *Our Common Future*; Butterworth Heinemann: Oxford, UK, 1987; Volume 8.
18. Kibert, C. *Final Session of First International Conference of CIB TG 16 on Sustainable Construction*; University of Florida: Gainesville, FL, USA, 1994.
19. Hill, R.C.; Bowen, P.A. Sustainable construction: Principles and a framework for attainment. *Constr. Manag. Econ.* **1997**, *15*, 223–239. [CrossRef]
20. Wolstenholme, A.; Austin, S.A.; Bairstow, M.; Blumenthal, A.; Lorimer, J.; McGuckin, S.; Rhys Jones, S.; Ward, D.; Whysall, D.; Le Grand, Z. *Never Waste a Good Crisis: A Review of Progress Since Rethinking Construction and Thoughts for Our Future*; Constructing Excellence: London, UK, 2009.
21. Abidin, N.Z.; Pasquire, C.L. Revolutionize value management: A mode towards sustainability. *Int. J. Proj. Manag.* **2007**, *25*, 275–282. [CrossRef]
22. SAVE. *Value Methodology Standard*; SAVE Inteernational: Mount Royal, NJ, USA, 2007.
23. Shen, Q. *A Knowledge Based Structure for Implementing Value Management in the Design of Office Buildings*; University of Salford: Greater Manchester, UK, 1993.
24. Thiry, M. *Value Management Practice. Newtown Square, PA: Project*; Management Institute: Newtown Square, PA, USA, 1997.
25. Al-Saleh, Y.M.; Taleb, H.M. The integration of sustainability within value management practices: A study of experienced value managers in the GCC countries. *Proj. Manag. J.* **2010**, *41*, 50–59. [CrossRef]
26. Parker, D.E. *Value Engineering Theory*, rev. ed.; The Lawrence D. Miles Value Foundation: Washington, DC, USA, 1998.
27. Alattyih, W.; Haider, H.; Boussabaine, H. Development of Value Creation Drivers for Sustainable Design of Green Buildings in Saudi Arabia. *Sustainability* **2019**, *11*, 5867. [CrossRef]
28. Rosłon, J.; Książek-Nowak, M.; Nowak, P. Schedules Optimization with the Use of Value Engineering and NPV Maximization. *Sustainability* **2020**, *12*, 7454. [CrossRef]
29. Kim, S.-Y.; Lee, Y.-S.; Nguyen, V.T. Barriers to applying value management in the Vietnamese construction industry. *J. Constr. Dev. Ctries.* **2016**, *21*, 55. [CrossRef]
30. Abdelghany, M.; Rachwan, R.; Abotaleb, I.; Albughdadi, A. Value engineering applications to improve value in residential projects. In Proceedings of the Annual Conference–Canadian Society for Civil Engineering, Edmonton, AB, Canada, 27–30 May 2015; pp. 27–30.
31. Aboelmaged, M. The drivers of sustainable manufacturing practices in Egyptian SMEs and their impact on competitive capabilities: A PLS-SEM model. *J. Clean. Prod.* **2018**, *175*, 207–221. [CrossRef]
32. Daoud, A.O.; Othman, A.; Robinson, H.; Bayati, A. Towards a green materials procurement: Investigating the Egyptian green pyramid rating system. In Proceedings of the 3rd International Green Heritage Conference, Porto, Italy, 10–12 October 2019.
33. Othman, I.; Kineber, A.; Oke, A.; Khalil, N.; Buniya, M. Drivers of Value Management Implementation in Building Projects in Developing Countries. *J. Phys.* **2020**, *1529*, 042083.
34. Aghimien, D.O.; Oke, A.E.; Aigbavboa, C.O. Barriers to the adoption of value management in developing countries. *Eng. Constr. Archit. Manag.* **2018**, *25*, 818–834. [CrossRef]
35. Oke, A.; Aghimien, D.; Olatunji, S. Implementation of value management as an economic sustainability tool for building construction in Nigeria. *Int. J. Manag. Value Supply Chain.* **2015**, *6*, 55–64.
36. Aarseth, W.; Ahola, T.; Aaltonen, K.; Økland, A.; Andersen, B. Project sustainability strategies: A systematic literature review. *Int. J. Proj. Manag.* **2017**, *35*, 1071–1083. [CrossRef]
37. Martens, M.L.; Carvalho, M.M. Key factors of sustainability in project management context: A survey exploring the project managers' perspective. *Int. J. Proj. Manag.* **2017**, *35*, 1084–1102. [CrossRef]
38. Fewings, P.; Henjewele, C. *Construction Project Management: An Integrated Approach*; Routledge: London, UK, 2019.
39. Australia, S. *Australian Standard: Value Management (AS 4183-2007)*; Council of Standards Australia: Sydney, Australia, 2007.
40. Bowen, P.A.; Edwards, P.J.; Cattell, K. Value management practice in South Africa: The built environment professions compared. *Constr. Manag. Econ.* **2009**, *27*, 1039–1057. [CrossRef]
41. Ann, T.; Shen, Q.; Kelly, J.; Hunter, K. An empirical study of the variables affecting construction project briefing/architectural programming. *Int. J. Proj. Manag.* **2007**, *25*, 198–212.
42. Kelly, J.; Male, S. *Value Management in Design and Construction*; Routledge: London, UK, 2003.

43. Abidin, N.Z.; Pasquire, C.L. Delivering sustainability through value management. *Eng. Constr. Archit. Manag.* **2005**, *12*, 168–180. [CrossRef]
44. Scotland, B. Report on Value Management Workshops for Community Self Build Scotland, BRE Scotland. Available online: http://goo.gl/rHv5WB (accessed on 10 December 2013).
45. Stephenson, M. How Can Value Engineering Be Used to Enhance the Sustainability of Major Engineering Developments? Master's Thesis, University of Strathclyde, Glasgow, UK, 2003.
46. Hayles, C. The role of value management in the construction of sustainable communities. *Value Manag.* **2004**, *10*, 15–19.
47. Al-Yousefi, A.S. The synergy between value engineering and sustainable construction. In Proceedings of the CTBUH 8th World Congress, Dubai, UAE, 3–5 March 2008; pp. 3–5.
48. Kelly, J.; Male, S.; Graham, D. *Value Managent of Construction Project*; Blackwell: London, UK, 2004.
49. Phillips, M.R. Towards sustainability and consensus through value management: Case study. Managing Sustainable Values. In Proceedings of the International Conference of the Institute of Value Management, Hong Kong, China, 6–7 May 2009.
50. Davis, D.F.; Golicic, S.L.; Boerstler, C.N. Benefits and challenges of conducting multiple methods research in marketing. *J. Acad. Mark. Sci.* **2011**, *39*, 467–479. [CrossRef]
51. Buniya, M.K.; Othman, I.; Sunindijo, R.Y.; Kineber, A.F.; Mussi, E.; Ahmad, H. Barriers to safety program implementation in the construction industry. *Ain Shams Eng. J.* **2020**. [CrossRef]
52. Bazeley, P. The evolution of a project involving an integrated analysis of structured qualitative and quantitative data: From N3 to NVivo. *Int. J. Soc. Res. Methodol.* **2002**, *5*, 229–243. [CrossRef]
53. Mason, M. Sample size and saturation in PhD studies using qualitative interviews. *Forum Qual. Sozialforschung Forum Qual. Soc. Res.* **2010**, *11*, 3.
54. Othman, I.; Kamil, M.; Sunindijo, R.Y.; Alnsour, M.; Kineber, A.F. Critical success factors influencing construction safety program implementation in developing countries. In Proceedings of the 2nd Joint International Conference on Emerging Computing Technology and Sports (JICETS) 2019, Bandung, Indonesia, 25–27 November 2019; p. 042079.
55. Ochieng, E.G.; Price, A.D. Managing cross-cultural communication in multicultural construction project teams: The case of Kenya and UK. *Int. J. Proj. Manag.* **2010**, *28*, 449–460. [CrossRef]
56. Haig, B.D. An abductive theory of scientific method. In *Method Matters in Psychology*; Springer: Berlin/Heidelberg, Germany, 2018; pp. 35–64.
57. Dubois, A.; Gadde, L.-E. Systematic combining: An abductive approach to case research. *J. Bus. Res.* **2002**, *55*, 553–560. [CrossRef]
58. Saunders, M.; Lewis, P.; Thornhill, A. *Research Methods for Business Students*, 6th ed.; Pearson: Harlow, UK; New York, NY, USA, 2012.
59. Fellows, R.F.; Liu, A.M. *Research Methods for Construction*; John Wiley & Sons: New York, NY, USA, 2015.
60. Williams, B.; Onsman, A.; Brown, T. Exploratory factor analysis: A five-step guide for novices. *Aust. J. Paramed.* **2010**, *8*. [CrossRef]
61. Field, A. *Discovering Statistics Using SPSS*; Sage Publications: New York, NY, USA, 2009.
62. Thompson, B. *Exploratory and Confirmatory Factor Analysis: Understanding Concepts and Applications*; American Psychological Associatio: Washington, DC, USA, 2004.
63. Costello, A.B.; Osborne, J. Best practices in exploratory factor analysis: Four recommendations for getting the most from your analysis. *Pract. Assess. Res. Eval.* **2005**, *10*, 7.
64. Tabachnick, B.G.; Fidell, L.S.; Ullman, J.B. *Using Multivariate Statistics*; Pearson: Boston, MA, USA, 2007; Volume 5.
65. Kim, T.-H.; Lee, H.W.; Hong, S.-W. Value engineering for roadway expansion project over deep thick soft soils. *J. Constr. Eng. Manag.* **2016**, *142*, 05015014. [CrossRef]
66. Luvara, V.G.; Mwemezi, B. Obstacles against value management practice in building projects of Dar es Salaam Tanzania. *Int. J. Constr. Eng. Manag.* **2017**, *6*, 13–21.
67. Shen, Q.; Liu, G. Critical success factors for value management studies in construction. *J. Constr. Eng. Manag.* **2003**, *129*, 485–491. [CrossRef]
68. Kothari, C.R. *Research Methodology: Methods and Techniques*; New Age International: Delhi, India, 2004.

69. Adedokun, O.A.; Ibironke, O.T.; Olanipekun, A.O. Vulnerability of motivation schemes in enhancing site workers productivity for construction industry's sustainability in Nigeria. *Int. J. Sustain. Constr. Eng. Technol.* **2013**, *4*, 21–30.
70. Sushil, S.; Verma, N. Questionnaire validation made easy. *Eur. J. Sci. Res.* **2010**, *46*, 172–178.
71. Chileshe, N.; Rameezdeen, R.; Hosseini, M.R. Drivers for adopting reverse logistics in the construction industry: A qualitative study. *Eng. Constr. Archit. Manag.* **2016**, *32*, 134–157. [CrossRef]
72. Kavishe, N.; Chileshe, N. Identifying project management practices and principles for Public–Private partnerships in housing projects: The case of Tanzania. *Sustainability* **2018**, *10*, 4609. [CrossRef]
73. King, N.; Horrocks, C.; Brooks, J. *Interviews in Qualitative Research*; SAGE Publications Limited: Thousand Oaks, CA, USA, 2018.
74. Sharma, G. Pros and cons of different sampling techniques. *Int. J. Appl. Res.* **2017**, *3*, 749–752.
75. Ripollés, M.; Blesa, A. International new ventures as "small multinationals": The importance of marketing capabilities. *J. World Bus.* **2012**, *47*, 277–287. [CrossRef]
76. Pallant, J. *SPSS Survival Guide*; Allen & Unwin: Crows Nest, Australia, 2005.
77. Shen, Q. Value management in Hong Kong's construction industry: Lessons learned. In Proceedings of the SAVE International Conference Proceeding, Hong Kong, China, 16 December 1997; pp. 260–265.
78. Ahadzie, D.; Proverbs, D.; Olomolaiye, P. Critical success criteria for mass house building projects in developing countries. *Int. J. Proj. Manag.* **2008**, *26*, 675–687. [CrossRef]
79. Al-Yami, A.M. An Integrated Approach to Value Management and Sustainable Construction during Strategic Briefing in Saudi Construction Projects. Ph.D. Dissertation, Loughborough University, Loughborough, UK, 2008.
80. Lai, N.K. Value Management in Construction Industry. Master's Thesis, Universiti Teknologi Malaysia, Johor Bahru, Malaysia, 2006.
81. SAVE. *Value Managemen*; SAVE: Dayton, OH, USA, 2008.
82. Jaapar, A.; Maznan, N.A.; Zawawi, M. Implementation of value management in public projects. *Procedia Soc. Behav. Sci.* **2012**, *68*, 77–86. [CrossRef]
83. Hwang, B.-G.; Zhao, X.; Toh, L.P. Risk management in small construction projects in Singapore: Status, barriers and impact. *Int. J. Proj. Manag.* **2014**, *32*, 116–124. [CrossRef]
84. Mohamad Ramly, Z.; Shen, G.Q. Value management in Malaysia: Past, present and future. In Proceedings of the International Conference on Value Engineering and Management 'Innovation in Value Methodology', Hong Kong, China, 6–7 December 2012; pp. 105–110.
85. Byrne, B. *Structural Equation Modeling with AMOS: Basic Concepts, Applications, and Programming*, 2nd ed.; Taylor & Francis Group: New York, NY, USA, 2010.
86. Sharma, S. *Applied Multivariate Techniques*; John Wiley and Sons: New York, NY, USA, 1996.
87. Pallant, J. *SPSS Survival Manual*, 3rd ed.; McGrath Hill: Berkshire, UK, 2007; Volume 15.
88. Tavakol, M.; Dennick, R. Making sense of Cronbach's alpha. *Int. J. Med. Educ.* **2011**, *2*, 53. [CrossRef]
89. Nunnally, J.C. *Psychometric Theory*, 3rd ed.; Tata McGraw-Hill Education: New York, NY, USA, 1994.
90. Yu, A.T.W.; Javed, A.A.; Lam, T.I.; Shen, G.Q.; Sun, M. Integrating value management into sustainable construction projects in Hong Kong. *Eng. Constr. Archit. Manag.* **2018**, *25*, 1475–1500. [CrossRef]
91. Tanko, B.L.; Abdullah, F.; Ramly, Z.M.; Enegbuma, W.I. An implementation framework of value management in the Nigerian construction industry. *Built Environ. Proj. Asset Manag.* **2018**, *8*, 305–319. [CrossRef]
92. Tanko, B.L.; Abdullah, F.; Ramly, Z.M.; Enegbuma, W.I. Confirmatory factor analysis of value management current practice in the Nigerian construction industry. *Pernabit Akademia Baru J. Adv. Res. Appl. Sci. Eng. Technol.* **2017**, *9*, 32–41.
93. British Standard EN 12973. Available online: http://bsonline.techindex.co.uk (accessed on 28 October 2020).
94. Zainul-Abidin, N. Achieving sustainability through value management: A passing opportunity? *Int. J. Constr. Manag.* **2008**, *8*, 79–91. [CrossRef]
95. Leung, M.; Liu, A.M. Developing a value management model–by value-goal system approach. In Proceedings of the 14th Annual Conference of Association of Researchers in Construction Management (ARCOM 98), Reading, UK, 9–11 September 1998; pp. 496–505.
96. Dallas, M. Revolutionising the Way We Build, Managing Sustainable Values. In Proceedings of the International Conference of the Institute of Value Management, Hong Kong, China, 6–7 May 1999; pp. 6–7.
97. Chougule, A.; Gupta, A.; Patil, S. Application of value engineering technique to A residential Building–Case study. *Int. J. Innov. Res. Adv. Eng. IJIRAE* **2014**, *1*.

98. Chen, W.T.; Liao, S.L. A job-plan based performance evaluation for construction value engineering study. *J. Chin. Inst. Eng.* **2010**, *33*, 317–333. [CrossRef]
99. Kelly, J.; Male, S.; Graham, D. *Value Management of Construction Projects*; John Wiley & Sons: New York, NY, USA, 2014.
100. Oke, A.E.; Aigbavboa, C.O. *Sustainable Value Management for Construction Projects*; Springer: Berlin/Heidelberg, Germany, 2017.
101. Kam Shadan, P. *Construction Project Management Handbook*; Federal Transit Administration: Washington, DC, USA, 2012; Volume 14, p. 2016.
102. Liu, G. A framework for Implementing Value Management in China's Construction Industry. Ph.D. Thesis, The Hong Kong Polytechnic University, Hong Kong, China, 2003.
103. Munyasya, B.M.; Chileshe, N. Towards sustainable infrastructure development: Drivers, barriers, strategies, and coping mechanisms. *Sustainability* **2018**, *10*, 4341. [CrossRef]
104. Švajlenka, J.; Kozlovská, M. Perception of user criteria in the context of sustainability of modern methods of construction based on wood. *Sustainability* **2018**, *10*, 116. [CrossRef]
105. Švajlenka, J.; Kozlovská, M.; Pošiváková, T. Analysis of selected building constructions used in industrial construction in terms of sustainability benefits. *Sustainability* **2018**, *10*, 4394. [CrossRef]
106. Švajlenka, J.; Kozlovská, M. Evaluation of the efficiency and sustainability of timber-based construction. *J. Clean. Prod.* **2020**, *259*, 120835. [CrossRef]
107. Spychalska-Wojtkiewicz, M. The Relation between Sustainable Development Trends and Customer Value Management. *Sustainability* **2020**, *12*, 5496. [CrossRef]
108. Yates, A.; Yates, A. *Sustainable Buildings: Benefits for Designers*; Building Research Establishment: London, UK, 2003.
109. Lu, P.; Guo, S.; Qian, L.; He, P.; Xu, X. The effectiveness of contractual and relational governances in construction projects in China. *Int. J. Proj. Manag.* **2015**, *33*, 212–222. [CrossRef]
110. Dallas, M.F. *Value and Risk Management: A Guide to Best Practice*; John Wiley & Sons: New York, NY, USA, 2008.
111. Kelly, J.; Male, S. A Technique for Understanding the Customer's Project Value Criteria. In Proceedings of Proceedings of the Society of American Value Engineers Conference, Denver, Colorado, 5–8 May 2002.
112. Atabay, S.; Galipogullari, N. Application of value engineering in construction projects. *J. Traffic Transp. Eng.* **2013**, *1*, 39–48.
113. Tang, P.; Bittner, R.B. Use of value engineering to develop creative design solutions for marine construction projects. *Pract. Period. Struct. Des. Constr.* **2014**, *19*, 129–136. [CrossRef]
114. Lin, G.; Shen, G.Q.; Sun, M.; Kelly, J. Identification of key performance indicators for measuring the performance of value management studies in construction. *J. Constr. Eng. Manag.* **2011**, *137*, 698–706. [CrossRef]
115. Kaczmarek, J. The Mechanisms of Creating Value vs. Financial Security of Going Concern—Sustainable Management. *Sustainability* **2019**, *11*, 2278. [CrossRef]
116. Zainul-Abidin, N.; Pasquire, C. Moving towards sustainability through value management. In Proceedings of the Joint International Symposium of CIB Working Commissions W55 and W107, Singapore, 22–24 October 2003; Volume 2, pp. 258–268.
117. Lee, W.J.; Mwebaza, R. The Role of the Climate Technology Centre and Network as a Climate Technology and Innovation Matchmaker for Developing Countries. *Sustainability* **2020**, *12*, 7956. [CrossRef]
118. Obiora, S.C.; Bamisile, O.; Opoku-Mensah, E.; Kofi Frimpong, A.N. Impact of Banking and Financial Systems on Environmental Sustainability: An Overarching Study of Developing, Emerging, and Developed Economies. *Sustainability* **2020**, *12*, 8074. [CrossRef]
119. Figge, F.; Hahn, T. Sustainable value added—measuring corporate contributions to sustainability beyond eco-efficiency. *Ecol. Econ.* **2004**, *48*, 173–187. [CrossRef]
120. Jabłoński, A.; Jabłoński, M. Research on business models in their life cycle. *Sustainability* **2016**, *8*, 430. [CrossRef]

sustainability

Article

Application of Hybrid SWARA–BIM in Reducing Reworks of Building Construction Projects from the Perspective of Time

Hamidreza Khalesi [1], Amirhossein Balali [1], Alireza Valipour [1,*], Jurgita Antucheviciene [2], Darius Migilinskas [2] and Viaceslav Zigmund [2]

1 Department of Civil Engineering, Shiraz Branch, Islamic Azad University, Shiraz 5-71993, Iran; khalesi.hr@gmail.com (H.K.); a.shz.balali@gmail.com (A.B.)
2 Department of Construction Management and Real Estate, Vilnius Gediminas Technical University, 10221 Vilnius, Lithuania; jurgita.antucheviciene@vgtu.lt (J.A.); darius.migilinskas@vgtu.lt (D.M.); viaceslav.zigmund@vgtu.lt (V.Z.)
* Correspondence: vali@iaushiraz.ac.ir; Tel.: +98-9177914214

Received: 29 September 2020; Accepted: 24 October 2020; Published: 27 October 2020

Abstract: One of the major issues of the construction industry has been the "reworks" that affect the time, quality, and cost of projects. Therefore, reworks and the ineffective use of site resources and materials will always result in significant losses on projects. The development of information technology has led to the widespread use of Building Information Modelling (BIM) to enhance the delivery of more sustainable building construction projects. The purpose of this study is to combine the Step-wise Weight Assessment Ratio Analysis (SWARA) method and BIM technologies to identify and reduce time delays caused by reworks in construction projects. Firstly, 49 rework causes in residential buildings were identified and ranked. Then, BIM was generated and compared to the initial model. It was observed that working hours were reduced by 4.6%. Moreover, using an Earned Value Management (EVM) system, a 0.06 increase in Schedule Performance Index (SPI) factor was illustrated. Results obtained by this study provide an effective step in reducing a project's time in the construction industry.

Keywords: rework causes; BIM; SWARA method; time; project success

1. Introduction

Rework is regarded as a serious issue for construction industry projects [1]. Cost, schedule, performance, and productivity of construction projects are influenced by reworks [2]. Cost and schedule overruns often occur due to rework in construction projects [2,3]. According to previous studies, rework costs range from 5% to 20% in major civil engineering projects [4,5]. To manage rework, its roots and causes must be identified first [1,6]. Many studies have been conducted to identify such causes [1,6–10]. It is essential to reduce rework due to severe potential consequences. Thus, managers are highly recommended to identify factors which result in rework in the planning phase of projects [1].

There have been various definitions of rework given by different researchers. According to Josephson et al. (2002), rework is defined as dispensable output resulting due to mistakes during the construction project [7]. Love (2002) defines rework as an event or process which is caused due to quality accidents, unqualified quality problems, deviations, or faults [11]. Ye et al. (2015) define rework as redoing a process which has already been done, to satisfy the functional requirements of the project [2]. Forcada et al. (2017) mention that any additional work that has resulted from order changes, design errors and scope changes must also be regarded as rework [12]. Many researchers have attempted to identify factors of rework. It is crucial to analyze all factors and to use appropriate decision-making tools for clients and construction project managers. Research has shown strong attention to these complex management issues to improve the productivity of projects in the construction industry. Most of the researchers are focused on the identification of risk processes and factors to support managers and decision-makers in identifying problems for efficient risk management [13].

According to Hwang et al. (2019), these factors in the construction projects can generally be put into six groups including "Contractor", "Subcontractor", "Supplier", "Manufacturer", "Designer" and "Client" [1]. Fayek et al. (2003) developed a fishbone diagram to illustrate the actual and potential causes of rework. They concluded that "Poor workmanship of prefabricated material", "Lack of inspection", and "Consistency not insured before issued for construction" are the major reasons for rework [6]. Rework can affect a projects' performance [1], thus, it seems necessary to identify and prevent them.

There have been various definitions for Building Information Modelling (BIM). For instance, Penttilä (2006) defines BIM as

> "A set of interrelating policies, processes and technologies that generate a systematic approach to managing the critical information for building design and project data in digital format throughout the life cycle of a building" [14].

The U.S. National BIM Standard also defines BIM as

> "A shared knowledge resource for information about a facility forming a reliable basis for decisions during its life-cycle; defined as existing from earliest conception to demolition" [15].

Most importantly in the use of BIM technologies for construction projects is to have reliable information at any construction project implementation stage and to make correct decisions [16].

There are different dimensions in BIM which are used in construction projects according to the complexity and requirements of such projects. The dimensions in BIM are known as 3D, 4D, 5D, 6D and 7D [17]. The third dimension of BIM (3D) represents the three geographical dimensions of a structure, commonly known as x, y, and z, which stand for length, width, and height of a structure, respectively [13,17–19]. The fourth dimension of BIM (4D) adds time and scheduling to the 3D BIM by simulating the construction process, which enables the project to be visualized at any point in time [19]. The fifth dimension of BIM (5D) integrates the 4D BIM and the project costs. This way, changes of the economic situation of a project can be observed at any phase of the construction project, which is a valuable feature, especially for an estimation of the initial budget forecast [20] and for the management of actual expenses. The sixth dimension of BIM (6D) considers sustainability, and more specifically energy, by estimating energy consumption in all phases of the project. The last dimension (7D) adds a facility management feature for a structure including its status, technical specifications, warranty information, and maintenance/operation manuals for owners and managers [17].

Using BIM technologies in construction has numerous positive effects [21]. For example, probable construction clashes can be identified and prevented using the model [22]. According to other studies, the overall performance of a project and project information management can be improved dramatically by using BIM besides other strategic innovations [23,24]. Non-value adding activities and their resulting wastes can be investigated in BIM-based project delivery [25]. The impact of various factors on delays can also be analyzed using BIM [26]. It is necessary to emphasize, that the use of BIM technologies is not limited with the construction of new buildings but can also be used in the reconstruction of heritage buildings. There are different ways of using BIM, and this effective support is not limited to the 3D modelling, but also uses photogrammetry, 3D scanning and other tools for existing buildings [27].

Lu et al. (2018) illustrated that construction errors on site can be decreased by sharing design information with site workers [28]. Decreasing the causes of rework, including design errors and defects, has been an aim for many researchers. For instance, Kwon et al. (2014) explored a defective management system by integrating BIM, image-machining and augmented reality to automatically identify and omit defects [29]. Moreover, according to different research, defect data were proposed to be shared using a BIM-integrated network [30]. Bryde et al. (2013) investigated the advantages and disadvantages of BIM use in projects and concluded that the advantages of using BIM are much greater in comparison to its drawbacks, challenges, and limitations [16,31]. However, the direct use of BIM technologies in reducing rework needs more investigation.

This paper aimed to reduce rework in the construction projects from the perspective of time using Building Information Modelling (BIM). As mentioned in the previous paragraphs, most of the research focuses on investigating the effect of BIM technologies on reducing rework. Thus, the consequences of using BIM technologies in a projects' schedule is considered as a gap in the body of knowledge. The novelty of this paper is that time effects of reworks are specifically investigated by using BIM technologies. A Step-wise Weight Assessment Ratio Analysis (SWARA) method was employed to weigh and rank the identified rework factors. Autodesk Revit software and Autodesk Navisworks software were used as the BIM tools in order to evaluate the benefits of using BIM in comparison to the traditional method, using the most important identified rework causes that are able to be simulated in BIM. Status-Curves (S-Curves) and an Earned Value Management (EVM) system were also used to calculate factors which illustrated the improvement of using BIM technologies. The findings of this study will illustrate the benefits of combining one of the decision-making tools, the SWARA method, with BIM technologies in order to identify and decrease reworks, and consequently their effects on the construction industry projects, and to ultimately enhance projects' sustainability.

2. Research Methodology

At first, rework causes in building construction projects were identified using literature including books, papers, documents, and online databases, as well as asking expert's opinions and also conducting field investigations. In this process, factors which had the most effect on increasing the projects' time (fourth dimension) were identified and ranked by the SWARA method. These causes were then illustrated on a fishbone diagram.

In the second stage, a building was selected as a case study and analyzed in terms of construction time using two different scenarios. The first scenario analyzed the construction process using the traditional system. On the other hand, the second scenario analyzed the abovementioned process according to BIM, in which clashes and therefore reworks could be diagnosed and managed at the beginning of the project.

In the next stage, the effects of utilizing BIM technologies on reducing reworks were investigated by comparing the two abovementioned scenarios in the previous stage. To do so, construction times of both scenarios were analyzed using S-Curves in Microsoft Project Software, and the effects of using BIM on reducing delays in the construction time were investigated.

Finally, the last stage focused on analyzing the benefits of using BIM technologies on reducing delays in the construction project by using the EVM system. These stages are demonstrated in Figure 1.

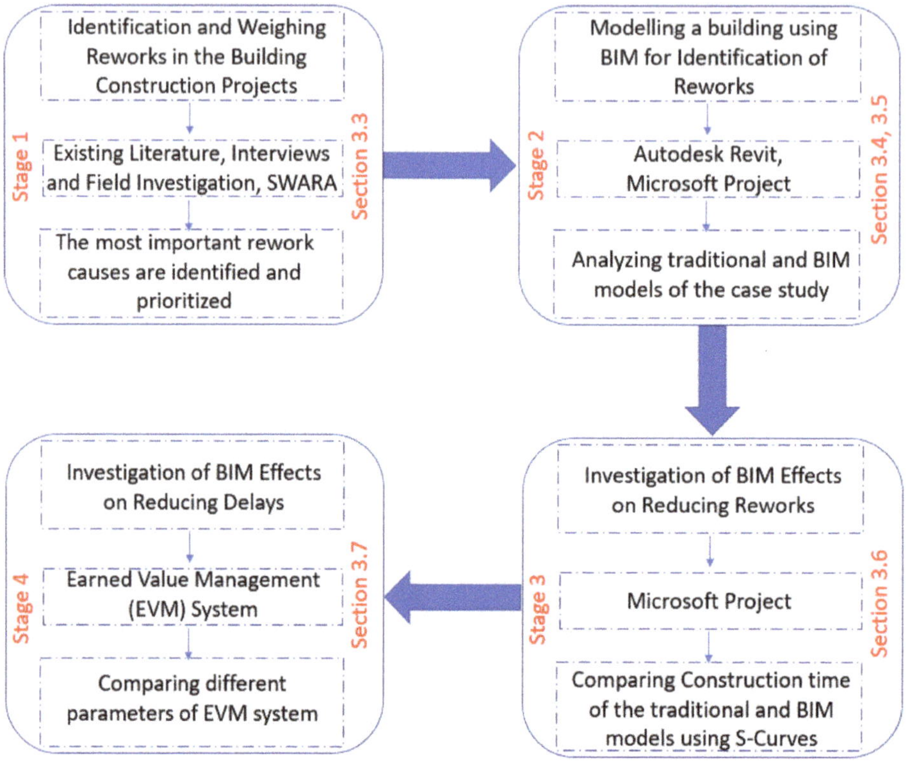

Figure 1. Research methodology.

2.1. Questionnaire

In the current study, a questionnaire was designed to weigh and rank rework causes. Designing the abovementioned questionnaire was conducted carefully, and experts played a significant role in making the final version. The final questionnaire included three sections. In the first section, there were some questions regarding general information such as occupational experience. In the second section, respondents were asked to weigh rework causes by considering the identified selection criteria. To attain this goal, a 5-point Likert scale was used in which 1 was the least importance, while 5 was defined as the most importance. Lastly, in the third part of the questionnaire, respondents were asked to mention any other rework causes or points about the topic. The information gained by this questionnaire was then analyzed by the SWARA method.

Cronbach's alpha is a coefficient calculated to check internal consistency. Therefore, it was a suitable coefficient to illustrate the reliability of the questionnaires. Cronbach's alpha values range from 0 to 1, where 0 means that all items are independent, while 1 means that items are perfectly correlated [32]. In this range, in terms of reliability, values above 0.9, 0.8, 0.7, 0.6, 0.5 are considered excellent, good, acceptable, questionable, and poor, respectively. Therefore, values below 0.5 are regarded as unacceptable [33,34]. There are two ways to calculate Cronbach's alpha. It can be calculated manually, according to the formula below [32]:

$$\alpha = \frac{j}{j-1}(1 - \frac{S_j^2}{S^2}) \quad (1)$$

where the number of items, variance of the jth criteria and variance of the total score are shown by j, S_j^2 and S^2, respectively. The second way to calculate Cronbach's alpha is by using software programs such as SPSS. Due to the complexity and difficulty of manual calculations, this way is usually preferred. In this study, Cronbach's alpha was computed using SPSS software.

2.2. SWARA (Step-Wise Weight Assessment Ratio Analysis) Method

Various Multi-Criteria Decision Making (MCDM) methods can be used in different cases, for example, they can be used for efficient application based on sustainability assessment tool efficiency, cost analysis and renewable energy evaluation [35,36]. Keršuliene et al. introduced the SWARA method in 2010 [37], and, in comparison to the other MCDM methods, SWARA is easier to employ in decision-making problems due to its understandable concept and analysis procedure [38].

The SWARA method is usually used for weighing decision criteria, which are the basis of assessing and prioritizing various alternatives [36]. To do so, knowledge, experience and opinions of experts are considered [39]. SWARA has been used in different topics by numerous researchers. For instance, the SWARA method was selected in an Iranian study to assess selection criteria for choosing the best passive energy reduction measures in Iran [40]. Balki et al. (2020) determined optimal operating parameters in Turkey and criteria were weighed using the SWARA method [38]. Rani et al. (2020) provided Solar Panel Selection [41], while Jafarzadeh Ghoushchi et al. (2020) ranked failures in Solar Panel Systems using the abovementioned method [42]. Akcan et al. (2019) aimed to reduce ecological risk factors by evaluating green suppliers and employed the SWARA method as a part of this evaluation [43]. Zarbakhshnia et al. (2018) evaluated and selected sustainable reverse logistic providers and used the abovementioned method in their analysis [44]. Chalekaee et al. (2019) applied SWARA when analyzing construction delay change response problems [45]. Morkunaite et al. (2019) evaluated the significance of criteria in contractor selections for the refurbishment of heritage buildings [46]. There is a lot more research which has utilized the SWARA method [47–58].

In this study, SWARA was used to weigh rework causes using the questionnaire. The procedure of using the SWARA method is illustrated below:

A. Selection criteria are identified and considered.
B. The abovementioned criteria are prioritized using experts' attitudes and then they are sorted from the most important to the least important.
C. Each criterion is compared to the upper criteria and the comparative average value of importance, s_j, is calculated.
D. Comparative importance (k_j) is computed as follows:

$$k_j = \begin{cases} 1 & j = 1 \\ s_j + 1 & j > 1 \end{cases} \quad (2)$$

E. Recalculated weights (q_j) are obtained using the formula below:

$$q_j = \begin{cases} 1 & j = 1 \\ \frac{q_{j-1}}{k_j} & j > 1 \end{cases} \quad (3)$$

F. Final weights (w_j) are calculated as follows:

$$w_j = \frac{q_j}{\sum_{m=1}^{n} q_m} \quad (4)$$

2.3. *Case Study*

Shiraz is one of the cities in Fars province, Iran, which is located in the southwest of the country. It is surrounded by various mountains and has a temperate climate [59]. A building which is located in Shiraz was considered as a case study for this research. The building is a 4-story steel structure in the western section of Shiraz, with an infrastructure area of 1100 square meters.

2.4. *Formation of the BIM Output*

Three kinds of BIM software were used as the key tools of this study. To model the case study building's elements, Autodesk Revit software and Autodesk Navisworks software were used. Autodesk Navisworks software and Microsoft Project software were also applied in the time management section of the research. Details about the usage of BIM software are shown in Table 1.

Table 1. Building Information Modelling (BIM) software used in the process of modelling.

Stage	BIM Software
Initial idea modelling	Autodesk Revit Architecture
Identifying the ability to build the idea	Autodesk Navisworks Manage
Modelling of the structure	Autodesk Revit Structure
Modelling the electrical and mechanical installation	Autodesk Revit MEP (Mechanical, Electrical, and Plumbing)
Identifying architectural and structural clashes	Autodesk Navisworks Manage
Identifying the change of results	Autodesk Revit Architecture and Structure
Identifying the construction schedule	Autodesk Navisworks Manage and Microsoft Project

2.4.1. Modelling the Architectural and Structural Information of the Building

Architectural and installation information of the building was modelled as follows:

- Step 1: Introducing the number of stories, as well as the height of each story according to the architectural plans.
- Step 2: Introducing and modelling the major elements of the building such as walls, roofs, and stairs.
- Step 3: Introducing and modelling the minor elements of the building such as doors and windows.
- Step 4: Adding supplementary details of the building such as stepped ceilings and parapets.
- Step 5: Modelling the building's risers and ducts, where installation components are located.
- Structural information of the building was modelled as follows:
- Step 1: Introducing the number of stories, as well as the height of each story according to the structural plans.
- Step 2: Introducing and modelling the major elements of the building such as the foundation, columns, and beams.
- Step 3: Modelling lateral bracing system.
- Step 4: Introducing and adding roofs and diaphragms.
- Step 5: Adding supplementary details such as roofs and connections.

2.4.2. Integrating and Simulating the Construction Process

In this study, Autodesk Revit was used in different parts. Autodesk provides the ability to integrate between the three versions of Revit, including Revit Architecture, Revit Structure and Revit MEP (mechanical, electrical, and plumbing) [60].

2.5. Data Extraction of the Traditional and BIM Methods

2.5.1. Traditional Method

In this part of the study, the abovementioned building case study was considered. All the construction documents of the buildings were investigated carefully to obtain reworks and gain the total construction time of the project.

2.5.2. BIM Method

As mentioned in the previous section, all the elements of the building were modelled in the BIM tools. Thus, all the clashes and reworks were identified and prevented at the beginning. This way, a massive amount of time and budget was saved.

2.6. Investigating the Ability of BIM Tools in the Identification of Clashes and Reworks

2.6.1. Automatic Identification of Errors Within the Process of Modelling

Using Autodesk Revit software, all the errors including clashes between architectural, structural and installation elements of the building were identified automatically, and a solution was given by the software. This feature gives engineers the ability to observe and prevent a large number of clashes and reworks at the initial level. Figure 2 illustrates one of the clashes that was identified by the software.

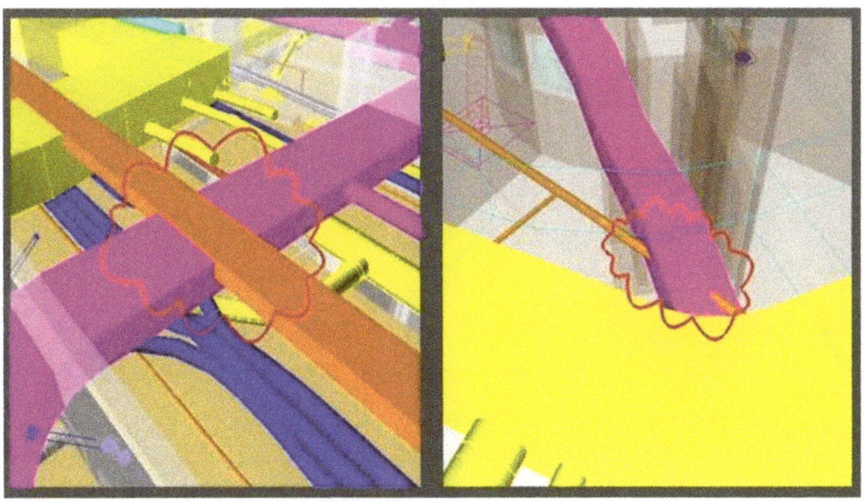

Figure 2. Automatic identification of errors within the modelling process.

2.6.2. Identification of Errors after the Process of Modelling

In this stage, BIM outputs and simulations were integrated, and all the clashes and errors were checked again using Autodesk Navisworks software. Data were imported directly from Autodesk Revit to Autodesk Navisworks. After this step, the probability of any clashes occurring during the construction process becomes almost zero. One of the identified clashes after the process of modelling is shown in Figure 3.

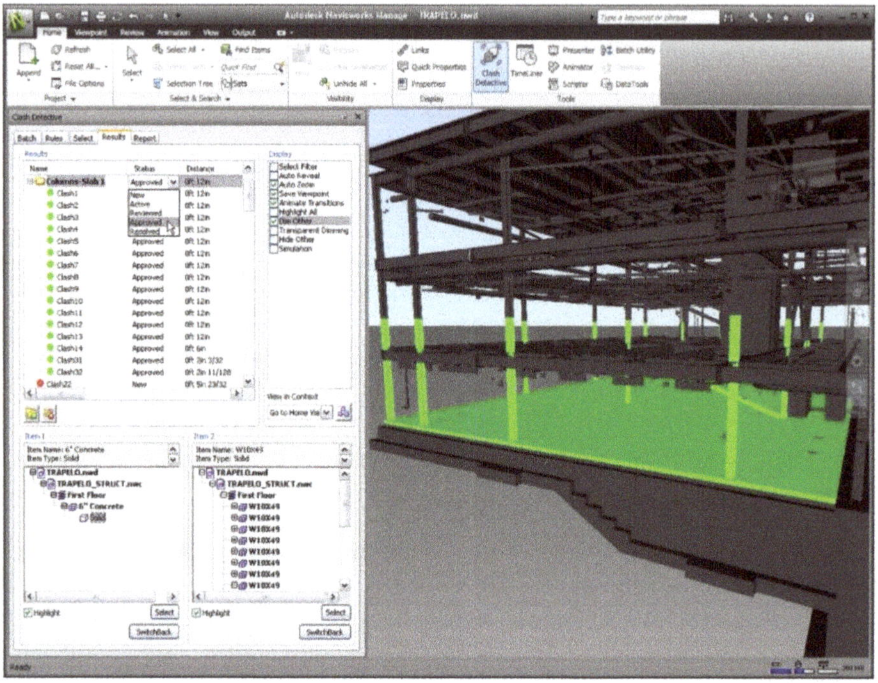

Figure 3. Identification of errors after the modelling process.

2.7. Earned Value Management (EVM)

The Earned Value Management (EVM) method is used to measure time, cost, and scope, for predicting the performance of projects [61,62]. The EVM method can consider the time passed as well as the cost, for calculating the value of work done [63]. In this stage, the aim was to observe the benefits of using BIM technologies in reducing delays in construction of the building. Thus, the procedure of this observation is illustrated as follows:

A. Calculating the Schedule Variance (SV) as follows:

$$SV = EV - PV \tag{5}$$

where EV and PV stand for Earned Value (budgeted cost of work performed) and Actual Cost (actual cost of work performed), respectively.

B. Calculation of Schedule Performance Index (SPI):

$$SPI = EV/PV \tag{6}$$

3. Results and Discussion

3.1. Analyzing the Sample Size

135 experts who have been involved in various sectors of the construction industry, including both private and governmental sectors, were the selected sample size of this study. In order to identify a sufficient number of respondents, Cochran's sample size formula was used. This formula is shown below [64]:

$$n = \frac{Nt^2pq}{Nd^2 + t^2pq} \tag{7}$$

where n, N, t, p, q and d stand for the sample size, population, confidence level value, probability of success, probability of failure and acceptable margin of error, respectively. In order to gain the required number of respondents, N, t, p, q and d were considered to be 135, 1.96, 0.5, 0.5 and 0.05, respectively. Therefore, n was computed as follows:

$$= \frac{135 * 1.96^2 * 0.5 * 0.5}{135 * 0.05^2 + 1.96^2 * 0.5 * 0.5} = 99.89.$$

Therefore, at least 100 respondents were needed to fill the questionnaire out. The abovementioned 135 questionnaires were distributed using the Internet (75 questionnaires) and postal system (60 questionnaires). 115 questionnaires were returned, which meant an 85.1% return rate: that was considered to be acceptable. Among the abovementioned returned questionnaires, six of them were unverified. Therefore, 109 questionnaires were analyzed which was more than required. Information regarding the questionnaires is shown in Table 2. Information about the experts who completed the verified questionnaires is illustrated in Table 3.

Table 2. Questionnaires return rate.

Questionnaire	Number	Percentage
Total distributed	135	100%
Total Returned	115	85.1%
Unreturned	20	14.8%
Unverified returned	6	4.4%
Verified returned	109	80.7%

Table 3. Information about the respondents.

Classification	Classification	Number	Percentage
Working background	Construction Engineer	29	26.6%
	Technical director	43	39.4%
	Project manager	12	11%
	Installation engineer	15	13.8%
	Employer	10	9.2%
Qualification	Bachelor	50	45.9%
	Master	36	33%
	PhD	23	21.1%
Working Experience	Less than 10 years	27	24.8%
	Between 10 and 20 years	32	29.3%
	More than 20 years	50	45.9%

3.2. Reliability of the Questionnaires

To assess the reliability of the questionnaires, Cronbach's alpha was calculated. Rework causes were categorized into seven groups according to their properties (Section 3.3). Thus, for each of the questionnaires, Cronbach's alpha value was computed separately. This value was calculated in order to check the reliability of questionnaires, and ranges from 0 to 1, where 0 and 1 stand for complete independency and complete dependency, respectively [32]. These computed values are demonstrated in Table 4.

Table 4. Values of Cronbach's alpha.

Questionnaire	Categorization of Rework Causes	Number of Rework Causes	Cronbach's Alpha
A	Engineering and Reviews	8	0.783
B	Implementation of Project	10	0.803
C	Material and Equipment Supply	6	0.776
D	Human Resource Capability	8	0.803
E	Construction Planning and Scheduling	4	0.735
F	Leadership and Communication	5	0.721
G	Effective External Causes	8	0.758

3.3. *Identification and Prioritization of Rework Causes*

The first step in this section was identifying rework causes. A comprehensive investigation took place in various literature resources. Then, experts were interviewed to add any other rework causes to the identified ones. Information regarding the experts is shown in Table 5. 42 of the identified causes were extracted from the literature [1,6] and the other factors were introduced by experts. Table 6 illustrates the final categorization of identified rework causes. Figure 4 also shows the abovementioned causes using a fishbone diagram. In the abovementioned figure, rework causes are separated into seven different groups including Engineering and Reviews (A), Implementation of Project (B), Material and Equipment Supply (C), Human Resource Capability (D), Construction Planning and Scheduling (E), Leadership and Communication (F), Effective External Causes (G). Each category also constitutes several subsections, which are illustrated in Figure 4.

Table 5. Information regarding experts.

Category	Classification	Number
Occupation	Academia	8
	Manager	9
	Contractor	8
	Technical expert	12
Sex	Male	20
	Female	17
Experience (years)	<5	7
	5–10	9
	10–15	10
	>15	11

Table 6. Categorization of rework causes.

Categorization of Rework Causes	Rework Causes	Sign
Engineering and Reviews	Design Errors	A1
	Scope Changes	A2
	Late Design Changes	A3
	Poor Document Control	A4
	Design Changes	A5
	Poor Supervision and Control	A6
	Poor Knowledge of Designer	A7
	Lack of Using Modern Design Tools	A8
Implementation of Project	Lack of Using Modern Implementation Systems	B1
	Changes in Work Volume	B2
	Difference among Plans and Operational Specifications	B3
	Incoherence of Structural Implementations	B4
	Unspecified Essential Operations	B5
	Lack of Operational Standards	B6
	Lack of Using Appropriate Appliances	B7
	Lack of Supervision in Controlling Quality	B8
	Poor Experience of Contractors	B9
	Poor Quality of Implementations	B10
Material and Equipment Supply	Non-compliance with Specifications	C1
	Materials not in the Right Place When Needed	C2
	Untimely Deliveries	C3
	Structural Non-compliance	C4
	Poor Quality of Materials	C5
	Lack of Suppliers' Information Regarding the Status of Project	C6
Human Resource Capability	Insufficient Skill Level	D1
	Lack of Knowledge in Occupational Planning	D2
	Unclear Instructions to Workers	D3
	Excessive Overtime	D4
	Lack of Occupational Security	D5
	Inadequate Control of Human Resource	D6
	Lack of Workers' Responsibility	D7
	Inadequate Training of Human Resource	D8
Construction Planning and Scheduling	Unrealistic Schedules	E1
	Insufficient Turnover and Commissioning Resourcing	E2
	Late Designer Input	E3
	Constructability Problems	E4
Leadership and Communication	Ineffective Management of Project Team	F1
	Lack of Operations	F2
	Lack of Safety and QA/QC Commitment	F3
	Poor Communication	F4
	Poor Attendance of Stakeholders	F5
Effective External Causes	Governmental Changes in Law	G1
	Economic Fluctuations	G2
	Social Contradictions	G3
	High Cost of Modern Technologies	G4
	Lack of Stakeholders' Training	G5
	Physical and Infrastructural Circumstances	G6
	Geographical Hazards	G7
	Political Circumstances and Sanctions	G8

After identifying the causes of rework, the next step focused on weighing the abovementioned causes. To do so, experts' opinions were extracted using a questionnaire. In the questionnaire, experts were asked to score the importance of causes from 1 to 5, where 1 and 5 meant the least and most effective, respectively. Mean values of the questionnaires were calculated after the questionnaires were returned. Then, using the SWARA method, rework causes were weighed and ranked in their specific groups. As defined above, rework causes were categorized into seven groups. Tables 7–13 show the results of the SWARA method for each group. The most important rework causes are also demonstrated in Table 14.

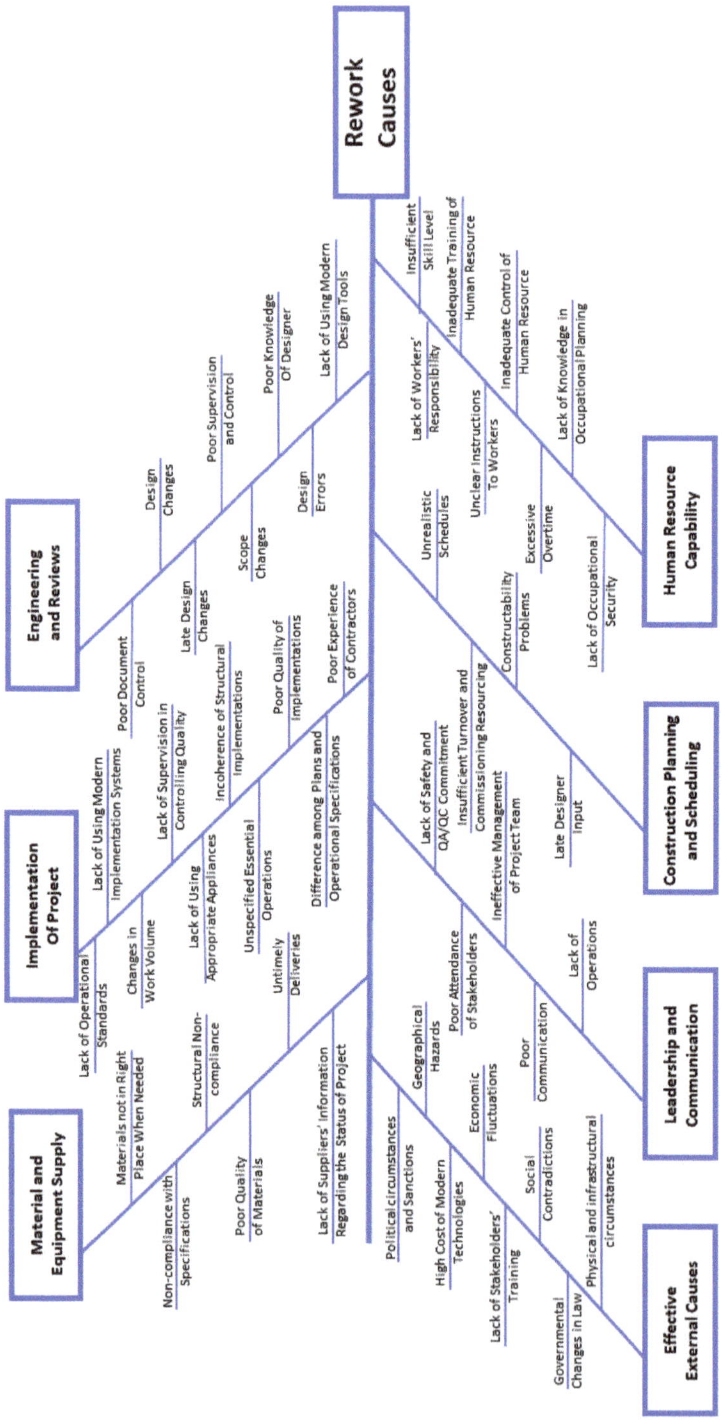

Figure 4. Fishbone diagram of rework causes.

Table 7. Weights of rework causes categorized in the "Engineering and Reviews" group.

Rework Cause	S_j	K_j=s_j+1	q_j	w_j	Rank
A1	—	1	1	0.48925	1
A5	0.902	1.902	0.52576	0.25723	2
A3	0.885	1.885	0.27892	0.13646	3
A8	0.831	1.831	0.15233	0.07453	4
A2	0.752	1.752	0.08695	0.04254	5
A6	0.712	1.712	0.05079	0.02485	6
A7	0.61	1.61	0.03154	0.01543	7
A4	0.585	1.585	0.01990	0.00974	8

Table 8. Weights of rework causes categorized in the "Implementation of Project" group.

Rework Cause	S_j	K_j=s_j+1	q_j	w_j	Rank
B3	—	1	1	0.48250	1
B10	0.896	1.896	0.52743	0.25448	2
B1	0.865	1.865	0.28280	0.13645	3
B5	0.712	1.712	0.16519	0.07970	4
B7	0.701	1.701	0.09711	0.04686	5
B8	0.618	1.618	0.06002	0.02896	6
B2	0.583	1.583	0.03792	0.01829	7
B6	0.524	1.524	0.02488	0.012000	8
B4	0.512	1.512	0.01645	0.00794	9
B9	0.493	1.493	0.01102	0.00532	10

Table 9. Weights of rework causes categorized in the "Material and Equipment Supply" group.

Rework Cause	S_j	K_j=s_j+1	q_j	w_j	Rank
C1	—	1	1	0.48299	1
C5	0.884	1.884	0.53079	0.25636	2
C6	0.842	1.842	0.28816	0.13918	3
C2	0.821	1.821	0.15824	0.07643	4
C4	0.697	1.697	0.09325	0.04504	5
C3	0.658	1.658	0.05624	0.02716	6

Table 10. Weights of rework causes categorized in the "Human Resource Capability" group.

Rework Cause	S_j	K_j=s_j+1	q_j	w_j	Rank
D1	—	1	1	0.48364	1
D3	0.893	1.893	0.52826	0.25549	2
D4	0.823	1.823	0.28978	0.14015	3
D7	0.808	1.808	0.16027	0.07752	4
D5	0.794	1.794	0.08939	0.04321	5
D8	0.769	1.769	0.05050	0.02443	6
D2	0.742	1.742	0.02899	0.01402	7
D6	0.717	1.717	0.01688	0.00817	8

Table 11. Weights of rework causes categorized in the "Construction Planning and Scheduling" group.

Rework Cause	S_j	K_j=s_j+1	q_j	w_j	Rank
E1	—	1	1	0.493960	1
E2	0.946	1.946	0.51387	0.25383	2
E3	0.873	1.873	0.27436	0.13552	3
E4	0.829	1.829	0.15000	0.07410	4

Table 12. Weights of rework causes categorized in the "Leadership and Communication" group.

Rework Cause	S_j	K_j=s_j+1	q_j	w_j	Rank
F4	—	1	1	0.47721	1
F5	0.864	1.864	0.53648	0.25602	2
F1	0.822	1.822	0.29445	0.14051	3
F2	0.754	1.754	0.16787	0.08011	4
F3	0.736	1.736	0.09670	0.4615	5

Table 13. Weights of rework causes categorized in the "Effective External Causes" group.

Rework Cause	S_j	K_j=s_j+1	q_j	w_j	Rank
G2	—	1	1	046900	1
G8	0.832	1.832	0.54585	0.25600	2
G4	0.768	1.768	0.30874	0.14480	3
G6	0.759	1.759	0.17552	0.08232	4
G1	0.719	1.719	0.10211	0.04789	5
G3	0.624	1.624	0.62870	0.02949	6
G5	0.607	1.607	0.03912	0.01835	7
G7	0.573	1.573	0.02484	0.01167	8

Table 14. The most important rework causes.

Sign	Rework Cause
A1	Design Errors
A5	Design Changes
B3	Difference among Plans and Operational Specifications
B10	Poor Quality of Implementations
C1	Non-compliance with Specifications
C5	Poor Quality of Materials
D1	Insufficient Skill Level
D3	Unclear Instructions to Workers
E1	Unrealistic Schedules
E2	Insufficient Turnover and Commissioning Resourcing
F4	Poor Communication
F5	Poor Attendance of Stakeholders
G2	Economic Fluctuations
G8	Political Circumstances and Sanctions

As it is clearly illustrated, the most three crucial rework causes are Design Errors (A1), Design Changes (A5), and Difference among Plans and Operational Specifications (B3). By considering the identified rework causes, and specially the most important causes, they will be reduced effectively or even prevented. Therefore, it can be very useful for different parties of the building industry. Then, through BIM, these effects were analyzed for the case study building.

3.4. *Investigating the Traditional Method of the Case Study Building's Construction*

As it was mentioned in the previous part of the paper, the case study of this project was a 4-story steel building located in the western section of Shiraz, Iran. Initial documents of the project's schedule were investigated carefully to find the predicted construction time. Using a Work Breakdown Structure (WBS), the project was divided into 25 levels, and the planned construction time was 348 days. However, after investigating the final documents of project, it was shown that the project had experienced a 176 days delay, and the project was finished after 524 days. Table 15 illustrates the project's WBS, anticipated, and implemented construction time.

Table 15. The case study building's Work Breakdown Structure (WBS), anticipated, and implemented construction time.

No.	WBS Level	Anticipated Time (Days)	Implemented Time (Days)
1	Delivery of Site	1	1
2	Site Preparation	12	19
3	Implementation of Foundation	26	42
4	Implementation of Building's Structure	43	99
5	Initial Flooring	46	107
6	Implementation of Stairs' Foundation	5	9
7	Implementation of Roofs and Internal Walls	84	99
8	Implementation of Stories' Foundation	33	38
9	Implementation of External Walls	9	11
10	Moving Frames and Doors to their Places	1	2
11	Implementation of Windows	18	24
12	Implementation of Stairs	2	2
13	Flooring the Stories	33	45
14	Moving Electrical Appliances to their Places	2	3
15	Installation	16	34
16	Installation of Frames	13	32
17	Implementing Toilets	1	5
18	Initial Joinery of the Floors	19	36
19	Final Flooring	24	36
20	Installation of Cornices	6	12
21	Final Joinery of the Floors	30	44
22	Installation of Floors' Appliances	6	12
23	Implementation of Facade	18	43
24	Painting	20	33
25	Delivery of Project	4	7
	Total Time	348	524

3.5. *Investigating BIM Output of the Case Study Building's Construction*

In this stage, BIM was generated using the initial documents of the project, including the most important identified rework causes. Thus, many of the errors and rework causes could have been found and their effects could have been diminished. Various errors were identified during the modelling and a summary of the errors is illustrated in Table 16.

Table 16. Summary of the identified errors and their effects on schedule.

No.	Error	Delay (Days)	Time Saved Using BIM (Days)
1	Interference among Structural and Architectural Elements	16	10
2	Designing Errors	37	20
3	Interference among Structural Elements, Openings, and Installation	13	9

3.6. *Investigation and Comparison of Delays between Traditional and BIM Assessments*

In this stage, cumulative working times of the project were calculated and S-Curves (Status-Curves) of the project's anticipated, actual and BIM output were drawn using Microsoft Project software. S-Curves showed that the project's anticipated, actual and BIM cumulative working times were 4336, 6936 and 6617 h, respectively. Results show that the project's cumulative working time would have been reduced by 319 h (4.6%) using BIM technologies, which is regarded as a valuable step in reducing projects' delays. It is necessary to mention that some delay and rework causes were not able to be modelled, such as political sanctions or geographical hazards, as they are not predictable. Figure 5 illustrates the S-Curves of the three abovementioned error circumstances. According to Figure 5, BIM would have reduced the actual time by 319 h, which could be regarded as a great improvement. Therefore, if BIM was used, the project would have been finished in 6617 h as opposed to 6936 h.

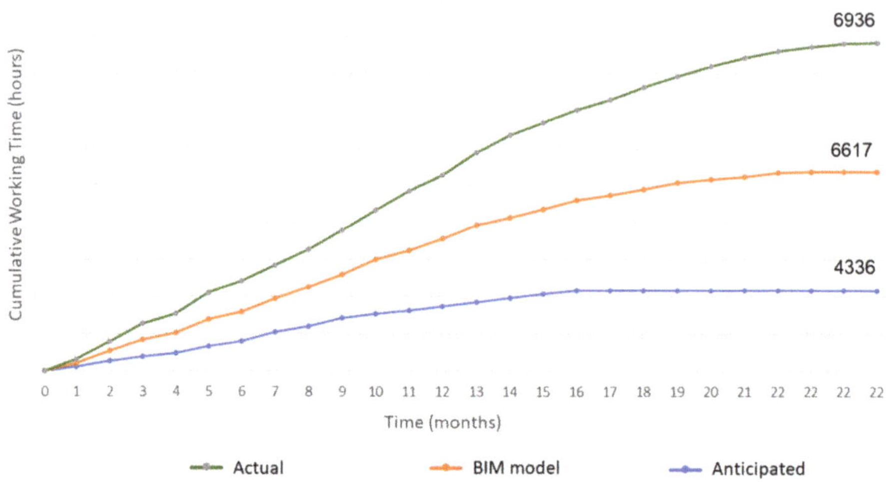

Figure 5. Status-Curves of the anticipated, actual and BIM cumulative working time.

3.7. Analyzing Time Using the Earned Value Management (EVM) System

To analyze the earned values and the status changes of this study, the parameters SPI and SV for the actual implementation and BIM output were calculated and are presented in Table 17. As can be seen, the computed values are still different from the actual values of the case study, although the results show the advantages of using BIM in a case study model. Moreover, a reduction in the SV value proves that using BIM would be successful.

Table 17. Earned values for the case study building's actual implementation and BIM prediction.

No.	Mode	SV	SPI
1	Actual Implementation	−0.18	0.83
2	BIM	−0.11	0.89

4. Conclusions

This study was conducted to identify rework causes in building construction projects and to analyze the benefits of using BIM technologies to process and predict them. Firstly, 49 rework causes were identified and categorized into seven groups including "Engineering and Reviews", "Implementation of Project", "Material and Equipment Supply", "Human Resource Capability", "Construction Planning and Scheduling", "Leadership and Communication" and "Effective External Causes". Then, the SWARA method was employed to weigh and rank the abovementioned rework causes into their own groups. Results showed that the most important causes in the abovementioned groups were "Design Errors (A1)", "Difference among Plans and Operational Specifications (B3)", "Non-compliance with Specifications (C1)", "Insufficient Skill Level (D1)", "Unrealistic Schedules (E1)", "Poor Communication (F4)" and "Economic Fluctuations (G2)". This was followed by Building Information Modelling for the selected case study using BIM software. In the next stage, anticipated, actual and BIM cumulative working times were calculated and illustrated using S-Curves. It was shown that BIM resulted in a 4.6% decrease in the working time. Finally, an EVM system was utilized to compute the positive effect of BIM technologies, and showed a 0.06 increase in the SPI value. The obtained results of this study could be very useful for different parties in the building industry who live in similar climatic and economic conditions. This study's method could be valuable for future studies, and could be very effective in other research as well. One of the limitations of this study was that only residential buildings were investigated. Although these results are highly beneficial for the residential construction sector, it is suggested that other types of buildings should be investigated in further studies. Prospective future researchers are suggested to investigate other dimensions of BIM, such as time. The authors suggest using BIM technologies at any stage of the construction project life cycle, for both new or refurbished buildings, and also recommend its benefits for sustainable construction.

Author Contributions: H.K.: Methodology, Software, Investigation; A.B.: Conceptualization, Methodology, Software, Investigation, Writing—Original draft; A.V.: Methodology, Visualization, Validation, Investigation, Writing—Reviewing and Editing, Supervision; J.A.: Writing-Reviewing and Editing, Supervision; D.M.: Investigation, Writing—Reviewing and Editing, V.Z.: Investigation, Writing—Reviewing and Editing. All authors have read and agreed to the published version of the manuscript.

Funding: This research received no external funding.

Conflicts of Interest: The authors declare no conflict of interest. The funders had no role in the design of the study; in the collection, analyses, or interpretation of data; in the writing of the manuscript; or in the decision to publish the results.

Abbreviations

Abbreviation	Meaning
BIM	Building Information Modelling
SWARA	Step-wise Weight Assessment Ratio Analysis
EVM	Earned Value Management
SPI	Schedule Performance Index
S-Curve	Status-Curve
MCDM	Multi-Criteria Decision Making
WBS	Work Breakdown Structure
SV	Scheduled Variance

References

1. Hwang, B.-G.; Thomas, S.R.; Haas, C.T.; Caldas, C.H. Measuring the impact of rework on construction cost performance. *J. Constr. Eng. Manag.* **2009**, *135*, 187–198. [CrossRef]
2. Ye, G.; Jin, Z.; Xia, B.; Skitmore, M. Analyzing causes for reworks in construction projects in China. *J. Manag. Eng.* **2015**, *31*, 040140972015. [CrossRef]
3. Love, P.E.; Irani, Z.; Edwards, D.J. A rework reduction model for construction projects. *IEEE Trans. Eng. Manag.* **2004**, *51*, 426–440. [CrossRef]
4. Burati, J.L., Jr.; Farrington, J.J.; Ledbetter, W.B. Causes of quality deviations in design and construction. *J. Constr. Eng. Manag.* **1992**, *118*, 34–49. [CrossRef]
5. Barber, P.; Graves, A.; Hall, M.; Sheath, D.; Tomkins, C. Quality failure costs in civil engineering projects. *Int. J. Qual. Reliab. Manag.* **2000**, *17*, 479–492. [CrossRef]
6. Fayek, A.R.; Dissanayake, M.; Campero, O.; Construction Owners Association of Alberta (COAA). Measuring and classifying construction field rework: A pilot study. *Res. Rep.* **2003**. Available online: https://www.coaa.ab.ca/ (accessed on 22 July 2020).
7. Josephson, P.-E.; Larsson, B.; Li, H. Illustrative benchmarking rework and rework costs in Swedish construction industry. *J. Manag. Eng.* **2002**, *18*, 76–83. [CrossRef]
8. Mastenbroek, Y. Reducing Rework Costs in Construction Projects. Bachelor's Thesis, University of Twente, Enschede, The Netherlands, 2010.
9. Palaneeswaran, E.; Love, P.E.; Kumaraswamy, M.M.; Ng, T.S. Mapping rework causes and effects using artificial neural networks. *Build. Res. Inf.* **2008**, *36*, 450–465. [CrossRef]
10. Love, P.E.; Edwards, D.J.; Watson, H.; Davis, P. Rework in civil infrastructure projects: Determination of cost predictors. *J. Constr. Eng. Manag.* **2010**, *136*, 275–282. [CrossRef]
11. Love, P.E. Influence of project type and procurement method on rework costs in building construction projects. *J. Constr. Eng. Manag.* **2002**, *128*, 18–29. [CrossRef]

12. Forcada, N.; Gangolells, M.; Casals, M.; Macarulla, M. Factors affecting rework costs in construction. *J. Constr. Eng. Manag.* **2017**, *143*, 040170322017. [CrossRef]
13. Migilinskas, D.; Ustinovičius, L. Methodology of risk and uncertainty management in construction's technological and economical problems. In Proceedings of the 25th International Symposium on Automation and Robotics in Construction, Vilnius, Lithuania, 26–29 June 2008; pp. 789–795.
14. Penttilä, H. Describing the changes in architectural information technology to understand design complexity and free-form architectural expression. *J. Inf. Technol. Constr.* **2006**, *11*, 395–408.
15. National Institute of Building Science. National BIM Standard-United States. Available online: https://www.nationalbimstandard.org/faqs#faq1 (accessed on 3 January 2020).
16. Pavlovskis, M.; Antucheviciene, J.; Migilinskas, D. Assessment of buildings redevelopment possibilities using MCDM and BIM techniques. *Procedia Eng.* **2017**, *172*, 846–850. [CrossRef]
17. United BIM. What are BIM Dimensions. Available online: https://www.united-bim.com/what-are-bim-dimensions-3d-4d-5d-6d-7d-bim-explained-definition-benefits/ (accessed on 28 May 2020).
18. Tarar, M.; Dang, D.T.P. Impact of 4D Modeling on Construction Planning Process. Master's Thesis, Chalmers University of Technology, Göteborg, Sweden, 2012.
19. Eastman, C.M.; Eastman, C.; Teicholz, P.; Sacks, R.; Liston, K. *BIM Handbook: A Guide to Building Information Modeling for Owners, Managers, Designers, Engineers and Contractors*; John Wiley & Sons: Hoboken, NJ, USA, 2011.
20. Kymmell, W. *Building Information Modeling: Planning and Managing Construction Projects with 4D CAD and Simulations*; The McGraw-Hill Companies, Inc.: New York, NY, USA, 2008.
21. Ham, N.; Moon, S.; Kim, J.-H.; Kim, J.-J. Economic analysis of design errors in BIM-based high-rise construction projects: Case study of Haeundae L project. *J. Constr. Eng. Manag.* **2018**, *144*, 05018006. [CrossRef]
22. Chau, K.; Anson, M.; Zhang, J. Four-dimensional visualization of construction scheduling and site utilization. *J. Constr. Eng. Manag.* **2004**, *130*, 598–606. [CrossRef]
23. Love, P.E.; Edwards, D.J.; Han, S.; Goh, Y.M. Design error reduction: Toward the effective utilization of building information modeling. *Res. Eng. Des.* **2011**, *22*, 173–187. [CrossRef]
24. Olawumi, T.O.; Chan, D.W. Building information modelling and project information management framework for construction projects. *J. Civ. Eng. Manag.* **2019**, *25*, 53–75. [CrossRef]
25. Liao, L.; Teo, E.A.L.; Chang, R.; Li, L. Investigating critical non-value adding activities and their resulting wastes in BIM-based project delivery. *Sustainability* **2020**, *12*, 355. [CrossRef]
26. Jang, S.; Lee, G. Impact of organizational factors on delays in BIM-based coordination from a decision-making view: A case study. *J. Civ. Eng. Manag.* **2018**, *24*, 19–30. [CrossRef]
27. Pavlovskis, M.; Migilinskas, D.; Antucheviciene, J.; Kutut, V. Ranking of heritage building conversion alternatives by applying BIM and MCDM: A case of Sapieha Palace in Vilnius. *Symmetry* **2019**, *11*, 973. [CrossRef]
28. Lu, Q.; Chen, L.; Lee, S.; Zhao, X. Activity theory-based analysis of BIM implementation in building O&M and first response. *Autom. Constr.* **2018**, *85*, 317–332.
29. Kwon, O.-S.; Park, C.-S.; Lim, C.-R. A defect management system for reinforced concrete work utilizing BIM, image-matching and augmented reality. *Autom. Constr.* **2014**, *46*, 74–81. [CrossRef]
30. Lee, D.-Y.; Chi, H.-l.; Wang, J.; Wang, X.; Park, C.-S. A linked data system framework for sharing construction defect information using ontologies and BIM environments. *Autom. Constr.* **2016**, *68*, 102–113. [CrossRef]
31. Bryde, D.; Broquetas, M.; Volm, J.M. The project benefits of building information modelling (BIM). *Int. J. Proj. Manag.* **2013**, *31*, 971–980. [CrossRef]
32. Bland, J.M.; Altman, D.G. Statistics notes: Cronbach's alpha. *BMJ* **1997**, *314*, 572. [CrossRef]
33. George, D. *SPSS for Windows Step by Step: A Simple Study Guide and Reference*, 17th ed.; Pearson Education India: New Delhi, India, 2011.
34. Balali, A.; Valipour, A. Identification and selection of building façade's smart materials according to sustainable development goals. *Sustain. Mater. Technol.* **2020**, *26*, e00213.
35. Vilutienė, T.; Migilinskas, D.; Bružas, A. Holistic approach to assess the sustainability and utility of refurbishment measures. *Procedia Eng.* **2015**, *122*, 137–142. [CrossRef]

36. Ghenai, C.; Albawab, M.; Bettayeb, M. Sustainability indicators for renewable energy systems using multi-criteria decision-making model and extended SWARA/ARAS hybrid method. *Renew. Energy* **2020**, *146*, 580–597. [CrossRef]
37. Keršuliene, V.; Zavadskas, E.K.; Turskis, Z. Selection of rational dispute resolution method by applying new step-wise weight assessment ratio analysis (SWARA). *J. Bus. Econ. Manag.* **2010**, *11*, 243–258. [CrossRef]
38. Mardani, A.; Nilashi, M.; Zakuan, N.; Loganathan, N.; Soheilirad, S.; Saman, M.Z.M.; Ibrahim, O. A systematic review and meta-analysis of SWARA and WASPAS methods: Theory and applications with recent fuzzy developments. *Appl. Soft Comput.* **2017**, *57*, 265–292. [CrossRef]
39. Balki, M.K.; Erdoğan, S.; Aydın, S.; Sayin, C. The optimization of engine operating parameters via SWARA and ARAS hybrid method in a small SI engine using alternative fuels. *J. Clean. Prod.* **2020**, *258*, 120685.
40. Balali, A.; Hakimelahi, A.; Valipour, A. Identification and prioritization of passive energy consumption optimization measures in the building industry: An Iranian case study. *J. Build. Eng.* **2020**, *30*, 101239. [CrossRef]
41. Rani, P.; Mishra, A.R.; Mardani, A.; Cavallaro, F.; Štreimikienė, D.; Khan, S.A.R. Pythagorean Fuzzy SWARA–VIKOR Framework for Performance Evaluation of Solar Panel Selection. *Sustainability* **2020**, *12*, 4278.
42. Jafarzadeh Ghoushchi, S.; Ab Rahman, M.N.; Raeisi, D.; Osgooei, E.; Jafarzadeh Ghoushji, M. Integrated Decision-Making Approach Based on SWARA and GRA Methods for the Prioritization of Failures in Solar Panel Systems under Z-Information. *Symmetry* **2020**, *12*, 310. [CrossRef]
43. Akcan, S.; Taş, M.A. Green supplier evaluation with SWARA-TOPSIS integrated method to reduce ecological risk factors. *Environ. Monit. Assess.* **2019**, *191*, 736. [CrossRef]
44. Zarbakhshnia, N.; Soleimani, H.; Ghaderi, H. Sustainable third-party reverse logistics provider evaluation and selection using fuzzy SWARA and developed fuzzy COPRAS in the presence of risk criteria. *Appl. Soft Comput.* **2018**, *65*, 307–319.
45. Chalekaee, A.; Turskis, Z.; Khanzadi, M.; Ghodrati Amiri, G.; Keršulienė, V. A new hybrid MCDM model with grey numbers for the construction delay change response problem. *Sustainability* **2019**, *11*, 776. [CrossRef]
46. Morkunaite, Z.; Bausys, R.; Zavadskas, E.K. Contractor Selection for Sgraffito Decoration of Cultural Heritage Buildings Using the WASPAS-SVNS Method. *Sustainability* **2019**, *11*, 6444. [CrossRef]
47. Akhanova, G.; Nadeem, A.; Kim, J.R.; Azhar, S. A multi-criteria decision-making framework for building sustainability assessment in Kazakhstan. *Sustain. Cities Soc.* **2020**, *52*, 101842. [CrossRef]
48. Prajapati, H.; Kant, R.; Shankar, R. Prioritizing the solutions of reverse logistics implementation to mitigate its barriers: A hybrid modified SWARA and WASPAS approach. *J. Clean. Prod.* **2019**, *240*, 118219. [CrossRef]
49. Valipour, A.; Yahaya, N.; Md Noor, N.; Antuchevičienė, J.; Tamošaitienė, J. Hybrid SWARA-COPRAS method for risk assessment in deep foundation excavation project: An Iranian case study. *J. Civ. Eng. Manag.* **2017**, *23*, 524–532. [CrossRef]
50. Maghsoodi, A.I.; Maghsoodi, A.I.; Poursoltan, P.; Antucheviciene, J.; Turskis, Z. Dam construction material selection by implementing the integrated SWARA–CODAS approach with target-based attributes. *Arch. Civ. Mech. Eng.* **2019**, *19*, 1194–1210. [CrossRef]
51. Jaber, A.Z. Assessment Risk in Construction Projects in Iraq Using COPRAS-SWARA Combined Method. *J. Southwest Jiaotong Univ.* **2019**, *54*. [CrossRef]
52. Alimardani, M.; Hashemkhani Zolfani, S.; Aghdaie, M.H.; Tamošaitienė, J. A novel hybrid SWARA and VIKOR methodology for supplier selection in an agile environment. *Technol. Econ. Dev. Econ.* **2013**, *19*, 533–548. [CrossRef]
53. Keshavarz Ghorabaee, M.; Amiri, M.; Zavadskas, E.K.; Antucheviciene, J. A new hybrid fuzzy MCDM approach for evaluation of construction equipment with sustainability considerations. *Arch. Civ. Mech. Eng.* **2018**, *18*, 32–49. [CrossRef]
54. Karabasevic, D.; Stanujkic, D.; Urosevic, S.; Maksimovic, M. Selection of candidates in the mining industry based on the application of the SWARA and the MULTIMOORA methods. *Acta Montan. Slovaca* **2015**, *20*, 116–124.
55. Hashemkhani Zolfani, S.; Pourhossein, M.; Yazdani, M.; Zavadskas, E.K. Evaluating construction projects of hotels based on environmental sustainability with MCDM framework. *Alex. Eng. J.* **2018**, *57*, 357–365. [CrossRef]

56. Hashemkhani Zolfani, S.; Chatterjee, P. Comparative evaluation of sustainable design based on Step-wise Weight Assessment Ratio Analysis (SWARA) and best worst method (BWM) methods: A perspective on household furnishing materials. *Symmetry* **2019**, *11*, 74. [CrossRef]
57. Valipour, A.; Yahaya, N.; Noor, N.M.; Valipour, I.; Tamošaitienė, J. A SWARA-COPRAS approach to the allocation of risk in water and sewerage public–private partnership projects in Malaysia. *Int. J. Strateg. Prop. Manag.* **2019**, *23*, 269–283. [CrossRef]
58. Balali, A.; Moehler, R.C.; Valipour, A. Ranking cost overrun factors in the mega hospital construction projects using Delphi-SWARA method: An Iranian case study. *Int. J. Constr. Manag.* **2020**, 1–9. [CrossRef]
59. Mohebbi, M.R.; Jashni, A.K.; Dehghani, M.; Hadad, K. Short-Term Prediction of Carbon Monoxide Concentration Using Artificial Neural Network (NARX) Without Traffic Data: Case Study: Shiraz City. *Iran. J. Sci. Technol. Trans. Civ. Eng.* **2019**, *43*, 533–540. [CrossRef]
60. Stine, D.J.; Hanson, J. *Autodesk Revit 2019 Architectural Command Reference*; SDC Publications: Mission, KS, USA, 2018.
61. Koke, B.; Moehler, R.C. Earned Green Value Management for Project Management: A systematic review. *J. Clean. Prod.* **2019**, *230*, 180–197. [CrossRef]
62. Bryde, D.; Unterhitzenberger, C.; Joby, R. Conditions of success for earned value analysis in projects. *Int. J. Proj. Manag.* **2018**, *36*, 474–484. [CrossRef]
63. Fleming, Q.W.; Koppelman, J.M. What's your project's real price tag? *Harv. Bus. Rev.* **2003**, *81*, 20.
64. Cochran, W.G. *Sampling Techniques*; John Wiley & Sons: Hoboken, NJ, USA, 2007.

sustainability

MDPI

Article

Generalized Resource-Constrained Critical Path Method to Improve Sustainability in Construction Project Scheduling

Kyunghwan Kim

Department of Architecture, Konkuk University, 120, Neungdong-ro, Gwangjin-gu, Seoul 05029, Korea; kykim@konkuk.ac.kr

Received: 21 September 2020; Accepted: 25 October 2020; Published: 27 October 2020

Abstract: Delays by limited supply of resources are common in many construction projects and may cause serious monetary disputes between project participants. Since the dispute resolution may require unnecessary additional time and cost, preventing delays in advance is an important goal in sustainable construction project management. To prevent delays, a feasible plan must be implemented, which reflects limited resources and provides reliable activity information. For this purpose, this study proposes a generalized resource-constrained critical path method (eRCPM). It consists of three steps to identify resource-dependent activity relationships (resource links) based on the result of resource-constrained scheduling (RCS) under multiple resource constraints. Compared to the existing resource-constrained critical path methods, the eRCPM has the advantage of identifying resource links irrespective of the applied RCS technique because it is based on the result rather than the RCS process. Further, this study presents a Microsoft (MS) Excel-based half-automated prototype system that is linked using file export and import functions to both P6 and MS Project software packages. The detailed process of the eRCPM algorithm and the operation process of the prototype system are described using an example schedule. Through a case study, it was demonstrated that eRCPM appropriately identifies the necessary resource links and provides reliable total floats.

Keywords: resource-dependent activity relationship; scheduling; scheduling software; Microsoft Excel Visual Basic for Applications (MS Excel VBA)

1. Introduction

The critical path method (CPM) is a representative scheduling technique used in construction project management. It provides important information necessary for managing construction projects and serves as a basis for analyzing the impact of delays in the construction process [1,2]. However, it has a limitation in that it assumes an infinite supply of resources available to perform an activity. Resources such as labor, equipment, and materials are highly limited in many projects, so that efficient resource allocation is essential for sustainable construction project management [3,4]. Delays are common in many construction projects and many delays are caused by limited supply of resources [5,6]. To make up for the delays, additional time and cost may be needed, which tends to cause disputes between project participants. Since the dispute resolution may require unnecessary additional time and cost, preventing delays in advance is an important goal in sustainable construction project management. To prevent delays, a feasible plan must be implemented, which requires reflection of limited resources. Many resource-constrained scheduling (RCS) techniques, such as the serial method, the parallel method, etc., have been developed to reflect limited resources to create more feasible schedules [7–10].

However, the RCS technique does not consider resource-dependent activity relationships (resource links) caused by resource constraints, resulting in loss of important information such as total floats and the critical path [10–13]. In addition, a regular schedule update to reflect changes during the construction progress can alter the work sequence after the data date, resulting in a plan that is different from the original [12]. The RCS technique itself alters the work sequence as it plans activities in the priority order set by the user when resource conflicts occur [7,9]. This work sequence change may make schedule control challenging after the data date. Analysis of the delay impact in RCS also becomes more problematic because of the unreliable total floats (also called phantom floats) and work sequence changes after schedule updates [14–16]. The same challenges occur in Primavera P6 and Microsoft (MS) Project, which are the two most frequently used scheduling software packages [17,18]. Consequently, several studies have been conducted to identify resource links considering limited resources [12,19–22].

Woodworth and Shanahan (1988) presents a method of identifying resource links in the RCS process for the first time among related studies [19]. This method applies the parallel RCS method and finds resource links based on the resource sequence label stored in an activity and an activity sequence list stored in a resource type. The case study applies a simple schedule in which one resource unit is allocated to one activity, so it is not sure whether it is applicable to multiple types of resources. Bowers (1995) proposes a method of finding resource links based on the resource utilization history generated in the process of the parallel RCS method [20]. However, the process of finding resource links is explained too briefly and the applied case study is also simple, so it is not clear if this will work for multiple resource constraints as well. Lu and Li (2003) presents detailed procedures for finding resource links by applying the serial RCS method [21]. This method considers multiple resources by finding resource activity interactions for every resource unit. However, in this process, the CPM relationships may not be considered, and unnecessary resource links may be identified [23,24]. Kim and de la Garza (2003) presents a step-by-step process of identifying resource links while applying the serial method [12]. This method identifies resource links in two steps. The first is to identify resource links of activities postponed due to resource constraints in the RCS process, and the second is to determine the additional resource links within the total float range of each activity after RCS. In the case study, a relatively complex example in which multiple resources have been freely allocated is applied. Nisar et al. (2013) applies the rank position weight method [25] as an RCS technique [22]. In this method, after applying the RCS method normally, the existing network is reversed, and the same RCS method is performed once more to identify each activity's essential time data such as early times, late times, floats, etc. By comparing these time data of each activity, resource links are identified. This heuristic process is not only too complicated to apply [24], but also has a limitation that it is only valid for one resource type.

These previous studies identified resource links in the RCS process and implemented a resource-constrained critical path method (RCPM) by applying a CPM process that considered the identified resource links and the existing technological activity relationships. The RCPM techniques used by these researches suggest methods to identify resource links using a specific technique or algorithm in the RCS process [23,24]. However, this approach has a limitation in that it is only valid in the currently applied RCS technique. That is, an RCPM system based on a certain RCS technique does not work for other RCS techniques. For example, a serial-method-based RCPM system that can be integrated with P6 is not valid for MS Project (MSP) that applies a parallel method [18]. Conversely, a parallel-method-based method that can be used in MSP is not effective in P6. Moreover, additional computer programming is required whenever different activity selection priorities are applied in RCS [26].

This study proposes a generalized RCPM (eRCPM) technique that can implement RCPM irrespective of the type of the RCS method applied under multiple resource constraints. In addition, it presents a half-automated prototype system that is linked using file export and import functions to both P6 and MSP to verify the validity and scalability of the proposed eRCPM. Since it does not depend on a specific RCS technique, eRCPM could be easily adopted in other scheduling software currently in use. This study can contribute not only to efficient project management but also to reduce unnecessary disputes by providing accurate time information, as eRCPM provides a more feasible schedule by reflecting resources and properly providing total floats, the critical path, and a stable work sequence.

2. Review of RCPM

2.1. Overview

This section reviews the process of identifying resource links by applying serial and parallel techniques [7,8] to the same example, based on the RCPM method proposed by Kim and de la Garza [12]. The method of identifying resource links is applied using the same concept in both RCS methods. In the RCS process, a resource link is generated when the start time of an activity is delayed owing to resource constraints or when the newly calculated total float after RCS does not fully exist owing to resource constraints. The detailed process is described with an example schedule in this section.

The example schedule is presented in Table 1; the maximum available units of resources A (R.A) and B (R.B) are 5 and 2, respectively. Early start time (EST), late start time (LST), and total float (TF) after performing CPM are also listed in this table. Figure 1 depicts a time-scaled network bar chart [27] reflecting this CPM result and indicates that R.A is exceeded in days 1 to 3 and R.B in days 1 to 2.

Table 1. Example Schedule Activity Data.

ID	Duration	Successors	Resource Usage		CPM Output		
			R.A	R.B	EST	LST	TF
A	4	F	2	2	1	5	4
B	5	E	2	0	1	1	0
C	2	F	0	2	1	7	6
D	3	-	3	0	1	8	7
E	3	F	0	1	6	6	0
F	2	-	3	1	9	9	0
Maximum resource units			5	2			

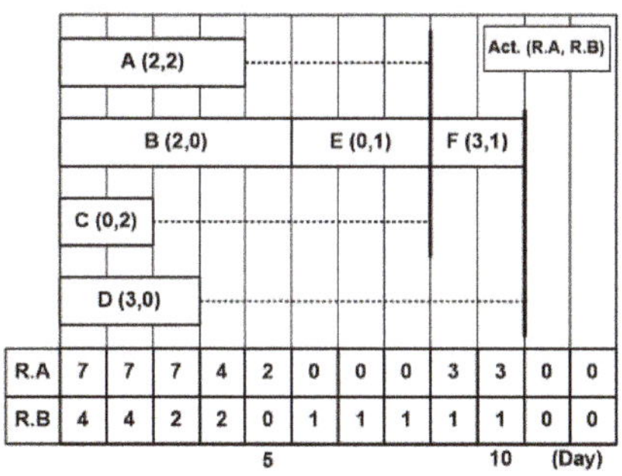

Figure 1. Initial schedule showing resource overruns.

During RCS, when resource conflicts occur between activities, the activity with higher priority is planned first, and the activity with lower priority is delayed until the required resources become available. The priority can be set in many ways, but in this study, the priority is set in the order of late start time, total float, and ID of each activity to achieve consistent results. When this criterion is applied to the CPM result of Table 1, the priority order is B, A, E, C, D, and F, which is applied in the same manner to both serial and parallel methods.

2.2. Serial RCS

In the serial method, activities are planned sequentially according to the priority. If an activity cannot be planned because of resource shortage owing to resource usage by already planned activities, the start time of the activity is delayed until the resources become available. In the example schedule, activities B, A, and E can be planned as the original CPM schedule, but the start time of activity C is delayed to day 9 because it cannot be planned on the original day 1 owing to lack of resources.

This delay of eight days is because planned activity A requires two R.Bs for days 1 to 5, and planned activity E requires one R.B for days 6 to 8. Activity C can start on day 9 using R.B released after the completion of activity E. In RCPM, resource links are created as illustrated in Figure 2a by reflecting this kind of resource-dependent activity relationship. In this study, the relationship by which the resources released at the end of an activity are immediately used for a subsequent activity without a time interval is defined as a direct resource link.

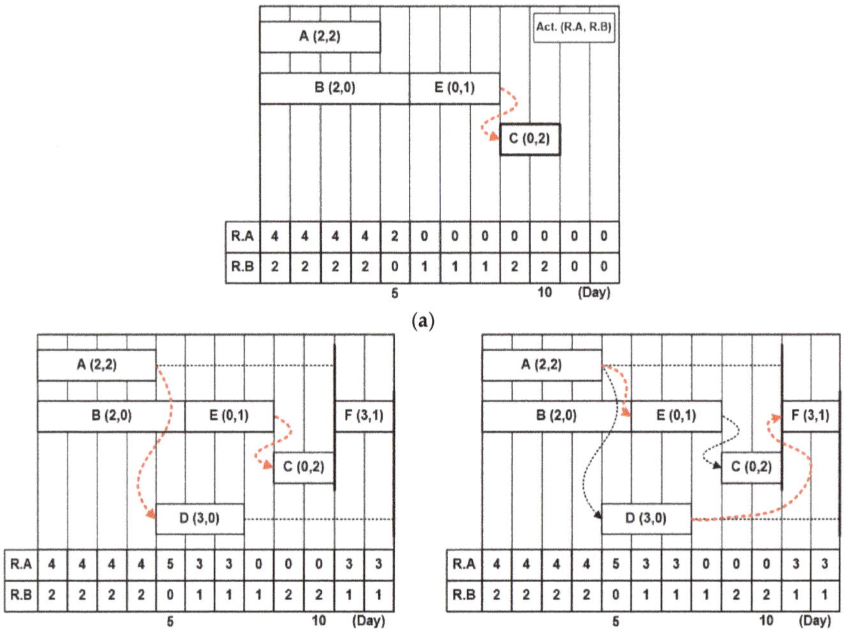

Figure 2. RCPM with serial method: (**a**) Resource link creation for a delayed activity owing to resource constraints; (**b**) RCS result with direct resource links; (**c**) All identified resource links including indirect resource links.

Activity D is also delayed to day 5 owing to the limitation of R.A, and it can start by receiving two R.As that are released at the end of activity A, creating a direct resource link between activities A and D. The last activity F can start after activity C without a delay, and the final serial RCS result is depicted in Figure 2b. Based on this result, if CPM is applied reflecting the existing technological relationship and resource links, the path through activities B, E, C, and F becomes the critical path, as illustrated in Figure 2b and activities A and D each have total float of five days. However, owing to resource constraints, activities A and D cannot have all five days of the total float.

Activities D and F cannot proceed simultaneously for the total float range of activity D owing to the constraints of R.A., and activities A and E cannot proceed simultaneously for the total float range of activity A owing to the constraints of R.B. To reflect this situation, RCPM identifies additional resource links between them, as illustrated in Figure 2c. When the resource links identified in this manner are reflected in CPM, the total floats of activities A and D become one and three days, respectively. In this study, the relationship identified within the total float range of an activity is defined as an indirect resource link.

2.3. Parallel RCS

The parallel method simultaneously or parallelly considers activities that can start in a time unit. If resource conflicts occur, activities are scheduled according to the activity priority order. Activities that cannot be scheduled owing to the lack of resources are delayed to the next time unit.

In the example schedule, the activity priority order is B, A, E, C, D, and F, as in the serial method. In the network conditions depicted in Figure 1, the activities that can be scheduled on the first day are A, B, C, and D, and they are considered in the order B, A, C, and D according to the priority. As displayed in Figure 3a, activities B and A can be scheduled to start on the first day. However, activities C and D cannot start on the first day because of resource constraints; hence, they are delayed to the next day.

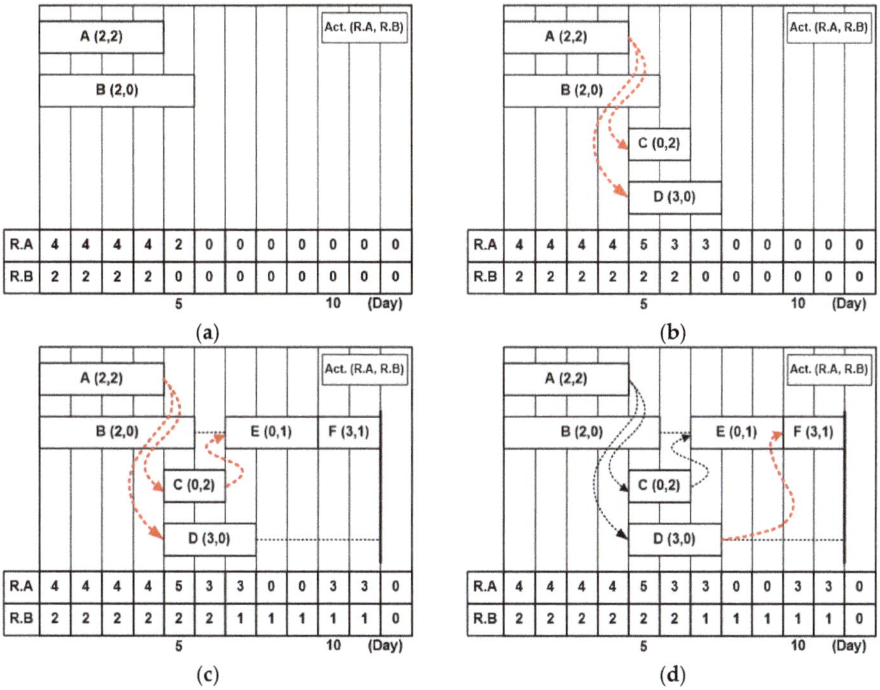

Figure 3. RCPM with parallel method: (**a**) Day 1 schedule; (**b**) Day 5 schedule with direct resource links; (**c**) All identified direct resource links; (**d**) All identified resource links including indirect resource links.

Activities that can start on days 2, 3, and 4 are C and D, but each day they are delayed to the next day owing to resource constraints. On day 5, activities that can start are still C and D, and as activity A ends on day 4, two R. As and R.Bs are released such that activities C and D can be scheduled to start. That is, activity A becomes the preceding activity of activities C and D, and direct resource links are created as illustrated in Figure 3b to reflect this condition. Because all activities that can start on day 5 are planned, the process moves to the next day. By the same process as for activities C and D, activity E is delayed by one day and can start on day 7, and a direct resource link is created between activities C and E. Finally, activity F is planned without a delay, as depicted in Figure 3c.

If CPM is performed by additionally reflecting all direct resource links identified in the parallel method, activities A, C, E, and F become critical, and the total floats of activities B and D are one and four days, respectively. Activity B can have a one day total float without resource conflict, but activity D cannot proceed in parallel with activity F owing to R.A restrictions. Reflecting this condition, an indirect resource link is created between activities D and F, as illustrated in Figure 3d, and the total float of activity D becomes two days.

2.4. Need for Improvement

When comparing the results of serial and parallel RCS methods, there may be differences in the activity sequence, project duration, and identified resource links. Because the processes are different, individual algorithms and coding processes are required to develop a computer program that executes RCPM. For example, P6 applying the serial method and MSP applying the parallel method require individual appropriate programs. Furthermore, there is a burden of executing additional programs suitable for each method for a new RCS technique that is different from the existing ones. Thus, this study proposes a generalized RCPM (eRCPM) that can be applied irrespective of the applied RCS techniques.

3. Generalized Resource-Constrained Critical Path Method

3.1. Overview

The process of identifying resource links in both serial and parallel RCS methods introduced in Sections 2.2 and 2.3 can be divided into three steps. The first step is to create a direct resource link. When a delayed activity can be started, a resource link is created with the activity that releases constrained resources with its completion immediately before the start time of the delayed activity. The second step is to perform CPM to identify the total float of each activity. At this time, CPM reflects the direct resource links identified in the first step besides the existing technological activity relationships. The third step is to create indirect resource links when the total float range of an activity is not fully available owing to resource conflicts with other activities. The second and third steps are applied in the same manner for serial and parallel methods and are the same as those applied in RCPM [12]. Therefore, this study proposes a method to identify direct resource links without depending on RCS techniques.

3.2. eRCPM Algorithm

Whereas RCPM identifies direct resource links in the process of RCS, eRCPM identifies direct resource links from the RCS result. For the example mentioned in the previous section, the result of serial RCS that does not reflect the resource links is depicted in Figure 4a. In this schedule, activities C and D are the activities whose start time is delayed out of CPM network logic owing to resource constraints. Activity F is also delayed, but this delay is because of a relationship with activity C, not because of resource constraints.

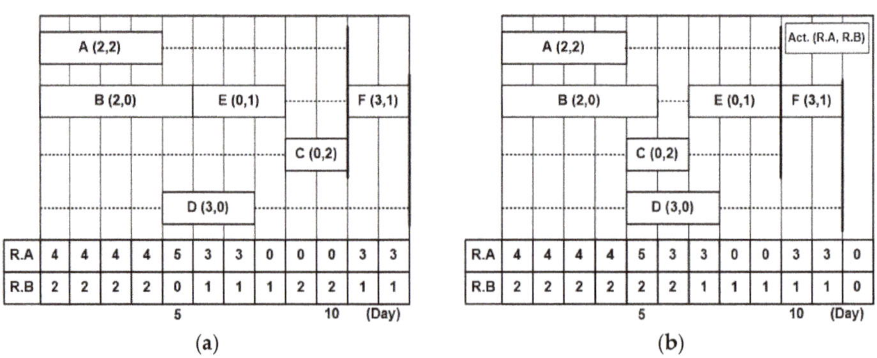

Figure 4. RCS results: (**a**) Serial method; (**b**) Parallel method.

Activity C deferred for lack of one R.B can start on day 9 because one required R.B is released when activity E ends on day 8. As one R.B is directly transferred from activities E to C, a direct resource link is required between them. In the same manner, deferred activity D requires a direct resource link with activity A because it can start on day 5 and activity A releases two R.As with its completion on day 4. Step 1 process is completed because no more activities are delayed owing to resource constraints. Subsequently, applying steps 2 and 3 yields the same result as displayed in Figure 2c.

The same eRCPM process can be applied to the RCS result of a parallel method. The result of parallel RCS of the given example schedule is illustrated in Figure 4b, in which activities delayed out of CPM logic are activities E, C, and D. Similar to the serial RCS result, it can be noted that direct resource transfers occur between activities C and E (one R.A), A and C (two R.Bs), and A and D (two R.As). All direct resource links identified in this manner are the same as those in Figure 3c. Subsequently, applying steps 2 and 3 yields the same result as depicted in Figure 3d.

As described here, eRCPM proposed in this study can identify the same resource links obtained by RCPM, based on the RCS result irrespective of the RCS techniques applied. System development based on this method also has the advantage that the eRCPM algorithm needs to be implemented only once.

4. eRCPM Prototype System

4.1. Overview

This study implements the prototype eRCPM system based on MS Excel, which is widely used for basic data management in various industries, including construction. Figure 5 shows the system development environment. As described earlier, the eRCPM system uses the RCS schedule data exported from P6 or MSP in an Excel format. The eRCPM system, an add-in program implemented with Excel Visual Basic for Applications (VBA), applies the eRCPM algorithm based on the exported schedule data to identify resource links and inserts the link information to the relevant Excel sheet. If the user imports this updated Excel file and executes CPM in P6 or MSP, the CPM output becomes the eRCPM output, which can consider resource constraints and provide reliable total floats, since the system has reflected resource links as well as technological activity relationships.

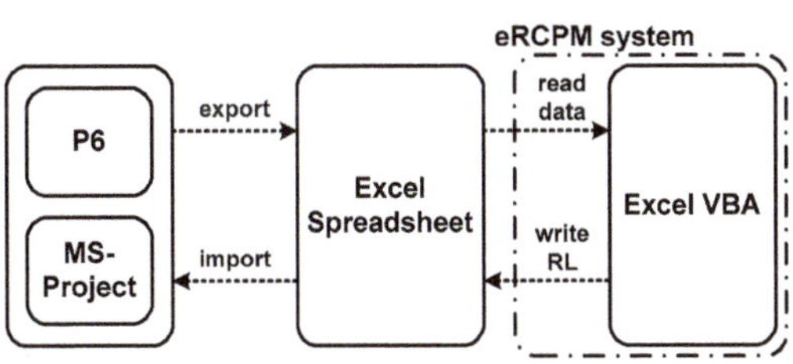

Figure 5. eRCPM system environment.

The main process of the eRCPM system is illustrated in Figure 6. Because the export and import characteristics of P6 and MSP are different, the eRCPM system requires functions to reflect the difference. The eRCPM system first checks whether the Excel data belong to P6 or MSP and reads the exported schedule data accordingly. Based on the data, the system executes the algorithm, step by step, to identify resource links and inserts resource links to the relevant Excel sheet. P6 or MSP will import this updated Excel file.

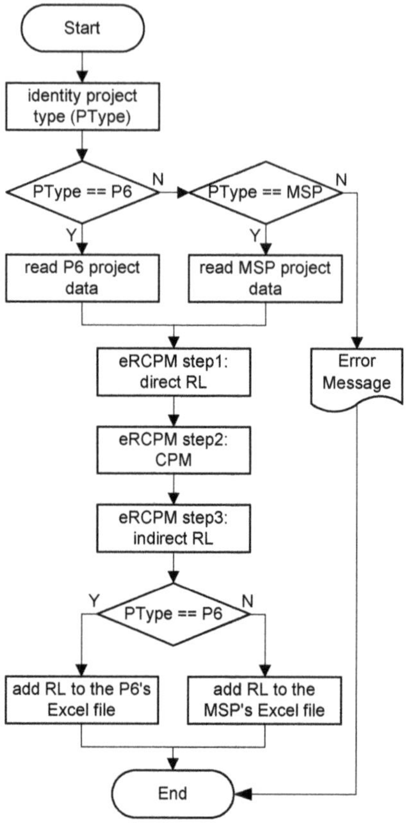

Figure 6. eRCPM system process overview.

4.2. eRCPM Key Processes

Among various processes of the eRCPM system, the following three main steps are explained in this section: input data processing prior to applying eRCPM, identifying direct resource links, and identifying indirect resource links.

4.2.1. Input Data Processing

Based on the exported Excel file, after RCS in P6 or MSP, the data necessary for the eRCPM operation are processed as shown in Figure 7. A brief description of each input data process in the eRCPM system is as follows.

1. Read ID, original duration (OD), early start time (EST), and early finish time (EFT) of every activity and store them in an array of activities.
2. Identify the project start and end times.
3. Read the relationship information of each activity and add it to the saved activity in the array of activities to establish the relationship between activities.
4. Read the ID, name, and maximum supply of every resource type and store them in an array of resources.
5. Read the amount of resources required for each activity and add them to the saved activity in the array of activities.

6. Create a daily resource table in a two-dimensional array according to the type of resource and the project period and record the demand for each resource for each date based on the time data and resource demand for each activity.
7. Identify start activities that have no predecessor and finish activities that have no successor and store their IDs in a separate array to apply forward and backward passes in steps 1 to 3.

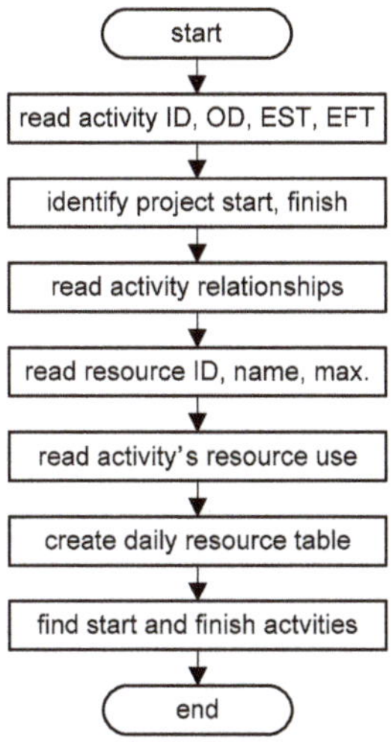

Figure 7. Reading and setting input data.

4.2.2. Step 1: Direct Resource Link

In the eRCPM process, after data creation, the same process is applied irrespective of P6 and MSP, as depicted in Figure 6. The process of identifying a direct resource link in step 1 is demonstrated in Figure 8a. Initiating from a start activity, a direct resource link is created when a delay occurred owing to resource constraints while visiting each activity in a forward pass method. When the start time of an activity in the RCS result is greater than EST of CPM, the activity is considered as a delayed activity owing to resource constraints.

For example, when the EST of a start activity, which has no predecessor, is greater than 1, that is, when the start activity is delayed as a result of RCS, a resource link can be obtained by referring to the values in the daily resource table that contains the total daily resource requirements per resource type and a list of activity IDs that require these resources. When identifying the resource link, the system stores the current step, the name of the resource that has caused the constraint, and the predecessor and the successor IDs of the resource link as a text log of the delayed activity. The forward pass, as depicted in Figure 8b, is applied as a recursive function and the entire network is searched to identify the required direct resource links.

Subsequently, the CPM backward pass computes the LST and LFT of each activity, considering both the technological relationships and the resource links identified in step 1. Based on the updated LST and LFT, the system calculates the total float of each activity. Because the resource links have been additionally reflected in the existing CPM relationships, the early times do not change, but the late times decrease owing to the resource links and the total float of the activity decreases accordingly.

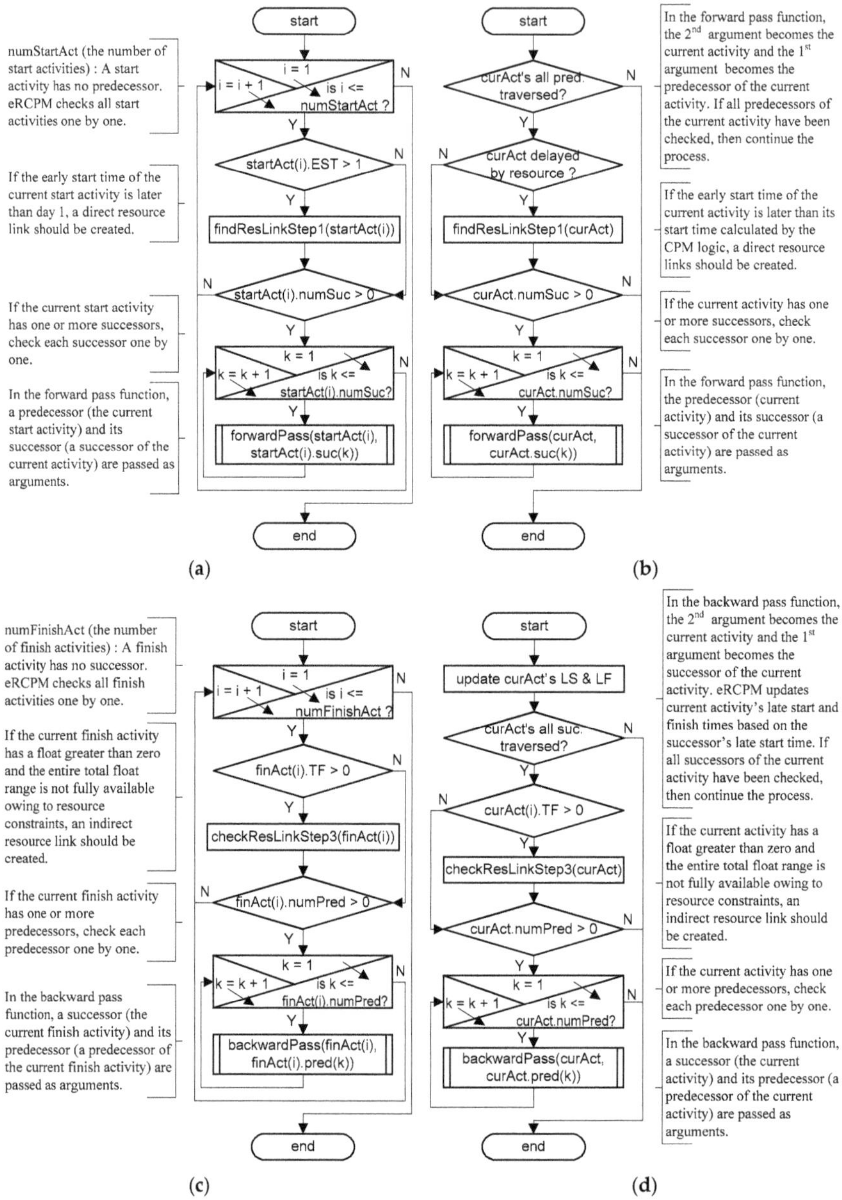

Figure 8. eRCPM processes of steps 1 and 3: (**a**) step 1; (**b**) forward pass; (**c**) step 3; (**d**) backward pass.

4.2.3. Step 3: Indirect Resource Link

Step 3 of eRCPM identifies indirect resource links by applying the backward pass starting from finish activities, as illustrated in Figure 8c. Starting from a finished activity, when the total float of an activity is greater than zero, the daily resource table is checked to determine whether this entire total float range is available without resource restrictions. When total floats are not fully available owing to resource constraints from a specific date, an indirect resource link is created by searching the daily resource table for the activity that starts on that date and has triggering resources. LFT and

LST should be updated by reflecting the newly identified resource link, and as a result, these values become smaller, and the total float becomes smaller accordingly.

As illustrated in Figure 8d, the backward pass checks all the activities on the network sequentially using a recursive function to identify necessary resource links. When a resource link is identified, the current step, triggered resource name, and predecessor and successor IDs are also stored in the text log. In this way, all resource links and text logs are created, and these are added to the appropriate Excel cells according to the format of P6 or MSP using Excel VBA code. P6 or MSP imports this Excel file and executes CPM to obtain an eRCPM output.

5. eRCPM System Demonstration

5.1. Export from P6 and MSP

The data required for the eRCPM execution are exported from P6 or MSP in an Excel format. P6 and MSP have different export characteristics, and users can select the specific data they require during each export process. Table 2 presents the data required to execute eRCPM. The data name itself makes sense, and certain terms have different expressions but the same properties. The biggest difference between P6 and MSP is the activity relationship. P6 exports the activity ID as a pair for a relationship; however, MSP does not export the activity relationships separately. Thus, for MSP data, the eRCPM system should refer directly to the predecessors and successors among the data of each activity. The RCPM log of activities is a user-defined text; it is set separately by the user in P6 and MSP and displays which resource link has been created by which resource at which step, as described earlier.

Table 2. Export data comparison between P6 and MSP.

Activities (Tasks) [1]		Activity Relationships		Resources		Resource Assignments (Assignments)	
P6	**MSP**	**P6**	**MSP**	**P6**	**MSP**	**P6**	**MSP**
Activity ID Activity Name Original Duration Planned Start Planned Finish Predecessors Successors RCPM Log Activity Status [2] WBS Code [2]	ID Name Duration Start Finish Predecessors Successors RCPM Log	Predecessor Successor Relationship Type [2] Predecessor Activity Status [2] Successor Activity Status [2]	N/A	Resource ID Resource Name Max Units/Time	ID Name Max Units	Activity ID Resource ID Budgeted Units/Time Activity Status [2] Role ID [2] Cost Account ID [2]	Task ID Resource ID Units

[1] The terms in parentheses refer to the MSP's export data type. [2] Values that must be included forcibly when exporting in P6.

5.2. P6 Example

The CPM result of P6 for the schedule presented in Table 1 is depicted in Figure 9a, and the RCS result is displayed in Figure 9b. Because of the application of RCS, the project completion time is extended by two days, from day 10 to 12; the float of each activity is also changed. Figure 9c demonstrates the result of CPM after exporting this RCS result in Excel format, applying eRCPM, and importing the Excel file. In the RCPM log column of Figure 9c, the step of identifying the resource link and the name of the resource that caused the constraint are presented; these resource links have been added as a predecessor and successor of the activity.

Notably, the total floats of RCS and eRCPM are identical to each other, as depicted in Figure 9b,c. The RCS output of P6 neither displays any additional activity relationships nor explains how these calculations resulted. It is assumed that the existence of phantom floats is recognized, and certain efforts are reflected. However, as explained by Franco-Duran and de la Garza [18] as well as this study, the total float after RCS is no longer reliable when the scale of the schedule becomes larger.

Figure 9. eRCPM with P6: (**a**) CPM output; (**b**) RCS output; (**c**) eRCPM output.

Figure 10 demonstrates the process of implementing eRCPM in Excel. Each Excel sheet in Figure 10a–d presents activities, activity relationships, resources, and resource assignments exported from P6, as presented in Table 2. The schedule of each activity is the same as the RCS result in Figure 9b; the activity relationships, resource demand, and maximum supply per resource are the same as those in Table 1.

Based on these data, if the eRCPM add-on program implemented in Excel VBA is executed, the identified resource links are recorded in the RCPM log, as depicted in Figure 10e. These resource links have been added as a predecessor and successor pair, as depicted in Figure 10f. P6 imports this updated Excel file and performs CPM. Figure 9c displays the eRCPM results.

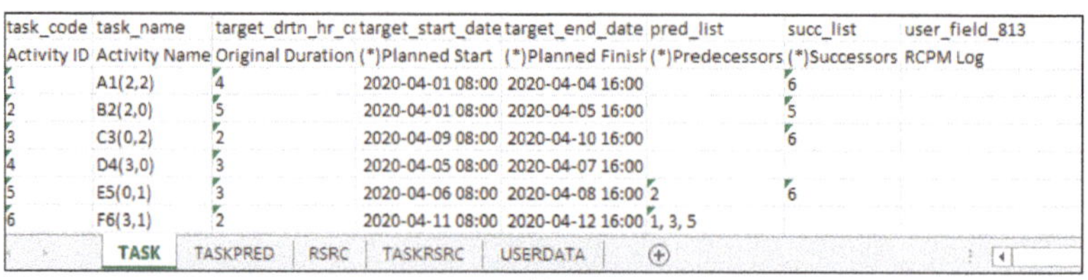

task_code	task_name	target_drtn_hr_c	target_start_date	target_end_date	pred_list	succ_list	user_field_813
Activity ID	Activity Name	Original Duration	(*)Planned Start	(*)Planned Finish	(*)Predecessors	(*)Successors	RCPM Log
1	A1(2,2)	4	2020-04-01 08:00	2020-04-04 16:00		6	
2	B2(2,0)	5	2020-04-01 08:00	2020-04-05 16:00		5	
3	C3(0,2)	2	2020-04-09 08:00	2020-04-10 16:00		6	
4	D4(3,0)	3	2020-04-05 08:00	2020-04-07 16:00			
5	E5(0,1)	3	2020-04-06 08:00	2020-04-08 16:00	2	6	
6	F6(3,1)	2	2020-04-11 08:00	2020-04-12 16:00	1, 3, 5		

TASK | TASKPRED | RSRC | TASKRSRC | USERDATA

(**a**)

Figure 10. *Cont.*

pred_task_id	task_id	pred_type
Predecessor	Successor	Relationship Type
2	5	FS
1	6	FS
3	6	FS
5	6	FS

TASK TASKPRED RSRC TASKRSRC USERDATA

(**b**)

rsrc_short_name	rsrc_name	max_qty_per_hr
Resource ID	Resource Name	(*)Max Units/Time(h/h)
RCPM_Ra	Ra	5
RCPM_Rb	Rb	2

TASK TASKPRED RSRC TASKRSRC USERDATA

(**c**)

task_id	rsrc_id	target_qty_per_hr
Activity ID	Resource ID	(*)Budgeted Units / Time(h/h)
1	RCPM_Ra	2
1	RCPM_Rb	2
2	RCPM_Ra	2
3	RCPM_Rb	2
4	RCPM_Ra	3
5	RCPM_Rb	1
6	RCPM_Ra	3
6	RCPM_Rb	1

TASK TASKPRED RSRC TASKRSRC USERDATA

(**d**)

task_code	task_name	target_drtn_hr_c	target_start_date	target_end_date	pred_list	succ_list	user_field_813
Activity ID	Activity Name	Original Duration	(*)Planned Start	(*)Planned Finish	(*)Predecessors	(*)Successors	RCPM Log
1	A1(2,2)	4	2020-04-01 08:00	2020-04-04 16:00		6	Step 1: RL 1 to 4 by Ra Step 3: RL 1 to 5 by Rb
2	B2(2,0)	5	2020-04-01 08:00	2020-04-05 16:00		5	
3	C3(0,2)	2	2020-04-09 08:00	2020-04-10 16:00		6	
4	D4(3,0)	3	2020-04-05 08:00	2020-04-07 16:00			Step 3: RL 4 to 6 by Ra
5	E5(0,1)	3	2020-04-06 08:00	2020-04-08 16:00	2	6	Step 1: RL 5 to 3 by Rb
6	F6(3,1)	2	2020-04-11 08:00	2020-04-12 16:00	1, 3, 5		

TASK TASKPRED RSRC TASKRSRC USERDATA

(**e**)

pred_task_id	task_id	pred_type
Predecessor	Successor	Relationship Type
2	5	FS
1	6	FS
3	6	FS
5	6	FS
1	4	
1	5	
4	6	
5	3	

TASK TASKPRED RSRC TASKRSRC USERDATA

(**f**)

Figure 10. P6's eRCPM data in Excel sheets: (**a**) Exported activity data; (**b**) Exported activity relationship; (**c**) Exported resource data; (**d**) Exported activity resource assignment; (**e**) RCPM Log after eRCPM; (**f**) Activity relationships after eRCPM.

5.3. MSP Example

MSP also proceeds in the same processes as P6. The CPM result is displayed in Figure 11a, and the RCS result is displayed in Figure 11b. In the RCS result, the critical path is broken because there is no critical activity on day 6, and the total floats are unreliable because the schedule does not reflect the resource dependency between activities. Each Excel sheet in Figure 11c–e presents the activity, resource, and resource assignment exported from MSP, as presented in Table 2. The result of executing the Excel add-on eRCPM based on this data is illustrated in Figure 11f, where the successor column presents the identified resource links and the RCPM log column presents the property of each identified resource link. Figure 11g depicts the result of executing CPM after MSP imports this Excel file. The critical path is continuous, and the total floats are reliable.

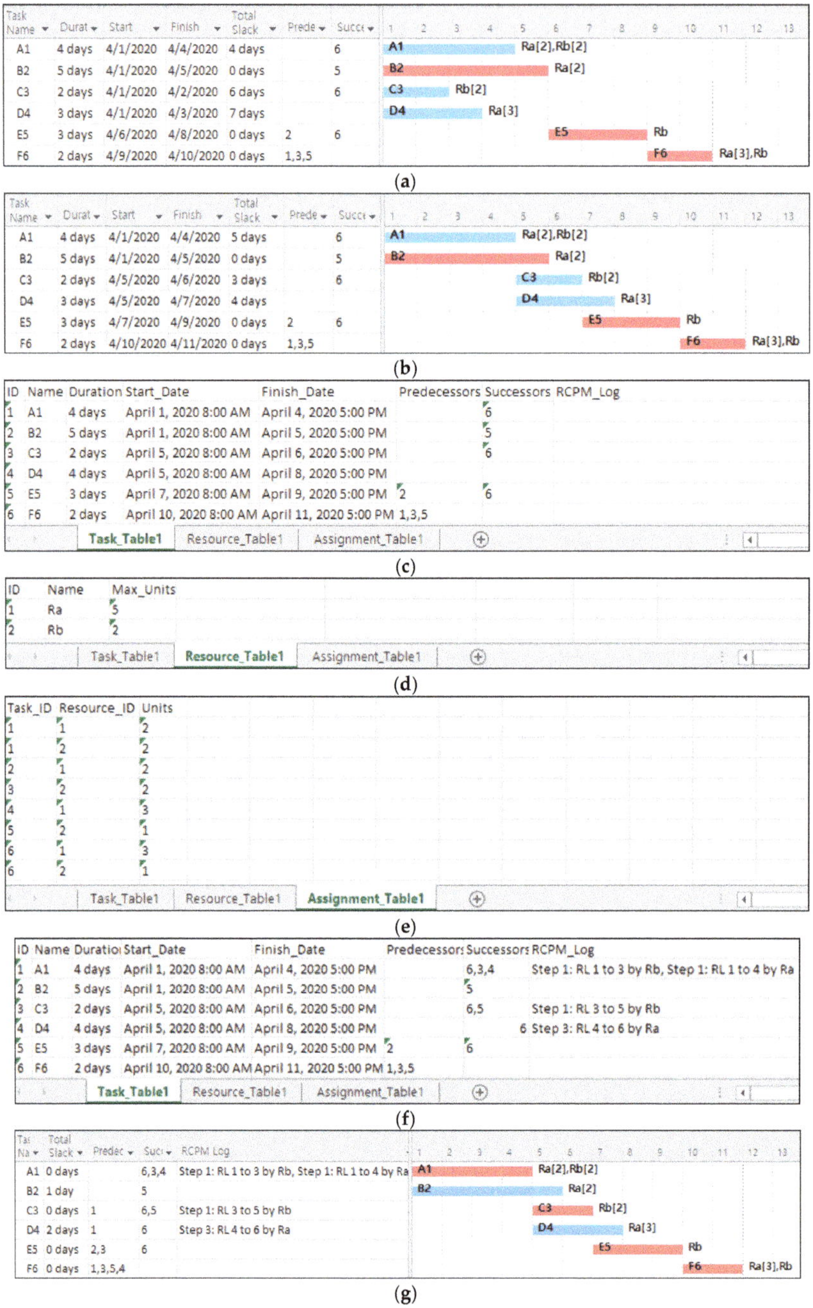

Figure 11. eRCPM process with MSP: (**a**) CPM output; (**b**) RCS output; (**c**) Exported activity data; (**d**) Exported resource data; (**e**) Exported activity resource assignment; (**f**) RCPM Log and inserted successors after eRCPM; (**g**) eRCPM output in MSP.

6. Case Study

This section validates eRCPM using a larger and more complex schedule than the previous example. This schedule originated from Fondahl (1991) [11] and was used as a case study in two previous studies [12,18]. In addition to the basic information of this schedule, Table 3 presents CPM, RCS, and eRCPM results for both P6 and MSP. The schedule information provides the ID, duration, successors, and required number of resources for each activity. There are three types of resources: carpenters (R1), iron workers (R2), and common laborers (R3), each of which has a maximum daily supply of four people. The CPM results are the same for P6 and MSP with a project duration of 27 days. The start time and total float of each activity are listed in Table 3.

Table 3. Case study schedule with CPM, RCS, and RCPM.

ID	Dur.	Successors	Resource			CPM		RCS				eRCPM			
			Carp (R1)	Iron (R2)	Lab (R3)	Start	TF	Start		TF		RL		TF	
								P6	MSP	P6	MSP	P6	MSP	P6	MSP
1	1	2				1	0	1	1	−5	0			0	0
2	3	3, 4, 5, 6, 7, 8				2	0	2	2	−5	0			0	0
3	2	30	4			5	19	25	25	0	0	29	29	0	0
4	2	20	2			5	13	7	15	9	1	5	8	0	0
5	4	16	4			5	9	9	9	1	3			0	0
6	1	9, 10			4	5	0	5	5	−5	3			0	0
7	2	25	2			5	17	19	20	4	5	23	23	1	0
8	3	23	4			5	14	16	17	2	5	24, 7	24, 7	1	0
9	1	11	4	2	2	6	0	6	6	−5	3	4		0	0
10	1	15				6	5	6	6	0	8			5	5
11	1	12	1		2	7	0	7	7	−5	3			0	0
12	1	13	1		1	8	0	8	8	−5	3	5	5	0	0
13	3	14, 15				9	0	9	9	−5	3			0	0
14	1	16, 19		4		12	2	12	12	1	3			0	0
15	2	17			2	12	0	12	12	−5	3			0	0
16	1	18		4		13	5	13	13	3	3	17		0	0
17	1	19	2	2		14	0	14	14	−5	3		4	0	0
18	1	20	2	2		14	5	14	14	3	3	8	4	2	0
19	1	21	1		2	15	0	15	15	−1	3	8		0	0
20	1	22		4		15	5	15	17	6	1			3	1
21	6	23, 30				16	0	16	16	−1	3			0	0
22	1	24		4		16	5	16	18	6	1			3	1
23	2	25, 26	4			22	0	22	22	−1	3			0	0
24	2	27	2			17	5	19	20	4	1	23	23	1	0
25	1	28	2			24	0	24	24	3	3	3	3	0	0
26	1	28			2	24	0	24	24	3	3			3	3
27	1	29			2	19	5	21	22	5	1			5	4
28	2	31				25	0	25	25	3	3			3	3
29	2	31	2			20	5	27	27	0	1	30	30	0	0
30	1	31	4			22	4	29	29	0	0			0	0
31	1				4	27	0	30	30	0	0			0	0
	Maximum units		4	4	4										

When the serial (P6) and parallel (MSP) techniques are applied to reflect the resource constraints, the start of several activities has been delayed, and the project durations of both methods are 30 days each, three days longer than CPM. However, owing to the difference in applied techniques, the start times of several activities are different, and both results provide unreliable total float values. Especially, if RCS is applied without selecting "preserve scheduled early and late dates" option in P6, an improbable result with negative total float values is obtained, as presented in Table 3 (P6's total float in RCS). In MSP, activities with a total float of zero are 1, 2, 3, 30, and 31, but as indicated by the start time in the table, the critical path is broken because it does not proceed continuously between activities 2 and 3 and between activities 3 and 30.

If the resource links identified by eRCPM are reflected, as depicted in Figure 12, the critical path becomes continuous from the project start to the end, and the total float values accurately reflect the resource constraints. Notably, as the RCPM log explains, most of the resource links were caused by the constraints of CARP (R1) in steps 1 and 3 of the eRCPM process.

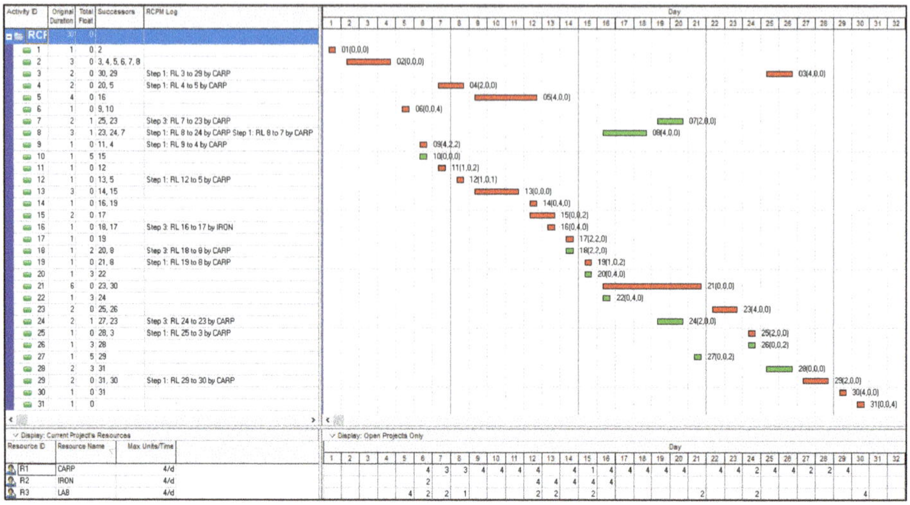

(a)

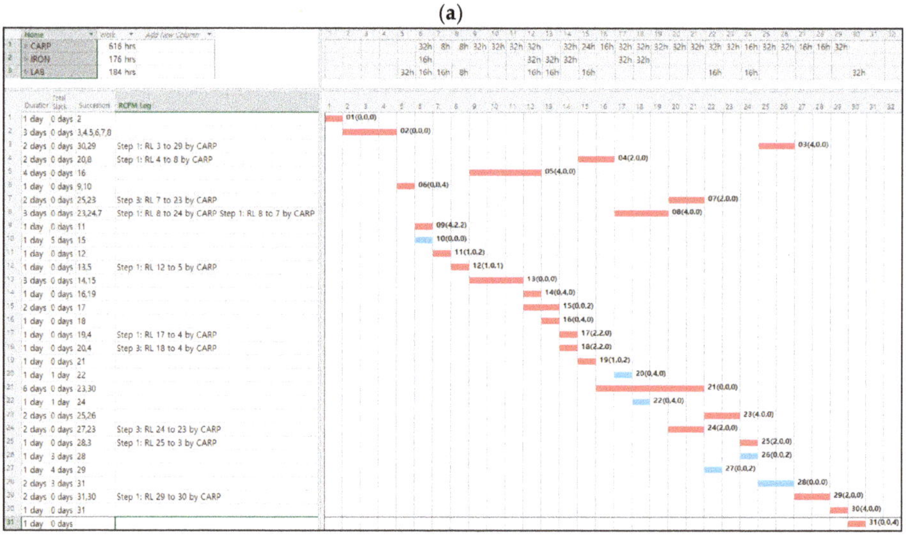

(b)

Figure 12. eRCPM output: (**a**) eRCPM in P6; (**b**) eRCPM in MSP.

When comparing the results of eRCPM with a recent previous study of Franco-Duran and de la Garza (2019) [18], P6's results were observed to be completely identical, but in MSP, there are slight differences in the identified resource links for activities 4, 17, 18, and 19. As presented in Table 3 and Figure 12b, activity 4 has been delayed from day 5 to day 15 owing to R1 resource constraints. To reflect this condition in step 1, the activities that released R1 with their completion on day 14 were identified as activities 17 and 18. These two activities released identical resources (two R1s and two R2s); hence, from the outlook of activity 4, these two activities were under equal conditions. In this case, eRCPM selects an activity with a smaller activity ID as a predecessor, so that it created resource links between activities 17 and 4. However, the manual application of RCPM in the previous study selected activity 18 as its predecessor. This difference in predecessor selection causes resource link differences for the other activities. Thus, in terms of considering resource constraints and creating a resource-dependent activity relationship, there is no difference between eRCPM and the previous study.

Through this case study, it can be noted that eRCPM appropriately identifies the necessary resource links based on the RCS results under multiple resource constraints irrespective of the method used and provides reliable total floats.

7. Conclusions

Resources are limited in most construction projects. To consider these resource restrictions, RCS can be applied based on the CPM schedule. However, after applying the RCS technique, important project management information, such as total float and the critical path, are lost. In addition, the activity sequence may be altered after a regular schedule update which increases the difficulty in project control and delay impact analysis. Accordingly, several studies related to RCPM were conducted that additionally reflect the resource-dependent activity relationships on the existing CPM network. The results of these studies were partially reflected in P6 as demonstrated with an example. However, because the existing RCPM techniques recognize resource links in a specific RCS process such as serial and parallel, separate algorithm development and system implementation are required according to the applied technique. Consequently, this study proposes eRCPM to implement the RCPM concept irrespective of the applied RCS technique under multiple resource constraints and develops a prototype of the eRCPM system with Excel VBA. The main achievements, applications, and future research of this study are summarized as follows.

- This study applied the serial method of P6 and the parallel method of MSP. However, it can be extended to other RCS results such as genetic algorithms or other optimization methods [26,28].
- The role and meaning of the resource link were specified by dividing it into a direct resource link identified in step 1 and an indirect resource link identified in step 3.
- This study developed an eRCPM prototype system with Excel VBA. The system automatically identifies resource links by linking it using the export and import functions of P6 and MSP. There are differences in input/output data types in the two systems, but the eRCPM process is executed in one process. Therefore, if required data input/output functions are added, it can be expanded to other types of systems or other RCS results. In addition, other activity-related project management functions not currently provided by P6 or MSP, such as schedule validation, procurement, quality, safety, etc., may be implemented in conjunction with their spreadsheet software through a similar procedure introduced in this study.
- Because the step that identifies the resource link and the resource that has caused the constraint are managed as a log, the system can be used as an additional planning technique, such as adjusting the amount of supplying resources.
- In the future, by adding functions for calendars, PDM (Precedence Diagramming Method) relationships, time constraints, progressed schedules, etc., the system will become a more practical tool for project schedule management.

In many construction projects, inaccurate information causes disputes between project participants, and these disputes are often resolved based on inaccurate data [11]. RCPM will provide more accurate information after RCS, and the eRCPM of this study will expand it to a wider area. These attempts will contribute to the implementation of sustainable construction project management by reducing unnecessary disputes as well as improving the efficiency of project management.

Funding: This paper was supported by Konkuk University in 2017.

Conflicts of Interest: The author declares no conflict of interest.

References

1. O'Brien, J.; Plotnick, F. *CPM in Construction Management*, 8th ed.; McGraw-Hill Education: New York, NY, USA, 2016; pp. 635–700.
2. De la Garza, J.M.; Franco-Duran, D.M.; Buckley, B. CPM Benefits in Estimating, Bidding Reported in Survey. 2017. Available online: https://www.enr.com/articles/43666-cpm-benefits-in-estimating-bidding-reported-in-survey (accessed on 15 September 2020).
3. Jo, S.-H.; Lee, E.-B.; Pyo, K.-Y. Integrating a Procurement Management Process into Critical Chain Project Management (CCPM): A Case-Study on Oil and Gas Projects, the Piping Process. *Sustainability* **2018**, *10*, 1817. [CrossRef]
4. Dasović, B.; Galić, M.; Klanšek, U. A Survey on Integration of Optimization and Project Management Tools for Sustainable Construction Scheduling. *Sustainability* **2020**, *12*, 3405. [CrossRef]
5. Khatib, B.A.; Poh, Y.S.; El-Shafie, A. Delay Factors Management and Ranking for Reconstruction and Rehabilitation Projects Based on the Relative Importance Index (RII). *Sustainability* **2020**, *12*, 6171. [CrossRef]
6. Leicht, D.; Castro-Fresno, D.; Dìaz, J.; Baier, C. Multidimensional Construction Planning and Agile Organized Project Execution—The 5D-PROMPT Method. *Sustainability* **2020**, *12*, 6340. [CrossRef]
7. Kelley, J.E. The critical-path method: Resource planning and scheduling. In *Industrial Scheduling;* Prentice Hall: Upper Saddle River, NJ, USA, 1963; pp. 347–365.
8. Moder, J.J.; Phillips, C.R.; Davis, E.W. *Project Management with CPM, PERT, and Precedence Diagramming*, 3rd ed.; Van Nostrand Reinhold: New York, NY, USA, 1983; pp. 203–204.
9. Demeulemeester, E.L.; Herroelen, W. *Project Scheduling: A Research Handbook*; Kluwer Academic Publishers: New York, NY, USA, 2002; pp. 203–343.
10. Wiest, J.D. Some Properties of Schedules for Large Projects with Limited Resources. *Oper. Res.* **1964**, *12*, 395–418. [CrossRef]
11. Fondahl, J.W. The development of the construction engineer: Past progress and future problems. *J. Constr. Eng. Manag.* **1991**, *117*, 380–392. [CrossRef]
12. Kim, K.; de la Garza, J.M. Phantom float. *J. Constr. Eng. Manag.* **2003**, *129*, 507–517. [CrossRef]
13. Pantouvakis, J.; Manoliadis, O.G. A practical approach to resource-constrained project scheduling. *Oper. Res. Int. J.* **2006**, *6*, 299–309. [CrossRef]
14. Ibbs, W.; Nguyen, L.D. Schedule analysis under the effect of resource allocation. *J. Constr. Eng. Manag.* **2007**, *133*, 131–138. [CrossRef]
15. Kim, K. Delay analysis in resource-constrained schedules. *Can. J. Civ. Eng.* **2009**, *36*, 295–303. [CrossRef]
16. Braimah, N. Construction Delay Analysis Techniques—A Review of Application Issues and Improvement Needs. *Buildings* **2013**, *3*, 506–531. [CrossRef]
17. Liberatore, M.J.; Pollack-Johnson, B.; Smith, C.A. Project management in construction: Software use and research directions. *J. Constr. Eng. Manag.* **2001**, *127*, 101–107. [CrossRef]
18. Franco-Duran, D.M.; de la Garza, J.M. Phantom float in commercial scheduling software. *Autom. Constr.* **2019**, *103*, 291–299. [CrossRef]
19. Woodworth, B.M.; Shanahan, S. Identifying the critical sequence in a resource constrained project. *Int. J. Proj. Manag.* **1988**, *6*, 89–96. [CrossRef]
20. Bowers, J.A. Criticality in resource-constrained networks. *J. Oper. Res. Soc.* **1995**, *46*, 80–91. [CrossRef]
21. Lu, M.; Li, H. Resource-activity critical-path method for construction planning. *J. Constr. Eng. Manag.* **2003**, *129*, 412–420. [CrossRef]
22. Nisar, S.A.; Yamamoto, K.; Suzuki, K. Resource-dependent critical path method for identifying the critical path and the 'Real floats' in resource-constrained project scheduling. *J. Jpn. Soc. Civ. Eng.* **2013**, *69*, 97–107. [CrossRef]
23. Kim, K.; de la Garza, J.M. Evaluation of the resource-constrained critical path method algorithms. *J. Constr. Eng. Manag.* **2005**, *131*, 522–532. [CrossRef]
24. Franco-Duran, D.M.; de la Garza, J.M. Review of Resource-Constrained Scheduling Algorithms. *J. Constr. Eng. Manag.* **2019**, *145*, 1–16. [CrossRef]
25. Tam, P.W.M.; Pujitha, B.G.D. Construction project scheduling by ranked positional weight method. *Can. J. Civ. Eng.* **1998**, *25*, 424–436. [CrossRef]

26. Lu, M.; Lam, H.; Dai, F. Resource-constrained critical path analysis based on discrete event simulation and particle swarm optimization. *Autom. Constr.* **2008**, *17*, 670–681. [CrossRef]
27. Melin, J.W.; Whiteaker, B. Fencing a bar chart. *J. Constr. Div.* **1981**, *107*, 497–507.
28. Chen, P.H.; Shahandashti, S.M. Hybrid of genetic algorithm and simulated annealing for multiple project scheduling with multiple resource constraints. *Autom. Constr.* **2009**, *18*, 434–443. [CrossRef]

Publisher's Note: MDPI stays neutral with regard to jurisdictional claims in published maps and institutional affiliations.

sustainability

MDPI

Article

Key Indicators for Linguistic Action Perspective in the Last Planner® System

Luis A. Salazar [1,2,*], Paz Arroyo [3] and Luis F. Alarcón [1]

1 School of Engineering, Department of Construction Engineering and Management, Pontificia Universidad, Católica de Chile, Avda, Vicuña Mackenna, Macul, Santiago 4860, Chile; lalarcon@ing.puc.cl
2 Faculty of Engineering, Construction Engineering, Universidad Andres Bello, Sazié, Santiago 2119, Chile
3 DPR Construction, San Francisco, CA 94111, USA; paza@dpr.com
* Correspondence: lasalaza@uc.cl

Received: 23 August 2020; Accepted: 19 October 2020; Published: 21 October 2020

Abstract: Since 2001, a link has been established between the Last Planner® System (LPS) and Linguistic Action Perspective (LAP). However, to date, it has not been studied in sufficient depth. This research developed a system of indicators to measure and control the management of commitments, through the Design Science Research (DSR) methodology, and thus contribute to the development of the social dimension of sustainability that is often neglected in construction management research. The main contributions of this paper are a proposal of five main activities to apply the DSR method, a checklist to analyze the engagement of meeting participants, a notebook for last planners, delve into the variations that can occur to the basic movements of LAP, and the creation of a system of indicators hence updating the Percent Plan Complete (PPC) with a reliability indicator. The main limitation of this research is that the system was only validated in two South American countries that implemented LPS. In future studies, we propose to apply case studies in weekly planning meetings in other industries worldwide and to determine the recommended values to improve communication and achieve the proper implementation of LAP with LPS and without LPS.

Keywords: linguistic action indicators; last planner system; linguistic action perspective

1. Introduction

1.1. Context

Traditional construction systems, mostly spread worldwide, are based on the concept of transforming raw materials (input) into a product result (output), through an established production process, not distinguishing between activities that add value and those that do not add value to the final product [1], which has generated a worldwide issue in construction productivity, since it adds costs to construction projects without really adding value [2]. As a result, in the last 50 years, productivity in the Architecture, Engineering and Construction (AEC) industry has dropped by almost 20%, while the productivity in non-farm business enterprises has grown by over 150% [3]. Therefore, according to González et al. [4], to improve project performance, it is necessary to increase the reliability of the planning of commitments at the operational level, through a production control and planning system.

1.2. Need and Relevance of Research

In response to the low productivity, starting in the 1950s, a production system called "Lean Production" was developed and led by engineers Taiichi Ohno and Eiji Toyoda, in the implementation of concepts, methods, and tools applied in the Toyota car manufacturing company in Japan [5], this production system later became part of the construction industry with the name of "Lean Construction" [6]. This system's main objective is to increase production efficiency by reducing losses or waste and satisfying customer requirements through the delivery of a product or service with higher value [5].

Regarding sustainability ("meet present needs without compromising the ability of future generations to meet their needs" [7]) the United Nations (UN) has defined 12 areas to be developed around sustainability. Of these, number 12 specifies "Ensure sustainable consumption and production patterns" [8], thus determining the need to ensure sustainability in industries such as construction. The high impact that building construction has on the environment has been demonstrated [9]. Therefore, improving the productivity of the construction industry is a fundamental need for a more sustainable society. Various authors have indicated a useful link between Lean Construction and Sustainability [10–12], as it is necessary to ensure survival in a constantly evolving business environment, through a sustainable environment driven by Lean, with significant untapped potential [11]. Furthermore, organizations have recognized the need for an approach that contributes not only to the production of buildings but primarily to the delivery process and improvement in product quality as a whole [11]. Thus, Lean Thinking and sustainability serve as complements given that Lean helps to increase efficient production of construction projects [11].

However, considering the interconnection of the three sustainability components: environmental, economic and social [9], the latter component should not be neglected. According to the Salem and al. [13] discoveries, it is necessary to generate changes in behavior through the use of Lean tools and concepts, mainly because construction structure is different from manufacturing' due to its greater complexity and uncertainty [14]. Under those circumstances, the Last Planner® System (LPS), developed by Glenn Ballard and Greg Howell in the 1990s [15], has led the introduction of Lean Construction concepts and principles [16]. The LPS is a system of planning and control of commitments, which is based on the principles of the Lean production philosophy and which directs its efforts to increase the reliability of planning and levels of performance [17,18]. Moreover, with the appropriate use of this system a reduction of the uncertainty and variability of the projects is achieved and hereby the coordination of the work participants is improved. This coordination is regarded as an internal social aspect of the project.

It is important to note that, according to studies carried out by Goldratt and Cox [19], the reliability of production is affected by the effectiveness of the control of dependencies and fluctuations between the different activities of the project; for example, a measure of reliability is variability [20], which is understood as the possible changes that can be generated in the execution time or the duration of the processes [21]. On the other hand, the uncertainty is due to the existence of variables that are not considered, such as availability of labor, administrative problems and availability of suppliers, among others. [22]. Consequently, it is necessary to achieve an adequate management of commitments to reduce the uncertainty and variability of the projects by strengthening the commitment management system in the weekly planning meetings, through coordinated action by a complex network of requests and promises, as this could be the only viable method of coordination under dynamic conditions [18].

Howell et al. [23] propose Linguistic Action Perspective (LAP) as a suitable framework for understanding the operation and effectiveness of LPS [24]. This perspective developed by Flores [25] is basically an application of the Speech Act Theory of Austin [26] and Searle [27] to organizational management. Flores [25] says that conversations do not simply precede action but constitute actions in themselves through the commitments that arise. In this way, language can be seen as the main means to create a common future for the coordination of human action, cooperation [25]. As we mentioned earlier, this idea builds on Austin's work [26] and the notion of illocutionary acts (the actions we carry out when we say certain words). For example, by saying, "I promise to do it," I am changing my environment, due to the actions that I take and those that other people take waiting for me to do what I promised to do [24]. It is important to note that this idea was later developed by Searle [27], who proposed a taxonomy of speech acts.

Flores [25] proposes a fundamental and universal structure for the coordination of actions in his book "Conversations for action," based on the completion of four essential speech acts: (1) request or offer, (2) promise or acceptance, (3) declaration of compliance and (4) declaration of satisfaction. These speech acts, according to Searle's taxonomy [28], correspond respectively to directives (request), commissives (offer, promise, and acceptance) and declaratives (statement of compliance and declaration of satisfaction), these acts modify the possibilities of action in the future, or in other words, they change the state of affairs through words [28].

According to Flores [25], four stages can be defined that generate a network or chain of commitments: (1) preparation of a request, (2) negotiation and agreements, (3) execution and declaration of compliance and (4) acceptance and declaration of satisfaction (for more details see Figure 1, based on Flores [24,25]).

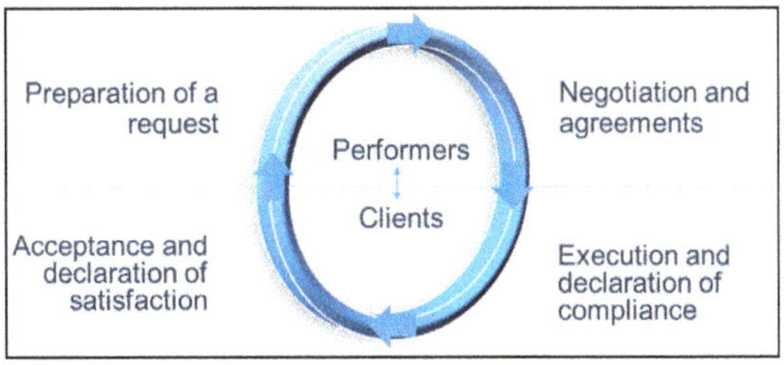

Figure 1. Network or Chain of Commitments.

In 2001 Ballard and Howell [15] established the first link between Last Planner® System and Linguistic Action Perspective. However, despite this, a system of quantitative instruments or methodologies has not been developed to carry out a specific management of the quality of commitments. Furthermore, recent studies show how the application of Lean Construction has focused on social responsibility (external) and engagement of the work environment (internal), filling gaps that sustainable practices alone cannot fill [11], also demonstrating that the application of Lean tools has a direct positive impact on the sustainability of projects [12].

1.3. State of the Art and Practice

Regarding state of the art, the authors carried out a thematic search in Web of Science, with the keywords "Language Action" or "Linguistic Action," obtaining, as a result, 217 documents. We consider it essential to clarify that in the literature, "Language Action" and "Linguistic Action" are used interchangeably. However, we respect both terms; we consider "Linguistic Action" more appropriate because "Linguistic," among many definitions, is the scientific study of language [29].

The authors carried out a preliminary analysis of co-occurrences, or joint appearances, to identify the conceptual and thematic structure of the scientific domain, through an analysis of the term co-occurrence based on text data, regarding the title and abstract fields of the 217 documents, creating Figure 2 through VOSviewer software. By means of this simple analysis, the co-occurrence of the key terms reported by the existing literature can be visually understood in order to obtain highly cited documents, to identify the most active lines of research in this domain of knowledge.

In Figure 2, each rounded rectangle represents a term, while the size of each figure indicates the number of publications in which that term appears in the title or abstract of the document. VOSviewer locates the terms that have greater coexistence close to each other in the visualization and defines the color according to the year of publication (see Figure 2 symbology, in the lower right corner). Determining with this, that the most published terms (language, study and system) are related to the area of linguistics, there is a gap in terms of construction, project, or indicators (our area of interest).

After this initial approach, the authors filtered the search, to find literature regarding construction projects and found only three publications [30–32]. The first paper, "The Role of Commitments in the Management of Construction Make-to-Order Supply Chains" [30], regards the commitment loop, network of commitments, and the number of citations for the class of problem (rebar supply chain and elevator supply chain). The second paper is based on "Understanding the theory behind the Last Planner System using the Language-Action Perspective: two case studies" [31] where they develop two case studies demonstrating the explicit representations in the commitment flows, through a proposal of symbols used for mapping the network of commitments (mapping of materials and equipment) and also an evaluation of the percentage of each type of activity performed during planning meetings. Finally, the third paper called "Variability propagation in the production planning and control mechanism of construction projects" [32] analyzes the variability propagation and planning and control function, structure and behavior of a complex adaptive system and the baseline mechanism and conversations over the LPS structure.

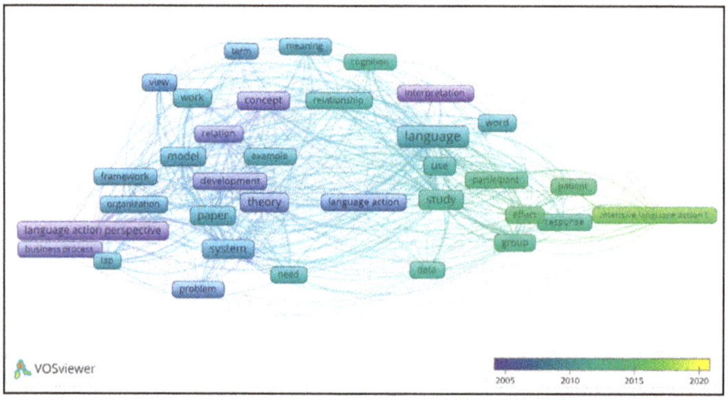

Figure 2. Term co-occurrence map created with VOSviewer.

Moreover, the authors review papers from the International Group for Lean Construction (IGLC) conference, where LAP has been proposed as a suitable framework to understand the effectiveness of LPS [23,33], but to date, the implications have not been fully understood. A first effort to understand the usefulness of LAP in LPS is the work of Viana, Formoso and Isatto [34], where based on a case study, they propose a descriptive model of the engagement networks in LPS, as well as a detailed analysis of the planning meetings. Otherwise, Vrijhoef, Koskela and Howell [35] studied construction supply chains from a LAP theoretical point of view, analyzing organizations as networks of commitments, determining that this perspective provides a plausible explanation for many of the root causes of problems in construction supply chains. Likewise, Ballard and Howell [15] argue that LAP represents an essential contribution to project management theory [33]. Macomber and Howell [33] pose five key aspects LAP: coordination, evaluation, speech (narration), trust, and moods, stating that these aspects influence how people work together and how this affects the results of the projects.

Nevertheless, the analysis offered by the publications mentioned above and all those that have been carried out to date in the IGLC does not explain the relationship between the way in which commitments are established, and the fulfillment of those commitments, except for the preliminary proposal of indicators carried out by these authors in the years 2018 and 2019 [24,36].

At last, the authors reviewed the database of this journal, mainly finding aspects of Social Networks, where the importance of understanding the relation between project dynamic interactions of the different project stakeholders and its performance has been mentioned [37]. Still, it has not been addressed through LAP.

1.4. Research Purpose and Why It Is New Knowledge

Based on the principles that view management as a process of opening, listening, obtaining, articulation and activation of the network of commitments produced mainly through promises and requests (allowing the autonomy of the productive unit), [38] and given that there are currently only three publications on the Web of Science about LAP in construction, and finally because these demonstrate the importance of this perspective in managing commitments [30–32], a vital element of the Last Planner® System is that the authors have worked on the development of a system of indicators to measure and control commitments, requirements, promises, and reliability, generating an extension and updated and corrected version of their preliminary work published in the IGLC conferences [24,36], through the Design Science Research (DSR) methodology.

Thus, the purpose of this document is to respond to the need of measuring the main elements of LAP, to control and improve the management of commitments in the weekly LPS meetings, and thus contribute to the development of an LPS 2.0 that centers its focus on the social (internal) dimension in the context of Lean Construction [39], since this dimension of sustainability [40] has been neglected in construction management approaches [41].

2. Research Method

2.1. Design Science Research

The authors used the Design Science Research (DSR) because this method manages to solve practical problems and produce artifacts as results [42]. These artifacts can be models, methods, indicators or any designed object in which a research contribution is incorporated into the design [43], thus solving problems found in the real world, and in this way also, contribute theoretically to the discipline in which it is applied [44].

According to Alan Hevner's proposal [45], DSR bridges the gaps among the contextual environment of the research project (people, organizational systems, technical systems, problems and opportunities), design science research (artifacts and processes), and the knowledge base of scientific foundations (scientific theories, methods, experience and expertise), iterating between the activities of construction and evaluation of research design artifacts and processes [45].

Figure 3 shows the research approach based on the updated DSR model proposed by Robert Briggs and Gerhard Schwabe [46] because the purpose of this work is to perform an Applied Science/Engineering (AS/E) to produce an artifact, faced with a phenomenon that varies according to time, contexts and application conditions [46]. This figure shows the main elements of the investigation, highlighting a relevancy cycle between discovery and design activities, and a design cycle between design and validation activities so that any DSR activity can be extracted or added to the knowledge base [46].

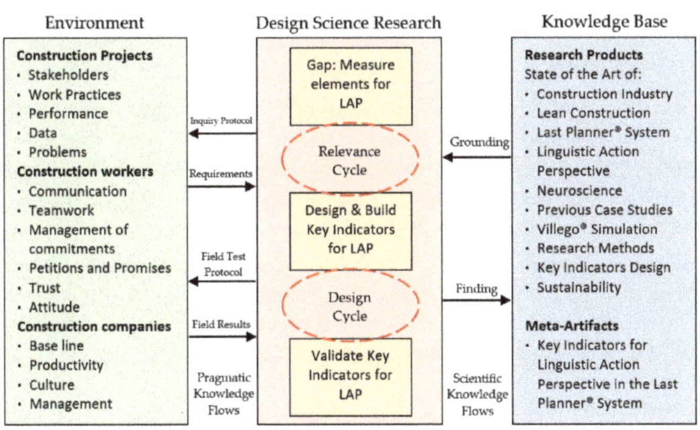

Figure 3. Design Science Research, Cycles Model.

This research consists of 5 main activities that have been adapted from the proposals of Lukka [44] and Peffers et al. [43], regarding the DSR method:

1. Discovery of problems and opportunities, through exhaustive analysis of the context.

2. Deep understanding of the subject, state of the art and practice.
3. Design and construction of artifact (indicators system), through constant iteration.
4. Evaluation of the solution to find a satisfactory solution (which fulfills its function).
5. Validation of artifact, through practical application and analysis of results.

Figure 4 shows how the artifact was developed through 3 cycles of continuous improvement based on Design Science Research (DSR). The first version of the artifact was developed from the identification of LAP elements, later it was evaluated by a panel of international experts, and it was applied in the Villego® Simulation (a game that allows teaching LPS by the construction of a house with Lego). Then, the second version of the artifact was developed from the elements of LAP applied in LPS, later it was evaluated through measurement in four construction projects in Chile and validated by means of comparison and analysis of results concerning artifact 1.0. The final version of the artifact was developed from the analysis of the LPS implementation in different countries. Later it was evaluated through measurement in two construction projects in Chile and two construction projects in Colombia, to be finally validated from the comparison and analysis of results (functionality) with respect to artifact 2.0.

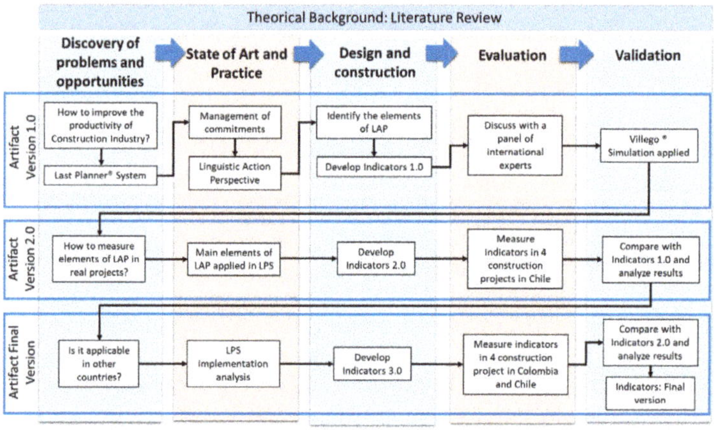

Figure 4. Research Method based on Design Science Research (DSR).

2.2. *Identification of the Problem*

Although the implementation of the Last Planner® System indeed increases the reliability of planning and performance levels through the management of commitments, the conversations during which commitments are made at weekly planning meetings have not been discussed and analyzed in sufficient depth [24]. Being a fundamental problem the lack of indicators to control this aspect, there is only the Percent Plan Complete (PPC) indicator, which is the number of planned activities completed, divided by the total number of planned activities and expressed as a percentage [47]; Hence, the authors set the goal of developing a measurement system that allows quantifying the fundamental elements of LAP present in LPS, since by measuring the management of commitments, we can enrich LPS and advance in an improved and updated version of it.

2.3. *Creation of Indicators: Version 1.0*

For the creation of the first version of indicators, the researchers carried out five key steps, after having discovered the problem and research opportunity, as detailed by Salazar et al. [24]:

1. To study the Linguistic Action Perspective, to generate a knowledge base, based mainly on Flores [25].

2. To identify the elements of this perspective that were potentially quantifiable, creating a list of concepts and data to be measured. Mainly aspects of the commitments, requests, promises and foundations of trust.
3. To develop indicators that could measure and control the previously identified elements, to generate the Design Science Research. For more details, see Table 1, which borrows the Indicators System found in Salazar et al. 2018 [24].
4. To discuss with a panel of international experts (Linguistic Action in Last Planner System Group) the feasibility of measuring and controlling these indicators, which allows improving the initial design. This panel consisted of academics from University of California-Berkeley, PhD students, a Master with a degree in linguistics, and Senior Lean consultants.
5. To validate proposed indicators, verifying the feasibility of observing these indicators utilizing a Villego® Simulation applied to a group of students as a pilot test to validate them through the Environment in a controlled situation.

These indicators are divided into three groups according to the study objective: (1) measurement and control of commitments: the primary data to be evaluated are network or chain of commitments, roles and responsibilities of the performers, declaration of the importance of the commitment, and the availability of the performers; (2) measurement and control of petitions and promises: the relevant data to be evaluated are specification of the deadline, unnecessary requests, and incomplete promises; and (3) measurement and control of fundamentals of trust: among the data to be evaluated are competence of the performer, reliability and engaged participants [24].

It is important to emphasize that these indicators are designed to analyze the management of commitments in weekly planning meetings, so the frequency of measurement should be every seven days [24]. Notwithstanding the preceding, it necessary to measure at least two weekly meetings to complete the network or chain of commitments, since the first two movements will be seen in the first meeting: (1) preparation of a request and (2) negotiation and agreements; and in the second meeting, the last two: (3) execution and declaration of compliance, and (4) acceptance and declaration of satisfaction [24]. If the client does not accept and declare satisfaction, it is understood that the committee did not comply with the agreed terms and therefore, the network or chain of commitments must be started again with the preparation of the request until the commitment is closed (when the client accepts and declares satisfaction).

2.4. Pilot Application-Villego® Simulation

To analyze and validate the proposed indicators system, we decided to verify the feasibility of observing and measuring these indicators employing the Villego® Simulation, which is a hands-on LPS experience where participants build a house with Lego bricks. For this purpose, 11 volunteer students of the sixth semester of Civil Engineering, at Pontificia Universidad Católica de Chile, were asked to perform the simulation. The authors video-recorded the two rounds of the simulation (first round simulates a traditional planning process and the second round simulates a process with LPS) to be able to analyze the proposed indicators system [24].

2.4.1. First Round Villego® Simulation

Next explaining the general rules of the simulation, we asked the students to define the roles and responsibilities that each one would assume in this round, establishing roles in the following areas: administration, quality, technical inspection, security, warehouse and several subcontractors who were identified with different colors; gray, blue, white, yellow, green and red [24].

Table 1. Indicators for observing elements of Linguistic Action Perspective in Last Planner® System.

Aim	Measure Name	Description	Formula	Means of Verification
Measurement and Control of Commitments	% of compliance network or chain of commitments	KPI measures the percentage of compliance with the chain of commitments; that is to say, that the 4 movements for the coordination are fulfilled	(Number of commitments in which the 4 movements for coordination are fulfilled)/(Total number of commitments) × 100	That the 4 movements for coordination (LAP movements presented in Figure 1) are fulfilled
	% of definition of roles and responsibilities of the performers	KPI measures the percentage of commitments that define roles and responsibilities of performers	(Number of commitments with defined roles and responsibilities)/(Total number of commitments) × 100	Roles (Who): Client and Performer Responsibilities (What): Definition of the promise of which the performer takes charge
	% of fulfillment of roles and responsibilities of performers	KPI measures the percentage of commitments in which the roles and responsibilities of previously defined performers are met	(Number of commitments that fulfilled previously defined roles and responsibilities)/(Total number of commitments) × 100	That the performer (not another) fulfills the promise and declares compliance to the client
	% of declaration of the importance of commitment	KPI measures the percentage of commitments that declare the importance (priority) of this, explicitly	(Number of commitments declaring importance)/(Total number of commitments) × 100	Declare the importance (priority) of commitment in the first 2 movements for coordination (LAP movements presented in Figure 1)
	% of compliance with priority commitments	KPI measures the percentage of commitments that were declared important (priority) and that are effectively met	(Number of priority commitments fulfilled)/(Total number of priority commitments) × 100	Review of priority commitment agreed previously
	% of verification of availability of performers in agreements	KPI measures the percentage of commitments that verify the availability of performers in the negotiation stage and agreements	(Number of commitments that verify availability of performers in agreements)/(Total number of commitments) × 100	Verification of the availability of performers in the negotiation stage and agreements. The executor's (workers) agenda can be requested from the foreman
	% of verification of availability of performers in execution	KPI measures the percentage of commitments that verify the availability of performers in the execution stage	(Number of commitments that verify availability of performers in execution)/(Total number of commitments) × 100	Compliance with the verification of the availability of performers in the execution stage

Table 1. *Cont.*

Aim	Measure Name	Description	Formula	Means of Verification
Measurement and Control of Petitions and Promises	% of specified deadlines	KPI measures the percentage of commitments that specify the deadline	(Number of commitments that specify the deadline)/(Total number of commitments) × 100	Specific deadline: date and time (AM, PM)
	% of unnecessary requests	KPI measures the percentage of commitments that make unnecessary requests	(Number of commitments that make unnecessary requests)/(Total number of commitments) × 100	When the client declares that the deadline specified in the request does not correspond to the last responsible moment and/or requested something that was not necessary (does not add value)
	% of incomplete requests and promises	KPI measures the percentage of requests and promises that do not comply with explicit conditions of satisfaction, background of obviousness and/or specific term	(Number of commitments that make requests and incomplete promises)/(Total number of commitments) × 100	Explicit conditions of satisfaction, background of obviousness and specific deadline
Measurement and Control of Fundamentals of Trust	% of compliance of the performer's competence	KPI measures the percentage of commitments where the performer is able to perform in the required domain	(Number of commitments where performer is competent)/(Total number of commitments) × 100	Performer is able to perform in the required domain (recurring performance according to accepted standards)
	% of reliability compliance (complementary to PPC)	KPI measures the percentage of commitments where the performer is able to perform reliably and timely in the required domain	(Number of commitments fulfilled + number of commitments revoked + number of counteroffers)/(Total number of commitments) × 100	Performer keeps his promises on time (PPC), counteroffer or revokes
	% of engaged participants	KPI measures the percentage of meeting participants who are engaged to it	(Number of participants engaged to the meeting)/(Total number of attendees) × 100	Participant attends the meeting, arrives at the time and remains in an attitude that suggests concentration (does not interact with the cell phone, looks at the speaker, takes notes, etc.). (Failing to comply with any previous aspect, justify it before or during)

Table 1. borrows the Indicators System found in Salazar et al. 2018 [24]

2.4.2. Second Round Villego® Simulation

After giving the new rules of the simulation, we asked students to redefine the roles and responsibilities of each member, according to the lessons learned from the initial round [24].

2.5. Indicators Validation: Version 1.0

Once the simulation ended, we analyzed the videos to determine the feasibility of observing the proposed indicators system. With this, not only did we manage to validate our indicators system, but we also analyzed the main differences between the results obtained in the Villego® Simulation and the expected results (according to preliminary field studies) in a real planning meeting. Given the nature of this simulation, we could only able to verify the indicators of commitment: compliance network or chain of commitments, definition of roles and responsibilities of the performers, fulfillment of the roles and responsibilities of the performers, specification of the deadline, compliance of the performer's competence and engaged participants. On the other hand, indicators of declaration of the importance of the commitment, compliance with priority commitments, verification of the availability of performers in agreements, verification of the availability of performers in execution, unnecessary requests, incomplete promises and promises and reliability compliance, were under verification process in construction projects in Chile. In the end, the authors consider that this first generation of validated Key Indicators is a useful tool to measure, control and improve the management of commitments in planning meetings, as they provide fast and specific feedback on these aspects, which undoubtedly enriches Last Planner® System. More details are provided in Salazar et al. [24].

2.6. Creation of Indicators: Version 2.0

After the publication of the first version of the Indicators System, the authors decided to evaluate the feasibility of its application, but this time in real construction projects in the field. The purpose of this was to observe which elements of LAP are intrinsically applied in LPS and which ones need to be taught and incorporated in order to improve the management of commitments in weekly planning meetings.

Furthermore, we realized that we had to delve into the explanation of the LAP movements, explaining that variations can occur in the basic movements shown in Figure 1. Specifically, if we observe the third movement, "Execution and declaration of compliance," we can renegotiate (generate a new commitment), revoke (performer suspends), and cancel (client suspends). These do not decrease reliability; it is increased [24,25]. Due to the fact that it is preferable that the performer or client suspends the original commitment and then they generate a new agreement in the field instead of what is currently happening, which is that they wait for the next weekly planning meeting to report that the initial agreement was not fulfilled. See Figure 5.

For all the above, we decided to carry out the study in Chile, due to the access to projects through the Collaborative Group of the Center of Excellence in Production Management (GEPUC), and also because we found previous studies of LPS in which it has been demonstrated that the incorporation of more actors in the planning process generates less variability, more reliable promises and higher productivity [48].

For the creation of the second version of Indicators System, we carried out five key steps, after having asked ourselves how to measure elements of LAP in real projects, as detailed in Salazar et al. [36]:

1. Deepen the study of Linguistic Action Perspective and Last Planner® System to generate a Knowledge Base mainly on Flores [25] and Ballard [47].
2. Identify the main elements of this Perspective that could be quantifiable and applied in LPS, taking the Indicators 1.0 as reference.
3. Develop Indicators System 2.0 from the improvement of version 1.0 through the Design Science Research.

4. Measure the proposed indicators in four construction projects in Chile to evaluate them through the Environment in a typical construction situation. Researchers recorder analyzed videotapes of weekly meetings and interviewed participants when necessary to assess each commitment for each proposed indicator.
5. Validate the proposed Indicators System, comparing with version 1.0, and analyzing the results.

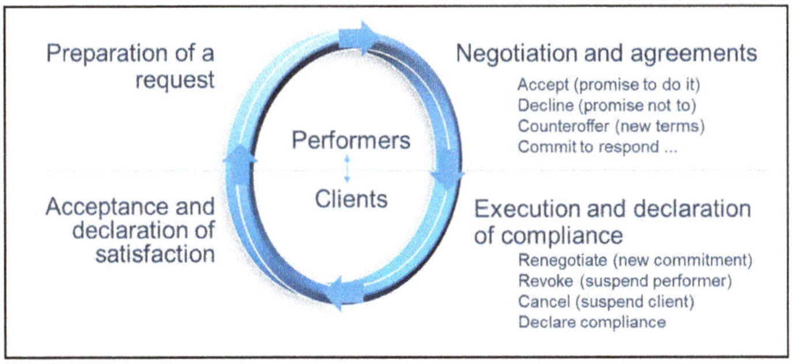

Figure 5. Network or Chain of Commitments with advanced movements.

In this second version of the Indicators System, the authors found deficiencies in functionality and in its inherent qualities (ease of use) that limited its usefulness in practice [45], given that they proposed a new set of indicators measured in the field according to the LAP to measure and control the fundamental aspects of the commitments, requests, promises and foundations of trust [36], by replacing some of the indicators and adding new ones based on the analysis of four case studies in construction projects in Santiago, Chile. We quote the results indicated in Salazar et al [36]:

1. Proposal to eliminate indicators:
    - The authors proposed not to measure the percentage of verification of the availability of performers in execution because most of the foremen verify the availability of their workers after the weekly meeting and in the field huddle, and these indicators are designed to be measured exclusively in weekly planning meetings.
    - We proposed to eliminate the percentage of incomplete requests and promises because it is confusing to measure it in the field.
    - Finally, we proposed to eliminate the percentage of compliance of the performer's competence because it is associated with the worker's curriculum vitae, and it is not possible to measure in the weekly meeting. It can only be associated with the correct fulfillment of each commitment or PPC (Percent Plan Complete).
2. Proposal to change the indicators:
    - The authors proposed modifying the percentage of declaration of the importance of each commitment because they consider it more appropriate to use the word "priority," so the indicator should be renamed as percentage of declaration of the priority of commitment. This change is proposed because it is necessary to deepen the conditions of satisfaction of the most relevant commitments. For more details, see Table 2, which borrows the Indicators System found in Salazar et al. 2019 [36].
    - Further, we proposed modification of the percentage of reliability compliance because we found a point of confusion in the formula of the indicator regarding the concept of counteroffers since counteroffers occur in the same meeting, whereas the concept after the meeting is "renegotiation." Additionally, we add "cancel" a commitment. See Table 2.
3. Measurement of original indicators:

- Table 2 shows the average results of the indicators in the four projects measured during three weeks, which incorporate the changes that we mentioned in version 1.0.

4. Newly proposed indicators:

- Finally, we proposed seven new indicators, which complement version 1.0. See Table 3, which borrows the Indicators System found in Salazar et al. [36].

This new Indicators System seeks to analyze the management of commitments in weekly planning meetings, so the measurement frequency is always every seven days. However, it is necessary to hold at least two weekly meetings to analyze the results, as mentioned in the version 1.0.

2.7. Application in Construction Projects

The strategy for selecting the case studies was based on "information-oriented selection" to establish "extreme cases/deviations" Flyvbjerg [49]. The units of the analysis were three multi-story building projects and an extension housing project with the LPS implemented with different degrees of maturity, in the Metropolitan Region of Santiago, Chile. The number of four projects were determined according to the recommendation of Hernández et al. [50], who recommend a maximum of eight cases when a multiple in-depth study is carried out [51], since the survey does not represent a "sample," as if an experiment does since it is a validation method.

Regarding the work methodology, the researchers used the "information-oriented selection" selecting four projects, one for each company, according to the preferences of the managers, recording availability and the degree of implementation maturity of the Last Planner® System due to the feasibility of research with construction companies and projects that belong to the Collaborative Group of GEPUC. As a result, projects with a primary degree of maturity (short-term planning system) up to projects with an advanced level were obtained (systematic use of an Workable Backlog) [52]. Through video recording of three consecutive weekly planning meetings, the researchers carried out the LAP in LPS practices survey, analyzing the indicators of the second and third meeting, since the objective was not to modify the "behavior" of the participants in the studio (we observed "strange" behavior by the team in the first meeting, probably because they knew they were being videotaped).

2.8. Indicators Validation: Version 2.0

To validate this new Indicators System, the authors measured the application of LAP in four construction projects, comparing and analyzing the results of version 1.0. We proposed improvements to the Indicators System: Version 1.0, eliminating indicators (percentage of verification of the availability of performers in execution, percentage of incomplete requests and promises, and % of compliance of the performer's competence) changing indicators (percentage of declaration of the importance of commitment and percentage of reliability compliance) and proposing new indicators for the measurement and control of the management of commitments in construction projects. Additionally, the researchers consulted contractors who participated in the weekly planning meetings about their perceptions. They stated that these measurements improved the ability to provide reliable promises since they understood the importance of speech acts, satisfaction conditions and trust in the management of commitments [36]. In this second version, we created new metrics to make the percentage compliance network or chain of commitments (four movements for coordination) and percentage reliability compliance (complementary to PPC) clearer, defining the latter as:

$$\text{Reliability} = \text{PPC} + \text{Revoke} + \text{Renegotiate} + \text{Cancel}, \quad (1)$$

Besides, we added Figure 5 and the results of the indicators: % of revoked commitments, percentage of renegotiated commitments, and percentage of canceled commitments.

Table 2. Results of indicators from the Linguistic Action Perspective in Last Planner®System.

Aim	Measure Name	Description	Formula	Results	General Comments
Measurement and Control of Commitments	% of compliance network or chain of commitments	KPI measures the percentage of compliance with the chain of commitments; that is to say, that the 4 movements for the coordination are fulfilled	(Number of commitments in which the 4 movements for coordination are fulfilled)/(Total number of commitments) × 100	0%	• The preparation of the petition is observed • There is no negotiation process, but rather an imposition by the client • The declaration of compliance is verified • The declaration of satisfaction is not observed
	% of definition of roles and responsibilities of the performers	KPI measures the percentage of commitments that define roles and responsibilities of performers	(Number of commitments with defined roles and responsibilities)/(Total number of commitments) × 100	83%	• In general, roles are defined intrinsically: client requests and performers agree. Regarding responsibilities, the scope of the commitment is not always clearly established
	% of fulfillment of roles and responsibilities of performers	KPI measures the percentage of commitments in which the roles and responsibilities of previously defined performers are met	(Number of commitments that fulfilled previously defined roles and responsibilities)/(Total number of completed commitments) × 100	15%	• In general, in the construction works the performer does not commit, the one who commits is the head of the performer (foreman) • Performer is engaged in administrative aspects (management team)
	% of declaration of the priority of commitment	KPI measures the percentage of commitments that declare the priority (importance) of this, explicitly	(Number of commitments declaring priority)/(Total number of commitments) × 100	10%	• In general, the priority of the commitments is not declared. This does not allow the foremen to carry out an adequate planning regarding the execution order of the assumed commitments
	% of compliance with priority commitments	KPI measures the percentage of commitments that were declared priority and that are effectively met	(Number of priority commitments fulfilled)/(Total number of priority commitments) × 100	100%	• The few commitments that were declared a priority were completed. The foregoing demonstrates the importance of making the priority statement
	% of verification of availability of performers in agreements	KPI measures the percentage of commitments that verify the availability of performers in the negotiation stage and agreements	(Number of commitments that verify availability of performers in agreements)/(Total number of commitments) × 100	18%	• There is a low percentage of verification of the availability of performers in the stage of negotiation and agreements

Table 2. *Cont.*

Aim	Measure Name	Description	Formula	Results	General Comments
+ M. and C. PP	% of specified deadlines	KPI measures the percentage of commitments that specify the deadline	(Number of commitments that specify the deadline)/(Total number of commitments) × 100	10%	• In general, only the date is specified, but it is not scheduled in detail, or if it will be completed in the morning or in the afternoon
	% of unnecessary requests	KPI measures the percentage of commitments that make unnecessary requests	(Number of commitments that make unnecessary requests)/(Total number of commitments) × 100	3%	• Low percentage in weekly meetings • According to the workers, the foremen often make unnecessary requests on the field
Measurement and Control of Fundamentals of Trust	% of reliability compliance (complementary to PPC)	KPI measures the percentage of commitments where the performer is able to perform reliably and timely in the required domain	(Number of commitments fulfilled + number of commitments revoked + number of counteroffers)/(Total number of commitments) × 100	81%	• It must always be a percentage equal to or greater than the PPC • Complements the PPC with additional movements, which happen after the initial agreement
	% of engaged participants	KPI measures the percentage of meeting participants who are engaged in it	(Number of participants engaged to the meeting)/(Total number of attendees) × 100	48%	• High degree of participation (only 10% left the meeting) • No meeting started at the agreed time • A lot of interaction with the cell phone during the meeting (calls, chat and e-mail) • Interruptions by radio • In some moments two or more people spoke at the same time • 60% of the team takes note (everyone should take note) • Non-verbal language indicates fatigue and lack of attention

Table 2. Borrows the Indicators System 2.0 found in Salazar et al. 2019 [36]. **+ M. and C. PP:** Measurement and Control of Petitions and Promises.

Table 3. Results of new indicators from the Linguistic Action Perspective in Last Planner®System.

Aim	Measure Name	Description	Formula	Results	General Comments
Measurement and Control of Commitments	% of fulfillment of a request	KPI measures the compliance percentage of the first movement; preparation of a request by the client	(Number of commitments in which the petition is prepared)/(Total number of commitments) × 100	100%	• Client is clear about the request (what) and to whom it will be entrusted (performer)
	% of compliance negotiation and agreements	KPI measures the compliance percentage of the second movement; negotiation and agreements	(Number of commitments in which a negotiation and agreement is made)/(Total number of commitments) × 100	20%	• In general, there is no negotiation before the agreement. The performer assumes the order established by the client. Sometimes he does not answer if he can or cannot comply with the agreement
	% of declaration of compliance with the commitment	KPI measures the percentage compliance of the third movement; execution and declaration of compliance with the commitment by the performer	(Number of commitments in which compliance is declared)/(Total number of commitments completed) × 100	78%	• It is verified by questions to clients and performers before the weekly meeting that there is a high percentage of declarations of compliance with the commitments. However, there are performers who do not inform clients that they finished with the assigned task
	% of fulfillment declaration of satisfaction	KPI measures the percentage of compliance of the fourth movement; acceptance and declaration of satisfaction by the client	(Number of commitments in which satisfaction is accepted and declared)/(Total number of commitments completed) × 100	5%	• There is a low percentage of commitments in which satisfaction is declared by the client. In general, it is only indicated if the commitment is fulfilled or not, without giving feedback to the performer
++ M. and C. FT	% of revoked commitments	KPI measures the percentage of commitments revoked	(Number of commitments revoked)/(Total number of commitments) × 100	4%	• Minor percentage of commitments are revoked after the weekly meeting
	% of renegotiated commitments	KPI measures the percentage of renegotiated commitments	(Number of renegotiated commitments)/(Total number of commitments) × 100	2%	• Practically no renegotiation of commitments after the weekly meeting
	% of canceled commitments	KPI measures the percentage of canceled commitments	(Number of canceled commitments)/(Total number of commitments) × 100	1%	• Practically no cancellation of commitments after the weekly meeting

Table 3. Borrows the Indicators System 2.0 found in Salazar et al. 2019 [36]. **++ M. and C. FT:** Measurement and Control of Fundamentals of Trust.

Finally, we conclude that this second generation of key indicators measured in the field (eliminating, changing and proposing the Indicators System from the first generation) generate a powerful tool to measure, control and improve the management of commitments in weekly planning meetings since they enable quick feedback that undoubtedly enriches the Last Planner® System [36].

2.9. Creation of Indicators: Final Version

After the publication of the Indicators System: Version 2.0, the authors decided to carry out the last iteration to achieve continuous improvement (Kaizen) and analyze the feasibility of application in another context, thus determining the scope of this artifact.

Ergo, we measured these indicators in two new projects in Santiago, Chile, thanks to the Collaborative Group of GEPUC and two projects in Bogotá, Colombia, thanks to the Engineering and Construction Management Research Group (INGECO).

For the creation of the final version of the Indicators System, we carried out five key steps, after having asked ourselves: Is it applicable in other countries?

1. Deepen the analysis of the implementation of the Last Planner® System and the Linguistic Action Perspective to generate a Knowledge Base.
2. Identify the key elements of this Perspective and which ones are redundant, taking the Indicators of version 2.0 as a reference.
3. Develop Indicators System 3.0 from the improvements of version 2.0 through the Design Science Research.
4. Measure the proposed indicators in four construction projects, two in Chile (a project of 24 houses and 22-story building) and two in Colombia (a 21-story building and a 23-story building), to evaluate them through the different construction Environment. For the above, the authors recorded and then analyzed videotapes of weekly meetings and interviewed participants when necessary to evaluate each commitment for each proposed indicator.
5. Validate the proposed Indicators System, comparing with version 2.0, and analyzing the results.

In this version, the authors found limitations in functionality and in its inherent qualities (ease of use) in practice [45] that must be optimized, so we propose the elimination of some of the indicators and the expansion of others based on the analysis of the four construction projects in Chile (2) and Colombia (2), which carry out weekly meetings using LPS.

This field test led to the following changes:

1. Proposal to eliminate indicators:

    - The authors propose not to measure the percentage of definition of roles and responsibilities of the performers because we realized that the roles (Who) are intrinsically established in the LPS weekly meeting structure. Concerning the responsibilities (What), this should be part of the correct fulfillment of the request, so it is incorporated into the % of fulfillment of a request.
    - Plus, we propose to eliminate the percentage of fulfillment of roles and responsibilities of performers, where the performer and not another one fulfills the promise and declares compliance to the client, because it is not possible to measure it directly. In general, in construction works, the performer does not commit, the one who commits is the head of the performer (foreman) [36].
    - Finally, we propose to eliminate the percentage of verification of the availability of performers in agreements because this verification should be carried out in the stage of negotiation and agreements. The client must request the agenda of the executor (workers) form the foreman. Therefore, it is incorporated into percentage of compliance negotiations and agreements.

2. Proposal to expand indicators:

- The authors propose to clarify the indicator percentage of engaged participants, providing that in the previous publications it was not defined precisely what would be considered as "engaged." Thence, we want to mention the key aspects to consider in this indicator, participants must:
    - Arrive on time (maximum 5 min late);
    - Avoid interaction with the cell phone (do not check, speak or ring the cell phone);
    - Remain in the room;
    - Avoid interacting with a walkie-talkie (not talking, hopefully off or at a volume that does not interrupt others);
    - Intervene in the meeting;
    - Take notes;
    - Look at the person who is speaking.

To see the final proposal of Indicators System to analyze the management of commitments in weekly planning meetings, see Table 4, while to measure the percentage of engaged participants, see checklist in Appendix A.

2.10. Application in Another Country

The units of the analysis were three multi-story building projects and an extension housing project with the LPS implemented with different degrees of maturity, in different countries, Chile and Colombia. We determined four projects according to the recommendation of Hernández et al. [36,50]. We used the work methodology based on the "information-oriented selection," selecting four projects, one for each company, according to the agreements established with the managers, availability of recording, and the degree of implementation maturity of Last Planner® System, as in version 2.0.

To measure the proposed indicator system, the principal investigator attended and videotaped three consecutive weekly planning meetings. In these meetings, the people who participated were the last planners, whose positions range from construction managers to foremen. We had a variable participation in each meeting: minimum 13, maximum 43 and an average of 25 participants. The researcher took notes of the meeting through the checklist (see in Appendix A) to measure the percentage of engaged participants and note down what was observed during the meeting, such as observations regarding the implementation of the room, topics discussed, adequate space, visual management, observed moods, among other comments that allowed feedback to the leaders and thus improve the implementation of LPS and LAP like a synergy. As well, after the meeting, the researchers analyzed the videos to complete the system of indicators, based on the means of verification and the proposed formula. For example, to determine the percentage of specified deadlines, according to the means of verification the number of commitments that specify the deadline are those in which date and time are set (AM, PM); For that reason, it is necessary to determine the number of commitments that meet the above, divide it by the total number of commitments and multiply it by 100, to obtain the result (see Table 4).

Table 4. Key Indicators for Linguistic Action Perspective in the Last Planner®System.

Aim	Measure Name	Description	Formula	Means of Verification
Measurement and Control of Commitments	% of compliance network or chain of commitments	KPI measures the percentage of compliance with the chain of commitments; that is to say, that the 4 movements for the coordination are fulfilled	(Number of commitments in which the 4 movements for coordination are fulfilled)/(Total number of commitments) × 100	That the 4 movements for coordination are fulfilled
	% of fulfillment of a request	KPI measures the compliance percentage of the first movement; preparation of a request by the client	(Number of commitments in which the petition is prepared)/(Total number of commitments) × 100	To determine that the petition was prepared, it must be verified that the following are established: • Roles (Who): Client and Performer • Responsibilities (What): Definition of the promise of which the performer takes charge
	% of compliance negotiation and agreements	KPI measures the compliance percentage of the second movement; negotiation and agreements	(Number of commitments in which a negotiation and agreement is made)/(Total number of commitments) × 100	- There must be an agreement and not an imposition by the client (Conversations for action are generated) - In addition, the client must consult the executor's (workers) agenda with the foreman
	% of declarations of compliance with the commitment	KPI measures the percentage compliance of the third movement; execution and declaration of compliance with the commitment by the performer	(Number of commitments in which compliance is declared)/(Total number of commitments completed) × 100	- Performers execute and inform clients that they are done with the assigned task (immediately) in the field huddle - Verification is done at the next weekly meeting
	% of fulfillment declaration of satisfaction	KPI measures the percentage of compliance of the fourth movement; acceptance and declaration of satisfaction by the client	(Number of commitments in which satisfaction is accepted and declared)/(Total number of commitments completed) × 100	- Client verifies compliance and reports if the commitment is satisfactory in the field huddle - Verification is done at the next weekly meeting
	% of declaration of the priority of commitment	KPI measures the percentage of commitments that declare the priority (importance) of this, explicitly	(Number of commitments declaring priority)/(Total number of commitments) × 100	Declare the priority (importance) of commitment in the first two movements: Request or Negotiation and agreements
	% of compliance with priority commitments	KPI measures the percentage of commitments that were declared priority and that are effectively met	(Number of priority commitments fulfilled)/(Total number of priority commitments) × 100	Review of priority commitment agreed at the previous weekly meeting

Table 4. *Cont.*

Aim	Measure Name	Description	Formula	Means of Verification
+ M. and C. PP	% of specified deadlines	KPI measures the percentage of commitments that specify the deadline	(Number of commitments that specify the deadline)/(Total number of commitments) × 100	Specific deadline: date and time (AM, PM)
	% of unnecessary requests	KPI measures the percentage of commitments that make unnecessary requests	(Number of commitments that make unnecessary requests)/(Total number of commitments) × 100	When the client declares that the deadline specified in the request does not correspond to the last responsible moment and/or requested something that was not necessary (does not add value)
Measurement and Control of Fundamentals of Trust	% of reliability compliance (complementary to PPC)	KPI measures the percentage of commitments where the performer is able to perform reliably and timely in the required domain	(Number of commitments fulfilled + number of commitments revoked + number of renegotiations + number of commitments canceled)/(Total number of commitments) × 100	Indicator is the sum expressed in the formula. It is essential to measure it because it provides a higher degree of reliability than the current PPC (Percent Plan Complete)
	% of revoked commitments	KPI measures the percentage of commitments revoked	(Number of commitments revoked)/(Total number of commitments) × 100	Performer informs the client immediately after the meeting (ASAP) that he will not be able to fulfill his commitments
	% of renegotiated commitments	KPI measures the percentage of renegotiated commitments	(Number of renegotiated commitments)/(Total number of commitments) × 100	Client and/or performer wishes to change the satisfaction conditions immediately after the meeting, a new agreement is generated in the field huddle (between weekly meetings)
	% of canceled commitments	KPI measures the percentage of canceled commitments	(Number of canceled commitments)/(Total number of commitments) × 100	Client informs the performer immediately after the meeting (ASAP) that the commitment made is no longer necessary
	% of engaged participants	KPI measures the percentage of meeting participants who are engaged to it	(Number of participants engaged to the meeting)/(Total number of attendees) × 100	- Participants must arrive on time (max 5 min late) - Avoid interaction with the cell phone and walkie-talkie - Remain in the room and intervene in the meeting - Take notes and look at the person who is speaking

Table 4. Own elaboration, based on Salazar et al [24,36]. **+ M. and C. PP:** Measurement and Control of Petitions and Promises.

As we did with Version 2.0, the researchers presented the average results of the indicators measured during the three weeks, this time in a list and not in a table:

- 25% of compliance network or chain of commitments: (1) The preparation of the petition is observed. (2) Although a negotiation process is observed, it must be improved. (3) There is a declaration of compliance although it must be measured in the field. (4) The declaration of satisfaction must be worked out by the team.
- 93% of fulfillment of a request: In general, roles are defined intrinsically: client requests and performers agree. Regarding responsibilities, the scope of the commitment is not always clearly established.
- 55% of compliance negotiation and agreements: The generation of agreements must be strengthened and the imposition by the client avoided. Currently, the performer assumes the order established by the client. We did not delve into the fact that the client must consult the executor's (workers) agenda with the foreman.
- 66% of declarations of compliance with the commitment: Before each weekly meeting, the investigators verified that there was a considerable percentage of declaration of compliance with commitments, through several questions to the clients and performers. However, there were performers who did not inform clients that they had finished with the assigned task.
- 48% of fulfillment declaration of satisfaction: We consider that work should be done on this indicator, since we observed that in general, it only indicated if the commitment was fulfilled or not, without giving feedback to the performer.
- 6% of declaration of the priority of commitment: In general, the priority of the commitments was not declared. This must be worked on to allow the foremen to carry out adequate planning regarding the order of execution of the assumed commitments.
- 20% of compliance with priority commitments: We consider this result exceptional because it is very important to comply with the commitments declared as priorities. In this case, what happened was that a contractor did not comply with the commitments, despite the fact that they had been informed as priorities (a contractor who presented delays throughout the project and had problems with the management). It is very important to note that we expect this percentage to always be close to 100%.
- 66% of specified deadlines: In general, only the date is specified, but not if it will be completed in the morning or in the afternoon.
- 0% of unnecessary requests: No unnecessary requests on the meetings. According to the workers, the foremen often make unnecessary requests on the field.
- 68% of reliability compliance: This indicator complements the PPC with additional movements, which occur after the initial deal. In this case we consider that there is important room for improvement.
- 0% of revoked commitments: No commitments revoked after the weekly meetings.
- 1% of renegotiated commitments: Practically no renegotiation of commitments after the weekly meetings.
- 0% of canceled commitments: No commitments cancelled after the weekly meetings.
- 61% of engaged participants: Regarding this indicator, we can detail by the average percentages obtained with the Check List—Meeting participants (see Appendix A): 75% of the participants arrived on time, 18% checked the cell phone, 2% of them had their walkie-talkie make sounds, 19% left the meeting room, 22% did not intervene, 50% made notes and 100% looked at the person who was speaking.

2.11. Indicators Validation: Final Version

This final version of indicators was validated by applying this LAP in LPS Indicator System in four construction projects, two in Chile and two in Colombia. Comparing and analyzing the results of version 2.0, proposing improvements to this system in order to achieve a refined system of key indicators of LAP in LPS.

Whence, we eliminated an indicator that could not be measured directly in the weekly meeting, percentage of fulfillment of roles and responsibilities of performers. We mix two indicators: (1) percentage of definition of roles and responsibilities of the performers, and (2) % of verification of the availability of performers in agreements; Because the first can be measured in % of fulfillment of a request, and the second can be measured in percentage of compliance negotiations and agreements. As well, we explained in detail the indicator percentage of engaged participants, because we consider that it was not evident in the previous versions and we consider it fundamental for the improvement of LAP. Finally, we want to mention that this Indicators System has metrics with the high expected value and low expected value; In other words, it is desirable that most indicators have high values (close to 100%, for example, percentage of compliance network or chain of commitments). In contrast, only some indicators like % of declaration of the priority of commitment, percentage of unnecessary requests, percentage of revoked commitments, percentage of renegotiated commitments, and percentage of canceled commitments, should have low values (close to 0%).

We decided to incorporate only the average results obtained during the three weeks, because the objective was to improve the proposed Indicators System through the optimization in the measurement and means of verification of each indicator. It is relevant to mention that we did not observe differences in the measurement system, but we did in the organizational culture (sociocultural of both countries), which will be analyzed in another document.

3. Analysis of Results

According to the parsimony of theory; that is, the number of constructions and statements that it requires to achieve its explanatory power [53], we have developed this "artifact" (Indicators System) trying to increase the explanatory power, but with fewer constructions or statements to contribute to the theoretical investigation (optimization).

The main criterion for this new contribution to Applied Science/Engineering (AS/E) knowledge is its practical utility. Because this artifact not only contributes theoretically but it is an AS/E contribution because it is original, generalizable, and validated [46]. Originality can be established by comparing contributions to state of the art, while a generalization can be established through the demonstration of the applicability of the artifact to a variety of contexts (different projects in different countries); in the end, the validity may be justified by the evaluation of the results (comparison of the three versions of indicators) [54]. Thus, researchers have reinforced efforts to justify these elements, including pilot tests in natural environments [55] (Villego® Simulation, projects in Chile and projects in Colombia), expert evaluations (including the creator of LPS) and feedback provided by the scientific community (two IGLC conferences).

Else, all DSR activities have been carried out to make these findings scientifically rigorous, meaning that there is no need for a separate Rigor Cycle [46] since all DSR activities have the potential to contribute to the Knowledge Base (theoretical contribution). Finally, since stakeholders participated in all DSR activities, we established a constant interaction with the Environment.

4. Discussion

4.1. Comments on the Results Obtained

We consider that our scientific research design was framed in a precise application context that not only managed to provide the requirements for the research (opportunity and/or problems) but also defined the acceptance criteria for the final evaluation of the results of the investigation [45].

Furthermore, through the DSR methodology, we demonstrated the usefulness of the proposed Indicators System, measuring and controlling the management of commitments in different situations: simulation with students, construction projects in a country, different construction projects (extension and height) and in different countries.

Through the effort made in the indicator design process (42 iterations in total, creation of two preliminary versions and a final proposal), we were able to meet the initial objective of this research, in terms of responding to the need to measure the elements of LAP, to control and improve the management of commitments in the weekly LPS meetings, and thus give the kick-off to the development of an LPS 2.0 that focuses on the social dimension (people).

4.2. Lessons Learned from the Methodology Used

The authors consider that the Design Science Research (DSR) methodology used was adequate because it allowed us to solve the problem of measuring and controlling the management of commitments in different situations, managing to produce an Indicators System (artifact) reliable (tested) as an outcome. For us, it was essential to use the updated model proposed by Briggs and Schwabe [46], since our study phenomenon varies according to time, contexts and conditions of application; for the same reason, we carry out a constant iteration applying Kaizen (continuous improvement) in each of the versions, based on the relevance cycle between discovery and design activities, and the design cycle between design and validation activities, always considering the knowledge base as the central axis of the research.

We believe that the five main activities that we proposed and developed applying the DSR method, are a contribution to the knowledge and updating of this method in constant evolution and not yet fully understood by the entire scientific community. Furthermore, we believe it is necessary to mention that as we implemented each system in new scenarios, we obtained more feedback, achieving an increasingly optimal result, although not perfect.

At the end of our research, we realized, thanks to the feedback from other researchers, that the process of having to analyze our indicators through video recording was a very slow and inefficient process, except for the "% of engaged participants" that we had already resolved with our checklist proposal. So we decided to propose a first version of a "Notebook for Last Planners" (see Appendix B) that allows the same last planners to write down their commitments and fill in the necessary data to directly measure seven of our indicators percentage of compliance network or chain of commitments, percentage of fulfillment of a request, percentage of compliance negotiation and agreements, percentage of declaration of compliance with the commitment, percentage of fulfillment declaration of satisfaction, % percentage of declaration of the priority of commitment and percentage of specified deadlines) and indirectly 2 of our indicators (percentage of compliance with priority commitments and percentage of unnecessary requests), leaving only the indicator pending "% of reliability compliance" (and the metrics that depend on it: percentage of revoked commitments, percentage of renegotiated commitments and percentage of canceled commitments) since as these actions occur in the third movement of "Execution and declaration of compliance" in the field huddle. It is recommended that it is discussed openly in the meeting where the PPC and the Reason for Missed Commitment (RMC) of the previous week are analyzed.

4.3. Scope of the Research

With respect to the scope of the proposed artifact, we can mention that it has already been validated in Villego® Simulation, in extension and height construction projects in Chile, and in height projects in Colombia that used LPS. Therefore, we consider that it could be generalizable to this type of construction projects all over the world since the culture may be different and the degree of maturity of LPS as well, but the methodology should be the same unless LPS is not implemented correctly.

Moreover, the research was focused on creating this Indicators System for projects that had LPS implemented and thus optimizing it. Still, we consider that adapting the indicators, according to the Linguistic Action Perspective, could measure and control aspects of commitment management in any construction project, without necessarily having LPS implemented, having as a requirement that the project has a formal meeting structure (ideally weekly) and that this meeting is based on the management of commitments and planning activities. According to the experience of the researchers, this system should not be applied in technical meetings, where work problems are solved without a defined structure, since we tried to measure in these types of meetings and it was very confusing to know who committed. Since even if solutions were given, the person in charge or the deadline was not specified.

As a result, the entire community linked to the construction industry is invited to use the proposed indicators to compare with the "location dimension" (Flyvbjerg 2006). The differences and similarities among different projects around the world, with the objective of determining the effect of the culture of the people and organization in the management of commitments and the general performance of construction projects. Because a theory is more useful if it explains more variations in a phenomenon in more contexts.

5. Conclusions

5.1. Summary

In 2001, Ballard and Howell, creators of LPS, raised a link between the Last Planner® System and Linguistic Action Perspective. However, to date, a system of quantitative instruments or methodologies had not been developed to carry out adequate management of the commitments.

Consequently, the authors wanted to respond to the need of measuring the main elements of LAP, in order to control and improve the management of commitments in the weekly LPS meetings, and thus contribute to the development of an LPS 2.0 that focuses its attention on the social dimension (people) in the context of Lean Construction, creating an Indicators System to measure and control commitments, requirements, promises, and reliability, through the Design Science Research methodology.

The authors used the Design Science Research (DSR) because this method manages to solve practical problems and produce artifacts as results, thus solving problems found in the real world, and in this way it also contributes theoretically in the discipline in which it is applied. Hence closing the gaps among the contextual environment of the research project (people, organizational systems, technical systems, problems and opportunities), design science research (artifacts and processes) and the knowledge base of scientific foundations (scientific theories, methods, experience and expertise), iterating between the activities of construction and evaluation of research design artifacts and processes.

5.2. Contributions

The main contributions of this document are: on the one hand, the proposal of five main activities where the DSR method can be applied, which allowed us to carry out an artifact (indicators system), managing to update this research method, which is continuously evolving, but unfortunately still is not fully understood by the entire scientific community.

Otherwise, we developed a checklist to analyze the engagement of meeting participants, a proposal of notebook for last planners, to simplify the measurement of this Indicators System, ergo avoiding the use of video recordings that require the informed consent of the participants and that can often be invasive.

Additionally, we delved into the Linguistic Action Perspective by creating a figure that details the variations that can occur to basic movements, which do not decrease reliability but increase it. It is important to note that these additional movements in the execution phase (revocation, renegotiation and cancellation of commitments) must be carried out as soon as possible after the commitment is established (once the meeting is over) because if the last planner notifies the client that the commitment will not be fulfilled one day before the meeting, it is no longer considered reliable.

Finally, our main contribution is the creation of an Indicators System that allows to measure and control the main aspects of Linguistic Action Perspective in Last Planner® System, updating the PPC

(Percent Plan Complete) with an updated indicator of Reliability (PPC + Revoke + Renegotiate + Cancel) and a detail of the indicators of engagement of meeting participants.

5.3. Limitations

The main limitations of this research are in two main aspects. The first aspect is that the Indicators System was validated only in projects that had the Last Planner® System implemented, which implies that they are not validated nor are they completely suitable for construction projects that do not have LPS implemented. However, we believe that they can be adapted. The second aspect is that the Indicators System was only validated in two South American countries, which could mean that it is not generalizable to everyone. Still, because LPS has a defined implementation methodology, if it is correctly implemented, differences in the result of the indicators according to the socio-cultural aspects of each country will probably be found. Still, the measurement and verification system should be the same.

At last, this Indicators System seeks to analyze the management of commitments in weekly planning meetings, so the measurement frequency was always every seven days. Nonetheless, it is necessary to hold at least two weekly meetings to analyze the results, since the first two movements will be seen in the first meeting: (1) preparation of a request and (2) negotiation and agreements; and in the second meeting, the last two: (3) execution and declaration of compliance and (4) acceptance and declaration of satisfaction.

5.4. Future Research

The authors see an opportunity to carry out case studies in different types of construction projects and different countries, to carry out a management benchmarking of project commitments with this new Indicators System, measuring and comparing different indicators of planning, production, term, cost, productivity, quality and safety of works, among others.

As well, in future studies, the authors propose to apply case studies in weekly planning meetings in other industries worldwide and to determine the recommended values to improve communication and achieve the proper implementation of LAP with LPS and without LPS (other planning systems).

Author Contributions: Conceptualization, L.A.S.; methodology, L.A.S.; validation, L.A.S.; formal analysis, L.A.S.; investigation, L.A.S.; resources, L.A.S.; data curation, L.A.S.; writing—original draft preparation, L.A.S.; writing—review and editing, P.A. and L.F.A.; visualization, L.A.S.; supervision, P.A. and L.F.A. All authors have read and agreed to the published version of the manuscript.

Funding: Luis A. Salazar acknowledges the financial support for his PhD studies from the Scholarship Programme of National Agency for Research and Development of Chile (ANID-PCHA/Doctorado Nacional/2016-21160819).

Acknowledgments: The authors wish to thank GEPUC and INGECO for their support in this research and all the organizations that participated in this study, as well as the experts for the insight provided.

Conflicts of Interest: The authors declare no conflict of interest.

Appendix A

CHECKLIST - MEETING PARTICIPANS

Instructions:
In each item, detail the number of people who perform the described action

	Item	Number of people	Observations
	Arrive on time		
	Person is late		
*	Check the cell phone		
*	Cell phone rings		
*	Talk by cell phone		
*	Leaves the room		
*	Walkie-talkie sounds		
*	Talk on walkie-talkie		
	Does not intervene in the meeting (does not speak)		
	Take notes		
	Does not look at the person who is speaking		

* Record in column "observations":
Cases in which the same person is persistent. Indicate approximate number of repetitions and other comments.

Other comments:

Figure A1. Checklist to analyze engagement of meeting participants.

Appendix B

NOTEBOOK FOR LAST PLANNERS												
Name:				Measurement start date:			___/___/2020		Measurement end date:		___/___/2020	
Position:				Symbology		-- (Very low)		- (Low)	0 (Mean)	+ (High)	++ (Very High)	
Company:						W: Well		N: Normal			P: Poor	
Week 1								Week 2				
Who asks for it	Activity and / or Task	(%)	Sector	Day (AM or PM)	Task Priority	Clarity in the Petition (request)	Negotiation and Agreement	% Completed	PPC	Declaration of compiance	Declaration of satisfaction	Comments (CNC)
Name and / or Position					--, -, 0, +, ++	W - N - P	W - N - P	%	Does it comply?	Yes - No	Yes - No	
Administrator	Floor slab installation -1	80%	Quadrant A and B	Tuesday (AM)	-	W	N	70%	NO	YES	NO	bla bla bla
Week 2								Week 3				
Who asks for it	Activity and / or Task	(%)	Sector	Day (AM or PM)	Task Priority	Clarity in the Petition (request)	Negotiation and Agreement	% Completed	PPC	Declaration of compiance	Declaration of satisfaction	Comments (CNC)
Name and / or Position					--, -, 0, +, ++	W - N - P	W - N - P	%	Does it comply?	Yes - No	Yes - No	

Figure A2. Proposal of Notebook for Last Planners.

References

1. Koskela, L. *An Exploration towards a Production Theory and its Application to Construction*; Helsinki University of Technology: Espoo, Finland, 2000.
2. Kapelko, M.; Abbott, M. Productivity Growth and Business Cycles: Case Study of the Spanish Construction Industry. *J. Constr. Eng. Manag.* **2017**, *143*, 05016026. [CrossRef]
3. World Economic Forum Shaping the Future of Construction: A Breakthrough in Mindset and Technology. Available online: http://www3.weforum.org/docs/WEF_Shaping_the_Future_of_Construction_full_report__.pdf (accessed on 16 October 2020).
4. González, V.; Alarcón, L.F.; Maturana, S.; Mundaca, F.; Bustamante, J. Improving planning reliability and project performance using the reliable commitment model. *J. Constr. Eng. Manag.* **2010**, *136*, 1129–1139. [CrossRef]
5. Womack, J.P.; Jones, D.T. *Lean Thinking: Banish Waste and Create Wealth in your Corporation*; Free Press: New York, NY, USA, 2003.
6. Koskela, L. *Application of the New Production Philosophy to Construction*; Stanford University: Stanford, CA, USA, 1992; Volume 72.
7. World Commission on Environment and Development (WCED). *Our Common Future*; Oxford University: Oxford, UK, 1987; ISBN 9780192820808.
8. Department of Economic and Social Affairs. United Nations Goal 12. Available online: https://sdgs.un.org/goals/goal12 (accessed on 14 October 2020).
9. Arroyo, P.; Tommelein, I.D.; Ballard, G. Comparing Multi-Criteria Decision-Making Methods to Select Sustainable Alternatives in the AEC Industry. In Proceedings of the International Conference on Sustainable Design, Engineering, and Construction 2012, Fort Worth, TX, USA, 7–9 November 2012; pp. 869–876.
10. Tasdemir, C.; Gazo, R. A Systematic Literature Review for Better Understanding of Lean Driven Sustainability. *Sustainability* **2018**, *10*, 2544. [CrossRef]
11. de Carvalho, A.; Granja, A.; da Silva, V. A Systematic Literature Review on Integrative Lean and Sustainability Synergies over a Building's Lifecycle. *Sustainability* **2017**, *9*, 1156. [CrossRef]
12. Zhang, B.; Niu, Z.; Liu, C. Lean Tools, Knowledge Management, and Lean Sustainability: The Moderating Effects of Study Conventions. *Sustainability* **2020**, *12*, 956. [CrossRef]
13. Salem, O.; Solomon, J.; Genaidy, A.; Minkarah, I. Lean construction: From theory to implementation. *J. Manag. Eng.* **2006**, *22*, 168–175. [CrossRef]
14. Ballard, G.; Howell, G.A. Toward construction JIT. In *Lean Construction*; Alarcón, L., Ed.; Balkema Publishers: Rotterdam, The Netherlands, 1997; pp. 291–300. ISBN 9054106484.
15. Ballard, G.; Howell, G.A. An Update on Last Planner. In Proceedings of the 11th Annual Conference of the International Group for Lean Construction, Blacksburg, VA, USA, 22–24 July 2003; pp. 1–13.
16. Daniel, E.I.; Pasquire, C.; Dickens, G. Exploring the implementation of the last planner® system through iglc community: Twenty one years of experience. In Proceedings of the IGLC 23rd Annual Conference of the International Group for Lean Construction, Perth, Australia, 29–31 July 2015; pp. 153–162.
17. Ballard, G. The Last Planner. In Proceedings of the Northern California Construction Institute Spring Conference, Monterey, CA, USA, 26–28 July 1994.
18. Ballard, G.; Tommelein, I. Current Process Benchmark for the Last Planner(R) System. *Lean Constr. J.* **2016**, *89*, 57–89.
19. Goldratt, E.M.; Cox, J. *La meta, un Proceso de Mejora Continua*; Ediciones Granica, S.A., Ed.; Tercera: Mexico, 2013; ISBN 9789506418069.
20. O'Brien, W.J.; Formoso, C.T.; Vrijhoef, R.; London, K.A. *Construction Supply Chain Management Handbook*; CRC Press: Boca Raton, FL, USA, 2008.
21. Alves, T.C.L.; Tommelein, I.D. Buffering and batching practices in the HVAC industry. In Proceedings of the 11th Annual Conference of the International Group for Lean Construction (IGLC), Blacksburg, VA, USA, 22–24 July 2003.
22. Rodríguez, A.D.; Alarcón, L.F.; Pellicer, E. La gestión de la obra desde la perspectiva del último planificador. *Rev. Obras Públicas* **2011**, *158*, 35–44.

23. Howell, G.A.; Macomber, H.; Koskela, L.; Draper, J. Leadership and Project Management: Time for a Shift from Fayol to Flores. In Proceedings of the 12th Annual Conference of the International Group for Lean Construction, Helsingør, Denmark, 3–5 August 2004.
24. Salazar, L.A.; Ballard, G.; Arroyo, P.; Alarcón, L.F. Indicators for Observing Elements of Linguistic Action Perspective in Last Planner® System. In Proceedings of the 26th Annual Conference of the International. Group for Lean Construction (IGLC), Chennai, India, 18–20 July 2018; pp. 402–411.
25. Flores, F. *Conversaciones Para la Acción: Inculcando una Cultura de Compromiso en Nuestras Relaciones de Trabajo (Conversations for Action and Collected Essays: Instilling a Culture of Commitment in Working Relationships)*; Flores, M., Ed.; Primera; Lemoine Editores: Bogotá, Colombia, 2015; ISBN 978-958-98664-9-8.
26. Austin, J.L. *How to Do Things with Words*, 2nd ed.; Oxford University Press: New York, NY, USA, 1975; ISBN 9780198245537.
27. Searle, J.R. *Speech Acts: An Essay in the Philosophy of Language*; Cambridge University Press: Cambridge, UK, 1969; Volume 626.
28. Searle, J.R. *A Taxonomy of Illocutionary Acts*; University Minnesota Press: Minneapolis, MN, USA, 1975; Volume 7, pp. 344–369.
29. Shahhoseiny, H. Differences between language and linguistic in the ELT classroom. *Theory Pract. Lang. Stud.* **2013**, *3*, 2234–2239. [CrossRef]
30. Isatto, E.L.; Azambuja, M.; Formoso, C.T. The Role of Commitments in the Management of Construction Make-to-Order Supply Chains. *J. Manag. Eng.* **2015**, *31*, 04014053. [CrossRef]
31. Viana, D.D.; Formoso, C.T.; Isatto, E.L. Understanding the theory behind the Last Planner System using the Language-Action Perspective: Two case studies. *Prod. Plan. Control* **2017**, *28*, 177–189. [CrossRef]
32. Zegarra, O.; Alarcon, L.F. Variability propagation in the production planning and control mechanism of construction projects. *Prod. Plan. Control* **2017**, *28*, 707–726. [CrossRef]
33. Macomber, H.; Howell, G.A. Linguistic Action: Contributing to the theory of lean construction. In Proceedings of the 11th Annual Meeting of the International Group for Lean Construction, Blacksburg, VA, USA, 22–24 July 2003.
34. Viana, D.D.; Formoso, C.T.; Isatto, E.L. Modelling the network of commitments in the last planner system. In Proceedings of the 19th Annual Conference of the International Group for Lean Construction, Lean Construction Institute, Lima, Peru, 13–15 July 2011; Volume 2011.
35. Vrijhoef, R.; Koskela, L.; Howell, G.A. Understanding construction supply chains: An alternative interpretation. In Proceedings of the 9th Annual Conference of the International Group for Lean Construction, Singapore, 6–8 August 2001; pp. 1–15.
36. Salazar, L.A.; Retamal, F.; Ballard, G.; Arroyo, P.; Alarcón, L.F. Results of indicators from the Linguistic Action Perspective in the Last Planner(r) System. In Proceedings of the 27th Annual Conference of the International Group for Lean Construction (IGLC), Dublin, Ireland, 3–5 July 2019; pp. 1241–1250.
37. Nunes, M.; Abreu, A. Applying Social Network Analysis to Identify Project Critical Success Factors. *Sustainability* **2020**, *12*, 1503. [CrossRef]
38. Winograd, T.; Flores, F. *Understanding Computers and Cognition: A New Foundation for Design*; Intellect Books: Bristol, UK, 1986.
39. Francis, A.; Thomas, A. Exploring the relationship between lean construction and environmental sustainability: A review of existing literature to decipher broader dimensions. *J. Clean. Prod.* **2020**, *252*, 119913. [CrossRef]
40. Zarta Ávila, P. La sustentabilidad o sostenibilidad: Un concepto poderoso para la humanidad. *Tabula Rasa* **2018**, *28*, 409–423. [CrossRef]
41. Arroyo, P.; Gonzalez, V. Rethinking Waste Definition to Account for Environmental and Social Impacts. In Proceedings of the 24th Annual Conference of the International Group for Lean Construction, Boston, MA, USA, 20–22 July 2016.
42. Holmström, J.; Ketokivi, M.; Hameri, A.-P. Bridging Practice and Theory: A Design Science Approach. *Decis. Sci.* **2009**, *40*, 65–87. [CrossRef]
43. Peffers, K.; Tuunanen, T.; Rothenberger, M.A.; Chatterjee, S. A Design Science Research Methodology for Information Systems Research. *J. Manag. Inf. Syst.* **2007**, *24*, 45–77. [CrossRef]
44. Lukka, K. The constructive research approach. In *Case Study Research in Logistics*; Ojala, L., Hilmola, O.-P., Eds.; Publications of the Turku School of Economics and Business Administration: Turku, Finland, 2003; pp. 83–101.
45. Hevner, A. A Three Cycle View of Design Science Research. *Scand. J. Inf. Syst.* **2007**, *19*, 87–92.

46. Briggs, R.O.; Schwabe, G. On Expanding the Scope of Design Science in IS Research. In *Proceedings of the Service-Oriented Perspectives in Design Science Research, Milwaukee, WI, USA, 5–6 May 2011*; Jain, H., Sinha, A.P., Vitharana, P., Eds.; Springer: Berlin/Heidelberg, Germany, 2011; pp. 92–106.
47. Ballard, H.G. The Last Planner System of Production Control. Ph.D. Thesis, The University of Birmingham, Birmingham, UK, 2000.
48. Alarcón, L.F.; Diethelm, S.; Rojo, Ó. Collaborative implementation of lean planning systems in Chilean construction companies. In Proceedings of the 10th Annual Conference of the International Group for Lean Construction, Gramado, Brazil, 6–8 August 2002.
49. Flyvbjerg, B. Five Misunderstandings About Case-Study Research. *Qual. Inq.* **2006**, *12*, 219–245. [CrossRef]
50. Hernandez Sampieri, R.; Fernández, C.; Baptista, P. Capítulo 4. Estudios de caso (Centro de recursos en línea). In *Metodología de la Investigación*; 2014; p. 31. ISBN 9781456223960. Available online: https://www.researchgate.net/publication/308385754_Robert_K_Yin_2014_Case_Study_Research_Design_and_Methods_5th_ed_Thousand_Oaks_CA_Sage_282_pages (accessed on 15 October 2020).
51. Yin, R.K. Case Study Research: Design and Methods. In *Applied Social Research Methods*, 3rd ed.; SAGE: Thousand Oaks, CA, USA, 2003; Volume 5.
52. Lagos, C.; Salazar, L.A.; Alarcón, L.F. Análisis de la relación entre el nivel de implementación de Last Planner System(R) y el desempeño de proyectos de construcción. In Proceedings of the 1er Congreso Latinoamericano de Ingeniería, Entre Ríos, Argentina, 13–15 September 2017.
53. Popper, K.R. *The Logic of Scientific Discovery*; ISSR library Routledge: Abingdon, UK, 2002; ISBN 9780415278447.
54. Frank, U. Die Konstruktion möglicher Welten als Chance und Herausforderung der Wirtschaftsinformatik. In *Wissenschaftstheorie und Gestaltungsorientierte Wirtschaftsinformatik*; Physica-Verlag HD: Heidleburg, Germany, 2009; pp. 161–173. [CrossRef]
55. Schwabe, G.; Krcmar, H. Piloting a SocioTechnical Innovation. In Proceedings of the 8th European Conference on Information Systems ECIS, Wienna, Austria, 3–5 July 2000; pp. 132–139.

Publisher's Note: MDPI stays neutral with regard to jurisdictional claims in published maps and institutional affiliations.

sustainability

MDPI

Article

Sensitivity Analysis in Probabilistic Structural Design: A Comparison of Selected Techniques

Zdeněk Kala

Department of Structural Mechanics, Faculty of Civil Engineering, Brno University of Technology, 602 00 Brno, Czech Republic; kala.z@fce.vutbr.cz

Received: 10 May 2020; Accepted: 5 June 2020; Published: 11 June 2020

Abstract: Although more and more reliability-oriented sensitivity analysis (ROSA) techniques are now available, review and comparison articles of ROSA are absent. In civil engineering, many of the latest indices have never been used to analyse structural reliability for very small failure probability. This article aims to analyse and compare different sensitivity analysis (SA) techniques and discusses their strengths and weaknesses. For this purpose, eight selected sensitivity indices are first described and then applied in two different test cases. Four ROSA type indices are directly oriented on the failure probability or reliability index beta, and four other indices (of a different type) are oriented on the output of the limit state function. The case study and results correspond to cases under common engineering assumptions, where only two independent input variables with Gaussian distribution of the load action and the resistance are applied in the ultimate limit state. The last section of the article is dedicated to the analysis of the different results. Large differences between first-order sensitivity indices and very strong interaction effects obtained from ROSA are observed for very low values of failure probability. The obtained numerical results show that ROSA methods lack a common platform that clearly interprets the relationship of indices to their information value. This paper can help orientate in the selection of which sensitivity measure to use.

Keywords: sensitivity analysis; uncertainty modelling; load action; resistance; limit states; stochastic simulation; failure probability; structural reliability; correlations

1. Introduction

Evaluating the reliability of building structures is a problem whose final goal remains a decision-making process [1]. In a probabilistic framework, the basic characteristic of engineering reliability is the probability of failure P_f, which represents the key quantity of interest in decision-making processes [2]. It is recommended by the best practices that such a report is supplemented with sensitivity analysis (SA), which describes the effect of changes in model inputs on the measure of reliability [3].

A classical measure of change in P_f is the derivative $\partial P_f/\partial \mu_{xi}$ with respect to the mean value μ of input variable X_i [4–7]. A drawback of the derivative-based SA is that it cannot detect interactions between input variables. Since only one μ_{xi} is varied at a time while others are fixed, it can be labelled as the One-At-a-Time (OAT) method or local SA (at point μ_{xi}). The aforementioned drawback can partially be overcome by using the factorial experiment, where SA is computed using two-level changes of μ_{xi} for all X_i in combinations, which permit the computation of interaction effects [8]. However, only absolute change of the distribution parameter μ_{xi} on P_f is investigated, not the relative influence of the random variability of X_i on P_f. For structural reliability, it is better to prefer such SA types that can compute the effects of the random variabilities of input variables and their interactions on P_f and not just changes in distribution parameters.

Compared with the local SA, global SA [9] can measure the effect of input variables on the model output in their entire distribution ranges and provide the interaction effect among different input variables. In the literature, many global SA techniques, such as the non-parametric techniques [10], screening approaches [11], Sobol's variance-based (ANOVA) methods [12,13] and moment-independent methods [14,15], can be found, among which the variance-based method has gained the most attention. Variance is an important component of reliability analysis, but is insufficient on its own for the analysis of structural reliability; see, for e.g., [16]. A more general approach, which generalized Sobol's sensitivity indices, was introduced by Fort et al. [17]. These indices are generally applicable (goal-oriented) because they can analyze various key quantities of interest, including P_f.

The selection of SA methods that focus on P_f or design quantiles is usually based on a stochastic model with binary output failure/nonfailure, 1/0 [18], but it is not a necessity. In first-order reliability method (FORM), P_f can be replaced by reliability index β [19], which is computed using the first two moments of resistance, and load action and can be applied as an alternative measure of reliability; see Figure 1. However, global SA of β has not yet been developed.

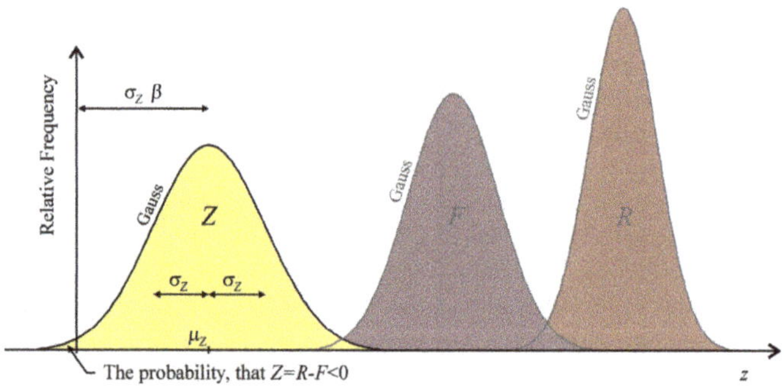

Figure 1. Illustration of the reliability index β.

From a computational point of view, probabilistic sensitivity measures have been comprehensively studied; however, from a decision-analytic point of view, they remain much less understood [20]. Their relationship to information value has not yet been particularly established. Linking the information value, SA, and forecasting with scoring rules remains a subject of research [20]. For a particular reliability task, it is necessary to look for means to select the most appropriate sensitivity measure with common rationale for this selection.

In civil and construction engineering, the scientific community uses SA in structural mechanics [21,22], geotechnics [23,24], landscape water management [25], building performance analysis [26], multi-criteria decision making (MCDM) [27], sustainable development of the building sector [28] or sensitivity audits to assess sustainability [29], but with a lower publication frequency than in basic sciences, such as chemistry, economics or mathematics [30]. In structural reliability, research deals with limit states [31] or the verification of partial safety factors of Eurocode standards [32] using various types of global SA based, for example, on the variance of model outputs [33,34].

The term "sensitivity analysis" can be understood differently in civil engineering than in basic sciences, where local and global SA types with random inputs are well established. For example, very specific (non-stochastic) SA methods based on advanced non-linear models are sometimes used for structures susceptible to buckling when the subject of interest is the stability (or potential energy) of structures [35,36] or imperfection sensitivity [37,38], whereby the main objective of these methods is to increase the stability limits of the structures through the variation of suitable design variables. In stochastic systems, stability often means insensitivity or low sensitivity of the output characteristics to the shapes of some input distributions [39]. In construction engineering, it is necessary to focus more on cooperation and integration of SA development [3] with reliability analysis tools [31,40] and decision-making processes [41,42].

This paper compares several existing sensitivity measures in the context of structural reliability in civil engineering. For this purpose, eight selected sensitivity indices are first described and then applied in two different test cases.

Four indices are oriented on the probability of failure or reliability index β, another four on the distribution or some moments of the output from the ultimate limit state function Equation (2). The reason for the inclusion of the second group of indices is their common use in the analysis of limit states, despite being only sensitive to reliability; see, for e.g., [43,44]. Correlations present a typical example; however, sensitivity techniques based on fuzzy probability analysis of constructions [45,46] are no exception. These alternative types of SA are not directly focused on the probability of failure, but they provide basic insight into the behaviour of computational models, their structures and their reactions to changes in model inputs.

The presented article deals with four ROSA type SA and four other SA, which are empathetic to reliability in civil engineering. Special attention is paid to small failure probabilities, which are relevant for assessing the engineering reliability of structures using design reliability conditions.

2. Design Reliability Conditions

In limit state design, the resistance of a structure R must be greater than the load action F with a predetermined probability [19]. Structural reliability can also be assessed by comparing the lower quantile of R with the upper quantile of F [19], where the quantiles represent alternative key quantities of interest. The decision-maker who develops or implements stochastic models is expected to provide a forecast of structural reliability, which can be performed by estimating the failure probability, quantiles or other computational statistics related to limit states of structure.

Let the reliability of building structures be a one-dimensional random variable Z, which is a function of random variables.

$$Z = g(X) = g(X_1, X_2, \ldots, X_M). \tag{1}$$

The reliability assessment of load-bearing structures is based on a semi-probabilistic approach of standard [19], which falls into the category of FORM methods [47]. Structural reliability is often expressed as a limit state function of random resistance R and random load action F:

$$Z = R - F \geq 0, \tag{2}$$

where R and F are statistically independent variables for which Gauss probability density functions (pdfs) are assumed with mean values μ_R, μ_F and standard deviations σ_R, σ_F. If R and F have Gauss pdfs, then Z has a Gauss pdf with mean value μ_Z and standard deviation σ_Z:

$$\mu_{\mathrm{Z}} = \mu_{\mathrm{R}} - \mu_{\mathrm{F}}, \tag{3}$$

$$\sigma_{\mathrm{Z}} = \sqrt{\sigma_{\mathrm{R}}^2 + \sigma_{\mathrm{F}}^2}. \tag{4}$$

The transformation of Z into a normalized Gaussian pdf of U with mean value $\mu_U = 0$, and standard deviation $\sigma_U = 1$ is written as

$$U = \frac{Z - \mu_{\mathrm{Z}}}{\sigma_{\mathrm{Z}}}. \tag{5}$$

The probability of failure (key quantity of interest) can be expressed as

$$P_{\mathrm{f}} = P(Z < 0) = P\left(U < -\frac{\mu_{\mathrm{Z}}}{\sigma_{\mathrm{Z}}}\right) = P(U < -\beta) = \Phi_U(-\beta), \tag{6}$$

where $\Phi_U(\bullet)$ is the cumulative distribution function of normalized Gaussian pdf and μ_Z/σ_Z is the so-called reliability index β; see Equation (7) and Figure 1.

It is assumed that $\beta > 0$. Standard [19] verifies reliability by comparing the obtained reliability index β with the target reliability index β_d.

$$\beta = \frac{\mu_Z}{\sigma_Z} \geq \beta_\mathrm{d}, \tag{7}$$

For instance, the reliability index has a target value of $\beta_d = 3.8$ ($P_{fd} = 7.2{\cdot}10^{-5}$), provided that we consider the ultimate limit state for common design situations within the reference period of 50 years; see Table C2 in [19] or [48]. Equation (8) can be written to obtain σ_Z as

$$\sigma_Z = \sqrt{\sigma_R^2 + \sigma_F^2} = \frac{\sigma_R^2 + \sigma_F^2}{\sqrt{\sigma_R^2 + \sigma_F^2}} = \frac{\sigma_R}{\sqrt{\sigma_R^2 + \sigma_F^2}}\sigma_R + \frac{\sigma_F}{\sqrt{\sigma_R^2 + \sigma_F^2}}\sigma_F = \alpha_R\sigma_R + \alpha_F\sigma_F, \tag{8}$$

where α_F, α_R are values of sensitivity coefficients (weight factors) according to the FORM method, which [19] introduces with constant values $\alpha_F = 0.7$, $\alpha_R = 0.8$. Substituting Equations (3) and (8) into Equation (7), we can write

$$\beta = \frac{\mu_R - \mu_F}{\alpha_R\sigma_R + \alpha_F\sigma_F} \geq \beta_d. \tag{9}$$

Equation (9) is the design reliability condition with formally separated random variables that can be expressed as

$$\mu_F + \alpha_F\beta_d\sigma_F \leq \mu_R - \alpha_R\beta_d\sigma_R. \tag{10}$$

where the left-hand side represents the design load F_d and the right-hand side the design resistance R_d; see Figure 2. The basic reliability targets for design values in the ultimate limit state recommended in [19] are based on the semi-probabilistic approach in Figure 2, with the target value of reliability index $\beta_d = 3.80$ for a 50 years reference period [48,49]. For $\beta_d = 3.8$, R_d can be approximately computed as 0.1 percentile [40]. Standard [19] enables the determination of design values F_d, R_d not only from a Gauss pdf but also from a two- or three-parameter lognormal (for resistance) or Gumbel or Gama (for load) pdfs. The probability of failure for non-Gaussian R and F can be estimated using Monte Carlo (or quasi-Monte Carlo) methods.

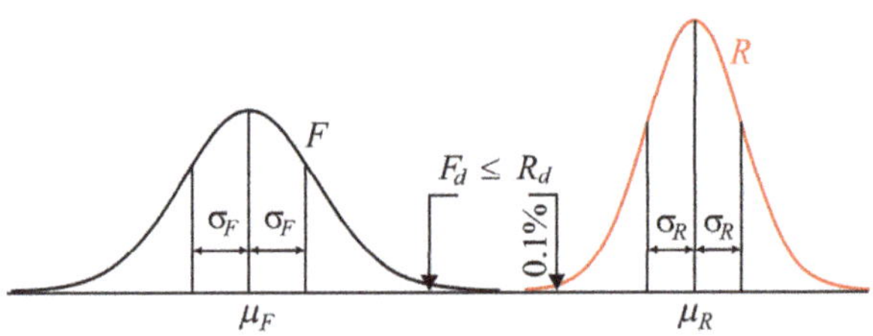

Figure 2. Illustration of the design condition of reliability.

3. Selected Types of Sensitivity Analysis Methods

In reliability engineering, SA methods quantify the effects of input variables on the failure probability, reliability index β or design quantiles. However, other statistical model-based inferences sensitive to reliability are often used. In this chapter, we present selected formulae of selected types of sensitivity measures in forms that are adapted to structural reliability analysis.

Cramér–von Mises indices [50]. Input random variables in Equation (1) are assumed to be statistically independent. Let Φ_Z be the distribution function of Z:

$$\Phi_Z(t) = P(Z \leq t) = E(1_{Z \leq t}) \text{ for } t \in R, \tag{11}$$

and Φ_Z^i is the conditional distribution function of Z conditionally on X_i:

$$\Phi_Z^i(t) = P(Z \le t|X_i) = E(1_{Z\le t}|X_i) \text{ for } t \in R. \tag{12}$$

The first-order Cramér–von Mises index G_i is based on measuring the distance between probability $\Phi_Z(t)$ and conditional probability $\Phi_Z^i(t)$ when an input is fixed [50].

$$G_i = \int_R \frac{E\left[\left(\Phi_Z(t) - \Phi_Z^i(t)\right)^2\right]}{\Phi_Z(t)(1-\Phi_Z(t))} \mathrm{d}\Phi_Z(t). \tag{13}$$

The second-order Cramér–von Mises index G_i can be expressed, on the basis of [50], as

$$G_{ij} = \int_R \frac{E\left[\left(\Phi_Z(t) - \Phi_Z^{ij}(t)\right)^2\right]}{\Phi_Z(t)(1-\Phi_Z(t))} \mathrm{d}\Phi_Z(t) - G_i - G_j, \tag{14}$$

where Φ_Z^{ij} is the conditional distribution function of Z conditionally on X_i, X_j, for $i < j$:

$$\Phi_Z^{ij}(t) = P\left(Z \le t\middle|X_i, X_j\right) = E\left(1_{Z\le t}\middle|X_i, X_j\right) \text{ for } t \in R. \tag{15}$$

Integration Equation (13) and Equation (14) respect t. Equation (13) is not oriented to one failure probability value P_f, but, depending on t, integrates the averages of squared values from the differences of all probabilities Equations (11) and (12) normalized by $F(t)(1 - F(t))$. The same applies to other higher-order indices [50]. Indices G_i, G_{ij}, etc., are based on Hoeffding decomposition; therefore, the sum of all indices is equal to 1 [50]. It can be noted that Cramér–von Mises indices can be formulated in copula theory framework [51].

Sensitivity indices subordinated to contrasts associated with probability [17] (*in short, Contrast P_f indices*). These indices measure the distance between probability P_f and the conditional probability $P_f|X_i$ using the contrast function in Equation (16). The input random variables in Equation (1) are assumed to be statistically independent.

$$\psi(\theta) = E(\psi(Z,\theta)) = E(1_{Z<0} - \theta)^2. \tag{16}$$

The first-order probability contrast index C_i is defined as Equation (17), where the contrast $\min_\theta \psi(\theta)$ is computed for probability estimator $\theta^* = \text{Argmin}\ \psi(\theta) = P_f$.

$$C_i = \frac{\min_\theta \psi(\theta) - E\left(\min_\theta E\ (\psi(Z,\theta)|X_i)\right)}{\min_\theta \psi(\theta)}. \tag{17}$$

The second term in the numerator in Equation (17) is computed as the average value of the conditional contrast functions whose probability estimator is $P_f|X_i$. The second-order probability contrast index C_{ij} can be expressed as

$$C_{ij} = \frac{\min_\theta \psi(\theta) - E\left(\min_\theta E\ \left(\psi(Z,\theta)\middle|X_iX_j\right)\right)}{\min_\theta \psi(\theta)} - C_i - C_j, \tag{18}$$

where $i < j$. Indices of the third and higher orders are computed similarly [17]. Sensitivity indices subordinated to contrasts are based on decomposition; therefore, the sum of all indices must be equal to one. Examples of the computation of indices using the Latin Hypercube Sampling method (LHS) [52,53] in engineering applications are in [54,55].

Sensitivity indices subordinated to contrasts associated with α-quantile [17]. The contrast function ψ associated with α-quantile can be written with parameter θ as [17]:

$$\psi(\theta) = E(\psi(Z,\theta)) = E((Y-\theta)(\alpha - 1_{Y<\theta})), \tag{19}$$

where the input random variables in Equation (1) are assumed to be statistically independent. The first-order quantile contrast index Q_i is defined as

$$Q_i = \frac{\min\limits_{\theta} \psi(\theta) - E\left(\min\limits_{\theta} E\left(\psi(Y,\theta)|X_i\right)\right)}{\min\limits_{\theta} \psi(\theta)}, \tag{20}$$

where $\min\limits_{\theta} \psi(\theta)$ is the contrast computed for the estimator of α-quantile θ^* = Argmin $\psi(\theta)$.

$$Q_{ij} = \frac{\min\limits_{\theta} \psi(\theta) - E\left(\min\limits_{\theta} E\left(\psi(Y,\theta)|X_i\right)\right)}{\min\limits_{\theta} \psi(\theta)} - Q_i - Q_j. \tag{21}$$

The second-order quantile contrast index Q_{ij} is defined as Equation (21), where $i < j$. Indices of the third and higher orders are computed in a similar manner [17]. Sensitivity indices subordinated to contrasts are based on decomposition; therefore, the sum of all indices must be equal to one. In engineering applications, the random variable Y is, for example, the load action F or resistance R [56]; see Figure 1.

Borgonovo moment independent importance measure [14] (*in short, Borgonovo indices*). The sensitivity indices described in [14] are defined by introducing a moment-independent uncertainty indicator that looks at the entire input/output distribution and whose definition is well-posed also in the presence of correlations among the input parameters.

$$B_i = \frac{1}{2} E \int \left|\varphi_Z(z) - \varphi_{Z|X_i}(z)\right| dz, \tag{22}$$

where $\varphi_Z(z)$ is the pdf of Z and $\varphi_{Z|X_i}(z)$ is the conditional pdf of Z given that one of the parameters, X_i, assumes a fixed value [14]. Fixing pairs X_i, X_j, leads to the second-order index B_{ij}, where $i < j$. Fixing triplets X_i, X_j, X_k leads to the third-order index B_{ijk}, where $i < j < k$, etc. The sum of all indices is not equal to one. As a general rule, $0 \leq B_i \leq B_{ij} \leq .. \leq B_{1,2,\ldots,M} \leq 1$ [14].

Reliability sensitivity index defined by Xiao et al. [57] (*in short, Xiao indices*). All the input variables are independent of each other. The first-order index S_i measures the individual effect of X_i on P_f.

$$K_i = \frac{1}{2P_f} E\left(\left|P_f - P_f|X_i\right|\right), \tag{23}$$

where $|P_f - P_f|X_i|$ measures the absolute difference between the unconditional failure probability P_f and the conditional failure probability $P_f|X_i$. The second-order interaction indices K_{ij}, where $i \neq j$, are asymmetrical:

$$K_{ij} = \frac{1}{2} E\left(\left|\frac{P_f|X_i}{P_f} - \frac{P_f|X_i, X_j}{P_f|X_j}\right|\right). \tag{24}$$

K_{ij} may or may not be equal to K_{ji}. Third-order and higher-order indices are not defined in [57].

Reliability sensitivity index defined by Ling et al. [58] (*in short, Ling indices*). The first-order index is the same as in Equation (23) $L_i = K_i$. Fixing pairs X_i, X_j, leads to the second-order index L_{ij}, where $i < j$:

$$L_{ij} = \frac{1}{2P_f} E\left(\left|P_f - P_f|X_i, X_j\right|\right). \tag{25}$$

Fixing triplets X_i, X_j, X_k leads to the third-order index L_{ijk}, where $i < j < k$, etc. The sum of all indices defined by Ling et al. [58] is not equal to one. As a general rule [58], $0 \leq L_i \leq L_{ij} \leq ... \leq L_{1,2,\ldots,M} \leq 1$.

Sobol's sensitivity indices [12,13] (*in short, Sobol's indices*). Sobol's first-order sensitivity indices can be written in the form:

$$S_i = \frac{V(Z) - E(V(Z|X_i))}{V(Z)} = \frac{V(E(Z|X_i))}{V(Z)} = corr^2(Z, E(Z|X_i)), \tag{26}$$

where *corr* is Pearson correlation coefficient. Fixing pairs X_i, X_j, leads to the second-order index S_{ij}, where $i < j$. Fixing triplets X_i, X_j, X_k leads to the third-order index S_{ijk}, where $i < j < k$, etc.; see, for example [9]. The sum of all indices is equal to one. It can be noted that Sobol's indices present a special case of sensitivity indices subordinated to contrasts in which the contrast function is associated with variance $\psi(\theta) = E(Z - \theta)^2$ [17].

Omission sensitivity factor [59] (*in short, Madsen's factor*). The omission sensitivity factor O_i is defined as the ratio between the conditional reliability index $\beta|X_i = \mu_{xi}$ and the reliability index β (7).

$$O_i = \frac{\beta\left|\left(X_i = \mu_{X_i}\right)\right.}{\beta}. \tag{27}$$

Random variable X_i is fixed at its mean value μ_{xi} in the numerator in Equation (27), but the possibility of fixing at the characteristic value [60] or median [3] is also indicated.

The indices described above can be divided into two groups. The first group (Sobol, Borgonovo and Cramér–von Mises) focuses on the distribution or some moments of the output function Z, while the second group (Xiao, Ling, Contrast, Madsen's) considers P_f, β or quantiles as the quantity of interest and thus can be referred to as reliability analysis indices. The first group can be classified as global SA, while the second group can be classified as reliability-oriented sensitivity analysis (ROSA) [3], of which Xiao, Ling and Contrast indices can terminologically [54,57,58] be classified as global ROSA. It can be noted that Xiao, Ling and Contrast ROSA indices are typical examples of ambiguous "local–global" indices [3]. On one hand, they can be considered as global since they are based on changes of P_f with regard to the variability of the inputs over their entire distribution ranges and they provide the interaction effect between different input variables. On the other hand, they can be considered as local in the sense of regional SA since they are based on the frequency of failures from the random realization in "region" of pairs of large load actions and small resistances.

Correlations. The last SA methods used are the analysis of the correlation between the input X_i and output Z according to Pearson, Spearman and Kendal Tau.

4. Case Studies

Many new sensitivity indices have been developed, but their ability in applications has not yet been reliably demonstrated. In this article, the properties of the selected sensitivity indices mentioned in Chapter 3 are examined in a case study of the probabilistic analysis of the reliability of a steel bar under axial tension; see Figure 3. A static time-independent study is considered.

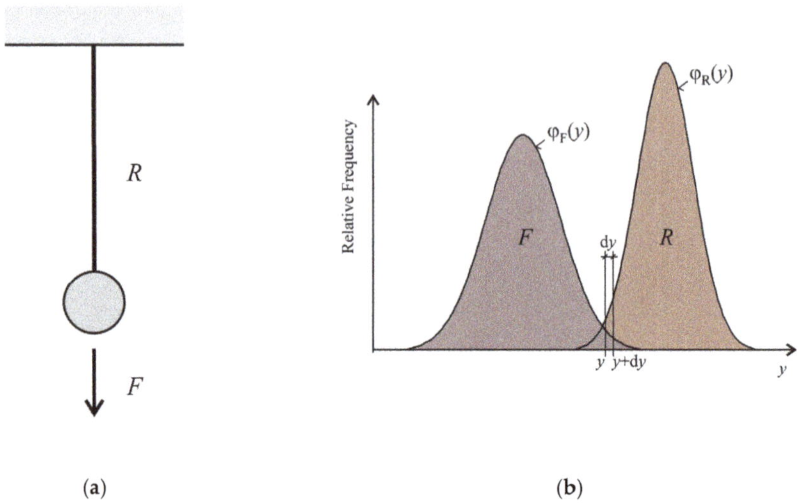

Figure 3. Static model: (**a**) bar under axial tension; (**b**) probability density functions of R and F.

In general, the random load action F and resistance R are usually described using appropriate types of distribution functions $\Phi_F(y)$, $\Phi_R(y)$ and corresponding pdfs $\varphi_F(y)$, $\varphi_R(y)$, where y denotes a general point of the observed variable (force with the unit of Newton), through which both variables F and R are expressed; see right part of Figure 3. It is assumed that F and R are statistically independent of each other with mean values μ_F, μ_R and standard deviations σ_F, σ_R.

The probability of failure $P_f = P(Z < 0) = P(R < F)$ can be computed as the integral:

$$P_f = \int_{-\infty}^{\infty} \Phi_R(y)\varphi_F(y)\mathrm{d}y. \quad (28)$$

In the case studies, integration in Equation (28) is performed numerically by Simpson's rule, using more than ten thousand integration steps over the interval $[\mu_Z - 10\sigma_Z, \mu_Z + 10\sigma_Z]$.

Reliability can be assessed by comparing the computed P_f in Equation (6) with the target value of P_f, where target values for design cases are listed in standard EN1990 [19]. Target values of P_f in Table 1 are taken from Table B2 in [19]. Table 1 lists the minimum values of P_f (the reliability index β) for ultimate limit state and 50 years reference period. The description of subsequent classes RC1, RC2, and RC3 with examples of building and civil engineering works are in [19,48].

Table 1. Recommended minimum values of β and related P_f.

Reliability Class	β	P_f
RC3	4.3	$8.5 \cdot 10^{-6}$
RC2	3.8	$7.2 \cdot 10^{-5}$
RC1	3.3	$4.8 \cdot 10^{-4}$

The aim of the presented study is the SA of the influence of input factors R, F on the output P_f using different types of sensitivity indices and the subsequent comparison of obtained results. Resistance R is the input random variable X_1, and load action F is the input random variable X_2.

4.1. Computation of Sensitivity Indices

This section includes a description of numerical methods for computing the size of sensitivity indices based on numerical integration methods in combination with sampling-based methods or analytical computation. Sensitivity indices were computed for eight SA types.

Contrast P_f indices [17] (ROSA). The contrast function Equation (16) is minimum if $\theta^* = P_f$. By substituting P_f into Equation (16), we can write first-order index in Equation (17) using $\min_{\theta} \psi(\theta) = P_f(1 - P_f)$, and similarly for $P_f|X_i$, we can write $\min_{\theta} E(\psi(Z,\theta)|X_i) = (P_f|X_i)(1 - (P_f|X_i))$.

$$C_i = \frac{P_f(1-P_f) - E\big((P_f|X_i)(1-P_f|X_i)\big)}{P_f(1-P_f)}. \quad (29)$$

By substituting $P_f(1 - P_f)$ and $(P_f|X_i)(1 - (P_f|X_i))$ into Equation (17), we can derive Equation (29) for practical use. C_i measures, on average, the effect of fixing X_i on P_f. The estimate of P_f is computed as the integral Equation (28). In the first loop, the estimate of $P_f|X_i = P((Z|X_i) < 0)$ is computed by numerical integration across $z\in[\mu_Z - 10\sigma_Z, \mu_Z + 10\sigma_Z]$. In the second loop, E[•] is computed by numerical integration of the pdf of X_i with a small step Δx_i taken over $[\mu_{Xi} - 10\sigma_{Xi}, \mu_{Xi} + 10\sigma_{Xi}]$. Since the second term in the numerator in Equation (18) is always equal to zero ($P_f|X_1,X_2$ is always equal to zero or one), $C_{12} = 1 - C_1 - C_2$.

Xiao indices [57] (ROSA). Indices K_1, K_2 are estimated from Equation (23) using double-nested-loop computation. In the outer loop, E[•] is computed by numerical integration of the pdf of X_i with a small step Δx_i taken over $[\mu_{Xi} - 10\sigma_{Xi}, \mu_{Xi} + 10\sigma_{Xi}]$. Note: the estimate E[•] obtained using the LHS method would be inaccurate because it requires an extremely high number of runs for small values of P_f. In the nested loop, estimates of P_f and $P_f|X_i$ are computed by integrating according to Equation (28). Indices K_{12} and K_{21} defined in Equation (24) are computed in a similar manner.

Ling indices [58] (ROSA). By definition, $L_1 = K_1$, $L_2 = K_2$. The computation of L_{12} includes an estimate of E[•], which is based on double numerical integration. In the outer loop, the pdf of X_2 is numerically integrated with a small step Δx_2 taken over $[\mu_{X2} - 10\sigma_{X2}, \mu_{X2} + 10\sigma_{X2}]$. In the inner loop, the pdf of X_1 is numerically integrated with a small step Δx_1 taken over $[\mu_{X1} - 10\sigma_{X1}, \mu_{X1} + 10\sigma_{X1}]$. During integration, the term $P_f|X_1,X_2$ can only have a value of 0 or 1.

Madsen factor [59] (ROSA). Indices O_1, O_2 are computed using one million LHS runs.

Cramér–von Mises indices [50]. Indices G_1 and G_2 are computed using Equation (13). Three nested loops are applied. In the first (outer) loop, numerical integration is computed with a small step $\Delta t = t_{l+1} - t_l$, where $t = (t_{l+1} + t_l)/2$, $t\in[\mu_Z - 10\sigma_Z, \mu_Z + 10\sigma_Z]$, $l = 1, 2, \dots, 10000$. To each Δt belongs $d\Phi(t)\approx P(t_l \le Z \le t_{l+1})$ and $\Phi(t)\approx P(Z \le (t_{l+1} + t_l)/2)$. In the second loop, E[•] in the numerator in Equation (13) is computed by numerical integration of the pdf of X_i with a small step Δx_i taken over $[\mu_{Xi}-10\sigma_{Xi}, \mu_{Xi}+10\sigma_{Xi}]$. Note: The LHS estimation of E[•] would be numerically very challenging but is possible. In the third (deep) loop, $\Phi^i(t)\approx P(Z \le (t_{l+1} + t_l)/2|X_i = \xi_i)$ is computed by numerical integration for fixed ξ_i, where ξ_i is the middle of interval Δx_i from the second loop. The index G_{12} is computed on the basis of Equation (14) in a similar manner.

Borgonovo indices [14]. Indices B_1, B_2 are estimated from Equation (13) using double-nested-loop computation. In the outer loop, 0.5·E[•] is computed using one million runs of the LHS method. In the nested loop, numerical integration $|\varphi_Z(z) - \varphi_{Z|Xi}(z)|$ is taken over $[\mu_Z - 10\sigma_Z, \mu_Z + 10\sigma_Z]$ using ten thousand runs. $B_{12} = 1$ in all case studies.

Sobol's indices [12,13]. Sobol's sensitivity indices are included only for comparison; these indices analyse the influence of the variance of R or F on the variance of Z, but not the influence on P_f. Sobol's indices are computed analytically as $S_1 = \sigma_R^2/(\sigma_R^2 + \sigma_F^2)$, $S_2 = \sigma_F^2/(\sigma_R^2 + \sigma_F^2)$, $S_{12} = 0$. It can be noted that Sobol's first-order indices are equal to the squares of the sensitivity coefficients (weight factors) in Equation (8): $S_1 = \alpha_R^2$, $S_2 = \alpha_F^2$.

Correlations. Correlation corr(X_1, Z) and corr(X_2, Z) are evaluated even though direct sensitivity to P_f is not measured by correlation. Pearson, Spearman and Kendal Tau correlation coefficients between input X_i and output Z are computed using one hundred thousand runs of the LHS method.

All E[•] are computed by numerical integration with the exception of Borgonovo indices and correlation. It can be noted that E[•] in the formulae in chapter 3 can also be numerically computed using Monte Carlo- (or quasi-Monte Carlo-) type simulation methods; however, the repeated computation of small values of P_f requires extremely high numbers of simulation runs and is numerically challenging.

4.2. Case Study 1

The aim of SA is to assess the influence of R and F on P_f. Let R (resistance) and F (load action) be statistically independent variables X_1, X_2 with Gauss pdfs, where μ_R = 412.54 kN, σ_R = 34.132 kN, σ_F = 34.132 kN, and mean value μ_F is the parameter; see Figure 4. Let parameter μ_F change with the step $\Delta\mu_F$ = 10 kN and gradually attain the values 92.54 kN, 102.54 kN,..., 722.54 kN. The sensitivity indices are plotted in dependence on P_f, where $P_f = \Phi_U(-(412.54 - \mu_F)/(2^{0.5} \cdot 34.132))$ is a function only of parameter μ_F. If μ_F decreases, then P_f decreases.

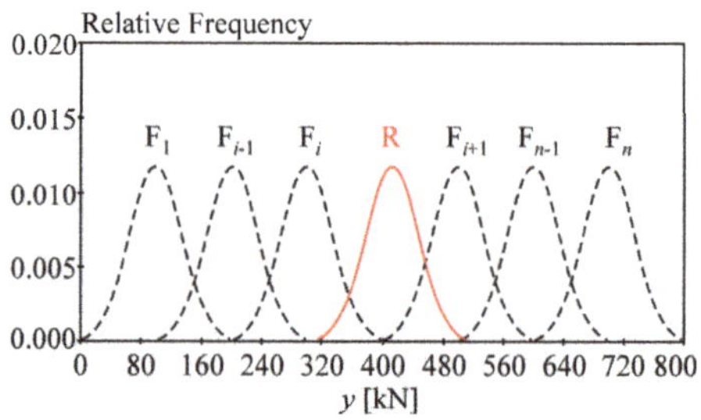

Figure 4. Case study 1, probability density functions of R and F.

First-order indices are depicted in Figure 5 and the second-order indices are depicted on the left part of Figure 6. Correlation coefficients corr(X_1, Z) and corr(X_2, Z) are added to Figure 5. Only indices within the interval [0, 1] are depicted. Madsen's factors are not plotted because they have a constant value $O_1 = O_2 = 1.415$ for all P_f (μ_F).

Changing P_f (μ_F) influences only indices C_1, C_2, C_{12}, indices K_1, K_2, K_{12}, K_{21} and indices L_1, L_2, L_{12}; see Figure 5 and the left part of Figure 6. For the other five types of SA, it was observed that two variables that have a different influence on the output have the same indices. This demonstrates properties of sensitivity indices that will prove useful in the interpretation of the result.

Ling and Xiao indices are the only indices with asymmetric plots and decrease with increasing P_f. Xiao asymmetrical interaction indices are identical: $K_{12} = K_{21}$. Contrast-based sensitivity indices have values of $C_1 = C_2 = C_{12} = 0.33$ for $P_f = 0.5$, but, otherwise, decrease with absolute distance from $P_f = 0.5$. Approaching $P_f \to 0$ or $P_f \to 1$ leads to $C_1 = C_2 \to 0$ and $C_{12} \to 1$. Change in mean value μ_F has no influence on the values of Sobol's indices, which are functions of only the variance and therefore remain constant $S_1 = S_2 = 0.5$, $S_{12} = 0$. Kendall's tau coefficient is approximately equal to 0.5 for all P_f (μ_F). Spearman's and Pearson coefficients confirm the dependence between the inputs R, F and the output Z. Borgonovo and Cramér–von Mises first-order indices have approximately the same value $B_1 = 0.306$, $G_1 = 0.286$, while the second-order indices are $B_{12} = 1.0$ and $G_{12} = 1.0 - G_1 - G_2 = 0.428$.

For common design situations, building constructions are considered reliable if $P_f < 7.2 \cdot 10^{-5}$ (RC2 in Table 1). For this case study, it occurs approximately for $\mu_F < 0.55 \cdot \mu_R$. Detail of the plots of sensitivity indices for $P_f < 1 \cdot 10^{-4}$ are depicted on the right part of Figure 6.

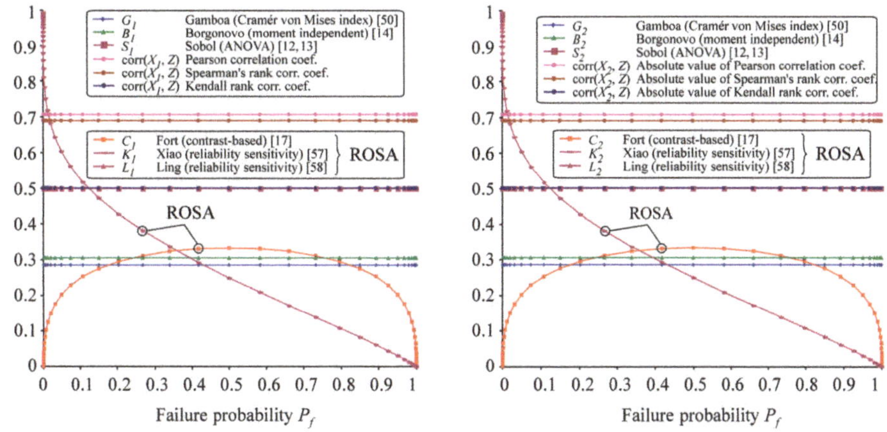

(**a**) (**b**)

Figure 5. Case study 1, first-order sensitivity indices of (**a**) resistance *R*; (**b**) load Action *F*.

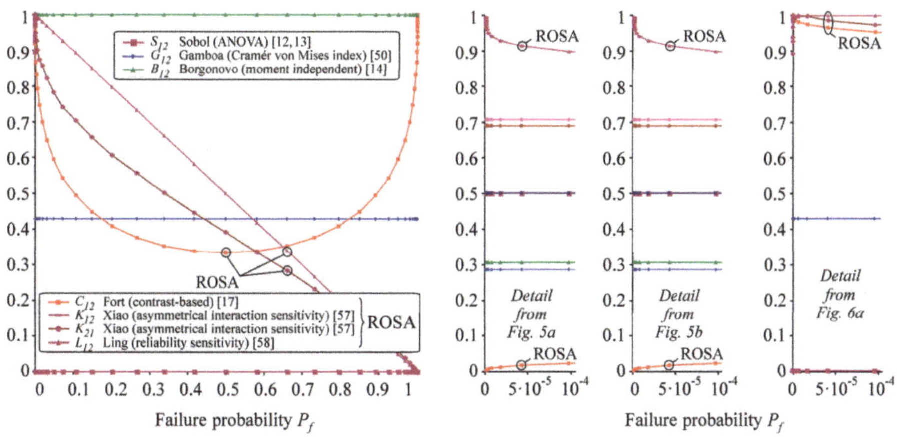

(**a**) (**b**)

Figure 6. Case study 1: (**a**) second-order sensitivity indices of *R*, *F*; (**b**) sensitivity analysis (SA) results related to structural reliability.

4.3. Case Study 2

Let *R* (resistance) and *F* (load action) be statistically independent variables X_1, X_2 with Gauss pdfs, where μ_R = 412.54 kN, σ_R = 34.132 kN and mean value μ_F is the parameter, while variation coefficient of *F* is constant $v_F = v_R$ = 34.132/412.54 = 0.0827 and thus $\sigma_F = v_F \cdot \mu_F$; see Figure 7. Let parameter μ_F change with the step $\Delta\mu_F$ = 12.89 kN and gradually attain values of 0.06 kN, 12.95 kN,..., 902.36 kN.

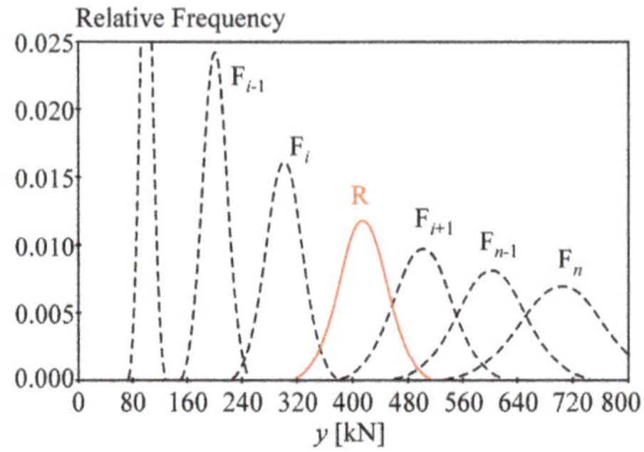

Figure 7. Case study 2, probability density functions of R and F.

The sensitivity indices are plotted in dependence on P_f, where $P_f = \Phi_U(-(412.54 - \mu_F)/(34.132^2 + (\mu_F \cdot 34.132/412.54)^2)^{0.5})$ in Equation (6) is a function of only parameter μ_F, where P_f decreases if μ_F decreases.

The first-order indices are depicted in Figure 8 and the second-order indices are depicted on the left part of Figure 9. Correlation coefficients corr(X_1, Z) and corr(X_2, Z) are added to Figure 8. Only indices within the interval [0, 1] are depicted. Omission sensitivity indices are not plotted because they have a value greater than one. The curves meet the expectation that small P_f (due to small μ_F and small σ_F) are less sensitive to F and more sensitive to R.

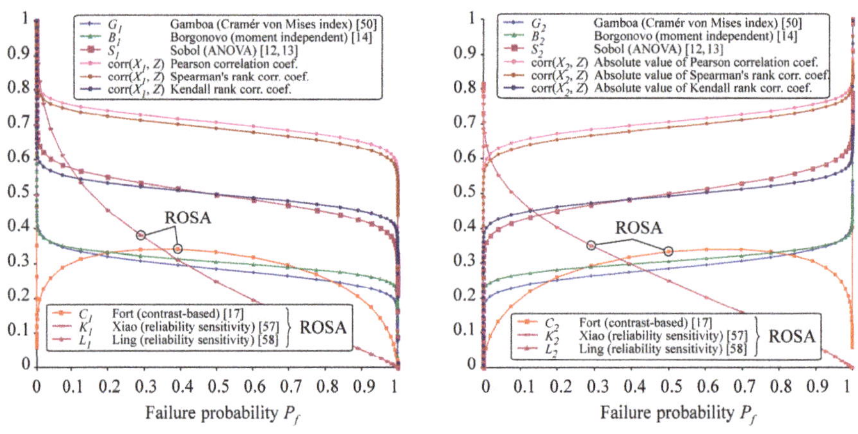

(a) (b)

Figure 8. Case study 2, first-order sensitivity indices of (**a**) Resistance R; (**b**) Load Action F.

The parametric change of P_f (μ_F) influences the values of all indices. All first-order indices of variable X_1 (R) are axially symmetrical to indices X_2 (F) along the vertical axis $P_f = 0.5$ with the exception of Ling and Xiao indices. Xiao asymmetrical interaction indices are $K_{12} \neq K_{21}$ with the exception of $P_f = 0.5$ where $K_{12} = K_{21}$. The plots of Borgonovo and Cramér–von Mises first-order indices are similar; the second-order indices are $B_{12} = 1.0$ and $G_{12} = 1.0 - G_1 - G_{12}$. The plots of Kendall's tau coefficient and plots of Sobol's indices are similar. The plots of Spearman's and Pearson coefficients are also similar. On the left side of the graphs, contrast P_f indices reach their extreme at point $P_f = 3.216 \cdot 10^{-10}$, $C_1 = 0.06$, $C_{12} = 0.94$, but no extreme on C_2. On the right side of the graphs, the extreme is at point $P_f = 1 - 3.216 \cdot 10^{-10}$, $C_2 = 0.06$, $C_{12} = 0.94$, but no extreme on C_1. Ling and Xiao indices have an extreme at $P_f = 3.216 \cdot 10^{-10}$, $K_2 = L_2 = 0.82$, $K_{12} = 0.94$; other extremes of Ling and Xiao indices are difficult to identify numerically (compared to other indices) because they quickly attain relatively small or large values for large or small P_f.

For standard design case $P_f = 7.2 \cdot 10^{-5}$ we obtain $K_1 = L_1 = 0.97$, $K_2 = L_2 = 0.76$, $K_{12} = 0.91$, $K_{12} = 0.994$, $L_{12} = 0.99993$. A detail of the plot of sensitivity indices for $P_f < 1 \cdot 10^{-4}$ is depicted on the right part of Figure 9. For instance, for $P_f = 7.2 \cdot 10^{-5}$, we obtain $O_1 = 1.88$, $O_2 = 1.18$, or for $P_f = 8.5 \cdot 10^{-6}$, we obtain $O_1 = 1.99$, $O_2 = 1.16$.

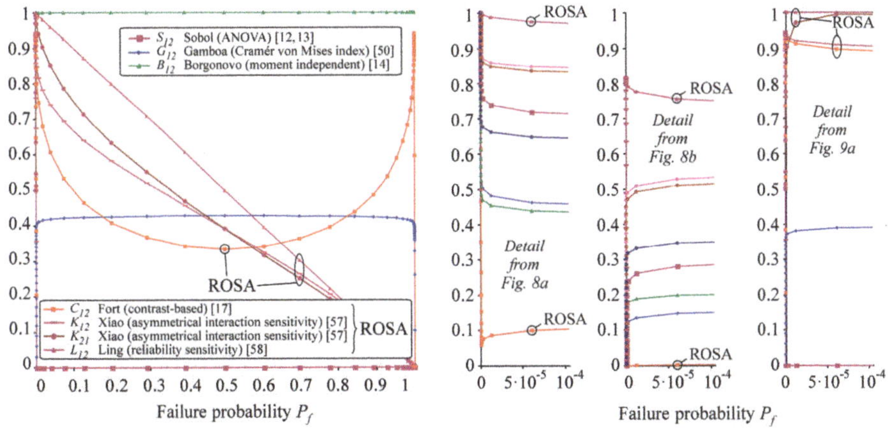

(a) (b)

Figure 9. Case study 2: (**a**) Second-order sensitivity indices of *R*, *F*; (**b**) SA results related to structural reliability.

5. Observation, Discussion and Questions

Reliability-oriented sensitivity indices (Xiao, Ling, Contrast and Madsen's) are computed together with the global indices (Sobol, Borgonovo and Cramér–von Mises). The effects of *R* and *F* on structural reliability are first analysed separately for each of the eight selected sensitivity indices.

Contrast P_f indices have relatively small values of first-order indices and high values of second-order indices for small P_f. The numerical results of reliability engineering tasks [54,55] with five input random variables have shown that the smaller P_f is, the smaller the values of first-order indices and the higher the values of higher-order indices. Change in the mean value or standard deviation of the dominant variables had a clear effect on P_f, confirming the rationality of the contrast indices applied in [54]. The sum of all indices is equal to one, which makes it easier to compare SA results for different P_f associated with different reliabilities, for e.g., different design conditions, different stages of the structural life or different loading conditions. For very small values of P_f, Equation (29) can be written approximately as:

$$C_i \approx \frac{P_f - E\left(\left(P_f | X_i\right)\right)}{P_f}, \tag{30}$$

and similarly

$$C_{ij} \approx \frac{P_f - E\left(\left(P_f | X_i, X_j\right)\right)}{P_f} - C_i - C_j. \tag{31}$$

The clear addressability to P_f is evident from Equation (30) and Equation (31). If the binary random variable $1_{Z<0}$ is considered, Equation (30) can then be written as:

$$C_i \approx \frac{E(1_{Z<0}) - E(E(1_{Z<0} | X_i))}{E(1_{Z<0})} = corr^2(1_{Z<0}, E(1_{Z<0} | X_i)), \tag{32}$$

where *corr* is Pearson correlation coefficient. SA based on contrast functions yields the same (symmetrical) results for unreliability (P_f) and reliability ($1 - P_f$) because C_i in Equation (29) is computed from the values of the contrast functions $P_f(1 - P_f)$ and $(P_f|X_i)(1 - (P_f|X_i))$. If $1_{Z<0}$ is rare, the evaluation of Equation (32) using Monte Carlo type methods requires an extreme number of samples.

The sum of all Ling or Xiao indices is not equal to one. For small P_f, the values of the first-order indices are relatively high; moreover, the second-order index is always greater than the first-order index, which complicates the comparison of the influence on P_f. The advantage of computing Ling indices is that the computation of the higher-order indices does not depend on the computation (accuracy) of lower-order indices; therefore, their computation may be performed parallelly on multiple processor cores.

Cramér–von Mises indices have the sum of all indices equal to one. However, there is no addressability of indices to the P_f level because Equations (13) and (14) are integrated over all $d\Phi_Z(t)$, i.e., over all t (which means over all P_f). This is also the reason that, at intervals relevant to reliability, index values are not extremely high or low. The advantage is that a zero value of the Cramér-von Mises index clearly means that the input is not important. Triple-nested-loop computation makes these indices very numerically challenging. Nevertheless, numerous effective approaches to reduce this computational complexity already exist; see, for e.g., [61].

Borgonovo first-order indices yield reasonable values in intervals relevant to reliability, similar to Cramér–von Mises indices. The advantage of these indices is the transparency and the clear interpretation of the influence of input uncertainty on the entire output distribution regardless of the specific moment of the output (moment independence). Moreover, the indices can be computed even in the presence of correlation between input variables. The computational complexity of indices is not high. The disadvantage is that the sum of all indices is not equal to one and the indices are not directly addressable to P_f.

Sobol's indices are functions of only the variance, which, although important, is not enough for SA of reliability. The computation of Sobol's indices is based on the double-nested-loop computation and can be numerically very challenging for engineering tasks. However, if we consider the binary random variable $1_{Z<0}$ as the quantity of interest, Sobol's indices can be an interesting reliability-oriented sensitivity technique [62].

Madsen's factor can be applied as a computationally undemanding (simple) SA in engineering tasks, but with a number of limitations. Madsen's factor can only be applied for $P_f < 0.5$. The disadvantage of Madsen's factor is that factor O_i can have values significantly greater than 1. For example, in Case study 2, for $P_f = 6.14 \cdot 10^{-32}$ we obtain $O_1 = 32$. A model with one random variable would theoretically lead to $O_1 = \infty$. A significant computational problem can occur in non-linear problems when fixing to the mean value X_i leads to the limit case of a given physical phenomenon. For example, the amplitude of the axial curvature of a slender bar subjected to buckling has a mean value equal to zero, which means a perfectly straight bar [63]. The resistance of such a perfectly straight (unrealistic) bar is always higher than the resistance of a bar with any non-zero imperfection [64], and thus the mean value of zero is not suitable for fixing in reliability analysis or SA. The modification of Equation (27) to the form $E(\beta|X_i)/\beta$ can be discussed, but with the proviso that fixing X_i must not lead to negative values of β. So far there is no global SA based on β, and it is questionable whether the first two statistical moments are sufficient to describe the influence on reliability.

The correlation coefficients are not directly addressable to P_f but can be used as sensitivity indicators if the output (Z) is monotonically dependent on the input variables R and F. Correlation points to dependence, but the opposite is not true. The advantage of correlation coefficients is their availability in computer software and they are relatively computationally inexpensive in simulation approaches.

ROSA-type indices have a different explanatory power than those of other types. Similar results cannot be expected from these two types (ROSA vs. non-ROSA) of indices as an input variable could be influential on P_f but not on the distribution of Z and conversely. Nevertheless, there is relatively good agreement between contrast Cramér–von Mises indices, Borgonovo indices and P_f indices in the interval of approximately $P_f \in (0.1, 0.9)$. However, common building structures are designed with a reliability of $P_f < 4.8 \cdot 10^{-4}$; see Table 1. For such small P_f, only Cramér–von Mises indices and Borgonovo indices have similar values. The values of the other sensitivity indices are considerably different.

It can be concluded from all the obtained numerical results that if $\sigma_F < \sigma_R$ ($\sigma_F > \sigma_R$), the sensitivity of P_f to R is higher (lower) than the sensitivity of P_f to F. It was confirmed that only ROSA-type indices are suitable for probability-based reliability assessment. The results of Case study 1 showed that change in μ_F together with σ_F = const. changes only the values of contrast P_f indices and Ling and Xiao indices, the other indices remain unchanged. Furthermore, change in σ_F together with μ_F = const. changes the values of all indices, except of course $B_{12} = 1$ and $S_{12} = 0$. The results of Case study 2 showed that changes in σ_F and μ_F with the condition $v_F = \sigma_F/\mu_F$ = const. causes changes in the values of all indices; therefore, none of these indices is a pure indicator of the influence of v_F on P_f.

It can be noted that resistance is generally a random variable that is a product of random variables such as yield strength (material characteristic), cross-sectional area (geometric characteristic), etc. Material and geometric characteristics usually do not have perfect Gauss pdf due to small skewness and kurtosis observed in histograms from real experiments [65,66]. The dead load can be roughly approximated using Gauss pdf, but other load types (wind, snow, traffic, long-term load action) have pdfs significantly different from Gauss pdf. For more complex structures, numerous load conditions and tens to hundreds of random variables with different pdfs can be expected. Furthermore, input random variables may have mutual correlations, which are implemented in beams [67] in more detail than in systems where each beam is represented by a smaller number of random variables independent of another beam; see for e.g., [68].

The question is, to what extent can different types of sensitivity indices oriented to P_f be influenced by the skewness and kurtosis values of the input variables or by correlations between them, and what is its importance for the analysis of reliability? For instance, the values of Sobol's indices change when the kurtosis changes, but not when the skewness changes [34]. Of the SA types presented here, only Borgonovo indices [14] have the ability to have correlations between input variables. This ability must also be sought in other indices suitable for structural reliability analysis.

A generally accepted measure of reliability is P_f; therefore, P_f should be the overall objective of SA. However, the concept of Eurocodes [19] assesses reliability according to the limit states using the so-called semi-probabilistic method, which compares the design values (quantiles) of resistance and load. Because probabilistic reliability analysis would be too expensive in common engineering practice, design values are usually computed deterministically according to design standards. These design values can be verified using the lower quantity of resistance (for e.g., 0.1 percentile) and upper quantity of load, where resistance and load are functions of other random variables; see Figure 2. Another useful property of SA could be that sensitivity indices oriented to P_f and design quantiles form pairs based on the same theoretical basis. For example, global SA subordinated to contrasts can be associated with both P_f and quantities R_d and F_d; see Figure 2. However, the question is whether there is a link between indices Equations (17), (18) and (20), (21) when the contrast functions Equations (16) and (19) are different. Preliminary studies show that partial similarity can be expected between the total indices.

6. Conclusions

The presented case studies have shown that the numerical results of reliability-oriented sensitivity analysis (ROSA) are inconsistent. ROSA was evaluated using Contrast, Xiao, Ling and Madsen's indices. For structural reliability, the key quantity of interest is failure probability P_f, which is lower than $4.8 \cdot 10^{-4}$.

Contrast P_f indices have relatively small values of first-order indices and high values of second-order indices for small P_f. Ling or Xiao indices have relatively high values of first-order indices, but also high values of second-order indices for small P_f. For instance, in the first case study, if $P_f \to 0$ then $C_1 = 0$, $C_2 = 0$, $C_{12} = 1$ and $L_1 = 1$, $L_2 = 1$, $L_{12} = 1$. In civil engineering, P_f is generally very small, and extreme values of sensitivity indices estimated by ROSA can be expected. The advantage of contrast P_f indices is that the sum of all indices is equal to one. The sum of all Ling or Xiao indices is not equal to one. The Madsen factor values were significantly greater than 1 and therefore cannot be compared in size with Contrast, Xiao and Ling indices. Madsen's factor does not reflect change in the mean value of input variables, although this change causes a change in P_f.

In contrast, Xiao, Ling and Madsen's indices have correctly identified the order of importance of input random variables to P_f; however, this observation only applies to the presented case studies and cannot be generalized. In the case studies, it is not possible to determine, even approximately, the percentage by which the dominant variable is more influential than the others, so this conclusion is true for each type of SA. Structural reliability lacks a common platform of SA that provides a clear interpretation of the size of sensitivity indices and defines their information value.

The other indices (Sobol, Borgonovo and Cramér–von Mises) and correlation coefficients are not directly addressable to P_f and therefore are not generally suitable for the analysis of reliability. As expected, these (out of ROSA type) sensitivity indices do not reflect the change in mean value of input variables, although this change causes a change in P_f. This means that the two variables that have a different influence on the reliability may have the same indices. In relation to the reliability of structures, the information value of these indices is not unambiguous. In the case of ROSA, Xiao and Ling indices, no two different P_f values exist for which the same sets of sensitivity indices exist, but contrast P_f indices have the same or similar values for unreliability (P_f) and reliability ($1 - P_f$).

There are many engineering reliability assessments in which non-ROSA indices are applied, although the connection with reliability is mentioned. The reason for these applications may be the simplicity of evaluating indices as well as the experience that known indices have at least partial sensitivity for reliability, which, along with other experience, is sufficient for basic decision-making. With the development of ROSA, a gradual transition to new types of reliability-oriented indices can be expected.

In connection with sustainable reliability, it is possible to discuss which type of ROSA should be applied and which key quantities of interest ROSA should be oriented to in particular. The Eurocode standards for structural design assess reliability using a so-called semi-probabilistic approach, which is based on design quantiles. The question remains whether P_f can be adequately replaced by design quantities, reliability index β or other model-based inferences so that the information value of SA results in relation to reliability is approximately maintained. Design quantiles are an important part of reliability analysis, and SA of the design quantiles may be required to provide results consistent with P_f.

In general, ROSA directly addressable to P_f may be preferred rather than focusing on the reliability index β or quantiles. Indices with the sum of one and a clear addressability to P_f present one SA, an advantage that facilitates the comparison of the results of different probability models. Contrast functions are a more general tool for estimating various parameters associated with probability distributions, and thus the partial consistency of requirements could perhaps be sought on the basis of contrasts. These and other tasks need to be addressed in order to make SA of structural reliability a useful and practical tool.

Funding: The work has been supported and prepared within the project namely "Probability oriented global sensitivity measures of structural reliability" of The Czech Science Foundation (GACR, https://gacr.cz/) no. 20-01734S, Czechia.

Conflicts of Interest: The author declares no conflict of interest.

References

1. Benjamin, J.R.; Cornell, C.A. *Probability, Statistics, and Decision for Civil Engineers*; McGraw-Hill: New York, NY, USA, 1970.
2. Au, S.-K.; Wang, Y. *Engineering Risk Assessment with Subset Simulation*; Wiley: New York, NY, USA, 2014.
3. Chabridon, V. Reliability-Oriented Sensitivity Analysis under Probabilistic Model Uncertainty—Application to Aerospace Systems. Ph.D. Thesis, University Clermont Auvergne, Clermont-Ferrand, France, 2018.

4. Melchers, R.E.; Ahammed, M. A fast-approximate method for parameter sensitivity estimation in Monte Carlo structural reliability. *Comput. Struct.* **2004**, *82*, 55–61. [CrossRef]
5. Wang, P.; Lu, Z.; Tang, Z. A derivative based sensitivity measure of failure probability in the presence of epistemic and aleatory uncertainties. *Comput. Math. Appl.* **2013**, *65*, 89–101. [CrossRef]
6. Jensen, H.A.; Mayorga, F.; Papadimitriou, C. Reliability sensitivity analysis of stochastic finite element models. *Comput. Methods. Appl. Mech. Eng.* **2015**, *296*, 327–351. [CrossRef]
7. MiarNaeimi, F.; Azizyan, G.; Rashki, M. Reliability sensitivity analysis method based on subset simulation hybrid techniques. *Appl. Math. Model.* **2019**, *75*, 607–626. [CrossRef]
8. Kala, Z.; Valeš, J. Sensitivity assessment and lateral-torsional buckling design of I-beams using solid finite elements. *J. Constr. Steel. Res.* **2017**, *139*, 110–122. [CrossRef]
9. Saltelli, A.; Ratto, M.; Andres, T.; Campolongo, F.; Cariboni, J.; Gatelli, D.; Saisana, M.; Tarantola, S. *Global Sensitivity Analysis: The Primer*; John Wiley & Sons: West Sussex, UK, 2008.
10. Saltelli, A.; Marivoet, J. Non-parametric statistics in sensitivity analysis for model output: A comparison of selected techniques. *Reliab. Eng. Syst. Saf.* **1990**, *28*, 229–253. [CrossRef]
11. Xiao, S.; Lu, Z.; Xu, L. A new effective screening design for structural sensitivity analysis of failure probability with the epistemic uncertainty. *Reliab. Eng. Syst. Saf.* **2016**, *156*, 1–14. [CrossRef]
12. Sobol, I.M. Sensitivity estimates for non-linear mathematical models. *Math. Model. Comput. Exp.* **1993**, *1*, 407–414.
13. Sobol, I.M. Global sensitivity indices for nonlinear mathematical models and their Monte Carlo estimates. *Math. Comput. Simul.* **2001**, *55*, 271–280. [CrossRef]
14. Borgonovo, E. A new uncertainty importance measure. *Reliab. Eng. Syst. Saf.* **2007**, *92*, 771–784. [CrossRef]
15. Borgonovo, E.; Castaings, W.; Tarantola, S. Moment independent importance measures: New results and analytical test cases. *Risk Anal.* **2011**, *31*, 404–428. [CrossRef] [PubMed]
16. Borgonovo, E. Measuring uncertainty importance: Investigation and comparison of alternative approaches. *Risk Anal.* **2006**, *26*, 1349–1361. [CrossRef] [PubMed]
17. Fort, J.C.; Klein, T.; Rachdi, N. New sensitivity analysis subordinated to a contrast. *Commun. Stat. Theory Methods* **2016**, *45*, 4349–4364. [CrossRef]
18. Wang, Y.; Xiao, S.; Lu, Z. An efficient method based on Bayes' theorem to estimate thefailure-probability-based sensitivity measure. *Mech. Syst. Signal. Process.* **2019**, *115*, 607–620. [CrossRef]
19. European Committee for Standardization. *EN 1990:2002: Eurocode—Basis of Structural Design*; European Committee for Standardization: Brussels, Belgium, 2002.
20. Borgonovo, E.; Hazen, G.; Plischke, E. A Common rationale for global sensitivity measures and their estimation. *Comput. Struct.* **2016**, *36*, 1871–1895. [CrossRef]
21. Greegar, G.; Manohar, G.S. Global response sensitivity analysis using probability distance measures and generalization of Sobol's analysis. *Probabilistic Eng. Mech.* **2015**, *41*, 21–33. [CrossRef]
22. Kotełko, M.; Lis, P.; Macdonald, M. Load capacity probabilistic sensitivity analysis of thin-walled beams. *Thin Walled Struct.* **2017**, *115*, 142–153. [CrossRef]
23. Štefaňák, J.; Kala, Z.; Miča, L.; Norkus, A. Global sensitivity analysis for transformation of Hoek-Brown failure criterion for rock mass. *J. Civ. Eng. Manag.* **2018**, *24*, 390–398. [CrossRef]
24. Mahmoudi, E.; Hölter, R.; Georgieva, R.; König, M.; Schanz, T. On the global sensitivity analysis methods in geotechnical engineering: A comparative study on a rock salt energy storage. *Int. J. Civ. Eng.* **2019**, *17*, 131–143. [CrossRef]
25. Tate, E.; Munoz, C.; Suchan, J. Uncertainty and sensitivity analysis of the HAZUS-MH flood model. *Nat. Hazards Rev.* **2014**, *16*, 04014030. [CrossRef]
26. Pang, Z.; O'Neill, Z.; Li, Y.; Niu, F. The role of sensitivity analysis in the building performance analysis: A critical review. *Energy Build.* **2020**, *209*, 109659. [CrossRef]
27. Su, L.; Wang, T.; Li, H.; Cao, Y.; Wang, L. Multi-criteria decision making for identification of unbalanced bidding. *J. Civ. Eng. Manag.* **2020**, *26*, 43–52. [CrossRef]
28. Salimi, S.; Hammad, A. Sensitivity analysis of probabilistic occupancy prediction model using big data. *Build. Environ.* **2020**, *172*, 106729. [CrossRef]
29. Saltelli, A.; Benini, L.; Funtowicz, S.; Giampietro, M.; Kaiser, M.; Reinert, E.; Van der Sluijs, J.P. The technique is never neutral. How methodological choices condition the generation of narratives for sustainability. *Environ. Sci. Policy* **2020**, *106*, 87–98. [CrossRef]

30. Ferretti, F.; Saltelli, A.; Tarantola, S. Trends in sensitivity analysis practice in the last decade. *Sci. Total Environ.* **2016**, *568*, 666–670. [CrossRef] [PubMed]
31. Joint Committee on Structural Safety (JCSS). Probabilistic Model Code. Available online: https://www.jcss-lc.org/ (accessed on 15 May 2020).
32. Sedlacek, G.; Kraus, O. Use of safety factors for the design of steel structures according to the Eurocodes. *Eng. Fail. Anal.* **2007**, *14*, 434–441. [CrossRef]
33. Kala, Z. Geometrically non-linear finite element reliability analysis of steel plane frames with initial imperfections. *J. Civ. Eng. Manag.* **2012**, *18*, 81–90. [CrossRef]
34. Kala, Z.; Valeš, J. Global sensitivity analysis of lateral-torsional buckling resistance based on finite element simulations. *Eng. Struct.* **2017**, *134*, 37–47. [CrossRef]
35. Szymczak, C. Sensitivity analysis of thin-walled members, problems and applications. *Thin Walled Struct.* **2003**, *41*, 271–290. [CrossRef]
36. Mang, H.A.; Höfinger, G.; Jia, X. On the interdependency of primary and initial secondary equilibrium paths in sensitivity analysis of elastic structures. *Comput. Methods Appl. Mech. Eng.* **2011**, *200*, 1558–1567. [CrossRef]
37. Hassan, M.S.; Salawdeh, S.; Goggins, J. Determination of geometrical imperfection models in finite element analysis of structural steel hollow sections under cyclic axial loading. *J. Constr. Steel Res.* **2018**, *141*, 189–203. [CrossRef]
38. Wu, J.; Zhu, J.; Dong, Y.; Zhang, Q. Nonlinear stability analysis of steel cooling towers considering imperfection sensitivity. *Thin Walled Struct.* **2020**, *146*, 106448. [CrossRef]
39. Rykov, V.; Zaripova, E.; Ivanova, N.; Shorgin, S. On sensitivity analysis of steady state probabilities of double redundant renewable system with Marshall-Olkin failure model. *Commun. Comput. Inf. Sci.* **2018**, *919*, 234–245.
40. Jönsson, J.; Müller, M.S.; Gamst, C.; Valeš, J.; Kala, Z. Investigation of European flexural and lateral torsional buckling interaction. *J. Constr. Steel Res.* **2019**, *156*, 105–121. [CrossRef]
41. Antucheviciene, J.; Kala, Z.; Marzouk, M.; Vaidogas, E.R. Solving civil engineering problems by means of fuzzy and stochastic MCDM methods: Current state and future research. *Math. Probl. Eng.* **2015**, *2015*, 362579. [CrossRef]
42. Zavadskas, E.K.; Antucheviciene, J.; Vilutiene, T.; Adeli, H. Sustainable decision-making in civil engineering, construction and building technology. *Sustainability* **2018**, *10*, 14. [CrossRef]
43. Lehký, D.; Pan, L.; Novák, D.; Cao, M.; Šomodíková, M.; Slowik, O. A comparison of sensitivity analyses for selected prestressed concrete structures. *Struct. Concr.* **2019**, *20*, 38–51. [CrossRef]
44. Novák, L.; Novák, D. On the possibility of utilizing Wiener-Hermite polynomial chaos expansion for global sensitivity analysis based on Cramér-von Mises distance. In Proceedings of the 2019 International Conference on Quality, Reliability, Risk, Maintenance, and Safety Engineering, QR2MSE 2019, Zhangjiajie, China, 6–9 August 2019; pp. 646–654.
45. Wen, Z.; Xia, Y.; Ji, Y.; Liu, Y.; Xiong, Z.; Lu, H. Study on risk control of water inrush in tunnel construction period considering uncertainty. *J. Civ. Eng. Manag.* **2019**, *25*, 757–772. [CrossRef]
46. Karasan, A.; Zavadskas, E.K.; Kahraman, C.; Keshavarz-Ghorabaee, M. Residential construction site selection through interval-valued hesitant fuzzy CODAS method. *Informatica* **2019**, *30*, 689–710. [CrossRef]
47. Zhao, Y.G.; Ono, T. A general procedure for first/second-order reliability method (FORM/SORM). *Struct. Saf.* **1999**, *21*, 95–112. [CrossRef]
48. Kala, Z. Reliability analysis of the lateral torsional buckling resistance and the ultimate limit state of steel beams with random imperfections. *J. Civ. Eng. Manag.* **2015**, *21*, 902–911. [CrossRef]
49. Sedlacek, G.; Müller, C. The European standard family and its basis. *J. Constr. Steel Res.* **2006**, *62*, 522–548. [CrossRef]
50. Gamboa, F.; Klein, T.; Lagnoux, A. Sensitivity analysis based on Cramér-von Mises distance. *SIAM ASA J. Uncertain. Quantif.* **2018**, *6*, 522–548. [CrossRef]
51. Plischke, E.; Borgonovo, E. Copula theory and probabilistic sensitivity analysis: Is there a connection? *Eur. J. Oper. Res.* **2019**, *277*, 1046–1059. [CrossRef]
52. McKey, M.D.; Beckman, R.J.; Conover, W.J. Comparison of the three methods for selecting values of input variables in the analysis of output from a computer code. *Technometrics* **1979**, *21*, 239–245.

53. Iman, R.C.; Conover, W.J. Small sample sensitivity analysis techniques for computer models with an application to risk assessment. *Commun. Stat. Theory Methods* **1980**, *9*, 1749–1842. [CrossRef]
54. Kala, Z. Global sensitivity analysis of reliability of structural bridge system. *Eng. Struct.* **2019**, *194*, 36–45. [CrossRef]
55. Kala, Z. Estimating probability of fatigue failure of steel structures. *Acta Comment. Univ. Tartu. Math.* **2019**, *23*, 245–254. [CrossRef]
56. Kala, Z. Quantile-oriented global sensitivity analysis of design resistance. *J. Civ. Eng. Manag.* **2019**, *25*, 297–305. [CrossRef]
57. Xiao, S.; Lu, Z. Structural reliability sensitivity analysis based on classification of model output. *Aerosp. Sci. Technol.* **2017**, *71*, 52–61. [CrossRef]
58. Ling, C.; Lu, Z.; Cheng, K.; Sun, B. An efficient method for estimating global reliability sensitivity indices. *Probabilistic Eng. Mech.* **2019**, *56*, 35–49. [CrossRef]
59. Madsen, H.O. Omission sensitivity factor. *Struct. Saf.* **1988**, *5*, 35–45. [CrossRef]
60. Leander, J.; Al-Emrani, M. Reliability-based fatigue assessment of steel bridges using LEFM—A sensitivity analysis. *Int. J. Fatigue* **2016**, *93*, 82–91. [CrossRef]
61. Sudret, B. Global sensitivity analysis using polynomial chaos expansions. *Reliab. Eng. Syst. Saf.* **2008**, *93*, 964–979. [CrossRef]
62. Wei, P.; Lu, Z.; Hao, W.; Feng, J.; Wang, B. Efficient sampling methods for global reliability sensitivity analysis. *Comput. Phys. Commun.* **2012**, *183*, 1728–1743. [CrossRef]
63. Kala, Z. Sensitivity assessment of steel members under compression. *Eng. Struct.* **2009**, *31*, 1344–1348. [CrossRef]
64. Mercier, C.; Khelil, A.; Khamisi, A.; Al Mahmoud, F.; Boissiere, R.; Pamies, A. Analysis of the global and local imperfection of structural members and frames. *J. Civ. Eng. Manag.* **2019**, *25*, 805–818. [CrossRef]
65. Melcher, J.; Kala, Z.; Holický, M.; Fajkus, M.; Rozlívka, L. Design characteristics of structural steels based on statistical analysis of metallurgical products. *J. Constr. Steel Res.* **2004**, *60*, 795–808. [CrossRef]
66. Kala, Z.; Melcher, J.; Puklický, L. Material and geometrical characteristics of structural steels based on statistical analysis of metallurgical products. *J. Civ. Eng. Manag.* **2009**, *15*, 299–307. [CrossRef]
67. Kala, Z.; Valeš, J.; Jönsson, L. Random fields of initial out of straightness leading to column buckling. *J. Civ. Eng. Manag.* **2017**, *23*, 902–913. [CrossRef]
68. Kala, Z. Sensitivity analysis of steel plane frames with initial imperfections. *Eng. Struct.* **2011**, *33*, 2342–2349. [CrossRef]

sustainability

MDPI

Article

Sustainable Risk Assessment through the Analysis of Financial Losses from Third-Party Damage in Bridge Construction

Sungjin Ahn, Taehui Kim and Ji-Myong Kim *

Department of Architectural Engineering, Mokpo National University, Mokpo 58554, Korea; sunahn@mokpo.ac.kr (S.A.); thkim@mokpo.ac.kr (T.K.)
* Correspondence: jimy@mokpo.ac.kr; Tel.: +82-61-450-2457

Received: 13 March 2020; Accepted: 17 April 2020; Published: 23 April 2020

Abstract: Due to the recent introduction of innovative construction methods and technologies, construction projects increasingly require sustainability in their high degrees of specialization and complex work processes. This is due to a wide variety of new risk factors associated with construction projects that can lead to extensive and severe damage. When an accident occurs during a construction project, it can cause material, property, or bodily damage not only within the actual construction site but also outside, affecting third parties. This study analyzed the record of such third-party damage and the subsequent financial losses in bridge construction management, to identify the objective and quantified relationship of risk indicators related to the damage and losses. In order to assess the actual losses in construction projects, we adopted the loss claim payout data as recorded and provided by a major Korean insurance company, and conducted a multiple regression analysis to identify the loss indicators and to develop a loss estimation model. In this study, the analysis of the data indicated that the superstructure type, the foundation type, floods, and company ranking by the amount of the contract were the four statistically significant risk indicators that affected financial losses from third-party damage, among the nine variables used as independent variables, which included the superstructure type, foundation type, superstructure construction method, maximum span length, floods, typhoons, total construction cost, total construction period, and company ranking. As this study focused on identifying the risk factors and producing a loss assessment model quantified in numerical values, the results provide important references for assessing and minimizing the risks to third parties and the consequential financial losses in bridge construction, while promoting sustainability objectives.

Keywords: bridge construction; risk analysis; loss assessment model; third-party damage; insurance

1. Introduction

1.1. Background and Purpose of the Study

Bridges are important infrastructure and industrial facilities for urban growth and economic development, as they allow a large volume of logistics and transportation by connecting rivers, canyons, islands, and lands. Due to the recent introduction of innovative construction methods and technologies, construction projects increasingly require sustainability in their high degrees of specialization and complex work processes. This is due to a wide variety of new risk factors associated with construction projects, which can lead to extensive and severe damage. The high level of expertise and technology also highlights the need for a sustainable and systematic management of finances, as well-organized fiscal management and fund execution, which can contribute to the sustainable practices of safety and risk management.

It is also important to reinforce the reliability of information regarding risks in construction projects and to use this information as a means for rational decision-making and effective strategy-making for sustainable risk management. However, there are limits to relying solely on government-led risk and safety management systems, which tend to be too broad and general, and thus the role of the private sector is becoming more emphasized [1]. More specifically, the current trend of risk management in construction projects should be more numerical, quantified, and thus more objective, as opposed to the empirical custom of the past, which relied on the individual experiences of experts and judgments from previous cases [2,3].

Risk management is defined in the *Principles for Risk Management* of the International Organization for Standardization (ISO) as: (1) a part of decision-making; (2) should clearly address uncertainty; (3) should be based on the best information available; (4) should be adapted to a specific purpose or method; and (5) should be comprehensive and obvious [4]. Risk management in construction is commonly focused on reducing risk factors, as well as transferring risk through purchasing construction insurances. For this reason, a risk management approach that recognizes internal and external risk factors in advance and analyzes the extent of possible losses, in order to share the risks according to the causes, is strongly required [5]. In other words, construction management and loss estimation models require a more sophisticated and scientific methodology. Furthermore, it is also crucial that such models consider the comprehensive aspects of losses, by including not only material damage within the site but also third-party damage, which is directly and indirectly due to the influence of construction activities or accidents.

In this study, a "third party" is defined as an outside entity not aligned with any of the stakeholders (including the workers, subcontractors, or general contractors) of a construction project. Therefore, "third-party damage" refers to the damage caused by construction activities or accidents to the third party's bodies and/or properties, including damage to, physical injury to, loss of, or destruction of tangible property [6]. On the insurance record data, which this study collected and analyzed, the "loss claim payout" refers to the financial amount spent to indemnify the third-party damage.

Previous studies on safety accidents and risk management in construction have been mostly limited to the workers and structures within the construction site. Analyses and studies on third-party damage beyond the construction sites are much rarer [7–10]. Examples of third parties can include, among others, pedestrians and agricultural and commercial workers around the site. The occurrence of third-party damage can give rise to many problems, including cost compensation, suspension of construction work, and administrative punishment, resulting in reduced work productivity, economic losses, and harm to the reputation of the company in question. This, in turn, can lead to unexpected secondary losses despite many efforts to improve the productivity and profitability, such as cost-cutting, construction period reducing, and so on. Therefore, at a time when many construction companies seek to develop and strengthen more advanced risk assessment and management methods to minimize potential losses, a more comprehensive and sustainable management system needs to be established by taking into account the risk factors beyond the construction site, rather than relying on the existing risk management, focused on accidents taking place within the boundaries of construction sites [11].

The purpose of this study is to provide a loss assessment model for third-party damage that can contribute to minimizing risks in a more systematic and evidence-based way. In other words, this study aims to identify the statistically significant risk factors in bridge construction projects from onset to completion and to present a risk prediction model, while reflecting the actual record of damage that occurred in bridge construction projects. Both aims are ultimately directed at the achievement of sustainable risk management.

1.2. Method and Scope of Research

This study analyzed the record of damage incurred to the third parties and subsequent financial losses to indemnify the damage in actual bridge construction projects, in order to identify the correlation between the risk factors of the damage and financial losses. To ascertain the actual losses in construction projects, we adopted the loss claim payout data recorded and provided by a major Korean insurance company as the data in this study, for the purpose of achieving the quantification of third-party damage in numerical values. More specifically, the loss claim payout record was used due to its clarity and objectivity [12], which can represent the cost of financial losses in construction work well. This quantified representation of insurance records can be especially useful, because of the specific and detailed information about each case of damage, which allows engineers, insurance underwriters, etc., to accurately and logically review, examine, and determine the loss in construction management.

We conducted risk analysis in order to find valid factors based on the accumulated past data and statistics. In the insurance industry, generalized linear models (GLMs) are commonly used to support critical decisions [13]; more specifically, the choice of variables in the model first considers all variables used in the calculation of insurance premium and filters out statistically meaningless variables through a statistical analysis process. Adopting this method, in this study, variables were selected based on the insurance claim payout record and the correlations with quantified losses were analyzed.

Based on the data, the financial loss incurred to compensate the third-party damage was then referred to as the "loss from third-party damage" in this study, and the term and concept of the "loss ratio" was established as the dependent variable. To identify the risk factors and the relationships between the loss and the factors, first, the dependent variable was defined using the term, "loss ratio." The loss ratio is defined as the amount of financial loss incurred to indemnify third-party damage, divided by the total cost of the construction project. We used the term and concept of loss ratio as it seemed reasonable to consider the fact that, even when the loss amount was relatively little in a project, if the size of the project was rather small, the extent of the loss could be more detrimental and severe, and the contrary fact that even when the loss amount was relatively large, if the size of the project was rather big, the extent of the loss could be considered somewhat benign and insignificant.

Second, the key risk factors in bridge constructions were selected based on previous studies and past insurance compensation records. The risk factors regarding the characteristics of bridge constructions were based on the project base information, and the occurrence of natural disasters were entered as the independent variables. Therefore, a total of eight variables (superstructure types, maximum span length, superstructure construction method, flood, typhoon, total construction duration, and company rank) were selected for the analysis. Third, a multiple regression analysis using the stepwise variable selection method was conducted to identify and verify the statistically significant risk factors as well as the correlation between the variables and to develop a loss assessment model. Figure 1.

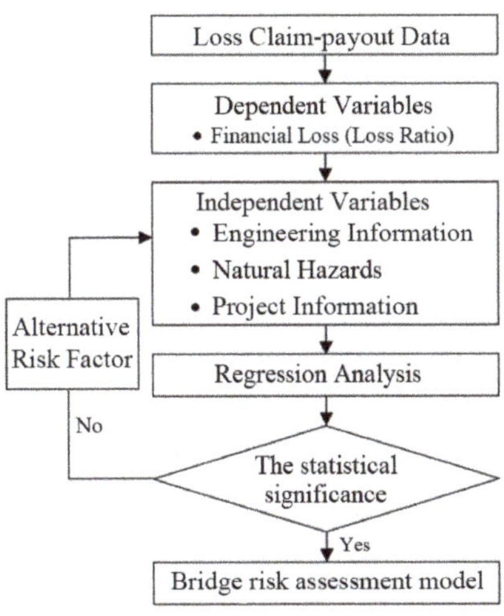

Figure 1. Our study procedure.

2. Review of Literature

2.1. Construction Insurance

Construction insurance covers the damage caused by unexpected accidents and consequential losses occurred during construction and civil engineering works, such as temporary construction work, main construction, and damage to the construction materials. The compensation for such damage can be divided into two types: first, the compensation for construction materials within the construction sites, and second, the compensation for third-party physical and property losses incurred beyond the construction site due to accidents involving the construction work [14].

The regulations on the compensation not only for construction materials but also for third-party damage have recently become more stringent. For example, it has become obligatory for construction projects ordered by the Korean central or local governments to purchase construction insurance, carried out mainly through a turn-key base contract or alternative bidding, mostly by large construction companies. The projects requiring pre-qualification are bridges over 50 meters in span. These regulations also require that the insurance purchased in such cases must guarantee the collateral damage for materials and third-party liability (Article 55 of the government's bidding and contract enforcement standards). The Korean Public Procurement Service is also seeking to reduce the loss burden on construction companies by applying construction insurance, which includes third-party liability, to the public building corporation in downtown areas starting from August 2019.

This study used the loss claim payout data from Contractor's All Risk (CAR) insurance. As mentioned earlier, construction insurance is designed to cover the entire range of unexpected losses at all stages of construction projects. If an accident occurs during a construction project, this could cause damage to the existing property of the contractors and to those involved in the construction around the construction site, and also to a third-party that is not necessarily directly related to the construction activities. In addition to property and material damage, damage such as death, physical injuries and/or any bodily harm can occur to third parties. For example, the *Korea National Environment Dispute Resolution Commission* recently ordered hundreds of millions of KRW (South Korean Won) in compensation for damage, such as damage to crops due to sunlight interruption from highway bridges and damage from mass death in fish farms due to the noise and water pollution from bridge construction. Such construction events relating to third-party damage, in turn, can cause losses to the owners, including the suspension of work or delay of completion.

2.2. Loss from Third-Party Damage

Many leading companies involved in construction projects make a great deal of effort to estimate the possible losses during construction work. In particular, insurance and re-insurance companies have developed their own loss assessment models to predict and prepare for the potential losses [15]. These loss estimation models in fact help the insurers and their customers to understand and estimate the potential risks in construction projects. However, there is a disadvantage that these models can only be accessed and used by certain insurers or limited customers, which makes it difficult for the others who are also involved in construction work to utilize the models. Therefore, it is quite difficult for general operators or public institutions to gain access to such models and to estimate the losses incurred during construction work.

Many vendors, including EQECAT, Risk Management Solution, and Applied Insurance Research, have designed and provided risk estimation models to evaluate the construction risks [16,17] to be generally used, but applying these models over a wide range of regions and countries has some limitations. This is because differences in the size and frequency of local vulnerabilities in construction projects and natural disasters can increase the uncertainty in forecasting losses, and models that do not adequately reflect the characteristics of a region or country may result in errors. This, in turn, can increase the differences or errors, especially in identifying third-party losses, as third-party losses tend to be heavily affected by the surrounding environmental conditions. More specifically, because third-party losses incur to people and places beyond the actual sites of construction, particular regional differences easily arise due to a variety of surrounding environmental factors and the construction related regulations and laws.

Therefore, for users, not only are the loss assessment models developed by insurance and re-insurance companies difficult to understand due to their black box type algorithms but also these models cannot easily be adjusted to reflect the important characteristics particular to specific regions and countries. For this reason, there is a need for studies to analyze the risk factors for third-party damage and also the losses from third-party damage, and to develop a loss assessment model that evaluates the losses from third-party damage, which can be applied to a wide range of regions with their particular details regarding the environment and laws.

2.3. Risks in Bridge Construction Projects

Existing studies on the risks in bridge construction have also been conducted with a wide range of approaches, including technical, social, environmental, and economic points of views, as risk is a combination of factors, such as disaster, vulnerability, and exposure, and is not solely or independently determined by a single factor [18]. Hastak and Baim proposed risk factors that influenced the cost-effectiveness of management and the operation and maintenance of urban infrastructures, such as bridges, highways and subway stations. They determined that the risk factors specific to bridges include the training of inspection personnel and deicing salts [19]. Wang and Elhag performed a comparative study that analyzed bridge risk modeling, focusing on safety, functionality, substantiality, and the environment by comparing multiple regression analysis and neural network analysis methods [20].

Cho and Kim proposed a probabilistic risk assessment in a virtual construction to evaluate the risks of erection control and the main cable wires' fracture during the construction phases [21]. Hashemi et al. identified the key bridge construction risk factors, i.e., delayed payment on the contract and extras, a shortage of labor, materials, and equipment, construction permitting issues, and poor relationship among parties [22]. Li et al. pointed out the risk factors for bridge construction using factor analysis. They categorized seven risk factors: economic, contract and law, building technology, design, environment, staff, and materials and equipment [23]. Choudhry et al. identified the critical risk factors related to bridge construction projects, including the financial risks, external risks, design risks, management risks, construction risks, contractual risks, and health and safety risks. They determined that financial risks were the key factor that affected the costs and schedule of bridge construction projects [24].

Various studies have been conducted on risk assessment methods and losses in bridge construction, and these studies contributed to recognizing certain critical risk factors. In existing studies, however, two major gaps can be pointed out. First, studies that identify significant risk factors through objective and statistical analysis using quantified data are still insufficient. Second, most of the previous studies on bridge construction risks did not distinguish the material damage that occurred within the construction sites from third-party damage, as it is difficult to exclusively distinguish and evaluate the losses from third-party damage. Thus, for a scientific and objective analysis of losses caused by third-party damage, and for the development of the loss assessment models, the use of data with quantified risk, as well as statistical analysis and verification, is crucial and necessary.

3. Data Collection and Analysis

3.1. Data Collection

For quantitative risk analysis and evaluation, in this study, we collected the data of 296 loss cases of third-party damage claim payouts from construction insurance coverages of actual bridge construction projects between 1999 and 2016. The collected data included various information, such as the date and place of accidents, structure component types, construction period, loss details, the amount of insurance coverage, and the loss amount in the bridge construction project. The detailed information in the data was then classified into three groups, for the sake of convenience: (1) accident information (accident date, site address, and accident details); (2) characteristic information of the bridge construction projects (superstructure types, maximum span length, and superstructure construction method); and (3) the project scale, represented by the total construction duration and company rank. Not all information was used to select the independent variables. However, the above information was used to form sets of independent variables, which will be introduced in the next section. For example, the accident information itself was not selected as an independent variable, but based on such information, such as the accident date, site address, and accident details, the indicators of natural hazards (floods and typhoons) were determined and used as independent variables.

3.2. *Comparison of Material Loss and the 3rd Party Loss*

Table 1 shows a descriptive statistical comparison between the losses from material damage and the losses from third-party damage from the collected data. The frequency of third-party damage was not very high, as it was estimated to be about half of the material damage. However, the average amount of the financial loss incurred to indemnify the third-party damage was approximately 20% higher than the loss incurred to indemnify the material damage.

Furthermore, a statistical comparison for a clearer verification was performed and is represented in Table 2. The table illustrates the results of analyzing the differences between the two groups. As can be seen, there is no significant difference in the average between the losses from the materials and from the third-party, which evidently indicates that the loss from third-party damage can and should be an important part of the consideration for construction loss analysis, and requires an equivalent and immediate level of management with the loss from the material damage.

Table 1. Comparison with materials and third-party losses.

Category	Frequency (%)	Avg. Loss (Mil. KRW)	Max. (Mil. KRW)	Min (Mil. KRW)	Std. Deviation
Material	66.8	84.98	1915	1.15	328.22
Third-party	33.2	95.34	841	1.03	176.41

Table 2. ANOVA test for materials and third-party losses.

Category	Sum of Squares	Mean Square	F-Value	Sig.
Between Groups	5.079e+15	5.079e+15	0.147	0.833
Within Groups	5.054e+18	4.458e+16		
Total	5.059e+18			

3.3. *Multiple Regression*

3.3.1. Dependent Variable

In this study, multiple regression analysis was used to define the relationship between the loss ratio and loss indicators to develop a loss estimation model. More specifically, through multiple regression analysis, we identified the loss indicators in quantitative, numerical values through analyzing the third-party losses, and determined the significant loss indicators among them. Based on the data, the term and concept of the "loss ratio" were established. This is the amount of the loss incurred to indemnify the third-party damage, divided by the size of the construction project. The size of the construction project in this study was represented by the total sum insured (TSI), which reflects the total bridge construction cost. This can be expressed in an equation as follows in Equation (1):

$$\mathrm{LR} = \frac{CP}{TSI}. \quad (1)$$

where:

LR: Loss Ratio
CP: Claim-Payout
TSI: Total Sum Insured

In each case, the loss from third-party damage was relatively small, compared to the TSI, and most loss ratios (LR) were inclined toward zero when expressed by Equation (1). For this reason, the dependent variables used in the regression analysis were converted by natural logarithms in order to meet the normal distribution. The value of the dependent variable used in the regression analysis is shown in Equation (2):

$$\text{Transformed Loss Ratio} = \mathrm{Ln(LR)}. \quad (2)$$

As mentioned, the normality test for the dependent variable was performed for the regression analysis. The LR tended to be excessively left-leaning, which required conversion to a normal distribution. As shown in Equation (2), the dependent variable was log transformed, and then the normality was tested by histogram, Q-Q plot, and Shapiro–Wilk tests (see Table 3 and Figure 2). As the p-value of the Shapiro–Wilk test was greater than 0.05, the dependent variable data can be interpreted as being normally distributed.

Table 3. Normality test of the dependent value.

	Shapiro–Wilk Test						
	Statistic	**df**	**sig.**		**Statistic**	**df**	**sig.**
LR	0.386	296	0.000	Ln (LR)	0.965	296	0.084

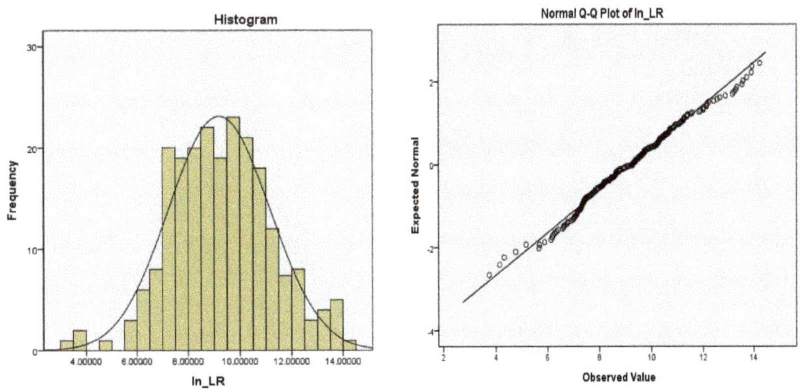

(**a**) Histogram of Ln(LR) (**b**) Q-Q plot of Ln(LR)

Figure 2. Normality test by histogram and Q-Q plot.

3.3.2. Independent Variables

This section introduces the independent variables in detail, focusing on how they were selected. This study included eight independent variables, which were categorized into three groups. The first is technical components, including the types of superstructures, types of foundations, construction methods, and lengths of bridges. The second group concerns natural disasters, such as floods and typhoons. The last group includes information regarding construction projects, e.g., the total duration of construction work and the sizes of the construction companies (company ranking by amount of contract). Numerical construction information indicates the complexity, scale, and task level of the construction projects and can reflect the risk of the work [25].

Technical Components

The types of superstructures of bridges refer to the structure above the bridge's abutment and bridge deck, and are determined by the composite assessment of constructability, economic feasibility, and safety along with topographical and environmental conditions. The Korean Public Procurement Service assesses the previous performance record by classifying them into A to D grades according to the difficulty of the construction in the contractor pre-qualification criteria for bridge construction. Researchers have suggested that optimizing the selection of the superstructure types can be done on the basis of such factors as the number and length of bridge spans and the approximate construction cost [26,27]. Differentiation according to the characteristics of the superstructure types can be considered as a risk factor in bridge construction. This study classified the types of superstructures into the order of the ordinal scale of arch bridges, pre-cast concrete (PSC) beam bridges and cable-stayed bridges, by the distribution of the average loss claim payout.

In addition, for long spanned bridges, the difficulty of construction can increase, and the risks can also increase due to the influence of the increased period and the cost, as well as severe wind speed risks [28–30]. In this study, ordinal scales were classified on the basis of 50, 100, and 500 m.

The foundation of the bridge is an important structural factor in the construction of a bridge and bears important risk factors and uncertainties, which necessitate specific management of the hazards [31]. In addition to this, increased flow rates and water speeds resulting from floods and typhoons can bring about scour, which can cause unforeseen damage and the collapse of bridges [32]. This, in turn, can trigger material and third-party damage. As such, the occurrence of scour can be a fatal hazard to the life and stability of a bridge. The data collected regarding the foundation of the bridge can be used as a major risk factor, and based on the distribution of the average loss claim payouts, the classification criteria of an ordinal scale, including pre-cast concrete pile, cast in place, and open-caisson type, were established.

The superstructure construction method is determined based on economic feasibility, construction speed, and overhead clearance (valid height from the bottom of the bridge body to the surface of the water or road). Previous studies analyzed the structural safety and optimal design methods, according to bridge construction methods [33–36]. These studies found that differences existed between the influence of loads according to the construction methods and the economic design methods. For this, risk analyses based on the characteristics of the different construction methods and their classifications are required. Kim and Cho developed a rough estimation model for the construction costs according to the classification of the typical construction methods of bridge superstructures and found that the classification of the superstructure construction methods was necessary [37]. In this study, the criteria for classification of ILM (Incremental Launching Method), FCM (Free Cantilever Method), and MSS (Movable Scaffolding System) were established on the basis of the average loss claim payout through classification of the construction method.

Natural Disasters

Losses from natural disasters were calculated through indicators, e.g., typhoons and floods, using the reinsurer's Natural Disaster Assessment Network (NATHAN). The map of world natural disasters uses the existing natural disaster occurrence data to index the risks of disasters in specific areas, e.g., floods, typhoons, and earthquakes. We used the risks, such as typhoon and flood, to represent the indicators of natural disasters. Natural disaster hazards were collected using the location information (address) from each construction project site. Meteorological accidents have been considered as a key factor that affects construction risks [38,39]. Construction sites located in major hurricane-prone areas face frequent construction delays due to hurricane-induced gusts and heavy rains, and there is a direct link to construction risks according to the risk levels of natural disasters [40–42].

Information Regarding Construction Project

In previous studies, the total duration of the project was used as a measure of the construction project risk and could be a key indicator in estimating the loss [43–45]. We found that, in these previous studies, in general building construction, the loss ratio became lower as the total period of the project became longer.

In addition, studies have suggested that the differences in the sizes of construction companies can entail notable differences in the awareness and support for safety accident prevention and risk management, which indicates that company sizes could be used as a major measure of risk analysis [46,47]. In this study, we used the company ranking by the number of contracts to indicate the size of the construction companies.

Table 4 shows the categories of the bridge project risk indicators.

Table 4. The categories of bridge project risk indicators.

	Engineering				Natural Hazard		Project	
Factor	Super-structure	Span Length	Construction Method	Foundation	Flood	Typhoon	Total Duration	Company Rank
Unit	Ordinal scale	meters	Ordinal scale	Ordinal scale	Ordinal scale (Zone)	Ordinal scale (Zone)	Month	Number
Description	1: Arch 2: PSC beam 3: Cable-stayed	Max. span length	1: ILM 2: FCM 3: MSS	1: Precast conc' pile 2: Cast in place 3: Open caisson	Occurrence per year (times) Zone 1: 1 Zone 2: 2 Zone 3: 3 Zone 4: 4 Zone 5: 5	Max. speed (Km/h) 1: 76–141 2: 142–184 3: 185–212 4: 213–251 5: 252–290 6: 300–	Total duration of the project	Company rank by contract amount

4. Results

Tables 5 and 6 show the descriptive statistics of the variables and the results of the regression analysis, respectively. Eight variables were used as independent variables, including the superstructure type, foundation type, superstructure construction method, maximum span length, floods, typhoons, total construction duration, and company rank by amount of contract. The analysis found that four independent variables were significant indicators that affected the loss ratio from third-party damage in the bridge construction. These are the superstructure type, foundation type, flood, and company rank. The p-value of the F test was less than the significance level (0.05).

Table 5. Descriptive statistics of the variables.

Variables	Min.	Max	Mean	Std. Deviation
Dependent variable				
Ln (Loss Ratio)	1.14	18.41	9.27	7.14
Independent variables				
Superstructure	1.00	2.00	1.24	0.65
Foundation	1.00	3.00	2.64	0.71
Flood	1.00	5.00	3.52	1.83
Company Rank	1.00	84.00	42.59	14.42

Table 6. Regression analyses results.

Variables	Coef.	Beta Coef.	p > |z|	VIF
Engineering factor				
Superstructure	4.243	0.708	0.008	1.032
Foundation	2.275	0.454	0.024	1.051
Natural Hazard factor				
Flood	1.075	0.353	0.021	1.072
Project factor				
Company rank	−0.092	−0.582	0.018	1.043
Number of Observations = 296 F value = 16.341 Adj-R^2 = 0.361				

In addition, the independent variables, such as the superstructure construction method, maximum span length, typhoons, total construction cost, and total construction period were found to be non-significant to the dependent variable and the loss ratio, as their levels of significance were greater than 0.05. In the regression analysis of the loss ratio of the bridge construction, as illustrated in Table 5, the revised R^2 value was 0.361, which indicated that 36.1% of the loss ratio could be explained by the regression model. In addition, the variance inflation factor (VIF) value range shown is from 1.032 to 1.072, reflecting no multicollinearity among the variables.

The standardized coefficient (beta value) is the value of the relationship between the independent and the dependent variables that compares the influence of the independent variable on the dependent variable. In other words, the standardized regression coefficient indicates the importance of the regression coefficient, and the higher the value of the beta coefficient of the variable, the greater the effect on the dependent variable. When prioritized by the absolute value of the beta coefficient, in order to understand each indicator's impact on the loss ratio, the variables were organized as follows: (1) the superstructure type (beta coefficient = 0.708); (2) the company ranking (beta coefficient = −0.582); (3) the foundation type (beta coefficient = 0.454); and (4) floods (beta coefficient = 0.353).

According to the analysis values, when the value of the superstructure type, one of the independent variables, was changed to the ordinal scale of 1 unit, the amount of change in the loss ratio increased by 0.708. In other words, it is possible to predict that the loss from third-party damage will increase when the superstructure type of the bridge is constructed in a cable-stayed bridge rather than a PSC bridge. In terms of the impact of the construction company ranking, the results show that companies with a greater number of contracts tended to have less losses from third-party damage. Regarding the foundation types, the selection of precast concrete pile, cast in place, and open-caisson type had an influence of 0.454 on the loss ratio. When the risk level of flood increased by one level, the influence of 0.353 arose on the loss ratio. This indicates that the higher the risk rating of flood, the higher the loss ratio due to loss from third-party damage. Figure 3 shows the scatter plot comparing the results of the log transformed loss ratio and predicted log transformed loss ratio. The plot proves that both values are well matched and verified that the values were consistent.

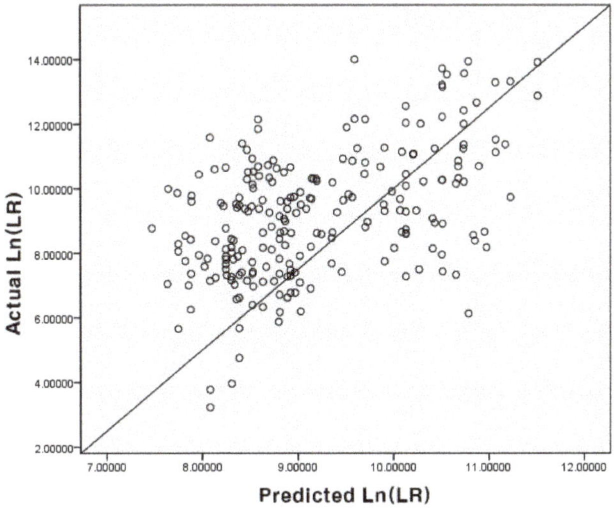

Figure 3. The scatter plot of actual Ln(LR) and predicted Ln(LR).

5. Conclusions and Discussion

Bridge construction projects increasingly require more thorough and sustainable risk management due to the scattering risks in the projects, such as heavy-duty construction equipment and high-altitude work under unstable environmental conditions. Consequently, in the fields of construction management, the processes and practices of risk assessment that align with sustainability objectives are all the more necessary. For this reason, studies that identify more objective and quantified risk factors at construction sites and develop a loss assessment models are called for.

In response to these needs, this study analyzed the financial losses that incurred to indemnify third-party damage in actual bridge construction projects and developed a loss assessment model through a multi-linear regression analysis. This was based on the data of loss claim payouts for third-party damage in bridge construction, as recorded by a Korean insurance company. The analysis found that the four independent variables, i.e., superstructure types, foundation types, floods, and construction company rank, were the significant risk factors that demonstrated actual relationships with the financial losses incurred to indemnify the third-party damage.

Some of the findings of this study demonstrated consistency with those of the previous studies. As regards to structure type, Gurcanli et al. [45] and Kim et al. [48] found that the type of structure is a key factor to be considered in construction risk analysis. In this study, we specified the types as the superstructure type, foundation type, span length, and construction method in our analysis. We found that superstructure type was a significant factor in bridge construction risk analysis.

Previous studies have shown that natural hazards played a main role in defining the risks in construction projects [38,39]. Among the natural hazards, this study found that the occurrence of flooding was proportional to the increase in third-party losses in bridge construction, which is consistent with the findings of previous studies that revealed that the elements of flooding are also key risk factors in construction projects [44]. In addition, as Kim et al. [49] contended, the results of this study also indicate that the size of a construction project affects the potential loss. This study classified the size of a construction project into two different elements, the construction duration and rank of the construction company and verified that the rank of the company is the more significant factor of the two.

This study sought to statistically analyze and evaluate the practical effectiveness of secured information in risk assessment in a situation where it is necessary to predict the future risks of bridge construction with minimum information. Therefore, the units of ordinal scale applied to the risk analysis of superstructures and foundation types, as well as construction method, do not indicate or evaluate the difficulty of the construction or the level of safety. Rather, this study identifies the statistically significant kinds and extent of the risk factors in bridge construction as reflected in the collected data. In this study, the derived risk factors had a correlation with the incurred financial losses but did not objectively explain a causal relation between them.

For example, a correlation was identified where the higher the construction company rank, the less loss from third-party damage, while the causal relationship between the two, such as whether the reason for the decrease in the loss ratio was due to decent risk management or disaster prevention measures within the company, is still unexplainable. Therefore, further studies are required in order to obtain more effective risk assessment measures through additional investigation and analysis to identify the causal relationship in the future. In addition, this study used loss claim payout data from only one Korean insurance company. Further research is needed to support this study and its reliability based on data from many other insurers and their loss data to reflect a wide range of characteristics in construction projects.

As this study verified the quantified risk factors that reflected the actual loss claim payout record and presented the framework used for deriving the factors and developing the loss assessment model, the results can be used as important references for the government, construction companies, insurance and re-insurance companies, and others related to bridge construction projects, all of whom aim to manage and minimize the risks and consequential financial losses in bridge construction management. In particular, insurance and reinsurance companies can refer to this study to reconstruct their own loss assessment model of measuring the potential risks for particular bridge construction projects and to adapt the results from the model in estimating the rates of premiums.

Construction companies can benefit from this study, as we provided a substantial risk assessment for third-party loss in bridge construction based on the predicted total value of property, structure type, and the foundation type of the bridge. By doing so, such information can be reflected in predicting the life cycles and consequential management costs for third parties as well as for the bridge. This is because having risk factors for third party damage in bridge construction projects are quantified and proportioned helps in managing, responding to, and ultimately reducing potential loss.

Thus, the framework and the findings of this study will be able to contribute to improving the bridge construction companies' judgment in risk management. In addition, for the central and local governments and government-invested institutions, the utilization of the findings of this study is also expected to contribute to the establishment of effective measures and regulations of risk management and prevention that reflect the characteristics of actual damage and losses in bridge construction projects. This study is also expected to be applicable to other regions and countries where the construction environment and the natural disasters are similar to Korea. Taking particular consideration of the issues surrounding bridge superstructure types, foundations, floods, and company ranks, which were the four significant risk factors found in this study, will allow various resources to be efficiently distributed and allocated, ultimately contributing to robust and sustainable risk management in bridge construction.

Overall, this study is novel as it provided quantified and therefore measurable risk factors in bridge construction. In this study, we derived the risk factors based on the data of the actual accidents that occurred in bridge construction projects that had been carried out and developed a loss prediction model. These two outcomes of this study are believed to be useful for risk management in the construction industry, especially in the sense that this study was based on the numerical and tangible records of construction accidents, which were then perceptibly represented in financial values.

Author Contributions: Conceptualization, S.A.; Data curation, S.A., J.-M.K.; Funding acquisition, T.K.; Investigation, S.A.; Methodology, S.A.; Project administration, T.K.; Software, J.-M.K.; Supervision, T.K.; Validation, S.A.; Writing—original draft, J.-M.K.; Writing—review & editing, S.A. All authors have read and agreed to the published version of the manuscript.

Funding: This research was supported by a grant (NRF-2019R1A2C1009398) from the National Research Foundation of Korea by Ministry of Science, ICT and Future Planning.

Conflicts of Interest: The authors declare no conflict of interest.

References

1. Paudel, Y. A Comparative study of public–private catastrophe insurance systems: Lessons from current practices. *Geneva Pap. Risk Ins.* **2012**, *37*, 257–285. [CrossRef]
2. Wood, G.; Ellis, R.C.T. Risk management practices of leading UK cost consultants. *Eng. Constr. Archit. Manag.* **2003**, *10*, 254–262. [CrossRef]
3. Dikmen, I.; Birgonul, M.T.; Arikan, A.E. A Critical Review of Risk Management Support Tools. In Proceedings of the 20th Annual Conference of Association of Researchers in Construction Management, Edinburgh, UK, 1–3 September 2004; pp. 1145–1154.
4. ISO. *ISO31000, "31000: 2009 Risk Management–Principles and Guidelines"*; International Organization for Standardization: Geneva, Switzerland, 2009.
5. Adeleke, A.Q.; Bahaudin, A.Y.; Kamaruddeen, A.M.; Bamgbade, J.A.; Salimon, M.G.; Khan, M.W.A.; Sorooshian, S. The Influence of Organizational External Factors on Construction Risk Management among Nigerian Construction Companies. *Saf. Health Work.* **2018**, *9*, 115–124. [CrossRef]
6. Dewar, J. *International Project Finance: Law and Practice*; Oxford University Press: Oxford, UK, 2011.
7. Lee, H.S.; Kim, H.; Park, M.; Ai Lin Teo, E.; Lee, K.P. Construction risk assessment using site influence factors. *J. Comput. Civil Eng.* **2012**, *26*, 319–330. [CrossRef]
8. Fung, I.W.; Tam, V.W.; Lo, T.Y.; Lu, L.L. Developing a risk assessment model for construction safety. *Int. J. Proj. Manag.* **2010**, *28*, 593–600. [CrossRef]
9. Xiang, P.; Jia, F.; Li, X. Critical Behavioral Risk Factors among Principal Participants in the Chinese Construction Industry. *Sustainability* **2018**, *10*, 3158. [CrossRef]
10. Taroun, A. Towards a better modelling and assessment of construction risk: Insights from a literature review. *Int. J. Proj. Manag.* **2014**, *32*, 101–115. [CrossRef]
11. Brockett, P.L.; Golden, L.L.; Betak, J. Different Market Methods for Transferring Financial Risks in Construction. In *Risk Management in Construction Projects*; IntechOpen: London, UK, 2019.
12. Odeyinka, H.A. An evaluation of the use of insurance in managing construction risks. *Constr. Manag. Econ.* **2000**, *18*, 519–524. [CrossRef]
13. De Jong, P.; Heller, G.Z. *Generalized Linear Models for Insurance Data*; Cambridge University Press: New York, NY, USA, 2008.
14. Bunni, N.G. *Risk and Insurance in Construction*, 2nd ed.; Spon Press: London, UK, 2003.
15. Kim, J.M.; Kim, T.; Son, K.; Yum, S.G.; Ahn, S. Measuring Vulnerability of Typhoon in Residential Facilities: Focusing on Typhoon Maemi in South Korea. *Sustainability* **2019**, *11*, 2768. [CrossRef]
16. Sanders, D.E.; Brix, A.; Duffy, P.; Forster, W.; Hartington, T.; Jones, G.; Levi, C.; Paddam, P.; Papachristou, D.; Perry, G.; et al. *The Management of Losses Arising from Extreme Events*; Convention General Insurance Study Group GIRO: London, UK, 2002.
17. Kunreuther, H.; Meyer, R.; Van den Bulte, C.; Chapman, R.E. *Risk Analysis for Extreme Events: Economic Incentives for Reducing Future Losses*; US Department of Commerce; Technology Administration; National Institute of Standards and Technology: Gaithersburg, MD, USA, 2004.
18. Crichton, D. The Risk Triangle. In *Natural Disaster Management*; Ingleton, J., Ed.; Tudor Rose: London, UK, 1999; pp. 102–103.
19. Hastak, M.; Baim, E.J. Risk factors affecting management and maintenance cost of urban infrastructure. *J. Infrastruct. Syst.* **2001**, *7*, 7–76. [CrossRef]
20. Wang, Y.-M.; Elhag, T.M.S. A comparison of neural network, evidential reasoning and multiple regression analysis in modeling bridge risks. *Expert Syst. Appl.* **2007**, *32*, 336–348. [CrossRef]

21. Cho, T.; Kim, T.S. Probabilistic risk assessment for the construction phases of a bridge construction based on finite element analysis. *Finite Elem. Anal. Des.* **2008**, *44*, 383–400. [CrossRef]
22. Hashemi, H.; Mousavi, S.M.; Mojtahedi, S.M. Bootstrap technique for risk analysis with interval numbers in bridge construction projects. *J. Constr. Eng. Manag.* **2011**, *137*, 600–608. [CrossRef]
23. Li, Q.F.; Li, Z.X.; Niu, J. Application of factor analysis to risk evaluation of bridge construction. *Adv. Mater. Res.* **2011**, *243*, 1848–1853. [CrossRef]
24. Choudhry, R.M.; Aslam, M.A.; Hinze, J.W.; Arain, F.M. Cost and schedule risk analysis of bridge construction in Pakistan: Establishing risk guidelines. *J. Constr. Eng. Manag.* **2014**, *140*, 04014020. [CrossRef]
25. Nevada DOT. Risk Management and Risk-based Cost Estimation Guidelines. 2012. Available online: https://www.nevadadot.com/home/showdocument?id=4518 (accessed on 22 April 2020).
26. Yun, S.Y.; Kim, C.H.; Kang, L.S. Development of model for selecting superstructure type of small size bridge using dual classification method. *J. Korean Soc. Civ. Eng.* **2015**, *35*, 1413–1420. [CrossRef]
27. Malekly, H.; Mousavi, S.M.; Hashemi, H. A fuzzy integrated methodology for evaluating conceptual bridge design. *Expert Syst. Appl.* **2010**, *37*, 4910–4920. [CrossRef]
28. Koh, H.M.; Kim, H.J.; Lim, J.H.; Kang, S.C.; Choo, J.F. Lifetime Design of Cable-Supported Super-Long-Span Bridges. In Proceedings of the 5th International Conference on Bridge Maintenance, Safety and Management, Philadelphia, PA, USA, 11–15 July 2010; pp. 26–42.
29. Jo, B.W.; Park, J.C.; Kim, C.H. Wind characteristics of existing long span bridge based on measured data. *J. Civ. Eng.* **2005**, *9*, 219–224. [CrossRef]
30. Yang, Y.; Gang, Y.; Wei, F.; Qin, W. Buffeting performance of long-span suspension bridge based on measured wind data in a mountainous region. *J. Vibroeng.* **2018**, *20*, 621–635. [CrossRef]
31. Giroux, R.P. Relevance of Roebling. *J. Perform. Constr. Facil.* **2009**, *23*, 2–4. [CrossRef]
32. Elsaid, A.; Seracino, R. Rapid assessment of foundation scour using the dynamic features of bridge superstructure. *Constr. Build. Mater.* **2014**, *50*, 42–49. [CrossRef]
33. Won, J.H.; Kim, G.Y. Structural Safety Analysis of a Long Span Cable-stayed Bridge with a Partially Earth Anchored Cable System on Dynamic Loads during Construction. *J. Korean Soc. Saf.* **2016**, *31*, 104–110. [CrossRef]
34. Sampaio, A.Z.; Martins, O.P. The application of virtual reality technology in the construction of bridge: The cantilever and incremental launching methods. *Autom. Constr.* **2014**, *37*, 58–67. [CrossRef]
35. Kwak, H.G.; Son, J.K. Determination of design moments in bridges constructed with a movable scaffolding system (MSS). *Comput. Struct.* **2006**, *84*, 2141–2150. [CrossRef]
36. Pan, N.F. Fuzzy AHP approach for selecting the suitable bridge construction method. *Autom. Constr.* **2008**, *17*, 958–965. [CrossRef]
37. Kim, S.B.; Cho, J.H. Development of the approximate cost estimating model for PSC box girder bridge based on the breakdown of standard work. *J. Korean Soc. Civ. Eng.* **2013**, *33*, 791–800. [CrossRef]
38. Choi, H.H.; Mahadevan, S. Construction project risk assessment using existing database and project-specific information. *J. Constr. Eng. Manag.* **2008**, *134*, 894–903. [CrossRef]
39. Kuo, Y.C.; Lu, S.T. Using fuzzy multiple criteria decision making approach to enhance risk assessment for metropolitan construction projects. *Int. J. Proj. Manag.* **2013**, *31*, 602–614. [CrossRef]
40. Carr, V.; Tah, J.H.M. A fuzzy approach to construction project risk assessment and analysis: Construction project risk management system. *Adv. Eng. Softw.* **2001**, *32*, 847–857. [CrossRef]
41. El-Sayegh, S.M. Risk assessment and allocation in the UAE construction industry. *Int. J. Proj. Manag.* **2008**, *26*, 431–438. [CrossRef]
42. Chan, D.W.; Chan, A.P.; Lam, P.T.; Yeung, J.F.; Chan, J.H. Risk ranking and analysis in target cost contracts: Empirical evidence from the construction industry. *Int. J. Proj. Manag.* **2011**, *29*, 751–763. [CrossRef]
43. Kim, J.M.; Kim, T.; Bae, J.; Son, K.; Ahn, S. Analysis of plant construction accidents and loss estimation using insurance loss records. *J. Asian Archit. Build. Eng.* **2019**, *18*, 507–516. [CrossRef]
44. Ryu, H.; Son, K.; Kim, J.M. Loss prediction model for building construction projects using insurance claim payout. *J. Asian Archit. Build. Eng.* **2016**, *15*, 441–446. [CrossRef]
45. Gurcanli, G.E.; Bilir, S.; Sevim, M. Activity based risk assessment and safety cost estimation for residential building construction projects. *Saf. Sci.* **2015**, *80*, 1–12. [CrossRef]
46. Lowery, J.T.; Borgerding, J.A.; Zhen, B.; Glazner, J.E.; Bondy, J.; Kreiss, K. Risk factors for injury among construction workers at Denver International Airport. *Am. J. Ind. Med.* **1998**, *34*, 113–120. [CrossRef]

47. Cañamares, M.S.; Escribano, B.V.; García, M.G.; Barriuso, A.R.; Sáiz, A.R. Occupational risk-prevention diagnosis: A study of construction SMEs in Spain. *Saf. Sci.* **2017**, *92*, 104–115. [CrossRef]
48. Kim, J.M.; Kim, T.; Son, K.; Bae, J.; Son, S. A quantitative risk assessment development using risk indicators for predicting economic damages in construction sites of South Korea. *J. Asian Archit. Build. Eng.* **2019**, *18*, 472–478. [CrossRef]
49. Kim, J.M.; Woods, P.K.; Park, Y.J.; Kim, T.; Son, K. Predicting Hurricane Wind Damage by Claim Payout Based on Hurricane Ike in Texas. *Geomat. Nat. Haz. Risk* **2015**, *7*, 1513–1525. [CrossRef]

sustainability

MDPI

Article

Prediction of Risk Delay in Construction Projects Using a Hybrid Artificial Intelligence Model

Zaher Mundher Yaseen [1], Zainab Hasan Ali [2], Sinan Q. Salih [3,*] and Nadhir Al-Ansari [4,*]

1. Sustainable Developments in Civil Engineering Research Group, Faculty of Civil Engineering, Ton Duc Thang University, Ho Chi Minh City, Vietnam; yaseen@tdtu.edu.vn
2. Civil Engineering Department, College of Engineering, University of Diyala, Baquba 32001, Iraq; zainabhasan222@gmail.com
3. Institute of Research and Development, Duy Tan University, Da Nang 550000, Vietnam
4. Civil, Environmental and Natural Resources Engineering, Lulea University of Technology, 97187 Lulea, Sweden

* Correspondence: sinanq.salih@duytan.edu.vn (S.Q.S.); nadhir.alansari@ltu.se (N.A.-A.)

Received: 22 December 2019; Accepted: 14 February 2020; Published: 18 February 2020

Abstract: Project delays are the major problems tackled by the construction sector owing to the associated complexity and uncertainty in the construction activities. Artificial Intelligence (AI) models have evidenced their capacity to solve dynamic, uncertain and complex tasks. The aim of this current study is to develop a hybrid artificial intelligence model called integrative Random Forest classifier with Genetic Algorithm optimization (RF-GA) for delay problem prediction. At first, related sources and factors of delay problems are identified. A questionnaire is adopted to quantify the impact of delay sources on project performance. The developed hybrid model is trained using the collected data of the previous construction projects. The proposed RF-GA is validated against the classical version of an RF model using statistical performance measure indices. The achieved results of the developed hybrid RF-GA model revealed a good resultant performance in terms of accuracy, kappa and classification error. Based on the measured accuracy, kappa and classification error, RF-GA attained 91.67%, 87% and 8.33%, respectively. Overall, the proposed methodology indicated a robust and reliable technique for project delay prediction that is contributing to the construction project management monitoring and sustainability.

Keywords: delay sources; risk management; random forest-genetic algorithm; computer aid; construction project

1. Introduction

1.1. Research Background

The construction sector has a crucial role in improving the economics of developed countries [1]. The success of this sector is measured by time, cost and quality performance of construction projects. Prediction of construction durations represents a problem for both researchers and project managers. The construction process is subject to many factors and unpredicted variables that result from many sources. These sources prevent the completion of projects within the specified time and lead to a delay risk in the construction process [2,3].

Delay risk is considered as one of the major challenges tackled by construction firms [3,4]. Delay can be defined as an action or event that extends the time required to complete the project identified in a contract [3]. Project delay has an adverse effect on the project performance, which leads to cost overruns and productivity reduction. Its effect extends to include the owner, consultant and contractor in terms of litigation, dispute and arbitration [5]. Delays are caused by many sources and factors such as the owner [6,7], designer [3,8], contractor [4,7], materials [4,7], project [7,8], labor [9] and external factors [3,10].

1.2. Literature Review

The prediction of project delay based on internal and external sources can help project managers to provide an accurate forecast of the project schedule, and this can assist a proactive management approach in the construction project [11]. Construction projects are dynamic and complex, included a huge number of project stockholders, feedback processes and non-linear relationships [12]. The existence of a delay problem is related to interdependent factors that affect the construction project and the complexity and uncertainty of construction activities. Thus, providing of an efficient tool for analyzing delay factors is key for estimating an accurate duration in construction projects [11].

By recalling previous studies, Chan (2001) used regression analysis to identify time–cost relationships for building projects in Malaysia [2]. This approach was developed for managers and owners to estimate the average time that is required for project delivery. Chan and Chan (2004) performed multiple regression exercises to analyze data related to the time performance of construction projects [13]. The results indicated that multiple regression was used as a useful method to predict time performance in construction projects. Rezaie et al. (2007) used Monte Carlo analysis to investigate the effects of uncertainties on the schedule performance [14]. The results revealed that this method is a good tool to simulate the relationship of uncertainties of construction projects, and represents many dimensions of the utility function. Hammad et al. (2008) developed a statistical model and sample tests to predict the cost and duration of construction projects in Jordan [15]. The proposed model showed reliable statistical results in the prediction of the time and cost performance of construction projects. Mohamed et al. (2009) introduced Analytical Hierarchy Processes (AHP) to predict time contingency in construction projects [16]. The model provided a reliable result in the estimation process. In another study, Abu Hammad et al. (2010) used statistical analysis to measure the risk impact on the time and cost of public building projects [17]. Dursun and Stoy (2011) modeled construction project duration by testing the statistical efficiency in classifying projects with respect to their locations [18]. The results demonstrated that the model was valid to select populations. Kokkaew and Wipulanusat (2014) presented a stochastic-based Monte Carlo approach for managing delay risk in construction projects [19]. This approach has been applied to two case studies, a commercial building and a build–operate–transfer (BOT) road. The results showed that this approach may enhance risk management in construction projects that are surrounded with uncertainties. However, although there have been several models introduced for delay risk simulation, these methods cannot explain and clarify the uncertainty and complexity of the construction process very well, and this can affect the prediction process of project duration.

Recently, artificial intelligence (AI) models have been widely applied in different fields of engineering and science [20]. Several studies used AI models in their research on construction project management. Elazouni (2006) Used an Artificial Neural Network (ANN) during the prequalification process before awarding a contract [21]. The model was used to classify contractors into several groups depending on their performance. The model demonstrated reliable result in representing contractors in four-dimensional space. Chao and Chien (2009) developed a neural network model to estimate the S-curve in a construction project, which represented by polynomial parameter [22]. The results displayed that the model can be useful for contractors and owners in the early planning phase of a construction project. Desai and Joshi (2010) used a decision tree to analyze labor productivity in construction projects [23]. Shin (2011) proposed the AdaBoost algorithm for the selection of a framework system in construction projects [24]. The algorithm was compared with an ANN model and the results revealed that the AdaBoost algorithm performed with better accuracy than the ANN model. In another study, Chou and Lin (2012) applied an ensemble method for disputing problems in public–private partnership projects [25]. The study used four regression and classification trees and two statistical techniques to compare the performance of the utilized method. The study revealed that the ensemble models provided better accuracy that the individual models. Rudžianskaitė-Kvaraciejienė et al. (2015) Integrated a Random Forest algorithm with the modeling of public–private partnerships in infrastructure projects [26]. The method performed with good accuracy in the prediction process for public and private projects.

Heravi and Eslamdoost (2015) investigated the potential of an ANN model for the prediction of labor productivity in construction projects [27]. The results discovered that the ANN model showed better modeling of labor productivity. Gerassis et al. (2016) applied Bayesian networks to analyze the causes of accidents in embankment construction [28]. The study revealed that this method provided an accurate identification of embankment stability in civil engineering projects. By recalling the related literature review studies, AI model application is still a new methodology in the field of construction management research and delay risk prediction [29]. Few studies used AI models in risk prediction and classification.

Asadi et al. (2015) used a decision tree and a Naive Bayes model based on a questionnaire survey to predict delay in construction logistics. The authors evidenced the capacity of the decision tree has higher accuracy by 79.41% over the Naive Bayes model, which showed a lower accuracy value of 73.52% [30]. Naji et al. (2018) used a Bayesian decision tree model to predict the impact of contract changes on the time and quality performance of construction projects [31]. The model performed with good accuracy in the prediction process and caused an improvement in the project performance. Gondia et al. (2020) utilized Naive Bayes and decision tree models to predict the delay risk in construction projects. The study revealed the power of AI models in delay risk prediction and improving risk management strategies [11]. Based on the reported studies in the literature, the current research is established with the aim of providing a reliable methodology for delay risk prediction that will contribute to the baseline knowledge of construction management. Owing to the fact that standalone AI models experienced some limitations on tuning their internal parameters for an optimal learning process [32], the current study is adopted based on the integration of a nature-inspired optimization algorithm called Genetic Algorithm (GA) with a Random Forest (RF) model. The GA optimization approach was demonstrated as a reliable technique in tuning AI models for multiple engineering applications and thus it was selected for the current study [33–35].

1.3. Research Objectives

In the current study, the authors aim to explore and develop an effective tool to predict delay risk problems based on delay sources using previous construction projects data. The main contribution of the current investigation is to provide an accurate methodology that can assist in the prediction of future durations and monitor risk levels, based on these projects. This work can enhance a proactive approach in risk management. To achieve this aim, sources and factors of delay risk are extracted from the literature, and then related data to the delay risk problems are collected from previous construction projects. A questionnaire survey is adopted to measure the impact of various sources on the delay level in construction projects. Based on the complex nature of the construction process and the associated uncertainties of the delay sources, a hybrid model based on the integration of the Random Forest and Genetic Algorithm (RF-GA) is developed in order to analyze the data of completed previous projects. The performance of the developed model is studied statistically and discussed comprehensively. The potential of the proposed RF-GA model is validated against the classic RF model.

2. Research Methodology

2.1. Random Forest Model

The RF model was first developed by Breiman (2001) based on the combination of decision tree classifiers [36]. Each tree provides a prediction for the class label and the algorithm selects the classes that have the most choices. Random Forest is a very popular tool that uses the bootstrapping method to train dataset samples and construct multiple random trees [37]. The algorithm gained significant important because it is invariant under scaling and it is robust to the inclusion of irrelevant features [38]. Several studies examined the application of Random Forest in engineering applications and demonstrated its feasibility in prediction processes [26,34,35]. Under the bootstrapping method, the data during the training phase are selected randomly and independently to develop an RF model, and the data that are not involved in the selection process are named "out-of-bag" [39]. The capacity of the random forest has been approved by several engineering problems such as [40,41]. During this process, the out-of-bag data are changed and the prediction error is measured to estimate the importance of input variables [41,42]. In the RF algorithm, overfitting does not occur due to large numbers of trees and the choice of the right type of random variables leads to accurate classification. Random Forests contain several parameters that need to be optimized, such as number of trees, minimum gain and maximum tree depth. In this study, these parameters were optimized by a genetic algorithm.

2.2. The Hybrid RF-GA Model

Genetic algorithm is a popular technique used to optimize problems in complex systems based on natural selection [43]. To solve a problem in a GA algorithm, random solutions are generated, and then the selection of the population is done to develop the model. The new solutions are developed by using selection, crossover and mutation. A string of bits or chromosomes is used to represent the solutions in the GA model. The position of bits is called the gene, and the gene contains many values that are named alleles. GA has been widely applied in different fields, such as image processing, pattern recognition and controlling systems [44]. The GA model was used in different research in construction management such as resources optimization [45], project scheduling [46,47], optimizing of time and cost in construction projects [48] and dispute classification [49]. In the present study, a GA model is presented to optimize parameters of the RF model, including number of trees, minimum gain and maximum tree depth. The description of the hybrid FR-GA model is illustrated in Figure 1.

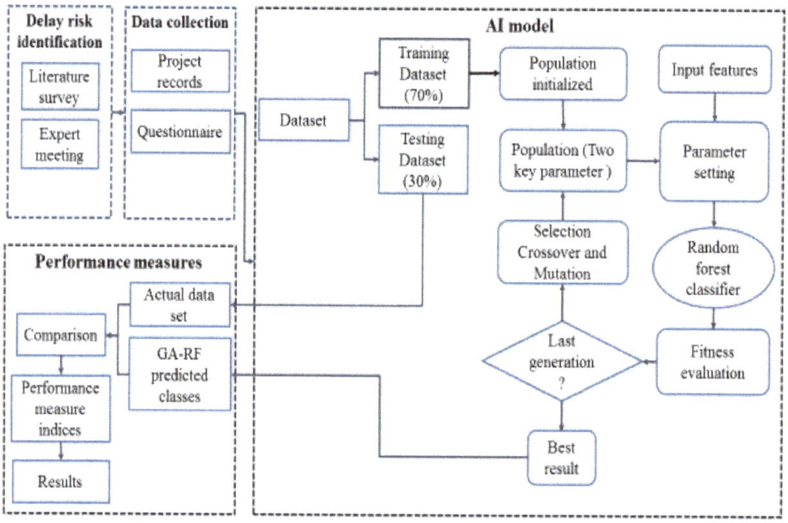

Figure 1. Hybrid artificial intelligence model Random Forest and Genetic Algorithm (RF-GA) structure.

2.3. Identification of Delay Sources and Factors in Construction Projects

The most important factors that affect delays in construction projects were identified from a literature survey, and these factors were categorized into different sources. These sources included owner, designer, contractor, project, material, equipment, labor and external factors. To obtain more information on the delay problems and their factors in construction projects, interviews were held with

15 experts in construction work [11,50]. By this interview, the identified sources and factors and their relevance to the construction industry were confirmed. Based on the reviews and literature review, the most important delay factors and their sources were identified as shown in Table 1.

Table 1. Identified factors and sources in construction projects.

Delay Source	Delay Factors
1. Owner	1.1 Owner financial problems
	1.2 Payment delay by the owner
	1.3 Choosing of inefficient design team
	1.4 Inadequate experience of the owner
	1.5 Issuing of change orders by the owner
	1.6 Delay in location delivery to the contractor
	1.7 Choice of inefficient contractor
	1.8 Delay in decision making procedure
2. Designer	2.1 Inadequate experience of design members
	2.2 Delay in the preparation of design documents
	2.3 Defects in the design and ambiguity of design drawings
3. Contractor	3.1 Ineffective project planning
	3.2 Financial contractor difficulties
	3.3 Inadequacy of contractor
	3.4 Rework due to defects in executed work
	3.5 Ineffective supervision and site management
	3.6 Many changes in subcontractor parties
	3.7 Poor communication between contractor and project parties
4. Project	4.1 Awarding the contract to an inadequate contractor
	4.2 Disputes between project parties
	4.3 Period of contract is very short
	4.4 Errors in contract documents
5. Material	5.1 Deficiency of materials in the market
	5.2 Delay in supplying materials
	5.3 Ineffective quality of materials
	5.4 Poor storage of materials
6. Equipment	6.1 Poor efficiency of equipment
	6.2 Unsuitable type of equipment
7. Labor	7.1 Poor labor productivity
	7.2 Inadequacy of workforce
	7.3 Lack of labor
8. External factors	8.1 Political situation and terrorism
	8.2 Inflation
	8.3 Legislation changes in the country
	8.4 Unpredicted surface conditions
	8.5 Neighbor problems
	8.6 Bad weather conditions

2.4. Data Collection

The compiled data included 40 completed projects that had different degrees of time overrun. These projects were executed in Diyala city, Iraq. The collected data included historical documents of previous projects that were investigated to extract the measure of risk delay in construction projects.

These documents included contract documents, specifications, change orders records and schedule baselines. To complete data collection, a questionnaire survey was arranged and constructed. Each questionnaire form contained a construction project and another nine variables. The first variable represented the delay level, and the other eight variables referred to the risk delay sources in the construction project. Each risk source was given scores depending on two scales. The first scale was the probability of risk to occur in the construction project and the second related to the impact of sources on the delay of the construction project, as shown in Table 2. The overall risk impact was evaluated by multiplying the two scales [3,43,44].

Table 2. Scales of probability and impact of risk delay in construction projects.

Scale	Probability	Impact
Very low	0.1	0.05
Low	0.3	0.1
Medium	0.5	0.2
High	0.7	0.4
Very high	0.9	0.8

The probability and impact of the variables were measured by using a five-point Likert scale with measures form very low to very high level [51]. The input variables were classified as: very low, low, medium, high and very high. The output variable (delay level) was also classified into three class measures. This method resulted in three categories of delay level that reduced the bias during the execution of the artificial intelligence model. Delay level was categorized as: <50% delay, 50%–100% delay and >100% delay. The questionnaire was allocated to a pilot study to measure the questionnaire reliability and to investigate the problems and determine the items that are more confusing than the others. The authors selected 40 parties for the pilot study as the size of the study was ranged between 30 and 50 parties [52]. To confirm the questionnaire reliability, Cronbach's alpha was adopted, and in this study the value of the alpha coefficient is 91.8%. The result of Cronbach's alpha confirms the reliability of the questionnaire.

2.5. Model Development Procedure

The questionnaire was distributed to 300 experts who worked in the collected projects. The experts were involved in different parties, which include client, engineer and the other experts of these projects. The collected projects were divided into two phases: 70% of the total projects (28 projects) were used for the training phase and 30% (12 projects) for the testing phase of the hybrid intelligence model. The genetic algorithm was applied to optimize the Random Forest classifier, and the hybrid RF-GA was used to predict the time performance in construction projects based on their risk levels by dividing the projects into a class label describing the predicted time delay. At first, the model describes the data with a set of records and variable values. The rows represent the individual projects and the columns represent the value of each project. The input variables included owner, designer, contractor, project, material, equipment, labor and external factors. The hybrid model was used for learning the dataset in order to predict the time delay at different levels of time overrun. The structure of the developed model is described in Figure 2.

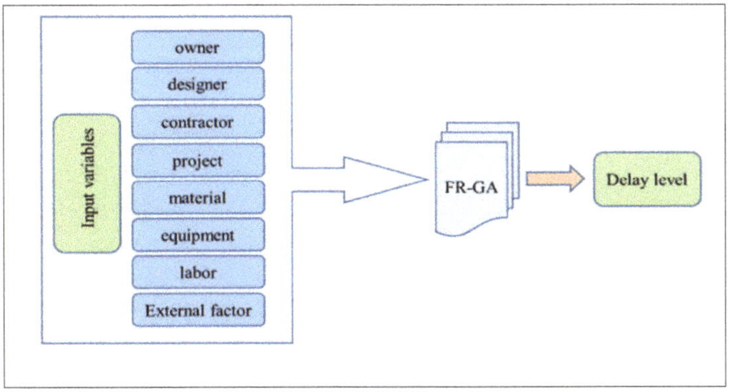

Figure 2. Structure of the delay prediction model.

2.6. Model Performance Measures

The performance of the predicted model was evaluated by using class performance and overall performance measures. Class performance was measured by precision, sensitivity and specificity [53,54]. The overall performance of the predicted model was evaluated by accuracy, classification error and kappa statistics. The kappa coefficient (k) was used in statistics to measure the quality of an item based on inter classifier agreement [55,56]. The equations of performance measures are explained as follows:

$$Precision = \frac{TP}{TP+FP} \quad (1)$$

$$Sensitivity = \frac{TP}{TP+FN} \quad (2)$$

$$Specificity = \frac{TN}{TN+FP} \quad (3)$$

$$Accuracy = \frac{TP+TN}{TP+FP+TN+FN} \quad (4)$$

$$Classification\ error = 1 - accuracy \quad (5)$$

$$k = \frac{p_o - p_e}{1 - p_e} \quad (6)$$

where:

TP means the number of positive classes that are correctly recognized by the algorithm;
FP represents the number of positive classes that are incorrectly classified by the algorithm;
TN means the number of negative classes that are correctly predicted by the algorithm;
FN represents the number of negative classed that are incorrectly recognized by the algorithm;
P_o means the observed agreement between rates; and
P_e represents the probability of chance agreement.

3. Results and Discussion

Analysis of collected data based on 40 projects was conducted to identify the sources of delay problems effectively. The properties of the complied data and the distribution of delay sources among the construction project are presented in Figures 3 and 4.

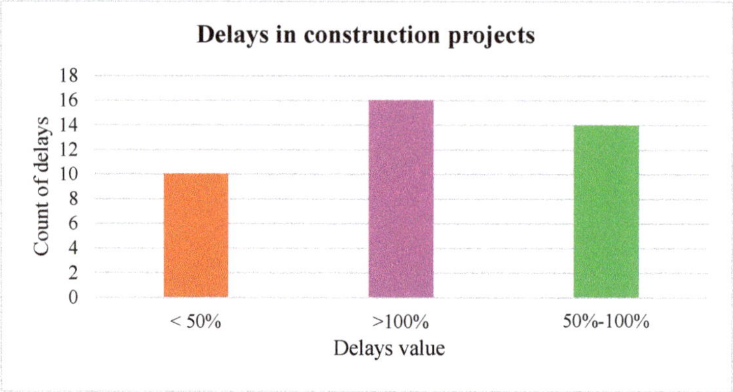

Figure 3. Frequency count of delays in construction projects.

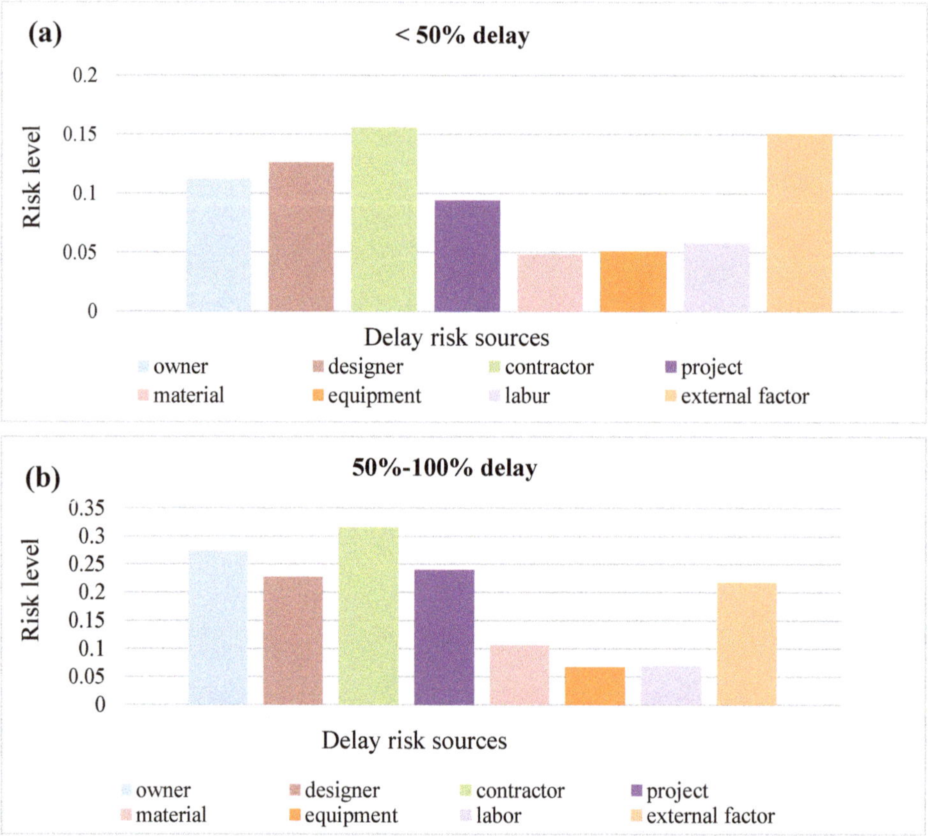

Figure 4. *Cont.*

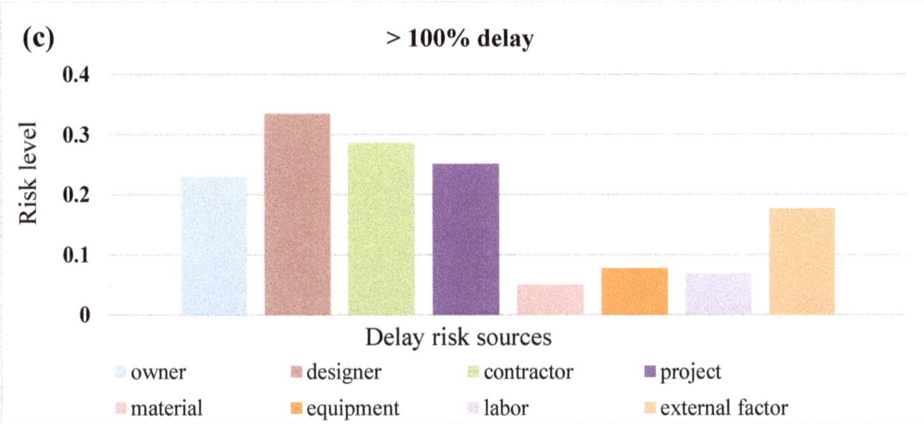

Figure 4. (**a**) Distribution of delay sources among projects with class <50% delay, (**b**) distribution of delay sources among projects with class 50%–100% delay, (**c**) distribution of delay sources among projects with class >100% delay.

Based on the reported results, Figure 3 shows the counts of projects with a <50% delay, 50%–100% delay and >100% delay were 10 (25%), 14 (35%) and 16 (40%), respectively. It can be seen that a high percentage of projects belongs to the class of >100% delay. Figure 4 demonstrates the distribution of delay sources among each class of delay problem, which was obtained from the historical records, pilot study and distributed questionnaire. These outcomes revealed the delay sources values of contractor, owner, designer, project and external factors have a higher impact that the other delay sources. Owner, designer, contractor and project are represented as the internal risk sources that have an impact on the project delay. External factors can be discussed by the special circumstances that are experienced in the studied region "Iraq" in a manner that severely affected the construction industry. These conditions have an enormous impact on the project stockholder and project performance. These conditions resulted in the stumbling and failure of many projects in the construction sector. On the other hand, the application of a robust predictive model can contribute to estimating an accurate duration in construction projects and analyzing delay risk sources that arise from the complex and dynamic nature of construction sector.

The statistical performance of the training and testing datasets of the proposed hybrid RF-GA model were evaluated based on the model performance measures against the classical Random Forest classifier. The performance measure metrics were evaluated based on the confusion matrix of the two classifiers. The confusion matrix is described in the performance of the classification model. The confusion matrix of the RF and RF-GA are displayed in Tables 3 and 4.

Table 3. Confusion matrix from RF classifier.

		Actual Class		
Predicted Class	**<50%**	**50%–100%**	**>100%**	**Total**
<50%	2	0	1	3
50%–100%	0	0	3	3
>100%	1	4	1	6
Total	3	4	5	

Table 4. Confusion matrix from RF-GA classifier.

	Actual Class			
Predicted Class	**<50%**	**50%–100%**	**>100%**	**Total**
<50%	2	0	0	2
50%–100%	1	5	0	6
>100%	0	0	4	4
Total	3	5	4	

The columns in the confusion matrix represent the actual classification within each class, while the rows correspond to the number of the predicted class. The correct predictors are located on the diagonal of the matrix. The confusion matrix of a high-performance model contains large numbers in its diagonal and the zero numbers outside the diagonal. The performance of the hybrid RF-GA and RF models during the training and testing phases was evaluated. Precision, sensitivity, specificity, accuracy, classification error and kappa statistics were computed and are presented in Tables 5 and 6.

Table 5. Comparison of two classifiers based on performance measures for the training phase (70%).

	RF			RF-GA		
Performance Index	**<50% Delay**	**50%–100% Delay**	**>100% Delay**	**<50% Delay**	**50%–100% Delay**	**>100% Delay**
Precision	87.5	100	90	87.5	100	100
Sensitivity	87.5	83.33	90	100	91.67	90
Specificity	95	100	94.44	95.2	100	100
Accuracy		92.86			96.43	
Classification error		7.41			3.57	
Kappa		89.2			94.6	

Table 6. Comparison of two classifiers based on performance measures for the testing phase (30%).

	RF			RF-GA		
Performance Index	**<50% Delay**	**50%–100% Delay**	**>100% Delay**	**<50% Delay**	**50%–100% Delay**	**>100% Delay**
Precision	66.67	100	90	66.67	83.33	100
Sensitivity	50	50	90	80	100	80
Specificity	87.5	100	94.44	71.4	85.7	100
Accuracy		75			91.67	
Classification error		25			8.33	
Kappa		62.5			87	

Tables 5 and 6 demonstrate the comparison of the RF-GA and RF models based on performance measures for the overall and class performance. Based on these results, the attained values of accuracy, classification error, and Kappa for RF-GA were 96.43%, 3.57% and 94.6%, respectively for the training phase; whereas the RF model provided an accuracy value of 92.86%, a classification error of 7.41% and a Kappa statistics value of 89.2%. It can be noticed that the both models gave good results for the training phase. Based on the results of testing phase, RF-GA revealed a better performance than the RF model. The provided values of accuracy, classification error and Kappa of RF-GA are 91.67%, 8.3%, and 87%, respectively.

With regards to performance measures, the RF-GA model exhibited a good performance in the prediction of delay in the construction sector. Based on the training phase, RF-GA achieved the minimum values of precision, sensitivity and specificity of 87.5, 90 and 95.2, respectively. The lowest values of RF in terms of precision, sensitivity and specificity were 87.5, 83.33 and 94.44, respectively. Based on the comparison between the two classifiers, it can be concluded that the RF-GA model outperformed the feasibility of the classical RF model in both the training and testing performances. Tables 5 and 6 revealed the superiority of the RF-GA classifier in terms of accuracy, classification error and Kappa statistics. This can be explained as due to the potential of the integration of the nature-inspired optimization algorithm (i.e., GA) that assisted in providing reliable hyperparameters optimization and thus attained a reliable learning process. The RF-GA model also provided higher values of precision, sensitivity and specificity in comparison with the RF model.

The RF-GA classifier showed an impressive performance in terms of overall and class measure indices. These results can be discussed by the ability of the genetic algorithm in solving optimization problems depending on the chromosome approach, and its capacity to solve the problems while dealing with multiple solutions [57]. It is even better to validate the current research results with the reported research over the literature. As compared with the previous results, it can be inferred that the RF-GA model demonstrated remarkable prediction superiority in comparison with the previous established studies as reported in Table 7. The capacity of the RF-GA model was compared with the best outcomes. The RF-GA model exceeded all of the reported related literature.

Table 7. Validation of the current research results against the reported related literature studies.

Author	Methods	Results
[30]	Questionnaire survey, decision tree and Naive Bayes	Accuracy of decision tree 79.41% is higher than Naive Bayes by 5.81%
[31]	Questionnaire and Bayesian decision tree	Bayesian decision tree gained an accuracy of 86.7%
[11]	Records of construction projects, meeting with experts, decision tree and Naive Bayes	Accuracy of Naive Bayes, 51.2%, is higher than decision tree by 4%
The current study	Records of construction projects, meetings and questionnaire survey, classical Random Forest, hybrid genetic Random Forest	Accuracy of genetic Random Forest, 91.76%, is higher than classical Random Forest by 16.67%

To summarize, a proactive management approach involves the identification of new risk delay sources and the monitoring of the sources that arise during the project lifecycle. As a result, the proposition of a reliable and robust methodology as an analysis tool that is able to mimic and comprehend the dynamic input variables is highly needed for this purpose. Hence, and based on the established methodology of the current research, the potential of the RF-GA model to be modified and set up for project duration prediction though the project lifecycle was evidenced. The RF-GA model was successfully developed for the investigated dynamic project delay risk prediction.

4. Conclusions

In this present study, an analysis tool that is capable of predicting the delay level in construction projects based on delay sources was proposed. To meet this goal, two approaches were adopted in this study. First, delay sources and factors were collected from a literature review and identified by an expert meeting. Data that are related to delay levels were compiled from 40 construction projects that are located in Diyala city, Iraq. The collected data included historical records of previous projects that were investigated, and in order to extract the measure of delay risk in construction projects a questionnaire was prepared and distributed to 300 experts so as to extract the information about delay sources in construction projects. Risk sources were measured by computing the probability and the impact of each source. An analysis of data results and distribution of delay sources among the collected previous projects was implemented in order to better understand delay factors in the construction sector. Secondly, a hybrid RF-GA model was developed to deal with the complex and dynamic nature of data in the construction sector. The RF-GA model was evaluated by performance measure indices and compared with the classical RF model. Based on the analysis results, RF-GA revealed a better performance than the RF model. The RF-GA provided values of accuracy, classification error and Kappa were 91.67%, 8.3%, and 87%, respectively. These results reflect the ability of the model to handle the nonlinearity and complexity of data in the construction sector. The results also revealed the capability of the genetic algorithm in solving problems with multiple solutions.

Author Contributions: Conceptualization, Z.M.Y. and Z.H.A.; Data curation, Z.M.Y.; Formal analysis, Z.M.Y. and N.A.-A.; Funding acquisition, N.A.-A.; Investigation, Z.M.Y., Z.H.A. and S.Q.S.; Methodology, Z.H.A. and S.Q.S.; Project administration, S.Q.S.; Resources, Z.H.A.; Software, Z.H.A. and S.Q.S.; Supervision, N.A.-A.; Validation, Z.H.A. and S.Q.S.; Visualization, Z.H.A. and S.Q.S.; Writing—original draft, Z.M.Y., Z.H.A., S.Q.S. and N.A.-A.; Writing—review and editing, Z.M.Y. and N.A.-A. All authors have read and agreed to the published version of the manuscript.

Funding: This research received no external funding.

Conflicts of Interest: The authors declare no conflict of interest.

References

1. Vorakulpipat, C.; Rezgui, Y.; Hopfe, C.J. Value creating construction virtual teams: A case study in the construction sector. *Autom. Constr.* **2010**, *19*, 142–147. [CrossRef]
2. Chan, A.P. Time-cost relationship of public sector projects in Malaysia. *Int. J. Proj. Manag.* **2001**, *19*, 223–229. [CrossRef]
3. Assaf, S.A.; Al-Hejji, S. Causes of delay in large construction projects. *Int. J. Proj. Manag.* **2006**, *24*, 349–357. [CrossRef]
4. Sambasivan, M.; Soon, Y.W. Causes and effects of delays in Malaysian construction industry. *Int. J. Proj. Manag.* **2007**, *25*, 517–526. [CrossRef]
5. Aibinu, A.; Jagboro, G. The effects of construction delays on project delivery in Nigerian construction industry. *Int. J. Proj. Manag.* **2002**, *20*, 593–599. [CrossRef]
6. Odeh, A.M.; Battaineh, H.T. Causes of construction delay: traditional contracts. *Int. J. Proj. Manag.* **2002**, *20*, 67–73. [CrossRef]
7. Fugar, F.D.; Agyakwah-Baah, A.B. Delays in Building Construction Projects in Ghana. *Australas. J. Constr. Econ. Build.* **2010**, *10*, 103–116. [CrossRef]
8. Aziz, R.F. Ranking of delay factors in construction projects after Egyptian revolution. *Alex. Eng. J.* **2013**, *52*, 387–406. [CrossRef]
9. Al-Momani, A.H. Construction delay: a quantitative analysis. *Int. J. Proj. Manag.* **2000**, *18*, 51–59. [CrossRef]
10. Jing, W.; Naji, H.I.; Zehawi, R.N.; Ali, Z.; Al-Ansari, N.; Yaseen, Z.M. System Dynamics Modeling Strategy for Civil Construction Projects: The Concept of Successive Legislation Periods. *Symmetry* **2019**, *11*, 677. [CrossRef]
11. Gondia, A.; Siam, A.; El-Dakhakhni, W.; Nassar, A.H. Machine Learning Algorithms for Construction Projects Delay Risk Prediction. *J. Constr. Eng. Manag.* **2020**, *146*, 04019085. [CrossRef]
12. Mahamid, I.; Bruland, A.; Dmaidi, N. Causes of delay in road construction projects. *J. Manag. Eng.* **2011**, *28*, 300–310. [CrossRef]
13. Chan, A.P.C.; Chan, D.W. Developing a benchmark model for project construction time performance in Hong Kong. *Build. Environ.* **2004**, *39*, 339–349. [CrossRef]
14. Rezaie, K.; Amalnik, M.; Gereie, A.; Ostadi, B.; Shakhseniaee, M. Using extended Monte Carlo simulation method for the improvement of risk management: Consideration of relationships between uncertainties. *Appl. Math. Comput.* **2007**, *190*, 1492–1501. [CrossRef]

15. Hammad, A.A.A.; Ali, S.M.A.; Sweis, G.J.; Bashir, A. Prediction model for construction cost and duration in Jordan. *Jordan J. Civ. Eng.* **2008**, *2*, 250–266.
16. Mohamed, D.; Srour, F.; Tabra, W.; Zayed, T. A prediction model for construction project time contingency. In *Construction Research Congress 2009: Building a Sustainable Future*; ASCE: Reston, VA, USA, 2009; pp. 736–745.
17. Abu Hammad, A.; Ali, S.M.A.; Sweis, G.J.; Sweis, R. Statistical Analysis on the Cost and Duration of Public Building Projects. *J. Manag. Eng.* **2010**, *26*, 105–112. [CrossRef]
18. Dursun, O.; Stoy, C. Time–cost relationship of building projects: statistical adequacy of categorization with respect to project location. *Constr. Manag. Econ.* **2011**, *29*, 97–106. [CrossRef]
19. Kokkaew, N.; Wipulanusat, W. Completion delay risk management: A dynamic risk insurance approach. *KSCE J. Civ. Eng.* **2014**, *18*, 1599–1608. [CrossRef]
20. Brunette, E.S.; Flemmer, R.C.; Flemmer, C.L. A review of artificial intelligence. In Proceedings of the 2009 4th International Conference on Autonomous Robots and Agents, Wellington, New Zealand, 10–12 February 2009; pp. 385–392.
21. Elazouni, A. Classifying Construction Contractors Using Unsupervised-Learning Neural Networks. *J. Constr. Eng. Manag.* **2006**, *132*, 1242–1253. [CrossRef]
22. Chao, L.-C.; Chien, C.-F. Estimating Project S-Curves Using Polynomial Function and Neural Networks. *J. Constr. Eng. Manag.* **2009**, *135*, 169–177. [CrossRef]
23. Desai, V.S.; Joshi, S. Application of decision tree technique to analyze construction project data. In *Communications in Computer and Information Science, Proceedings of the International Conference on Information Systems, Technology and Management, Bangkok, Thailand, 11–13 March 2010*; Springer: Piscataway, NJ, USA, 2010; pp. 304–313.
24. Shin, Y.-S. Formwork System Selection Model for Tall Building Construction Using the Adaboost Algorithm. *J. Korea Inst. Build. Constr.* **2011**, *11*, 523–529. [CrossRef]
25. Chou, J.-S.; Lin, C. Predicting disputes in public-private partnership projects: Classification and ensemble models. *J. Comput. Civ. Eng.* **2012**, *27*, 51–60. [CrossRef]
26. Rudžianskaitė–Kvaraciejienė, R.; Apanaviciene, R.; Gelzinis, A. modelling the effectiveness of ppp road infrastructure projects by applying random forests. *J. Civ. Eng. Manag.* **2015**, *21*, 290–299. [CrossRef]
27. Heravi, G.; Eslamdoost, E. Applying Artificial Neural Networks for Measuring and Predicting Construction-Labor Productivity. *J. Constr. Eng. Manag.* **2015**, *141*, 04015032. [CrossRef]
28. Gerassis, S.; Martín, J.E.; García, J.T.; Saavedra, A.; Taboada, J. Bayesian decision tool for the analysis of occupational accidents in the construction of embankments. *J. Constr. Eng. Manag.* **2016**, *143*, 4016093. [CrossRef]
29. Bilal, M.; Oyedele, L.; Qadir, J.; Munir, K.; Ajayi, S.; Akinade, O.; Owolabi, H.A.; Alaka, H.A.; Pasha, M. Big Data in the construction industry: A review of present status, opportunities, and future trends. *Adv. Eng. Inform.* **2016**, *30*, 500–521. [CrossRef]
30. Asadi, A.; Alsubaey, M.; Makatsoris, C. A machine learning approach for predicting delays in construction logistics. *Int. J. Adv. Logist.* **2015**, *4*, 115–130. [CrossRef]
31. Hassan, Z.; Ibrahim, A.M.; Naji, H. Evaluation of Legislation Adequacy in Managing Time and Quality Performance in Iraqi Construction Projects- a Bayesian Decision Tree Approach. *Civ. Eng. J.* **2018**, *4*, 993. [CrossRef]
32. Yaseen, Z.; Mohtar, W.H.M.W.; Ameen, A.M.S.; Ebtehaj, I.; Razali, S.F.M.; Bonakdari, H.; Salih, S.Q.; Al-Ansari, N.; Shahid, S. Implementation of univariate paradigm for streamflow simulation using hybrid data-driven model: Case study in tropical region: Implementation of univariate paradigm for streamflow simulation using hybrid data-driven model: Case study in tropical region. *IEEE Access* **2019**, *7*, 74471–74481. [CrossRef]
33. Chou, J.-S.; Pham, A.-D. Hybrid computational model for predicting bridge scour depth near piers and abutments. *Autom. Constr.* **2014**, *48*, 88–96. [CrossRef]
34. Yaseen, Z.M.; Ehteram, M.; Hossain, S.; Chow, M.F.; Koting, S.; Mohd, N.S.; Jaafar, W.B.; Afan, H.A.; Hin, L.S.; Zaini, N.; et al. A Novel Hybrid Evolutionary Data-Intelligence Algorithm for Irrigation and Power Production Management: Application to Multi-Purpose Reservoir Systems. *Sustainability* **2019**, *11*, 1953. [CrossRef]

35. Yaseen, Z.M.; Ebtehaj, I.; Kim, S.; Sanikhani, H.; Asadi, H.; Ghareb, M.I.; Bonakdari, H.; Mohtar, W.H.M.W.; Al-Ansari, N.; Shahid, S. Novel Hybrid Data-Intelligence Model for Forecasting Monthly Rainfall with Uncertainty Analysis. *Water* **2019**, *11*, 502. [CrossRef]
36. Breiman, L.; Cutler, A. State of the art of data mining using Random forest. In Proceedings of the Salford Data Mining Conference, San Diego, CA, USA, 24–25 May 2012.
37. Breiman, L. Random Forests. *Mach. Learn.* **2001**, *45*, 5–32. [CrossRef]
38. Ruppert, D. The Elements of Statistical Learning: Data Mining, Inference, and Prediction. *J. Am. Stat. Assoc.* **2004**, *99*, 567. [CrossRef]
39. Catani, F.; Lagomarsino, D.; Segoni, S.; Tofani, V. Landslide susceptibility estimation by random forests technique: sensitivity and scaling issues. *Nat. Hazards Earth Syst. Sci.* **2013**, *13*, 2815–2831. [CrossRef]
40. Naghibi, S.A.; Ahmadi, K.; Daneshi, A. Application of Support Vector Machine, Random Forest, and Genetic Algorithm Optimized Random Forest Models in Groundwater Potential Mapping. *Water Resour. Manag.* **2017**, *31*, 2761–2775. [CrossRef]
41. Alipour, M.; Harris, D.K.; Barnes, L.E.; Ozbulut, O.; Carroll, J. Load-Capacity Rating of Bridge Populations through Machine Learning: Application of Decision Trees and Random Forests. *J. Bridg. Eng.* **2017**, *22*, 04017076. [CrossRef]
42. Liaw, A.; Wiener, M. Classification and regression by randomforest. *R News* **2002**, *2*, 18–22.
43. Holland, J.H. *Adaptation in Natural and Artificial Systems: An Introductory Analysis with Applications to Biology, Control, and Artificial Intelligence*; MIT Press: Cambridge, MA, USA, 1992.
44. Azar, A.T.; Vaidyanathan, S. *Computational Intelligence Applications in Modeling and Control*; Springer: Piscataway, NJ, USA, 2015.
45. Kandil, A.; El-Rayes, K. Parallel Genetic Algorithms for Optimizing Resource Utilization in Large-Scale Construction Projects. *J. Constr. Eng. Manag.* **2006**, *132*, 491–498. [CrossRef]
46. Senouci, A.; Eldin, N.N. Use of Genetic Algorithms in Resource Scheduling of Construction Projects. *J. Constr. Eng. Manag.* **2004**, *130*, 869–877. [CrossRef]
47. Long, L.D.; Ohsato, A. A genetic algorithm-based method for scheduling repetitive construction projects. *Autom. Constr.* **2009**, *18*, 499–511. [CrossRef]
48. Rogalska, M.; Bozejko, W.; Hejducki, Z. Time/cost optimization using hybrid evolutionary algorithm in construction project scheduling. *Autom. Constr.* **2008**, *18*, 24–31. [CrossRef]
49. Chou, J.-S.; Cheng, M.-Y.; Wu, Y.-W.; Pham, A.-D. Optimizing parameters of support vector machine using fast messy genetic algorithm for dispute classification. *Expert Syst. Appl.* **2014**, *41*, 3955–3964. [CrossRef]
50. Xia, N.; Zhong, R.; Wu, C.; Wang, X.; Wang, S. Assessment of Stakeholder-Related Risks in Construction Projects: Integrated Analyses of Risk Attributes and Stakeholder Influences. *J. Constr. Eng. Manag.* **2017**, *143*, 04017030. [CrossRef]
51. Ismail, I.; Memon, A.H.; Rahman, I.A. Expert opinion on risk level for factors affecting time and cost overrun along the project lifecycle in Malaysian construction projects. *Int. J. Constr. Technol. Manag.* **2013**, *1*, 2289.
52. Thomas, S.J. *Using Web and Paper Questionnaires for Data-Based Decision Making: From Design to Interpretation of the Results*; Corwin Press: Thousand Oaks, CA, USA, 2004.
53. Helmer, G.; Wong, J.; Honavar, V.G.; Miller, L. Automated discovery of concise predictive rules for intrusion detection. *J. Syst. Softw.* **2002**, *60*, 165–175. [CrossRef]
54. Davis, J.; Goadrich, M. The relationship between Precision-Recall and ROC curves. In *Proceedings of the 23rd International Conference on Machine Learning*; ACM: New York, NY, USA, 2006; pp. 233–240.
55. Smeeton, N.C. Early history of the kappa statistic. *Biometrics* **1985**, *41*, 795.
56. Pontius, R.G.; Millones, M. Death to Kappa: birth of quantity disagreement and allocation disagreement for accuracy assessment. *Int. J. Remote Sens.* **2011**, *32*, 4407–4429. [CrossRef]
57. Tabassum, M.; Mathew, K. A genetic algorithm analysis towards optimization solutions. *Int. J. Digit. Inf. Wirel. Commun.* **2014**, *4*, 124–142. [CrossRef]

sustainability

MDPI

Article

A Comprehensive Analysis: Sustainable Trends and Awarded LEED 2009 Credits in Vietnam

Duy Hoang Pham [1], Byeol Kim [1], Joosung Lee [2,*], Abraham Chiwon Ahn [3] and Yonghan Ahn [4,*]

1 Department of Architectural Engineering, Hanyang University, 55 Hanyangdaehak-ro, Sangrok-gu, Ansan-si, Gyeonggi-do 15588, Korea; phamduyhoang@hanyang.ac.kr (D.H.P.); keemstars@naver.com (B.K.)
2 Innovative Durable Building and Infrastructure Research Center, Hanyang University, 55 Hanyangdaehak-ro, Sangrok-gu, Ansan-si, Gyeonggi-do 15588, Korea
3 Department of Architectural Design, Hanyang University, 55 Hanyangdaehak-ro, Sangrok-gu, Ansan-si, Gyeonggi-do 15588, Korea; abeahn@hanyang.ac.kr
4 School of Architecture and Architectural Engineering, Hanyang University, 55 Hanyangdaehak-ro, Sangrok-gu, Ansan-si, Gyeonggi-do 15588, Korea
* Correspondence: js4ever@hanyang.ac.kr (J.L.); yhahn@hanyang.ac.kr (Y.A.)

Received: 20 December 2019; Accepted: 20 January 2020; Published: 23 January 2020

Abstract: Established by the U.S. Green Building Council (USGBC), the Leadership in Energy and Environmental Design (LEED) became a sustainable leader of green building rating systems in American and many other countries. In Vietnam, LEED is expected as a potential solution in improving the sustainable quality of buildings for residents and solving the housing/infrastructure demand with a limit in resource consumption and minimizing negative environmental impacts. The study analyzed the awarded LEED 2009 credits by investigating the data of 36 of the total 42 LEED BC+D 2009 certified projects in Vietnam. The results of the investigation indicated the awarded credits were significantly implemented in Vietnam. These results were converted based on the summary updated on LEED version 4 of the USGBC report, to become a useful guideline for green building cost-efficiency strategies. Additionally, it also served as reference data for the Vietnamese public agency to update their green regulations based on the specific characteristics of Vietnam.

Keywords: sustainability; green building; LEED; Vietnam

1. Introduction

Besides the rapid economic growth, environmental pollution and the lack of housing demand, the research of sustainability has become a top concern for developing countries [1]. Especially in 2019, the study of Climate Central showed that a large area of Vietnam could be drowned due to rising sea levels in 2050, and a quarter of Vietnam's population will be profoundly affected by climate change [2]. Along with the limitations of developing countries such as seeking large budgets to address environmental issues, it is essential and necessary to find a way to promote sustainable development. Obviously, the construction industry not only accounts for a large proportion of Vietnam's production, but buildings also contribute a large part to the consumption of electricity, water, and other research resources [3]. Thus, green buildings had flourished in recent years after recognizing their potential social and economic benefits in solving many negative environmental issues and problems [4]. Moreover, the increase in electricity prices and recent urban environment incidents in Vietnam also became a strong motivation for real estate developers to choose green building certification for their future projects [1,3]. This trend can be reflected by the increase in the number of projects participating in green building certification in Vietnam in recent years, especially LEED certificates with notable numbers.

LEED and other green building certification systems provided technical guides and a framework to promote sustainable construction practical activities by balancing the elements of environmental protection, energy/water consumption, the benefits of user and community and operations, and maintenance solutions [3,5]. While promoting sustainable construction and bringing many benefits to building residents and the community, several problems related to the applying of LEED are also identified [6], such as encountering challenges for the project team and cost increasing [3]. Considerable challenges of implementing a LEED building project were the expensive quests, higher construction cost [7] and non-experience extra works for the construction team [5]. The additional charges might come from hiring LEED consultants, Commissioning Authority (CxA), air-testing and flush-out test contractors [8], which are entirely unfamiliar and expensive to the Vietnamese building project [3]. However, several studies indicated that LEED projects may have little incremental costs (2.5 to 9.4%) but will significantly reduce the operating costs (about 31%) of the building [9]. The study by D. Langdon (2007) and J.O. Choi (2015) also confirmed that this cost impact could be completely limited by their previous similar project experience and commitment during project implementation [5,10]. The research of Y.H. Ahn (2015) showed that it is crucial to apply the Integrated Construction Process (ICP) to collect the knowledge and ideas of all key stakeholders for successfully implementing green buildings with an acceptable cost [4].

In addition, C.P. Cheng's (2015) study mentioned that the better judgment of target credit selection and the appropriate green design technologies largely depend on the experience of managers from previous LEED projects [11]. However, construction projects are unique, and it is often difficult to participate in similar projects, especially when the number of certified LEED projects is very limited in Vietnam at present. Besides that, the new projects applying for LEED certification form 2018 have to comply with the unique requirements of LEED version 4 with many considerable changes, which might create several difficulties for the project team [3]. As a result, a useful guide for the green building project in selecting LEED credits that are appropriate to the project goals, from the synthesis and analysis of previous projects, would bring many significant benefits to the LEED projects and promote the sustainable construction. Thus, this study was conducted to create a useful guideline for the project and building developer for identifying the suitable project LEED credits/goals and its risk by investigating and analyzing all similar projects that have received LEED certification in Vietnam. This paper aims to:

(1) Identify the significantly awarded credits at LEED projects in Vietnam.
(2) Investigate the significant different credits between the different LEED certificate levels and provide the recommendations for credits required for future LEED projects.
(3) Review and summarize the significant differences between LEED 2009 and LEED v4 to combine with the analysis results and make more suitable guidelines and recommendations.

2. Literature Review

2.1. Background of Research

"The most widely used green building rating system in the world," "The framework to create healthy, highly efficient and cost-saving green buildings," and "The globally recognized symbol of sustainability achievement" are the definitions of the U.S. Green Building Council (USGBC) about the "Leadership in Energy and Environmental Design" (LEED) rating system [8]. Additionally, LEED is the meaningful attempt of USGBC to create a global rating system, which guides technically to design and deliver the green building. Thus, green building projects that comply with the LEED requirement not only promote sustainable construction but also increase the project life cycle cost-effective health benefits to users [3,6,12]. LEED is considered as the most globally popular green building rating tool because of its overlooked variations between different climatic and environmental conditions, rules, standards, and laws. The selection of LEED credits dramatically affects the results of the project, but it is more difficult due to the rising tide of cost, schedule, and LEED experience. Thus, understanding the relationships between decision-making factors and characteristics of LEED on various credits is essential for project stakeholders and regulatory authorities [3,6]. As a result, many pieces of research are conducted to guide better efficiency of the green building projects and encourage sustainable construction. In particular, a significant number of them were conducted to support the design process and identify appropriate LEED credits for the project during the early phase, when the design changes are much higher in efficiency and less costly [3,4].

Cheng and Ma (2015) studied the relationship between LEED credits and green building technologies/sustainability design strategies to improve the effectiveness of LEED credits selection [11]. M.A. Jun's paper (2015) presented a methodology for identifying the target LEED credits based on project information and climatic factors [13]. J. Ma's study (2016) analyzed LEED credit achievements in previous projects using data-driven techniques and provided useful instruction so as to help the project team to understand and identify the appropriate LEED credits [14]. Jack C.P. Cheng's paper (2015) proposed a case-based reasoning (CBR) approach to find out the suitable case study base on the similar previously certificated green building projects and suggestions on target LEED credits [11]. F. Jalaei and A. Jrade's study (2015) (BIM) explains how this integration would assist project teams in making sustainability-related decisions [15]. P.H. Chen and T.C. Nguyen developed an application by interacting web map service (WMS) technology and BIM to pre-calculate LEED awarded points of the "Location and transportation" category [16]. The research of S. Pushkar (2018) pointed out that several data analysis studies of previous LEED-certified projects have to consider the problem of pseudoreplication [17]. Recently, when Wu et al. (2017) [6] and Wu et al. (2018) [18] analyzed previous LEED certification project data, they used aggregated processes to achieve extremely low P-values. Da Silva and Ruwanpura (2009) investigated LEED Canada certificated building data and found out that material (MR) and energy and atmosphere (EA) criteria are not common at LEED projects in Canada [19]. Cell and Beata (2009) presented that the lack of consideration in the design phase of a green building project may have an adverse impact on the implementation phase [20]. J.-Y. Park's study (2017) developed an optimization algorithm that aims to determine the LEED awarded points that can be achieved with minimum cost [12]. Madanayake and Ruwanpura (2012) suggested a method for determining appropriate LEED credits based on cost, productivity, and environmental impact information [21]. Juan, Gao, and Wang (2010) addressed a selection of green building technology strategies by optimizing the trade-off between costs and achievements in different green building rating systems [22]. Castro-Lacouture, Sefair, Flórez, and Medaglia studied the factors influencing the selection of sustainable materials in the LEED project [23]. Boschmann and Gabriel proposed (2007) the implementation and development of green technologies directly related to a particular LEED credit [24]. Pushkar (2018) identified the significant differences in the choice of implementing LEED credits among several countries, including SS, WE, EA, EQ, and ID [25]. In other words, it is highly useful for analyzing the influence and development of green buildings in developed countries to benchmark the Green Building regulation system of developing countries [26].

In summary, the previous studies have provided the LEED credits choosing by using various methods of selection supporting the appropriate credit through the assistance of BIM, available previous LEED building data, or decision-making strategies technic and so on. However, there is no previous study which offers limited guidance for developing countries like Vietnam. Therefore, an analysis of certified LEED projects is needed to promote the development of sustainable construction in Vietnam by providing direction of goal-setting to designers and developers. In addition, the developers of the Vietnam Green Building Council also have grounds to adjust their policies more appropriately for promoting sustainable construction in Vietnam.

2.2. Comparison of LEED 2009 and LEED Version 4

Every few years, the USGBC offers a new version of LEED to capture new trends in the construction industry, and embraces new opportunities for encouraging sustainable construction. [27]. There are several versions of LEED being used in Vietnamese existing buildings, but the 2009 LEED version has occupied for the majority of current certified projects and is applied to projects registered before 2018. The previous versions of LEED, which are LEED v2.2 and LEED 2009, both have six main categories, including sustainable sites (SS), the water efficiency (WE), the energy and atmosphere (EA), the material and resources (MR), the indoor environmental quality (IEQ), and the innovation in design (ID) [6,25]. LEED 2009 consists of 6 major sets of criteria with 12 prerequisite credits and 42 optional credits, and credits also have different performance strategies [11]. The newer version is the LEED v4 and LEED v4.1 that was announced in 2014 and mandatorily applied to all newly registered projects from 2018 [8]. Currently, most current projects in Vietnam are certified LEED 2009, while new projects need to comply with LEED v4 requirements. To apply for future projects, the results of this analysis should be considered by checking the differences between LEED 2009 and LEED v4.

The updated version, LEED v4, still has the same total score (110 pts), but the SS (26 pts) criteria group is divided into two smaller criteria groups, which are "Location and Transportation (LT)" (16 pts) and SS (10 pts) [8,25,28]. The updated Location and Transportation criteria were published in LEED v4, including some Sustainable Site credits of LEED 2009, such as protection and green transportation parking, land usage and so on. Meanwhile, the new SS criteria group includes light pollution reduction, site usage and protection, rainwater, and heat island credits [28]. In fact, the assessment of SS issues is very similar to the assessment of LT+SS issues [25]. In LEED v4 EA, MR, EQ criteria, credits are still included as in LEED 2009 but there are some changes in the detailed requirements and more recommended green strategies [28]. Only one new credit, acoustic performance, was introduced in the LEED NCv4 [25]. According to the USGBC report [28], although there is not much change in the score distribution between categories, several credits have been added and some of its requirements have been majorly changed in the new version. Svetlana Pushkar's research evaluates that the majority of LEED version 4 is similar to the previous LEED 2009 and that it is feasible to modify the algorithm for LEED v4 extended credits [25].

3. Research Method

3.1. Scope of the Study

The objective of the study is investigating LEED-certified buildings data to find the association between each LEED credits compliance performance and LEED certificate targeted level of the project. From the results, we provided a guideline to the project team in making decisions on their green strategies and the LEED credits decision that will be more cost-effective. The study also aims to identify differences in green building projects in Vietnam and other developed countries as a basis for adjusting Vietnam's existing green building standards such as LOTUS (Vietnam Green Building Council) and TCVN 09-2017 (Vietnam Ministry of Construction). Therefore, the population of the study is all LEED projects in Vietnam, but the number of projects is minimal and mostly LEED NC projects. The target of this study is collecting and analyzing all existing LEED NC projects (November 2019). Furthermore, similar to the other countries' conditions, the number of LEED v4 projects in Vietnam is also minimal [3], so data of the research is obtained from LEED 2009 projects in Vietnam.

3.2. Data Collection

Project data was collected from the official website of the U.S, Green Building Council. In an effort to promote and disseminate the LEED standard system globally, the USGBC has published numerous documents as well as data on their successful projects. At the time of the study, a total of 42 LEED 2009 NC projects were certified, of which only 36/42 projects had complete data on the credits received. Despite the limited number of samples, the study was conducted over the entire population, and the statistical results of the data were accepted as a normal distribution according to the central limit principle [29] (p. 208).

3.3. Characteristics of the Vietnamese Green Building

Factors of project characteristics, location, and climate are critical to the selection of appropriate investment strategies. The average annual temperature in Vietnam is from 22 °C to 27 °C. Every year, there are about 100 rainy days with an average rainfall of 1500 mm to 2000 mm and the air humidity is around 80% [30]. The majority of projects (28/36) investigated in this study are located in the southern part of Vietnam, which has a humid tropical climate with a low amplitude of heat fluctuation during the year, with high temperature and humidity. Therefore, most LEED projects are not equipped with heating equipment. Furthermore, these projects are also mainly factory projects or office buildings, where the developer is responsible for operating their project, and the energy-saving strategies bring direct efficiency to their operating costs. Post-tropical humid weather with heavy rainfall and the diversity of native species also offer many advantages in developing LEED strategies related to greenery and irrigation. The sunshine hours in many areas of Vietnam are quite high, with the number of sunshine hours being about 1500–2000 h [30]; also, the average annual radiant heat is 100 kcal/cm^2 which provides many advantages for solar energy solutions. However, similar to many developing countries, the shortage of various green materials also makes it more difficult to implement LEED credits for material criteria.

3.4. Data Analysis

In this study, 2009 LEED credits are divided into two groups for analysis due to their characteristics. The first group is the credits that only awarded a maximum of 1 point, meaning that the project team can only decide to implement or not. The research will then investigate the credit of group 1 based on the frequency of awarded credits for LEED-certified projects in Vietnam. In the second group, the number of points awarded per credit is greater than 2, and the project may achieve different levels of compliance to request based on the project's conditions. For group 2, the credit achievement degree (CAD) will be revised from W. Peng's (2017) [6] study, and be calculated based on the following Equation (1):

$$CAD = \frac{PCO}{TPC} \times 100\% \quad (1)$$

where CO is the point that credit obtained and TC equals to total points of credits. As such, the CAD represents the level of fulfillment of LEED's requirements.

In this study, the statistical tests used to analyze LEED awarded credits from collected data include the Mean-Value Ranking test, One-sample *t*-test, Kolmogorov–Smirnov test, Wilcoxon–Mann–Whitney and non-parametric effect size index (Figure 1). At first, in order to identify common LEED credits, the study conducted the ranking techniques for first and second group credits. In the second step, the One-sample *t*-test was used to determine the significant LEED awarded credits in the previous Vietnamese LEED projects. The given hypothesis H0 is "the frequency of awarded credit is less than 50% in the Vietnamese LEED projects" with the confident index of 95%. In the third step, the normal distribution test is implemented because the number of samples of LEED silver, gold LEED, and LEED platinum projects is limited. Furthermore, the results of the Kolmogorov–Smirnov test showed that data is not a normal distribution, and LEED data are presented in ordinal scales [17]. To investigate the significant differences between the two groups with different LEED certificate levels of certification and consider the problem of pseudoreplication, the statistical tests used are based on suggestions of S. Pushkar's (2018) study [25]. Following the non-parametric tests of S. Pushkar's (2018) study, Cliff's δ and WMW tests [29] were used to compare the two unpaired groups. The data are presented as the median ± interquartile range (IQR, 25th–75th percentile) [17]. Cliff's δ is used to measure the substantive significance (effect size) between two non-paired groups. Cliff's δ [31] (p. 495) is calculated following Equation (2) where x1 and x2 are the points in 2 compared groups respectively; n1 and n2 are the sizes of the sample groups, and # is the index times. The WMW test was used to determine the statistical difference between the two unpaired groups. It should be noted that WMW tests can be applied in two forms: approximate or extracted. If the sample size is more than 9, an approximate WMW test has been used, and the extracted form for all else [32] (p. 56).

$$\delta = \#(x1 > x2) - \#(x1 < x2)/(n1n2) \quad (2)$$

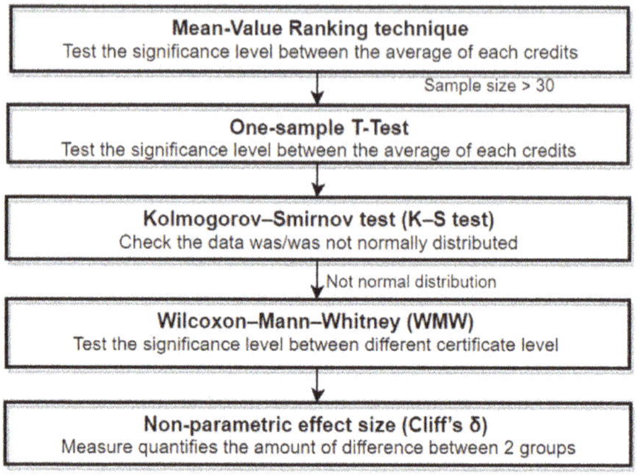

Figure 1. The summary of the data analysis test framework of the study.

Finally, the Neo-Fisherian significance assessments (NFSAs) are used to interpret the signs and magnitudes of the statistical effects [17]. Based on NFSAs, precise P-values were evaluated and shown according to a three-valued logic as follows:

- "Positive": The credit seems to be a difference between two compared groups.
- "Negative": The credit does not seem to be a difference between two compared groups.
- "Suspended": The credit regards the difference between two compared groups.

4. Results

4.1. The Popularly Awarded LEED Credits in the Vietnamese Projects

The critical objective of this study is to investigate the credits that are commonly implemented at LEED projects in Vietnam. As a result, the study suggested a list of LEED credits that should be considered for future projects, which were shown through the analyzed result of ranking Technic methods (in Table 1) and *t*-test (in Table 2). Please see the detailed description of the LEED credits and the corresponding symbol in Appendix A.

As mentioned above, the 1st group are the credits that the project can only achieve or not. Thus, the result of ranking technics of the 1st group, which were shown in Table 1, illustrated the ranking based on credits' frequency. If the credits' frequency is similar, the credits have a smaller standard deviation index that would be prioritized. Similar to the first group, Table 2 showed the ranking of the second group of LEED credits based on the CAD index; the average points the creditors have set are also presented. At LEED projects in Vietnam, the most popular 1st group credits include IDc2 (97.22%), EQc4.2 (94.44%), and SSc1 (91.64%) compared to the opposite side of rarely awarded credits such as SSc3 (0%), MRc6 (0%), and MRc7 (0%). In the second group of credits, the high popularity credits are SSc4.3 (100%), SSc4.4 (97.22%), MRc5 (94.44%), WEc3 (93.75%), and WEc2 (91.67%), and the low popularity credits were MRc1 (10.19%), and MRc3 (0%). We suggested that the inexperienced LEED project teams should select the credits to consider for their projects based on the rankings in Table 1.

Table 1. (**a**) Ranking the popularity of 1st group LEED credits in Vietnam; (**b**) ranking the popularity of 2nd group LEED credits in Vietnam.

Rank	ID	Points	Frequency	STD	Rank	ID	Points	Frequency	STD
1	IDc2	1	97.22	16.67	16	EQc3.2	1	36.11	48.71
2	EQc4.2	1	94.44	23.23	17	SSc5.1	1	33.33	47.81
3	SSc1	1	91.67	28.03	18	SSc6.2	1	33.33	47.81
4	EQc3.1	1	88.89	31.87	19	EQc4.4	1	30.56	46.72
5	SSc7.1	1	86.11	35.07	20	EQc7.1	1	30.56	46.72
6	SSc4.2	1	83.33	37.80	21	EQc7.2	1	27.78	45.43
7	SSc7.2	1	80.56	40.14	22	SSc6.1	1	25.00	43.92
8	EQc4.3	1	77.78	42.16	23	EQc1	1	22.22	42.16
9	EQc4.1	1	72.22	45.43	24	EQc5	1	19.44	40.14
10	SSc5.2	1	69.44	46.72	25	EQc6.2	1	5.56	23.23
11	EQc2	1	63.89	48.71	26	MRc1.2	1	2.78	16.67
12	EQc6.1	1	52.78	50.63	27	SSc3	1	0.00	0.00
13	EQc8.1	1	50.00	50.71	28	MRc6	1	0.00	0.00
14	SSc8	1	41.67	50.00	29	MRc7	1	0.00	0.00
15	EQc8.2	1	38.89	49.44					

(**a**)

Rank	ID	Max Points	CAD Mean	Ave. Points	STD	Rank	ID	Max Points	CAD Mean	Ave. Points	STD
1	SSc4.3	3	100.00	3.00	0.00	11	IDc1	3	72.22	2.17	22.57
2	SSc4.4	2	97.22	1.94	16.67	12	EAc3	2	69.44	1.39	46.72
3	MRc5	2	94.44	1.89	23.23	13	EAc4	2	69.44	1.39	46.72
4	WEc3	4	93.75	3.75	15.09	14	EAc1	19	53.36	10.14	30.85
5	WEc2	2	91.67	1.83	28.03	15	SSc2	5	27.78	1.39	45.43
6	WEc1	4	87.50	3.50	21.96	16	EAc2	7	25.00	1.75	41.03
7	MRc2	2	80.56	1.61	38.32	17	EAc6	2	19.44	0.39	40.14
8	MRc4	2	80.56	1.61	38.32	18	MRc1.1	3	10.19	0.31	29.62
9	SSc4.1	6	80.56	4.83	40.14	19	MRc3	2	0.00	0.00	0.00
10	EAc5	3	80.56	2.42	40.14						

(**b**)

Table 2. The results of the normal distribution test and one-sample *t*-test.

1st Group					2nd Group				
ID	N	Frequency	*p*-Value	Sig. Value	ID	N	CAD Mean Value	*p*-Value	Sig. Value
SSc1 *	36	91.67	0.000 [a]	0.000 [c]	SSc2	36	27.78	0.000 [a]	0.000 [c]
SSc3	36	0	- [b]	- [b]	SSc4.1	36	80.56	0.000 [a]	0.412
SSc4.2 *	36	83.33	0.000 [a]	0.000 [c]	SSc4.3 *	36	1000.00	- [b]	- [b]
SSc5.1	36	33.33	0.000 [a]	0.044	SSc4.4 *	36	97.22	0.000 [a]	0.000 [c]
SSc5.2 *	36	69.44	0.000 [a]	0.017 [c]	WEc1 *	36	87.50	0.000 [a]	0.002 [c]
SSc6.1	36	25	0.000 [a]	0.002 [c]	WEc2 *	36	91.67	0.000 [a]	0.001 [c]
SSc6.2	36	33.33	0.000 [a]	0.044 [c]	WEc3 *	36	93.75	0.000 [a]	0.000 [c]
SSc7.1 *	36	86.11	0.000 [a]	0.000 [c]	EAc1 *	36	53.36	0.063	0.000 [c]
SSc7.2 *	36	80.56	0.000 [a]	0.000 [c]	EAc2	36	250.00	0.000 [a]	0.000 [c]
SSc8	36	41.67	0.000 [a]	.324	EAc3	36	69.44	0.000 [a]	0.480
MRc1.2	36	2.78	0.000 [a]	0.000 [c]	EAc4	36	69.44	0.000 [a]	0.480
MRc6	36	0	- [b]	- [b]	EAc5	36	80.56	0.000 [a]	0.412
MRc7	36	0	- [b]	- [b]	EAc6	36	19.44	0.000 [a]	0.000 [c]
EQc1	36	22.22	0.000 [a]	0.000 [c]	MRc1.1	36	10.19	0.000 [a]	0.000 [c]
EQc2	36	63.89	0.000 [a]	0.096	MRc2	36	80.56	0.000 [a]	0.390
EQc3.1 *	36	88.89	0.000 [a]	0.000 [c]	MRc3	36	00.00	- [b]	- [b]
EQc3.2	36	36.11	0.000 [a]	0.096	MRc4	36	80.56	0.000 [a]	0.390
EQc4.1 *	36	72.22	0.000 [a]	0.006 [c]	MRc5 *	36	94.44	0.000 [a]	0.000 [c]

Table 2. *Cont.*

1st Group					2nd Group				
ID	N	Frequency	*p*-Value	Sig. Value	ID	N	CAD Mean Value	*p*-Value	Sig. Value
EQc4.2 *	36	94.44	0.000 [a]	0.000 [c]	IDc1	36	72.22	0.004 [a]	0.465
EQc4.3 *	36	77.78	0.000 [a]	0.000 [c]	SSc2	36	27.78	0.000 [a]	0.000 [c]
EQc4.4	36	30.56	0.000 [a]	0.017 [c]	SSc4.1	36	80.56	0.000 [a]	0.412
EQc5	36	19.44	0.000 [a]	0.000 [c]	SSc4.3 *	36	1000.00	- [b]	- [b]
EQc6.1	36	52.78	0.000 [a]	0.744	SSc4.4	36	97.22	0.000 [a]	0.000 [c]
EQc6.2	36	5.56	0.000 [a]	0.000 [c]	WEc1	36	87.50	0.000 [a]	0.002 [c]
EQc7.1	36	30.56	0.000 [a]	0.017 [c]	WEc2	36	91.67	0.000 [a]	0.001 [c]
EQc7.2	36	27.78	0.000 [a]	0.006 [c]					
EQc8.1	36	50	0.000 [a]	10.000					
EQc8.2	36	38.89	0.000 [a]	0.186					
IDc2 *	36	97.22	0.000 [a]	0.000 [c]					

[a] The Kolmogorov–Smirnov test *p*-value is less than 0.05. [b] The Standard deviation = 0, the Kolmogorov–Smirnov test and *t*-test could not be calculated. [c] The *t*-test *p*-value is less than 0.05. * The awarded credits with a significant popular rate in Vietnam.

As shown in Table 2, the results of the Kolmogorov–Smirnow test determine that most of the data collected are not a normal distribution. Therefore, the *t*-test is only performed for the whole sample (number of samples >30), and its results are also presented in Table 2. The *t*-test was conducted to identify the significantly awarded credits at LEED projects in Vietnam with a comparative value of 50%, and 95% of the confident index was set. The results of significantly awarded credits at LEED projects in Vietnam are:

- First group: SSc1, SSc4.2, SSc5.2, SSc7.1, SSc7.2, EQc3.1, EQc4.1, EQc4.2, EQc4.3, IDc2.
- Secord group: SSc4.3, SSc4.4, WEc1, WEc2, WEc3, EAc1, MRc5.

4.2. The Difference between LEED Project Target Goal Levels

USGBC classifies buildings by four levels of LEED certification, including LEED certificated, LEED Silver, LEED Gold, and LEED Platinum. The higher the certification level, the more the project has to comply with LEED's strict requirements, which means they have to spend more on sustainable investments and additional work. Investigation of LEED credits that differ between these different project groups on project goals is based on the results of the Wilcoxon–Mann–Whitney test and the Cliff's delta values. The approximate WMW test results and Cliff's delta values of LEED credits between the groups of silver LEED and gold LEED projects (n = 12) are shown in Tables 3a and 4a. Similarly, Tables 3b and 4b presented the results of the extract WMW test and Cliff's delta between the groups of the LEED gold and LEED platinum projects (n = 7).

Table 3. The results of the Wilcoxon–Mann–Whitney test and Cliff's delta indexes for the 1st group credits. (**a**) Between the groups of LEED Silver and LEED Gold certification projects; (**b**) between the groups of LEED Gold and LEED Platinum certification projects.

(a) LEED Silver vs. LEED Gold					(b) LEED Gold vs. LEED Platinum				
ID	**n**	**Sig.**	**Cliff's Delta**	**NFSAs**	**ID**	**n**	**Sig.**	**Cliff's Delta**	**NFSAs**
SSc1	12	0.317	0.083	Negative	SSc1	7	0.530	0.143	Negative
SSc3	12	1.000	0.000	Negative	SSc3	7	1.000	0.000	Negative
SSc4.2	12	0.356	0.167	Negative	SSc4.2	7	0.023	−0.571	Positive
SSc5.1	12	0.356	−0.167	Negative	SSc5.1	7	0.122	−0.429	Suspended
SSc5.2	12	.688	−0.083	Negative	SSc5.2	7	0.060	−0.429	Suspended
SSc6.1	12	0.356	0.167	Negative	SSc6.1	7	0.530	0.143	Negative
SSc6.2	12	1.000	0.000	Negative	SSc6.2	7	0.530	−0.143	Negative
SSc7.1	12	0.284	−0.167	Negative	SSc7.1	7	1.000	0.000	Negative
SSc7.2	12	0.028	0.417	Positive	SSc7.2	7	0.023	−0.571	Positive
SSc8	12	0.105	−0.333	Suspended	SSc8	7	0.591	0.143	Negative
MRc1.2	12	0.317	−0.083	Negative	MRc1.2	7	1.000	0.000	Negative
MRc6	12	1.000	0.000	Negative	MRc6	7	1.000	0.000	Negative
MRc7	12	1.000	0.000	Negative	MRc7	7	1.000	0.000	Negative
EQc1	12	1.000	0.000	Negative	EQc1	7	0.141	−0.286	Suspended
EQc2	12	1.000	0.000	Negative	EQc2	7	0.591	−0.143	Negative
EQc3.1	12	0.070	0.250	Suspended	EQc3.1	7	1.000	0.000	Negative
EQc3.2	12	0.105	0.333	Suspended	EQc3.2	7	0.530	−0.143	Negative
EQc4.1	12	0.028	−0.417	Positive	EQc4.1	7	1.000	0.000	Negative
EQc4.2	12	0.148	0.167	Suspended	EQc4.2	7	1.000	0.000	Negative
EQc4.3	12	0.140	−0.250	Suspended	EQc4.3	7	1.000	0.000	Negative
EQc4.4	12	0.660	0.083	Negative	EQc4.4	7	0.606	−0.143	Negative
EQc5	12	0.032	0.333	Positive	EQc5	7	0.141	−0.286	Suspended
EQc6.1	12	0.418	−0.167	Negative	EQc6.1	7	0.107	−0.429	Suspended
EQc6.2	12	0.317	0.083	Negative	EQc6.2	7	0.317	−0.143	Negative
EQc7.1	12	0.187	0.250	Suspended	EQc7.1	7	0.141	−0.286	Suspended
EQc7.2	12	0.356	0.167	Negative	EQc7.2	7	0.141	−0.286	Suspended
EQc8.1	12	0.000	−0.750	Positive	EQc8.1	7	0.060	0.429	Suspended
EQc8.2	12	0.216	−0.250	Negative	EQc8.2	7	0.254	0.286	Negative
IDc2	12	0.317	−0.083	Negative	IDc2		1.000	0.000	Negative

Table 4. The results of the Wilcoxon–Mann–Whitney test and Cliff's delta indexes for the 2nd group credits. (**a**) Between the groups of LEED Silver and LEED Gold certification projects; (**b**) between the groups of LEED Gold and LEED Platinum certification projects.

(a) LEED Silver vs. LEED Gold					(b) LEED Gold vs. LEED Platinum				
ID	**n**	**Sig.**	**Cliff's Delta**	**NFSAs**	**ID**	**n**	**Sig.**	**Cliff's Delta**	**NFSAs**
SSc2	12	0.356	−0.167	Negative	SSc2	7	0.591	−0.143	Negative
SSc4.1	12	0.623	−0.083	Negative	SSc4.1	7	1.000	0.000	Negative
SSc4.3	12	1.000	0.000	Negative	SSc4.3	7	1.000	0.000	Negative
SSc4.4	12	0.317	0.083	Negative	SSc4.4	7	1.000	0.000	Negative
WEc1	12	0.065	−0.333	Suspended	WEc1	7	0.317	0.143	Negative
WEc2	12	0.070	−0.250	Suspended	WEc2	7	1.000	0.000	Negative
WEc3	12	0.654	−0.076	Negative	WEc3	7	0.142	−0.286	Suspended
EAc1	12	0.011	−0.611	Positive	EAc1	7	0.020	−0.735	Positive
EAc2	12	0.776	0.049	Negative	EAc2	7	0.001	−1.000	Positive
EAc3	12	0.680	0.083	Negative	EAc3	7	0.141	−0.286	Suspended
EAc4	12	1.000	0.000	Negative	EAc4	7	0.530	0.143	Negative
EAc5	12	0.356	−0.167	Negative	EAc5	7	0.317	−0.143	Negative
EAc6	12	0.284	−0.167	Negative	EAc6	7	0.530	−0.143	Negative
MRc1.1	12	0.482	−0.097	Negative	MRc1.1	7	0.317	0.143	Negative
MRc2	12	0.683	−0.069	Negative	MRc2	7	0.872	−0.041	Negative
MRc3	12	1.000	0.000	Negative	MRc3	7	1.000	0.000	Negative
MRc4	12	0.638	−0.090	Negative	MRc4	7	0.142	−0.286	Suspended
MRc5	12	1.000	0.000	Negative	MRc5	7	0.317	−0.143	Negative
IDc1	12	0.738	0.076	Negative	IDc1	7	0.030	−0.633	Positive

The summary of results of Tables 3a and 4a, which were the result of comparing two groups of LEED Silver projects and the group of LEED Gold projects showed the five credits had significant differences, including SSc7.2, EQc4.1, EQc 5, EQc 8.1, and EAc1. In addition, other credits have the "suspended" status, such as SSc8, EQc3. 1, EQc3.2, EQc4.2, EQc4.3, etc., can be considered as a secondary choice for getting the LEED Gold certificate. When comparing the difference in credits achievement between the LEED Gold projects group and the LEED Platinum projects group, Tables 3b and 4b illustrated the five credits which were identified with significant differences, including SSc4.2, SSc7.2, EAc1, EAc2, and IDc1. The credits SSc5.1, SSc.2, EQc1, EQc5, etc., were also considered for the second choice for getting the LEED Platinum certificate. When comparing Tables 3 and 4, there were some duplicated credits such as EAc1, EQc5, EQc7.1, EQc7.1, EQc8.1, and SSc7.2. These credits are all part of the second set of credits, so project teams need to consider the different levels of achievement of these credits to achieve higher levels of credits.

5. Discussion

5.1. Comparison with the Well-Developed Countries

Table 5 is referenced from Jae-Yong Park's study (2017), and compared with the results of this study, the following points should be noted. The popular LEED credits of Vietnamese projects were focused on Costless-Easy and Cost-Easy credits groups. Additionally, several Cost-Hard group credits are popular such as WEc2, EAc1, EAc2, EQc4.3, with the reason being that these credits have a large percentage of points in LEED and most of the LEED projects in Vietnam are LEED silver or gold. A lot of Costless-Hard credits have not been implemented, reflecting the limited experience of LEED consultants in Vietnam. Therefore, some of the credits that are costly but still implemented, such as SSc5.2, SSc7.2, WEc1, WEc3, EQc4.1, EQc4.2, need to be carefully studied and judged before being applied to the projected practice.

Table 5. Classification table of costs and difficulty level of LEED credits 2009 by Jae-Yong Park (2017) [12].

Costless-Easy	Costless-Hard	Cost-Easy	Cost-Hard
SSc1 *	SSc3	SSc4.2	SSc6.1
SSc2	SSc7.1 *	SSc5.1	SSc6.2
SSc4.1	SSc8	SSc5.2 *	WEc2 *
SSc4.3 *	EAc4	SSc7.2 *	EAc1 *
SSc4.4 *	MRc1.1	WEc1 *	EAc2 *
MRc2	MRc1.2	WEc3 *	EAc3
MRc4	MRc3	EAc6	EAc5
MRc5 *	EQc2	MRc7	MRc6
EQc7.1	EQc3.1 *	EQc3.2	EQc1
EQc7.2	EQc8.1	EQc4.1 *	EQc4.3 *
	EQc8.2	EQc4.2 *	EQc4.4
		EQc6.1	EQc5
			EQc6.2

* are the significant popular credits at LEED projects in Vietnam.

In addition, LEED specialists need to study more and understand clearly the costless and uncommon credits group to offer appropriate solutions. For example, SSc2 and SSc4.1 credits relate to community connectivity planning, and it may not be feasible for many projects because of the current urban conditions in Vietnam. The construction waste management, MRc2, is currently only implemented by well-known contractors, so the Vietnamese government should have stricter regulations to promote them. As a result of uncommon materials with sustainability certificates, the credits for materials are not offered in the Vietnamese LEED project. The credits of SSc8, EQc8, EQc2, EAc4, and SSc3 had not been implemented in many projects, and this predicted that the idea contribution meetings did not work well. Therefore, the roles of LEED experts and the integrative design process need to be addressed in the future.

5.2. Converting Results to LEED Version 4

Besides the significant changes, as mentioned in the literature review section, some of the credits are combined or renamed, which included the following changes, such as the SSc7.1 and SSc7.2 credits are combined into SSc5; the EQc4.1, EQc4.2, EQc4.3 credits are grouped into EQc2; and most credits have changed names. Not only changing the names and aggregation of credits, but the content of these credits also had specific changes that the project team needs to refer to before deciding on the selection of sustainable credits for their LEED project in the future. Thus, after comparing and converting common LEED credits of LEED 2009 projects in Vietnam, the list of LEED v4 credits proposed by this study includes:

1. SSc1—Site Selection. (1 point)
2. LTc6—Bicycle facilities. (1 point)
3. SSc3—Open space. (1 point)
4. SSc5—Heat island reduction. (1 point)
5. EQc3—Construction indoor air quality management plan. (1 point)
6. EQc2—Low emitting materials. (2.75 points)
7. INc2—LEED Accredited Professional. (1 point)
8. LTc8—Green vehicles. (3.00 point)
9. LTc7—Reduced parking footprint. (1.95 points)
10. WEc1—Water Efficient Landscaping. (3.51 points)
11. WEc3—Water Use Reduction. (3.76 points)
12. WEc2—Innovative Wastewater Technologies. (1.84 points)
13. EAc2—Optimize energy performance. (10.38 points)
14. MRc3—Building product disclosure and optimization-sourcing of raw materials. (3.62 points)

6. Conclusions

The important contribution of LEED in driving green building in Vietnam is evidenced through the increase in the number of recent LEED projects registered in recent years. More studying and a deeper understanding of project objectives are needed for the development of green buildings without much additional cost. Understanding the development characteristics of green buildings in Vietnam is also significant for the Vietnamese government to establish appropriate Green Building guidelines and regulations. Therefore, the main contribution of this study to the investigation of the credits earned by LEED buildings in Vietnam, provide benchmarks for future LEED projects to improve. In summary, the key points are:

➢ The results of significantly awarded credits at LEED projects in Vietnam are identified and the recommended LEED v4 credits are SSc1—Site Selection; LTc6—Bicycle facilities; SSc3—Open space; SSc5—Heat island reduction; EQc3—Construction indoor air quality management plan; EQc2—Low emitting materials; INc2—LEED Accredited Professional; LTc8—Green vehicles; LTc7—Reduced parking footprint; WEc1—Water Efficient Landscaping; WEc3—Water Use Reduction; WEc2—Innovative Wastewater Technologies; EAc2—Optimize energy performance; and MRc3—Building product disclosure and optimization—sourcing of raw materials.

➢ The significant different credits between the LEED Silver and the LEED Gold projects are SSc5 (*SSc7.2*), ECc2 (*EQc4.1*), EQc3 (*EQc5*), EQc7 (*EQc8.1*), and EAc2 (*EAc1*).

➢ The significant different credits between the LEED Gold and the LEED Platinum projects are SSc1 (*SSc4.2*), SSc5 (*SSc7.2*), EAc2 (*EAc1*), EAc5 (*EAc2*), and INc1 (*IDc1*).

In addition, many limitations of research need to be mentioned. First, the number of LEED certificated projects in Vietnam is minimal, and detailed statistical analyses for each type of credit, or buildings in each region, are also not conducted. Thus, project teams need to have deeply researched similar projects because of the uniqueness of construction works. Research data was also collected from LEED v2009, and further research is needed based on data from new LEED v4 projects when the number of LEED v4 projects in Vietnam is large enough.

Author Contributions: Conceptualization draft, D.H.P. and B.K.; writing—original draft preparation, D.H.P.; writing—review and editing, D.H.P., A.C.A and J.L.; supervision, Y.A. and A.C.A. All authors have read and agreed to the published version of the manuscript.

Funding: This research received no external funding.

Acknowledgments: This research was supported by the Basic Science Research Program through the National Research Foundation of Korea (NRF) funded by the Ministry of Education (2019R1I1A1A01063207).

Conflicts of Interest: The authors declare no conflict of interest.

Appendix A

Table A1. Credits and Symbols of Credits in LEED 2009 and LEED v4.

LEED 2009			LEED v4		
ID	**Description**	**Max Points**	**ID**	**Description**	**Max Points**
Sustainable sites		26	**IPc1**	Integrative process	1
SSc1	Site selection	1	**Location & transportation**		**32**
SSc2	Development density and community connectivity	5	LTc1	LEED for Neighborhood Development location	16
SSc3	Brownfield redevelopment	1	LTc2	Credit Sensitive land protection	1
SSc4.1	Alternative transportation—public transportation access	6	LTc3	Credit High priority site	2
SSc4.2	Alternative transportation—bicycle storage and changing rooms	1	LTc4	Credit Surrounding density and diverse uses	5
SSc4.3	Alternative transportation—low-emitting and fuel-efficient vehicles	3	LTc5	Credit Access to quality transit	5
SSc4.4	Alternative transportation—parking capacity	2	LTc6	Credit Bicycle facilities	1
SSc5.1	Site development—protect or restore habitat	1	LTc7	Credit Reduced parking footprint	1
SSc5.2	Site development—maximize open space	1	LTc8	Credit Green vehicles	1
SSc6.1	Stormwater design—quantity control	1	**Sustainable sites**		10
SSc6.2	Stormwater design—quality control	1	SSc1	Site assessment	1
SSc7.1	Heat island effect—non-roof	1	SSc2	Site development—protect or restore habitat	2
SSc7.2	Heat island effect—roof	1	SSc3	Open space	1
SSc8	Light pollution reduction	1	SSc4	Rainwater Mgmt	3
Water efficiency		10	SSc5	Heat island reduction	2
WEc1	Water efficient landscaping	4	SSc6	Light pollution reduction	1
WEc2	Innovative wastewater technologies	2	**Water efficiency**		11
WEc3	Water use reduction	4	Wc1	Cooling tower water use	2

Table A1. *Cont.*

LEED 2009			LEED v4		
ID	Description	Max Points	ID	Description	Max Points
Energy and atmosphere		35	Wc2	Water metering	1
EAc1	Optimize energy performance	19	Wc3	Outdoor water use reduction	2
EAc2	On-site renewable energy	7	Wc4	Indoor water use reduction	6
EAc3	Enhanced commissioning	2	**Energy and atmosphere**		**33**
EAc4	Enhanced refrigerant management	2	EAc1	Enhanced commissioning	6
EAc5	Measurement and verification	3	EAc2	Advanced energy metering	1
EAc6	Green power	2	EAc3	Demand response	2
Material and resources		14	EAc4	Renewable energy production	3
MRc1.1	Building reuse—maintain existing walls, floors and roof	3	EAc5	Enhanced refrigerant Mgmt	1
MRc1.2	Building reuse—maintain interior nonstructural elements	1	EAc6	Green power and carbon offsets	2
MRc2	Construction waste management	2	EAc7	Optimize energy performance	18
MRc3	Materials reuse	2	**Material and resources**		**13**
MRc4	Recycled content	2	Mrc1	Building life-cycle impact reduction	5
MRc5	Regional materials	2	Mrc2	Building product disclosure and optimization—environm-ental product declarations	2
MRc6	Rapidly renewable materials	1			
MRc7	Certified wood	1	Mrc3	Building product disclosure and optimization—sourcing of raw materials	2
Indoor environmental quality		15			
EQc1	Outdoor air delivery monitoring	1	Mrc4	Building product disclosure and optimization—material ingredients	2
EQc2	Increased ventilation	1			
EQc3.1	Construction IAQ management plan—during construction	1	Mrc5	Construction and demolition waste Mgmt	2
EQc3.2	Construction IAQ management plan—before occupancy	1	**Indoor environmental quality**		**16**
EQc4.1	Low-emitting materials—adhesives and sealants	1	IEQc1	Enhanced IAQ strategies	2
EQc4.2	Low-emitting materials—paints and coatings	1	IEQc2	Low-emitting materials	3
EQc4.3	Low-emitting materials—flooring systems	1	IEQc3	Construction IAQ Mgmt plan	1
EQc4.4	Low-emitting materials—composite wood and agrifiber products	1	IEQc4	IAQ assessment	2
EQc5	Indoor chemical and pollutant source control	1	IEQc5	Thermal comfort	1
EQc6.1	Controllability of systems—lighting	1	IEQc6	Interior lighting	2
EQc6.2	Controllability of systems—thermal comfort	1	IEQc7	Daylight	3
EQc7.1	Thermal comfort—design	1	IEQc8	Quality views	1
EQc7.2	Thermal comfort—verification	1	IEQc9	Acoustic performance	1
EQc8.1	Daylight and views—daylight	1	**Innovation**		**6**
EQc8.2	Daylight and views—views	1	INc1	Innovation	5
Innovation		8	INc2	LEED Accredited Professional	1
IDc1	Innovation in design	3	**Regional priority credits**		**4**
IDc2	LEED Accredited Professional	1			
Regional priority credits		**4**			

References

1. Nguyen, H.-T.; Skitmore, M.; Gray, M.; Zhang, X.; Olanipekun, A.O.J. Will green building development take off? An exploratory study of barriers to green building in Vietnam. *Resour. Conserv. Recycl.* **2017**, *127*, 8–20. [CrossRef]
2. Kulp, S.A.; Strauss, B.H.J. New elevation data triple estimates of global vulnerability to sea-level rise and coastal flooding. *Nat. Commun.* **2019**, *10*, 4844. [CrossRef] [PubMed]
3. Pham, D.H.; Lee, J.; Ahn, Y.J. Implementing LEED v4 BD+ C Projects in Vietnam: Contributions and Challenges for General Contractor. *Sustainability* **2019**, *11*, 5449. [CrossRef]

4. Ahn, Y.H.; Jung, C.W.; Suh, M.; Jeon, M.H.J. Integrated construction process for green building. *Procedia Eng.* **2016**, *145*, 670–676. [CrossRef]
5. Choi, J.; Bhatla, A.; Stoppel, C.; Shane, J.J.S. LEED credit review system and optimization model for pursuing LEED certification. *Sustainability* **2015**, *7*, 13351–13377. [CrossRef]
6. Wu, P.; Song, Y.; Shou, W.; Chi, H.; Chong, H.-Y.; Sutrisna, M.J.R. A comprehensive analysis of the credits obtained by LEED 2009 certified green buildings. *Renew. Sustain. Energy Rev.* **2017**, *68*, 370–379. [CrossRef]
7. Dwaikat, L.N.; Ali, K.N.J. Green buildings cost premium: A review of empirical evidence. *Energy Build.* **2016**, *110*, 396–403. [CrossRef]
8. U.S. Green Building Council. *LEED v4 for Building Design and Construction*; U.S. Green Building Council: Washington, DC, USA, 2014.
9. Nyikos, D.M.; Thal, A.E.; Hicks, M.J.; Leach, S.E. To LEED or not to LEED: Analysis of cost premiums associated with sustainable facility design. *Eng. Manag. J.* **2012**, *24*, 50–62. [CrossRef]
10. Morris, P.; Langdon, D.J. What does green really cost. *PREA Q.* **2007**, 55–60.
11. Cheng, J.C.; Ma, L.J. A non-linear case-based reasoning approach for retrieval of similar cases and selection of target credits in LEED projects. *Build. Environ.* **2015**, *93*, 349–361. [CrossRef]
12. Park, J.-Y.; Choi, S.-G.; Kim, D.-K.; Jeong, M.-C.; Kong, J.-S. Credit optimization algorithm for calculating LEED costs. *Sustainability* **2017**, *9*, 1607. [CrossRef]
13. Jun, M.; Cheng, J.C. Selection of target LEED credits based on project information and climatic factors using data mining techniques. *Adv. Eng. Inform.* **2017**, *32*, 224–236. [CrossRef]
14. Ma, J.; Cheng, J.C. Data-driven study on the achievement of LEED credits using percentage of average score and association rule analysis. *Build. Environ.* **2016**, *98*, 121–132. [CrossRef]
15. Jalaei, F.; Jrade, A. Integrating building information modeling (BIM) and LEED system at the conceptual design stage of sustainable buildings. *Sustain. Cities Soc.* **2015**, *18*, 95–107. [CrossRef]
16. Chen, P.-H.; Nguyen, T.C. Integrating web map service and building information modeling for location and transportation analysis in green building certification process. *Autom. Constr.* **2017**, *77*, 52–66. [CrossRef]
17. Pushkar, S. Sacrificial Pseudoreplication in LEED Cross-Certification Strategy Assessment: Sampling Structures. *Sustainability* **2018**, *10*, 1353. [CrossRef]
18. Wu, P.; Song, Y.; Wang, J.; Wang, X.; Zhao, X.; He, Q. Regional variations of credits obtained by LEED 2009 certified green buildings—A country level analysis. *Sustainability* **2018**, *10*, 20. [CrossRef]
19. Da Silva, L.; Ruwanpura, J.Y. Review of the LEED points obtained by Canadian building projects. *J. Archit. Eng.* **2009**, *15*, 38–54. [CrossRef]
20. Cidell, J.; Beata, A. Spatial variation among green building certification categories: Does place matter? *Landsc. Urban Plan.* **2009**, *91*, 142–151. [CrossRef]
21. Madanayake, H.S.P.; Ruwanpura, J.Y. Multi-criteria decision making to improve performance in construction projects with LEED certification. In Proceedings of the Construction Research Congress 2012: Construction Challenges in a Flat World, West Lafayette, IN, USA, 21–23 May 2012; pp. 1899–1909.
22. Juan, Y.-K.; Gao, P.; Wang, J. A hybrid decision support system for sustainable office building renovation and energy performance improvement. *Energy Build.* **2010**, *42*, 290–297. [CrossRef]
23. Castro-Lacouture, D.; Sefair, J.A.; Flórez, L.; Medaglia, A.L. Optimization model for the selection of materials using a LEED-based green building rating system in Colombia. *Build. Environ.* **2009**, *44*, 1162–1170. [CrossRef]
24. Zimmerman, A.; Kibert, C.J. Informing LEED's next generation with The Natural Step. *Build. Res. Inf.* **2007**, *35*, 681–689. [CrossRef]
25. Pushkar, S. A Comparative Analysis of Gold Leadership in Energy and Environmental Design for New Construction 2009 Certified Projects in Finland, Sweden, Turkey, and Spain. *Appl. Sci.* **2018**, *8*, 1496. [CrossRef]
26. Potbhare, V.; Syal, M.; Arif, M.; Khalfan, M.M.; Egbu, C. Emergence of green building guidelines in developed countries and their impact on India. *J. Eng. Des. Technol.* **2009**, *7*, 99–121. [CrossRef]
27. Kubba, S. *LEED Practices, Certification, and Accreditation Handbook*; Butterworth-Heinemann: Oxford, UK, 2009.
28. USGBC. *Summary of Changes: LEED 2009 to v4—BD+C*; USGBC: Washington, DC, USA, 2009.
29. Hogg, R.V.; Tanis, E.A.; Zimmerman, D.L. *Probability and Statistical Inference*; Pearson: London, UK, 2015.
30. Tourism, G.D.o. Socialist Republic of Vietnam. Available online: http://asosai14.vn/80-1/viet-nam-dat-buoc-con-nguoi.sav (accessed on 20 January 2020).

31. Cliff, N. Dominance statistics: Ordinal analyses to answer ordinal questions. *Psychol. Bull.* **1993**, *114*, 494. [CrossRef]
32. Marx, A.; Backes, C.; Meese, E.; Lenhof, H.-P.; Keller, A. EDISON-WMW: Exact dynamic programing solution of the Wilcoxon–Mann–Whitney test. *Genom. Proteom. Bioinform.* **2016**, *14*, 55–61. [CrossRef]

sustainability

MDPI

Article

Integrated Design Process for Modular Construction Projects to Reduce Rework

Hosang Hyun [1], Hyunsoo Kim [2], Hyun-Soo Lee [3], Moonseo Park [3] and Jeonghoon Lee [4,*]

1 Institute of Engineering Research, Seoul National University, Gwanak-ro 1, Gwanak-gu, Seoul 08826, Korea; hhs0518@snu.ac.kr

2 Department of Architectural Engineering, Dankook University, 152 Jukjeon-ro, Suji-gu, Yongin-si, Gyeonggi-do 16890, Korea; hkim13@dankook.ac.kr

3 Department of Architecture and Architectural Engineering, Seoul National University, Gwanak-ro 1, Gwanak-gu, Seoul 08826, Korea; hyunslee@snu.ac.kr (H.-S.L.); mspark@snu.ac.kr (M.P.)

4 Technical Policy Research Team, Construction Engineer Policy Institute of Korea, Eonju-ro 650, Gangnam-gu, Seoul 06098, Korea

* Correspondence: jhlee@cepik.re.kr; Tel.: +82-10-8896-5956

Received: 26 November 2019; Accepted: 7 January 2020; Published: 10 January 2020

Abstract: In modular construction projects, unit production and onsite work are conducted concurrently, enabling shorter duration, lower cost, and improved quality. Because of the nature of the work, building design details should be determined early in the design phase, which requires information from participants. However, the design process for stick-built construction does not include such information, which leads to errors in design, such as omissions and conflicts of information from participants, causing reworking in the design phase. To reduce errors, an information flow should be identified representing when/what/how the information should be shared, and with whom. This paper proposes an integrated design process based on the information flow. To identify the flow, a precedence relationship between activities is represented using a dependency structure matrix (DSM). Then, the order of activities is rearranged using a partitioning algorithm. In this manner, unnecessary feedback and reverse information flow, which are related to errors, are reduced. Finally, the rearranged activities are proposed as an integrated design process. To validate the impact of the proposed process and methodology, interviews with experts were conducted. The validation results suggest that the project delivery method should also be considered in the early project phase in practical application.

Keywords: modular construction; rework; integrated design process; dependency structure matrix (DSM); process optimization

1. Introduction

1.1. Background

The global construction industry suffers from various issues, such as labor shortage, low productivity, and environmental pollution [1,2]. Modular construction has gained attention as an effective approach for overcoming these problems because of its inherent benefits, such as improved productivity, shorter project duration, higher quality, and better occupational safety, owing to the controlled work environment of a factory [3–6]. Moreover, modular construction reduces waste by at least 70% in comparison with the industry average [7,8]. Thus, it is a more sustainable construction method [9–11].

In modular construction, a significant proportion of work in a project is conducted in a factory; the produced units are transported and then assembled onsite [5,11]. Because production and onsite work are conducted concurrently, project duration is reduced significantly [12]. Owing to the nature of the work, design details, such as unit installation and the connection method, should be determined early in the design phase to reduce errors such as omissions and conflicts and to start unit production and onsite work simultaneously based on the agreed design [13]. The decision-making process for finalizing the details requires integration of project participants in the early design phase; hence, the complexity of information flows between the participants increase in comparison with the conventional stick-built construction [13,14]. This complexity due to multilateral information exchange is a source for errors resulting from omissions and conflicts in the design, which in serious cases may even lead to redoing of the whole design process for rectification [13,15]. The feedback in the phase comprises communication between participants to include the information into the design. However, when the necessary participants for providing the information are not identified in the information flow before the phase, omissions or conflicts may occur when the work planning is established after the design phase, which may necessitate redoing of the whole design process for rectification [7,12,15–21]. For example, if the Road Traffic Act regulations of an area are not considered in a related design step, the size of units may need to be reworked for transportation, which can cause schedule and cost overruns.

In modular construction, the complexity involved makes identification of the information flows difficult [14]. To facilitate the integration of participants and identification of the information flows in the design phase of a modular project, an integrated design process is required to ensure that the project is managed considering the entire project process, including design, unit production, transportation, erection, and performance [22,23]. However, designs are still established based on conventional stick-built construction methods, which are unsuitable for modular construction, and lack of expertise and knowledge in the marketplace make it difficult to recognize the need for the integrated design process [18,22,24]. This causes uncertainty in the design process, leading to unnecessary and inappropriate feedback, which may affect project performance [12,18,19,25]. Consequently, the lower performance caused by reworking results in a negative stigma and barrier in the usage of modular construction [18].

1.2. Research Methodology

To address this problem, this paper proposes an integrated design process for modular construction projects based on identified information flows between activities in the design and work planning phases. The scope of the design process in this paper includes activities related to architectural, mechanical, structural, and electrical design of buildings and site analysis in the early design phase. It also includes selected activities from the related work planning phase that should be integrated into the design phase to reduce rework. This is because the inability to make changes onsite is the biggest barrier to using modular construction methods as it is costlier to rework than conventional construction methods [2,6,12,22,26,27]. To select the activities, major issues causing rework in modular construction are identified by analyzing daily reports of modular projects and reviewing existing literature. Then, strategies for mitigating reworking are identified by reviewing descriptions of the report and literatures. These strategies consist of activities that prevent the causes of errors, such as omissions or conflicts in design, which should be included to reduce rework in the field such as production and on-site if not considered at the design stage.

In suggesting an integrated design process, information flows and the relationship between activities are first identified based on the precedence relationship. Then, as a methodology to facilitate information flows, a dependency structure matrix (DSM) is derived. DSM has been used to represent complex processes in systems and also used to alleviate the complexity by representing the information flow using a matrix [28]. Based on the DSM, the sequence of activities is rearranged using the partitioning algorithm. Through this rearrangement, information flows are optimized to reduce feedback (or reverse information flows) between activities; the rearranged activity sequence can be used as the integrated planning process because the process provides information to project participants: (1) when the mitigation activities should be conducted, and (2) what information should be shared with whom in the design phase. Finally, the rearranged activities are briefly described. It is expected that adoption of the process outlined in this study will help reduce engineering rework (rework to rectify the design), as well as field rework (rework to rebuild or reassemble building components).

2. Preliminary Study

2.1. Rework in Modular Construction Projects

Rework is simply defined as the unnecessary effort of redoing a process or activity that was incorrectly implemented [17]. Rework significantly affects the cost, schedule, and quality of construction projects. The direct costs alone are estimated to be 2–12% of the total construction cost; hence, rework should be managed effectively [15,17,29–31]. The causal relationship and effects of rework are usually analyzed to reduce rework in construction [15,16,20,21,31–33]. Typically, a significant proportion of rework is caused by errors made during the design process [15,17]. These errors usually appear downstream, and therefore have a negative impact on a project's performance [15,17,20,21]. The various causes of errors in the design phase include omissions from the project brief; ineffective or lack of coordination; inaccurate, incomplete, or conflicting documents; or an unrealistic design schedule [15]. To reduce errors in the design phase, it is necessary to integrate, and effectively coordinate among, the participants in the early design phase [16,17,29].

In modular construction, it is necessary for the project participants to be integrated in the early design phase, not only to reduce rework, but also to determine design details, such as unit installation and connection method [13]. Therefore, information from the participants, such as unit production and onsite work manager, should be included and applied to the related design steps [18]. However, using conventional stick-built construction design processes for integrating the participants may also lead to rework in itself, although there may be no errors in the design [18,24]. This is because the stick-built design processes do not suggest when/what/how the information from the other participants is applied. Hence, issues with identification of information flow lead to unnecessary reverse information flows and feedback, increasing the chances of omissions and conflicts between the completed design and requirements from participants. For example, if a modular building structure is designed using the stick-built construction process, and the deformation of units during lifting is not considered in the design, damage to internal or external finishes caused by deformation may occur because the self-load increases during lifting. The damaged finish would need to be reworked onsite [24].

In summary, (1) rework can occur without error in the design document that was established using the stick-built construction design process; (2) information regarding both unit production and onsite work should be applied to the related design step. Moreover, there are various details such as installation and connection method for unit assembly that should be determined, for which multilateral information exchange is required, which in turn increases the complexity of information flows in the design processes, as depicted in Figure 1 [2,22]. Figure 1 shows the extra intermediate process such as unit production, transportation compared to stick-built construction and feedbacks between participants in early phase. This means increased project constraints and considerations related to the extra process, which requires multilateral interactive communication [14]. The increased communication results in increased complexity [34,35].

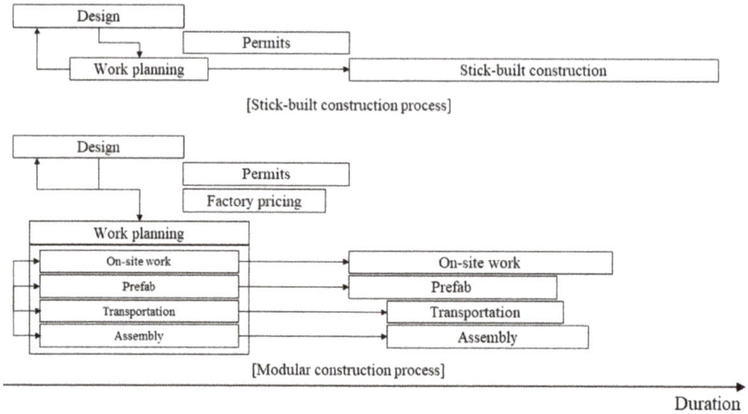

Figure 1. Increased complexity caused by interposed processes.

Figure 2 depicts the rework in structural components caused by omissions and conflict in design [15,16]. The errors causing the rework were not detected before conducting the works, and this proves that the information related to unit production and lifting were not applied to the design. Owing to the nature of modular construction work, any changes made onsite are not only difficult but also prohibitive in terms of cost [12,13,36]. In other words, if the design phase is not coordinated properly, modular construction can cost much more than stick-built construction [2,6,12,22,26,27].

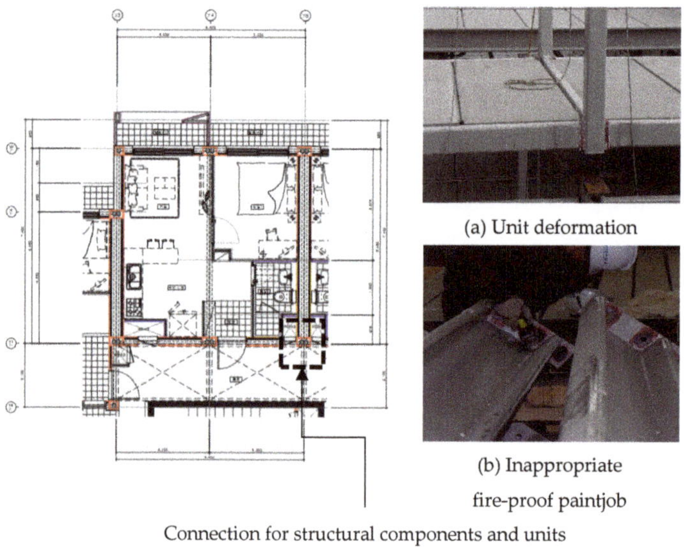

Figure 2. Examples of errors caused by omission in design.

In addition to the absence of an integrated process, there is currently limited expertise in modular construction in the marketplace; hence, the approach to design is still largely based on traditional methods [18,37,38]. This may lead to difficulty in identifying errors and omissions in the design, especially with complex information flows [15,17]. Figure 3 depicts the impact of undetected omissions in each design step. To alleviate the issue of lack of expertise, an integrated design process is required.

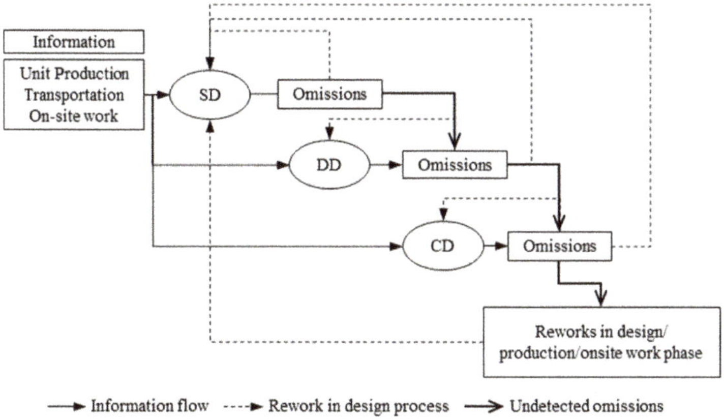

Figure 3. Impact of undetected omissions in each design step.

To support the design phase of modular constructions, several studies have been conducted previously [36–40]. Nahangi et al., proposed an automated approach for monitoring the defection and damage of modular units using laser scanned data [36]. Han et al., developed an optimization model for tower crane operation that can be used in the planning phase [39]. Lei et al., suggested an automated method for checking crane path, and the method could be used to prevent interference during tower crane operation in the planning phase [40]. Olearczyk et al. proposed a methodology for selecting tower crane capacity and location that can be used in the early planning phase [41]. To facilitate the decision making process in the design phase, Sharafi et al., developed a unified matrix approach for the automated design of multi-story modular buildings to effectively compare the building design alternatives in the early design phase considering architectural, structural, and constructional aspects. By using the approach, design alternatives can be suggested considering multidimensional relationship between units and spatial design that are able to achieve minimum cost, maximum regularity, efficient building energy consumption [37]. Then, Sharafi et al., suggested a list of critical decision making criteria that can be used for feasibility studies to select construction methods, such as the modular construction and stick-built construction methods, and developed a decision support system to facilitate the decision making process. By using the process, an appropriate construction system and the level of modularity can be determined [38]. These studies proved the importance of decision making in the early design phase and provided considerations that should be included for efficient modular building design. Smith identified the major issues; suggested guidelines for design, unit production, transportation, and onsite work planning; and analyzed the effect of design and planning change in the early stages [12]. However, these studies did not analyze the information relationship between activities in the design phase or an integrated design process. In summary, the project participants should be integrated to reduce errors in the design phase. However, the errors cannot be reduced only by participation and without the integrated design process. To reduce the errors, this paper proposes an integrated design process to identify and facilitate the information flow between participants.

2.2. Information Flow in the Design Process

To use the proposed process in the design phase, it should be established beforehand, and information from participants should be applied to the process. Therefore, what information is required at each design step and which activity generates the information needs to be identified. The information may include considerations for unit production, transportation, onsite work, or the whole project, which should be applied to the design. In addition to this, activities in each design step require information in advance, and information from each participant requires preceding activities. Activities that generate information have different precedence relationships, and feedback occurs between participants. This multilateral information exchange increases the complexity of the information flows, which acts as a source of errors—as identified earlier, omissions and conflicts. It has usually been difficult to handle this complexity because of unawareness and limited expertise in information flow management [18,22,24]. To alleviate the issue, information flows should be first identified, and then feedback in the flow should be managed to facilitate the design process.

To identify the information flow in a process, network analysis and bar chart techniques for planning construction work have been used extensively, but they are not capable of dealing with iterations in the planning process, such as feedback processes between activities [42]. To represent and solve the problem caused by information flow in a process, IDEF0 is used. In the IDEF0 model, an activity can be divided into sub-activities. Moreover, the flow representing input or output data of activities can be identified in the model, which is useful in modeling the system processes based on the information relationship between activities [43]. However, IDEF0 also does not have the capability to manage the feedback process and iterations in the system because the involved activity amount is too large, and the graph in the model will become unclear [43–45]. DSM has been used in modular building design [37,46]. Sharafi, Samali et al., used DSM for representing the connection type between modules and 3-dimensional cost optimized building design can be suggested using the unified matrix method in the paper. In the DSM, various interaction types are used for representing connection types between components [37]. DSM has also been used to improve the workflow of the modular unit production process based on the precedence relationship between activities on the production line [46]. It has been used to identify information flows and iterations within a process and to schedule activities with the objective of optimizing the task order. After the identification of information flow, DSM rearranges the order of activities to reduce unnecessary feedbacks and reverse information flows, thereby reducing engineering rework in the planning phase [42,47]. Therefore, DSM can be used to identify and facilitate information flows in the integrated design process for modular construction and to alleviate the complexity of the flow. In DSM, the relationship among activities can be characterized by three fundamental building blocks: parallel (independent), sequential (dependent), and coupled (interdependent), as shown in Figure 4. After the relationship is identified based on the relationship type, process optimization is conducted to reduce feedbacks and iterations in the process. There are various algorithms for process optimization, such as partitioning, tearing, clustering; an algorithm is selected based on the optimization objective. In this paper, a partitioning algorithm, which can help reduce feedback processes based on the information flow by reallocating the activity order, is used [42]. The partitioning algorithm manipulates the rows and columns in DSM to minimize feedbacks in the rearranged DSM. In the partitioning process, the activity that can be conducted without input from other activities is identified and placed at the top of the matrix; this process is iteratively conducted until no such activity remains. After the previous step, the activity delivering no information to other activity is identified and placed at the bottom of matrix and this process also repeated. Completion of these process means the DSM is partitioned [28]. Then, to manage the remaining feedback or reverse information flow, a clustering algorithm is used. The objective of clustering is to find subsets of DSM elements that are called clusters. The clusters should contain most of the interaction between activities internally and the clusters are generated based on the rule that the link or interaction between the clusters should be eliminated or minimized [28]. Using the clustering algorithm, the number of clusters can be minimized and it means a reduced number of feedback loops in the process. In modular projects, the reduced number of clusters means the project participants groups for sharing information can also be reduced and thus, improvement in the efficiency of information flow management.

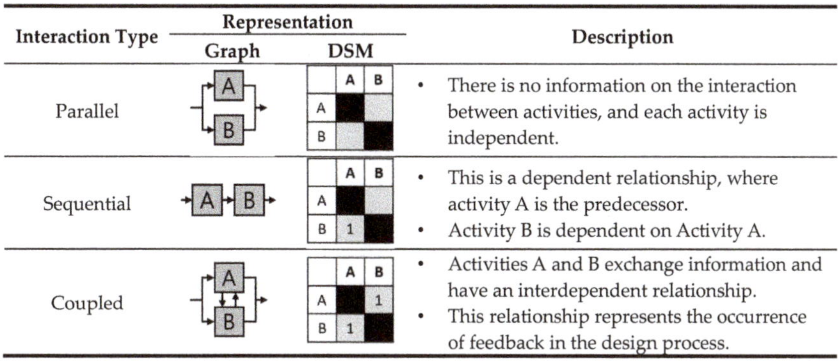

Figure 4. Information relationships in a DSM.

3. Information Flow Identification

3.1. Activities in Design Phase

To conceptualize an integrated design process, the information flow of the design process was first identified. This identification was based on the precedence relationship of each activity in the design phase. The precedence relationship includes activities that generate information, which is in turn required to conduct activities in the design steps. Based on the identified information flow, a DSM was established. It was then optimized using a partitioning algorithm to suggest an integrated design process [46,47]. The process consists of activities involved in the design process of the conventional stick-built construction. The scope of the process includes schematic design (SD), design development (DD), and construction documentation (CD). The American Institute of Architecture (AIA) has defined the typical building design process [48]. The activities in each design step were adopted from the process of the AIA and limited to those activities related to the determination of architectural, structural, electrical, and mechanical design. In this paper, the adopted activities were subdivided to represent the interaction between architecture and architectural engineers, such as mechanical–structural engineer. For example, article 3.2.5 in the AIA standard form agreement describes SD documents as follows: "Preliminary selections of major building systems and construction materials shall be noted on the drawings or described in writing". This article describes building systems and materials that should be provided in the SD documentation. However, the same comprehensive description cannot be used for the proposed integrated design process because the activities in the description, such as selecting building system and material, require different information to determine the specifications. Therefore, the description is subdivided, and the subdivided activities related to, inter alia, the building system and materials are listed in Table 1 from ID 5 to 8 [12,13,24,39].

All of the subdivided activities in each design step are listed in Table 1. The subdivided activities are conducted similar to activities in the design process for a conventional stick-built construction. In the SD phase, site layout planning is developed based on the site analysis results in the pre-design phase. Then, materials and a structural system are applied to each building design alternative. The alternatives are prepared to be selected for the DD phase, where the building design is selected among design alternatives and, finally, the selected design is developed. In this phase, materials that will be applied to the design are selected. Each project participant reviews the electrical, mechanical, and structural systems through a feedback process between participants. In the CD phase, details of the building components; mechanical, electrical, and plumbing (MEP); and structural system are documented. The document should be crosschecked by various participants because undetected errors in this phase directly lead to rework in the unit production and onsite work.

Table 1. Activities in the design phase.

Design Phase	ID	Activities
SD	1	Site analysis
	2	Site layout planning
	3	Establish design direction in terms of plan, section, and elevation
	4	Building core planning
	5	Explore interior and exterior material
	6	Review the strengths of materials, structural design criteria, and design load
	7	Review alternatives of structural system design, such as size of components (rough estimation of unit weight)
	8	MEP planning and MEP space requirement review
DD	9	Develop and modify the schematic design
	10	Determine interior and exterior materials
	11	Determine structural system design
	12	Structural design analysis and structural calculation documentation
	13	Draft the locations and sizes of structural components
	14	Bar arrangement drawing documentation
	15	Determine MEP system
	16	Review MEP and structural component interference
CD	17	Prepare construction document of the design
	18	Structural and MEP system adjustment and documentation
	19	Architectural detail, specifications, structural calculation documentation and incorporate subcontractor's documentation
	20	Principal structural parts finishing propriety review
	21	Check errors and omissions in documentation and constructability review

3.2. *Causes of Rework in Modular Constructions*

As mentioned earlier, rework can be classified into two categories: engineering rework and field rework [29,49,50]. The proposed integrated design process intends to reduce engineering rework, which is rework made to rectify errors or omissions in the design phase. In this paper, it is assumed that engineering rework in modular projects occurs when information from participants is omitted in the design phase. To identify such information, many works were reviewed, focusing on modular construction planning [12,13,24,36,39–41,51–56]. However, rework is also performed to rectify defects after the design phase, such as the unit production, transportation, or onsite work phases. Rework conducted in these phases is called field rework in this paper. Field rework also affects project performance. A significant portion of the causes of field rework are from errors and omissions that are not detected in the design phase [15,17]. An efficient measure to reduce field rework is to develop improved design procedures by incorporating work planning to check causes [49]. In order to include activities that should be integrated into the design process to reduce rework, causes of rework that are not included in the existing literature are identified by analyzing daily reports of modular projects, which document details of work activities and issues. These cases and their causes of rework are collected from daily reports and other sources in the literature, as presented in Table 2. Then, activities to mitigate rework are established. The mitigation plans related to the causes are derived from the literature [12,13,24,36,39–41,51–56]. The mitigation plans are described in Table 3. Each mitigation plan is an activity in the work plan for off/on-site work that should be integrated into the design process. Activities in the mitigation plan consist of activities for reviewing or checking the causes, and errors in design can be reduced by conducting the activities in the design phase. For example, in Figure 2, rework was caused because a fireproof paint was applied to a unit assembly connection. Fireproof paint should not be applied to such connections as the structural performance of the painted connection may decrease. Therefore, the paint had to be removed from the connection and the units had to be reassembled. This implies that the interference between fireproof painting and connection was not marked in the shop drawing for unit manufacturing. To prevent rework, an activity whereby errors and omissions in the design are checked is included, to be conducted while preparing shop drawings in the manufacturing planning phase.

Table 2. Cases and causes of rework in modular construction projects.

Phase	Cause	Related Research
Manufacturing	Interference between MEP and structural components	[19,24,54]
	Fireproof paint application to connection	[24], daily report
	Rework caused by error of activity order on manufacturing line	[19,24], daily report
	Excess manufacturing tolerance	[36]
	Quality deterioration in manufacturing process	[24,36]
Transportation	Damage and deformation in transportation	[19,24,55]
	Revision of design caused by omission of reviewing Road Traffic Act in the design process	[12,13,24]
Onsite work	Interference between MEP and structural components, such as concrete foundations	[19,24,54], daily report
	Damage to unit caused by onsite work interference	[24,39–41,51,52]
	Shortage of passage space for workers in PIT	[24], daily report
	Occurrence of component tolerance errors	[24,56], daily report
	Occurrence of tolerance error in MEP	[24,54], daily report
	Unit deformation caused by unit lifting	[12,24]
	Damage caused by weather conditions on un-proofed components	[24], daily report

Table 3. Mitigation plan for reducing rework.

Phase	ID	Activities in Rework Mitigation Plan
Manufacturing	22	(Production plan) Determine and review the work activities for unit production in the factory (determine factory work)
	23	(Production plan) Prepare shop drawings and check interference of unit production
	24	(Production plan) Prepare unit production line design
	25	(Quality management plan) Prepare manufacturing tolerance management plan
	26	(Quality management plan) Prepare quality management plan
Transportation	27	(Transportation plan) Prepare management plan for reducing deformation and damage
	28	(Transportation plan) Review the Road Traffic Act regulations (check the weight and size of unit)
Onsite Work	29	(Onsite work plan) Determine onsite work activities
	30	(Onsite work plan) Select tower crane location
	31	(Onsite work plan) Select tower crane specification
	32	(Onsite work plan) Prepare shop drawings and check for interference
	33	(Onsite work plan) Review constructability for onsite work
	34	(Quality management plan) Prepare onsite work tolerance management plan
	35	(Quality management plan) Prepare deformation management plan in unit lifting process
	36	(Quality management plan) Prepare unit proofing plan to reduce damage from weather conditions

3.3. *Identification of Information Flow for DSM*

To develop the proposed integrated design process, including the rework mitigation plan, the information flows between activities should be identified. Through this identification, the activities in the process are rearranged to facilitate information flows using the partitioning algorithm. Through the rearrangement, activities in the mitigation plan are allocated in the design process, and the type and time of information from project participants to be shared are provided. Therefore, the integrated process shows the procedure of the design process including the activities in the mitigation plan. In Table 4, the information flow between activities in the designing and planning phases has been identified [12,13,24,36,39–41,51–56]. Each activity in Table 4 has predecessors. The predecessor column lists the IDs of the activities that must be conducted before each activity. In other words, the activity can only start with the information obtained by completing the predecessor activities. For example, site layout planning should be preceded by activities, such as site analysis, reviewing the Road Traffic Act regulations, and tower crane location selection. However, when selecting the tower crane location, the site layout planning is considered to be a predecessor, which implies that there is a feedback process between the site layout planning and tower crane location selection. Therefore, when conducting activities, activity information, such as building and crane location, should be exchanged and shared. Moreover, the Road Traffic Act regulation review follows the activity of site layout planning while also providing the site layout planning activity with information. The reverse information flow here indicates that there is a possibility of engineering rework in the design phase. Feedback and reverse information flows can be reduced by rearranging the activity order during the optimization of the design process. The order is rearranged based on information flows.

Table 4. Activity relationship based on information flows.

Phase	ID	Activities	Predecessor
Schematic design	1	Site analysis	—
	2	Site layout planning	1, 28, 30
	3	Establish design direction in terms of plan, section, and elevation	2, 4, 28
	4	Building core planning	3
	5	Explore interior and exterior material	3
	6	Reviewing the strength of material, structural design criteria, and design load	1, 3
	7	Reviewing alternatives of structural system designs, such as sizes of component (rough estimation of unit weight)	3, 6, 28
	8	MEP planning and MEP space requirement review	1, 3, 4
Design development	9	Develop and modify the schematic design	3, 10, 11, 13, 15, 16
	10	Determine interior and exterior materials	9
	11	Determine structural system design	6, 7, 9, 12
	12	Structural design analysis and structural calculation documentation	11, 35
	13	Draft the locations and sizes of structural components	9, 11, 12
	14	Bar arrangement drawing documentation	13
	15	Determine MEP system	1, 8, 9
	16	Reviewing MEP and structural component interference	9, 13, 15
Construction documentation	17	Prepare construction document of the design	9, 18, 20, 21
	18	Structural and MEP system adjustment and documentation	11, 15, 17, 21
	19	Architectural details, specifications, and structural calculation documentation and incorporate subcontractor documentation	17, 18, 20
	20	Principal structural parts finishing and propriety review	21
	21	Check the errors and omissions in documentation and constructability review	17, 20
Manufacturing	22	Determine and review the work activities for unit production in factory (determine factory work)	1, 17, 27, 28, 29
	23	Prepare shop drawings and check interference in unit production	22
	24	Prepare unit production line design	23
	25	Prepare manufacturing tolerance management plan	17, 22, 23, 26
	26	Prepare quality management plan	22, 25
Transportation	27	Prepare management plan for reducing deformation and damage	1, 22
	28	Review the Road Traffic Act regulations (weight and size of unit)	1
Onsite Work	29	Determine onsite work activities	22
	30	Select tower crane location	1, 2, 28
	31	Select tower crane specification	17, 22, 30
	32	Prepare shop drawings and check for interference	29, 33
	33	Review constructability for onsite work	29, 32
	34	Prepare onsite work tolerance management plan	29
	35	Prepare deformation management plan in unit lifting process	12
	36	Prepare unit proofing plan to reduce damage from weather conditions	17, 22

For the rearrangement, a DSM of the modular construction design process was developed based on the information flow relationship in Table 4. Figure 5 illustrates the DSM. The marks above the diagonal represent the reverse information flows. The optimization objective of the DSM is to move as many marks as possible to below the diagonal [47]. Figure 6 depicts the rearranged activities and information flow in the DSM after applying the partitioning algorithm. Then, activities with strong interdependencies were grouped, and each group has feedback and reverse information flow between activities. The blue boxes in Figure 6 indicate the groups, and to constitute the blue box, a clustering algorithm was used. Although the number of feedback processes were reduced by the optimization, some still remain. Activity groups A, B, C, and D require information flow management, because the feedback processes are concentrated within the groups. As a quantitative result, the total number of reverse information flows in the process was reduced from 23 to 18 after rearrangement of the activity order. For example, the Road Traffic Act regulations review in the transportation planning phase was reallocated to the SD phase. The reverse information flow of reviewing the regulation was reduced from 4 to 0. This reduced reverse flow implies that the regulation review affects other activities in the process. By allocating the regulation review to the early design phase, engineering rework in the design and planning phases can be reduced. Given that the results of the SD phase affect the following activities, reallocation in the early phase implies that the potential rework in the following phase can also be reduced.

Element Name	ID	1	2	3	4	5	6	7	8	9	10	11	12	13	14	15	16	17	18	19	20	21	22	23	24	25	26	27	28	29	30	31	32	33	34	35	36
Site analysis	1	1																																			
Site layout planning	2	1	2																										1		1						
Establish design direction in terms of plan, section, and elevation	3		1	3	1																								1								
Building core planning	4			1	4																																
Explore interior and exterior material	5			1		5																															
Reviewing the strength of material, structural design criteria, design load	6	1		1			6																														
Reviewing alternatives of structural system design such as size of component (rough estimation of unit weight)	7			1			1	7																					1								
MEP planning and MEP space requirement review	8	1		1	1				8																												
Develop and modify the schematic design	9			1						9	1	1		1		1	1																				
Determine interior and exterior material	10									1	10																										
Determine structural system design	11						1	1		1		11	1																								
Structural design analysis and structural calculation documentation	12											1	12																							1	
Draft the location and size of structural component	13									1		1	1	13																							
Bar arrangement drawing documentation	14													1	14																						
Determine MEP system	15	1							1	1						15																					
Reviewing MEP and structural component interference	16									1				1		1	16																				
Prepare construction document of the design	17									1								17	1		1	1															
Structural and MEP system adjustment and documentation	18											1				1		1	18			1															
Architectural detail, specifications, structural calculation documentation and incorporate subcontractor's documentation	19																	1	1	19	1																
Principal structural parts finishing propriety review	20																				20	1															
Check the errors and omissions in documentation and constructability review	21																	1			1	21															
Determine and review the work activities for unit production in factory (Determine factory work)	22	1																1					22					1	1	1							
Prepare shop drawing and check the interference for unit production	23																						1	23													
Prepare unit production line design	24																							1	24												
Prepare manufacturing tolerance management plan	25																	1					1	1		25	1										
Prepare quality management plan	26																						1			1	26										
Prepare management Plan for reducing deformation and damage	27	1																					1					27									
Review the Road Traffic Act regulation (weight and size of unit)	28	1																											28								
Determine the on-site work activities	29																						1							29							
Select tower crane location	30	1	1																										1		30						
Select tower crane specification	31																	1					1								1	31					
Prepare shop drawing and check the interference	32																													1			32	1			
Review constructability for on-site work	33																													1			1	33			
Prepare on-site work tolerance management plan	34																													1					34		
Prepare deformation management plan in unit lifting process	35												1																							35	
Prepare unit proof plan to reduce the damage from weather condition	36																	1					1														36

Figure 5. Information flow identification using DSM.

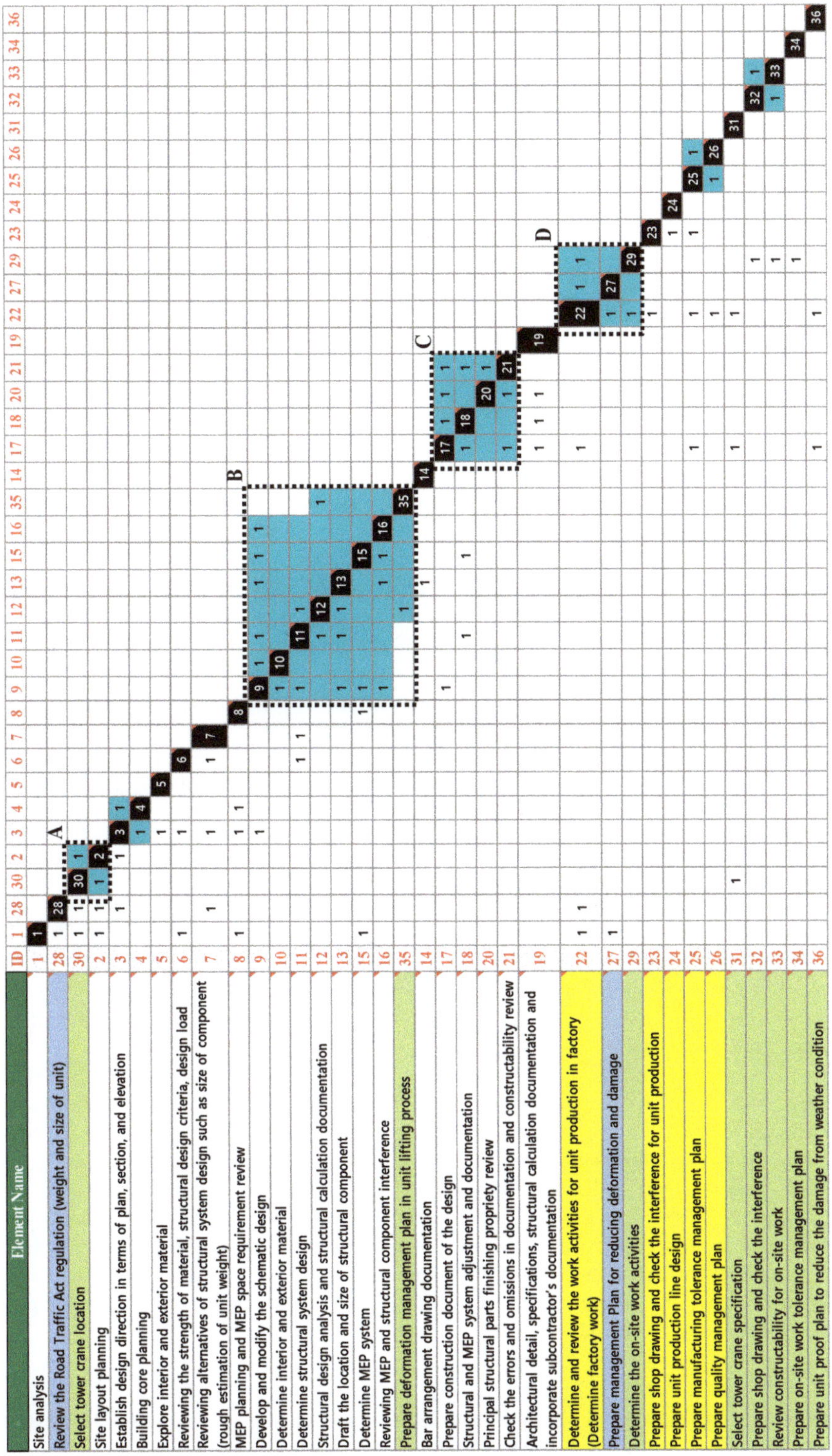

Figure 6. Rearranged activities after optimization.

4. Integrated Design Process for Modular Construction

4.1. Schematic Design Phase

Through DSM optimization, activities in the design phase and rework mitigation plan were rearranged and an integrated design process was suggested. In the process, activities are conducted sequentially from above. The marks on the matrix refer to the relationship described in Figure 4 and the marks on the diagonal line indicate feedback or reverse information flow. Using this process, it is expected that reworks in modular projects will be reduced. In this section, each design step in the design phase is explained, focusing on the rearranged activities and groups of activities. In the schematic design step, the Road Traffic Act regulations review in transportation planning and tower crane location selection in onsite work planning were included. When developing building design alternatives, it is necessary to review the Road Traffic Act regulations [12,55]. If the design does not meet the regulations, the building design is reworked because modular units must be transported to the site.

In this step, activity group A indicates the feedback process between the site layout planning and tower crane location selection. In the planning phase of the onsite work, the selection and positioning of tower cranes on the construction site are essential, because heavy units must be lifted [39,51,52]. In the tower crane operational plan, capacity is estimated based on a combination of the maximum distance that the tower crane must reach and the weight of the material. By reducing the maximum distance, the tower crane's operational cost can be reduced. For example, in the SD phase, after estimating the approximate weight of the unit, tower crane location alternatives are suggested considering the alternative site layout planning. Then, the tower crane capacity can be estimated on the basis of the combination of alternatives. If a tower crane with the required capacity cannot be obtained from a crane rental company or if the operational cost is uneconomical, the weight of the unit, the location of the crane, or the site layout plan are modified to meet the project objectives. After the design process—along with the regulation review and tower crane operation—is completed, a building design is selected from among the design alternatives by considering the feasibility and project objectives.

4.2. Design Development Phase

In this phase, the selected design from the schematic design phase is developed and the building systems—such as architectural, structural, and MEP systems—are determined. Activity group B shows the feedback process between activities to determine the building systems. To facilitate information interchange between project participants, information flow management is required. In this phase, when developing the MEP and structural design, interference between these activities often occurs [55,56]. Therefore, to reduce rework caused by interference, an accurate design review between activities is required, which implies that there will be feedback in this design process.

In this phase, onsite work planning is included to prevent the deformation of units [12,24]. When lifting a unit onsite, deformation caused by self-load of the unit can occur. This can affect the quality of unit and cause difficulties during the onsite unit assembly phase. To prevent this deformation, a balance beam can be used; however, deformation may still occur for heavy units. There is also an approved deformation range in the assembly phase. However, when the deformation exceeds the approved range, rework to revise the deformation should be conducted. The higher the quality and precision standards of the modular units, the higher the probability of running into common problems when using less precise components onsite or when the precision of the unit decreases [18]. Therefore, when selecting the structural system and conducting structural analysis of units, it is necessary to plan for preventing deformation induced by self-load.

4.3. Construction Documentation Phase

In this phase, the results of the previous phase are developed, and details are determined. Then, construction documents, such as drawings for details and specifications are prepared for manufacturing and onsite work. In this phase, the participants in the design planning phase crosscheck the design documents to rectify errors and omissions. The design document is used to prepare shop drawings for unit production and onsite work. The quality of shop drawings is directly related to project quality. Hence, interference between components or activities should be checked and rectified. Activity group C shows the feedback process in this phase. Here, constructability, information interchange, and rectification should be facilitated between participants [12,13,24].

4.4. Manufacturing, Transportation, and Onsite Work Planning Phase

In this phase, manufacturing, transportation, and onsite work plans, which are not included in the previous design phases, are finalized based on the results of the CD phase. To improve the efficiency of modular construction, the largest number of activities possible must be conducted in the manufacturing process. After the manufacturing process, units are transported to the construction site. It is necessary to establish the work plans according to the site environment and road conditions. Therefore, the work activities conducted in the manufacturing process are selected considering the unit deformation in transportation, damage to units, and site environment. Then, the remaining activities to complete the project are conducted onsite. Given that the work proportions are determined depending on the project characteristics, the proportions are flexible. This work activity distribution is shown by the activity group D in Figure 6. In this group, manufacturing, transportation, and onsite work plans are cross-checked and rectified by the participants. Therefore, it involves feedback processes between activities, and therefore, cooperation is required between participants. Moreover, in addition to work planning related to rework, other work planning for activities is also included in this phase.

The proposed design process can contribute to reducing rework in modular construction projects by integrating the rework mitigation plans and work planning. To employ this planning process, modular construction must be considered in the early project phase by the project client or in the early design stage. Then, to facilitate the process and include project participants in the early design phase, project delivery methods, such as integrated project delivery (IPD), should also be considered [12]. However, there are many hindrances to choosing modular construction, such as early design freeze, limited experience, short overall project timescale, and lack of availability of advice in the early phase [22]. Moreover, follow-on projects cannot use the same processes as previous projects, which is a constraint caused by a lack of experience and knowledge [22]. To overcome this, a standardized modular construction process is required. The proposed design process can be used to overcome the constraints and reduce rework in modular construction.

5. Discussion

To facilitate information flow in the design process, an integrated design process was suggested using DSM. To validate the process and effect on reducing errors in the design phase, an expert interview was conducted. The expert group consisted of researchers and architects; experts with over 5 years of experience in modular construction. In the interview, the experts agreed that an integrated design process is required to reduce errors in the design phase, and that the suggested process could reduce the errors. Since each participant in the process is a non-professional in all of the other fields (for example, a manager in a transportation company without knowledge of design processes), they do not know to whom or at what stage they should provide relevant information. Therefore, error can be reduced by informing them of what information should be provided to other participants and when. However, they also mentioned that further investigation into the information relationship between activities is required to find the unidentified relationship in this paper. For example, when planning tower crane operation, the crane may be located near the building core to attach the crane to the core. This means that the increased load caused by tower crane operation should be included in the structural design of the building core, because of the heavy weight of the unit and crane. Therefore, this unidentified relationship should be further investigated and included in the suggested design process. Moreover, the above-mentioned experts who are to use the process should consider project delivery methods at the beginning of the project, because some methods such as design-bid-building do not allow early integration of participants. The expert interview results show the effect and necessity of the process. However, there are limitations. In real projects, various decision-making criteria, such as level of modularity and integration strategies, are applied in the design phase that affect the design process, but these have not been considered in this paper, as they are outside the scope of study. In this paper, the design process of the AIA was applied to the suggested process, but when applied to other countries, legal considerations need to be reviewed. Moreover, to represent the information flow more clearly, various types of relationship can be used, representing the type of information or subject of information exchange, but the DSM in this paper used only 3 types of relationship. Thus, it can only represent information flow as a binary system. Finally, to ensure the robustness of this research, further validation of the process through application to a real project is required.

6. Conclusions

An integrated design process was developed using DSM and optimized using a partitioning algorithm. Through process optimization, feedback and reverse information flows were reduced, and thus, the complexity of the design process was alleviated. It is expected that rework in modular projects can be reduced by using the proposed process. Additionally, by using the information flow identification method proposed in this paper, other considerations for modular projects can also be included in the process. However, the process has limitations in that (1) application of the process to modular projects is required for validation; (2) the cases and causes of rework used are limited to only a few cases in the daily report and literature; and (3) the process is not able to suggest information about participants, i.e., who should be included for each activity in the planning process; and (4) although many criteria affect the decision-making process in the design phase, such as quality, safety, cost and constructability, the suggested design process in this paper focused on the information relationship between activities, and thus, the effect of the suggested process on the criteria were not considered. To overcome these limitations, other criteria affecting the design process should be included in the integrated design process. Accordingly, a case study including more cases of rework and validation will be conducted in future research.

Author Contributions: Conceptualization, H.H. and J.L.; methodology, H.H.; formal analysis, H.K. and H.-S.L.; investigation, H.H.; resources, H.H.; writing—original draft preparation, H.H. and J.L.; writing—review and editing, M.P., H.K, and H.-S.L.; project administration, H.-S.L.; funding acquisition, H.-S.L. All authors have contributed substantially to the work reported. All authors have read and agreed to the published version of the manuscript.

Funding: This work was supported by the Housing Environment Research Program funded by Ministry of Land, Infrastructure and Transport of Government of Korea (Grant number 20RERP-B082884-07).

Acknowledgments: The authors would also like thanks to Ministry of Land, Infrastructure and Transport of Government of Korea, Housing Environment Research Program, Korea Agency for Infrastructure Technology Advancement for supporting this research.

Conflicts of Interest: The authors declare no conflict of interest.

References

1. Chen, Y.; Okudan, G.E.; Riley, D.R. Sustainable performance criteria for construction method selection in concrete buildings. *Automa. Constr.* **2010**, *19*, 235–244. [CrossRef]
2. Lu, N. The current use of offsite construction techniques in the United States construction industry. In Proceedings of the Construction Research Congress 2009, Seattle, WA, USA, 5–7 April 2009; p. 96.
3. Boafo, F.E.; Kim, J.-H.; Kim, J.-T. Performance of modular prefabricated architecture: Case study-based review and future pathways. *Sustainability* **2016**, *8*, 558. [CrossRef]
4. Eastman, C.M.; Sacks, R. Relative productivity in the AEC industries in the United States for on-site and off-site activities. *J. Constr. Eng. Manag.* **2008**, *134*, 517–526. [CrossRef]
5. Mullens, M.A. *Factory Design for Modular Homebuilding: Equipping the Modular Factory for Success;* Constructability Press: Winter Park, FL, USA, 2011.

6. Shaked, O.; Warszawski, A. CONSCHED: Expert system for scheduling of modular construction projects. *J. Constr. Eng. Manag.* **1992**, *118*, 488–506. [CrossRef]
7. Lawson, R.M.; Ogden, R.G.; Bergin, R. Application of modular construction in high-rise buildings. *J. Archit. Eng.* **2011**, *18*, 148–154. [CrossRef]
8. Shen, K.; Cheng, C.; Li, X.; Zhang, Z. Environmental Cost-Benefit Analysis of Prefabricated Public Housing in Beijing. *Sustainability* **2019**, *11*, 207. [CrossRef]
9. Whole Building Design Guide Sustainable (WBDG). Available online: http://www.wbdg.org/design-objectives/sustainable (accessed on 11 February 2019).
10. Jiang, Y.; Zhao, D.; Wang, D.; Xing, Y. Sustainable Performance of Buildings through Modular Prefabrication in the Construction Phase: A Comparative Study. *Sustainability* **2019**, *11*, 5458. [CrossRef]
11. Lee, J.-H.; Kim, J.-S.; Lee, H.-J.; Lee, Y.-M.; Kim, H.-G. Small-Scale Public Rental Housing Development Using Modular Construction—Lessons learned from Case Studies in Seoul, Korea. *Sustainability* **2019**, *11*, 1120. [CrossRef]
12. Smith, R.E. *Prefab Architecture: A Guide to Modular Design and Construction*; John Wiley & Sons: Hoboken, NJ, USA, 2011.
13. Lawson, M.; Ogden, R.; Goodier, C. *Design in Modular Construction*; CRC Press: Boca Raton, FL, USA, 2014.
14. Alvanchi, A.; Azimi, R.; Lee, S.; AbouRizk, S.M.; Zubick, P. Off-site construction planning using discrete event simulation. *J. Archit. Eng.* **2011**, *18*, 114–122. [CrossRef]
15. Love, P.E.; Mandal, P.; Li, H. Determining the causal structure of rework influences in construction. *Constr. Manag. Econ.* **1999**, *17*, 505–517. [CrossRef]
16. Love, P.E.; Li, H.; Mandal, P. Rework: A symptom of a dysfunctional supply-chain. *Eur. J. Purch. Supply Manag.* **1999**, *5*, 1–11. [CrossRef]
17. Love, P.E.; Li, H. Quantifying the causes and costs of rework in construction. *Constr. Manag. Econ.* **2000**, *18*, 479–490. [CrossRef]
18. Blismas, N.; Wakefield, R. Drivers, constraints and the future of offsite manufacture in Australia. *Constr. Innov.* **2009**, *9*, 72–83. [CrossRef]
19. Johnsson, H.; Meiling, J.H. Defects in offsite construction: Timber module prefabrication. *Constr. Manag. Econ.* **2009**, *27*, 667–681. [CrossRef]
20. Love, P.E.; Holt, G.D.; Shen, L.Y.; Li, H.; Irani, Z. Using systems dynamics to better understand change and rework in construction project management systems. *Int. J. Proj. Manag.* **2002**, *20*, 425–436. [CrossRef]
21. Love, P.E.; Edwards, D.J. Determinants of rework in building construction projects. *Eng. Constr. Archit. Manag.* **2004**, *11*, 259–274. [CrossRef]
22. Blismas, N.G.; Pendlebury, M.; Gibb, A.; Pasquire, C. Constraints to the use of off-site production on construction projects. *Archit. Eng. Des. Manag.* **2005**, *1*, 153–162. [CrossRef]
23. Jiang, L.; Li, Z.; Li, L.; Gao, Y. Constraints on the promotion of prefabricated construction in China. *Sustainability* **2018**, *10*, 2516. [CrossRef]
24. Park, H.K.; Ock, J.-H. Unit modular in-fill construction method for high-rise buildings. *KSCE J. Civ. Eng.* **2016**, *20*, 1201–1210. [CrossRef]
25. Pasquire, C.L.; Gibb, A.G. Considerations for assessing the benefits of standardisation and pre-assembly in construction. *J. Financ. Manag. Prop. Constr.* **2002**, *7*, 151–161.
26. Al-Bazi, A.; Dawood, N. Developing crew allocation system for the precast industry using genetic algorithms. *Comput. Aided Civ. Infrastruct. Eng.* **2010**, *25*, 581–595. [CrossRef]
27. Arif, M.; Espinal, D.; Broadway, R.S. Estimating, Planning and Controlling Labor in the Industrialized Housing Factory. In *IIE Annual Conference. Proceedings*; Institute of Industrial Engineers-Publisher: Orlando, FL, USA, 2002; p. 1.
28. Yassine, A.; Braha, D. Complex concurrent engineering and the design structure matrix method. *Concurr. Eng.* **2003**, *1*, 165–176. [CrossRef]
29. Hwang, B.-G.; Thomas, S.R.; Haas, C.T.; Caldas, C.H. Measuring the impact of rework on construction cost performance. *J. Constr. Eng. Manag.* **2009**, *135*, 187–198. [CrossRef]
30. Josephson, P.-E.; Hammarlund, Y. The causes and costs of defects in construction: A study of seven building projects. *Autom. Constr.* **1999**, *8*, 681–687. [CrossRef]
31. Smith, G.; Jirik, T. *Making Zero Rework a Reality: A Comparison of Zero Accident Methodology to Zero Rework and Quality Management*; Research Report; Construction Industry Institute: Austin, TX, USA, 2006; pp. 203–211.

32. Burati, J.L., Jr.; Farrington, J.J.; Ledbetter, W.B. Causes of quality deviations in design and construction. *J. Constr. Eng. Manag.* **1992**, *118*, 34–49. [CrossRef]
33. Rahmandad, H.; Hu, K. Modeling the rework cycle: Capturing multiple defects per task. *Syst. Dyn. Rev.* **2010**, *26*, 291–315. [CrossRef]
34. Bruns, T.; Stalker, G. *The Management of Innovation*; Tavistock: London, UK, 1961; pp. 120–122.
35. Gidado, K. Project complexity: The focal point of construction production planning. *Constr. Manag. Econ.* **1996**, *14*, 213–225. [CrossRef]
36. Nahangi, M.; Safa, M.; Shahi, A.; Haas, C.T. Automated registration of 3D point clouds with 3D CAD models for remote assessment of staged fabrication. In Proceedings of the Construction Research Congress 2014: Construction in a Global Network, Atlanta, GA, USA, 19–21 May 2014; pp. 1004–1013.
37. Sharafi, P.; Samali, B.; Ronagh, H.; Ghodrat, M. Automated spatial design of multi-story modular buildings using a unified matrix method. *Autom. Constr.* **2017**, *82*, 31–42. [CrossRef]
38. Sharafi, P.; Rashidi, M.; Samali, B.; Ronagh, H. Identification of factors and decision analysis of the level of modularization in building construction. *J. Archit. Eng.* **2018**, *24*, 04018010. [CrossRef]
39. Han, S.H.; Hasan, S.; Bouferguène, A.; Al-Hussein, M.; Kosa, J. Utilization of 3D visualization of mobile crane operations for modular construction on-site assembly. *J. Manag. Eng.* **2014**, *31*, 04014080. [CrossRef]
40. Lei, Z.; Taghaddos, H.; Olearczyk, J.; Al-Hussein, M.; Hermann, U. Automated method for checking crane paths for heavy lifts in industrial projects. *J. Constr. Eng. Manag.* **2013**, *139*, 04013011. [CrossRef]
41. Olearczyk, J.; Al-Hussein, M.; Bouferguène, A. Evolution of the crane selection and on-site utilization process for modular construction multilifts. *Autom. Constr.* **2014**, *43*, 59–72. [CrossRef]
42. Austin, S.; Baldwin, A.; Li, B.; Waskett, P. Analytical design planning technique ADePT): A dependency structure matrix tool to schedule the building design process. *Constr. Manag. Econ.* **2000**, *18*, 173–182. [CrossRef]
43. Giaglis, G.M. A taxonomy of business process modeling and information systems modeling techniques. *Int. J. Flex. Manuf. Syst.* **2001**, *13*, 209–228. [CrossRef]
44. Mayer, R.J.; Benjamin, P.C.; Caraway, B.E.; Painter, M.K. A framework and a suite of methods for business process reengineering. *Bus. Process Reeng. Manag. Perspect.* **1995**, *3*, 245–290.
45. Wei, H.-Q. Concurrent design process analysis and optimization for aluminum profile extrusion product development. *Int. J. Adv. Manuf. Technol.* **2007**, *33*, 652–661. [CrossRef]
46. Lee, J.; Park, M.; Lee, H.-S.; Kim, T.; Kim, S.; Hyun, H. Workflow dependency approach for modular building construction manufacturing process using Dependency Structure Matrix (DSM). *KSCE J. Civ. Eng.* **2017**, *21*, 1525–1535. [CrossRef]
47. Oloufa, A.A.; Hosni, Y.A.; Fayez, M.; Axelsson, P. Using DSM for modeling information flow in construction design projects. *Civ. Eng. Environ. Syst.* **2004**, *21*, 105–125. [CrossRef]
48. AIA. *The American Institute of Architects Document B101-Standard Form of Agreement between Owner and Architect*; AIA: Washington, DC, USA, 2017.
49. Fayek, A.R.; Dissanayake, M.; Campero, O. Developing a standard methodology for measuring and classifying construction field rework. *Can. J. Civ. Eng.* **2004**, *31*, 1077–1089. [CrossRef]
50. O'connor, J.T.; Tucker, R.L. Industrial project constructability improvement. *J. Constr. Eng. Manag.* **1986**, *112*, 69–82. [CrossRef]
51. Al-Hussein, M.; Alkass, S.; Moselhi, O. An algorithm for mobile crane selection and location on construction sites. *Constr. Innov.* **2001**, *1*, 91–105. [CrossRef]
52. Han, S.; Al-Hussein, M.; Hasan, S.; Gökçe, K.U.; Bouferguene, A. Simulation of mobile crane operations in 3D space. In Proceedings of the 2012 Winter Simulation Conference (WSC), Berlin, Germany, 9–12 December 2012; pp. 1–12.
53. Han, S.; Bouferguene, A.; Al-Hussein, M.; Hermann, U. 3D-Based Crane Evaluation System for Mobile Crane Operation Selection on Modular-Based Heavy Construction Sites. *J. Constr. Eng. Manag.* **2017**, *143*, 04017060. [CrossRef]
54. Kalasapudi, V.S.; Tang, P.; Zhang, C.; Diosdado, J.; Ganapathy, R. Adaptive 3D imaging and tolerance analysis of prefabricated components for accelerated construction. *Procedia Eng.* **2015**, *118*, 1060–1067. [CrossRef]
55. Hwang, B.-G.; Shan, M.; Looi, K.-Y. Key constraints and mitigation strategies for prefabricated prefinished volumetric construction. *J. Clean. Prod.* **2018**, *183*, 183–193. [CrossRef]

56. Shahtaheri, Y.; Rausch, C.; West, J.; Haas, C.; Nahangi, M. Managing risk in modular construction using dimensional and geometric tolerance strategies. *Autom. Constr.* **2017**, *83*, 303–315. [CrossRef]

sustainability

MDPI

Article

The Modelling of Roof Installation Projects Using Decision Trees and the AHP Method

Augustinas Maceika [1,†,‡], Andrej Bugajev [2,*,†,‡] and Olga R. Šostak [2,†,‡]

1 The Faculty of Mechanics, Vilnius Gediminas Technical University, Sauletekio ave. 11, LT-10223 Vilnius, Lithuania; augustinas.maceika@vgtu.lt
2 The Faculty of Fundamental Sciences, Vilnius Gediminas Technical University, Sauletekio ave. 11, LT-10223 Vilnius, Lithuania; olgaolgaregina@yahoo.com
* Correspondence: zvex77777@gmail.com
† Current address: Sauletekio ave. 11, LT-10223 Vilnius, Lithuania.
‡ These authors contributed equally to this work.

Received: 15 November 2019; Accepted: 14 December 2019; Published: 19 December 2019

Abstract: In this work, the process of roofing projects' execution is considered. The proper analysis of this process is important to optimise the behaviour of a project's participants and to perform risk evaluation. The main result of this work is methodology, which can be used to optimise a project owner's decisions and potentially can be applied for risk control or integrated into expert systems. This methodology includes the application of a decision tree and AHP (analytic hierarchy process) method to perform the modelling for roof installation project selection. In the proposed approach, a decision tree describes the process with nodes representing the states of a project. The tree includes the decision on whether to sell the project results or not, which requires the estimation of the subjective opinion of the project owner. These subjective values are used in the decision tree leaves. We propose to perform this estimation with the AHP method and describe how to do it in this paper. A particular example was considered. The proposed methodology was applied to that case, and all details of the process and results are provided. Using the proposed methodology, the adapted version of a specific, current situation model of project participants' behaviours can be formed, allowing one to make the most efficient decisions in the light of the existing constraints. The application of results can increase the investor protection and contribute to the general sustainability of investments.

Keywords: decision tree; analytic hierarchy process; dynamic programming; sustainable investment; project participants' behaviour; roof installation projects

1. Introduction

1.1. The Background of the Research

In this research we consider roof installation projects, which can be defined as the organised, temporary processes of constructing, renovating, refurbishing, etc., a roof. We investigate such projects from the point of view of project owner (investor); thus, the profit is the goal function in optimisation problems that arise as a part of the this research. The results and the methodology presented in this work can be applied to some other similar building projects; however, we focus on this particular case to overview the complete set of the properties that are typical for such types of projects.

The construction business is often associated with risky and contentious situations and their solutions. Investors should avoid risky situations and the damage that they can cause. Many construction projects include roof installation, replacement or repair processes. There is a need for tools that will give a clear picture of the projects' implementation situation and find out about the results of the various options. One of the tools, selected to investigate by the authors of this article, is the decision tree, which can be used to analyse the behaviours of the participants in the selection and implementation of the roof installation projects and to make the most efficient decisions. The second tool is the analytic hierarchy process, which is used to provide a decision tree based on subjective opinion data from the roof installation project owner.

Human beings need to make decisions based on several different criteria in order to solve many problems around themselves [1]. When assessing roofing projects, it is important to choose a set of criteria that will be relevant to decision-making, taking into account the project life cycle stages. One of the important criteria is length of construction time. The results of cost performance (deviation) investigation with respect to the length of construction time implies that cost performance is project specific and it is difficult to conclude based on certain attributes; the most important thing is to consistently evaluate and have a closer look on cost development and establish a learning platform to overcome unexpected changes [2].

Other criteria can be related to the social costs of construction. Attempts to investigate the social costs of building constructions in urban residential areas are still insufficient due to probable difficulties and complexities of including the third parties; however, in the decision making process for construction projects, apart from economic sense of it, decision-makers need to be provided with other useful information: analytical and procedural assessments to comment on the convenience of the expected environmental impacts once the proposed project is implemented, and a social assessment to find out if the consequences of developing the proposed project are socially acceptable [3].

Additional costs may arise from conflicts with third parties. To analyse this situation, stochastic dynamic programming can be used. That enables a broad analysis of the dependencies between the optimal investor's strategy and the probabilities that third parties will select a certain strategy [4]. One of the implementations of dynamic programming is the decision tree approach, which is useful because this method enables the analysis of the step-by-step project execution process, including evaluation of the behaviour of the participants. It is also suitable for analysing the possible choice alternatives.

For project evaluation, it is important to select appropriate methodology. Advancements (framework and methodology for evaluating project competencies and project key performance indicators (KPIs); relationship determined between the different project competencies and grouping them using factor analysis; applying of advanced modelling techniques through the joint application of prioritised fuzzy aggregation, factor analysis, and fuzzy neural networks (FNNs) to identify the relationship between the different project competencies and project KPIs) should allow construction practitioners to identify and monitor the relationships between evaluation criteria of project competencies' and project KPIs throughout the project lifecycle to ensure better project performance [5].

One of proposed multiple criteria techniques is based on the decision tree for determining the project, using various types of criteria for comparisons of alternate strategies [6]. When a single strategy is the best with respect to all criteria, the problem is trivial; however, in most situations it is faced with conflicting criteria. A common approach to address such a problem consists of two phases: determining a set of efficient strategies and selecting the best of those strategies [6].

1.2. *The Relationship to the Agency, Stakeholder and Stewardship Theories*

Next, in order to establish the relationships between the main participants of the roof installation projects, agency, stakeholder and stewardship topics will be overviewed.

For decision making it is important to find interaction the mechanisms between the project owner (principal) and the managers, consultants and supervisors (agents), taking into account environmental conditions.

A social psychological perspective on agency relations and solution mechanisms reveals that each agency problem (each solution of an agency problem) becomes a problem of social power to the principal (agent) if she (he) changes her (his) beliefs, attitudes, or behaviour as a result of the action, or presence of the agent (principal) [7]. An analysis of one agency problem after the other revealed that in all situations in which an agency problem exists, the principal possesses quantitatively more bases of power than the agent; in the hidden intentions and hidden action-situations, she additionally possesses those power bases with a broader range; agency theory assumes implicitly an asymmetry in power in favour of the principal [7].

The nature of relationship between the client and agent has been presented as the model of "one servant of two masters" based on to the triangular relationship between the client, the construction company and its project manager [8].

In this field of research, an agent-based simulation that conceptualises insights from behavioural economics to increase understanding of price formations in housing markets was used, and it was found that the model produces the most favourable results when there are agents with extremely myopic expectations in the market and when other agents mimic them [9].

Analysis of stakeholders' impacts on the creation and successful implementation of roof installation projects is the part of decision tree development. For this reason, a number of works about the methodology and results devoted to stakeholder theory were analysed.

In the work of this field, through the application of focus group study, 13 stakeholder groups were identified across the lifecycles of net zero energy homes, including home buyers, sales personnel, financial institutions, developers, designers and drafting personnel, estimators, project managers/coordinators, regulators, superintendents, inspectors, trades/suppliers, net zero energy home occupants and warranty staff [10]. According to another source, eleven groups of stakeholders were identified; namely, employees, customers, shareholders, creditors, suppliers and partners, environment and resources agencies, local communities, government, competitors and non-governmental organisations [11]. In this work, corporate social responsibility (CSR) indicators were extracted for each performance issue by analysing environment, health and safety, human resources, supply chain management, customers and communities, governance and ethics aspects of construction enterprises [11].

The value of good stakeholders' management is visible from the work, where the findings from the interviews emphasised the need for a "proactive" stakeholder management approach which takes into account both the views of primary and secondary stakeholders, and there, through building internal capabilities for secondary stakeholder management, organisations have to recognise the importance of creating the right vision for major public infrastructure and construction projects and delivering not just assets but bringing extra value either at national, regional or local level [12].

The effects of underlying collaboration between designers and contractors were examined and the results show that the best means to promote the collaboration is "reducing collaboration costs", followed by "increasing the collaboration benefits" and "decreasing the loss caused by a lack of collaboration" [13].

In one article, off-site manufacturing stakeholders' business information was analysed and 59 sustainability perceptions were identified, covering the social, environmental and economic sustainability dimensions. Among them, "high quality" and a "customer-focused approach and customisation" were most valued [14].

In the other article, relevant governance criteria are introduced which can be used to judge on extents to which stakeholder participation effectively contributes to increased urban sustainability. They include legitimacy, accountability, representation, responsibility and transparency; moreover, a literature review was conducted, from which outcomes are presented across three categories of stakeholder participation—stakeholder based initiatives, government based initiatives and science based initiatives, outcomes which have briefly been assessed with respect to the governance criteria [15].

For our investigation, the model for analysing stakeholder conflicts in urban redevelopment projects is important too, because some roofing projects can be related to renovation. The proposed stakeholder salience theory and Pawlak's conflict theory are useful, and generated an action scheme that mitigates stakeholder conflicts and maximises project benefits [16]—which can be taken into account during roofing project developments.

The people themselves and the environment they live in, their homes and neighbourhoods, if located around the building construction zones, are exposed to adverse impacts of the construction activities [3]. In return, people react via altering their daily routine to resolve or alleviate the exposed disruptions to their common life patterns and the cost of this reaction is defined as the social costs associated with building construction projects. As the definition of social cost implies, there are costs caused by constructions that are to be paid by the third parties [3].

The cradle to cradle (C2C) theory propounds that environmental impact reduction can provide a positive economic impulse to stakeholders; current sustainability strategies focus on reducing the negative environmental impacts of buildings. The systems theory of C2C, however, aims at a positive impact; this could suggest that the state-of-the-art becomes inadequate when adopting C2C as a strategy for improvement, focusing on closed or continuous materials, and energy and water cycles [17].

Stewardship theory concept provides that the managers of different organisations participating in roof installation projects must act responsibly, taking into account stakeholders needs.

In the field of stewardship theory, an opinion given is that understanding the institutional framing, underpinnings and logic of mega projects can provide the key to successful delivery of solutions in water, transportation, energy, communications, health, education and a variety of related sectors, through the development of best practices for building social, organisational and political legitimacy that can enhance the security and stability of the role that such projects play in an increasingly interconnected world [18].

Organisational support, customer pressure and regulatory pressure have significant positive effects on the adoption of environmental practices in construction projects; moreover, environmental practices have a positive effect on the environmental and economic performances of construction firms. This result may be even more important for managers of construction firms to convince stakeholders that investments in environmental practices will not reduce competitiveness and may even increase profitability while simultaneously benefiting the environment [19].

Local governments, community-based organisations, foundations, neighbourhood and other advocacy groups, construction companies, investors, commercial banks, tenants and their brokers, ecologists, media and unions, all are the participants of the city development and city stakeholders, and they should be able to create feasible projects which generate benefits and reduce the risk involved in urban development [20].

Following the analysis of the literature and the existing construction practice situation, the main parties involved in the selection and implementation of roof installation projects were identified and are presented in Figure 1.

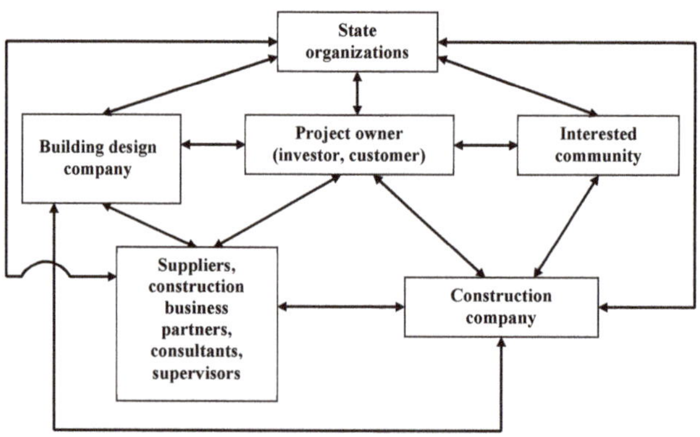

Figure 1. The main participants in the selection and implementation of the roof installation projects.

The project owner (investor, customer) is connected with other participants who are interested in the results of a roofing project's implementation. For successful project creation and implementation, the project owner must initiate a number of processes to ensure that the communication network between the participants in the process will be developed.

1.3. The Importance of Roofing Projects

As new construction develops and the renovation process is under way, one of the most complex and responsible building elements—the roof—requires additional attention and effort to achieve a more sustainable state. In studies of whole-building lifecycle assessments, it has been mentioned that the most significant components added during renovation are the roof access floors, and new windows, while the new construction scenario was overwhelmingly burdened by manufacturing intensive structural (concrete and steel) and envelope components (brick and terracotta walls) [21]. The roof renovation process can be important for energy-saving changes and the social network or professionals can influence the decision to implementing the energy renovation measures [22]. As measures on both occupant behaviour and physical improvement have influences on energy saving to various extents, effective renovation strategies should be developed by combining both building technologies and behavioural changes [23].

1.4. The Problem Statement

We discussed the information about connections between participants, possible events along with their probabilities and monetary and time costs during those events. However, in raw form, such information is not suitable for direct analysis. Thus, there is a need for software layer between the human and raw data.

The end goal of the investor's activity is profit. The process during which the profit is achieved can be described as Markov decision process. Such a process can be described as a decision tree where the nodes represent states (events) which are connected between by edges representing the transitions between those states. It might look like some classical approaches, such as typical Markov decision process modelling methods, might solve that problem and was already developed a long time ago (see, for example, [24]). However, to apply classical Markov decision process modelling, the values for all states must be given. In the considered case the values describing the final states depend on:

- The costs of all previous states.
- The accumulated costs due to time, passed during some of the previous states; i.e., some time-dependent project costs.
- The evaluation of the project value that must be properly addressed using multi-criteria evaluation techniques.

The goal of this research was to provide a methodological approach for roofing projects evaluation. When there are decision nodes, first, an optimal investor behaviour strategy must be formed. Thus, this brings forth the optimisation problem that is needed to be addressed. In this article, it is done by taking the profit value as a goal function. Two abstraction levels of the problem can be considered:

- Evaluating a single project.
- Choosing the best project out of all possible alternatives.

Choosing a project out of all possible alternatives might be a non-trivial task, which depends on many circumstances. For example, there are individual risks thresholds that can be taken for different investors—larger companies can take larger risks if they are part of a bigger expected profit.

If the investor chooses between different projects using some trivial technique, for example, by comparing the expected profit values of different alternatives, then the choice can be easily automated as well; i.e., the choice between projects becomes a part of the investor's strategy. Thus, in this work we consider the most general case when different projects are evaluated and are followed by the choice between alternatives.

If the there is a need to apply some non-trivial (risk evaluation) technique for comparison of different projects, the approach provided is suitable to process every single project separately and extract the profit distribution—the information that can be applied for further analysis.

As it was mentioned already, roofing project evaluation involves the estimation of the subjective value of the construction object. Thus, the modelling of such a process must include some kind of the evaluation method, such as AHP. The existing solutions [25] for modelling of similar processes do not support such evaluation. Moreover, some process costs occur as additional costs with some probabilities; however, the occurrences of these costs do not change the structure of the tree. So including these costs into a decision tree would require duplicating some of branches of a tree. That should not be done by the user, since it can be done automatically.

1.5. The Methods of the Research

As it was mentioned before, the modelled process is Markov decision process, so a decision tree is a natural representation of it. However, differently from a typical Markov process, the current model lacks of information which must be derived (preprocessed); thus, we developed solutions in the most flexible and elegant form to process tree data structure—recursive algorithms. It is important to note that preprocessing does not introduce any additional computational errors. Thus, the final result is the exact solution due to the Bellman [26] principle which can be read as the optimal solution is the best out of optimal subsolutions, also known as dynamic programming.

The AHP method is suitable for application to the current problem due to a small number of evaluated parameters; it will be shown later. Also, the number of alternatives is small. It is equal to two. We will show that in conjunction with a decision tree, we can isolate the multi-criteria part of the problem to a single project evaluation. Two alternatives will be considered: to sell a project or not. Moreover, instead of direct usage of AHP method we will solve the inverse problem: the value of the price attribute will be chosen so that it will fit the threshold after which the decision change. i.e., the AHP method is directly used for the price evaluation instead of a typical usage to perform a decision. It will affect the decision via decision tree. Thus, we need to fit a single parameter for a monotonic function (for it to be equal to 0.5). We derived an explicit formula for that.

It should be emphasised that in AHP method the criteria hierarchy and weight value allocations are important. When more criteria are taken into consideration, the interrelations between criteria can be altered, and alternative hierarchies may influence the weights allocation [27]. The simplicity and the relative ease of use of the AHP method do not mean that it provides poor results [28]. Moreover, AHP method is widely used and is well-studied, making the results robust and interesting for a wide range of researchers. This widespread use is due to its ease of applicability and the structure of AHP which follows the intuitive way in which managers solve problems [29].

This research is dedicated to covering the already mentioned drawbacks of the existing solutions and show how apply the proposed methodology to roofing projects. This work makes the following contributions:

1. We analysed the participants and other specificities of a roofing project and showed how to describe such a process as a decision tree describing a Markov decision process.
2. We propose a modelling approach to evaluate the roofing project. The model was based on another work in [25], where the modelling did not support the specificity of the roofing project.
3. We propose to deal with the subjective evaluation of the projects' values using the AHP method; the details of its application for such projects are provided.

This paper is organised as follows. In Section 2 we present an example of roofing project's execution process and provide the details on algorithmic formalisation to perform the modelling of such process. The application of the model and the solution of the case study problem are provided in

Section 3. At the end of the paper, we discuss the practical value and the conclusions of the research (Sections 4 and 5).

2. The Creation of Mathematical Model

2.1. The Description of the Studied Case and Construction of the Decision Tree

For modelling, the case of the situation where an investor can make decisions to carry out a roof installation project or not and to choose type of the project to select, depending on the risk assessment and the behaviour of participants, was investigated. Each situation, presented in decision tree, was related to the losses or incomes, and to the time used to resolve the situation, taking into account the risk as probability. Also the decisions about consultants and construction supervisor, building design company, project, construction company, suppliers and construction business partners selection were investigated. The decisions to sell the implemented project or not, in the model, were based on the market price of similar objects and the opinion of the project owner on how much the projects' results were worth for him in monetary units.

The model was provided with option that the project owner would be able to hire consultants which would receive a constant hourly wage, so the cost of consulting would depend on the duration of the project, were the consultants to be hired. It is also anticipated that the risk of project activities can be reduced by consulting.

The project owner has the opportunity to choose a building design company from two possibles. It is also foreseen that each building design company will prepare four roof installation projects, two expensive, and two of average prices, so in the model, eight projects were presented. Four of them were medium-priced roof projects: when an inexpensive roof with a simple loft was installed. The other four projects were expensive roof projects with a fully equipped loft and comfortable living space. It was taken into account that there can be losses related to the situations when additional investment is needed to correct a project, and alignment with the interested community and state organisations is necessary.

Another step was the choice of construction company. It was anticipated that for a single project it would be possible to select one from two construction companies.

Each building company was associated with the choice of two possible options of suppliers and construction business partners' organisations. Evaluation of losses due to additional investment if construction and supply organisation fails was included into the model.

The main decisions that must performed at the different project implementation stages are provided on the left part of the Figure 2. On the right part of the Figure 2 are what losses must be evaluated during the associated decisions.

The graph for the considered example is provided in Figure 3. Different types of nodes were used:

1. Decision nodes which are represented by rectangle shapes:
    - Of blue colour—the decision on whether the project owner should sell the product or not;
    - Of white colour with a yellow frame—the rest of decisions that should be made during the project.
2. End nodes that are represented by red rectangles.
3. Probability nodes represented by green ovals.

We also provide the zoomed-in fragment of the whole graph in Figure 4.

The detailed information about the data is provided in Appendix A and will be commented in more detail in the next section.

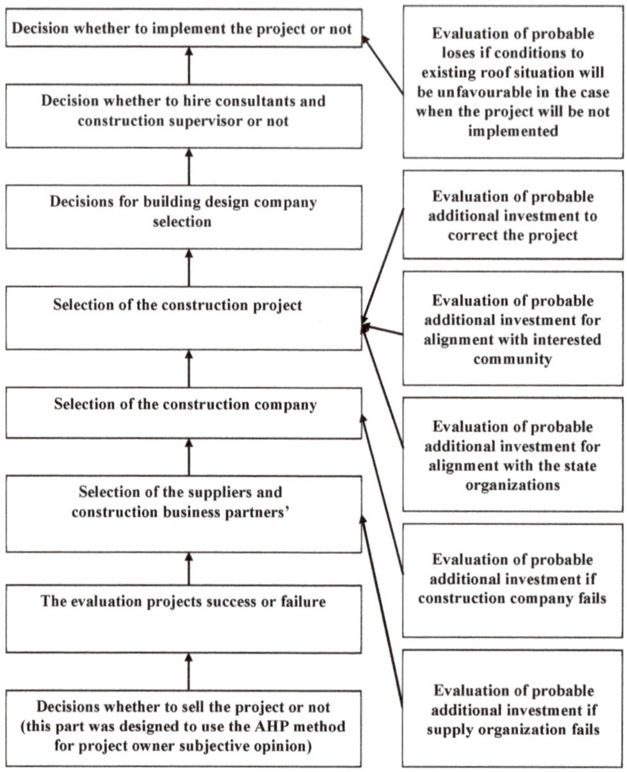

Figure 2. The main decisions at different stages of project's execution.

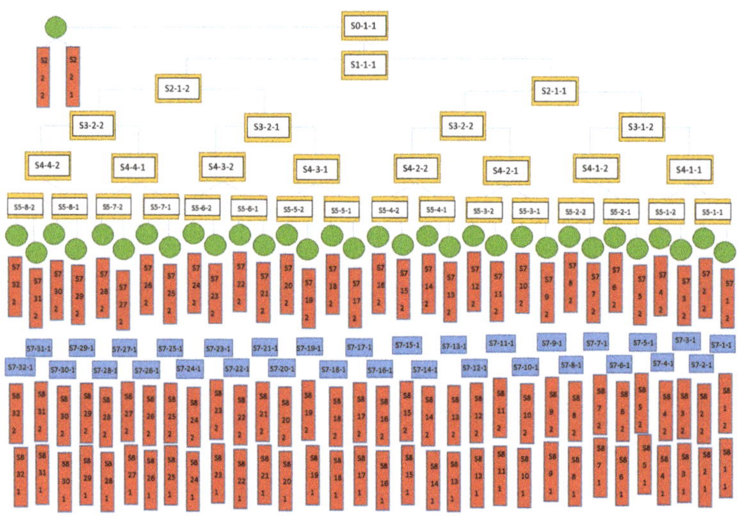

Figure 3. The graph for the example considered.

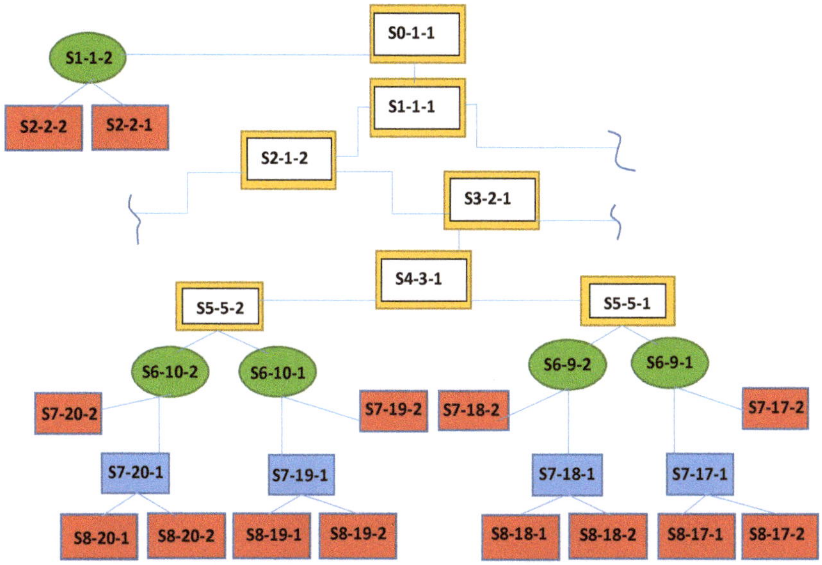

Figure 4. The zoomed-in fragment of the graph.

2.2. The Mathematical Model

The decision tree nodes represent the states which the project may be at. The transition between states will be referred to as an event. Which data will be stored at the node corresponds to the state the event transits to. Each task level consists of profit/loss values, which depend on the monetary value obtained at the previous state and probability of events or selection, optimising value (profit) for the customer. Since the decision tree has been divided into separate levels, the profit/loss is equal to the previous level's value (profit)/loss after evaluation of probability or decision made.

The objective is to maximise profit for the roof installation project owner, where a solution of decision strategy is defined by $X_{ijk} = 0$ or 1 (integer input variables), $i = 1, 2, 3; \ldots$ g (the number of the decision tree level); $j = 1, 2, 3, \ldots$ m (the number of the branch group in the decision tree level); $k = 1, 2, 3, \ldots$ n (the number of the branch in the decision tree branch group); $X_{ijk} \geq 0$; $\sum_{k=1}^{n} X_{ijk} = 1$; if $X_{ijk} = 1$– decision to take the action ijk; if $X_{ijk} = 0$—decision do not to take the action ijk; if P_{ijk} – probability of events, $P_{ijk} \geq 0$; $\sum_{k=1}^{n} P_{ijk} = 1$; $S_{(i-1)jk}$ value (profit)/losses received after evaluation; for each decision tree branch group if decision must to be done:

$$S_{(i-1)jk} = \sum_{k=1}^{n} S_{ijk} X_{ijk}; \tag{1}$$

or if there are probabilities of events:

$$S_{(i-1)jk} = \sum_{k=1}^{n} S_{ijk} P_{ijk}. \tag{2}$$

Also, each node in the tree has a duration parameter t_{ijk} (measured in days). And if in decision tree position was the probability of events P_{ijk} , the expected time was also counted. The calculations provided total time for each branch of the tree, and the time planed for consultants and construction supervisor activity, if they were hired.

The basic element of our data structure (a tree) is a node with connections to parent and children nodes (Figure 5).

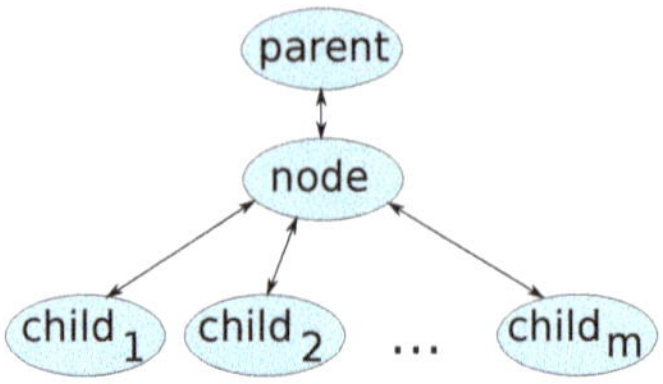

Figure 5. The relationships between a node and its parent and children nodes.

A more detailed description is presented in Algorithm 1, where the symbol "//" denotes comments that describe the corresponding fields.

Algorithm 1: Data structure.

```
struct {
    int type; //node type
    float p; //the probability for this node to be selected by parent
    float price; //the price of event
    float time; //the time before event starts
    float ap; //the probability of additional cost and duration
    float aprice; //the additional pricefloat atime; //the additional time
    float tariff; //the additional price per daynode* parent; //the pointer to parent node
    vector< node* > children; //the list of children
    float priceTotal; //accumulated price
    float timeTotal; //accumulated time
    float value; //node expected profit value
    float extime; //node expected time
} node;
```

There are 12 fields in the provided structure in total. First eight are known from the data directly, some notes on these fields:

1. Field *type* describes the type of a node which can be either equal to 1, which means the investor decision node, or equal to 0, which means all other nodes.
2. Field *tariff* denotes the cost which will be paid for every day of the project that after the node with this field is selected; i.e., the corresponding event activates the tariff mode, in our case primarily due to paying to consultations. Note, that the negative values of this field may stop the tariff mode period; however, for our particular case it will be not needed. I.e., it will be active until the project ends.
3. The pointers to the parent and the children nodes are defined by the graph edges.

As for the last four fields—they are computed by algorithms provided in this paper.

Note, that in all functions presented it is assumed that the arguments are passed by reference; i.e., the changes of arguments are seen outside of these functions. We propose the algorithm that consists of these steps:

1. Apply the AHP method to evaluate the values on leafs—we use field *price* with negative values for that.
2. Calculate total times (field *timeTotal*) and costs (field *priceTotal*) up to the moment when events are finished—this is implemented in the function *CalcPars* in Algorithm 2.
3. Evaluate end-node scenarios (calculate field *value*) and select optimal strategy—function *CalcValues* in Algorithm 3.
4. Create profit (field *value*) distribution by calculating different scenario probabilities. A simple implementation is provided in Algorithm 4.

Algorithm 2: Costs parameters calculation algorithm.

```
Function CalcPars(node)
    node.timeTotal = node.time + node.ap * node.atime
    node.priceTotal = node.price + node.ap * node.aprice
    if node.parent != NULL then
        node.timeTotal += node.parent.timeTotal
        node.priceTotal += node.parent.priceTotal
    end
    for each node t in node.children do
        CalcPars(t)
    end
end
```

Algorithm 3: Scenario profit evaluation and strategy selection algorithm.

```
Function CalcValues(node)
    for each node t in node.children do
        CalcValues(t)
    end
    if node.children = NULL then
        node−>extime = node−>timeTotal
    else
        if node.type = 1 then
            for each node t in node.children do
                t.p = 0
            end
            best = arg max_{t∈node.children} t.value
            best.p = 1
        end
        n = node.children.size
        node.value = Σ_{i=1}^{n} node.children[i].p · node.children[i].value
        node.extime = Σ_{i=1}^{n} node.children[i].p · node.children[i].extime
    end
end
```

Algorithm 4: Probabilities calculation algorithm with distribution extraction.

```
Function CalcProbs(node, distribution)
    if node.parent = NULL then
        node.probTotal = 1
    else
        node.probTotal = node.p * node.parent.probTotal
    end
    if node.children.size() > 0 then
        for each node t in node.children do
            CalcProbs(t)
        end
    else
        distribution.add(node.value, node.probTotal)
    end
end
```

The algorithms were based on the results presented in [25]. We additionally included the support of additional costs with probabilities. There are two main techniques that were applied:

1. Pre-order tree-traversal, when the calculations must be performed from top to the down of the tree; i.e., when each node must inherit the calculations part from the parent node.
2. Post-order tree-traversal, when the calculations must be performed from down to top; i.e., each node must collect the results from it's children and aggregate the result for itself—it is the implementation of a dynamic programming method.

Because the decision tree was limited to a monetary criterion, it was decided to search for additional approaches to take in consideration project owner's opinion, based on multi-criteria decision-making. In this field a lot of methods are created and used. For example, [30] proposed to use discrete stochastic multi-criteria decision-making method, based on distance from average solution, to handle many real-life decision-making problems.

We applied the AHP method to evaluate the roof installation projects taking into account the suitability for the chosen problem solution and the benefits visible from the works, where: subjective weightings were analysed by using the AHP computer software programme the Expert Choice 11 [31]; the AHP method was adopted in order to rank assessment themes and identify the priorities of the study's participating stakeholders [32]; a hierarchical structure for pavement alternatives was constructed and individual pairwise comparison was done by the experts [33].

It was found that the decision tree requires additional data about the value of the project, as it is subjectively perceived by the owner of the roof installation project. This data can be obtained by using the AHP method developed by T.L. Saaty [34].

The steps to use the AHP method to evaluate the price of the object:

1. Weighting of the criteria for AHP. For this step we need to form the matrix for pair-wise C_{lm} elements:

$$\begin{bmatrix} C_{11} & C_{12} & \dots & C_{1z} \\ C_{21} & C_{22} & \dots & C_{2z} \\ \dots & \dots & \dots & \dots \\ C_{z1} & C_{z2} & \dots & C_{zz} \end{bmatrix}; \tag{3}$$

each entry in the matrix is positive ($C_{lm} > 0$) and reciprocal ($C_{lm} = 1/C_{lm}, \forall, m = 1, 2, ...z$).

Then it is necessary to divide each element in the matrix by its column total:

$$Y_{lm} = \frac{C_{lm}}{\sum\limits_{l=1}^{z} C_{lm}}; \tag{4}$$

where z is the number of criteria.

Then, it is possible to generate normalised, pair-wise matrix:

$$\begin{bmatrix} Y_{11} & Y_{12} & \dots & Y_{1z} \\ Y_{21} & Y_{22} & \dots & Y_{2z} \\ \dots & \dots & \dots & \dots \\ Y_{z1} & Y_{z2} & \dots & Y_{zz} \end{bmatrix}. \tag{5}$$

If to divide the sum of normalised row of matrix by the number of criteria:

$$W_l = \frac{\sum\limits_{m=1}^{z} Y_{lm}}{z}; \tag{6}$$

we will have weighted matrix:

$$\begin{bmatrix} W_1 \\ W_2 \\ \ldots \\ W_z \end{bmatrix}, \tag{7}$$

the values of which represent different criteria weights.

To estimate the correctness of estimation (whether the evaluations for importance between different criteria are consistent), a consistency vector is calculated by dividing the weighted sum vector by criterion weight:

$$Cv_1 = \frac{1}{W_1}\left[C_{11}W_1 + C_{12}W_2 + \ldots + C_{1z}W_z\right];$$

$$Cv_2 = \frac{1}{W_2}\left[C_{21}W_1 + C_{22}W_2 + \ldots + C_{2z}W_z\right];$$

$$\ldots$$

$$Cv_z = \frac{1}{W_z}\left[C_{z1}W_1 + C_{z2}W_2 + \ldots + C_{zz}W_z\right]. \tag{8}$$

Eigen value λ_{max} is calculating by averaging the value of the consistency vector:

$$\lambda_{\max} = \frac{1}{z}\sum_{l=1}^{z} Cv_l, \tag{9}$$

a consistency index (CI):

$$CI = \frac{\lambda_{\max} - z}{z - 1}, \tag{10}$$

where $\lambda_{\max}$ is the maximal eigen value.

Consistency ratio:

$$CR = \frac{CI}{RI}, \tag{11}$$

where RI is the random index.

In practice, a CR of 0.1 or below is considered acceptable. Any higher value at any level indicates that the judgements warrant re-examination.

2. Evaluation of the criteria for the projects A–H. An overall assessment of alternatives (to sell the object or not) for each of the project was planned to carry out without criteria of compliance with economic logic weighed by scores, because this criterion would be found for the situation where both alternatives are equal.
3. Finding the scores of economic logic for selling and reserving, for the roof owner, results in the same value (solution values are equal to 0.5 weighed and normalised points, because in this case we have only two alternatives). Here we assume a solution to be the value of the choice for the first alternative. To derive a formula we can exploit the fact that there are only two alternatives and write the solution in the simplified form

$$F = \sum_{m=1}^{z} \frac{W_m V_m}{V_m + 1}; \tag{12}$$

here, V_m is the m-th criteria evaluation before the normalisation. Next, we express the score value of the first criteria

$$V_1 = \frac{a}{1-a}, a = \left(F - \sum_{m=2}^{z} \frac{W_m V_m}{V_m + 1}\right) / W_1. \quad (13)$$

Substituting the needed value $F = 0.5$ gives the desired result—the subjective price evaluation measured in score. In general the achieved result is not integer; therefore, we convert it into the price monetary value using linear interpolation of the values in table 2. Note that the aforementioned table points are presented in a typical-for-AHP way—integers from -9 to 9, excluding 0, so before interpolation. we must perform additional mapping and use the value

$$\bar{V}(V_1) = \begin{cases} -1/V_1, & \text{when } V_1 < 1 \\ V_1, & \text{otherwise.} \end{cases} \quad (14)$$

3. The Application of The Model

Compliance with the psychological and social needs, economics logic, strategic objectives and best location variant criteria were chosen. It was done after a practical case study of construction projects and analysis of literature sources, in which the number of bedrooms, bathrooms, outdoor spa, swimming pool, garden, parking spaces, state of repair, architecture style, socioeconomic profile of neighbourhood, energy efficiency, distance [35], situation in the labour market, labour supply cost, amenities and neighbourhood quality [36], appropriative control of the green building project development [37], distance from the city centre and quality of the site [38], structural attributes, accessibility, public and private service amenities [39], city structure and house location [40] and strategy to rent [41], were factors for house pricing and housing strategy development. The goal of the AHP method, its criteria and its alternatives are presented in the Figure 6. Here are some comments about the criteria:

- The content of the criterion of compliance with the psychological and social needs is related to the situation of neighbourhood, habitual place, status, way of life, appearance, romanticism and history.
- Compliance with the economic logic criterion is related to the price of the object. It may also be related to energy consumption, maintenance, resale value and other costs—but for simplicity we did not include those.
- Compliance with the strategic objectives criterion is related to the owner's plans—whether the object will be rented, sold or used for living.
- Compliance with best location variant criterion is related to the quality of the site, accessibility and public and private service amenities.

In order to determine the weights of the criteria, the project owner had to assess which criterion were more important, and how important, with a score from 1 to 9 granting the alternative to sell the project and from -9 to -1 does not allow sale. The weights obtained after the calculations are given in the Table 1.

Then the project owner had to determine the scale for valuation of eight projects from A to H, where points are related to the monetary value for him if not to sell the projects. The monetary value dependence on score is presented in linear form in Table 2.

Evaluation of the criteria was done by the project owner (decision to sell the object or not), and is presented in Table 3. Positions of compliance with economics logic were not evaluated by the project owner, because this number would be found by using the algorithm, whereby, knowing the non-linear dependency between the scores of the economic logic parameter (the score from 1 to 9 for alternative to sell the project and from -9 to -1 to not sell) and the evaluation of each project's alternatives without a compliance with economics logic parameter (on 0 to 1 scale), we can determine a economic logic parameter where both alternatives are the same; i.e., 0.5 weighed scores. This is clearly a subjective assessment. Knowing the scores, we can set the compliance with economic logic parameter from the scale system presented in Table 2. This is clearly a subjective assessment of the limit project value for owner if not to sell the roof for each project is presented in Table 4.

For comparison, the sale price for each project roof, based on analogues, is presented in Table 4. This data was incorporated into the decision tree branches.

The Algorithm 3 was applied to the tree, it was found that the best option for the consumer in economic terms would be to implement the project and sell the roof. The most effective way is from position S_{8-12-1} to S_{0-1-1}, with a value of 5891.58 Euros. This is an average price project variant C, with a market sale price of 28,468 Euros. The cost of the project implementation would be of 22,576.4 Euros. The risk and additional costs were reduced by the choice of consultants and construction supervisor, although for them it was necessary to spend a considerable amount of money. The other organisations participating in the project chose taking into account the minimum cost for project C implementation. The duration of the project was found—171.003 days, which takes into account the probabilities of probable time losses. The tariff for consulting was 6.25 Euros per day.

Expensive projects were not selected because the best option was D, with the value of 3345 Euros, which is less than for the project C. The results of modelling for these expensive projects showed that the owner does not sell the roof, but its installation costs are higher and the projects last longer.

Based on the results, if one does not to take the project activity, expected loses can be of 1000 Euros.

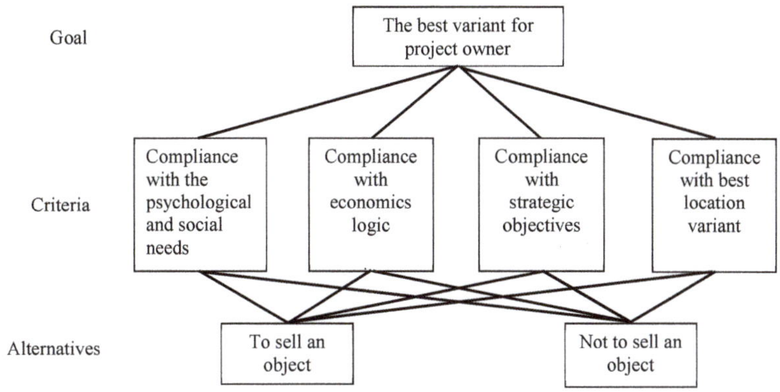

Figure 6. AHP hierarchy for roofing project.

Table 1. Weights for the criteria.

Criteria	Compliance with the Psychological and Social Needs	Compliance with Economics Logic Economics Logic	Compliance with Strategic Objectives	Compliance with Best Location Variant
Weights	0.1376	0.3935	0.3935	0.0754

Table 2. The scale to connect points system and monetary value of the project.

Points System	Projects A-H Value Scale for the Owner in Euros							
	A	B	C	D	E	F	G	H
9	16,694	34,800	17,694	36,000	15,194	32,300	16,194	33,500
8	17,694	37,300	18,694	38,500	16,194	34,800	17,194	36,000
7	18,694	39,800	19,694	41,000	17,194	37,300	18,194	38,500
6	19,694	42,300	20,694	43,500	18,194	39,800	19,194	41,000
5	20,694	44,800	21,694	46,000	19,194	42,300	20,194	43,500
4	21,694	47,300	22,694	48,500	20,194	44,800	21,194	46,000
3	22,694	49,800	23,694	51,000	21,194	47,300	22,194	48,500

Table 2. *Cont.*

Points System	Projects A–H Value Scale for the Owner in Euros							
	A	B	C	D	E	F	G	H
2	23,694	52,300	24,694	53,500	22,194	49,800	23,194	51,000
1	24,694	54,800	25,694	56,000	23,194	52,300	24,194	53,500
−1	24,694	54,800	25,694	56,000	23,194	52,300	24,194	53,500
−2	25,694	57,300	26,694	58,500	24,194	54,800	25,194	56,000
−3	26,694	59,800	27,694	61,000	25,194	57,300	26,194	58,500
−4	27,694	62,300	28,694	63,500	26,194	59,800	27,194	61,000
−5	28,694	64,800	29,694	66,000	27,194	62,300	28,194	63,500
−6	29,694	67,300	30,694	68,500	28,194	64,800	29,194	66,000
−7	30,694	69,800	31,694	71,000	29,194	67,300	30,194	68,500
−8	31,694	72,300	32,694	73,500	30,194	69,800	31,194	71,000
−9	32,694	74,800	33,694	76,000	31,194	72,300	32,194	73,500

Table 3. Evaluation of the criteria by the customer (decision to sell the object or not).

Criteria	Projects A-H Alternatives Evaluation in Points							
	A	B	C	D	E	F	G	H
Compliance with the psychological and social needs	−3	−5	−3	−5	−3	−5	−3	−5
Compliance with strategic objectives	−2	−2	−1	−1	−3	−3	−2	−2
Compliance with best location variant	2	2	2	2	2	2	2	2
Compliance with economics logic, points from AHP, when both alternatives are equal	2.60	3.02	1.25	1.41	4.14	5.05	2.60	3.02

Table 4. Final alternatives to sell or not data for entering into decision tree (to select best variant).

Criteria	Projects in Euros							
	A	B	C	D	E	F	G	H
Limit value if we reserve for ourselves from AHP	23,095	49,748	25,444	54,981	20,053	42,183	22,595	48,448

4. Discussion

The authors of the article proposed the methodology, which lets one model the behaviour of roof installation project participants, and they applied it to the example case. Application of the results is useful:

1. If it is necessary to select the economically best project from a set of possible alternatives.
2. If it is necessary to take into account the stages involved in the project and the needs of the main stakeholders, and review different scenarios.
3. If individual parameters (criteria) need to be assessed and transferred to monetary value.

We showed how to isolate the multi-criteria part of the problem and model it separately from the decision tree. Instead of including it into a tree processing logic, we derived the threshold value of the price which is on the edge between the solution to sell the roof or not, and we used this value as the subjective price evaluation which was added to the profit value—it affected the decision via single criteria Markov process modelling.

There were some assumptions made that might be revised; for example, in Algorithm 3 the formula $best = arg \max_{t \in node.children} t.value$ was used, which means a branch of a tree is evaluated by the expected profit value alone; however, for an investor the most important thing might be the profit per time unit. Still, the proposed approach is flexible enough and easy to enough modify to support such an evaluation as well; instead of *t.value* the value *t.value*/*t.extime* can be used. Some more advanced techniques could be used for risk analysis, etc. However, it might break the conditions which are necessary for the Bellman [26] principle leading to the loss of the optimality. But even without those conditions it might lead to a decent heuristic with a reasonably good results. That might be the subject of the future research.

The proper (advanced) risk evaluation techniques are out of scope of this research. The project selection from several alternatives might be done outside of the proposed methodology—it lets us evaluate different projects separately, which can be followed by processing with some advanced risk evaluation tools.

Regarding AHP sensitivity analysis, we donot apply the AHP directly. We solve the inverse problem to extract the threshold value for the object price. Next, we apply this price as a price evaluation in Markov process modelling; this is where the choice of the project is performed. This means, that even if the choice of the project is sensitive to some criteria evaluation, then the actual difference of profit values will be small. A sensitive choice would mean small difference in profit values. Therefore, it is not reasonable to apply the sensitivity analysis in the scope of the current research. Such analysis could be useful to determining the importance of different criteria; however, that is out of scope of the current research.

5. Conclusions

1. A combination of decision tree and AHP techniques covered the needs of this research; i.e., the decision tree directly describes the real process, and AHP lets one estimate some of the states of this process, which are not known initially. More specifically—it lets one perform a multi-criteria evaluation of the subjective prices of the projects (for each of them separately) according to the project owner's opinion.
2. The proposed approach lets one perform modelling with a large number of possible alternatives automatically. Therefore, the number of alternatives and the size of the tree itself depends on the user and is limited by a human factor alone. A limited number of possible alternatives can be reviewed, as the process of creating alternatives is rather labour-intensive and may take a lot of time.
3. The probabilistic parameters of the model are related to the market situation; however, the project decisions evaluated at the current moment and the situation during the project implementation can change. This risk can be controlled by extending the tree with possible scenario nodes; thus, it can be easily be performed by a user.
4. The estimation of the information that is necessary to fill the model by the data can be a non-trivial task. For example, in order to apply the proposed methodology, it is necessary to evaluate the costs of each project implementation. In addition, some project details may be unknown, and some data should be evaluated without exact information as well. However, the proposed methodology supports non-determinism, which partially solves the aforementioned problems. Technically, it means that instead of a single scenario, a few possible scenarios can be used with given probabilities.
5. Evaluation and implementation of the project is directly linked to the specific project owner; he can have personal reasons to consider why and how valuable the project is for him. The AHP method enables one to include this information into a model.
6. In this work, the computational part of the methodology is provided in the flexible form of general algorithm description through pseudo code. This lets one implement the technique easily, using general-purpose programming languages, such as Java, Python, C++, etc. Thus, the results provided here can be easily integrated into other systems, such as business management, insurance and expert systems.

Author Contributions: O.R.Š. and A.M. formulated the problem, and collected and analysed the data; A.B. created appropriate algorithms, implemented them in C++ and wrote pseudocode. All authors contributed to performing experiments, analysis of results and writing the paper equally.

Funding: This research received no external funding.

Conflicts of Interest: The authors declare no conflict of interest.

Appendix A

Table A1. Description of the transition from one state to another in the decision tree, part 1.

Marking of the Edges S_{i-1jk}-S_{ijk}	Description of Behavioural Options and Probable Events
S_{0-1-1}-S_{1-1-1}	Project owner decided to carry out the project
S_{0-1-1}-S_{1-1-2}	Project owner refuses the project
S_{1-1-1}-S_{2-1-1}	Project owner decided to hire the consultants and supervisor
S_{1-1-1}-S_{2-1-2}	Project owner decided do not to hire the consultants and supervisor
S_{1-1-2}-S_{2-2-1}	The loses if conditions to existing roof situation are unfavourable
S_{1-1-2}-S_{2-2-2}	The loses if conditions to existing roof situation are favourable
S_{2-1-1}-S_{3-1-1}	Project owner decided to select the building design company "A" (with consulting)
S_{2-1-1}-S_{3-1-2}	Project owner decided to select the building design company "B" (with consulting)
S_{2-1-2}-S_{3-2-1}	Project owner decided to select the building design company "A" (without consulting)
S_{2-1-2}-S_{3-2-2}	Project owner decided to select the building design company "B" (without consulting)
S_{3-1-1}-S_{4-1-1}; S_{3-1-1}-S_{4-1-2}; S_{3-1-2}-S_{4-2-1}; S_{3-1-2}-S_{4-2-2}; S_{3-2-1}-S_{4-3-1}; S_{3-2-1}-S_{4-3-2}; S_{3-2-2}-S_{4-4-1}; S_{3-2-2}-S_{4-4-2}	Project owner decided to select the medium price project "A" or expensive project "B" for building design company "A" (with consulting); "C" or "D" for "B" (with consulting); "E" or "F" for "A" (without consulting); "G" or "H" for "B" (without consulting)
S_{4-1-1}-S_{5-1-1}; S_{4-1-1}-S_{5-1-2}; S_{4-1-2}-S_{5-2-1}; S_{4-1-2}-S_{5-2-2}; S_{4-2-1}-S_{5-3-1}; S_{4-2-1}-S_{5-3-2}; S_{4-2-2}-S_{5-4-1}; S_{4-2-2}-S_{5-4-2}; S_{4-3-1}-S_{5-5-1}; S_{4-3-1}-S_{5-5-2}; S_{4-3-2}-S_{5-6-1}; S_{4-3-2}-S_{5-6-2}; S_{4-4-1}-S_{5-7-1}; S_{4-4-1}-S_{5-7-2}; S_{4-4-2}-S_{5-8-1}; S_{4-4-2}-S_{5-8-2}	Project owner decided to select the construction company "AA" or "AB" for project "A"; "BA" or "BB" for "B"; "CA" or "CB" for "C"; "DA" or "DB" for "D"; "EA" or "EB" for "E"; "FA" or "FB" for "F"; "GA" or "GB" for "G"; "HA" or "HB" for "H"
S_{5-1-1}-S_{6-1-1}; S_{5-1-1}-S_{6-1-2}; S_{5-1-2}-S_{6-2-1}; S_{5-1-2}-S_{6-2-2}; S_{5-2-1}-S_{6-3-1}; S_{5-2-1}-S_{6-3-2}; S_{5-2-2}-S_{6-4-1}; S_{5-2-2}-S_{6-4-2}; S_{5-3-1}-S_{6-5-1}; S_{5-3-1}-S_{6-5-2}; S_{5-3-2}-S_{6-6-1}; S_{5-3-2}-S_{6-6-2}; S_{5-4-1}-S_{6-7-1}; S_{5-4-1}-S_{6-7-2}; S_{5-4-2}-S_{6-8-1}; S_{5-4-2}-S_{6-8-2}; S_{5-5-1}-S_{6-9-1}; S_{5-5-1}-S_{6-9-2}; S_{5-5-2}-S_{6-10-1}; S_{5-5-2}-S_{6-10-2}; S_{5-6-1}-S_{6-11-1}; S_{5-6-1}-S_{6-11-2}; S_{5-6-2}-S_{6-12-1}; S_{5-6-2}-S_{6-12-2}; S_{5-7-1}-S_{6-13-1}; S_{5-7-1}-S_{6-13-2}; S_{5-7-2}-S_{6-14-1}; S_{5-7-2}-S_{6-14-2}; S_{5-8-1}-S_{6-15-1}; S_{5-8-1}-S_{6-15-2}; S_{5-8-2}-S_{6-16-1}; S_{5-8-2}-S_{6-16-2}	Project owner decided to select the supply company "AA" or "AB" for construction company "AA"; "AA" or "AB" for "AB"; "BA" or "BB" for "BA"; "BA" or "BB" for "BB"; "CA" or "CB" for "CA"; "CA" or "CB" for "CB"; "DA" or "DB" for "DA"; "DA" or "DB" for "DB"; "EA" or "EB" for "EA"; "EA" or "EB" for "EB"; "FA" or "FB" for "FA"; "FA" or "FB" for "FB"; "GA" or "GB" for "GA"; "GA" or "GB" for "GB"; "HA" or "HB" for "HA"; "HA" or "HB" for "HB

Table A2. Description of the transition from one state to another in the decision tree, part 2.

S_{6-1-1}-S_{7-1-1}; S_{6-1-2}-S_{7-2-1}; S_{6-2-1}-S_{7-3-1}; S_{6-2-2}-S_{7-4-1}; S_{6-3-1}-S_{7-5-1}; S_{6-3-2}-S_{7-6-1}; S_{6-4-1}-S_{7-7-1}; S_{6-4-2}-S_{7-8-1}; S_{6-5-1}-S_{7-9-1}; S_{6-5-2}-S_{7-10-1}; S_{6-6-1}-S_{7-11-1}; S_{6-6-2}-S_{7-12-1}; S_{6-7-1}-S_{7-13-1}; S_{6-7-2}-S_{7-14-1}; S_{6-8-1}-S_{7-15-1}; S_{6-8-2}-S_{7-16-1}; S_{6-9-1}-S_{7-17-1}; S_{6-9-2}-S_{7-18-1}; S_{6-10-1}-S_{7-19-1}; S_{6-10-2}-S_{7-20-1}; S_{6-11-1}-S_{7-21-1}; S_{6-11-2}-S_{7-22-1}; S_{6-12-1}-S_{7-23-1}; S_{6-12-2}-S_{7-24-1}; S_{6-13-1}-S_{7-25-1}; S_{6-13-2}-S_{7-26-1}; S_{6-14-1}-S_{7-27-1}; S_{6-14-2}-S_{7-28-1}; S_{6-15-1}-S_{7-29-1}; S_{6-15-2}-S_{7-30-1}; S_{6-16-1}-S_{7-31-1}; S_{6-16-2}-S_{7-32-1}	Variant of project implemented with success
S_{6-1-1}-S_{7-1-2}; S_{6-1-2}-S_{7-2-2}; S_{6-2-1}-S_{7-3-2}; S_{6-2-2}-S_{7-4-2}; S_{6-3-1}-S_{7-5-2}; S_{6-3-2}-S_{7-6-2}; S_{6-4-1}-S_{7-7-2}; S_{6-4-2}-S_{7-8-2}; S_{6-5-1}-S_{7-9-2}; S_{6-5-2}-S_{7-10-2}; S_{6-6-1}-S_{7-11-2}; S_{6-6-2}-S_{7-12-2}; S_{6-7-1}-S_{7-13-2}; S_{6-7-2}-S_{7-14-2}; S_{6-8-1}-S_{7-15-2}; S_{6-8-2}-S_{7-16-2}; S_{6-9-1}-S_{7-17-2}; S_{6-9-2}-S_{7-18-2}; S_{6-10-1}-S_{7-19-2}; S_{6-10-2}-S_{7-20-2}; S_{6-11-1}-S_{7-21-2}; S_{6-11-2}-S_{7-22-2}; S_{6-12-1}-S_{7-23-2}; S_{6-12-2}-S_{7-24-2}; S_{6-13-1}-S_{7-25-2}; S_{6-13-2}-S_{7-26-2}; S_{6-14-1}-S_{7-27-2}; S_{6-14-2}-S_{7-28-2}; S_{6-15-1}-S_{7-29-2}; S_{6-15-2}-S_{7-30-2}; S_{6-16-1}-S_{7-31-2}; S_{6-16-2}-S_{7-32-2}	Variant of project implemented with total fail
S_{7-1-1}-S_{8-1-1}; S_{7-2-1}-S_{8-2-1}; S_{7-3-1}-S_{8-3-1}; S_{7-4-1}-S_{8-4-1}; S_{7-5-1}-S_{8-5-1}; S_{7-6-1}-S_{8-6-1}; S_{7-7-1}-S_{8-7-1}; S_{7-8-1}-S_{8-8-1}; S_{7-9-1}-S_{8-9-1}; S_{7-10-1}-S_{8-10-1}; S_{7-11-1}-S_{8-11-1}; S_{7-12-1}-S_{8-12-1}; S_{7-13-1}-S_{8-13-1}; S_{7-14-1}-S_{8-14-1}; S_{7-15-1}-S_{8-15-1}; S_{7-16-1}-S_{8-16-1}; S_{7-17-1}-S_{8-17-1}; S_{7-18-1}-S_{8-18-1}; S_{7-19-1}-S_{8-19-1}; S_{7-20-1}-S_{8-20-1}; S_{7-21-1}-S_{8-21-1}; S_{7-22-1}-S_{8-22-1}; S_{7-23-1}-S_{8-23-1}; S_{7-24-1}-S_{8-24-1}; S_{7-25-1}-S_{8-25-1}; S_{7-26-1}-S_{8-26-1}; S_{7-27-1}-S_{8-27-1}; S_{7-28-1}-S_{8-28-1}; S_{7-29-1}-S_{8-29-1}; S_{7-30-1}-S_{8-30-1}; S_{7-31-1}-S_{8-31-1}; S_{7-32-1}-S_{8-32-1}	Project owner decided to sell the object and gain the profit or loses
S_{7-1-1}-S_{8-1-2}; S_{7-2-1}-S_{8-2-2}; S_{7-3-1}-S_{8-3-2}; S_{7-4-1}-S_{8-4-2}; S_{7-5-1}-S_{8-5-2}; S_{7-6-1}-S_{8-6-2}; S_{7-7-1}-S_{8-7-2}; S_{7-8-1}-S_{8-8-2}; S_{7-9-1}-S_{8-9-2}; S_{7-10-1}-S_{8-10-2}; S_{7-11-1}-S_{8-11-2}; S_{7-12-1}-S_{8-12-2}; S_{7-13-1}-S_{8-13-2}; S_{7-14-1}-S_{8-14-2}; S_{7-15-1}-S_{8-15-2}; S_{7-16-1}-S_{8-16-2}; S_{7-17-1}-S_{8-17-2}; S_{7-18-1}-S_{8-18-2}; S_{7-19-1}-S_{8-19-2}; S_{7-20-1}-S_{8-20-2}; S_{7-21-1}-S_{8-21-2}; S_{7-22-1}-S_{8-22-2}; S_{7-23-1}-S_{8-23-2}; S_{7-24-1}-S_{8-24-2}; S_{7-25-1}-S_{8-25-2}; S_{7-26-1}-S_{8-26-2}; S_{7-27-1}-S_{8-27-2}; S_{7-28-1}-S_{8-28-2}; S_{7-29-1}-S_{8-29-2}; S_{7-30-1}-S_{8-30-2}; S_{7-31-1}-S_{8-31-2}; S_{7-32-1}-S_{8-32-2}	Project owner decided not to sell the object and gain the value from holding

References

1. Dahooie, J.H.; Zavadskas, E.K.; Firoozfar, H.; Vanaki, A.; Mohammadi, N.; Brauers, W.K.M. An improved fuzzy MULTIMOORA approach for multi-criteria decision making based on objective weighting method (CCSD) and its application to technological forecasting method selection. *Eng. Appl. Artif. Intell.* **2019**, *79*, 114–128. [CrossRef]
2. Belay, A.M.; Torp, O. Do Longer Projects Have Larger Cost Deviation Than Shorter Construction Projects? *Procedia Eng.* **2017**, *196*, 262–269. [CrossRef]
3. Çelik, T.; Kamali, S.; Arayici, Y. Social cost in construction projects. *Environ. Impact Assess. Rev.* **2017**, *64*, 77–86. [CrossRef]
4. Šostak, O.R.; Vakrinien, S. Mathematical modelling of dispute proceedings between investors and third parties on allegedly violated third-party rights. *J. Civ. Eng. Manag.* **2011**, *17*, 126–136. [CrossRef]
5. Omar, M.N.; Fayek, A.R. Modeling and evaluating construction project competencies and their relationship to project performance. *Autom. Constr.* **2016**, *69*, 115–130. [CrossRef]
6. Nowak, M. Defining project approach using decision tree and quasi-hierarchical multiple criteria method. *Procedia Eng.* **2017**, *172*, 791–799. [CrossRef]
7. Saam, N.J. Asymmetry in information versus asymmetry in power: Implicit assumptions of agency theory? *J. Socio-Econ.* **2007**, *36*, 825–840. [CrossRef]
8. Sha, K. Incentive strategies for construction project manager: A common agency perspective. *Constr. Manag. Econ.* **2019**, *37*, 461–471. [CrossRef]
9. Ozbakan, T.A.; Kale, S.; Dikmen, I. Exploring House Price Dynamics: An Agent-Based Simulation with Behavioral Heterogeneity. *Comput. Econ.* **2019**, *54*, 783–807. [CrossRef]

10. Li, H.X.; Patel, D.; Al-Hussein, M.; Yu, H.; Gül, M. Stakeholder studies and the social networks of NetZero energy homes (NZEHs). *Sustain. Cities Soc.* **2018**, *38*, 9–17. [CrossRef]
11. Zhao, Z.Y.; Zhao, X.J.; Davidson, K.; Zuo, J. A corporate social responsibility indicator system for construction enterprises. *J. Clean. Prod.* **2012**, *29*, 277–289. [CrossRef]
12. Di Maddaloni, F.; Davis, K. Project manager's perception of the local communities' stakeholder in megaprojects. An empirical investigation in the UK. *Int. J. Proj. Manag.* **2018**, *36*, 542–565. [CrossRef]
13. Xu, X.; Li, C.Z.; Wang, J.; Huang, W. Collaboration between designers and contractors to improve building energy performance. *J. Clean. Prod.* **2019**, *219*, 20–32. [CrossRef]
14. Hu, X.; Chong, H.Y.; Wang, X. Sustainability perceptions of off-site manufacturing stakeholders in Australia. *J. Clean. Prod.* **2019**, *227*, 346–354. [CrossRef]
15. Soma, K.; Dijkshoorn-Dekker, M.; Polman, N. Stakeholder contributions through transitions towards urban sustainability. *Sustain. Cities Soc.* **2018**, *37*, 438–450. [CrossRef]
16. Yu, T.; Liang, X.; Shen, G.Q.; Shi, Q.; Wang, G. An optimization model for managing stakeholder conflicts in urban redevelopment projects in China. *J. Clean. Prod.* **2019**, *212*, 537–547. [CrossRef]
17. van Dijk, S.; Tenpierik, M.; van den Dobbelsteen, A. Continuing the building's cycles: A literature review and analysis of current systems theories in comparison with the theory of Cradle to Cradle. *Resour. Conserv. Recycl.* **2014**, *82*, 21–34. [CrossRef]
18. Biesenthal, C.; Clegg, S.; Mahalingam, A.; Sankaran, S. Applying institutional theories to managing megaprojects. *Int. J. Proj. Manag.* **2018**, *36*, 43–54. [CrossRef]
19. Yusof, N.; Awang, H.; Iranmanesh, M. Determinants and outcomes of environmental practices in Malaysian construction projects. *J. Clean. Prod.* **2017**, *156*, 345–354. [CrossRef]
20. Wojewnik-Filipkowska, A.; Węgrzyn, J. Understanding of Public–Private Partnership Stakeholders as a Condition of Sustainable Development. *Sustainability* **2019**, *11*, 1194. [CrossRef]
21. Hasik, V.; Escott, E.; Bates, R.; Carlisle, S.; Faircloth, B.; Bilec, M.M. Comparative whole building life cycle assessment of renovation and new construction. *Build. Environ.* **2019**, 106218. [CrossRef]
22. Broers, W.; Vasseur, V.; Kemp, R.; Abujidi, N.; Vroon, Z. Decided or divided? An empirical analysis of the decision-making process of Dutch homeowners for energy renovation measures. *Energy Res. Soc. Sci.* **2019**, *58*, 101284. [CrossRef]
23. Santangelo, A.; Yan, D.; Feng, X.; Tondelli, S. Renovation strategies for the Italian public housing stock: Applying building energy simulation and occupant behaviour modelling to support decision-making process. *Energy Build.* **2018**, *167*, 269–280. [CrossRef]
24. Stonebraker, J.S.; Kirkwood, C.W. Formulating and solving sequential decision analysis models with continuous variables. *IEEE Trans. Eng. Manag.* **1997**, *44*, 43–53. [CrossRef]
25. Bugajev, A.; Šostak, O. An algorithm for modelling the impact of the judicial conflict-resolution process on construction investment. *Sustainability* **2018**, *10*, 182. [CrossRef]
26. Bellman, R.E.; Dreyfus, S.E. *Applied Dynamic Programming*; Princeton University Press: Princeton, NJ, USA, 2015.
27. Si, J.; Marjanovic-Halburd, L.; Nasiri, F.; Bell, S. Assessment of building-integrated green technologies: A review and case study on applications of Multi-Criteria Decision Making (MCDM) method. *Sustain. Cities Soc.* **2016**, *27*, 106–115. [CrossRef]
28. Kokaraki, N.; Hopfe, C.J.; Robinson, E.; Nikolaidou, E. Testing the reliability of deterministic multi-criteria decision-making methods using building performance simulation. *Renew. Sustain. Energy Rev.* **2019**, *112*, 991–1007. [CrossRef]
29. Ishizaka, A.; Labib, A. Analytic hierarchy process and expert choice: Benefits and limitations. *Insight* **2009**, *22*, 201–220. [CrossRef]
30. Keshavarz Ghorabaee, M.; Amiri, M.; Zavadskas, E.K.; Turskis, Z.; Antucheviciene, J. Stochastic EDAS method for multi-criteria decision-making with normally distributed data. *J. Intell. Fuzzy Syst.* **2017**, *33*, 1627–1638. [CrossRef]
31. Khalil, N.; Kamaruzzaman, S.N.; Baharum, M.R. Ranking the indicators of building performance and the users' risk via Analytical Hierarchy Process (AHP): Case of Malaysia. *Ecol. Indic.* **2016**, *71*, 567–576. [CrossRef]

32. Kamaruzzaman, S.N.; Lou, E.C.W.; Wong, P.F.; Wood, R.; Che-Ani, A.I. Developing weighting system for refurbishment building assessment scheme in Malaysia through analytic hierarchy process (AHP) approach. *Energy Policy* **2018**, *112*, 280–290. [CrossRef]
33. Zheng, X.; Easa, S.M.; Yang, Z.; Ji, T.; Jiang, Z. Life-cycle sustainability assessment of pavement maintenance alternatives: Methodology and case study. *J. Clean. Prod.* **2019**, *213*, 659–672. [CrossRef]
34. Saaty, T.L. *Fundamentals of Decision Making and Priority Theory with the Snalytic Hierarchy Process*; RWS Publications: Pittsburgh, PA, USA; Volume 6, 2000.
35. Fuerst, F.; Warren-Myers, G. Does voluntary disclosure create a green lemon problem? Energy-efficiency ratings and house prices. *Energy Econ.* **2018**, *74*, 1–12. [CrossRef]
36. Picarelli, N. There is no free house. *J. Urban Econ.* **2019**, *111*, 35–52. [CrossRef]
37. Ahmad, T.; Aibinu, A.A.; Stephan, A.; Chan, A.P. Investigating associations among performance criteria in Green Building projects. *J. Clean. Prod.* **2019**. [CrossRef]
38. d'Acci, L. Quality of urban area, distance from city centre, and housing value. Case study on real estate values in Turin. *Cities* **2019**, *91*, 71–92. [CrossRef]
39. Li, H.; Wei, Y.D.; Wu, Y.; Tian, G. Analyzing housing prices in Shanghai with open data: Amenity, accessibility and urban structure. *Cities* **2019**, *91*, 165–179. [CrossRef]
40. Chen, C.S.; Chiu, Y.H.; Tsai, L. Evaluating the adaptive reuse of historic buildings through multicriteria decision-making. *Habitat Int.* **2018**, *81*, 12–23. [CrossRef]
41. Waltl, S.R. Variation across price segments and locations: A comprehensive quantile regression analysis of the Sydney housing market. *Real Estate Econ.* **2019**, *47*, 723–756. [CrossRef]

sustainability

Article

LDA-Based Model for Defect Management in Residential Buildings

Byeol Kim [1], Yonghan Ahn [1] and Sanghyo Lee [2,*]

1 Department of Architectural Engineering, Hanyang University, 55, Hanyangdaehak-ro, Sangrok-gu, Ansan-si, Gyeonggi-do 15588, Korea; keemstars@naver.com (B.K.); yhahn@hanyang.ac.kr (Y.A.)
2 Division of Architecture and Civil Engineering, Kangwon National University, 346, Jungang-ro, Samcheok-si, Gangwon-do 25913, Korea
* Correspondence: leesh0903@kangwon.ac.kr; Tel.: +82-33-570-6527

Received: 11 November 2019; Accepted: 13 December 2019; Published: 16 December 2019

Abstract: This study systematically analyzes various defect patterns that occur during the warranty period of residential buildings using the loss distribution approach (LDA). This paper examines 16,108 defects from 133 residential buildings where defect disputes occurred between 2008 and 2018 in South Korea. The analysis results showed that the defect losses were relatively high in reinforcement concrete (RC) work (3/5/10 years), waterproof work (5 years), and finish work (2 years). It is shown that RC work has a high frequency of defects, such as cracks in concrete in public spaces affected by external factors. In addition, it was analyzed that the type of defect needed high repair cost because the area where the defect—such as incorrect installation and missing task—occurred, needed construction again. According to the level of frequency and severity, losses were divided within four zones to provide detailed strategies (by period). This will effectively contribute to minimizing unnecessary losses from defects as quantifying the losses of defects.

Keywords: defects liability period; risk matrix; residential buildings; loss distribution approach

1. Introduction

The proportion of people living in cities is expected to reach 66% among the global population by 2050. For this trend, apartments are becoming a major residential type in many countries, including Vietnam, Singapore, South Korea, and the United States [1,2].

As the functionality of apartments becomes more complex and higher, the importance of quality management is increasing [3]. The quality control of apartments aims to maintain a high-quality construction environment so that many residents can live in a safe, comfortable, residential environment [4]. In particular, defect management is critical to maintaining the building performance [5]. Thus, defect management can contribute to the sustainability and durability of residential buildings.

However, the construction of apartments causes defects due to the complex interconnections of various work types, unexpected design mistakes, material defects, faults during construction, and environmental factors [6]. Various stakeholders suffer practical damage because significant costs are incurred to solve these defects [7]. Many efforts have been continuously made to ensure the quality and to prevent poor construction of housing [8]. However, conflicts and disputes between residents and contractors can arise due to differences in the interpretations and perceptions of defects that occur after the houses are transferred [9]. Many countries have institutionalized the defect liability period (DLP) to minimize the social losses caused by the defects [10–13]. In the case of the United Kingdom, the DLP is arbitrarily implemented, while in Japan, the DLP is mandatory. As such, DLP in each country operates at various standards.

The DLP obligates the contractor to repair any defects that occur within a set period; however, the obligation is terminated after the set period. Due to this characteristic of the DLP, the contractor should identify various defect risks, including the lack of knowledge of residents, daily wear and tear, and construction failures, while managing the defects efficiently during the warranty period. Furthermore, occupants need to recognize any defects that may occur within the DLP and secure their rights to minimize the defects [12].

Therefore, it is essential for stakeholders to recognize the defect occurrence pattern preemptively to take a rational decision about the defect repair for the DLP [14]. Many studies have attempted to analyze the causes of defect generation in residential buildings. However, these studies simply investigated the importance of defects by summarizing the frequency of the defects and their costs [15–18]. The defect frequency can be recognized as loss-occurring events and the repair cost of a defect can be interpreted as the weight of loss. This means that if the frequency and cost of defects are considered at the same time, the defect occurrence pattern can be analyzed in a more integrated manner.

This study aims to systematically analyze the patterns and profiles of various defects that occur during the DLP of residential buildings by using the loss distribution approach (LDA).

2. Literature Review

A defect is generally, "a state of a product that does not have the quality or performance that it must normally have. Defects of apartments include cracks, settlements, damage, swelling, leakage, or other flaws due to construction errors. This can result in problems in the safety, function, or aesthetics of the building or facilities" [19,20]. There are two main methods to minimize the loss resulting from defects. The first is to prevent defects in advance. Defects are caused by mutually complex factors throughout the process of design, construction, materials, and maintenance [21]. To prevent them, a fundamental analysis about the causes of defects that have occurred is required. In this respect, existing studies have examined the characteristics of defect generation by reclassifying the defect cases based on various criteria [13,16,22,23]. To prevent defects practically, the fundamental causes should be analyzed based on the types of recognized defects. In this context, various studies on the causes of defects have been conducted [18,24]. In order to identify the cause of the defects that occur in various forms, research has been conducted regarding the classification of the defects behavior [17,25–27]. Moreover, some studies analyzed defect occurrence patterns by thoroughly tracking the causal relations by considering such factors as objects, locations, and work types, as well as direct causes [28,29]. These studies suggested design errors, human errors, financial problems, pressing schedules, materials, and maintenance as the main causes of defects. However, defects are caused by complex nonlinear causal relations between various factors. The mechanism of these complex causes are major obstacles to managing the defect risks because they can change dynamically during time or depend on the project type. In other words, the methods to prevent the construction defects of apartments ultimately have limitations and defects inevitably occur [15,30].

The second method for minimizing the loss from defects is to find rational response measures to defects. Since most industrial products have a warranty system for the product's quality for a certain period, buildings also have a quality assurance system. It is common practice for a construction company, which is the business entity, to take responsibility for the quality assurance of the apartment in the event of defects for a certain period of time [11]. In this regard, existing studies revealed the measures to efficiently operate the warranty system [10,12,31]. Davey et al. (2006) derived the problems during the DLP and emphasized the need for a short-term processing plan to effectively respond to the defects that occur during the handover stage. To address this issue, they proposed measures such as improving the management during the DLP as well as improving the procedures through the cooperation of contractors, consultants, and customers [12]. Hopkin et al. (2016) surveyed UK major stakeholders in regard to the defect frequency to determine the most important types of defects and to identify which defects must be focused on after defects have occurred [32]. However, there are differences in the perspectives about defects that occur during DLP among the major stakeholders that were analyzed [9]. Hence, defect occurrence patterns need to be analyzed in consideration of the DLP.

Efficient defect resolution is difficult due to a fierce clash of opinions and conflicts about the adequacy of the incurred defect repair cost. Therefore, it is necessary to investigate the rationality of the criteria, the scope of the defect judgment, and the detailed criteria for estimating the defect costs. Mills et al. (2009) discussed the characteristics of major defects based on the defect repair costs for buildings constructed between 1977 and 1983 and quantified the effects of the types of contractors and buildings [15]. Furthermore, Hughes et al. (2000) researched the budget range that must be prepared to protect users and to complete the contractor's project by calculating the cost required to resolve the defects [31]. However, these studies were specifically based on an individual analysis without considering other detailed factors such as the tasks, components, and location in an integrated manner. From this background, the present study analyzes the various defect profiles that occur during the DLP to explore reasonable directions for efficient quality control and dispute reconciliation for residential buildings.

3. Materials and Method

The causes of defects that have occurred in residential buildings include: the neglect of supervision; pressing schedules; construction by inexperienced workers; defects in materials; and inadequate inspection. In other words, defects are operation risks mainly caused by humans or process errors. In the financial sector, which already has a high level of risk measurement and management systems, the loss distribution approach (LDA) is a representative method for measuring and managing rising operational risks. LDA is used mainly in sales organizations to assess objective loss estimates through statistical analysis based on existing loss data. As a method that finds the distribution characteristics of operational risks using a vast amount of data generated, the LDA is considered as the most sophisticated and accurate risk measurement method.

This paper establishes a process for measuring defect risks based on the LDA process used to measure the operational risks in the financial field (Figure 1). There are four steps of measurements when measuring the operational risks using the LDA. In step 1, a defect classification system is constructed in 3D. In the existing financial sector, the operating risk frequency distribution and severity distribution are estimated for each of the eight business areas and seven loss cases. That is, 56 areas (8 Business Line (BL) * 7 Loss event type LET) are combined with each other to generate a loss distribution. Then, the cost for responding to the operational risks is calculated using the loss distribution. At this time, the areas are reclassified according to the detailed measurement level and items of the measurement target. In this study, areas were classified based on 11 work types, 7 defect occurrence locations, and 11 defect types (11 * 7 * 11). The data is collected in step 2. The collected data was frequency and severity data corresponding to each cell. In step 3, the frequency and severity distribution of the defect are estimated based on the data. The loss distribution is calculated by combining the two estimated distributions using Monte Carlo simulation. The severity data was scaled according to the size of the apartments and defined as the defect cost versus the total floor area. Finally, in step 4, the risk profiles are organized by the service year based on the calculated loss distribution. Afterwards, the overall analysis is performed.

To estimate the loss frequency and severity distribution, which was step 3 in the LDA measurement process, the distribution of each item is set. For the frequency distribution, which is a discrete distribution, the most representative Poisson distribution is used. In contrast, the severity distribution is a continuous distribution, which is set as a model with a tail distribution with log-normal distribution, Weibull distribution, gamma distribution, and Pareto distribution. To set the severity distribution for each cell in this study, the best fit test was performed for each distribution. Then, the loss distribution was set by performing Monte Carlo simulation 100,000 times using the parameters of the frequency and severity distributions for each cell. After that, the confidence levels of all of the distributions were set ranging from 0% to 100% and the risk level was checked. The loss distribution for each cell was derived through this process and was combined to identify and analyze the defect risk profile of each item classified by the DLP, work type, or space.

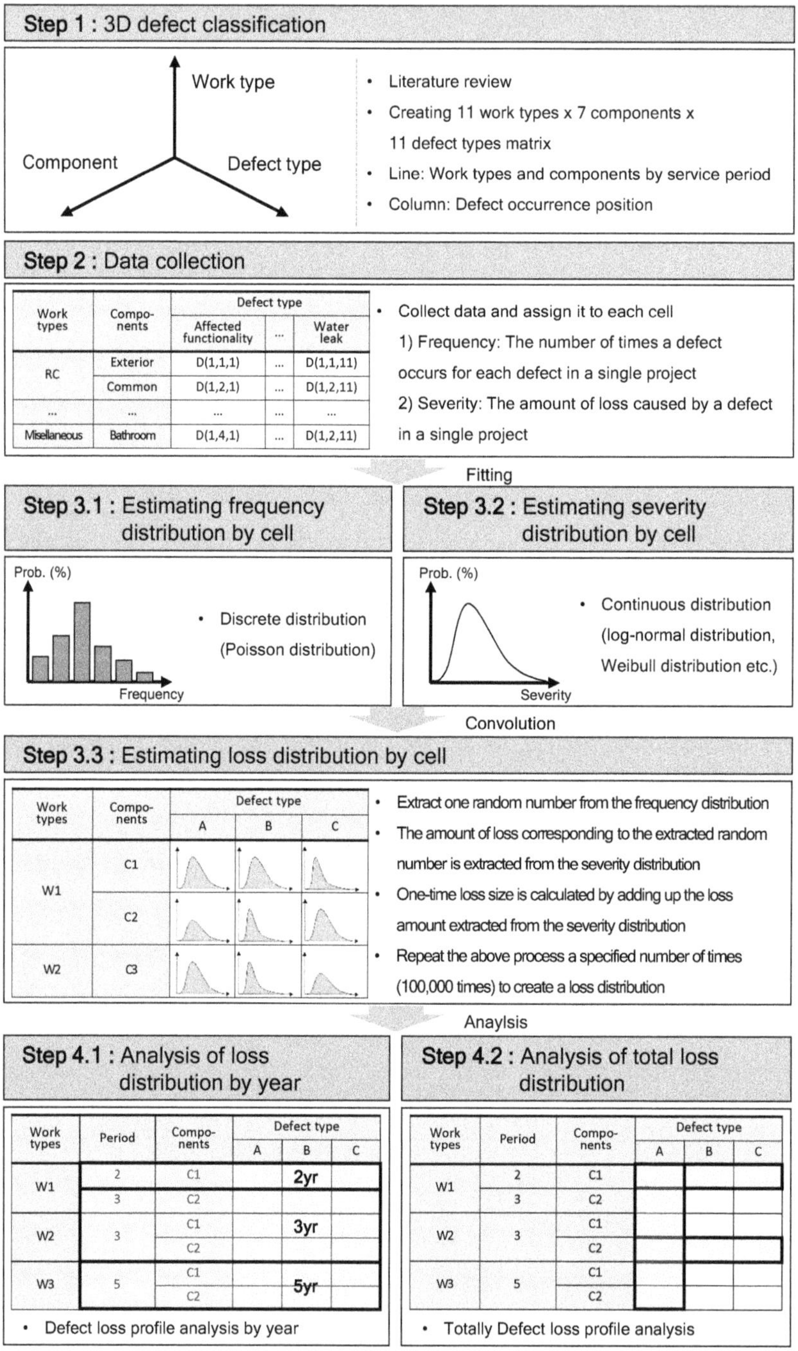

Figure 1. Conducting estimations with the loss distribution approach.

In the process of analyzing defect risks using the LDA, a lack of data may occur in each cell of the 3D defect matrix constructed in accordance with the defect classification system. In this case, the reliability may drop due to the neutrality of the parameter estimates. To prevent this problem, the minimum data corresponding to each cell should be obtained in the setting step of the frequency and severity distributions. Each cell must not be set too broadly or in too much detail. In addition, the range of cells should be classified in consideration of the classification system for the items to be measured. In consideration of this, Table 1 presents the work type-defects liability period, which was classified into 11 categories. These include: reinforced concrete (Rc) (3); Rc (5); Rc (10); masonry (5); finish (2); insulation (3); waterproof (5); mechanical, electrical and plumbing (MEP) (3); door and windows (3); furniture (2); and miscellaneous (3), as suggested by the Ministry of Land, Infrastructure, and Transport of South Korea. The defect locations were classified into seven categories: exterior; common area; garage; hall/corridor; balcony; room/entrance; and bathroom/kitchen. The types of defects were reclassified into 11 categories: affected functionality; broken; corrosion; crack; detachment; incorrect installation; missing task; surface appearance; excess moisture; entrapped water; and water leak (11 work types-liability periods* 7 locations *11 defects). As a result, each cell was classified as shown in Table 2.

Table 1. Defect classification.

Defect Classification	Description	Writer(s)
Affected functionality	Materials/components that must be replaced because its functionality is completely affected. Materials/components that must be repaired but not necessarily replaced because its functionality is partially affected.	Forcada et al. (2013), Forcada et al. (2013), Macarulla et al. (2013), Rotimi et al. (2015) [16,17,22,24]
Broken	Materials/components physically and forcibly separated into pieces or split.	Forcada et al. (2013), Forcada et al. (2013), Macarulla et al. (2013), Georgiou et al. (2010), Rotimi et al. (2015), Chong et al. (2006), Chong et al. (2005) [16,17,22–24,27,30]
Corrosion	All defects caused by living beings as molds. Corrosion of metals or the carbonation of concrete.	Macarulla et al. (2013), Rotimi et al. (2015) [22,24]
Crack	Cracks in construction elements.	Forcada et al. (2013), Macarulla et al. (2013), Rotimi et al. (2015), Chong et al. (2005) [16,22,24,30]
Detachment	Materials/components that are not fixed in their position.	Forcada et al. (2013), Forcada et al. (2013), Macarulla et al. (2013), Rotimi et al. (2015), Chong et al. (2006), Chong et al. (2005) [16,17,22,24,27,30]
Incorrect installation	Materials/components are not well positioned or do not satisfy project specifications or do not have the characteristics they should.	Forcada et al. (2013), Forcada et al. (2013), Macarulla et al. (2013), Rotimi et al. (2015) [16,17,22,24]
Missingtask	Works that are not completed/done, although the project or specifications are supposed to be collocated or completed/done.	Forcada et al. (2013), Forcada et al. (2013), Macarulla et al. (2013) [16,17,22]
Surface appearance	Protuberance on a level surface. Opposite effect to a bump. Surface uneven or uniform e.g., in shape or texture, an uneven color, uneven ground, uneven margins, wood with an uneven grain. The result of a collision or abrasion. Surface with a powdery deposit caused by the evaporation of water when there is a certain level of dissolved salts.	Forcada et al. (2013), Forcada et al. (2013), Macarulla et al. (2013), Rotimi et al. (2015), Chong et al. (2006), Chong et al. (2005) [16,17,22,24]
Excess moisture	Wetness caused by moisture, including rising dampness. Penetration damp and condensation.	Forcada et al. (2013), Forcada et al. (2013), Macarulla et al. (2013), [16,17,22]
Entrapped water	Water that does not drain.	Macarulla et al. (2013) [22]
Water leak	Defects related to water that seeps through walls, slabs, roofs, etc.	Forcada et al. (2013), Forcada et al. (2013), Macarulla et al. (2013), Chong et al. (2005) [16,17,22,30]

Table 2. Risk matrix of the different types of defects and work types.

Work Types [a] (Liability Period)	Locations [b]	Defects [c]										
		Af	Br	Co	Cr	De	Ii	Mt	Sa	Em	Ew	Wl
Rc (3)	Ex	d(1,1,1)	d(1,1,2)	d(1,1,3)	d(1,1,4)	d(1,1,5)	d(1,1,6)	d(1,1,7)	d(1,1,8)	d(1,1,9)	d(1,1,10)	d(1,1,11)
	Ca	d(1,2,1)	d(1,2,2)	d(1,2,3)	d(1,2,4)	d(1,2,5)	d(1,2,6)	d(1,2,7)	d(1,2,8)	d(1,2,9)	d(1,2,10)	d(1,2,11)
	Ga	d(1,3,1)	d(1,3,2)	d(1,3,3)	d(1,3,4)	d(1,3,5)	d(1,3,6)	d(1,3,7)	d(1,3,8)	d(1,3,9)	d(1,3,10)	d(1,3,11)
	Hc	d(1,4,1)	d(1,4,2)	d(1,4,3)	d(1,4,4)	d(1,4,5)	d(1,4,6)	d(1,4,7)	d(1,4,8)	d(1,4,9)	d(1,4,10)	d(1,4,11)
	Ba	d(1,5,1)	d(1,5,2)	d(1,5,3)	d(1,5,4)	d(1,5,5)	d(1,5,6)	d(1,5,7)	d(1,5,8)	d(1,5,9)	d(1,5,10)	d(1,5,11)
	Re	d(1,6,1)	d(1,6,2)	d(1,6,3)	d(1,6,4)	d(1,6,5)	d(1,6,6)	d(1,6,7)	d(1,6,8)	d(1,6,9)	d(1,6,10)	d(1,6,11)
	Bk	d(1,7,1)	d(1,7,2)	d(1,7,3)	d(1,7,4)	d(1,7,5)	d(1,7,6)	d(1,7,7)	d(1,7,8)	d(1,7,9)	d(1,7,10)	d(1,7,11)
Rc (5)	Ex	d(2,1,1)	d(2,1,2)	d(2,1,3)	d(2,1,4)	d(2,1,5)	d(2,1,6)	d(2,1,7)	d(2,1,8)	d(2,1,9)	d(2,1,10)	d(2,1,11)
	Ca	d(2,2,1)	d(2,2,2)	d(2,2,3)	d(2,2,4)	d(2,2,5)	d(2,2,6)	d(2,2,7)	d(2,2,8)	d(2,2,9)	d(2,2,10)	d(2,2,11)
	Ga	d(2,3,1)	d(2,3,2)	d(2,3,3)	d(2,3,4)	d(2,3,5)	d(2,3,6)	d(2,3,7)	d(2,3,8)	d(2,3,9)	d(2,3,10)	d(2,3,11)
	Hc	d(2,4,1)	d(2,4,2)	d(2,4,3)	d(2,4,4)	d(2,4,5)	d(2,4,6)	d(2,4,7)	d(2,4,8)	d(2,4,9)	d(2,4,10)	d(2,4,11)
	Ba	d(2,5,1)	d(2,5,2)	d(2,5,3)	d(2,5,4)	d(2,5,5)	d(2,5,6)	d(2,5,7)	d(2,5,8)	d(2,5,9)	d(2,5,10)	d(2,5,11)
	Re	d(2,6,1)	d(2,6,2)	d(2,6,3)	d(2,6,4)	d(2,6,5)	d(2,6,6)	d(2,6,7)	d(2,6,8)	d(2,6,9)	d(2,6,10)	d(2,6,11)
	Bk	d(2,7,1)	d(2,7,2)	d(2,7,3)	d(2,7,4)	d(2,7,5)	d(2,7,6)	d(2,7,7)	d(2,7,8)	d(2,7,9)	d(2,7,10)	d(2,7,11)
Rc (10)	Ex	d(3,1,1)	d(3,1,2)	d(3,1,3)	d(3,1,4)	d(3,1,5)	d(3,1,6)	d(3,1,7)	d(3,1,8)	d(3,1,9)	d(3,1,10)	d(3,1,11)
	Ca	d(3,2,1)	d(3,2,2)	d(3,2,3)	d(3,2,4)	d(3,2,5)	d(3,2,6)	d(3,2,7)	d(3,2,8)	d(3,2,9)	d(3,2,10)	d(3,2,11)
	Ga	d(3,3,1)	d(3,3,2)	d(3,3,3)	d(3,3,4)	d(3,3,5)	d(3,3,6)	d(3,3,7)	d(3,3,8)	d(3,3,9)	d(3,3,10)	d(3,3,11)
	Hc	d(3,4,1)	d(3,4,2)	d(3,4,3)	d(3,4,4)	d(3,4,5)	d(3,4,6)	d(3,4,7)	d(3,4,8)	d(3,4,9)	d(3,4,10)	d(3,4,11)
	Ba	d(3,5,1)	d(3,5,2)	d(3,5,3)	d(3,5,4)	d(3,5,5)	d(3,5,6)	d(3,5,7)	d(3,5,8)	d(3,5,9)	d(3,5,10)	d(3,5,11)
	Re	d(3,6,1)	d(3,6,2)	d(3,6,3)	d(3,6,4)	d(3,6,5)	d(3,6,6)	d(3,6,7)	d(3,6,8)	d(3,6,9)	d(3,6,10)	d(3,6,11)
	Bk	d(3,7,1)	d(3,7,2)	d(3,7,3)	d(3,7,4)	d(3,7,5)	d(3,7,6)	d(3,7,7)	d(3,7,8)	d(3,7,9)	d(3,7,10)	d(3,7,11)
Ma (5)	Ex	d(4,1,1)	d(4,1,2)	d(4,1,3)	d(4,1,4)	d(4,1,5)	d(4,1,6)	d(4,1,7)	d(4,1,8)	d(4,1,9)	d(4,1,10)	d(4,1,11)
	Ca	d(4,2,1)	d(4,2,2)	d(4,2,3)	d(4,2,4)	d(4,2,5)	d(4,2,6)	d(4,2,7)	d(4,2,8)	d(4,2,9)	d(4,2,10)	d(4,2,11)
	Ga	d(4,3,1)	d(4,3,2)	d(4,3,3)	d(4,3,4)	d(4,3,5)	d(4,3,6)	d(4,3,7)	d(4,3,8)	d(4,3,9)	d(4,3,10)	d(4,3,11)
	Hc	d(4,4,1)	d(4,4,2)	d(4,4,3)	d(4,4,4)	d(4,4,5)	d(4,4,6)	d(4,4,7)	d(4,4,8)	d(4,4,9)	d(4,4,10)	d(4,4,11)
	Ba	d(4,5,1)	d(4,5,2)	d(4,5,3)	d(4,5,4)	d(4,5,5)	d(4,5,6)	d(4,5,7)	d(4,5,8)	d(4,5,9)	d(4,5,10)	d(4,5,11)
	Re	d(4,6,1)	d(4,6,2)	d(4,6,3)	d(4,6,4)	d(4,6,5)	d(4,6,6)	d(4,6,7)	d(4,6,8)	d(4,6,9)	d(4,6,10)	d(4,6,11)
	Bk	d(4,7,1)	d(4,7,2)	d(4,7,3)	d(4,7,4)	d(4,7,5)	d(4,7,6)	d(4,7,7)	d(4,7,8)	d(4,7,9)	d(4,7,10)	d(4,7,11)
Fi (2)	Ex	d(5,1,1)	d(5,1,2)	d(5,1,3)	d(5,1,4)	d(5,1,5)	d(5,1,6)	d(5,1,7)	d(5,1,8)	d(5,1,9)	d(5,1,10)	d(5,1,11)
	Ca	d(5,2,1)	d(5,2,2)	d(5,2,3)	d(5,2,4)	d(5,2,5)	d(5,2,6)	d(5,2,7)	d(5,2,8)	d(5,2,9)	d(5,2,10)	d(5,2,11)
	Ga	d(5,3,1)	d(5,3,2)	d(5,3,3)	d(5,3,4)	d(5,3,5)	d(5,3,6)	d(5,3,7)	d(5,3,8)	d(5,3,9)	d(5,3,10)	d(5,3,11)
	Hc	d(5,4,1)	d(5,4,2)	d(5,4,3)	d(5,4,4)	d(5,4,5)	d(5,4,6)	d(5,4,7)	d(5,4,8)	d(5,4,9)	d(5,4,10)	d(5,4,11)
	Ba	d(5,5,1)	d(5,5,2)	d(5,5,3)	d(5,5,4)	d(5,5,5)	d(5,5,6)	d(5,5,7)	d(5,5,8)	d(5,5,9)	d(5,5,10)	d(5,5,11)
	Re	d(5,6,1)	d(5,6,2)	d(5,6,3)	d(5,6,4)	d(5,6,5)	d(5,6,6)	d(5,6,7)	d(5,6,8)	d(5,6,9)	d(5,6,10)	d(5,6,11)
	Bk	d(5,7,1)	d(5,7,2)	d(5,7,3)	d(5,7,4)	d(5,7,5)	d(5,7,6)	d(5,7,7)	d(5,7,8)	d(5,7,9)	d(5,7,10)	d(5,7,11)
In (3)	Ex	d(6,1,1)	d(6,1,2)	d(6,1,3)	d(6,1,4)	d(6,1,5)	d(6,1,6)	d(6,1,7)	d(6,1,8)	d(6,1,9)	d(6,1,10)	d(6,1,11)
	Ca	d(6,2,1)	d(6,2,2)	d(6,2,3)	d(6,2,4)	d(6,2,5)	d(6,2,6)	d(6,2,7)	d(6,2,8)	d(6,2,9)	d(6,2,10)	d(6,2,11)
	Ga	d(6,3,1)	d(6,3,2)	d(6,3,3)	d(6,3,4)	d(6,3,5)	d(6,3,6)	d(6,3,7)	d(6,3,8)	d(6,3,9)	d(6,3,10)	d(6,3,11)
	Hc	d(6,4,1)	d(6,4,2)	d(6,4,3)	d(6,4,4)	d(6,4,5)	d(6,4,6)	d(6,4,7)	d(6,4,8)	d(6,4,9)	d(6,4,10)	d(6,4,11)
	Ba	d(6,5,1)	d(6,5,2)	d(6,5,3)	d(6,5,4)	d(6,5,5)	d(6,5,6)	d(6,5,7)	d(6,5,8)	d(6,5,9)	d(6,5,10)	d(6,5,11)
	Re	d(6,6,1)	d(6,6,2)	d(6,6,3)	d(6,6,4)	d(6,6,5)	d(6,6,6)	d(6,6,7)	d(6,6,8)	d(6,6,9)	d(6,6,10)	d(6,6,11)
	Bk	d(6,7,1)	d(6,7,2)	d(6,7,3)	d(6,7,4)	d(6,7,5)	d(6,7,6)	d(6,7,7)	d(6,7,8)	d(6,7,9)	d(6,7,10)	d(6,7,11)
Wa (5)	Ex	d(7,1,1)	d(7,1,2)	d(7,1,3)	d(7,1,4)	d(7,1,5)	d(7,1,6)	d(7,1,7)	d(7,1,8)	d(7,1,9)	d(7,1,10)	d(7,1,11)
	Ca	d(7,2,1)	d(7,2,2)	d(7,2,3)	d(7,2,4)	d(7,2,5)	d(7,2,6)	d(7,2,7)	d(7,2,8)	d(7,2,9)	d(7,2,10)	d(7,2,11)
	Ga	d(7,3,1)	d(7,3,2)	d(7,3,3)	d(7,3,4)	d(7,3,5)	d(7,3,6)	d(7,3,7)	d(7,3,8)	d(7,3,9)	d(7,3,10)	d(7,3,11)
	Hc	d(7,4,1)	d(7,4,2)	d(7,4,3)	d(7,4,4)	d(7,4,5)	d(7,4,6)	d(7,4,7)	d(7,4,8)	d(7,4,9)	d(7,4,10)	d(7,4,11)
	Ba	d(7,5,1)	d(7,5,2)	d(7,5,3)	d(7,5,4)	d(7,5,5)	d(7,5,6)	d(7,5,7)	d(7,5,8)	d(7,5,9)	d(7,5,10)	d(7,5,11)
	Re	d(7,6,1)	d(7,6,2)	d(7,6,3)	d(7,6,4)	d(7,6,5)	d(7,6,6)	d(7,6,7)	d(7,6,8)	d(7,6,9)	d(7,6,10)	d(7,6,11)
	Bk	d(7,7,1)	d(7,7,2)	d(7,7,3)	d(7,7,4)	d(7,7,5)	d(7,7,6)	d(7,7,7)	d(7,7,8)	d(7,7,9)	d(7,7,10)	d(7,7,11)
Me (3)	Ex	d(8,1,1)	d(8,1,2)	d(8,1,3)	d(8,1,4)	d(8,1,5)	d(8,1,6)	d(8,1,7)	d(8,1,8)	d(8,1,9)	d(8,1,10)	d(8,1,11)
	Ca	d(8,2,1)	d(8,2,2)	d(8,2,3)	d(8,2,4)	d(8,2,5)	d(8,2,6)	d(8,2,7)	d(8,2,8)	d(8,2,9)	d(8,2,10)	d(8,2,11)
	Ga	d(8,3,1)	d(8,3,2)	d(8,3,3)	d(8,3,4)	d(8,3,5)	d(8,3,6)	d(8,3,7)	d(8,3,8)	d(8,3,9)	d(8,3,10)	d(8,3,11)
	Hc	d(8,4,1)	d(8,4,2)	d(8,4,3)	d(8,4,4)	d(8,4,5)	d(8,4,6)	d(8,4,7)	d(8,4,8)	d(8,4,9)	d(8,4,10)	d(8,4,11)
	Ba	d(8,5,1)	d(8,5,2)	d(8,5,3)	d(8,5,4)	d(8,5,5)	d(8,5,6)	d(8,5,7)	d(8,5,8)	d(8,5,9)	d(8,5,10)	d(8,5,11)
	Re	d(8,6,1)	d(8,6,2)	d(8,6,3)	d(8,6,4)	d(8,6,5)	d(8,6,6)	d(8,6,7)	d(8,6,8)	d(8,6,9)	d(8,6,10)	d(8,6,11)
	Bk	d(8,7,1)	d(8,7,2)	d(8,7,3)	d(8,7,4)	d(8,7,5)	d(8,7,6)	d(8,7,7)	d(8,7,8)	d(8,7,9)	d(8,7,10)	d(8,7,11)
Dw (3)	Ex	d(9,1,1)	d(9,1,2)	d(9,1,3)	d(9,1,4)	d(9,1,5)	d(9,1,6)	d(9,1,7)	d(9,1,8)	d(9,1,9)	d(9,1,10)	d(9,1,11)
	Ca	d(9,2,1)	d(9,2,2)	d(9,2,3)	d(9,2,4)	d(9,2,5)	d(9,2,6)	d(9,2,7)	d(9,2,8)	d(9,2,9)	d(9,2,10)	d(9,2,11)
	Ga	d(9,3,1)	d(9,3,2)	d(9,3,3)	d(9,3,4)	d(9,3,5)	d(9,3,6)	d(9,3,7)	d(9,3,8)	d(9,3,9)	d(9,3,10)	d(9,3,11)
	Hc	d(9,4,1)	d(9,4,2)	d(9,4,3)	d(9,4,4)	d(9,4,5)	d(9,4,6)	d(9,4,7)	d(9,4,8)	d(9,4,9)	d(9,4,10)	d(9,4,11)
	Ba	d(9,5,1)	d(9,5,2)	d(9,5,3)	d(9,5,4)	d(9,5,5)	d(9,5,6)	d(9,5,7)	d(9,5,8)	d(9,5,9)	d(9,5,10)	d(9,5,11)
	Re	d(9,6,1)	d(9,6,2)	d(9,6,3)	d(9,6,4)	d(9,6,5)	d(9,6,6)	d(9,6,7)	d(9,6,8)	d(9,6,9)	d(9,6,10)	d(9,6,11)
	Bk	d(9,7,1)	d(9,7,2)	d(9,7,3)	d(9,7,4)	d(9,7,5)	d(9,7,6)	d(9,7,7)	d(9,7,8)	d(9,7,9)	d(9,7,10)	d(9,7,11)
Fu (2)	Ex	d(10,1,1)	d(10,1,2)	d(10,1,3)	d(10,1,4)	d(10,1,5)	d(10,1,6)	d(10,1,7)	d(10,1,8)	d(10,1,9)	d(10,1,10)	d(10,1,11)
	Ca	d(10,2,1)	d(10,2,2)	d(10,2,3)	d(10,2,4)	d(10,2,5)	d(10,2,6)	d(10,2,7)	d(10,2,8)	d(10,2,9)	d(10,2,10)	d(10,2,11)
	Ga	d(10,3,1)	d(10,3,2)	d(10,3,3)	d(10,3,4)	d(10,3,5)	d(10,3,6)	d(10,3,7)	d(10,3,8)	d(10,3,9)	d(10,3,10)	d(10,3,11)
	Hc	d(10,4,1)	d(10,4,2)	d(10,4,3)	d(10,4,4)	d(10,4,5)	d(10,4,6)	d(10,4,7)	d(10,4,8)	d(10,4,9)	d(10,4,10)	d(10,4,11)
	Ba	d(10,5,1)	d(10,5,2)	d(10,5,3)	d(10,5,4)	d(10,5,5)	d(10,5,6)	d(10,5,7)	d(10,5,8)	d(10,5,9)	d(10,5,10)	d(10,5,11)
	Re	d(10,6,1)	d(10,6,2)	d(10,6,3)	d(10,6,4)	d(10,6,5)	d(10,6,6)	d(10,6,7)	d(10,6,8)	d(10,6,9)	d(10,6,10)	d(10,6,11)
	Bk	d(10,7,1)	d(10,7,2)	d(10,7,3)	d(10,7,4)	d(10,7,5)	d(10,7,6)	d(10,7,7)	d(10,7,8)	d(10,7,9)	d(10,7,10)	d(10,7,11)

Table 2. *Cont.*

Work Types [a] (Liability Period)	Locations [b]	Defects [c]										
		Af	Br	Co	Cr	De	Ii	Mt	Sa	Em	Ew	Wl
Mi (3)	Ex	d(11,1,1)	d(11,1,2)	d(11,1,3)	d(11,1,4)	d(11,1,5)	d(11,1,6)	d(11,1,7)	d(11,1,8)	d(11,1,9)	d(11,1,10)	d(11,1,11)
	Ca	d(11,2,1)	d(11,2,2)	d(11,2,3)	d(11,2,4)	d(11,2,5)	d(11,2,6)	d(11,2,7)	d(11,2,8)	d(11,2,9)	d(11,2,10)	d(11,2,11)
	Ga	d(11,3,1)	d(11,3,2)	d(11,3,3)	d(11,3,4)	d(11,3,5)	d(11,3,6)	d(11,3,7)	d(11,3,8)	d(11,3,9)	d(11,3,10)	d(11,3,11)
	Hc	d(11,4,1)	d(11,4,2)	d(11,4,3)	d(11,4,4)	d(11,4,5)	d(11,4,6)	d(11,4,7)	d(11,4,8)	d(11,4,9)	d(11,4,10)	d(11,4,11)
	Ba	d(11,5,1)	d(11,5,2)	d(11,5,3)	d(11,5,4)	d(11,5,5)	d(11,5,6)	d(11,5,7)	d(11,5,8)	d(11,5,9)	d(11,5,10)	d(11,5,11)
	Re	d(11,6,1)	d(11,6,2)	d(11,6,3)	d(11,6,4)	d(11,6,5)	d(11,6,6)	d(11,6,7)	d(11,6,8)	d(11,6,9)	d(11,6,10)	d(11,6,11)
	Bk	d(11,7,1)	d(11,7,2)	d(11,7,3)	d(11,7,4)	d(11,7,5)	d(11,7,6)	d(11,7,7)	d(11,7,8)	d(11,7,9)	d(11,7,10)	d(11,7,11)

[a] Rc = reinforced concrete; Ma = masonry; Fi = finish; In = insulation; Wa = waterproof; Me = MEP; Dw = door and windows; Fu = furniture; Mi = miscellaneous. [b] Ex = exterior; Ca = common area; Ga = garage; Hc = hall/corridor; Ba = balcony; Re = room/entrance; Bk = bathroom/kitchen. [c] Af = affected functionality; Br = broken; Co = corrosion; Cr = crack; De = detachment; Ii = incorrect installation; Mt = missing task; Sa = surface appearance; Em = excess moisture; Ew = entrapped water; Wl = water leak.

4. Analysis

4.1. Data Collection and Setting the 3D Defect Classification

This study has analyzed 16,108 defects from 133 residential buildings where defect disputes occurred in South Korea between 2008 and 2018. As shown by the frequency of each defect type in Table 3, cracks (28.50%) were the most frequent defect type, followed by missing tasks (14.23%), detachments (10.91%), affected functionalities (9.87%), and surface appearance (9.86%). Severity data was analyzed by correcting the defect costs by the gross floor area (GFA). Water leaks (30.30%) had the highest defect cost. Incorrect installation (27.18%), surface appearance (13.86%), missing tasks (11.19%), and cracks (6.16%) required relatively high defect costs.

Table 3. Defect frequency and the cost for each defect type.

Category	Defect Frequency (Number)	Defect Frequency Rate (%)	Defect Cost/G.F.A ($/m^2)	Defect Cost/G.F.A Rate (%)
Affected functionality	11.44	9.87	0.46	0.78
Broken	4.46	3.85	0.54	0.91
Corrosion	3.34	2.88	1.33	2.25
Crack	33.02	28.50	3.64	6.16
Detachment	12.64	10.91	2.92	4.94
Incorrect installation	9.16	7.91	16.07	27.18
Missing task	16.49	14.23	6.61	11.19
Surface appearance	11.42	9.86	8.20	13.86
Excess moisture	1.89	1.63	0.71	1.20
Entrapped water	3.22	2.78	0.72	1.22
Water leak	8.77	7.57	17.91	30.30
Total	115.85	100.00	59.11	100.00

Note: unit is number of projects.

4.2. Estimating Loss Distributions by the Cell

Each country has a warranty system to protect consumers from damage due to defects after completion of the construction and occupancy. The warranty system of South Korea requires the contractor to deposit 3% of the total construction cost for repairing defects to the public agency after completion. When defects occur, the occupants use the deposit for any necessary repairs for the DLP for each facility that was constricted. The contractor is responsible for the repair of the defect even if the repair cost exceeds the deposit. On the other hand, if the repair cost is less than the deposit, afterwards, the balance is returned to the contractor. In view of this, the profile of the defect risk for each DLP was analyzed and the step-by-step management measures are derived.

In this study, the frequency and severity distribution of each cell in the above risk matrix was set based on the classified data. The Poisson distribution was set as the frequency distribution because it is most widely used in discrete distribution. Furthermore, among the continuous distributions, including log-normal distribution, Weibull distribution, gamma distribution, and Pareto distribution, the distribution that has the best fit was set for the severity distribution of each cell. Then, the defect loss distribution of each cell was derived by performing a Monte Carlo simulation based on these frequency and severity distributions while considering the DLP of each cell.

First, a general analysis on the defect loss distribution was performed. Figure 2 illustrates the scatter chart for each cell based on the results of the frequency and severity distributions. Scatter chart analysis can be used to suggest defects risk profile analysis and application method. The finish work has a very high defect frequency by the missing task of public locations and was a typical case of HFLS (high-frequency, low-severity). The defects of finish work of buildings in public locations showed a high likelihood due to differences between the design drawings and the actual work or poor construction.

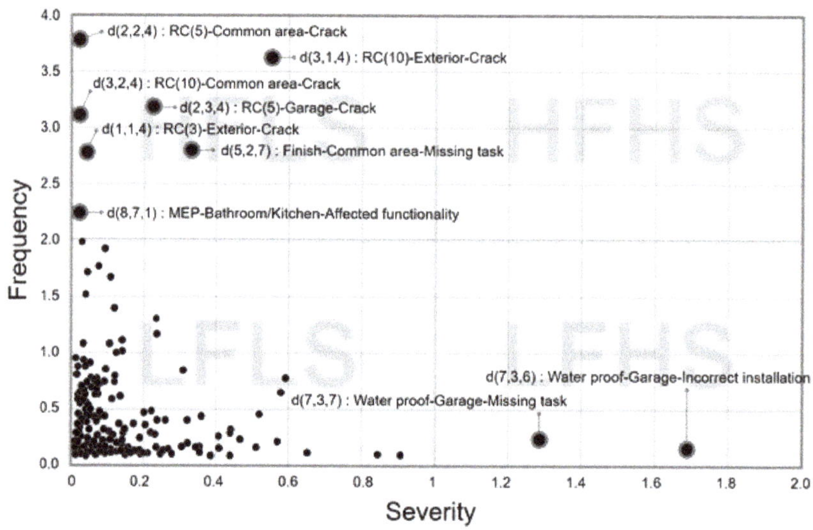

Figure 2. Frequency and severity scatter chart of the defects.

The RC work had a very high defect frequency due to cracks in public locations, which was found to be a typical case of HFLS (high-frequency, low-severity). Cracks in structures in public locations, such as the exterior, garage, and common areas, are natural phenomenon caused by aging. However, cracks can be regarded as serious defects. If cracks are not repaired, external factors accelerate the deterioration and aging of the defective parts and lead to continuous defects.

By contrast, the MEP work showed a very high defect frequency in private locations and was found to be a typical case of HFLS. Furthermore, the MEP work showed a high loss of malfunction defects. This defect is caused by a combination of construction factors such as missing equipment or modified construction, and mechanical factors such as poor product operation.

The waterproof work showed a very high defect cost in public locations, which was a typical case of LFHS (low-frequency high-severity). In particular, poor waterproof finishing in places exposed to the outside, such as underground parking lots and rooftops, can cause serious damage to the structure; thus, responsible quality control is required on the part of the contractor.

Next, a detailed analysis was performed of the defect loss distribution by the DLP. Table 4 lists the cells with 10 of the highest average values of the defect loss distribution for defects occurring within two years of the DLP. Figure 3 shows a scatter chart for the average values of the frequency/severity distributions for each cell that occurred within two years of the DLP. For the finish work, the losses of missing tasks, incorrect installations, and detachments occurred at all locations including shared/exclusive spaces. These defects are considered to have occurred due to the neglect of supervision at the site as well as work by inexperienced workers. Furthermore, the frequency of occurrence seems to be high because it is easier to recognize defects. Table 5 outlines the characteristics of the defect risks by defect type, location, and work type for defects occurring within two years of the DLP. This presents management measures by phase including the design, construction, handover, post-handover, and occupancy phase. Defects for finish work, which occur in private spaces, had higher defect losses due to problems associated with the contractor's management rather than misuse by the occupants. Furthermore, the finish process should be completed before the handover phase because the defects can be easily seen by the occupants or contractor.

Table 4. Ranking the defects loss occurring within two years of the defect liability period (DLP).

Ranking	Work Types	Location	Defects	Average
1	Finish	Bathroom/Kitchen	Incorrect installation	0.370
2	Finish	Hall/Corridor	Missing task	0.309
3	Finish	Bathroom/Kitchen	Missing task	0.279
4	Finish	Room/Entrance	Incorrect installation	0.208
5	Finish	Room/Entrance	Detachment	0.183
6	Finish	Room/Entrance	Broken	0.181
7	Finish	Bathroom/Kitchen	Detachment	0.174
8	Finish	Hall/Corridor	Incorrect installation	0.134
9	Finish	Common area	Missing task	0.131
10	Finish	Garage	Missing task	0.125

Note: average means the mean value of loss distribution statistics.

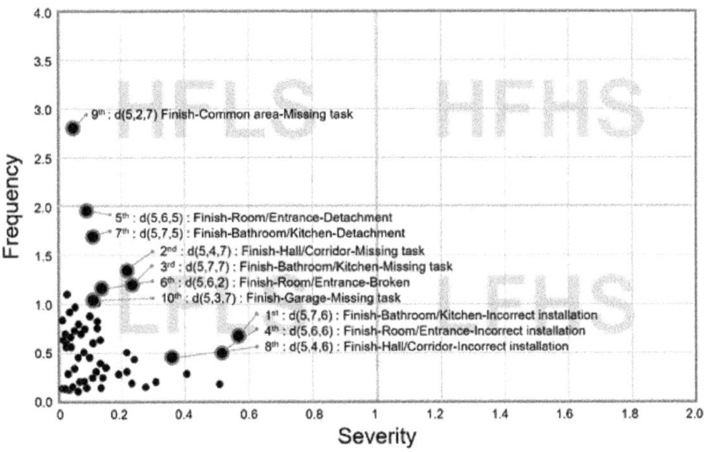

Figure 3. Frequency and severity scatter chart of a defect occurring within two years of the DLP.

Table 5. Defect risk management strategies occurring within two years of the DLP.

Risk Zone	Defect Characteristics		Defect Risk Management Strategies
	Cell No	Phase	
HFLS	d(5,2,7)	Construction	Construction of components is required after disassembly, because incorrect installations and missing tasks appear as major defects. Thorough management during construction is required because of a serious defect that prevents the building from performing its function and role at all, or demolishing is required simultaneously with construction depending on the field situation.
LFLS	d(5,7,6) d(5,4,7) d(5,7,7) d(5,6,6) d(5,4,6) d(5,3,7)		
LFLS	d(5,6,5) d(5,6,2) d(5,7,5)	Handover	Defects such as detachment and breakage need to be reviewed through visual inspection, along with careful management of the construction status by reviewing the design documents and design change documents during the inspection of use.

Table 6 lists the cells with 12 of the highest average values of the defect loss distribution for defects occurring after 2 years and before 3 years of the DLP. Figure 4 shows the scatter chart for the average frequency/severity distributions for each cell for defects that occurred after 2 years and before 3 years of the DLP. This analysis confirmed that the RC work was associated with relatively high losses due to cracks, surface appearance, and detachment defects in the exterior. It is a consequence of the exterior walls that are exposed to outside conditions and have a higher probability of occurrence of defects due to external factors such as typhoons, daily/seasonal temperature and humidity variations, sun exposure, etc. MEP work has a high loss of missing tasks and affected functionality defects. These defects occur during a series of construction of various components within the building because some elements are not constructed or due to poor connection between components. Furthermore, they can also be caused by the malfunction or unconformities of the product itself. In practice, MEP operations often occur in the early stages because they are easily recognizable by the user. However, the reason why it is significantly higher within three years compared to other periods is that the proportion of other defects falls significantly within the three years.

Table 6. Ranking of the defect loss occurring within three years of the DLP.

Ranking	Work Types	Location	Defects	Average
1	RC	Exterior	Crack	0.923
2	RC	Exterior	Surface appearance	0.428
3	RC	Exterior	Detachment	0.138
4	MEP	Common area	Missing task	0.134
5	RC	Exterior	Water leak	0.111
6	MEP	Balcony	Missing task	0.081
7	MEP	Bathroom/Kitchen	Missing task	0.076
8	Insulation	Bathroom/Kitchen	Incorrect installation	0.066
9	Insulation	Room/Entrance	Excess moisture	0.062
10	Door and Windows	Room/Entrance	Affected functionality	0.060
11	MEP	Bathroom/Kitchen	Affected functionality	0.057
12	MEP	Entrance	Affected functionality	0.056

Note: average means the mean value of loss distribution statistics.

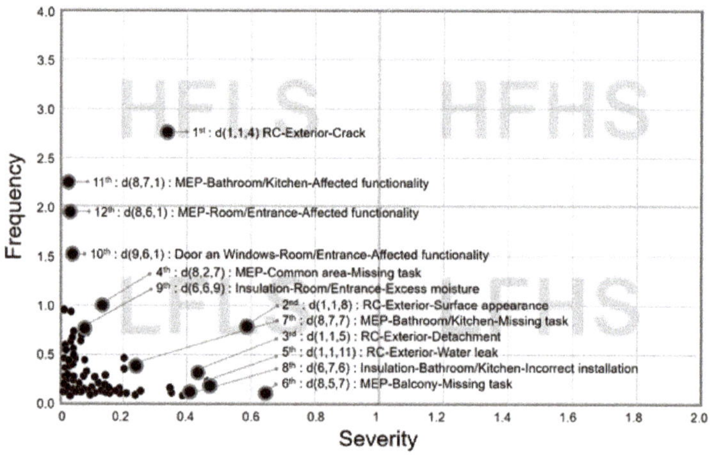

Figure 4. Frequency and severity scatter chart of a defect occurring within three years of the DLP.

These affected functionality defects also caused a high loss in doors and windows. Along with the MEP work, this is likely caused by defective or damaged furniture or windows installed in exclusive spaces. Focusing on the highest loss in doors and windows is not the defects that the user is responsible for.

Defects caused by poor quality insulation are also high. In the event of defects in the insulation work, the area was first removed and the area was constructed again.

If a crack or damage occurs in a structure at a public location such as an outer wall, a series of defects such as poor surface appearance or water leakage defects may occur. As a result, an immediate response is necessary to reduce the defect loss. Moreover, high defect losses caused by incorrect installations and missing tasks can be minimized through management and supervision in the construction phase. The affected functionality defect of MEP construction and the excess moisture defect of insulation work are caused by complex factors in all life cycles of construction. To prevent these in advance, the potential defects need to be examined from the design phase (Table 7).

Table 7. Defect risk management strategies occurring within three years of the DLP.

Risk Zone	Defect Characteristics		Defect Risk Management Strategies
	Cell No	Phase	
HFLS	d(1,1,4)	Post-Handover and Occupancy	Secondary defects can occur from cracks and damage in the main structure. An early response at the beginning of a defect is effective to reduce serial losses.
LFHS	d(1,1,8) d(1,1,5) d(1,1,11)		
LFLS	d(8,2,7) d(8,5,7) d(8,7,7) d(6,7,6)	Construction	Thorough management and supervision in the construction phase is required to reduce the losses of incorrect installations and missing tasks.
LFLS	d(6,6,9) d(9,6,1) d(8,7,1) d(8,6,1)	Design, Construction, and Handover	Management measures are required, such as constructability review in the design phase of MEP work and close collaboration between work types in the construction phase. Water leakage due to poor insulation work has a significant effect on the satisfaction of occupants, causing severe inconvenience. Hence, continuous management measures such as thorough design, appropriate method, and proper material selection are required.

Table 8 lists the cells with high average values for the defect loss distribution for defects that occurred after 3 years and before 5 years of the DLP. Figure 5 shows a scatter chart of the average values of the frequency/severity distributions for each cell for the defects that occurred after 3 years and before 5 years of the DLP. In particular, the RC work had relatively high losses of crack and water leak defects in the common area and garage. This is different from the pattern of 3 years of the DLP with a high loss of crack defects in the RC work due to internal and external factors. The structure of the apartments consists of concrete, which inevitably causes cracks because concrete is a construction material composite, made of different elements. In particular, cracks in underground parking lots cause water leaks, which can significantly affect the durability of the building. Therefore, a strategy is required to prevent defects that may occur in the future through early detection of potential defect sites through frame inspections during the finish work (Table 9).

Table 8. Ranking of the defects loss occurring within five years of the DLP.

Ranking	Work Types	Location	Defects	Average
1	RC	Garage	Crack	0.706
2	Water proof	Garage	Incorrect installation	0.319
3	Water proof	Garage	Missing task	0.307
4	RC	Exterior	Crack	0.167
5	Water proof	Balcony	Water leak	0.136
6	RC	Garage	Water leak	0.135
7	RC	Balcony	Crack	0.120
8	RC	Garage	Surface appearance	0.103
9	RC	Garage	Incorrect installation	0.101
10	Water proof	Exterior	Water leak	0.086
11	RC	Common area	Crack	0.073

Note: average means the mean value of loss distribution statistics.

Waterproof work featured high losses of incorrect installations and missing task defects in the garage. This appears to be linked to the crack and water leak defects in the RC work. The incorrect installations and missing tasks of waterproof work causes water tightness and surface defects. In reverse order, when cracks are generated, the vicious circle of water leaks, corrosion of the reinforcement materials, and crack enlargement are repeated. Thus, thorough management is required to reduce incorrect installation of waterproof work with LFHS characteristics. This is because such defects are caused by poor construction. Water leak defects in waterproof work with LFLS characteristics too are caused by poor construction. However, it is also caused by other defects such as cracks. It is related to places affected by external factors such as exterior or balcony. Therefore, an immediate response to the defect in the early stage is required to reduce the defect losses (Table 9).

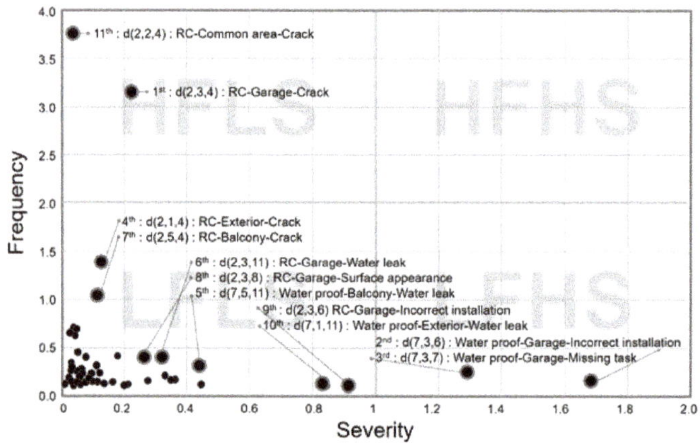

Figure 5. Frequency and severity scatter chart of a defect occurring within five years of the DLP.

Table 9. Defect risk management strategies occurring within five years of the DLP.

Risk Zone	Defect Characteristics		Defect Risk Management
	Cell No	Phase	Strategies
HFLS	d(2,3,4) d(2,2,4)	Handover, Post-Handover and Occupancy	Cracks/damage occurs in structures such as underground parking lots. RC work of the underground parking lot can be visually inspected. Thus, a strategy to discover potential defects early through thorough defect inspection in the handover phase is required. Crack defect can lead to a series of other defects such as surface defects and water leakage. Therefore, an immediate response in the early stage of a defect is required.
LFLS	d(2,1,4) d(2,3,11) d(2,5,4) d(2,3,8) d(2,3,6)		
LFLS	d(7,5,11) d(7,1,11)	Post-Handover and Occupancy	Crack/water leakage occurs in parts that have contact with the outside environment. This can lead to other defects such as crack enlargement and reinforcement corrosion. Thus, an early response to the defect is critical to reduce the defect losses.
LFHS	d(7,3,6) d(7,3,7)	Construction	Reconstruction is required because incorrect installation and missing tasks appear as the main defects.

Table 10 lists the cells with high average values of the defect loss distribution for defects that occurred after 5 years and before 10 years of the DLP. Figure 6 illustrates a scatter chart of the average values of the frequency/severity distributions for each cell for defects that occurred after 5 years and before 10 years of the DLP. The analysis results show that in the case of the RC work, various defects including crack, surface appearance, water leakage, detachment, and corrosion occurred in various exterior and interior parts such as halls/corridors and the garage. In particular, the crack defect losses were high for the outer wall. This is because the RC work, which is the largest type of construction work, has additional accessory processes and repair work in high places. These crack defects caused high losses in the halls/corridors, the garage, common areas, as well as the outer walls. Since this is a defect due to the negligence of the business entity, the business entity should perform the repair. Crack defects must be repaired because they are serious issues in the durability and functionality of the structure. Cracks on concrete generally define the length of width required for repair cracks as 0.3

to 0.4 mm, depending on structure, component, defects, etc. Even if the width of the crack is less than 0.3 mm, problems can occur from the perspective of the structure's function, safety, and durability as rebars are corroded and cracks spread due to penetration of rainwater.

Table 10. Ranking of the defects loss occurring within ten years of the DLP.

Ranking	Work Types	Location	Defects	Average
1	RC	Exterior	Crack	2.029
2	RC	Exterior	Surface appearance	0.262
3	RC	Hall/corridor	Crack	0.137
4	RC	Exterior	Water leak	0.123
5	RC	Garage	Crack	0.078
6	RC	Garage	Water leak	0.066
7	RC	Common area	Crack	0.060
8	RC	Exterior	Detachment	0.045
9	RC	Hall/corridor	Water leak	0.039
10	RC	Exterior	Corrosion	0.032

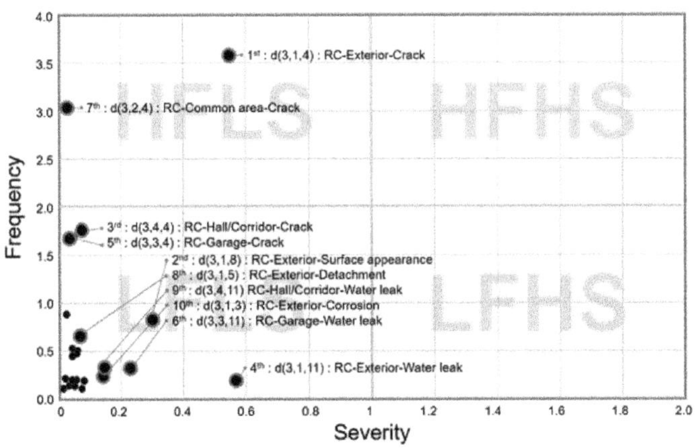

Figure 6. Frequency and severity scatter chart of a defect occurring within ten years of the DLP.

Due to the presence of a crack defect, various defects generated in RC work are due to errors made by the contractor such as inaccurate arrangements of reinforcements, poor curing after casting, and cold joints. Therefore, if the cause of the defects is clearly by the contractor, the contractor must perform the repair work even if the warranty period of the item has expired (Table 11).

Table 11. Defect risk management strategies occurring within ten years of the DLP.

Risk Zone	Defect Characteristics		Defect Risk Management
	Cell No	Phase	Strategies
HFLS	d(3,1,4) d(3,2,4)	Construction	Crack/damage occurs in the structure. Losses must be minimized through thorough management in the construction phase rather than response/repair after completion. Even if the defect liability period has expired, the business entity must take responsibility for defects that occurred a long time before the end of the durable period of the component (Only if the contractor has the cause of the defect).
LFLS	d(3,1,8) d(3,1,5) d(3,1,11) d(3,1,3) d(3,3,4) d(3,3,11) d(3,4,4) d(3,4,11)		

5. Discussion and Conclusions

The risk management of defect repairs for existing apartments is focused primarily on individual management methods such as the improvement of construction methods. As a result, it is difficult to identify the cause of defects resulting from complex mechanisms; hence, there is a limit in deriving the fundamental reduction measures. This study identified the problems of individual risk assessment methods and proposed a risk quantification technique that introduced the LDA (loss distribution approach). The main objective of this study is to construct a classification system based on the defect type, work type, and defect location, by applying the proposed evaluation method. Afterwards, the defect risk profile was systematically analyzed according to the defect liability period by simultaneously considering the frequency and cost. Using methods that have already been proven in the financial field is useful for analysis from a variety of perspectives.

The results of the analysis for the defect risk profile by the DLP in this study are described below. First, in the case of two years of DLP, the severity of the incorrect installation defect and the frequencies of the missing task and detachment were relatively high in the finish work. The analysis results revealed that the loss of defect types that require reconstruction, such as incorrect installations and missing tasks, is high overall in the all shared/exclusive areas. These defects are characterized by easy visual verification.

Second, in the case of 3 years of DLP, the loss of crack defects on the outer wall of RC was relatively high. It was found that cracks and damage defects occurred in the concrete due to the influence of external factors on the parts exposed to the outside. Furthermore, the MEP work had relatively high losses of missing tasks, which affected the functionality defects. The insulation work showed relatively high losses due to the incorrect installations and the excess moisture defects. The corresponding function was affected because the process of the corresponding defect did not work well with other various processes.

Third, in the case of 5 years of DLP, the defect losses of the RC work and waterproof work were relatively high. The RC work had a high defect frequency in the waterproof work. In particular, defects such as cracks in the structure of the underground parking lot along with poor waterproofing work, required reconstruction. These defects have a relatively high risk according to the influence of the groundwater.

Fourth, in the case of 10 years of DLP, high losses were caused by various types of defects such as cracks, surface appearance, water leakage, detachments, and corrosion of the RC work in all of the general public areas. The cracks of the RC work generated in the previous period have relatively high defect risks because they are accompanied by secondary defects such as surface appearance and water leakage.

This study proposes a defect risk management strategy reflecting the characteristics of defect risks analyzed above. In other words, based on the risk profile for each period, the major defects to be considered in the design, construction, handover, post-handover, and occupancy phases, and defect risk management strategies for each defect were derived.

The defect risk matrix proposed in this study has four dimensions: defect type, location, work type, and service life period. However, the defect risk matrix can be analyzed in more detail by adding the dimensions of the components. Another option is performing a multi-faceted analysis through item linkage and integration. If the defect risk matrix is further subdivided and sufficient data is available to produce meaningful results, it is possible to derive effective ways to manage the risk of the defect.

If the practical applications of the LDA in this study are expanded, this will effectively contribute to minimizing unnecessary losses caused by defects by quantifying the defect risks by the period. In other words, the average value of the probabilistically derived total distribution of losses can be calculated based on the amount of losses due to the defects. Then, the level of the defect cost can be identified using percentiles from the perspective of each party to the defect dispute, which includes construction companies and consumers. Consequently, it is expected to be useful for coordinating in the dispute reconciliation process. This loss distribution can be derived for each cell to enable multifaceted analysis, thus assisting in flexible decision making.

The LDA-based defect risk analysis method proposed in this study is a post evaluation method. However, to minimize the defect risks, it also needs measures to minimize the defects identified in this study in advance. In reality, defects occur due to various causes before the construction process. For example, a more detailed analysis is required to identify the obvious causes of cracks in concrete structures that are inevitably created. In this context, it necessary to analyze the mechanism of defects that has not been identified through defect package pattern analysis for the construction project lifecycle and to explore quality control measures that can prevent the defects in advance.

Author Contributions: B.K. developed the concept and drafted the manuscript. S.L. revised the manuscript and supervised the overall work. Y.A. reviewed the manuscript. All authors read and approved the manuscript.

Funding: This research was supported by a grant (19CTAP-C152020-01) from Technology Advancement Research Program (TARP) funded by Ministry of Land, Infrastructure and Transport of Korean government.

Conflicts of Interest: The authors declare no conflict of interest.

References

1. Raslanas, S.; Alchimoviene, J.; Banaitiene, N. Residential areas with apartment houses: Analysis of the condition of buildings, planning issues, retrofit strategies and scenarios. *Int. J. Strat. Property Manag.* **2011**, *15*, 158–172. [CrossRef]
2. Yuen, B.; Yeh, A.; Appold, S.J.; Earl, G.; Ting, J.; Kwee, L.K. High-rise living in Singapore public housing. *Urban Stud.* **2006**, *43*, 583–600. [CrossRef]
3. Sacks, R.; Goldin, M. Lean Management Model for Construction of High-Rise Apartment Buildings. *J. Construct. Eng. Manag.* **2007**, *133*, 374–384. [CrossRef]
4. Josephson, P.E.; Hammarlund, Y. The causes and costs of defects in construction: A study of seven building projects. *Autom. Construct.* **1999**, *8*, 681–687. [CrossRef]
5. Kim, J.H.; Go, S.S. Evaluation of defective risk for the finishing work of apartment house. *Korean J. Constr. Eng. Manag.* **2012**, *13*, 63–70. [CrossRef]
6. Kim, Y.S.; Oh, S.W.; Cho, Y.K.; Seo, J.W. A PDA and wireless web-integrated system for quality inspection and defect management of apartment housing projects. *Autom. Construct.* **2008**, *17*, 163–179. [CrossRef]
7. Jang, H.S.; Seo, C.H. A Study on the Improvement of Defect Information Management System of Apartment House. *J. Korea Inst. Build. Constr.* **2010**, *10*, 115–123. [CrossRef]
8. Kang, I.; Shu, D.S.; Ann, K.S. A study on the optimum defect management system of apartment house in Korea. *JAIK* **1997**, *13*, 343–353.
9. Craig, N.; Sommerville, J.; Auchterlounie, T. Customer satisfaction and snagging in the UK private house building sector. In Proceedings of the 26th Annual ARCOM Conference, Leeds, UK, 6–8 September 2010; Egbu, C., Ed.; Association of Researchers in Construction Management: Leeds, UK, 2010; pp. 1199–1208. Available online: http://www.arcom.ac.uk/-docs/proceedings/ar2010-1199-1208_Craig_Sommerville_and_Auchterlounie.pdf (accessed on 6 May 2019).
10. Cooper, I. Post-occupancy evaluation—Where are you? *Build. Res. Inform.* **2001**, *29*, 158–163. [CrossRef]
11. de Silva, N.; Dulaimi, M.F.; Ling, F.Y.Y.; Ofori, G. Improving the maintainability of buildings in Singapore. *Build. Environ.* **2004**, *39*, 1243–1251. [CrossRef]
12. Davey, C.L.; McDonald, J.; Lowe, D.; Duff, R.; Powell, J.A.; Powell, J.E. Defects liability management by design. *Build. Res. Inform.* **2006**, *34*, 145–153. [CrossRef]
13. Forcada, N.; Macarulla, M.; Fuertes, A.; Casals, M.; Gangolells, M.; Roca, X. Influence of building type on post-handover defects in housing. *J. Perform. Constr. Fac.* **2012**, *26*, 433–440. [CrossRef]
14. Aljassmi, H.; Han, S. Analysis of Causes of Construction Defects Using Fault Trees and Risk Importance Measures. *J. Construct. Eng. Manag.* **2013**, *139*, 870–880. [CrossRef]
15. Mills, A.; Love, P.E.D.; Williams, P. Defect costs in residential construction. *Korean J. Constr. Eng. Manag.* **2009**, *135*, 12–16. [CrossRef]
16. Forcada, N.; Macarulla, M.; Love, P.E.D. Assessment of residential defects at post-handover. *J. Construct. Eng. Manag.* **2013**, *139*, 372–378. [CrossRef]

17. Forcada, N.; Macarulla, M.; Gangolells, M.; Casals, M.; Fuertes, A.; Roca, X. Post-handover housing defects: Sources and origins. *J. Perform. Constr. Fac.* **2013**, *27*, 756–762. [CrossRef]
18. Pan, W.; Thomas, R. Defects and Their Influencing Factors of Post handover New-Build Homes. *J. Perform. Constr. Fac.* **2015**, *29*, 04014119. [CrossRef]
19. Watt, D.S. *Building Pathology: Principles and Practice*; Blackwell Science: Oxford UK, 1999.
20. Seo, D.S.; Park, J.M. Analysis of Consulting Reports on Defect Disputes in Apartment Building. *J. Korea Inst. Build. Constr.* **2013**, *13*, 498–505. [CrossRef]
21. Ishak, S.N.H.; Chohan, A.H.; Ramly, A. Implications of design deficiency on building maintenance at post-occupational stage. *J. Build. Apprais.* **2007**, *3*, 115–124. [CrossRef]
22. Macarulla, M.; Forcada, N.; Casals, M.; Gangolells, M.; Fuertes, A.; Roca, X. Standardizing Housing Defects: Classification, Validation, and Benefits. *J. Construct. Eng. Manag.* **2013**, *139*, 968–976. [CrossRef]
23. Georgiou, J. Verification of a building defect classification system for housing. *Struct. Surv.* **2010**, *28*, 370–383. [CrossRef]
24. Rotimi, F.E.; Tookey, J.; Rotimi, J.O. Evaluating Defect Reporting in New Residential Buildings in New Zealand. *Buildings* **2015**, *5*, 39–55. [CrossRef]
25. Atkinson, A.R. The pathology of building defects; a human error approach. *Eng. Construct. Architect. Manag.* **2002**, *9*, 53–61. [CrossRef]
26. Minato, T. Representing causal mechanism of defective designs: A system apoach considering human errors. *Construct. Manag. Econ.* **2003**, *21*, 297–305. [CrossRef]
27. Chong, W.K.; Low, S.P. Latent Building Defects: Causes and Design Strategies to Prevent Them. *J. Perform. Constr. Fac.* **2006**, *20*, 213–221. [CrossRef]
28. Bortolini, R.; Forcada, N. Building Inspection System for Evaluating the Technical Performance of Existing Buildings. *J. Perform. Constr. Fac.* **2018**, *32*, 04018073. [CrossRef]
29. Lee, J. Value Engineering for Defect Prevention on Building Façade. *J. Construct. Eng. Manag.* **2018**, *144*, 04018069. [CrossRef]
30. Chong, W.K.; Low, S.P. Assessment of defects at construction and occupancy stages. *J. Perform. Constr. Fac.* **2005**, *19*, 283–289. [CrossRef]
31. Hughes, W.; Hillebrandt, P.; Murdoch, J. The impact of contract duration on the cost of cash retention. *Construct. Manag. Econ.* **2000**, *18*, 11–14. [CrossRef]
32. Hopkin, T.; Lu, S.L.; Rogers, P.; Sexton, M. Detecting defects in the UK new-build housing sector: A learning perspective. *Construct. Manag. Econ.* **2016**, *34*, 35–45. [CrossRef]

sustainability

MDPI

Article

A Methodology for a Performance Information Model to Support Facility Management

Rossella Marmo [1,*], Maurizio Nicolella [1], Francesco Polverino [1] and Andrej Tibaut [2]

[1] Department of Civil, Architectural and Environmental Engineering of the University of Naples Federico II, 80125 Naples, Italy; nicolell@unina.it (M.N.); polverin@unina.it (F.P.)

[2] Faculty of Civil Engineering, Transportation Engineering and Architecture of the University of Maribor, 2000 Maribor, Slovenia; andrej.tibaut@um.si

* Correspondence: rossella.marmo@unina.it

Received: 10 October 2019; Accepted: 6 December 2019; Published: 8 December 2019

Abstract: Current facility management practice relies on different systems which require new technologies to integrate and manage information more easily. Building information modeling offers a good opportunity to improve facility information management by providing a unified platform for various data sources rather than an intuitive information interface. Although current research trends reveal that there is a continuously growing interest in facility management aided by building information modeling, an integrated model is still hard to obtain. This paper aims at developing a novel methodology based on building information modeling and facility management systems integration, underpinned by a performance information model. The implementation process of a performance information model is described, including information technologies involved, the data and process requirements, and the building performance assessment methods used. A first pilot case-study has been conducted with regards to surgery rooms in healthcare buildings. The proposal can support condition-based maintenance work schedule, as well as the achievement of organizational, environmental, and technical requirements. Among the practical implications found: Improved technological and environmental performances assessment; better visualization of building condition; improved decision-making process; facilitated maintenance tasks planning and maintenance records management.

Keywords: building information modeling (BIM); building performance assessment (BPA); key performance indicators (KPIs); facility management (FM); operation and maintenance (O&M); operating room (OR)

1. Introduction

The International Organization for Standardization (ISO) defines facility management (FM) as the "organizational function which integrates people, place and process within the built environment with the purpose of improving the quality of life of people and the productivity of the core business" [1].

Facility management has to support a wide range of activities (commonly referred to as non-core business) which enhance the work environment and well-being of people; enable the organization to deliver effective and responsive services; make the physical assets highly cost-effective, allowing also future changes; enhance the organization's image and culture [2] (pp. 36–37).

Particularly in the healthcare facilities sector, a facility manager has to consider many factors when making a strategic decision. The identification of a set of specific key performance indicators (KPIs) [3,4] helps the performance assessment and strategic planning. An integrated healthcare facility management model has been proposed [5] to hierarchically analyze healthcare FM core parameters, showing that an analytical quantitative model may significantly contribute to a better understanding of facility management performance.

Based on studies conducted by several authors [6–9], it is possible to identify different clusters of services and competencies within the FM domain. These tasks are carried out through different strategies (insourcing, total FM, public private partnership, etc.) but mostly through outsourcing [10].

Among the competency areas, the operation and maintenance (O&M) service plays a key role. It is meant to let the facility and its required systems function efficiently, reliably, safely, securely in a manner consistent with existing regulations and standards [9].

Building maintenance activities requires a comprehensive information system to capture and retrieve data related to building equipment [11]. The current FM practice relies on different systems (i.e., building energy management systems (BEMS), building automation systems (BAS), computerized maintenance management system (CMMS) [12], computer aided facilities management (CAFM), document management system (DMS)), which utilize new technologies to integrate and manage information more easily [13]. Studies on maintenance issues reveal that the most frequent problem is the information accessibility [14].

Building information modeling (BIM) can be considered as a tool or a method to face information management challenges throughout a building life cycle. It has been defined as the "use of a shared digital representation of a built asset to facilitate the design, construction and operation processes to form a reliable basis for decisions" [15]. BIM is semantically-based and object-oriented; it has 3D modeling capabilities and allows users to retrieve comprehensive information represented by objects and their attributes [16]. BIM provides a unified platform (Figure 1) for various data sources needed for daily O&M activities [11,17–19], so that data regarding technical specification, planned activities, and building performances (simulated or monitored) can be integrated to facilitate the decision-making process.

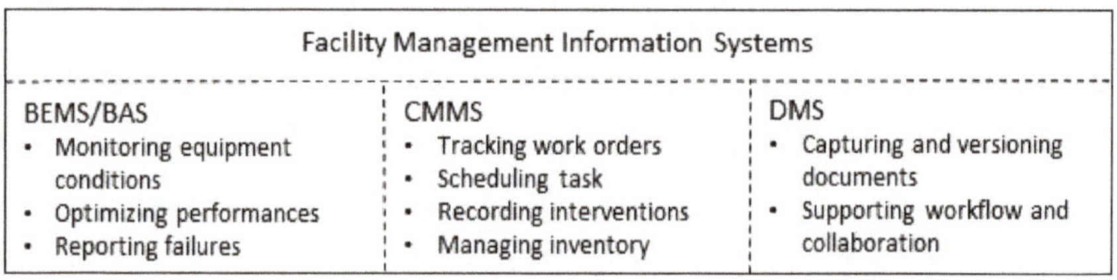

Figure 1. Facility management (FM) information systems.

Recent studies regarding BIM application in O&M activities indicates that energy management has been relatively analyzed by researchers, followed by emergency management and maintenance and repair [19].

Although current research trends reveal that there is a continuously growing interest in facilities information management using BIM [19–21], a seamless information process between BIM and FM systems does not exist yet [18]. Data exchange and interoperability remain problematic topics [18,19,22] while a lack of real-life examples of BIM-FM applications [17,23] and a lack of demonstrated benefits of BIM in FM [22] are among the key barriers to BIM implementation in FM.

Within the condition-based maintenance, monitoring of physical variables related to symptoms of failures is needed. The condition-based maintenance aids facility managers in identifying anomalies early enough to minimize the impact of operational interruptions [24]. Building performance assessment (BPA) provides for a better knowledge of an asset, so to make correct decisions at the right time. The performance assessment enriches BIM models with the purpose of evaluating the residual performances, so that coherent interventions can be selected [25]. For example, when certain spaces are performing under a certain threshold the integrated model can make suggestions regarding maintenance planning.

An improvement that is expected from BIM-FM integration is the systematic generation of information, such as KPIs [8]. Developing an FM benchmarking framework enables organizations to identify best practice and strategies improvement [26].

The aim of this paper is to propose a methodology that integrates FM systems, BIM, and BPA, to support maintenance planning and organizational requirements achievement.

To do so, the following research objectives are defined: (i) Analyzing FM and BPA processes and requirements; (ii) identifying relevant performance information which can be translated into KPIs; (iii) developing a framework for BIM-aided FM and BPA based on that information. To better understand the implications of BIM implementation in FM and performance assessment domain an overview of current BIM-FM integration is provided, with particular regard to information communication technologies (ICT) in FM, the impact of BIM in FM systems, and BPA methods.

A systematic literature review related to data and process requirements for BIM-FM integration was carried out via Scopus database with the following keywords: 'Building Information Modeling', 'BIM', 'Information Management', 'Facilit * Management', 'Operation and Maintenance', 'CMMS', 'CAFM', 'case study', 'Building Performance Assessment' in title/abstract/keywords. A total of 60 publications were examined in detail.

As evident, we were also interested in publications describing use cases to better understand the information exchange needs, the challenges to be faced and the expected results of BIM implementation in our research. For the scope of the paper the case studies analyzed concern BIM application in the O&M domain. Table 1 summarizes the analysis of the selected publications according to the following categories: The purpose of the case study; the BIM use purpose; information requirements; information references; information exchange supports; benefits achieved; challenges encountered. In the table, the 'BIM use purpose' is mapped according to [27] where a BIM use purpose is "the specific objective to be achieved when applying building information modeling during a facility's life.".

According to the Table 1, BIM is mostly appreciated for gathering (i.e., to capture, monitor, qualify), communicating (i.e., to visualize), and analyzing (i.e., to coordinate, validate, forecast) data and information. In few cases [28,29] the BIM model is integrated with a benchmarking system to report current performances, while it commonly contains maintenance activities records [13,30,31], asset characteristics and specification [13,32–42], and space management information [29,34,36].

Table 1. Building information modelling-operation and maintenance (BIM-O&M) integration case studies.

Case Study, Year, Ref.	Purpose of the Case Study	BIM Use Purpose	Information Requirements	Information References	Information Exchange Supports and Methods	Benefits	Challenges
Sydney Opera House 2007, [28]	Supporting building system alterations and asset management	To communicate and analyze	Properties of building elements; Building Condition Index	2D CAD drawings and Sydney Opera House specifications	IFC model (integrated data model)	Control of costs and environmental data; support to decision-making	Not discussed
US Coast Guard 2008, [29] (pp. 339–357)	Facilitating better decision-making for strategic planning and facility assessment	To gather	Facility Condition Index; Mission Dependency Index; Space Utilization Index	As-built documents (including 3D models); assessment team data; assessors' data; new BIM objects	Customized systems based on open standards (IFC, XML etc.)	Cost and time savings; standardizing processes and capturing knowledge digitally	BIM-based processes must support the integration of a variety of data and must be accessible to a wide range of users
A campus building 2010, [30]	Integrating facility maintenance data with BIM to support maintenance planning	To communicate and analyze	Maintenance activities information such as replacement, installation and status change	Not discussed except for the work order records	Manual integration of maintenance data into BIM model	Spatiotemporal analysis to optimize future interventions	Data capture and collection; updating the model and related information
Taiwan's school 2011, [31]	Creating a single repository of facility data for facilities maintenance	To communicate and gather	Schedule of planned tasks; results of maintenance works; facilities maintenance documents	3D CAD models; existing FM systems	Application Programming Interface and C# programming language	Improved information accessibility; enhancement of tasks planning and quality of inspection	Information exchange; updating information in BIM models
Norrtälje hospital 2011, [32]	Developing a customized life cycle management system to support proactive maintenance	To gather, communicate, and analyze	Geometrical model; material properties; environmental properties; condition assessment data; degradation model	2D CAD drawings; administrative documents; condition surveys	wrl. file, transformed from a dwg. file by the use of a third software	BIM-based tools serve as information repository for life cycle management; simplified build-up of information; enriched data	Needs for BIM integrated life cycle solution based on open standards
University of Chicago 2013, [33] (pp. 294–314)	Supporting maintenance activities	To communicate and gather	List of asset inventory information and data	Design and construction models; existing FM systems	Spreadsheet (modified version of COBie)	Improved data accuracy; streamlined data acquisition process	Handling with the variety of information resources; need for FM team information expertise

Table 1. *Cont.*

Case Study, Year, Ref.	Purpose of the Case Study	BIM Use Purpose	Information Requirements	Information References	Information Exchange Supports and Methods	Benefits	Challenges
Manchester City Council Town Hall 2014, [8,23]	Investigating the use of BIM in FM domain	Not discussed	Operation and maintenance information	Various FM systems; 3D building information and cloud-based repository for digital documents	Not discussed	Faster maintenance process and shorter service disruption	Need for FM team BIM expertise; limited software interoperability; unclear BIM FM requirements etc.
Kerr Hall, Ryerson University 2015, [34]	Testing how to overcome key challenges while developing 7D BIM	To communicate, gather, and analyze	Space allocation; lighting feasibility calculations; asbestos hazard map	Survey and reports; existing space management systems	Spreadsheet	Improved data updating and assessment of potential energy retrofit	Identify critical information; create/modify BIM models; information transfer; documentation uncertainty
Northumbria University's campus 2015, [13]	Investigating the value of BIM in space management	To communicate, gather, analyze, and generate	Asbestos properties, location, date of removal and survey documentation	DWG floor plans, scans of elevations, JPEG sections, Excel databases	Not discussed	Improved space management and geometric information record	Identifying necessary information; need for FM team information and BIM expertise; interoperability;
University of British Columbia 2015, [35]	Understanding the transition from a paper-based to a BIM-based approach in handover and FM	Not discussed	List of operation and maintenance information	Building management systems; facilities information systems; asset management systems	Not discussed	Not discussed	Methods and process changes
Terrassa Campus 2016, [36]	Investigating the benefits of the integration of maintenance management and BIM	To gather, communicate, and analyze	List of building characteristics, space management, maintenance and building monitoring data	Physical stock and intranet; building management system; maintenance management systems	Definition of a unique identifier (ID) for each object and space	Improved data consistency, intelligence in the model and reports generation; integration of facility systems	Correlating different kind of data sources; information exchanges
Laboratory and office building 2018, [37,38]	Developing more efficient data collection in post-occupancy facilities management	To gather and communicate	Mechanical and electrical asset data	As-built 2D drawings; project documents; asset data list	Comparison among different methods: manual; spreadsheet; CSV; IFC	Not discussed	Data transferring processes
Public University building 2018, [39]	Creating a central facility data repository to support FM tasks	To gather and generate	List of maintenance and equipment information	Owner's guidelines and handover products	COBie and IFC	Easier updating of CMMS thanks to handover BIM models	Data transfer and data quality control; needs for resources and collaboration among teams

Table 1. *Cont.*

Case Study, Year, Ref.	Purpose of the Case Study	BIM Use Purpose	Information Requirements	Information References	Information Exchange Supports and Methods	Benefits	Challenges
Melzo's school buildings, 2018, [40]	Developing a decision support model to define the priorities of refurbishment actions	To gather and analyze	List of information regarding accessibility; energy efficiency; acoustic performance	Legislation and technical standards; thermal simulations;	SQL and Dynamo	Semi-automatic evaluation of the level of compliance of existing buildings, with reduced time and costs	Lack of information suitable to perform a complete assessment in BIM models
Training center 2019, [41]	Merging BIM and Product-Service System to enhance maintenance operations	To gather and communicate	List of ordinary and extraordinary maintenance activities information	Maintenance reports and interviews with customers and suppliers	Not discussed	More effective management of maintenance activities, facilitated data record and tracking	Lack of knowledge and skills concerning the use of BIM tools
Hospitals in Scandinavia and Denmark 2019, [42] (cases A and B)	Investigating the enabling and constraining elements of digital FM in Scandinavia.	Not discussed	Building inventory	Design and construction information	Manual integration of as-built information in FM systems (case A); customized classification system (case B)	Not perceived (case A); time savings thanks to common project library shared by different design teams (case B)	Information exchange; few interests in ICT investment; lack of knowledge concerning ICT implementation; needs for digitalization strategy

1.1. Facility Management (FM) and Information Communication Technologies (ICT)

FM needs high level of interaction and information sharing, currently supported by ICT. Ranging from email documents to BIM, including computerized maintenance management system (CMMS), computer aided facilities management (CAFM), and BAS/BEMS, different tools have supported FM activities during the past decades [43].

The CMMS includes the creation and the management of asset records, bill of materials, and work orders; inventory control etc. [44], thus they support maintenance scheduling, facilities monitoring, and preventive maintenance [45].

The CAFM is a collection of tools used for organizing and managing various activities within the facilities such as space planning and management [45]. CAFM systems are largely based on the 'office suite' tools, including tabulated data, spreadsheet, and 2D drawings [43].

Both CMMS and CAFM have limited visualization capabilities, as traditional utilization of paper based or digital 2D plans limits the facility manager to identify the exact maintenance location context and the history of modifications [43].

The BEMS are regularly applied to the control of active systems, i.e., heating, ventilation, and air-conditioning (HVAC) system, determining their operating times. While sensors send feedback and alarms to these controlling systems, facility managers can monitor and change any benchmark or override the information [12].

In order to guarantee the required building operational performance, facility managers must check technical and environmental conditions. In this sense, building sensors and controllers can inform maintenance activities.

The complexity of BEMS can be integrated to CAFM, BIM and Maintenance Management System to control the operating equipment [45].

1.2. Benefits of Building Information Modelling-Facilities Managament (BIM-FM) Integration

BIM model might interact with the systems described above, as a source for data input, providing material/spatial data, reports or technical analyses, or as an interface for a repository, providing data capture, monitoring, processing, and transformation [34,46].

The main expected benefits from BIM-FM integration are cost reduction, thanks to ready to use data provided at the handover phase; performance improvement, it is to say more accessible FM data allows faster analysis and correction of problems; integration of several information technologies [33] (pp. 1,2).

BIM promises to provide a reliable database and integrated views across all facility systems [30] so that facility managers can base their decision on a more comprehensive knowledge of the building systems. BIM also provides 3D spatial information; therefore, it supports visualization and spatial analyses of various maintenance activities. Such analyses might not be easily performed with traditional databases [30].

Owners can use a BIM model to quickly populate an FM database [29] (pp. 108–111). As-built BIMs can enable the transfer of facility information from the design and construction phases to the operational phase. Retrieving necessary facility information from a BIM model and importing them into CMMS allows relevant costs savings, avoiding recapturing and transferring information by architects, engineers, and contractors [30].

Owners can also use a BIM model strategically and effectively to manage facility assets. They can evaluate the impact of retrofit or maintenance works or associate each building object with a condition assessment over time, supporting critical analyses [29] (pp. 108–111) such as maintenance planning and sustainability management [33] (pp. 11–13).

However, the BIM implementation in FM systems is not currently achieved without challenges, as described below.

1.3. Challenges of BIM-FM Integration

Three major category of issues can be defined [30]: Challenges encountered by the facility team or facility owners (i.e., lack of knowledge about how to use BIM in their practice); challenges encountered by the designers and contractors (i.e., lack of guidance about data requirements and delivery); technical issues (i.e., interoperability).

To connect BIM data to FM systems, FM teams can face the interoperability issue in several manners [37]. As different companies are involved either in the design or in the operational phase, common data exchange formats can be considered an effective strategy for exchanging data between BIM models and FM systems. Examples of open standards are the Construction Operations Building Information Exchange (COBie) [47] or the Industry Foundation Classes (IFC), in particular the FM Handover Model View definition [48]. They define standard structure and minimum data fields to support facility management [2].

Other BIM-FM linking approaches concern manual integration of data (i.e., through spreadsheets) and proprietary middleware [49].

Due to the simplicity of their inherent structure, spreadsheets are useful means of moving data (text and numbers) between software [28]. They are generally used in CAFM/CMMS or BAS, plus they are linkable to BIM objects. With a customized application, it is possible to read/write and import/extract data from a BIM based platform that also supports spreadsheet-based documents. For example, Dynamo, a tool for visual programming, which works within the Revit environment, can act as a bidirectional link from Revit to an Excel spreadsheet [38].

However, the transfer of data at the handover phase is commonly limited to graphical spatial information (i.e., room areas and attributes) and building inventory. Facility managers hardly update information from small projects, work orders, and major renovations in as-built BIM [33]. In order to enhance the maintenance planning there is a need of capturing information about maintenance and repair works during the operational phase. Retrieving this information facilitates project financial analysis and maintenance works prioritization [50].

In addition, an as-built model that is developed without early guidance is not effective for operational purposes [51]. In early project phases, designers and contractors have to know what information the FM team will need, as well as what organizational standard structure for information inventories is needed [52], which is not commonly known by the owners [53].

Defining the BIM-FM integration goals and developing the BIM-FM information collection and related information exchange process are necessary steps to effectively design the integration of BIM for FM [54]. The strategic identification of operational information is critical; thus facility managers need to detail and prioritize their information requirements [13,34], identifying by whom and when the data should be provided throughout the project life cycle [17]. This data will depend on specific user systems, organizational structure, and scope of the model.

In conclusion, owners might not be accustomed to the technological side of building management issues and not educated on BIM, how to request it, or how to adopt it to their practices. At the same time, few contractors are willing to perform BIM that does not directly benefit their daily work process without charging significant additional costs [55].

For these reasons, the cost of BIM-FM integration can be high, requiring investment in infrastructure, training, and new software and hardware [30].

1.4. Building Performance Assessment Methods

The performance evaluation of buildings and their components has always been a very complex and controversial topic. The problem arose when it was necessary to introduce assessments on the duration of the components, within the wider topic of scheduled maintenance. If we assume that every maintenance intervention must be associated with a performance threshold, and that the status of failure must be identified and coded also for those components for which performance is not measurable, then methods and tools to evaluate the performances are needed.

For the assessment of building performance, it is very important to evaluate whether it is necessary to intervene [56], because there is the risk that maintenance interventions may have high costs if not necessary or urgent [57].

In the last decades, however, the concept of performance evaluation has strongly moved towards those concerning the environment and sustainability [58–60]. In this sense, we can say that building performances have a strong integration with building users [61], and the reference methods have therefore become those that prefer aspects such as quality, health, safety, security, comfort, without neglecting others such as the social ones [62].

Among them, there are the POE (post occupancy evaluation), that is a method developed in the 1960s, conceived to measure the performances of buildings that have been built and occupied for a set time duration, and the BPE (building performance evaluation), that is a method conceived in the 1990s that was developed for upgrading POE and to improve the quality of decisions made at every phase of building life cycle [63–66]. These methodologies mostly tend to evaluate the performance of the whole building, and not those of the single components, essentially using analyses of users' satisfaction.

Similarly, KPIs propose performance assessments indirectly, mainly investigating the satisfaction of users and are detected through non-material indicators such as psychological or perceptive or performance indicators, defined in facility management contracts.

Many researchers have emphasized the importance of measuring building performances through KPIs [67], and many others have categorized them—at operational level—into the following ones: Technical, functional, behavioral, aesthetic, environmental [56].

Performance indicators are useful for measuring status and plan improvement activities and continuously assess changes over time [68]. Technical performance indicators are considered the most critical, and within this category structural resistance to fire and stability are two important indicators to be considered [69], while other researchers [70] identified asset failures and the severity of their consequences as an indicator.

It has been highlighted [71] that performance categories and examples of operational indicators, on the basis of studies conducted by several authors (such as Straub, 2003; Hovde and Moser, 2004; Lützkendorf and Lorenz, 2006; Preiser and Nasar, 2008; Yan et al., 2015), can be summarized as shown in Table 2.

Table 2. Categories of indicators.

Performance Category	Example of Indicator
Technical	Good lay-out of evacuation routes Structural condition [number of defects, severity]
Functional	Suitability of spaces [occupation/m^2] Air quality [CO_2 level]
Behavioral	Thermal comfort [number of complaints about temperature year]
Aesthetic	Façade appearance [peeling defects on façade/m^2]
Environmental	Waste generation [kg/year] Energy consumption [kWh/m^2/year]

From the publications reviewed in this research it is possible to demonstrate that building performance assessment could be limited to safety and efficiency, health and comfort, space functionality, and energy performance.

Some researchers [72] consider methods such as the BREEAM (Building Research Establishment Environmental Assessment Method) or approaches such as LEED (Leadership in Energy and Environmental Design) as real performance evaluation tools, although they refer to performances in a more than qualitative way but in any case codified by means of scores and/or attributes.

There are also possibilities, widely exploited, to use BIM as a source of data for a prediction of performance indicators of the planned building. That is for obtaining quantifiable predictions that can help in identifying strategies, tools, and methods to improve the overall building performances, and in particular the energy efficiency.

BIM-based computational analysis tools provide possibilities for integrating design and analysis process from the earliest stages of design and can also assist in design decision making [73]; some researchers [74] have emphasized the effectiveness of building performance-based design method compared to traditional method.

Recent publications have defined BIM-based workflows to compute and compare KPIs in order to make a qualitative assessment of the building and its parts [75] and to automate the monitoring of buildings during their regular operation [76]. In both cases the digital model becomes the mirror of the building and stores its actual performances to support facility managers in making decisions.

1.5. Interpretation and Discussion

Starting from the review of existing ICT tools for FM, the use cases analysis of BIM implementation for O&M purposes and the review of BPA methods it has been demonstrated that BIM can enhance different FM activities, even though with some challenges.

BIM as a repository tool, able to support different analysis, has been tested in several applications. For example, BIM can support proactive maintenance through gathering information about materials, environmental, and condition data so that a BIM-based life cycle management system can be developed [32]. The prioritization of refurbishment actions can be improved too, developing a decision support model based on accessibility, energy efficiency and acoustic performance information [40]. Three-dimensional data visualization allows analysis to optimize future interventions planning [8,13,23,30,31]. Faster maintenance processes and shorter periods of disruption have been proved [15,18] and tested merging BIM and product-service system [41].

At the same time, it appears that a lack of BIM expertise among the FM team and the owners [8,13,23,33,41,42] and the information exchange processes [8,13,23,31,32,37,38] are major challenges.

Furthermore, it can be deduced that a preliminary analysis of the FM process and policies, both currently adopted or expected, is necessary. In fact, the sources of required information for facility maintenance mostly involve the existing FM documentation, FM personnel's experience, and building management systems [19]. Interviews with the owner and the FM team allow to better understand the organization's information requirements. The analyzed case studies involved the owner and other relevant stakeholders in order to define data needs based on current and future goals of O&M activities [13,34–39,41].

Finally, the integration of operational conditions and performances in BIM models is a lesser-known topic, even if it can facilitate the decision making for facility planning and assessment. For this purpose, a specific set of information for a complete BIM-aided performance assessment must be defined [40], a wide variety of data and a wide range of users must be involved in BIM processes [29], and BIM-FM links must be based on open standards [32].

1.6. Research Contribution

The BIM-based approach presented in this paper includes the definition of FM and BPA information requirements, involves a wide range of data and users, and is based on open standard data format, in order to facilitate the analysis of current building performances, reduce corrective maintenance and emergency repairs.

The methodology is underpinned by a performance information model (PIM), which links facility management information systems, FM workflow repository and BIM's common data environment (CDE) (Figure 2).

The methodology addresses the interoperability issue with the latest IFC 4 × 2 data schema. Novelties in the schema like the entities IfcFacility and IfcFacilityPart along with IfcBuildingStorey and IfcSpace provide for detailed spatial breakdown of built facilities which can support location context information in the FM systems. The IFC schema can support also organizational requirements like IfcActionRequest for maintenance work orders through the IfcProjectOrder entity, which supports different maintenance types (corrective, condition based, planned corrective, scheduled). The IFC schema could support assessment of environmental conditions with the use of the property Pset_Condition, which determines the state or condition of a physically existent object like IfcElement (i.e., IfcSensor), at a particular point in time. Therefore, our methodology is intended to resource to BIM technologies for building performance assessment and facility maintenance information. Also, Business process modelling notation (BPMN) is considered for the FM workflow automation.

A pilot implementation of the methodology has been conducted in a use case, which includes surgery rooms of a healthcare facility. The case study concerns the environmental quality assessment, considered as a major performance feature to be monitored in such an environment. To easier control and quantify this performance feature a new KPI has been defined. Evaluation of the environmental performance indicator is related to technical and equipment conditions, consequently it can facilitate the identification of maintenance works eventually needed.

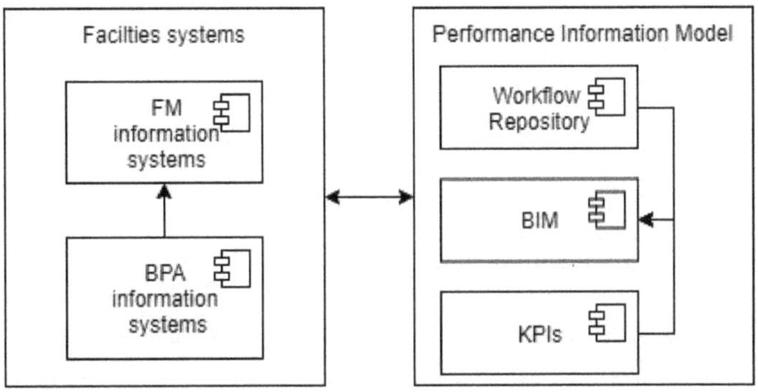

Figure 2. Organization of performance information model (PIM) elements.

2. Methodology for Developing a Performance Information Model (PIM)

In the context of the maintenance information modeling the BIM-aided BPA takes place. The PIM is achieved by the following workflow:

Identify building performances to be monitored and FM information requirements. To achieve a deep understanding of the required information to be gathered and managed through BIM we acquired and studied: Tender specifications about facilities management, in particular specification on O&M management and surgery units environmental condition assessment; reports on the monitoring process and results; adjustment plans; organizational documents on risk assessment. The collected data regard a public hospital in the province of Salerno, South of Italy. Interviews and focus groups were conducted among the FM personnel and the prevention and safety team. The interviews were carried out in person and involved two different Italian public healthcare authorities. These interviews resulted in deeper understanding of: the information needed to control performances and conditions; processes and systems in use to obtain and gather those information; the means by which to communicate the results. Furthermore, aimed by the contractors, we achieved and studied the CMMS database, in particular the history of corrective maintenance intervention, planned and preventive maintenance tasks and schedule, register of work orders. This process of information enrichment led to the definition of the stakeholder's information requirements, a crucial knowledge to inform the BIM model efficiently. Databases used in this research protect patients' privacy. They regard only maintenance and monitoring activities and patient-related information were not collected. Analyzed data are the outcome of maintenance activities carried out according to the Italian regulation, furthermore processes and environments analyzed in this paper were not modified for the scope of this study.

Establish methods of performance assessment. To identify how to achieve the performance assessment, procedures and systems currently in use among the healthcare facilities were analyzed. KPIs have been chosen as performance measurement tool because their functionality is already well-known within the organization involved in this study and they best facilitate the achievement of the BIM-aided BPA as they can be managed in form of objects parameters within the BIM platform. For each performance to be assessed a KPI must be set. In the context of this research an environmental condition index (ECI) was defined ex-novo. Such ECI is a weighted average of control parameters values. The weights are related to the criticality of the parameters in relation to the environmental quality. First, it was necessary to identify which parameters describe the KPI, then their weights. To do so the local health authority of Salerno and the University Hospital of Verona were involved. The list of measurable control parameters was achieved by two focus groups with expert panels from the above-mentioned healthcare authorities. To achieve an agreement on the criticality of each parameters referred to operating room air quality, a Delphi method was conducted. Then the criticality weights were obtained by a combination of Delphi method and analytical hierarchy process (AHP). The Delphi involved a panel of 17 experts including engineers and other technicians, doctors, nurses, and a chemist, all experts of maintenance and condition assessment of surgery rooms. The Delphi allowed to achieve the pair comparison matrix related to the control parameters, then the vector of the weights was calculated by AHP. The pair comparisons referred to the Saaty's values scale.

Link the monitored performances to preventive/corrective activities. According to certain performance values the interventions needed can be identified. From the environmental performance assessment, the performance of the technical systems can be deducted. The links were defined with the collaboration of mechanical systems and indoor air quality experts, taking into account the way by which the environmental quality is monitored and the type of installed plants. These relationships can be translated in a deterministic logic and then transposed in a BIM platform to inform and update the model utilizing IfcActionRequest (description of maintenance request), IfcApproval (approval of maintenance request) and IfcActor (person or organization(s) fulfilling the request such as a facilities manager or contractor.).

Define the BIM use purpose and PIM requirements. Establishing the potential value of BIM use on the project helps to identify the BIM implementation goals and the specific BIM uses. Once the BIM uses are identified then the model requirements can be defined, i.e., in terms of parameters to be inserted in the model, customized dataset, links to maintenance database, level of development required, implementation process needs, etc. Once the implementation process has been established then the information exchanges can be defined. The exchange files contain instances of a subset of entities compliant with the IFC schema, such as IfcActionRequest, which are addressing the PIM requirements. For consumption of the exchange files a customized software needs to be developed.

Implement the PIM. PIM input data come from facility information management systems, including the BPA process. The actual condition of the facility is also required, so that the model to which the FM attributes refer can be created. Monitoring information were pulled in the model in an automatized manner, creating a link between the model and the Excel spreadsheets used to handle the monitoring results. The output data are the required inspection tasks associated with the failed systems. They are visualized in the model in the form of text shared parameters, but they can be exported or linked to CMMS to inform future work orders.

3. Performance Information Model: Use Case of Hospital Surgery Rooms

Considering international guidelines [77–80] and literature analysis, a proposal of PIM process development is deducted (Figure 3).

The proposed process provides for the operational condition assessment of a building and its elements. The monitored data can be gathered in the model, then analyzed and translated in the form of performance indicators. Relevant information can be generated, such as the interventions needed to satisfy the necessary requirements. The KPIs, in form of objects properties, are visualized in the digital model, allowing further spatiotemporal analysis and supporting decision making tasks of subcontractors and FM team. According to the monitored conditions different performances of the building during its lifecycle can be assessed and managed (i.e., sustainability, affordability, energy consumption, safety, efficiency, environmental quality, etc.). Building information modeling can play a key role as it allows collaboration and information exchange among the operators, shows the building actual performances and related suggested interventions, supports the maintenance planning, and facilitates the asset inventory management.

The local health authority has recently provided new contractors for FM and prevention and safety activities for hospitals under its responsibility. No existing BIM models were held by the authority or the FM contractors, and the processes currently in use among them were not BIM-oriented. This is a common situation within the Italian built environment. Therefore, recent laws and regulations [81–83] regarding the digitalization of the information process in the construction sector require to face the digitalization of existing buildings and related services. In such context this case-study constitutes the first step taken to a BIM-aided FM.

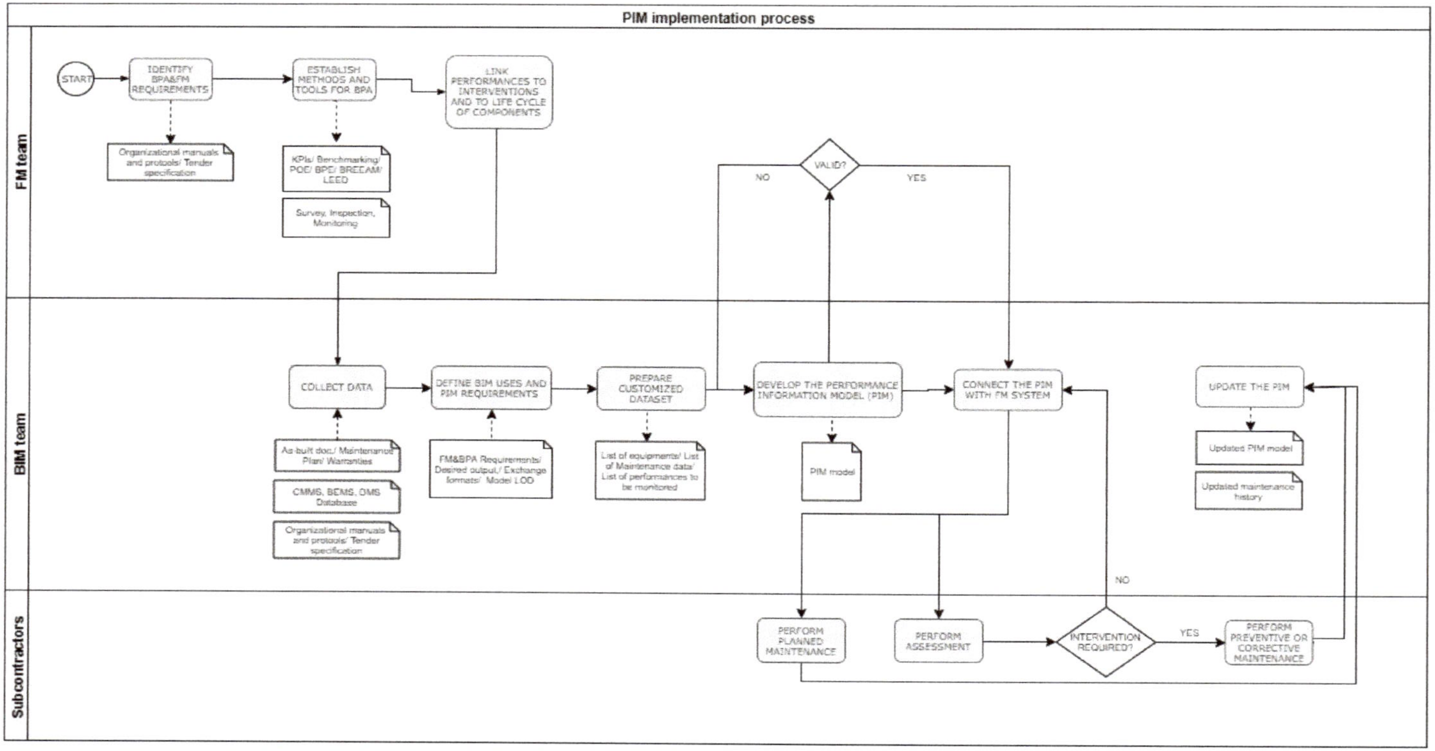

Figure 3. Performance information model (PIM) implementation process in business process modelling notation (BPMN) standard. The PIM approach has been validated by a surgery rooms BIM model, regarding a public healthcare building in Italy. A surgery room is a very complex system, it has to comply with a set of requirements [84] which can be transposed in performances to be assessed.

3.1. Building Performance Assessment (BPA) and FM Information Requirements

Especially when no BIM exists yet, it is crucial to do the prior analysis of stakeholders' information requirements to optimize geometric modeling and information handling effort. Prior to modelling, local health authority's specifications on surgery rooms (related to maintenance, risk management and work organization) have been examined.

The list of collected data related to preventive and corrective maintenance tasks is reported in Table 3.

Table 3. List of preventive and corrective maintenance information that can fill the model as objects properties.

Work Orders History	
WorkOrderID	Description
BuildingID	DateOfRegistration
DateOfCompletition	Duration
Requests for Intervention	
RequestID	Created by
Reported by	Description
Location	ContractualAuthority
SiteID	BuildingID
FloorID	UniteID
RoomID	Equipment
DateRequestCreated	UrgencyLevel
UrgencyTimeConstraints	ProblemType
InterventionType	ResolutionType
InsuranceDeductable	ExpectedCompletitionDate
MaintenanceCompany	DateOfCompletition
Notes	StatusID

Monitoring significant parameters related to the condition of hygienic/engineering/structural systems allows healthcare facilities to adopt preventive procedures. The risk management team, according to the performance assessment and risk evaluation results, informs the FM team in order to align findings to the required condition. The local health authority commissioned a specialized company to evaluate, analyze, and report the environmental and equipment condition of surgery areas. The hospital has no BAS, but the quality control is performed according to a planned schedule of activities. The list of information to be monitored is reported in Table 4.

Table 4. Monitored conditions in surgery rooms.

Parameter Name	
Particle concentration	Air volumes/air exchange
Microbiological concentration	Noise
Anesthetic gases	Recovery time
Microclimatic conditions	Water quality
Pressure gradient	Lighting intensity

3.2. Key Performace Indicator (KPI) for Surgery Rooms: Environmental Condition Index

Using a set of established KPIs simplifies the evaluation process and helps the management team to make strategic decisions towards the organization's mission.

We proposed a novel environmental condition index (ECI), which has the following Formula (1):

$$\frac{\sum_i^n P_i \times W_i}{\sum_i^n W_i} \tag{1}$$

where:

- ECI_{UAk} = environmental condition index referred to the surgery room k. It varies from 0 (best scenario) to 1 (worst scenario). As an example, the ECI for the orthopedic surgery room resulted to be 0.09, while for the general surgery room it was equal to 0.21. In both cases the 'noise' and the 'contamination at rest' controls were not satisfied, but in the latter case also the 'microclimatic condition in operational' was not fulfilled;
- P_i = value of each environmental quality factor. It is evaluated on a binary scale: P_i is equal to 0 if the control associated to it is fulfilled, otherwise it is equal to 1. As an example, for the orthopedic surgery room the value 1 was associated with the parameter's 'noise' and 'contamination at rest', while the value 0 was associated with the remainder;
- W_i = criticality weight of each factor. The sum of all the weights is 1 (100%). In the orthopedic room case, the sum of the products was 0.09 as the 'noise' and the 'contamination at rest' weigh respectively 4% and 5%.

The ECI has the following features:

- It eliminates overlapping and redundant information, as some parameters are grouped when depending on the same equipment element. Then the identification of the required intervention was simplified;
- It expresses each relevant aspect of the system assessed. The list presented in Table 3 was discussed in two focus groups to select seven parameters necessary and sufficient to evaluate a surgery room environmental quality;
- It provides for a wide applicability across the authority FM systems, as it is based upon their requirements;
- It is expressed by a number, which values can vary from 0 to 1. This is a consequence of two factors: The formula which expresses the KPI and the evaluation mechanisms.

The Delphi method provided for the pair comparison matrix (consistency ratio equal to 0.07). Each expert involved in the Delphi expressed his opinion regarding parameters confronted two by two with respect to their contribution to the environmental quality of operating rooms. The AHP was used to deduce the parameters priorities with respect to the above-mentioned criterion. These weights have been deducted by calculating, in the MATLAB programming platform, the main eigenvector associated to the main eigenvalue of the pair comparison matrix. The results are reported in Table 5.

Table 5. Pair comparison matrix and criticality weights.

		1	2	3	4	5	6	7	Weights %
Contamination at rest	1	1.00	0.19	0.64	0.40	0.26	0.15	0.93	4
Contamination in operational	2	5.27	1.00	4.57	0.73	0.44	0.27	4.78	14
Microclimatic conditions at rest	3	1.57	0.22	1.00	0.20	0.19	0.15	0.51	4
Microclimatic conditions in operational	4	2.48	1.37	4.94	1.00	0.26	0.21	2.64	12
Air volumes/Air exchanges/Recovery time	5	3.90	2.27	5.39	3.78	1.00	0.27	2.47	21
Anesthetic gases concentration	6	6.65	3.71	6.65	4.72	3.66	1.00	5.62	40
Noise	7	1.08	0.21	1.98	0.38	0.40	0.18	1.00	5

3.3. Correlation Between Environmental and Technical Performances

In order to integrate BPA with maintenance planning and to enhance the value of the ECI use, a link was established between control parameter values and the interventions required. These interventions are defined in terms of inspections and checks to be performed in order to verify possible failures or inadequate operational conditions within the technological and functional system. Table 6 proposes the correlations list.

Table 6. Links between environmental and technological system.

N° Parameter	Parameter Name	N° Task	Task Description
1	Contamination at rest	1.1	HEPA filters inspection
		1.2	HVAC pipes inspection
2	Contamination in operational	2.1	Behavioral protocols check
3–4	Microclimatic conditions at rest and in operational	3.1	Project condition check
		3.2	ATU supplied power control
5	Air exchanges/Recovery time	5.1	Filters inspection
		5.2	Load loss check
		5.3	Forced air volume calculation
		5.4	Mixing and ventilation efficiency control
6	Anesthetic gases concentration	6.1	Pipes fitting controls (high- and low-pressure systems)
		6.2	Gas evacuation system controls
7	Noise	7.1	Air-cooled inspection
		7.2	HVAC ducts inspection
		7.3	ATU inspection

3.4. *PIM Implementation*

The PIM methodology was tested on a healthcare building in the province of Salerno, south of Italy (Figure 4). In this case study BIM was used to gather information related to the environmental control, to communicate the monitoring results, and to analyze the condition assessment in terms of maintenance interventions required. The controls discussed in this study concern the risk management associated with surgery rooms activities. We accessed the database containing the surgery units' environmental controls, which regard air quality. Other factors and engineering devices were not monitored. The methods used to perform those tests respect the Italian regulations and are based on the Italian guidelines regarding the assessment of the efficiency of the preventive measures adopted by the prevention and safety department of healthcare organizations. The PIM described here has a basic geometric development (a BIM model with LOD 200) [85] but contains specific non-graphical information for facility management. The geometric model was created in Autodesk Revit 2019, starting from 2D CAD plans regarding the architectural and HVAC systems. Autodesk Revit allows different types of analysis when combined with other software, such as Green Building Studio to conduct energy performance simulation [86] or Dynamo, as illustrated below.

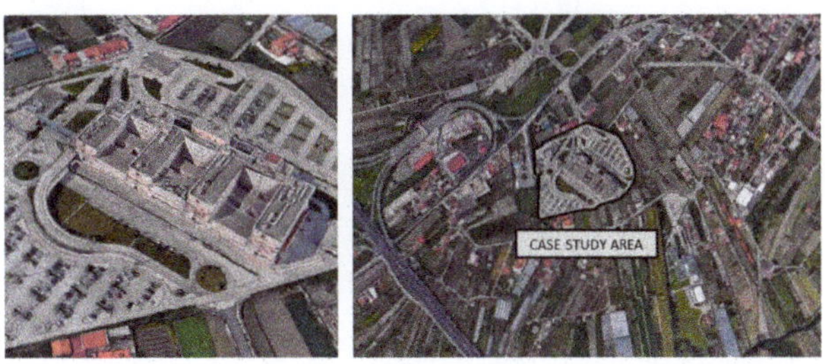

Figure 4. Location of the case study: building and surrounding.

The case study is focused on the environmental systems management, so it was enriched by the definition of rooms and related properties (i.e., environmental condition index).

We analyzed the database of the monitoring results related to one semester of activities (last semester of 2018 year) carried out in three operating rooms. In this database the results were individually presented for each type of test. The authors provided to translate them in a summarized Excel sheet to make them easier to read by Dynamo (Table 7). The results were translated to Boolean values to define the failure (1) or the fulfillment (0) of each control in each room.

Table 7. Monitoring results associated to each surgery rooms presented as Boolean values.

Orthopedic Surgery Room	1	0	0	0	0	0	1
General Surgery Room	1	0	0	1	0	0	1
Pediatric Surgery Room	1	0	0	0	0	0	1

The input data in Excel sheets (Figure 5) can be easily updated when BPA activities are conducted. The data concern all the results enabling to calculate the ECI for each surgery room (i.e., the value of control parameters, their respective weights, and the value of the resulted ECI).

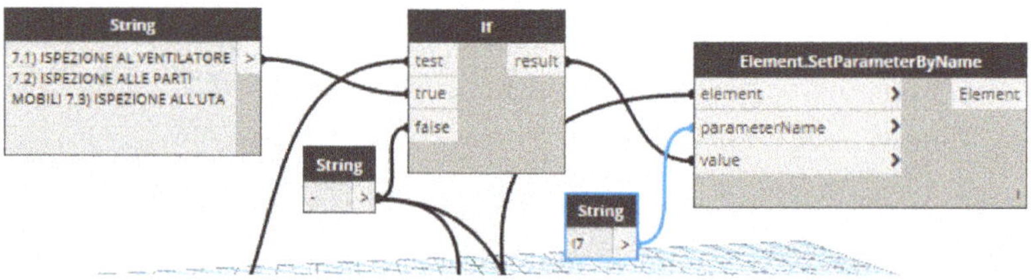

Figure 5. Conditional logic in Dynamo.

Dynamo was used to create bidirectional links between the model and external data, as systems integration tool. The Excel.ReadFromFile node was used to connect the BPA results spreadsheet-based with the model parameters. The 'If' statement was used to check the needs of intervention according to the monitoring activities results. The 'If' statement contains a Boolean statement so that the 'true' condition was associated with the failure of environmental controls. The results of the performance assessment were transposed in the model through the node Element.SetParameterByName (Figure 6). The BPA results and maintenance tasks needed are visualized in the model in the form of shared parameters, furthermore it is possible to visualize the performance assessment by thematic drawings. The Figure 7 shows the thematic plan of three surgery rooms and the properties associated to them, in terms of ECI, controls (I1, I2, etc.) and interventions required (1.1, 1.2, etc.). In this case, which regards the general surgery room, the controls I1, I4, and I7 are not fulfilled, so the corresponding required interventions are reported in the model (1.1 HEPA filters inspection, 1.2 HVAC pipes inspection; 3.1 Project condition check, 3.2 ATU supplied power control; 7.1 Air-cooled inspection, 7.2 HVAC ducts inspection, 7.3 ATU inspection).

The corresponding fragment of the IFC file shows an example of the custom IfcPropertySet 'Environmental Condition Index' for the use case.

```
#836 = IFCPROPERTYSET('3DF3k2$9z3HRdciX9bgXts',#59,'Environmental Condition Index',$,(#850,#854));
#850 = IFCPROPERTYSINGLEVALUE('Description',$,IFCLABEL('1.1 HEPA filters inspection, 1.2 HVAC pipes inspection; 3.1 Project condition check, 3.2 ATU supplied power control; 7.1 Air-cooled inspection, 7.2 HVAC ducts inspection, 7.3 ATU inspection'),$);
#854 = IFCPROPERTYSINGLEVALUE('Value',$,IFCREAL(0.217256),$);
```

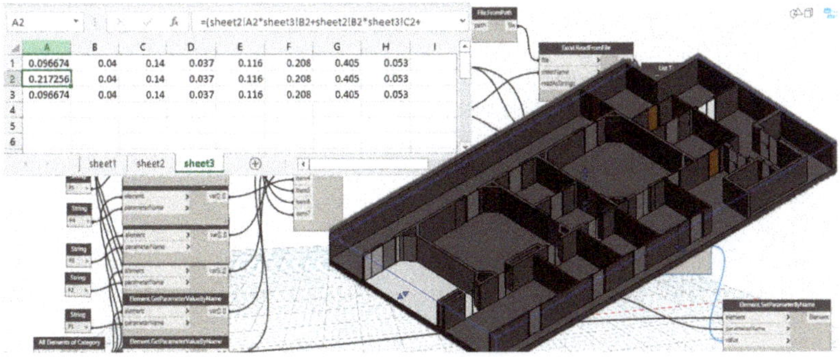

Figure 6. Excel-Dynamo-Revit links.

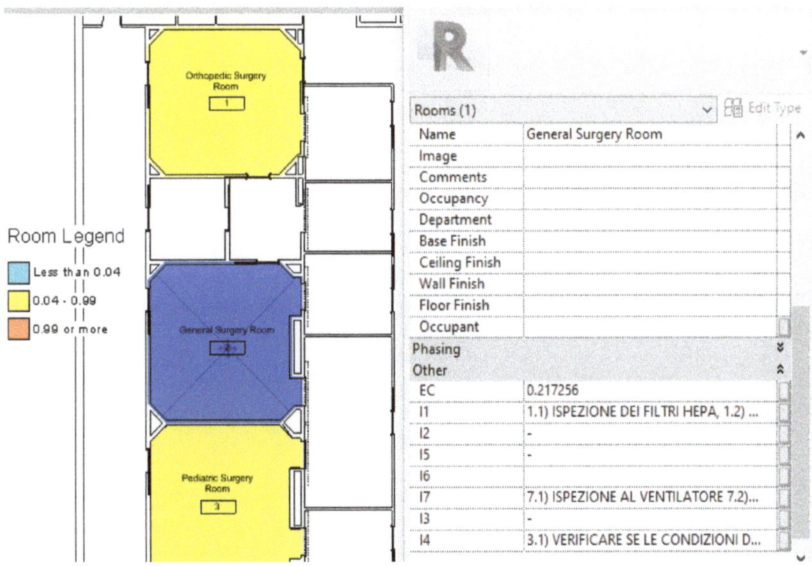

Figure 7. Thematic plans of orthopedic, pediatric and general surgery rooms. ECIndex and inspections required are reported in the model as rooms properties.

4. Analysis and Discussion

In this section, a synthesis of findings and their analysis is presented. The purpose of the synthesis is to assess the stability of results.

The performance information model tested on surgery rooms provides for:

1. Proposing a workflow for PIM implementation based on BPMN model (Figure 3),
2. Listing maintenance related information which can be added to the BIM model as the IFC object properties (Table 2),
3. Defining a new KPI for surgery rooms, measuring the environmental and functional performances (Formula (1)),
4. Correlating the measured performances to required maintenance intervention in terms of inspections and controls (Table 3),
5. Implementing a conditional logic and the information systems integration (Figure 6).

This application demonstrates the positive impact that BIM can have on analyzed FM and BPA processes.

The prevention and safety department is in charge of measuring the air quality of the surgery units, with the support of an external contractor. The results inform the maintenance department with a list of suggested interventions. The maintenance department is in charge of the maintenance planning and require the intervention, if needed, of the maintenance provider, an additional contractor.

A performance information model in such context can act as a tool to facilitate the analysis of building performances, through KPI definition (Figure 7), and as a system to speed-up the information exchange process among the different parties. A PIM can also support the maintenance works and assessment activities record, information that are rarely captured in BIM models nowadays. Figure 8 shows the instance of PIM implementation, starting from the already developed model (see Figure 3). The requests for intervention and related approvals are omitted in order to simplify the diagram interpretation.

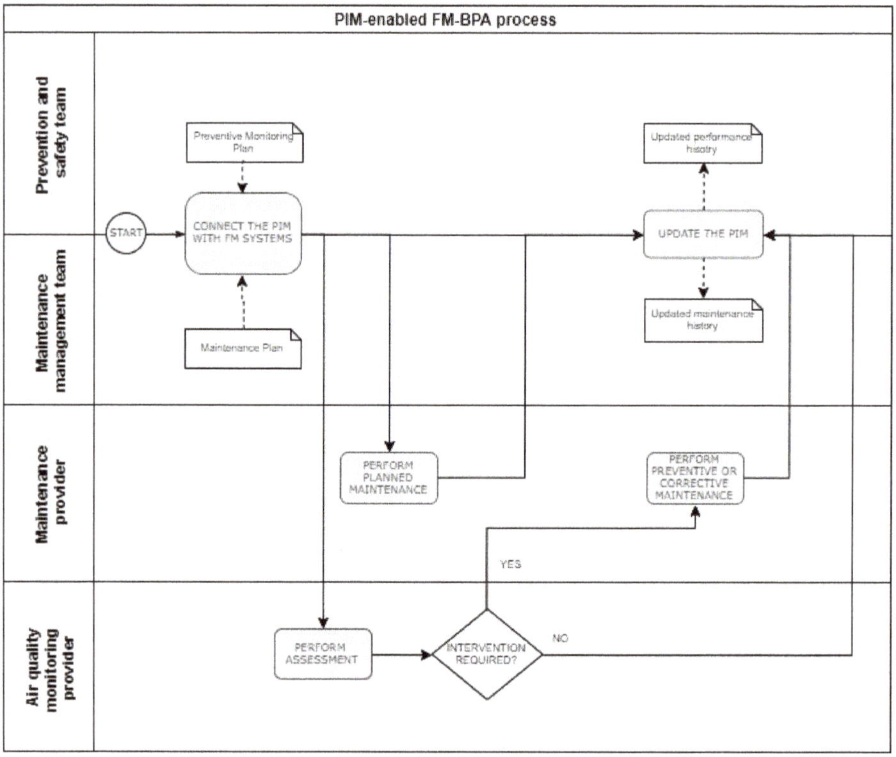

Figure 8. Instance of the PIM application process.

From the case study analysis, it is possible to deduct that a comprehensive PIM, extended to maintenance management and further systems control, has at least the following application benefits:

1. Improved technological and environmental performances assessment,
2. Integrated visualization of the operating condition of building and its elements,
3. Inventory management of building components, spaces, furniture and documents,
4. Automation of the quantity take-off,
5. Support to identify and schedule future maintenance interventions,
6. Management of historical data about previous maintenance.

5. Conclusions

Current research trends reveal that there is a continuously growing interest in facilities information management using BIM, which offers a good opportunity for a technical platform to integrate various data sources needed for daily O&M activities. An improvement that is expected from the BIM-FM integration is the systematic generation of information, such as KPIs to identify best practice and improvement strategies. Even though such an integration process has promising potential benefits, for example relating performance thresholds to maintenance planning, in very few cases the benchmarking of performance was tested.

Surveys aiming to develop the common BIM data requirements for O&M are limited and more focused surveys for specific building types and for specific O&M tasks have to be conducted [19].

Within this context, this article aimed at developing a methodology to integrate BIM, BPA and FM systems, supporting organizational, environmental and technical requirements achievement.

It contributes to the body of knowledge by defining a comprehensive approach to achieve this systems integration and has proposed a novel performance information model as decision-making support tool.

Furthermore, it explores in detail the healthcare buildings sector, with particular regard to operating rooms, reporting specific information requirements which can be used for future O&M tasks.

As no existing BIM models were held by the authority or the FM contractors, the starting point for the PIM development was the owner's and FM team's needs investigation, which resulted in modeling the environmental system and related attributes. As the BPA and FM process were not BIM oriented, we created a link between the existing FM systems and BIM model by customized application and commercial software tools.

At the moment the PIM implementation is limited to operating room environmental system. Other engineering systems (i.e., fire protection, electrical systems, energy consumption meters, weather station etc.) were not monitored.

Further research will be conducted to achieve a more comprehensive as-built model in order to enrich the PIM with maintenance information related to technological system elements (i.e., mechanical plant elements as ducts, pipes, filters, etc.; architectural elements as floor, ceiling, doors etc.; structural elements as pillars, columns, beams, etc.).

Further research and development of the proposed methodology concern automation of the information exchange, development of a prototype application in Python with an API that utilizes IfcOpenShell and a customized database, integrated with a BPMN automation support. In the development, preventive and corrective maintenance information from the Table 2 will be mapped to the corresponding objects from the asset BIM models using the standardized IFC entities identified as relevant for FM. The prototype application will be tested on several case studies to deeper analyze the proposed approach and for developing a more powerful and generally applicable BIM-based solution for FM and BPA activities.

Author Contributions: Writing—original draft preparation, R.M.; writing—review and editing, M.N. and A.T.; supervision, F.P.

Funding: This research received no external funding.

Acknowledgments: The authors want to acknowledge the local health authority of Salerno and the University Hospital of Verona for the support given to the development of the case study discussed in this research.

Conflicts of Interest: The authors declare no conflict of interest.

References

1. ISO 41011:2017. Facility Management—Vocabulary. Available online: https://www.iso.org/obp/ui/#iso:std:iso:41011:ed-1:v1:en (accessed on 4 August 2019).
2. Atkin, B.; Brooks, A. *Total Facility Management*, 4th ed.; John Wiley & Sons: Chichester, UK, 2015.

3. Shohet, I.M.; Lavy, S. Facility maintenance and management: A health care case study. *Int. J. Strateg. Prop. Manag.* **2017**, *21*, 170–182. [CrossRef]
4. Lavy, S.; Garcia, J.A.; Dixit, M.K. KPIs for facility's performance assessment, Part I: Identification and categorization of core indicators. *Facilities* **2014**, *32*, 256–274. [CrossRef]
5. Lavy, S.; Shohet, I.M. A strategic integrated healthcare facility management model. *Int. J. Strateg. Prop. Manag.* **2007**, *11*, 125–142. [CrossRef]
6. Chotipanich, S. Positioning facility management. *Facilities* **2004**, *22*, 364–372. [CrossRef]
7. Roper, K.; Payant, R. *The Facility Management Handbook*; AMACOM Division of American Management Association International: New York, NY, USA, 2014.
8. Kiviniemi, A.; Codinhoto, R. Challenges in the Implementation of BIM for FM—Case Manchester Town Hall Complex. In *Computing in Civil and Building Engineering*; Issa, R., Flood, I., Eds.; American Society of Civil Engineers (ASCE): Orlando, FL, USA, 2014; pp. 665–672.
9. International Facility Management Association. Competency Guide. Available online: https://www.fm.training/sites/collabstore/files/images/Competency_Guide_Context_Public_v1_7-12-18.pdf (accessed on 4 August 2019).
10. Ancarani, A.; Capaldo, G. Supporting decision-making process in facilities management services procurement: A methodological approach. *J. Purch. Supply Manag.* **2005**, *11*, 232–241. [CrossRef]
11. Motawa, I.; Almarshad, A. A knowledge-based BIM system for building maintenance. *Autom. Constr.* **2019**, *29*, 173–182. [CrossRef]
12. Shalabi, F.; Turkan, Y. IFC BIM-Based Facility Management Approach to Optimize Data Collection for Corrective Maintenance. *J. Perform. Constr. Facil.* **2016**, *31*. [CrossRef]
13. Kassem, M.; Kelly, G.; Dawood, N.; Serginson, M.; Lockley, S. BIM in facilities management applications: A case study of a large university complex. *Built Environ. Proj. Asset Manag.* **2015**, *5*, 261–277. [CrossRef]
14. Liu, R.; Issa, R.R.A. Survey: Common Knowledge in BIM for Facility Maintenance. *J. Perform. Constr. Facil.* **2015**, *30*. [CrossRef]
15. International Organization for Standardization. ISO 19650-1:2018—Organization and Digitization of Information about Buildings and Civil Engineering Works, Including Building Information Modelling (BIM)—Information Management Using Building Information Modelling. Part 1: Concepts and Principles. Available online: https://www.iso.org/obp/ui/#iso:std:iso:19650:-1:ed-1:v1:en (accessed on 9 October 2019).
16. Jeong, W.S.; Kim, K.H. A performance evaluation of the BIM-based object-oriented physical modeling technique for building thermal simulations: A comparative case study. *Sustainability* **2016**, *8*, 648. [CrossRef]
17. Becerik-Gerber, B.; Jazizadeh, F.; Li, N.; Calis, G. Application Areas and Data Requirements for BIM-Enabled Facilities Management. *J. Constr. Eng. Manag.* **2012**, *138*, 431–442. [CrossRef]
18. Matarneh, S.T.; Danso-Amoako, M.; Al-Bizri, S.; Gaterell, M.; Matarneh, R. Building information modeling for facilities management: A literature review and future research directions. *J. Build. Eng.* **2019**, *24*. [CrossRef]
19. Gao, X.; Pishdad-Bozorgi, P. BIM-enabled facilities operation and maintenance: A review. *Adv. Eng. Inform.* **2019**, *39*, 227–247. [CrossRef]
20. Ilter, D.; Ergen, E. BIM for building refurbishment and maintenance: Current status and research directions. *Struct. Surv.* **2015**, *33*, 228–256. [CrossRef]
21. Edirisinghe, R.; London, K.A.; Kalutara, P.; Aranda-Mena, G. Building information modelling for facility management: Are we there yet? *Eng. Constr. Archit. Manag.* **2017**, *24*, 1119–1154. [CrossRef]
22. Pärn, E.A.; Edwards, D.J.; Sing, M.C.P. The building information modelling trajectory in facilities management: A review. *Autom. Constr.* **2017**, *75*, 45–55. [CrossRef]
23. Codinhoto, R.; Kiviniemi, A. BIM for FM: A Case Support for Business Life Cycle. In *Product Lifecycle Management for a Global Market*; Fukuda, S., Bernard, A., Gurumoorthy, B., Bouras, A., Eds.; Springer: Berlin/Heidelberg, Germany, 2014; pp. 64–74.
24. Hao, Q.; Xue, Y.; Shen, W.; Jones, B.; Zhu, J. A Decision Support System for Integrating Corrective Maintenance, Preventive Maintenance and Condition-based Maintenance. In Proceedings of the Construction Research Congress 2010, Banff, AB, Canada, 8–10 May 2010.
25. Bruno, S.; De Fino, M.; Fatiguso, F. Historic Building Information Modelling: Performance assessment for diagnosis-aided information modelling and management. *Autom. Constr.* **2018**, *86*, 256–276. [CrossRef]
26. Kensek, K. BIM Guidelines Inform Facilities Management Databases: A Case Study over Time. *Buildings* **2010**, *5*, 899–916. [CrossRef]

27. Kreider, R.G.; Messner, J.I. *The Uses of BIM: Classifying and Selecting BIM Uses*; The Pennsylvania State University: State College, PA, USA, 2013; pp. 6–8.
28. CRC. *Adopting BIM for Facilities Management: Solutions for Managing the Sydney Opera House*; Cooperative Research Centre (CRC) for Construction Innovation: Brisbane, Australia, 2007.
29. Eastman, C.M.; Teicholz, P.; Sacks, R.; Liston, K. *BIM Handbook: A Guide to Building Information Modeling for Owners, Managers, Designers, Engineers, and Contractors*; John Wiley & Sons: Hoboken, NJ, USA, 2008.
30. Akcamete, A.; Akinci, B.; Garrett, J.H. Potential utilization of building information models for planning maintenance activities. In Proceedings of the International Conference on Computing in Civil and Building Engineering, Clearwater Beach, FL, USA, 17–20 June 2012.
31. Su, Y.C.; Lee, Y.C.; Lin, Y.C. Enhancing Maintenance Management Using Building Information Modeling in Facilities Management. In Proceedings of the 28th International Symposium on Automation and Robotics in Construction (ISARC 2011), Seoul, Korea, 29 June 2011; pp. 752–757. [CrossRef]
32. Hallberg, D.; Tarandi, V. On the use of open bim and 4D visualisation in a predictive life cycle management system for construction works. *Electron. J. Inf. Technol. Constr.* **2011**, *16*, 445–466.
33. Teicholz, P. *BIM for Facility Managers*; John Wiley & Sons: Hoboken, NJ, USA, 2013.
34. McArthur, J.J. A Building Information Management (BIM) Framework and Supporting Case Study for Existing Building Operations, Maintenance and Sustainability. *Procedia Eng.* **2015**, *118*, 1104–1111. [CrossRef]
35. Cavka, H.; Staub-French, S.; Pottinger, R. Evaluating the Alignment of Organizational and Project Contexts for BIM Adoption: A Case Study of a Large Owner Organization. *Buildings* **2015**, *5*, 1265–1300. [CrossRef]
36. Bortolini, R.; Forcada, N.; Macarulla, M. BIM for the integration of Building Maintenance Management: A case study of a university campus. In Proceedings of the 11th European Conference on Product & Process Modelling (ECPPM), Limassol, Cyprus, 7–9 September 2016.
37. Thabet, W.; Lucas, J. Asset Data Handover for a Large Educational Institution: Case-Study Approach. *J. Constr. Eng. Manag.* **2017**, *143*. [CrossRef]
38. Lucas, J.; Thabet, W. Using a Case-Study Approach to Explore Methods for Transferring BIM-Based Asset Data to Facility Management Systems. In Proceedings of the Construction Research Congress (CRC 2018), New Orleans, LA, USA, 2–4 April 2018. [CrossRef]
39. Pishdad-Bozorgi, P.; Gao, X.; Eastman, C.; Self, A.P. Planning and developing facility management-enabled building information model (FM-enabled BIM). *Autom. Constr.* **2018**, *87*, 22–38. [CrossRef]
40. Carbonari, A.; Corneli, A.; Di Giuda, G.; Ridolfi, L.; Villa, V. BIM-Based Decision Support System for the Mangement of Large Building Stocks. In Proceedings of the 35th International Symposium on Automation and Robotics in Construction (ISARC), Berlin, Germany, 20–25 July 2018. [CrossRef]
41. Fargnoli, M.; Lleshaj, A.; Lombardi, M.; Sciarretta, N.; Di Gravio, G. A BIM-based PSS approach for the management of maintenance operations of building equipment. *Buildings* **2019**, *9*, 139. [CrossRef]
42. Koch, C.; Hansen, G.K.; Jacobsen, K. Missed opportunities: Two case studies of digitalization of FM in hospitals. *Facilities* **2019**, *37*, 381–394. [CrossRef]
43. Aziz, N.D.; Nawawi, A.H.; Ariff, N.R.M. ICT Evolution in Facilities Management (FM): Building Information Modelling (BIM) as the Latest Technology. *Procedia—Soc. Behav. Sci.* **2016**, *234*, 363–371. [CrossRef]
44. Márquez, A.C.; De León, P.M.; Fernández, J.F.G.; Márquez, C.P.; Campos, M.L. The maintenance management framework: A practical view to maintenance management. *J. Qual. Maint. Eng.* **2009**, *15*, 167–178. [CrossRef]
45. Mohanta, A.; Das, S. ICT-Based Facilities Management Tools for Buildings. In Proceedings of the International Conference on ICT for Sustainable Development, Ahmedabad, India, 3–4 July 2015; pp. 125–133.
46. Volk, R.; Stengel, J.; Schultmann, F. Building Information Modeling (BIM) for existing buildings—Literature review and future needs. *Autom. Constr.* **2014**, *38*, 109–127. [CrossRef]
47. What Is COBie? Available online: https://www.thenbs.com/knowledge/what-is-cobie (accessed on 9 October 2019).
48. MVD Database. Available online: https://technical.buildingsmart.org/standards/mvd/mvd-database/ (accessed on 9 October 2019).
49. Ibrahim, K.F.; Abanda, F.H.; Vidalakis, C.; Woods, G. BIM for FM: Input versus Output Data. In Proceedings of the 33rd CIB W78 Conference, Brisbane, Australia, 31 October–2 November 2016.
50. Klammt, F. Financial Management for Facility Managers. In *Facility Design and Management Handbook*; McGraw-Hill Companies Inc.: New York, NY, USA, 2001; pp. 510–537.

51. Lui, R.; Zettersten, G. Facility Sustainment Management System Automated Population from Building Information Models. In Proceedings of the Construction Research Congress, San Juan, Puerto Rico, 31 May–2 June 2016; American Society of Civil Engineers (ASCE): Reston, VA, USA, 2016.
52. Mayo, G.; Giel, B.; Issa, R. BIM use and requirements among building owners. In *Computing in Civil Engineering*; ASCE: Reston, VA, USA, 2012; pp. 349–356.
53. Keady, R. *Equipment Inventories*; Wiley: Hoboken, NJ, USA, 2013.
54. Lin, Y.-C.; Chen, Y.-P.; Huang, W.-T.; Hong, C.-C. Development of BIM Execution Plan for BIM Model Management during the Pre-Operation Phase: A Case Study. *Buildings* **2016**, *6*, 8. [CrossRef]
55. Gleason, D. Getting to a facility management BIM. In *Lake Constance 5D-Conference*; Trimble Navigation: Constance, Germany, 2013.
56. Talon, A.; Boissier, D.; Chevalier, J.L.; Hans, J. Temporal quantification method of degradation scenarios based on FMEA. In Proceedings of the 10th International conference on durability of building materials and components, Lyon, France, 17–20 April 2005.
57. Silva, A.; de Brito, J.; Gaspar, P.L. *Methodologies for Service Life Prediction of Buildings*; Springer International Publishing: Basel, Switzerland, 2016. [CrossRef]
58. Isaac, A.; Meir, I.A.; Garb, Y.; Jiao, D.; Cicelsky, A. Post-Occupancy Evaluation: An Inevitable Step Toward Sustainability. *Adv. Build. Energy Res.* **2019**, *3*, 189–220. [CrossRef]
59. Built to Last: Measuring the Life Cycle of a Facility. Available online: http://www.greenerbuildings.com/feature/2008/04/10/built-last-measuring-life-cycle-a-facility (accessed on 9 October 2019).
60. Meir, I.A.; Motzafi-Haller, W.; Krüger, E.L.; Morhayim, L.; Fundaminsky, S.; Oshry-Frenkel, L. Towards a comprehensive methodology for Post Occupancy Evaluation (POE): A hot dry climate case study. In Proceedings of the 2nd PALENC and 28th AIVC Conference, Crete, Greece, 27–29 September 2007; pp. 644–653.
61. Wahab, A.M.; Kamaruzzaman, S.N. Building Performance and Evaluation Methods: A Preliminary Review. In Proceedings of the 2nd International Conference on Project and Facilities Management, Kuala Lumpur, Malaysia, 16 September 2011.
62. Vischer, J.C. Applying knowledge on building performance: From evidence to intelligence. *Intell. Build. Int.* **2009**, *1*, 239–248. [CrossRef]
63. Preiser, W.F.E. Post-occupancy evaluation: How to make buildings work better. *Facilities* **1995**, *13*, 19–28. [CrossRef]
64. Preiser, W.F.E. Building performance assessment- from POE to BPE: A personnel perspective (post occupancy evaluation and building performance evaluation). *Archit. Sci. Rev.* **2005**, *48*, 201–204. [CrossRef]
65. Sanni-Anniber, M.O.; Hassanain, M.A.; Al Hammad, A.-M. POE of Housing Facilities Overview & Summary of Methods. *J. Perform. Constr. Facil.* **2016**, *30*, 1–9. [CrossRef]
66. Menon, R.; Gilbert, J.; Bridgestock, M. Development of Post Occupancy Evaluation for Evaluation of Innovative Low Carbon Social Housing Projects. Available online: http://www.cicstart.org/fs03.htm (accessed on 9 October 2019).
67. Sinou, M.; Kyvelou, S. Present and future of building performance assessment tools. *Manag. Environ. Qual.* **2006**, *17*, 570–586. [CrossRef]
68. Talamo, C.; Bonanomi, M. *Knowledge Management and Information Tools for Building Maintenance and Facility Management*; Springer International Publishing: Basel, Switzerland, 2015. [CrossRef]
69. Lützkendorf, T. A comparison of international classifications for performance requirements and building performance categories used in evaluation methods. In Proceedings of the 11th Joint BIM International Symposium, Helsinki, Finland, 13–16 June 2005.
70. Weber, A.; Thomas, R. *Key Performance Indicators—Measuring and Managing the Maintenance*; Iavara Corporation: Burlington, ON, Canada, 2005.
71. Bortolini, R.; Forcada, N. Facility managers' perceptions on Building Performance Assessment. *Front. Eng. Manag.* **2018**, *5*, 324–333. [CrossRef]
72. Amasuomo, T.T.; Atanda, J.; Baird, G. Development of a building performance assessment and design tool for residential buildings in Nigeria. *Procedia Eng.* **2017**, *180*, 221–230. [CrossRef]
73. Aksamija, A. BIM-based building performance analysis: Evaluation and simulation of design decisions. In Proceedings of the 2012 ACEEE Summer Study on Energy Efficiency in Buildings, Pacific Grove, CA, USA, 12–17 August 2012.

74. Aksamija, A. Analysis and Computation: Sustainable Design in Practice. *Des. Princ. Pract. Int. J.* **2010**, *4*, 291–314. [CrossRef]
75. Re Cecconi, F.; Maltese, S.; Dejaco, M.C. Leveraging BIM for digital built environment asset management. *Innov. Infrastruct. Solut.* **2017**, *2*, 14. [CrossRef]
76. Bonci, A.; Carbonari, A.; Cucchiarelli, A.; Messi, L.; Pirani, M.; Vaccarini, M. A cyber-physical system approach for building efficiency monitoring. *Autom. Constr.* **2019**, *102*, 68–85. [CrossRef]
77. CIC Research Group. *BIM Project Execution Planning Guide—Version 2.1*; The Pennsylvania State University: University Park, PA, USA, 2011.
78. Common BIM Requirements. Series 12. In *Use of Models in Facility Management*; BuildingSMART Finland: Helsinki, Finland, 2012.
79. PAS 1192-3:2014. *Specification for Information Management for the Operational Phase of Assets Using Building Information Modelling*; BSI Standards Limited: London, UK, 2014; ISBN 978-0-580-86674-6.
80. U.S. General Service Administration Public Buildings Office of Design and Construction. *GSA Building Information Modeling Guide Series: 08—GSA BIM Guide for Facility Management*; General Services Administration: Washington, DC, USA, 2011.
81. Minister of Transport and Infrastructure. Ministerial Decree n.560/2017. Available online: http://www.mit.gov.it/normativa/decreto-ministeriale-numero-560-del-01122017 (accessed on 9 October 2019).
82. UNI 11337/1-4:2017. Building and Civil Engineering Works—Digital Management of the Informative Processes. Available online: http://www.uni.com (accessed on 9 October 2019).
83. President of Italian Republic. Legislative Decree n.50/2016. Gazzetta Ufficiale della Repubblica Italiana. 2016. 91, 1–378. Available online: https://www.anticorruzione.it/portal/rest/jcr/repository/collaboration/Digital%20Assets/anacdocs/MenuServizio/English%20section/ITALIAN_PUBLIC_CONTRACT_CODE%2015%20giugno%202018_sito%20(2).pdf (accessed on 7 December 2019).
84. President of Italian Republic. Presidential Decree 14/01/1997. *G.U.* **1997**, *42*, 1–70.
85. AIA. *AIA Document G202–2013, Project Building Information Modeling Protocol Form*; American Institute of Architects: Washington, DC, USA, 2013.
86. Ozarisoy, B.; Altan, H. Low Energy Design Strategies for Retrofitting Existing Residential Buildings in Cyprus. In *Proceedings of the Institution of Civil Engineers–Engineering Sustainability*; ICE Publishing: London, UK, 2018; pp. 1–15. ISSN 1478 4629. E-ISSN 1751 7680. [CrossRef]

sustainability

MDPI

Article

Exploring On-Site Safety Knowledge Transfer in the Construction Industry

Ying-Hua Huang [1,*] and Tzung-Ru Yang [2]

[1] Department of Civil and Construction Engineering, National Yunlin University of Science and Technology, Yunlin 640, Taiwan

[2] Engineering Department, Yuan Tay Construction Co., Ltd., Kaohsiung 807, Taiwan

* Correspondence: huangyh@yuntech.edu.tw

Received: 21 October 2019; Accepted: 13 November 2019; Published: 15 November 2019

Abstract: A primary cause of occupational accidents is on-site workers not having proper or even adequate safety knowledge and awareness, leading to them failing to employ safety measures, equipment, or behavior to protect themselves. The complexity of construction projects and changes in organizational personnel complicate the safety knowledge transfer process. Therefore, to reduce occupational accidents in the construction industry, this study explored the on-site safety knowledge transfer process as well as its relationship with a safe working environment; it did this to understand the associations between various constructs in the process, which could be used as a reference for management personnel to promote on-site safety education and behaviors. This would allow safety knowledge to be learned and practiced by on-site workers, changing their unsafe behaviors and creating a safe on-site work environment. This study used structural equation modeling to empirically study the relationship between various constructs during safety knowledge transfer on a construction site. The results revealed that an excellent safety knowledge transfer environment can lead to favorable safety behavior as well as safety knowledge application and inspiration of on-site workers, which would affect their safety behaviors. More satisfactory safety behaviors of on-site workers could produce a safer working environment on the construction site. Moreover, although safety application and inspiration do not directly affect the safety of a work environment, they do so indirectly through safety behaviors.

Keywords: knowledge transfer; safety behavior; safety management; construction site; structural equation modeling (SEM)

1. Introduction

The construction industry is characterized by one-time products, multi-level subcontracting, and complex staffing. Different construction stages usually have different staffing. Labor safety issues occur easily because of the uncertain labor source demands. According to statistics compiled by Taiwan's Ministry of Labor, the occupational injury death rate per thousand was 0.113 (excluding traffic accidents) for the construction industry in Taiwan in 2018, which was the highest amongst all industries; furthermore, the occupational fatal injury rate in the construction industry per thousand (including morbidity, disability, and death) was 9.385, which was also the highest amongst all industries [1]. Over the years, occupational accident analysis of Taiwan's construction industry has shown that one of the most common accidents in public works is workers falling without personal protective equipment. In private construction projects, accidents are caused by employers not providing personal protective equipment for workers or workers not using it or ignored warning signs when preparing, repairing, or finishing work on a ladder, platform, floor surface, or steel component [2]. Occupational accidents are mostly caused by noncomplying facilities, improper personal protective equipment, unsafe guardrails and covers, or the absence of protective measures [3]. Relevant studies on construction site safety performance and assessment have noted that workers have low safety awareness [4,5], inadequate safety knowledge and training [6,7], and unsafe practices [8]. A review of major occupational accidents and relevant studies reveals some of the main causes of occupational accidents are workers on construction sites lacking proper personal safety knowledge and failing to take appropriate safety measures, equipment, or actions to protect themselves.

The causes of occupational accidents are complicated by numerous events with causal relationships. However, if on-site workers do not possess correct safety concepts, the number of accidents in the construction industry cannot be reduced. Approximately 90% of accidents are caused by unsafe environments and behaviors [8–10]. Person, behavior, and environment are three dynamic and interactive factors, and changes in any one of them will affect the other two [11]. How to ensure on-site workers learn and practice safety knowledge to correct unsafe behaviors and create a safe construction environment forms the basis for reducing occupational accidents in the construction industry.

Knowledge involves commitment and action [12]. The knowledge transfer conceptual model developed by Gilbert and Cordey-Hayes [13] emphasized that knowledge transfer is a dynamic process that must be assimilated into members of an organization. That is, workers are considered to truly acquire relevant knowledge after they apply it to the daily activities of their organization, which is reflected in changes in personal cognition and behavior. According to the high occupational injury rates in the construction industry, a gap exists between the education of on-site safety knowledge and the safety awareness and behaviors of on-site workers. Because of complex construction projects and changes in organizational personnel, the process of transferring safety knowledge becomes more complex.

This study aimed to identify both of the on-site safety knowledge transfer processes and the relationship between safety knowledge transfer and the safety of on-site work environments. The results of this study can be used as a good reference when the constructors plan to establish specific on-site safety education and to promote essential working environment safety. This study has the following two objectives:

1. To develop a safety knowledge transfer model and to identify on-site safety knowledge transfer mechanisms.
2. To explore the correlation between on-site safety knowledge transfer and the safe working environment.

The remainder of this paper is organized as follows. Section 2 presents a literature review on constructs required for construction worker's safety knowledge transfer. Section 3 presents the research model and research hypotheses. Section 4 presents the research methodology. Section 5 explains the design of the questionnaire and survey process. Sections 6 and 7 describe how structural equation modeling (SEM) was used to verify the relevance of knowledge transfer to a safe working environment and present results and discussion. Finally, Section 8 presents the study's conclusions.

2. Literature Review

2.1. Knowledge Transfer

Knowledge is the result of information processing, which involves rearranging, quantifying, qualifying, grouping, and learning [14]; these processes allow people to make meaningful connections between information and actions based on responses to information [15]. Knowledge is not only stored in documents and storage systems for an organization, but also in the processes, implementation, and specifications of daily work [16]. Knowledge can be divided into tacit knowledge and explicit knowledge. Tacit knowledge refers to subjective and substantive knowledge that cannot be expressed in words or sentences. It is personal, difficult to formalize, and deeply embedded in individual actions and experiences, personal ideals, values, and inner feelings; by contrast, explicit knowledge refers to objective and metaphysical knowledge that can be expressed in words and numbers. It can be conveyed in formal and institutionalized language, and easily communicated and shared using specific materials, scientific formulas, standardized procedures, and general principles [17,18]. Both tacit and explicit knowledge are learned through the continuous flow of knowledge, such as drawing on the personal experiences of friends and family, observing and imitating the states of others, knowledge being passed from teachers to students in schools, knowledge being imparted from senior staff to junior staff at a company, and companies holding meetings to announce changes in internal rules. All of these are knowledge transfer processes.

Szulanski [19] defined knowledge transfer as the flow of knowledge among organizational members; that is, the routine exchange of knowledge between knowledge providers and recipients for the operation of individual skills and social systems. Davenport and Prusak [16] emphasized that this process involves transmission and absorption, where transmission refers to knowledge being transmitted to potential recipients and absorption refers to people absorbing accepted knowledge. Knowledge transfer fails if the knowledge is not absorbed by recipients; this reflects the importance of interaction between knowledge providers and recipients, because knowledge can only be transferred when both parties are willing to share and receive the knowledge, respectively [20]. In addition, Wiig [21] defined knowledge transfer from a systematic point of view. He asserted that knowledge transfer includes actions such as knowledge acquisition, organization, structural reconstruction, storage, memory, and reassembly for deployment and dissemination. Organizations must provide basic construction capabilities that support knowledge transfer and create incentives to motivate employees, teams, departments, and business units to work together toward mutual goals.

According to a literature review and actual surveys, Gilbert and Cordey-Hayes [13] compiled their knowledge transfer conceptual framework into a five-step model to illustrate the connotations of knowledge transfer.

1. Acquisition: The first step in knowledge transfer requires knowledge acquisition, which can be achieved through acquiring knowledge from past experiences of work processes or environments, and discussions, as well as through obtaining new knowledge from personal sharing.
2. Communication: Organizations must develop communication mechanisms in written and verbal forms, and must be aware of and eliminate obstacles that hinder communication within the organization to ensure that knowledge can be effectively transferred.
3. Application: The application of acquired and communicated knowledge can ensure it is transferred and retained within the organization; knowledge application enables all members of an organization to learn, apply, and pass on knowledge.
4. Acceptance: The most essential key to knowledge transfer is acceptance. Once knowledge is applied, it must be accepted by individuals for consistent application as well as be assimilated into core daily activities.
5. Assimilation: Another crucial key to knowledge transfer is assimilation, which is the result of the acceptance of knowledge application. The essence of assimilation refers to the process of knowledge creation, including the cumulative learning process, which demonstrates individuals' changes in cognition and behavior that enable them to practically apply the knowledge they have acquired to daily activities of their organization.

2.2. *Safety Knowledge*

In terms of construction-safety knowledge, explicit knowledge exists in the form of accident records, mandated safety rules and regulations, and best practices [22], whereas tacit knowledge mainly relies on workers' personal experience of safety engineering [23]. Safety climate research has suggested that safety knowledge is a critical determinant of safety behavior or may act as a mediator between safety climate and safety performance [24]. Improving the safety knowledge of workers helps to reduce the risk of accidents [25].

Safety knowledge, whether explicit or tacit, is mostly scattered and fragmented. Thus, relevant research has focused more on safety knowledge management and learning technologies. Zhang et al. [22] proposed a construction-safety knowledge ontology to formalize safety management knowledge, as well as explored its connection with building information modeling to enable more effective safety-knowledge inquiries. Taher et al. [26] integrated developed earthwork safety regulation knowledge with an earthwork ontology to bridge the gap between high-level safety regulations and task-level instructions. Le et al. [27] proposed an online social virtual reality system framework, which allowed students to perform role-playing, dialogic learning, and social interactions for construction health and safety education. Li et al. [25] presented Internet of things applications for knowledge exchange and safety awareness in construction. Although researchers have mentioned the importance of safety knowledge in safety performance, relatively little is known about the mechanism of safety knowledge transformation.

3. Research Model and Hypotheses

3.1. Constructing a Safety Knowledge Transfer Model

This study referenced the knowledge transfer concept proposed by Gilbert and Cordey-Hayes [13] and summarized relevant studies on construction safety to establish a model for developing on-site safety knowledge transfer, as well as investigated the correlation between safety knowledge transfer and safe work environments. Following the concepts proposed by Gilbert and Cordey-Hayes [13], the on-site safety knowledge transfer model is described as follows:

1. Acquisition of safety knowledge:

According to the channels of learning and acquiring safety knowledge proposed by scholars, this study summarized the concepts of safety knowledge acquisition into safety training, safety leadership, and self-learning. Safety training equips workers with the skills required to recognize and manage hazards [28]. It can improve the safety climate and its effect on enhancing worker's safety attitudes, perceptions, and behaviors. Organizations can influence the safety performance of their workers by teaching and training them in safety behaviors and attitudes [24,29]. Blair [30] emphasized that supervisors must provide education, training, and resources to ensure that workers are effectively developed and prepared to contribute to safety. Hofmann and Morgeson [31] argued that the interaction between supervisors and workers can influence an organization's safety communication and worker accidents, which they defined as safety leadership. Conducting safety leadership will drive workers to demonstrate safe behavior [32]. Through enabling workers to understand policies and rules in the workplace, promoting an open and trustworthy relationship with workers, increasing worker rights and responsibilities for safety performance, and taking the initiative to conduct safety-related activities, supervisors can establish a safety paradigm, thereby enhancing workers' safety in the workplace [33]. In addition, informal knowledge of industry professions, such as rules of thumb and skills, must be observed by workers in the work environment so they acquire knowledge through comprehending empirical knowledge.

2. Communication of safety knowledge:

Communication is the process by which a supervisor or worker communicates opinions expressed by one party to another through a formal or informal communication channel. In addition to long-term accumulated knowledge in an organization, organization members are also required to contribute their knowledge for the sake of accumulation [34]. The sharing, interaction, and communication of members' safety knowledge within the organization will contribute to the flow of safety knowledge and learning applications. Supervisors conducting safety communication will also drive workers to demonstrate safety behaviors [32], improving their safety behaviors through effective communication [35,36]. Consistent sharing of safety information among employees helps to prevent on-site accidents.

3. Application of safety knowledge:

This study named this concept safety knowledge application and inspiration. This means the application of safety knowledge (acquired through the acquisition and communication of safety knowledge) that enables workers to confirm the acquired knowledge in practical operations; and enables other personnel in the organization to acquire or receive the safety knowledge. Attempts by workers to apply their acquired knowledge within the organization allow the continuous dissemination of safety knowledge.

4. Acceptance of safety knowledge:

This study named this concept safety knowledge acceptance. When workers apply and accept the transferred knowledge, they obtain the knowledge required to prevent and eliminate accidents. Workers will clearly know what they should do as well as where and why when accidents occur, reducing the number of accidents caused by uncertainty [37].

5. Assimilation of safety knowledge:

The assimilation of safety knowledge is an individual's change in safety awareness and behaviors. This process allows an individual to be aware of the importance of safety, strengthens his or her awareness of safety, and thereby change his or her safety behaviors. This study named this concept safety concern and safety awareness.

Through conducting a literature review and extending Gilbert and Cordey-Hayes' knowledge transfer model [13], this study developed eight subconstructs, namely safety training, safety leadership, self-learning, safety knowledge communication, safety knowledge application and inspiration, safety knowledge acceptance, safety concern, and safety awareness, to indicate safety knowledge transfer (as shown in the dotted box in Figure 1). In the first four subconstructs (safety training, safety leadership, self-learning, and safety knowledge communication), an organization plays a crucial role in providing a proper environment for workers to acquire safety knowledge as well as appropriate channels for safety knowledge communication. These actions enable workers to apply the safety knowledge they acquire through the acquisition and communication of safety knowledge and inspire each other (safety knowledge application and inspiration). Additionally, these actions drive the acceptance and assimilation of safety knowledge (safety knowledge acceptance, safety concern, and safety awareness) for presenting specific safety behaviors. Therefore, this study defined the first four subconstructs as a safe knowledge transfer environment, emphasizing the importance of organizations providing an appropriate environment for initiating the transfer of safety knowledge. In addition, the latter three subconstructs were defined as safety behaviors.

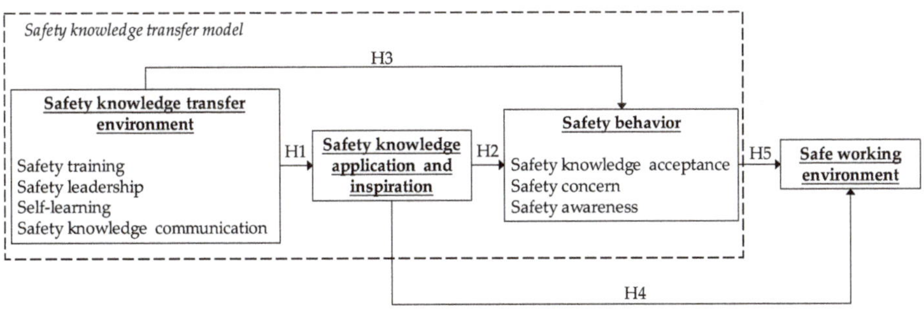

Figure 1. Diagram of the research hypotheses.

3.2. Research Framework

This study explored the constructs of a safe working environment and safety knowledge transfer model (safety knowledge transfer environment, safety knowledge application and inspiration, and safety behavior) presented in the previous section. A safe working environment is a place where all construction activities are conducted safely and workers have the working conditions they require to complete tasks safely. To prevent the occurrence of any accidents, a good on-site working environment is an essential ingredient. This study employed safety training, safety leadership, self-learning, and safety knowledge communication in response to the safety knowledge transfer environment, and safety knowledge acceptance, safety concern, and safety awareness in response to safety behaviors. In addition, to explore the effects of on-site safety knowledge transfer in a safe working environment, this study hypothesized that a safety knowledge transfer environment will affect safety behaviors and safety knowledge application and inspiration, which could lead to a safe working environment. Figure 1 presents the research hypotheses. According to the abovementioned discussion, this study proposed the following five hypotheses as shown in Table 1.

Table 1. Research hypotheses.

Item	Hypothesis
H1	A safety knowledge transfer environment has a positive effect on safety knowledge application and inspiration.
H2	Safety knowledge application and inspiration have a positive effect on safety behavior.
H3	A safety knowledge transfer environment has a positive effect on safety behavior.
H4	Safety knowledge application and inspiration have a positive effect on a safe working environment.
H5	Safety behavior has a positive effect on a safe working environment.

4. Methodology

According to the research objectives and literature review, this study used statistical methods and SEM as research tools and applied empirical analysis to explore the correlation between construction-safety knowledge transfer and a safe working environment. On the basis of the literature review and an actual situation of on-site safety, this study drafted questionnaire items for each construct, discussed the definition of each item with practical experts, and confirmed the wording of sentences. A pretest of 50 samples [38] was performed prior to the issuance of the formal questionnaire to determine the reliability and goodness-of-fit of the questionnaire items.

Analysis results of the pretest were processed in four stages: (1) The correlation coefficient matrix of all items was calculated, and one of the two items with high correlation (the correlation coefficient exceeded 0.9) was eliminated. (2) An internal consistency indicator was used, and the sum of all the pretest samples was sorted in order; the difference between the top 27% (high-score group) and bottom 27% (low-score group) was used as the basis for item discrimination, and nondiscriminatory items were removed [39]. (3) Factor analysis was used to remove items with factor loadings less than 0.5 [40] to ensure the reliability and validity of the items. (4) The internal consistency of each construct was tested using the Cronbach's α value, where higher coefficient values indicate higher reliability.

This study used SEM and an empirical analysis to build an on-site safety knowledge transfer model for construction projects, and explored the relationship between on-site safety knowledge transfer and a safe working environment. SEM was proposed by scholars in the 1970s, and it combines the concept of path analysis with latent variables and factor analysis. In short, SEM integrates latent variables and observed variables, which forms the measurement model of SEM, and the structural model uses a path analysis model to study the causal relationship between latent variables; the integration of both models forms SEM.

When establishing SEM, the goodness-of-fit coefficient of the theoretical model should be judged. The higher the model's goodness-of-fit is, the more satisfactory the display model and the more meaningful the estimation parameters become. This study used the indexes listed in Table 2 as the measurement indicators, and composite reliability (CR) and average variance extracted (AVE) for testing [40–45]. According to Chin and Todd [45] and Hair et al. [40], the value of chi-square/degree of freedom (χ^2/df) should not exceed 3 in rigorous research; thus, the judging criterion of χ^2/df should be between 1 to 3. Doll et al. [44] indicated that goodness-of-fit index (GFI) and adjusted GFI (AGFI) range from 0.80 to 0.89, which represent a reasonable goodness-of-fit of the model; therefore, the judging criteria of GFI and AGFI in this study had to exceed 0.9. The remaining indicators of normed fit index (NFI) > 0.90, comparative fit index (CFI) > 0.9, root mean square error of approximation (RMSEA) < 0.08, and root mean square residual (RMR) < 0.05 are often used as references for goodness-of-fit, which also served as the criteria for this study. Hair et al. [40] proposed that CR can be used to measure the internal consistency of indicator items of latent variables; the higher the CR value, the higher the consistency of the indicator items, with the suggested CR value being > 0.7. The AVE is the average explanatory power of the latent variables for each observation number. Bagozzi and Yi [41] proposed that the internal quality of a model is more favorable when the AVE is > 0.5.

Table 2. Model fit indexes.

Index	Recommended Standards
Chi-square/degree of freedom (χ^2/df)	<3
Goodness-of-fit index (GFI)	>0.9
Root mean square error of approximation (RMSEA)	<0.08
Root mean square residual (RMR)	<0.05
Adjusted GFI (AGFI)	>0.9
Normed fit index (NFI)	>0.9
Comparative fit index (CFI)	>0.9

5. Questionnaire Development and Distribution

This study referred to relevant literature and held discussions with industry experts, such as construction site directors, occupational safety and health personnel, and on-site supervisors of architectural firms, to develop various construct items including safety training, safety leadership, safety communication, self-learning, safety knowledge application and inspiration, safety knowledge acceptance, safety concern, safety awareness, and a safe working environment. The initial questionnaire featured a total of 66 items. To measure participants' level of agreement on each item, they were measured using a five-point Likert scale (5 = strongly agree, 4 = agree, 3 = neutral, 2 = disagree, and 1 = strongly disagree; the points for reversed items were from 1 (strongly agree) to 5 points (strongly disagree)).

Analysis results of the pretest questionnaires showed that the correlation coefficient values between the items were lower than 0.9, indicating no similar items; thus, no items were removed. Regarding discriminant analysis, seven items that failed the analysis at a significance level of 10% were removed. Subsequently, factor analysis was performed for each construct, and nine items with a load factor less than 0.5 were eliminated. The formal questionnaire had a total of 50 items, as shown in Table 3. The Cronbach's α values of all constructs were greater than 0.7, indicating that each construct had satisfactory reliability.

Participants in this study were on-site personnel in the construction industry in western Taiwan. A total of 311 questionnaires were distributed to construction site management personnel and workers, with a response rate of 78.14% (243 copies) and valid response rate of 66.24% (206 copies). Table 4 presents the basic information of the samples. The majority of the on-site personnel were aged between 26 and 35 years (38.3%), followed by 36 to 45 years (33.5%). A total of 62.6% of the on-site personnel had an undergraduate degree as their highest educational attainment. The majority had worked in the industry for fewer than 5 years (36.4%), followed by 6 to 10 years (19.4%) and 20 years and above (17%). Most of the on-site personnel attended at least one safety training session per year (83%), and most worked for a Level A construction company (73.3%). In addition, most of the on-site personnel were engineers (51.5%), followed by site directors (18.9%). Finally, most of the on-site personnel had not encountered occupational accidents on site within the last 3 years (67.5%).

Table 3. Factor analysis results and Cronbach's α values of each construct.

Construct (Cronbach's α)	Item	Factor Load
Safety Training (0.803)	T1: Current on-site supervisors often organize routine occupational health and safety training courses.	0.662
	T2: Current on-site supervisors often train workers on how to work safely and comply with safety regulations.	0.826
	T3: Current on-site supervisors rarely train workers on proper safety behaviors and attitudes.	0.646
	T4: Current on-site supervisors often discuss with workers to review and improve dangerous workplace situations.	0.763
	T5: Current on-site supervisors often encourage workers to express ideas and recommendations on safety issues.	0.746
	T6: Current on-site supervisors regularly ask colleagues to pay more attention to the safety awareness of new workers.	0.697
Safety Leadership (0.899)	L1: Current on-site supervisors often guide workers to be cautious and strictly follow safety procedures.	0.648
	L2: Current on-site supervisors emphasize the company's belief that safety and work are equally crucial.	0.591
	L3: Current on-site supervisors promptly remind workers who have incorrect concepts regarding safety.	0.743
	L4: Current on-site supervisors often remind me of safety issues that should be noted at work.	0.665
	L5: My supervisor often leads by example to actively share safety knowledge.	0.688
	L6: Current on-site supervisors immediately correct workers who do not comply with safety regulations.	0.682
Self-learning (0.808)	SL1: I often try to apply safety knowledge to the construction site.	0.818
	SL2: I often share my safety knowledge.	0.754
	SL3: I often learn work safety from the contents of the health and safety management section of the construction plan.	0.638
Safety Knowledge Communication (0.752)	C1: On-site supervisors can provide appropriate guidance and solutions to colleagues who face any safety problems.	0.622
	C2: On-site supervisors value proposals or recommendations related to accident prevention.	0.678
	C3: On-site supervisors often raise safety-related issues during meetings.	0.704
	C4: I often convert safety knowledge into text and store it in a shared folder in an electronic file.	0.541
	C5: I often learn about work safety by chatting with colleagues.	0.553
Safety Knowledge Application and Inspiration (0.828)	AI1: I inform my superiors when I notice any unsafe behavior.	0.624
	AI2: I inform everyone when I receive a new company safety policy.	0.753
	AI3: I often pass on safety knowledge to other people at the scene.	0.782
	AI4: I do not often share accident cases with other people at the scene.	0.534
	AI5: I often solve the safety concerns of my colleagues or workers.	0.776
	AI6: I think that individuals affect each other in terms of safety behavior.	0.909
Safety Knowledge Acceptance (0.722)	A1: I understand the accidents that I may encounter at work.	0.538
	A2: I usually use tools, utensils, and materials correctly.	0.572
	A3: I understand all approaches and regulations on the use of personal protective equipment.	0.576
	A4: I understand the safety risks posed by unsafe actions.	0.511
	A5: I always remember construction-safety knowledge.	0.686
Safety Concern (0.790)	SC1: I check and confirm that it is safe to start work.	0.691
	SC2: When I notice any unsafe behavior during work, I stop work and eliminate it.	0.781
	SC3: I correct my colleagues or workers immediately if I notice they are behaving unsafely.	0.784
	SC4: I not only pay attention to my own safety but also to that of my colleagues.	0.742

Table 3. *Cont.*

Construct (Cronbach's α)	Item	Factor Load
Safety Awareness (0.726)	SA1: I restore or maintain dismantled safety protection equipment.	0.502
	SA2: I still continue to work when I fall sick.	0.804
	SA3: I pay special attention to construction safety when the project is rushed.	0.606
	SA4: I do not propose discussions when I have any safety-related suggestions.	0.723
Safe Working Environment (0.916)	E1: The site has an excellent safety management system.	0.705
	E2: Safety warning signs can be found everywhere on the construction site.	0.771
	E3: The on-site personnel is always alert in the work environment at all times.	0.703
	E4: A guardrail will be added to areas in a workplace where it is easy to fall.	0.768
	E5: All holes in the workplace will be covered.	0.690
	E6: Illumination of the workplace is extremely inadequate.	0.805
	E7: All fire safety equipment in the workplace is adequately set up.	0.768
	E8: Workers have appropriate safety equipment when working at heights.	0.781
	E9: Water is sprinkled when there is excessive dust.	0.661
	E10: Everyone wears protective equipment when there is excessive dust.	0.618
	E11: Safety of the site is reviewed daily before work.	0.637

Table 4. Basic information distribution of the samples.

Variable	Category	Number	Percentage
Gender	Male	180	87.4%
	Female	26	12.6%
Age	Below 25 years	14	6.8%
	26–35 years	79	38.3%
	36–45 years	69	33.5%
	46–55 years	39	18.9%
	56 years and above	5	2.4%
Highest educational level obtained	Below junior high school	6	2.9%
	High school	34	16.5%
	Undergraduate	129	62.6%
	Postgraduate and above	37	18.0%
Years of working experience in the construction industry	Below 5 years	75	36.4%
	6–10 years	40	19.4%
	11–15 years	26	12.6%
	16–20 years	30	14.6%
	20 years and above	35	17.0%
Number of safety training events attended in the past year	0	35	17.0%
	1	83	40.3%
	2	34	16.5%
	3	23	11.2%
	3 and above	31	15.0%
Grade of the general construction business	Level A	151	73.3%
	Level B	16	7.8%
	Level C	14	6.8%
	Others	25	12.1%
On-site position	Worker	25	12.1%
	Engineer	106	51.5%
	Management personnel	24	11.7%
	Site director	39	18.9%
	Others	12	5.8%

6. Results and Analysis

6.1. Measurement Model Test Results

This section describes how this study used SEM to verify the causal relationships between variables. The correlations between knowledge transfer environment (safety training, safety leadership, safety knowledge communication, and self-learning), safety knowledge application and inspiration, and safety behaviors (safety knowledge acceptance, safety concern, and safety awareness) were explored.

This study first analyzed the measurement model and verified whether the goodness-of-fit of each construct of the model was acceptable. If it was not acceptable, the items were removed to modify the model. The evaluation indicators of χ^2/df, GFI, AGFI, RMSEA, RMR, NFI, and CFI were used as the basis for judging whether the measurement model had a good fit, and the AVE and CR were used for testing. After eliminating T1, T2, L1, L2, C4, SA1, SA3, A4, AI4, AI5, E1, E3, E4, E6, and E11, each construct met all fit indicators with the recommended standards in Table 2 (Table 5). As shown in Table 6, the CR of each construct was > 0.7, indicating their adequate reliability. The AVE values of each construct were all > 0.5, demonstrating that each had adequate convergent validity.

Table 5. Goodness-of-fit of the measurement model and the structural model.

Index	Construct of the Measurement Model				Structural Model
	Safety Knowledge Transfer Environment	Safety Knowledge Application and Inspiration	Safety Behavior	Safe Working Environment	
χ^2/df	1.223	1.039	1.648	2.241	1.134
GFI	0.941	0.990	0.953	0.940	0.927
RMSEA	0.033	0.014	0.056	0.078	0.026
RMR	0.021	0.013	0.029	0.034	0.260
AGFI	0.910	0.969	0.918	0.900	0.902
NFI	0.945	0.985	0.930	0.922	0.919
CFI	0.989	0.999	0.971	0.955	0.990

Table 6. CR and AVE for each construct of the measurement model.

Construct	CR	AVE
Safety knowledge transfer environment	0.9364	0.7873
Safety knowledge application and inspiration	0.837	0.5677
Safety behavior	0.7885	0.5790
Safe working environment	0.8573	0.5083

6.2. Structural Model Test Results

According to the proposed research hypotheses, the structural model between each construct was established and a path analysis was performed. Figure 2 illustrates the final structural model of this study. The index of the adjusted structural model achieved the following results; χ^2/df = 1.134, GFI = 0.927, RMSEA = 0.026, RMR = 0.260, AGFI = 0.902, NFI = 0.919, and CFI = 0.990, as shown in Table 5. The overall fit statistics indicated an excellent fit for the model; thus, the results supported the hypothesized relationship.

Hypothesis testing was performed according to the results of models empirically presented in this study. H1 was that a knowledge transfer environment has a positive and significant effect on safety knowledge application and inspiration. The path coefficient of knowledge transfer environment to safety knowledge application and inspiration was 0.874 ($p < 0.001$); thus, H1 was supported, indicating that an improved knowledge transfer environment can result in greater safety knowledge application and inspiration of on-site workers. H3 posited that a knowledge transfer environment has a positive and significant effect on safety behavior. The path coefficient of knowledge transfer environment to safety behavior was 0.587 ($p < 0.001$); thus, H3 was supported, indicating that a more satisfactory knowledge transfer environment will drive on-site workers to demonstrate higher levels of safety behavior. H2 posited that safety knowledge application and inspiration have a positive and significant effect on safety behavior. The path coefficient of safety knowledge application and inspiration to safety behavior was 0.440 ($p < 0.01$); thus, H2 was supported, signifying that greater safety knowledge application and inspiration of on-site workers can lead to improved safety behavior. Moreover, this showed that, in addition to directly affecting safety behaviors, the knowledge transfer environment indirectly affects safety behaviors through safety knowledge application and inspiration. H5 posited that safety behavior has a positive and significant effect on a safe working environment. The path coefficient of personal safety behavior to safe working environment was 0.795 ($p < 0.001$); thus, H5 was supported, indicating that the working environment of a construction site is safer when on-site workers exhibit more safety behaviors. In the case of a safe working environment, a higher chance exists for preventing on-site accidents.

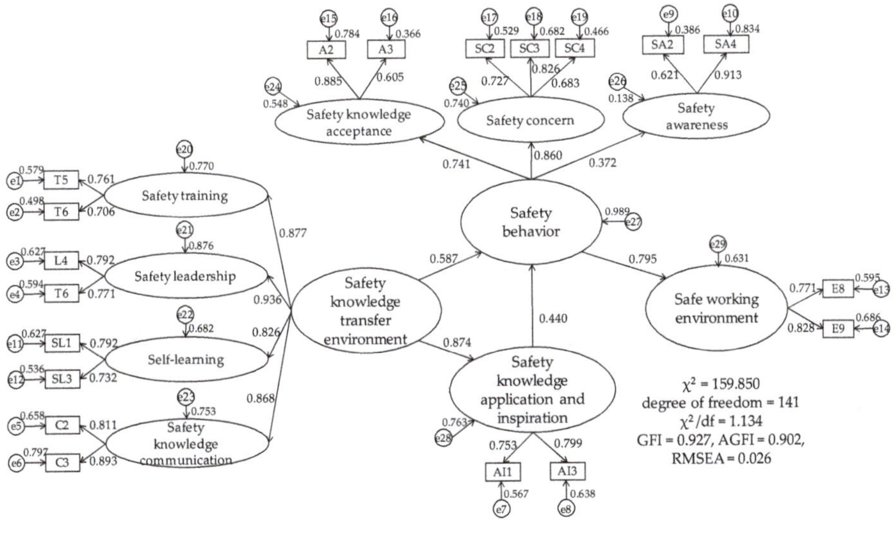

Figure 2. Overall structural model.

H4 posited that safety knowledge application and inspiration have a positive and significant effect on a safe working environment; however, as shown in Figure 2, safety knowledge application and inspiration did not directly affect safe working environment. By contrast, because H2 and H5 were supported, safety knowledge application and inspiration indirectly affect safe working environment through safety behaviors; this means that H4 should be supported because safety knowledge application and inspiration have a positive and significant effect on safe working environment. Table 7 presents the direct and indirect effects of latent independent variables of the overall model on latent dependent variables. Among the total effects, the knowledge transfer environment had the greatest effect on safety behavior, and safety knowledge application and inspiration had the smallest effect on safe working environment. Path values in the structural model were positive, indicating that the various constructs are positively related. Knowledge transfer environment and safety knowledge application and inspiration indirectly affected a safe working environment through safety behavior in a positive and increasing trend.

Table 7. Effects of the overall model.

Latent Independent Variable	Latent Dependent Variable	Direct Effect	Indirect Effect	Total Effect	Hypothesis Supported?
Safety knowledge transfer environment	Safety knowledge application and inspiration	0.874	-	0.874	H1 was supported
	Safety behavior	0.587	0.874 × 0.44 = 0.385	0.972	H3 was supported
Safety knowledge application and inspiration	Safety behavior	0.440	-	0.440	H2 was supported
	Safe working environment	-	0.44 × 0.795 = 0.350	0.350	H4 was supported
Safety behavior	Safe working environment	0.795	-	0.795	H5 was supported

7. Discussion

The knowledge transfer environment constructs included four subconstructs: safety training, safety leadership, safety knowledge communication, and self-learning. Among them, safety leadership had the highest factor loading (0.936), which was the most vital factor in the knowledge transfer environment, followed by safety training (0.877), safety knowledge communication (0.868), and self-learning (0.826). On-site supervisors often remind workers to pay attention to safety and immediately correct unsafe behaviors on the spot, resulting in the effect of safety leadership. Safety training is improved when on-site supervisors encourage workers to express safety concerns or regularly ask their colleagues to pay more attention to the safety of new workers. Furthermore, safety knowledge communication will be enhanced when on-site supervisors emphasize proposals or suggestions for accident prevention, as well as frequently discuss safety-related topics. In addition, applying safety knowledge to the on-site construction environment for confirming knowledge and understanding safety hazards at work through an organization's health and safety management plan can be regarded as self-learning performance. In addition to safety leadership, on-site supervisors play an essential role in safety training and safety knowledge communication. This shows that the safety concerns of on-site supervisors are vital for establishing workers' safety knowledge. The results of this study indicated that safety knowledge acquisition and self-learning orientation of Taiwan's construction workers are relatively passive, where the workers acquire safety knowledge through safety training and communication provided by their on-site supervisors as opposed to them learning them on their own.

Safety behavior constructs included safety knowledge acceptance, safety concern, and safety awareness. Among them, safety concern had the highest factor loading (0.860), indicating that it was the most crucial factor affecting safety behaviors (i.e., individuals who are more concerned with safety tend to exhibit higher levels of safety behavior), followed by safety knowledge acceptance (0.741), with the lowest being safety awareness (0.372). This result indicated that on-site workers should strengthen links between safety awareness and safety behavior. In terms of the acceptance of safety knowledge, most of the on-site workers could use tools, equipment, or materials correctly, and could understand the use and regulations of personal protective equipment, indicating that they possessed a certain amount of safety knowledge. Safety concern is reflected in on-site workers stopping work to eliminate the unsafe behaviors of others, indicating that they do not only pay attention to their own safety at work. On-site workers' safety awareness can be improved by enhancing their perceptions of their bodies, having them understand that they should stop working whenever they experience physical discomfort, and having them raise their own concerns about on-site safety. Safety knowledge application and inspiration can affect safety behaviors. Therefore, applying acquired safety knowledge to unsafe behaviors at work and informing management, or passing safety knowledge on to others, can directly affect the safety behaviors of on-site workers and indirectly make the work environment safer.

According to the aforementioned analysis, effective improvements of safety knowledge transfer are necessary for a safe working environment, which can be achieved through safety training, safety leadership, safety knowledge communication, and self-learning. On-site supervisors should encourage workers to express their thoughts and suggestions on safety issues, regularly ask their colleagues to pay more attention to new workers' safety awareness, and counsel and encourage workers to participate in safety issues to improve the effects of safety knowledge transfer. Furthermore, the leadership method of on-site management personnel has a great effect on knowledge transfer; if they emphasize safety, they will deliver safety knowledge to others. For example, on-site supervisors should often remind workers of safety issues to pay attention to at work. Moreover, supervisors should immediately correct workers' behavior in accordance with safety regulations. Such an active and positive leadership method can effectively transfer knowledge. Pu et al. (2013) noted that leaders with high levels of care or control can effectively control the unsafe behaviors of workers; workers' safety behaviors can be developed easily if the leadership behavior of on-site personnel has a more apparent development.

Moreover, the communication of on-site management personnel is critical. For instance, on-site supervisors should emphasize accident prevention-related proposals or suggestions, regularly raise safety-related topics in meetings, and be willing to discuss safety prevention with other workers. Regardless of whether the content discussed will be accepted, results must be obtained from discussions and idea exchange between management personnel and workers. Furthermore, management personnel must provide warnings about the unsafe behavior they have noticed to improve safety knowledge communication and safety prevention.

On-site knowledge transfer is not limited to management personnel; workers must also acquire safety knowledge by themselves, such as attempting to apply safety knowledge to their construction site and learning work safety from the health and safety management plan. To acquire knowledge, workers must not wait for it to be imparted by others but also identify their own sources of knowledge through observing, imitating, and reading. Applying the knowledge acquired on-site that enables other workers to learn and follow, and teaching workers of accidents that may occur can achieve mutual vigilance and influence, thereby reducing accidents.

Successful knowledge transfer requires people to impart and receive knowledge. Knowledge possessed by each individual is not necessarily similar. Therefore, a knowledge consensus can be achieved through mutual exchange, presentation, and discussion, which means that everyone is a seed of knowledge, and effective knowledge dissemination can turn it into a normal cycle. Safety in on-site construction environments can only be achieved when knowledge is transferred continuously in a cycle.

8. Conclusions

This study constructed a safety knowledge transfer model and explored the relationship between safety knowledge transfer and safe working environments. The safety knowledge transfer model included knowledge transfer environment (safety training, safety leadership, safety knowledge communication, and self-learning), safety knowledge application and inspiration, and safety behaviors (safety knowledge acceptance, safety concern, and safety awareness) in response to the processes of acquisition, communication, application, acceptance, and assimilation of transferred knowledge. After the structural model was established using SEM, hypothesis testing was performed. With the significance level set at 95%, all hypotheses in Table 1 reached significance, indicating that the working environment at a construction site is safer when the degree of knowledge transfer is higher. The knowledge transfer environment affects the safety behaviors of on-site workers as well as their safety knowledge application and inspiration. Furthermore, safety behaviors are affected by the safety knowledge application and inspiration of other personnel. The on-site working environment is directly and indirectly affected by safety behaviors and safety knowledge application and inspiration, respectively, improving the safety of the site.

The path values of each path could be determined after structural model analysis was performed. The total effects of each path value were knowledge transfer environment to safety behavior (0.972), knowledge transfer environment to safety knowledge application and inspiration (0.874), safety knowledge application and inspiration to safety behavior (0.440), safety knowledge application and inspiration to safe working environment (0.350), and safety behavior to safe working environment (0.795). Safety behaviors may change because of the knowledge transfer of on-site workers, such as on-site supervisors encouraging workers to express their ideas and suggestions on safety issues, immediately correcting workers' unsafe behaviors, and raising safety-related issues in meetings, as well as on-site workers applying safety knowledge to the construction site, which will result in safety knowledge transfer and changes in safety behaviors of the workers. Once safety behaviors are improved, on-site workers will pay attention to matters such as correctly using tools and protective equipment, stopping work and eliminating unsafe behaviors when they notice them, and stopping work when they experience physical discomfort. The improvement of safety behaviors can enhance safe on-site working environments.

In summary, safety-related knowledge transfer must be practiced to improve the safety of on-site work environments. Increasing the safety knowledge of on-site workers through meetings, on-site teaching, and safety-related issues can improve their appreciation of the importance of protective equipment. Moreover, on-site workers will pay more attention to the safety of themselves and others. This study's research model showed that safety behavior and safety knowledge application and inspiration mediated the effect of the knowledge transfer environment on the safe working environment, indicating the importance of on-site knowledge transfer. A safe working environment can be formed if on-site management personnel can transfer safety knowledge to on-site workers and cultivate it, provide guidance to them, ask for relevant safety opinions, always pay attention to and correct workers' safety problems, and correctly apply safety knowledge. Therefore, on-site management personnel must emphasize knowledge transfer, whether through education and training, on-site communication and guidance, and listening to the opinions of on-site workers. In addition, this study revealed that the linkage between the safety awareness and safety behavior of on-site workers was weak; therefore, strengthening it should be a focus in safety knowledge transfer.

According to the discussion in this study, several recommendations and future research directions were proposed. This study only used four subconstructs (i.e., safety training, safety leadership, safety communication, and self-learning) to represent a knowledge transfer environment, and assumed that each subconstruct was independent. Subsequent research can continue to explore the effectiveness of on-site workers' knowledge transfer and the correlation between each subconstruct, as well as develop a knowledge transfer effectiveness scale for on-site workers in the construction industry to understand the pros and cons of knowledge transfer, on the basis of which the knowledge transfer model can be revised. The participants of this study were on-site personnel in the construction industry. Future research can further investigate differences between the various levels of personnel, such as on-site construction workers, middle management personnel (on-site construction management personnel), and senior management personnel (supervisors). Because the cultures of different countries differ, further discussions and comparisons between the construction industries of different countries are warranted.

Author Contributions: Conceptualization, Y.-H.H. and T.-R.Y.; data curation, Y.-H.H.; formal analysis, T.-R.Y.; investigation, T.-R.Y.; methodology, Y.-H.H. and T.-R.Y.; project administration, Y.-H.H.; writing—original draft, Y.-H.H.; writing—review and editing, Y.-H.H.

Funding: This research received no external funding.

Acknowledgments: The authors would like to express their sincere gratitude to the anonymous reviewers who significantly enhanced the contents of the study with their insightful comments.

Conflicts of Interest: The authors declare no conflict of interest.

References

1. Occupational Safety and Health Administration. *2018 Labor Inspection Annual Report*; Occupational Safety and Health Administration: Ministry of Labor: Taipei, Taiwan, 2019. (In Chinese)
2. Cheng, C.W.; Leu, S.S.; Lin, C.C.; Fan, C. Characteristic analysis of occupational accidents at small construction enterprises. *Saf. Sci.* **2010**, *48*, 698–707. [CrossRef]

3. Chi, C.F.; Yang, H.S.; Chen, W.S.; Liu, K.C.; Chang, T.C.; Ting, H.I. In-depth analysis and prevention of fatal falls in construction industry. *J. Occup. Saf. Health* **2008**, *16*, 383–400. (In Chinese)
4. Musonda, I.; Smallwood, J. Health and safety (H&S) awareness and implementation in Botswana's construction industry. *J. Eng. Des. Technol.* **2008**, *6*, 81–90.
5. Chen, W.T.; Lu, C.S.; Huang, Y.H. Investigating the safety cognition of Taiwan's construction personnel. *J. Mar. Sci. Technol.* **2011**, *19*, 398–408.
6. Leu, S.S.; Chang, C.M. Bayesian-network-based safety risk assessment for steel construction projects. *Accid. Anal. Prev.* **2013**, *54*, 122–133. [CrossRef] [PubMed]
7. Nadhim, E.A.; Hon, C.K.; Xia, B.; Stewart, I.; Fang, D. Investigating the relationships between safety climate and safety performance indicators in retrofitting works. *Constr. Econ. Build.* **2018**, *18*, 110–129. [CrossRef]
8. Teo, E.A.L.; Ling, F.Y.Y.; Chong, A.F.W. Framework for project managers to manage construction safety. *Int. J. Proj. Manag.* **2005**, *23*, 329–341. [CrossRef]
9. Heinrich, H. *Industrial Accident Prevention*, 4th ed.; Mc Graw-Hill Book Company: New York, NY, USA, 1959.
10. Cheng, C.W.; Leu, S.S.; Cheng, Y.M.; Wu, T.C.; Lin, C.C. Applying data mining techniques to explore factors contributing to occupational injuries in Taiwan's construction industry. *Accid. Anal. Prev.* **2012**, *48*, 214–222. [CrossRef]
11. Geller, E.S. Ten principles for achieving a total safety culture. *Prof. Saf.* **1994**, *9*, 18–24.
12. Nonaka, I. A dynamic theory of organizational knowledge creation. *Organ. Sci.* **1994**, *5*, 14–37. [CrossRef]
13. Gilbert, M.; Cordey-Hayes, M. Understanding the process of knowledge transfer to achieve successful technological innovation. *Technovation* **1996**, *16*, 301–312. [CrossRef]
14. Spiegler, I. Knowledge management: A new idea or a recycled concept? *Commun. Assoc. Inf. Syst.* **2000**, *3*, 14. [CrossRef]
15. Nancy, M.D. *Common Knowledge: How Companies Thrive by Sharing What They Know*; Harvard Business School Press: Boston, MA, USA, 2000.
16. Davenport, T.H.; Prusak, L. *Working Knowledge: How Organizations Manage What They Know*; Harvard Business School Press: Boston, MA, USA, 1998.
17. Polanyi, M.; Sen, A. *The Tacit Dimension*; University of Chicago Press: Chicago, IL, USA, 2009.
18. Nonaka, I.; Takeuchi, H. *The Knowledge-Creating Company*; Oxford University Press: Oxford, UK, 1955.
19. Szulanski, G. Exploring internal stickiness: Impediments to the transfer of best practice within the firm. *Strateg. Manag. J.* **1996**, *17*, 27–43. [CrossRef]
20. Verkasalo, M.; Lappalainen, P. A method of measuring the efficiency of the knowledge utilization process. *IEEE Trans. Eng. Manag.* **1998**, *45*, 414–423. [CrossRef]
21. Wiig, K.M. Integrating intellectual capital and knowledge management. *Long Range Plan.* **1997**, *30*, 399–405. [CrossRef]
22. Zhang, S.; Boukamp, F.; Teizer, J. Ontology-based semantic modeling of construction safety knowledge: Towards automated safety planning for job hazard analysis (JHA). *Autom. Constr.* **2015**, *52*, 29–41. [CrossRef]
23. Hadikusumo, B.H.W.; Rowlinson, S. Capturing safety knowledge using design-for-safety-process tool. *J. Constr. Eng. Manag.* **2004**, *130*, 281–289. [CrossRef]
24. Neal, A.; Griffin, M.A.; Hart, P.M. The impact of organizational climate on safety climate and individual behavior. *Saf. Sci.* **2000**, *34*, 99–109. [CrossRef]
25. Li, R.Y.M.; Chau, K.W.; Lu, W.; Ho, D.C.W.; Shoaib, M.; Meng, L. Construction hazard awareness and construction safety knowledge sharing epistemology. In Proceedings of the International Conference on Smart Infrastructure and Construction 2019 (ICSIC): Driving Data-Informed Decision-Making, Churchill College, Cambridge, UK, 8–10 July 2019; pp. 283–290.
26. Taher, A.; Vahdatikhaki, F.; Hammad, A. Integrating earthwork ontology and safety regulations to enhance operations safety. In Proceedings of the 36th International Symposium on Automation and Robotics in Construction (ISARC), Banff, AB, Canada, 21–24 May 2019.
27. Le, Q.T.; Pedro, A.; Park, C.S. A social virtual reality based construction safety education system for experiential learning. *J. Intell. Robot. Syst.* **2015**, *79*, 487–506. [CrossRef]
28. Jeelani, I.; Albert, A.; Gambatese, J.A. Why do construction hazards remain unrecognized at the work interface? *J. Constr. Eng. Manag.* **2017**, *143*, 4016128. [CrossRef]
29. Shin, M.; Lee, H.-S.; Park, M.; Moon, M.; Han, S. A system dynamics approach for modeling construction workers' safety attitudes and behaviors. *Accid. Anal. Prev.* **2014**, *68*, 95–105. [CrossRef] [PubMed]

30. Blair, E. Culture and leadership: Seven key points for improved safety performance. *Prof. Saf.* **2003**, *48*, 18–22.
31. Hofmann, D.A.; Morgeson, P. Safety-related behavior as a social exchange: The role of perceived organizational support and leader-member exchange. *J. Appl. Psychol.* **1999**, *84*, 286–296. [CrossRef]
32. Tsaur, C.C.; Lin, S.H.; Kuo, C.C.; Gau, C.Y. The safety leadership and communication of supervisors and the safety behavior of workers: The moderating effect of safety management. *J. Occup. Saf. Health* **2012**, *20*, 1–33. (In Chinese)
33. O'Dea, A.; Flin, R. Site managers and safety leadership in the offshore oil and gas industry. *Saf. Sci.* **2001**, *37*, 39–57. [CrossRef]
34. Simon, M. *The Creative Marketer*; Butterworth-Heinemann Ltd.: London, UK, 1991.
35. Ojanen, K.; Seppälä, A.; Aaltonen, M. Measurement methodology for the effects of accident prevention programs. *Scand. J. Work Environ. Health* **1988**, *14*, 95–96.
36. Holt, A.S.J. *Principles of Construction Safety*; Blackwell Science: London, UK, 2001.
37. Smith-Crowe, K.; Burke, M.J.; Landis, R.S. Organizational climate as a moderator of safety knowledge safety performance relationships. *J. Organ. Behav.* **2003**, *24*, 861–876. [CrossRef]
38. Fowler, F.J., Jr. *Survey Research Methods*; Sage: Thousand Oaks, CA, USA, 2013.
39. Cureton, E.E. The upper and lower twenty-seven percent rule. *Psychometrika* **1957**, *22*, 293–296. [CrossRef]
40. Hair, J.F.; Black, B.; Brain, B.; Anderson, R.E.; Tatham, R.L. *Multivariate Data Analysis*, 6th ed.; Prentice Hall: Upper Saddle River, NJ, USA, 2006.
41. Bagozzi, R.P.; Yi, Y. On the evaluation for structural equation models. *J. Acad. Mark. Sci.* **1998**, *16*, 74–94. [CrossRef]
42. Bentler, P.M. Comparative fit indexes in structural models. *Psychol. Bull.* **1990**, *107*, 238–246. [CrossRef]
43. Bentler, P.M. On the fit of models to covariances and methodology to the Bulletin. *Psychol. Bull.* **1992**, *112*, 400–404. [CrossRef] [PubMed]
44. Doll, W.J.; Xia, W.; Torkzadeh, G. A confirmatory factor analysis of the end-user computing satisfaction instrument. *MIS Q.* **1994**, *18*, 453–461. [CrossRef]
45. Chin, W.W.; Todd, P.A. On the use, usefulness, and ease of use of structural equation modeling in MIS research: A note of caution. *MIS Q.* **1995**, *19*, 237–246. [CrossRef]

sustainability

MDPI

Article

Causes and Mitigation Strategies of Delay in Power Construction Projects: Gaps between Owners and Contractors in Successful and Unsuccessful Projects

Edwin Thomas Banobi and Wooyong Jung *

Department of Nuclear Power Plant Engineering, KEPCO International Nuclear Graduate School, Ulsan 45014, Korea; banjun85@gmail.com

* Correspondence: wooyong@kings.ac.kr; Tel.: +82-52-712-7120; Fax: +82-52-712-7261

Received: 30 September 2019; Accepted: 24 October 2019; Published: 27 October 2019

Abstract: Few studies have verified the different causes of project delays between the owner and contractor perspectives. This article's goal is to find what the causes of delay are and how to mitigate this delay depending on project performance. Thus, this study investigated 82 owner-side experts and 106 contractor-side experts in Tanzanian power construction projects. In successful projects (less than 10% time delay), the owners and contractors weighted similar causes such as vandalism and permits from authorities. They suggested similar mitigation strategies such as close project supervision, capacity building training, and proper logistics management. While in unsuccessful projects (more than 10% time delay), they exhibited many different responses. In particular, contractors weighted the causes incurred by changes in scope, owner's poor supervision, delays in approval, failure in planning and designing risk more than contractors. Owners weighted the mitigation strategies such as top management support and timely procurement more than contractors. These findings will help project managers to understand owners' and contactors' different concerns and develop better solutions. This study mainly contributes to improving delay management in power construction projects in developing countries.

Keywords: time management; delay management; mitigation strategy; owner perspective; contractor perspective; power construction project; Tanzania

---

1. Introduction

A multitude of construction projects in many countries still suffer from project delays that lead to losses and claims on the part of both owners and contractors [1,2]. In the case of Tanzania, 32 among 39 power projects experienced on average six month delays compared to their planned completion date [3]. Therefore, many studies have investigated delay causes to achieve better construction project management [4–6]. Although these studies contribute to improved delay management, several aspects have not been studied well to date. First, previous studies usually analyzed the causes only from the owner's or contractor's perspective [4]. Even though some research compared the rank of causes from the owner and contractor, they did not verify the statistical significance of the difference between owner and contractor [7–9]. Besides, even if another research analyzed the gaps well between owners and contractors, their subjects focused on the risk, contract and conflict, which are relevant but some different issues with project delay [10–12]. Second, they did not consider the performance-oriented cause of delay. Depending on the project progress performance, the cause and mitigation strategies can be varied.

Project owners usually are responsible for the basic planning, funding, risk allocation, award criteria, payment rules, procurement of major items, licensing and design approval [10,13]. Contractors play a role in design details, detail procurement, risk management, the process and scheduling, productivity, labor, equipment, environment, and so on [10,12]. The failure to meet any of these responsibilities is an important reason that projects are delayed. However, owners usually underestimate their responsibilities and blame the contractors. In contrast, contractors insist that the delays frequently are attributable to the owner's mismanagement. Thus, the same delays can be understood differently depending on the owner and contractor's perspectives, and these differences can be the root causes of delays. Moreover, this problem of shifting blame is more severe when the project's performance does not meet the original plan [14]. Thus, depending on the project progress performance, the delay causes and mitigation strategies can be varied.

Therefore, this study compares owners and contractors' different evaluations of the same causes of delay according to the project's performance. Then, it suggests more comprehensive mitigation strategies to reduce construction project delays that consider both owner and contractor perspectives. To do so, this study follows the research process, as shown in Figure 1. First, the study reviews previous research that deals with the causes of construction project delays and mitigation strategies. Second, this study derives the 35 factors of delay causes in 5 groups and 15 factors for mitigation strategies. The authors designed a questionnaire based on these factors. Third, this research surveyed and collected the valid responses from 82 owner-related and 106 contractor-related experts in Tanzania. Fourth, this study analyzed the delay cause rank and gaps between owner and contractor in the successful and unsuccessful project respectively. Last, this study analyzed the mitigation strategy rank and gaps between owner and contractor in the successful and unsuccessful project respectively.

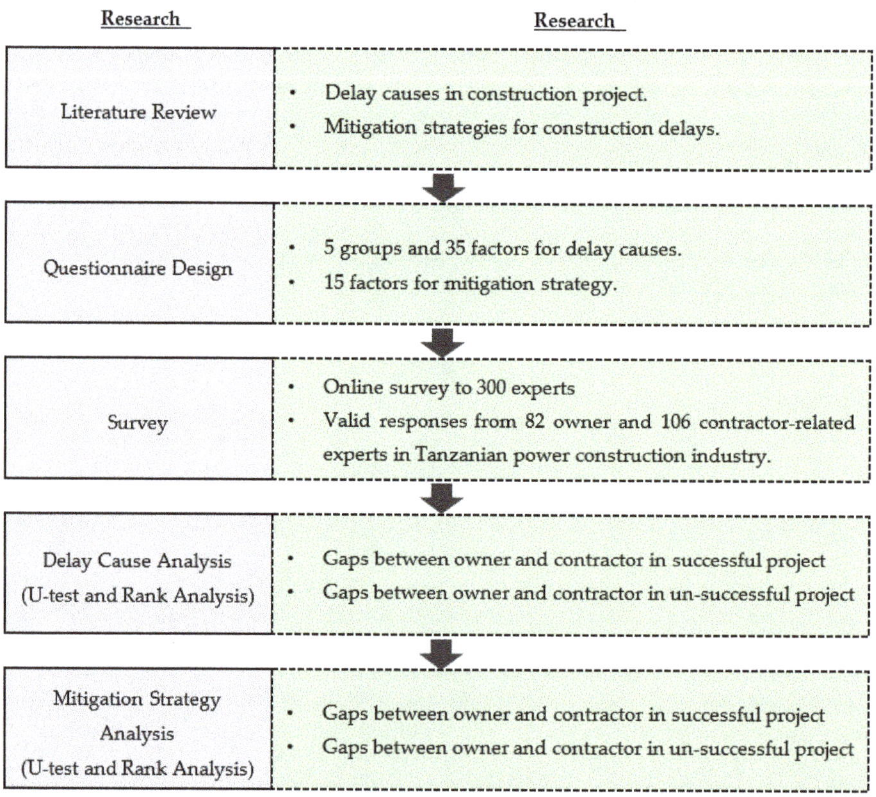

Figure 1. Research Procedure.

2. Literature Review

2.1. Causes of Construction Project Delays

Previous studies have largely identified causes of construction project delays based on literature reviews and expert interviews, and the delay factors were classified into various groups depending upon each paper's authors, as follows. Chan and Kumaraswamy [15] classified causes of delay into eight groups; project-, client-, design team-, and contractor-related, material, labor, equipment, and external factors. Odeh and Battaineh [16] studied causes of schedule delays in construction projects with traditional types of contracts and identified eight major groups of causes: client-, contractor-, consultant-, material-, labor and equipment-, contract-, and external-related factors. Aziz and Abdel-Hakam [17] used various categories to classify delay factors; project-, owner-, and contractor-related, financing, contract, design, site, labor, material, equipment, rules and regulations, scheduling and controlling, external, and contractual relationships. Alsuliman [18] investigated causes of delay according to stages in a public construction project, including before, during, and after the award, as well as general factors. Risk factors, which cause schedule delays in a construction project, can be categorized into various standards depending on the purpose of establishing the categorization [19]. This study reviews the literature related to causes of delay in power construction projects from five perspectives related to the: owner, contractor, design, infrastructure and social, and external factors.

2.1.1. Owner-Related Causes of Delay

The owner's insufficient project management capability affects delays significantly. Majid and McCaffer [20] suggested that inadequate fund allocation, insufficient communication among participants, and damaged materials and equipment are owner-related causes of delay, while Ogunlana et al. [21] suggested change orders and slow decision making. Long et al. [6] conducted a case study in Vietnam to identify common and general problems in large construction projects in developing countries. They suggested ten owner-related problems that cause delays, including lack of strategic management, construction requirements, improper project feasibility study, lack of a clear bidding process, excessive change orders, unclear responsibility, lack of capable representatives, owner's financial difficulties, poor contract management, and slow decision making. In addition, they specified delays attributable to participants' communication and coordination into seven factors. According to Frimpong et al.'s [22] study, participants, material procurement, and frequent breakdowns in the construction plant and equipment contributed to projects' schedule delays. Koushki et al. [23] studied the causes of time delays associated with the construction of private residential projects, particularly from the owner's and developer's perspectives. Their analysis of 450 questionnaires suggested three main causes of delays from the owner's perspective: change orders, financial constraints, and owner's lack of experience. Aziz and Abdel-Hakam [17] also identified a total of twenty highly frequent causes of delay in their literature review. Among them, owner-related causes were the owner's slow decisions, shop drawings and samples' slow preparation and approval, the owner's change orders during construction, owner's financial problems for the project, and owner's delay in contractors' progress payment. As owner-related causes of delay, Khatib et al. [24] suggested their financial problems and difficulties, change orders, delays and shortages of materials, poor site management and supervision, poor communication and coordination among construction parties, lowest bid awards, slow decisions, the contract type, delays in performing inspections and tests, and lack of clarity of the project's scope. Alsuliman [18] used a questionnaire to investigate the causes of delay according to stages of a public construction project. The results showed that the most significant group of causes of the delay was the factors associated with awarding tenders. In particular, the bid and award process, financial problems, approval delay, and owner's poor management capability were ranked among the top 20 causes of delays. Marques and Berg [10], showed that the lowest tender award system frequently fosters contractors to assume optimistic design and price estimation for winning awards.

This opportunism induces the changes in scope for increasing project contract price. Marques and Berg [11], also explained that the budget reduction by government frequently invokes the project delay.

Based on the factors reviewed above, this study selected seven owner-related causes of project delays: change in scope; owner's poor supervision; poor communication and coordination; approval delay; delay in procuring items, lowest bid tender award; owner's inadequate fund or budget allocation, and materials/equipment damaged during construction.

2.1.2. Contractor-Related Causes of Delay

Many delay factors are related to the contractor, who largely is responsible to execute a construction project and manage its schedule. Ogunlana et al. [21] identified material management problems, organizational deficiencies, planning, scheduling, and equipment allocation problems, financial difficulties, and inadequate site inspection as contractor-related causes of delay. Frimpong et al. [22] used a survey in their study of causes of delay in groundwater construction projects in developing countries. The survey showed that monthly payment difficulties, material procurement, and contractor's financial difficulties were ranked among the top 5 causes of delay. In addition, deficiencies in preparing cost estimates were ranked 10th in 25 delay factors. Long et al. [6] suggested 17 contractor-related causes of delay: improper planning and scheduling; inadequate experience; insufficient modern equipment; inaccurate time estimates; inaccurate cost estimates; poor site management; improper monitoring and control; poor labor and management relations; inappropriate construction methods; contractor's financial difficulties; incompetent project teams; poor contract management; severe overtime; material waste; lack of necessary skills; inadequate site inspection, and lack of competent subcontractors or suppliers. Koushki [23] argued that ensuring the delivery of materials and the contractor's capability are major factors that contribute to delays in construction projects in Kuwait. Aziz and Abdel-Hakam's [17] study also identified contractor-related factors, such as their poor site management and supervision, construction methods, ineffective project planning, and scheduling, financing during construction, and inadequate experience that caused errors as highly frequent causes of delay. Khatib et al. [24] suggested improper planning and scheduling, subcontractor's incompetence, contractor's lack of experience, discrepancies between drawings and specifications, construction mistakes and defective work, inaccurate estimates, inadequate tools and equipment, and price escalation.

Based upon this literature review, this study selected poor quality construction materials and equipment, poor cost management, poor project planning and scheduling, contractor's inadequate site supervision, additional work attributable to construction errors, misrepresentation of information before bid, poor cost estimation, late payments to suppliers or for contractor's work, late procurement orders for material and equipment, and changes in types and specifications, as contractor-related causes of delay.

2.1.3. Design-Related Causes of Delay

Poor design management causes delays in construction projects' schedules. Ogunlana et al. [21] suggested incomplete drawings and designers' slow responses are two major causes of delays related to design. In particular, nine of twelve cases they studied suffered from incomplete drawings. Razek et al. [5] also identified causes of delay in a construction project based on a questionnaire. In their study, these causes were the owner or his agent's design changes during construction, lack of a database to estimate activities' duration and resources, and designers' errors or incomplete designs. Khatib et al. [24] identified the following problems: design changes and modifications, errors, delays, shop drawings' slow preparation and approval, and erroneous sources of information. Furthermore, Alsuliman's [18] study indicated that variations in orders that occur during the project period and failure to determine quantities, specifications, and drawings accurately were the most frequent causes of delays related to the design of public construction projects.

Based on the literature review, five design-related delay factors were used in this study: design changes during construction; inappropriate data collection; errors and delays in providing design documents; a failure in planning and designing risk, and poor resource estimation and allocation.

2.1.4. Infrastructure and Socially Related Causes of Delay

The surrounding infrastructure and social environment also influence a project's schedule. Majid and McCaffer [20] suggested that labor-related causes of delay include workers' low morale/motivation and strikes, and poor workmanship. Ogunlana et al. [21] argued that problems with neighbors, government agencies' slow issuance of permits, resources' late delivery, shortage of site workers, and shortage of technical personnel lead to schedule delays in construction projects. Long et al. [6] found that slow government permits and unsatisfactory site compensation were the 4th and 10th most frequent problems that cause delays among a total of 62 delay factors in large construction projects. In particular, unsatisfactory site compensation was the 7th most influential problem that caused delays. Aziz and Abdel-Hakam [17] suggested other resource-related delay factors, including shortages in construction materials, equipment, and labor, slow delivery of materials, and low work productivity. They also pointed out that obtaining permits from municipalities is one of the main causes of delay in road construction projects in Egypt. Khatib et al. [24] suggested that delays are attributable to shortages of skilled workers and equipment, lack of qualified and experienced personnel, poor labor productivity, difficulties obtaining work permits, poor site conditions, and frequent interruptions from the public.

We extracted the following nine delay factors that affect schedule: workers' absenteeism, low motivation and morale, and strikes; poor working conditions; unskilled or inexperienced labor; late delivery of materials and equipment; delays in obtaining permits from authorities; conflicts with neighbors, and vandalism.

2.1.5. Externally Related Causes of Delay

Some literature has addressed uncontrollable external factors that delay construction projects, such as the host country's political climate and site's geological status. Long et al. [6] identified unforeseen ground conditions and inclement weather as environmental causes of delay. Frimpong et al. [22] confirmed such uncontrollable delay factors as ground problems and inclement weather, as well as unexpected geological conditions. Khatib et al. [24] listed unforeseen ground and weather conditions, and political insecurity and instability. In particular, multiple studies have identified weather conditions as one of the major delay factors. In their literature review, Aziz and Abdel-Hakam [17] found that 21 academic papers identified weather conditions as the most frequent delay factor. Koushki et al. [23] also analyzed inclement weather as the fifth important factor that owners in Kuwait reported caused delays. In Frimpong et al.'s [22] study, both owners and contractors ranked bad weather in the top 10 among 25 factors.

According to the literature reviewed, the authors identified three externally related delay factors that cause schedule delays, including force Majeure related to natural disasters, unexpected geological conditions, and political instability or control.

2.2. *Strategies to Mitigate Project Delays*

Proactive efforts to mitigate risks help achieve a project's objectives [25]. Previous studies have suggested various strategies to mitigate delays' adverse effects on project performance. Wang et al. [26] suggested optimal mitigation strategies that respond to various risk events and developed a risk management framework that suggests a proper mitigation strategy in accordance with country, market, and project risks, respectively. Kim et al. [27] also developed a risk assessment and mitigation model to support decision-making for investment in a steel-plant project. They selected six representative risk factors related to the target project and measured various mitigation strategies' effectiveness. Asadi, et al. [28], insisted that project risk management is a critical method to improve the cost, schedule and quality management. They introduced the risk management guideline and suggested the risk management tool using fuzzy model. The strategies that mitigate project delays include three main critical factors: project mission; top management's support, and project scheduling, all of which affect project performance during different phases of implementation. Tripathi and Jha [13], derived the six organizational success factors for project performance. Among the experience and performance, top management competence, project factor, supply chain and leadership, availability of resources and effective cost control measures, they ranked the top management competency highest. Guo, et al. [29], validated the project performance difference between non-supervised project and supervised project by engineer using evolutionary game theory. They recommended that the compulsory supervision is effective way to control project performance. Su et al. [30], emphasized on the accurate the time estimation skill for delay management. In particular, they suggested the solution of float ownership to prevent the delay and conflict between owner and contractor.

Critical factors are key issues or areas of activities in which favorable results are necessary for a project manager to achieve his/her target [31]. Project mission and schedule, top management's support, client consultations and acceptance, technical tasks, monitoring as well as feedback, communication, and troubleshooting are some the factors that influence project success [32]. Nguyen, et al. [33], recommended five possible factors that may be used to minimize project delays, these includes experienced project manager, satisfactory funds throughout project life cycle, experienced project team, project stakeholder's commitment, and resources availability. Moreover, Nguyen's another journal [34] emphasized the relationship between client satisfaction and the team behavior-related strategies such as project planning and organizing, coordination, contractor assurance and empowerment. Aibinu and Jagboro [35], suggested that the speeding up of site activities, and incidental stipend could be applied to reduce project delays. Odeh and Battaineh [16], proposed the following approaches creating and classifying human resources through appropriate training; consideration of capability and experience of contractor more than price during contract award, and adoption of design-build and construction management contracts. Li, et al. [36] reviewed the publications from 2005 to 2018 on dealing critical success factors for project performance. Then, they suggested the most frequently cited success factors such as communication and cooperation, effective project planning and controlling, owner's involvement and commitment and clear goals and objectives in order.

This study adopted the mitigation factors Pinto and Kharbanda [32] proposed, including the project's adequate financing and arrangement, previous work experience on similar projects, donors' influence, close project supervision, suitable time estimation skills, availability and quality of the workforce, and availability of materials and equipment. Furthermore, timely payments of completion certificates, good presentation of information during tendering, finishing the design on time, workers' motivation and morale, capacity building training, good logistic management (Transportation), top management's support, and site location were identified as strategies to mitigate project delays.

3. Methodology

This study derived the delay causes and mitigation strategy factors from the literature review, investigated the data from survey and further conducted the rank analysis and U-test. This study has several methodological strengths: (1) relatively large number of samples; (2) quantitative comparison analyses between owner and contractor, and between successful project and unsuccessful project. However, this study also has several methodological weaknesses: (1) lack of in-depth qualitative analyses; (2) limited investigation conducted in one country.

3.1. Questionnaire Design

This questionnaire consists of five sections: (1) respondent information; (2) project information; (3) causes of project delay; (4) consequences of project delay; (5) mitigation strategy for delay management. This study does not use the "consequences of project delay" section.

3.1.1. Delay Factors

This study derived the causes of delay in the literature review (Section 2.1) and grouped them in five categories as shown in Table 1. These delay factors were used in the questionnaire survey. The questionnaire asked the respondents to evaluate the importance of causes of delay based on their experience with projects. The importance levels were measured using five-point Likert scales: one point (less than 1-month delay); two points (approximately 1-month delay); three points (approximately 2-months delay); four points (approximately 3-month delay); five points (more than 3-months delay).

Table 1. Major Practices Causing Project Delays.

Group	Number	Delay Causes	References
Owner-related	O1	Change in scope	[6,10,11,17,18,20–24]
	O2	Owner's poor supervision	
	O3	Poor communication and coordination	
	O4	Delays in approval	
	O5	Delays in procuring materials	
	O6	Lowest bid tender award	
	O7	Owner's inadequate funds or budget allocation	
	O8	Damaging materials/equipment during construction	
Contractor-related	C1	Poor quality construction materials and equipment	[6,17,21–24]
	C2	Poor cost management	
	C3	Poor project planning and scheduling	
	C4	Contractor's poor site supervision	
	C5	Additional work attributable to mistakes	
	C6	Misrepresentation of information before bid	
	C7	Poor cost estimation	
	C8	Contractor's late payment to suppliers or works	
	C9	Late procurement order of material and equipment	
	C10	Change in types and specifications	
Design-related	D1	Design changes during construction	[5,18,21,24]
	D2	Inappropriate data collection	
	D3	Mistakes and delays in design documents	
	D4	Failure in planning and design risk	
	D5	Poor resource estimation and allocation	
Infrastructure and Socially related	I1	Worker's absenteeism	[6,17,20,21,24]
	I2	Workers' low motivation and morale	
	I3	Worker's strikes	
	I4	Poor working conditions	
	I5	Unskilled or inexperienced labour	
	I6	Late delivery of material and equipment	
	I7	Delay in obtaining permits from authorities	
	I8	Neighbor's conflicts	
	I9	Vandalism	
Externally related	E1	Force Majeure attributable to natural disaster	[6,17,22–24]
	E2	Unexpected geological condition	
	E3	Political instability or controls	

3.1.2. Mitigation Strategy Factors

The study summarizes the delay mitigation strategies found in the literature review (Section 2.2) that contribute to project success as shown in Table 2. The mitigation strategy factors were used in the questionnaire survey, in which the respondents weighted the importance using five-point Likert scales. The importance levels were measured using five-point Likert scales: one point (very low, approximately 0–20% contribution); two points (low, approximately 20–40% contribution); three points (medium, approximately 40–60% contribution); four points (high, approximately 60–80%); five points (very high, approximately 80–100%).

Table 2. Major Mitigation Strategies for Project Success.

Number	Mitigation Strategy Factors	References
M1	Proper planning of project financial arrangements	
M2	Use of skilled labors with and experience on similar projects	
M3	Consideration of Donor's Influence	
M4	Close project supervision	
M5	Use of suitable time estimation skills	
M6	Conducting capacity building training	
M7	Timely procurement and supply of materials and equipment	
M8	Timely payments of completion certificates	[13,16,25–28,30–33,35,36]
M9	Proper presentation of information during tendering	
M10	Finishing design on time	
M11	Timely site visits	
M12	Motivating workers to raise morale	
M13	Risk identification and assessment	
M14	Proper logistics management	
M15	Top management's support	

3.2. Survey

The authors circulated 300 questionnaires electronically to different professional project owners and contractors throughout Tanzania. Table 3 describes the respondents' profiles including the experience of the respondents. All respondents were asked to evaluate the causes of delay as well as mitigation strategies based on their project experiences. One-hundred ninety-nine responses were collected and 188 were confirmed valid. Eighty-two responses were collected from the project owner group and 106 from the project contractor group, as shown in Table 3. Figure 2 indicates the project profile to which the respondents referred to answer the questions. The distribution of projects planned and schedule performance varied and is relatively uniform. Project types are skewed slightly toward distribution rather than power generation and transmission projects.

Table 3. Respondents Profile.

	Owner		Contractor		Total	
	Number (Respondents)	Experience (Years)	Number (Respondents)	Experience (Years)	Number (Respondents)	Experience (Years)
Project Managers	37	8	27	2.5	64	13
Engineers	32	13	48	13	80	8
Technicians	9	2.5	20	8	29	2.5
Consultants	4	15	11	15	15	15
Total	82		106		188	

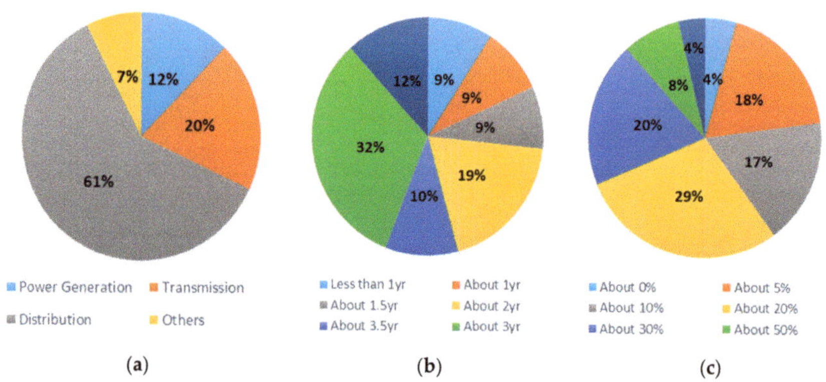

Figure 2. Project Profile: (**a**) Project Types; (**b**) Planned Project Duration; (**c**) Schedule Performance (Delay Schedule/Planned Schedule).

3.3. Analysis Method

To compare the owners' and contractors' different perspectives on the causes of delays and mitigation strategies, this study tested the data sample's normality first with the Shapiro–Wilk test; the result indicated that the sample was not distributed normally at $p < 0.05$ (average p-value = 0.015). Therefore, the study uses the Mann–Whitney U-test rather than the t-test with SPSS software. The Mann–Whitney U-test is a non-parametric statistical analysis that verifies the difference between two sample groups in cases of non-normal distributions [37]. If the U-test meets the significance level, the two groups compared differ significantly.

Furthermore, this study analyzes the differences from the successful project group's perspective and the unsuccessful group's perspective because the experts' responses and suggestions can vary depending on their experience with project performance. Thus, the sample data are divided into two groups. Successful projects are those that had less than a 10% increase in the originally planned project duration, while unsuccessful projects had more than a 10% increase in project duration.

4. Results

4.1. Causes of Project Delays

4.1.1. Causes of Delays in Successful Projects

Table 4 shows the causes of delay in successful power projects. In the group overall, the owner's inadequate funding or budget allocation (O7_1st Rank), vandalism (I9_2nd Rank), the contractor's late payment to suppliers or workers (C8_3rd Rank), late delivery of materials and equipment (I6_4th Rank), and delays in obtaining permits from authorities (I7_5th Rank) rank the highest. In the owner group, the owner's inadequate funds or budget allocation (O7_1st Rank), late delivery of materials and equipment (I6_2nd Rank), Unskilled or inexperienced labour (I5_3rd Rank), vandalism (I9_4th Rank), and changes in scope (O1_5th Rank) rank as the top five. In the contractor group, the contractor's late payment to suppliers or workers (C8_1st Rank), lowest bid tender award (O6_2nd Rank), delays in procuring materials (O5_3rd Rank), vandalism (I9_4th Rank), and poor communication and coordination (O3_5th Rank) rank as the top five. The owner and contractor groups ranked the top five delay factors somewhat differently.

Table 4. Delay Causes in Successful Projects.

Group	Number	Total		Owner		Contractor		U-test
		Mean	Rank	Mean	Rank	Mean	Rank	Sig.
Owner-related	O1	2.53	6	2.41	5	2.69	11	0.421
	O2	2.19	27	2.05	27	2.39	24	0.169
	O3	2.45	13	2.18	19	2.82	5	0.042
	O4	2.38	17	2.10	23	2.73	9	0.118
	O5	2.49	8	2.23	16	2.86	3	0.068
	O6	2.48	9	2.08	24	3.04	2	0.009
	O7	2.74	1	2.90	1	2.55	19	0.265
	O8	2.17	28	2.16	20	2.19	33	0.930
Contractor-related	C1	2.27	23	2.24	15	2.31	29	0.684
	C2	2.45	14	2.38	8	2.54	20	0.515
	C3	2.27	24	2.21	17	2.36	26	0.452
	C4	2.51	7	2.31	10	2.79	6	0.145
	C5	2.35	20	2.14	21	2.62	14	0.065
	C6	2.32	21	2.26	14	2.39	25	0.484
	C7	2.46	11	2.37	9	2.59	17	0.754
	C8	2.64	3	2.29	11	3.14	1	0.023
	C9	2.43	15	2.27	12	2.68	12	0.278
	C10	2.46	12	2.40	6	2.56	18	0.864

Table 4. *Cont.*

Group	Number	Total		Owner		Contractor		U-test
		Mean	Rank	Mean	Rank	Mean	Rank	Sig.
Design-related	D1	2.37	18	2.27	13	2.50	21	0.593
	D2	2.17	29	2.07	25	2.30	30	0.683
	D3	2.39	16	2.13	22	2.75	8	0.148
	D4	2.26	25	2.00	32	2.64	13	0.087
	D5	2.32	22	2.05	28	2.71	10	0.056
Infrastructure and Socially related	I1	1.87	34	1.74	35	2.03	34	0.466
	I2	2.13	32	2.05	29	2.23	32	0.608
	I3	1.8	35	1.86	33	1.71	35	0.317
	I4	2.21	26	2.05	26	2.41	23	0.178
	I5	2.47	10	2.50	3	2.43	22	0.909
	I6	2.56	4	2.54	2	2.59	16	0.936
	I7	2.55	5	2.39	7	2.77	7	0.346
	I8	2.02	33	1.82	34	2.29	31	0.193
	I9	2.65	2	2.50	4	2.85	4	0.430
Externally related	E1	2.16	30	2.02	31	2.36	27	0.664
	E2	2.15	31	2.03	30	2.32	28	0.441
	E3	2.37	19	2.19	18	2.61	15	0.218

In particular, the contractors ranked contractors' late payment to suppliers or workers (C8), lowest bid tender award (O6) and poor communication and coordination (O3) significantly more highly than did owners. Contractors' late payments to suppliers or workers are from the financial crises that contractors face. Furthermore, it is very common to find cases in which a contractor or subcontractor who has not been paid what s/he is due intimidates workers or suspends work under the contract until the balance is paid in full. The lowest bid tender award is a significant challenge to contractors and most often results in poor performance. Contractors may bid at the lowest price to obtain the award but ultimately may adopt low-quality techniques that can save cost. Hence, this factor has a greater effect on the contractor than the owner. Poor communication and coordination can result from work stress, poor communication skills on workers' part, unclear and inconsistent site information, and misinterpretation of instructions.

However, as Table 4 shows, there are not many significant differences between the owners and contractors in successful projects, compared to the unsuccessful projects in Table 5. If the project goes well, owners and contractors understand each other and reduce the gaps between their different views.

Table 5. Delay Causes in Unsuccessful Projects.

Group	Number	Total		Owner		Contractor		U-test
		Mean	Rank	Mean	Rank	Mean	Rank	Sig.
Owner-related	O1	3.09	26	2.60	32	3.36	15	0.000
	O2	2.94	33	2.34	35	3.25	23	0.000
	O3	2.87	35	2.74	28	2.94	35	0.389
	O4	3.28	16	2.97	24	3.44	10	0.018
	O5	3.00	32	2.76	27	3.13	29	0.107
	O6	3.16	21	3.03	20	3.23	24	0.321
	O7	3.48	9	3.61	5	3.42	11	0.228
	O8	3.05	29	3.03	21	3.06	31	0.902
Contractor-related	C1	3.17	20	2.95	25	3.29	21	0.370
	C2	3.59	6	3.50	8	3.64	5	0.700
	C3	3.37	10	3.38	10	3.37	14	0.472
	C4	3.31	12	3.24	14	3.34	17	0.777
	C5	3.32	11	3.35	12	3.31	19	0.767
	C6	3.08	27	2.54	33	3.37	13	0.000
	C7	3.86	1	3.76	3	3.92	2	0.937
	C8	3.54	7	3.53	7	3.54	7	0.421
	C9	3.61	4	3.84	2	3.49	8	0.082
	C10	3.13	23	3.05	19	3.17	28	0.818

Table 5. *Cont.*

Group	Number	Total		Owner		Contractor		U-test
		Mean	Rank	Mean	Rank	Mean	Rank	Sig.
Design-related	D1	3.12	25	3.21	16	3.07	30	0.207
	D2	3.03	31	2.66	30	3.22	25	0.009
	D3	2.88	34	2.63	31	3.01	34	0.071
	D4	3.28	17	2.92	26	3.47	9	0.002
	D5	3.29	14	3.18	17	3.35	16	0.689
Infrastructure and Socially related	I1	3.13	24	3.03	22	3.18	27	0.642
	I2	3.04	30	3.03	23	3.04	32	0.476
	I3	3.16	22	2.68	29	3.42	12	0.015
	I4	3.06	28	2.51	34	3.34	18	0.000
	I5	3.60	5	3.22	15	3.79	3	0.317
	I6	3.74	3	3.92	1	3.65	4	0.193
	I7	3.29	15	3.49	9	3.19	26	0.029
	I8	3.30	13	3.30	13	3.30	20	0.883
	I9	3.86	2	3.76	4	3.92	1	0.598
Externally related	E1	3.20	19	3.08	18	3.26	22	0.995
	E2	3.23	18	3.59	6	3.04	33	0.012
	E3	3.50	8	3.38	11	3.56	6	0.817

4.1.2. Causes of Delays in Unsuccessful Projects

Table 5 shows the causes of delay in unsuccessful power projects. Poor cost estimation (C7_1st Rank), vandalism (I9_2nd Rank), late delivery of material and equipment (I6_3rd Rank), late procurement orders for material and equipment (C9_4th Rank), and additional work attributable to errors (I5_5th Rank) rank as the top 5 in the group overall. In the owner group, late delivery of material and equipment (I6_1st Rank), late procurement orders for material and equipment (C9_2nd Rank), poor cost estimation (C7_3rd Rank), vandalism (I9_4th Rank), and the owner's inadequate funds or budget allocation (O7_5th Rank) rank as the top five. In the contractor group, vandalism (I9_1st Rank), poor cost estimation (C7_2nd Rank), unskilled or inexperienced labor (I5_3rd Rank), late delivery of material and equipment (I6_4th Rank), and poor cost management (C2_5th Rank) rank as the top five. As such, the owner and contractor groups ranked the top five delay factors similarly, which indicates that the two do not have significantly different perspectives on delay factors.

However, there are some significant differences between the owners and contractors at the rank level. Owners ranked delays in obtaining permits from authorities (I7) and unexpected geological conditions (E2) significantly higher than did contactors. In contrast, contractors ranked change in scope (O1), owner's poor supervision (O2), delays in approval (O4), misrepresentation of information before bid (C6), inappropriate data collection (D2), failure in planning and designing risk (D4), workers' strikes (I3), and poor working conditions (I4) significantly higher than did owners. These causes usually are attributable not to the contractor, but to owners or external factors.

4.2. *Mitigation Strategies*

4.2.1. Mitigation Strategies in Successful Projects

Table 6 shows the project delay mitigation strategies in successful power construction projects. Close project supervision (M4_1st Rank), conducting capacity building training (M6_2nd Rank), and proper logistics management (M14_3rd Rank) rank as the top three in the total group. In the owner group, close project supervision (M4_1st Rank), top management's support (M15_2nd Rank), and proper logistics management (M14_3rd Rank) rank as the top three, while in the contractor group, conducting capacity building training (M6_1st Rank), proper logistics management (M14_2nd Rank), and timely site visits (M11_3rd Rank) rank as the top three. The Mann–Whitney U-test found no significant differences between owners and contractors in all strategies as shown in Table 6, which implies that if project schedule goes well, owner and contractor have similar delay management strategy.

Table 6. Mitigation Strategy in Successful Projects.

Number	Total		Owner		Contractor		U-test
	Mean	Rank	Mean	Rank	Mean	Rank	Sig.
M1	3.29	14	3.24	12	3.35	14	0.683
M2	3.32	12	3.19	14	3.52	9	0.328
M3	3.16	15	2.95	15	3.48	11	0.093
M4	3.65	1	3.67	1	3.62	7	0.922
M5	3.54	5	3.43	6	3.69	4	0.296
M6	3.64	2	3.50	4	3.86	1	0.256
M7	3.41	9	3.33	10	3.54	8	0.430
M8	3.38	11	3.45	5	3.26	15	0.583
M9	3.42	8	3.37	9	3.50	10	0.580
M10	3.43	7	3.31	11	3.63	5	0.257
M11	3.53	6	3.40	8	3.71	3	0.248
M12	3.32	13	3.24	13	3.45	12	0.461
M13	3.39	10	3.40	7	3.36	13	0.796
M14	3.61	3	3.52	3	3.72	2	0.381
M15	3.59	4	3.57	2	3.63	6	0.975

Conducting capacity building training ranked highest as a mitigation strategy from the contractors' perspective. Capacity building training offers a good opportunity for any industry to enhance its workers' knowledge and skills, as well as teams' self-esteem. Some of the benefits of conducting capacity building training are improved worker performance, satisfaction, and retention, an increased number of qualified workers, and workers who are updated on technology changes. Top management's support also ranked high. This indicates that the owners believe that top management's support contributed significantly to reducing power projects' delay. Furthermore, management usually defines the project's scope, facilitates the provision of resources, and selects the project team. They also ensure appropriate project funding and make some very critical decisions, such as approving funding allocation, authorizing scope changes, and whether to allow schedule overruns.

4.2.2. Mitigation Strategies in Unsuccessful Projects

Table 7 presents the project delay mitigation strategies in unsuccessful power construction projects. The results are significantly different from those in Table 6, and eight of fifteen mitigation strategies differed significantly between owners and contractors, which implies that if the project does not go well, owners and contractors think different solutions. If these different strategies are not understood and integrated, the project delay is not difficult to be solved.

Timely payments of completion certificates (M8_1st Rank), proper planning of project financial arrangements (M1_2nd Rank), and consideration of Donor's Influence (M3_3rd Rank) rank as the top three in the group overall. In the owner group, timely procurement and supply of materials and equipment (M7_1st Rank), top management's support (M15_2nd Rank), and proper planning of project financial arrangements (M1_3rd Rank) rank as the top three, while in the contractor group, conducting capacity building training (M6_1st Rank), timely payments of completion certificates (M8_2nd Rank), and finishing the design on time (M10_3rd Rank) rank as the top three. The Mann–Whitney U-test found no significant differences between owners and contractors, as shown in Table 7.

The practice of effective and well-timed payment in construction projects is a major factor that contributes to a project's success. For example, if the employer makes a late payment to the contractor, the payment due to the subcontractors or suppliers who are bound contractually to supply goods or services also will be late. Various reasons for delayed payment include the client's poor financial management, delays in certification, and disagreements on the valuation of the work performed.

Table 7. Mitigation Strategy in Unsuccessful Projects.

Number	Total		Owner		Contractor		U-test
	Mean	Rank	Mean	Rank	Mean	Rank	Sig.
M1	4.09	2	4.48	3	3.87	4	0.001
M2	3.87	10	4.10	10	3.75	8	0.104
M3	3.95	6	4.28	7	3.78	7	0.001
M4	3.83	13	4.03	12	3.72	9	0.106
M5	3.95	7	4.38	5	3.71	10	0.001
M6	4.04	3	3.95	14	4.10	1	0.325
M7	3.98	5	4.58	1	3.65	12	0.000
M8	4.13	1	4.23	8	4.07	2	0.356
M9	4.03	4	4.38	6	3.84	5	0.005
M10	3.95	8	4.03	13	3.90	3	0.487
M11	3.77	14	4.21	9	3.53	14	0.001
M12	3.74	15	3.82	15	3.70	11	0.667
M13	3.86	12	4.40	4	3.56	13	0.000
M14	3.89	9	4.10	11	3.78	6	0.095
M15	3.87	11	4.53	2	3.51	15	0.000

5. Discussions

Many studies have dealt with delay causes and mitigation strategies in construction projects. However, so many construction projects are still frequently delaying, which results in poor project performance such as cost and time overruns, disputes, arbitration, litigation, and complete termination [35,38,39]. Thus, this study tried to find more specific knowledge to owners and contractors respectively because owners and contractors have different roles and capability to deal with delay management. In addition, depending on the project's difficulty and performance, the delay causes and mitigation strategy can vary. Thus, this study suggested the delay causes and mitigation strategies separately in successful project and unsuccessful projects.

If the project progress meets the planned schedule or delays less than 10% of planned schedule, the owners and contractors can refer to Tables 4 and 6. Owners and contractors are likely to suggest similar delay causes. They need to take care of inadequate funding or budget allocation, vandalism, the contractor's late payment to suppliers or workers, late delivery of materials and equipment, and delays in obtaining permits from authorities. They can establish mitigation plan such as close project supervision, conducting capacity building training, and proper logistics management.

However, if the project delay more than 10% of the planned schedule, the owner and contractor can refer to Tables 5 and 7. Owners and contractors have to scrutinize the delay cause and mitigation strategy. Owners and contractors are likely to transfer their poor performance to the counter party's responsibility. The owner needs to review the late delivery of material and equipment, late procurement orders for material and equipment, poor cost estimation, vandalism, and owner's inadequate funds or budget allocation. The contractor needs to check the vandalism, poor cost estimation, unskilled or inexperienced labor, late delivery of material and equipment, and poor cost management. After analyzing the delay causes, owner can establish mitigation strategies such as timely procurement and supply of materials and equipment, top management's support, and proper planning of project finance arrangements. Whereas, contractor can build mitigation strategies such as conducting capacity building training, timely payments of completion certificates, and finishing the design on time.

These results make a contribution not only to delay management but also to risk and conflict management. As a result, there are some gaps between owner and contractor for perceiving the delay causes and mitigation strategies differently. These differences frequently induce conflicts between them, which further delay the project schedule. Owners particularly when making decision, should build reasonable benefit-sharing mechanism and risk allocation of resource arrangement [14], which increase the contractor's trust in the owner. This trust builds a strong foundation for contractors to conduct their responsibility [12].

6. Conclusions

This study analyzed the delay causes and mitigation strategies between owner and contractor in successful and unsuccessful power construction project. This study found that the delay causes and mitigation strategies significantly varied depending on project progress performance as described below.

First, if the project progress meets the plan well, there are not many different gaps between owner and contractors. Owners and contractors can easily converge the delay causes and build mitigations strategy for their success. In particular, the owner should manage the funding well and control the budget, whereas the contractor should take care of late payments to suppliers or work.

Second, if the project progress delays much, there are serious different gaps between owners and contractors. Even more, owner and contractor differently evaluate the delay causes and mitigation strategies to catch up with the progress. Therefore, the decision makers should encourage the owner and contractor-side experts to perceive the various gaps and communicate each other. Then, together they should build the mitigation strategies. In particular, the owner should manage the monitoring of late delivery of material and equipment, reviewing the cost estimation, funding and budget control, whereas the contractor should take care of vandalism, poor cost estimation, unskilled and unexperienced labors.

Third, several causes and mitigation strategies are much related to decision makers. Lowest bid tender award and inadequate funds or budget allocation causes the project delay. Top management support ranks high in the mitigations strategy. These causes and mitigation strategies should be improved by the involvement of the decision maker.

Even though this study contributes to improving the delay management of construction projects, this has several limitations. First, the investigation of this study was conducted at a power construction project in Tanzania. Thus, if the practitioners use this study in other industries or country, they have to consider these specific conditions. Second, this study did not reflect the project size, detail types of project and experience levels of respondents. These attribute can affect the causes and mitigation strategies. Therefore, in the future, this study will analyze the delay causes and mitigation strategies according to the project size and experience levels of respondents.

Author Contributions: E.T.B. and W.J. conceived and designed the model. E.T.B. collected and analyzed data, and wrote the manuscript. W.J. developed the overall idea and revised the manuscript. All authors have read and approved the final manuscript.

Funding: This research was supported by the 2019 Research Fund of the KEPCO International Nuclear Graduate School (KINGS), the Republic of Korea.

Conflicts of Interest: The authors declare no conflict of interest.

References

1. Sambasivan, M.; Soon, Y.W. Causes and effects of delays in Malaysian construction industry. *Int. J. Proj. Manag.* **2007**, *25*, 517–526. [CrossRef]
2. Kikwasi, G.J. Causes and effects of delays and disruptions in construction projects in Tanzania. *Australas. J. Constr. Econ. Build.* **2012**, *1*, 52–59. [CrossRef]
3. *Tanzania Electric Supply Company Limited (TANESCO 2018)*, Power Project Report; Unpublished.
4. Shubhan, V. Causes of delay in project construction in developing countries. *Indian J. Commer. Manag. Stud.* **2013**, *4*, 24–29.
5. El-Razek, M.E.A.; Bassioni, H.A.; Mobarak, A.M. Causes of Delay in Building Construction Projects in Egypt. *J. Constr. Eng. Manag.* **2008**, *134*, 831–841. [CrossRef]
6. Long, N.D.; Ogunlana, S.; Quang, T.; Lam, K.C. Large construction projects in developing countries: A case study from Vietnam. *Int. J. Proj. Manag.* **2004**, *22*, 553–561. [CrossRef]
7. Assaf, S.A.; Al-Hejji, S. Causes of Delay in Large Building Construction Projects. *J. Manag. Eng.* **2006**, *1*, 45–50. [CrossRef]

8. Olajide, T.I.; Timo, O.O.; Onaopepo, A.; Idowu, V.E. Analysis of Non-Excusable Delay Factors Influencing Contractors' Performance in Lagos State Nigeria. *J. Constr. Dev. Ctries.* **2013**, *18*, 53–72.
9. Pillai, N.; Kannan, K.P. *Time and Cost Over-Runs of Power Projects in Kerala*; Working Paper No. 320; Centre for Development Studies: Thiruvananthapuram, India, 2001.
10. Marques, R.C.; Berg, S. Risks, Contracts, and Private-Sector Participation in Infrastructure. *J. Constr. Eng. Manag.* **2011**, *137*, 925–932. [CrossRef]
11. Marques, R.C.; Berg, S. Revisiting the Strengths and Limitations of Regulatory Contracts in Infrastructure Industries. *J. Infrastruct. Syst.* **2010**, *16*, 334–342. [CrossRef]
12. Xu, Z.; Yin, Y.; Li, D.; Browne, G.J. Owner's Risk Allocation and Contractor's Role Behavior in a Project: A Parallel-mediation Model. *Eng. Manag. J.* **2018**, *30*, 14–23. [CrossRef]
13. Tripathi, K.K.; Jha, K.N. Determining Success Factors for a Construction Organization: A Structural Equation Modeling Approach. *J. Manag. Eng.* **2018**, *34*, 04017050. [CrossRef]
14. Wu, G.; Wang, H.; Chang, R. A Decision Model Assessing the Owner and Contractor's Conflict Behaviors in Construction Projects. *Adv. Civ. Eng.* **2018**, *2018*, 1347914. [CrossRef]
15. Chan, D.W.; Kumaraswamy, M.M. A comparative study of causes of time overruns in Hong Kong construction projects. *Int. J. Proj. Manag.* **1997**, *15*, 55–63. [CrossRef]
16. Odeh, A.M.; Battaineh, H.T. Causes of construction delay: Traditional contracts. *Int. J. Proj. Manag.* **2002**, *20*, 67–73. [CrossRef]
17. Aziz, R.F.; Abdel-Hakam, A.A. Exploring delay causes of road construction projects in Egypt. *Alex. Eng. J.* **2016**, *55*, 1515–1539. [CrossRef]
18. Alsuliman, J.A. Causes of delay in Saudi public construction projects. *Alex. Eng. J.* **2019**, *58*, 801–808. [CrossRef]
19. Jung, W.; Han, S.H. Which Risk Management Is Most Crucial for Controlling Project Cost? *J. Manag. Eng.* **2017**, *33*, 04017029. [CrossRef]
20. Majid, M.Z.A.; McCaffer, R. Factors of Non-Excusable Delays That Influence Contractors' Performance. *J. Manag. Eng.* **1998**, *14*, 42–49. [CrossRef]
21. Ogunlana, S.; Promkuntong, K.; Jearkjirm, V. Construction delays in a fast-growing economy: Comparing Thailand with other economies. *Int. J. Proj. Manag.* **1996**, *14*, 37–45. [CrossRef]
22. Frimpong, Y.; Oluwoye, J.; Crawford, L. Causes of delay and cost overruns in construction of groundwater projects in a developing countries; Ghana as a case study. *Int. J. Proj. Manag.* **2003**, *21*, 321–326. [CrossRef]
23. Koushki, P.A.; Al-Rashid, K.; Kartam, N. Delays and cost increases in the construction of private residential projects in Kuwait. *Constr. Manag. Econ.* **2005**, *23*, 285–294. [CrossRef]
24. Khatib, B.; Poh, Y.S.; El-Shafie, A. Delay Factors in Reconstruction Projects: A Case Study of Mataf Expansion Project. *Sustainability* **2018**, *10*, 4772. [CrossRef]
25. Zhang, Y.; Fan, Z.P. An optimization method for selecting project risk response strategies. *Int. J. Proj. Manag.* **2014**, *32*, 412–422. [CrossRef]
26. Wang, S.Q.; Dulaimi, M.F.; Aguria, M.Y. Risk management framework for construction projects in developing countries. *Constr. Manag. Econ.* **2004**, *22*, 237–252. [CrossRef]
27. Kim, M.-S.; Lee, E.-B.; Jung, I.-H.; Alleman, D. Risk Assessment and Mitigation Model for Overseas Steel-Plant Project Investment with Analytic Hierarchy Process-Fuzzy Inference System. *Sustainability* **2018**, *10*, 4780. [CrossRef]
28. Asadi, P.; Zeidi, J.R.; Mojibi, T.; Yazdani-Chamzini, A.; Tamošaitienė, J. Project risk evaluation by using a new fuzzy model based on Elena guideline. *J. Civ. Eng. Manag.* **2018**, *24*, 284–300. [CrossRef]
29. Guo, S.; Zhang, P.; Yang, J. System dynamics model based on evolutionary game theory for quality supervision among construction stakeholders. *J. Civ. Eng. Manag.* **2018**, *24*, 318–330. [CrossRef]
30. Su, Y.; Lucko, G.; Thompson, R.C., Jr. Application of Voting Theory to the Float Ownership Problem. *JCEM* **2018**, *144*, 04017094. [CrossRef]
31. Rockart, J.F. The changing role of the information systems executive: A critical success factors perspective. *Sloan Manag. Rev.* **1982**, *24*, 3–13.
32. Pinto, J.K.; Kharbanda, O.P. *Successful Project Managers Leading Your Team to Success*; Van No Strand Reihnhold: New York, NY, USA, 1995.
33. Nguyen, L.D.; Ogunlana, S.O.; Lan, D.T.X. A study on project success factors in large construction projects in Vietnam. *Eng. Constr. Arch. Manag.* **2004**, *11*, 404–413. [CrossRef]

34. Nguyen, L.H. Relationships between Critical Factors Related to Team Behaviors and Client Satisfaction in Construction Project Organizations. *J. Constr. Eng. Manag.* **2019**, *145*, 04019002. [CrossRef]
35. Aibinu, A.; Jagboro, G. The effects of construction delays on project delivery in Nigerian construction industry. *Int. J. Proj. Manag.* **2002**, *20*, 593–599. [CrossRef]
36. Li, Y.; Song, H.; Sang, P.; Chen, P.-H.; Liu, X. Review of Critical Success Factors (CSFs) for green building projects. *Build. Environ.* **2019**, *158*, 182–191. [CrossRef]
37. Weiner, I.B.; Craighead, W.E. *Corsini Encyclopedia of Psychology*, 4th ed.; John Wiley & Sons, Inc.: Hoboken, NJ, USA, 2010; pp. 960–961.
38. Kassanga, E.J. Comparative Analysis of Cost Performance in Contracted Projects with Internally Done Projects in Tanzania. Master's Thesis, University of Dar es Salaam, Dar es Salaam, Tanzania, 2011.
39. Koushki, P.A.; Kartam, N. Impact of construction materials on project time and cost in Kuwait. *Eng. Constr. Arch. Manag.* **2004**, *11*, 126–132. [CrossRef]

sustainability

MDPI

Article

Sustainable Construction Project Management (SCPM) Evaluation—A Case Study of the Guangzhou Metro Line-7, PR China

Na Dong [1], Yanting Fu [1,*], Feng Xiong [1], Lujie Li [1,2], Yibin Ao [3] and Igor Martek [4]

1 College of Architecture and Environment, Sichuan University, Chengdu 610065, China; dongna@scu.edu.cn (N.D.); xiongfeng@scu.edu.cn (F.X.); 2018223050046@stu.scu.edu.cn (L.L.)

2 Avic Chengdu Aircraft Industrial (Group) Co., Ltd., Chengdu 610091, China

3 College of Environment and Civil Engineering, Chengdu University of Technology, Chengdu 610059, China; aoyibin10@mail.cdut.edu.cn

4 School of Architecture and Built Environment, Deakin University, Locked Bag 20001, Geelong, VIC 3220, Australia; igor@deakin.edu.au

* Correspondence: fuyanting@stu.scu.edu.cn

Received: 13 September 2019; Accepted: 14 October 2019; Published: 16 October 2019

Abstract: As a pillar industry of the Chinese national economy, the construction sector needs to improve its level of management to embrace sustainable development. Sustainable construction project management (SCPM) performance evaluation can help to raise the level of management. However, the existing evaluation system that takes into account both the sustainable development and the dimension of traditional project management is meager. In order to address this problem, this study sets out an integrated sustainable performance evaluation method for SCPM, along with a comprehensive analysis of both traditional and future management directions. Through literature review and enterprise data analysis of the relevant factors of finance, schedule, quality, and safety, etc., indicators are filtered and classified. In order to determine the strength of each indicator, a questionnaire is administered to construction professionals within a large construction enterprise (group). From the result of the weight with an improved Group-G1 (iG1) method (finance 0.206, schedule 0.206, quality 0.185, safety 0.134, informatization 0.134, and greenization 0.134), it indicates that finance, schedule and quality management are still top three important dimensions in SCPM. However, amazingly, the greenization and informatization management is as significant as safety management. Finally, based on set pair analysis, the Guangzhou Metro Line 7 project is used as a verifying case, affirming the validity of the sustainable performance evaluation model. The above SCPM evaluation model can not only provide a guideline for construction companies' sustainable management in China, but also serve as reference cases for other countries/regions to carry out relevant research work.

Keywords: sustainable construction project management (SCPM); sustainable performance evaluation; set pair analysis; informatization; greenization; Guangzhou metro; China

1. Introduction

In recent years, with the rapid growth of China's economy and the acceleration of urbanization, the construction industry has developed rapidly [1]. The total output value of China's construction industry in 2017 was 5569 billion yuan, nearly four times higher than a decade ago [2]. Against this background, the construction industry has significant economic, environmental, and social impacts on society [3]. For instance, the construction industry in China contributes 6.7% to the gross domestic product (GDP) and provides more than 5.5 million jobs [2]. However, problems such as low profit, large resource consumption, serious environmental pollution, and poor ability to utilize information technology within the construction industry have raised concerns [4]. Moreover, traditional management fails to adequately address problems that often exist in areas of quality, scheduling, and safety, etc. [5]. China's state-owned construction companies generate a profit margin of 3.5%, compared with 10% for foreign companies [2]. According to the World Business Council for Sustainable Development, the construction industry accounts for about 40% of total national energy consumption, and produces between 45% and 65% of the disposed waste in landfills [6]. At the same time, the informatization level of the construction industry is only higher than that of agriculture, ranking the second to last among all industries [7]. Simply, the high-speed construction industry development pattern of 'high consumption, high pollution, low informatization, and low profit' is a matter requiring urgent attention [8].

Consequently, how best to improve the economy, environment, and social equity, by way of improving the level of management, has become topical within the industry. The quality of management within construction enterprises, which directly impacts the value of projects, and return to stakeholders [9], also impacts the enterprises' sustainable development. Effective performance evaluation of CPM is an important way to improve this management level. Effective performance evaluation of CPM can help managers not only identify existing project problems in a timely manner, but also put forward relevant improvement measures and solutions, ultimately improving the overall performance and management efficiency of construction project managers. However, at present, studies on evaluating the CPM performance have mostly focused on the economic benefits [8], while the environmental problems and information management issues are largely ignored. Furthermore, the majority of construction enterprises in China do not yet have a complete and effective SCPM performance evaluation system [9].

In line with the discussions above, it is, thus, necessary to establish an SCPM performance evaluation system geared to improving both the 'greenization' of projects and 'informatization', ensuring the sustainable development of the construction industry. Therefore, this study seeks to address the following research questions:

1. The existing studies for CPM mostly consider the financial, quality and schedule management. However, these dimensions are insufficient for sustainability evaluation from the social, environmental, and economic perspective.
2. The complexity of processes results in difficulty in establishing a SCPM evaluation system because of the lack of an effective method. No coordination analysis was made in the existing sustainable evaluation systems, and the comprehensive optimal management for SCPM could not be obtained.
3. This is a problem of how to establish a unified dimensional evaluation model for SCPM form integration from the perspective of sustainable development.

To answer the above overall questions, the remainder of this research is structured into the following sections. Section 2 focuses on related literature review. The establishment of a sustainable performance index system and a brief introduction for the assessment model are provided in Section 3. Then, Section 4 confirms the weight of the indicators with an iG1 method. In Section 5, a set-pair analysis (SPA) is proposed to assess the sustainable performance in a case study of Guangzhou metro line-7. Finally, the major findings, conclusions, and future areas of research of this study are given.

2. Literature Review

2.1. Traditional CPM Performance

The PMBoK (project management body of knowledge) contains the globally recognized standards and guides for the project management profession [9]. It is the general description of knowledge, skills, and tools required for project management by the Project Management Institute (PMI) [10,11]. It identifies ten knowledge areas for organizing processes: Stakeholder, integration, scope, time, cost, risk, quality, human resources, procurement, and communication [12]. Certain studies have measured performance across some of these knowledge areas or revealed the impact of specific knowledge areas on performance [12]. The performance evaluation of CPM often involves some aspects of the project knowledge area. In terms of CPM, the majority of research studies focus on timely completion [13–16], under budget completion [17–20], safely completed works [21–23], and meeting quality criteria [24–26]. In light of previous studies, those mostly cited indicators will be retained for traditional CPM performance evaluation.

2.2. Sustainable Development

There have been extensive studies conducted on sustainable development over the last decade [27]. These studies have been undertaken in both developed countries and developing countries, indicating that this is a global concern [28]. The concept of sustainable development was first defined in 1987, in the report of the World Commission on Environmental and Development titled "Our Common Future" (the so-called Brundtland Report) [29]. Sustainable development is defined as development that meets the needs of the present without compromising the ability of future generations to meet their own needs [29]. In 1997, sustainable development which encompasses the 'triple bottom line' of economic, environmental, and social condition into account, was defined as a necessary strategy for China's construction modernization [30]. It is worth emphasizing that in the Brundtland report itself, there are many direct references to management sciences [29]. The phrase 'sustainable' appears in many different contexts, from such narrow, specialist approaches as farm management [31] or coastal zone management [32], to more general areas like economic management [33], international management [34], or even global management [35].

After more than 30 years of development, research on sustainable development in management science is still growing [28]. Many fields have been studied, such as resource consumption [36], garbage disposal [37], and sustainable technology innovation [38]. With such extensive research on the concept of sustainable development in the field of management, greenization has emerged as a widely accepted phenomenon necessary to the implementation of triple bottom-line development of buildings in the construction industry [39]. Many scholars have conducted research in this field. For example, Ding reviewed several widely used building and construction environmental assessment methods across different countries and concluded that current assessment methods did not adequately consider environmental impacts [40]. Whang and Kim developed an assessment of factors for sustainable construction projects in Korea, by concentrating on environmental issues as well as economic and social dimensions [41]. There is a consensus that environmental performance is highly characterized by greenization, such as improved energy, water efficiency, enhanced air quality of construction sites, and reduced environmental pollution [42]. For this reason, greenization is considered an essential ingredient in any CPM performance evaluation index.

2.3. SCPM Performance

Construction projects are increasingly more difficult and complex [43], where construction is a highly project-based industry in which various organizations must couple with each other through project-specific collaborative relationships [44]. In order to optimize, automate, and modernize the traditional processes of this industry, information management has been increasingly valued and is already changing the current systems used in construction projects [45]. The industry has witnessed a growing interest in using the concept of building information modeling (BIM) in conjunction with sustainability principles during the design and construction phases of building projects [46]. BIM also helps reduce miscommunication and errors arising among construction stakeholders [47].

Greenization [48] (the ability to implement 'green' construction) places emphasis on environmental issues while the informatization [49] (the ability to manage and utilize accrued construction information) stresses the importance of innovation technologies [43,45,50,51]. Both of them are indispensable and necessary conditions for the sustainable development of the construction industry [42]. However, few studies on performance evaluation take these two points into consideration simultaneously. In this study, SCPM is defined in line with the definition proposed by Silvius [48]. This refers to the comprehensive and harmonized assimilation of social, economic, and environmental principles into effective construction project management systems [52]. The SCPM has been widely investigated in recent years, and the strategic importance of sustainability for construction enterprises has increased [53]. Zheng evaluates the trend of sustainable development in construction industry based on factor-cluster analysis [54]. Banihashemi and Kiani look at the critical success factors affecting the integration of sustainability into project management practices of construction projects in developing countries [55,56]. Pham measures the performance of CPM in the effects of sustainable practices and managers' leadership competencies [57]. However, these studies only consider some factors driving the sustainability of construction companies and lack a specific and feasible sustainable evaluation system.

Bamgbage suggests that sustainable performance in the construction industry includes social well-being, environmental protection, and financial earnings—the three main objectives [58]. The SCPM should not only consider the above performance concerns, but also considers the current industry's need to move towards sustainable development.

3. Methodology

In view of the above literature review, this paper aims at establishing a set of sustainable performance evaluation indexes considering social well-being, environmental protection, and financial earnings, along with a set of feasible evaluation methods to ensure that reasonable performance evaluation can be realized within a project. Firstly, the comprehensive evaluation index system of SCPM is constructed. In the index system, the greenization and informatization are not considered in isolation, but their impact on traditional project management performance measures, such as finances and safety, is also reflected. Secondly, the ratings for the evaluation index that is modeled by the iG1, are given in linguistic terms. Given different kinds of professional knowledge and work experience, the evaluators' understanding of the evaluation index and the enterprises may be different. Thus, in aggregating linguistic ratings, the amount of calculation is small so as to reduce errors. Thirdly, the feasibility and practical application of the proposed approach is verified using a case study of the Guangzhou Metro Line 7 (GML7). It also provides a valuable reference for the implementation of the evaluation of SCPM for other construction enterprises.

3.1. *Index System of SCPM Performance*

3.1.1. The Dimension of the Index System

An important step in establishing a comprehensive performance evaluation system for SCPM is to identify key aspects. It is impossible to analyze all aspects of SCPM. Such an attempt might not only confuse the experts of interactive information but also lead to an inefficient evaluation process. The indicators of this research are mainly established through three channels: Literature review, enterprise database analysis, and the future development trend analysis of the construction industry. The framework of the performance evaluation index system establishment is shown in Figure 1. The first step is the selection of the dimension of the evaluation index system. The traditional dimension of management performance includes financial, safety, schedule, and quality. Others, the frontier dimension includes two aspects, which are informatization and greenization. The second step is classification and filtering the secondary indicator based on literature review and construction enterprise management database.

The "knowledge area" proposed by PMI has been used in traditional CPM performance research studies. Various studies have focused on reducing the integration management [59], construction cost [20,60], and schedule control [61,62]. Demirkesen and Ozorhon contributed to the project management body of knowledge in that it develops a conceptual framework consisting of specific components for integration CPM [63], whereas the dimensions of project management performance are time, cost, quality, safety, and client satisfaction. Tan and Xiong sorted out 50 domestic theses related to performance evaluation in China and finally listed the high-frequency performance indicators which were divided into different hierarchical structures [59]. In their review, the most frequent indicators focus on financial, quality and time management. The performance evaluation indicators of large-scale engineering projects were divided into five categories: Human, material, machine, method, and environment [64]. In this study, we chose the four most representative dimensions, which are finance [65,66], safety [67–69], schedule [70–72], quality [73–75], as the dimensions of the performance index system for SCPM.

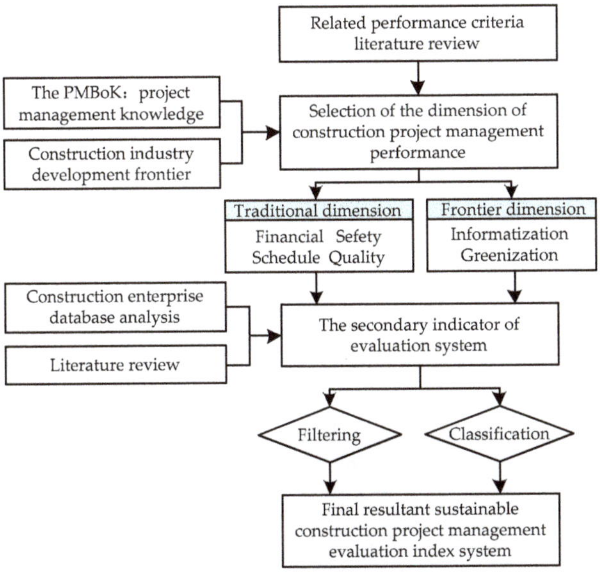

Figure 1. Conceptual framework of performance evaluation index system establishment.

In addition to the four dimensions of project management mentioned above, the SCPM performance should take the sustainable development of enterprises into account and encourage the application of cutting-edge technologies to adapt to the future development of the industry. Therefore, based on the traditional view and the perspective of future development, the final sustainable performance index system should consist of the six dimensions: Finance, safety, schedule, quality, informatization, and greenization management.

3.1.2. The Secondary Indicators of the Index System

The database established by this enterprise includes two parts: historical projects and actual needs of the Sichuan Huashi construction enterprise group (Huashi Group), which contains 18 sub-construction companies at home and abroad. The Huashi Group ranks 18th among China's top 60 contractors in ENR, with annual revenue from its main business exceeds 12.7 billion yuan, and stands for the overall development level of the construction industry.

The secondary indicators are filtered and classificated from the Huashi Group management database and literature review. This research looks for the 39 secondary indicators of the six dimensions, Figure 2 shows the final index system for the comprehensive performance evaluation of the SCPM, and then, the secondary indicators are described in the next chapter.

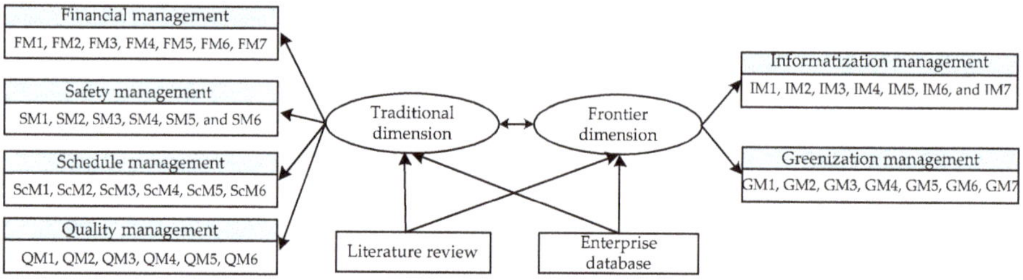

Figure 2. The comprehensive performance index system for the sustainable construction project management (SCPM).

Indicators for Financial Management

A company attaches much significance to financial management, which determines how much benefit the construction enterprise can get from a project. According to the enterprise database, the financial management contains three aspects, which are overall financial position, construction costs and financial management plans of the project. The overall financial situation of the project is measured by FM1 (project profit ratio), FM2 (unit profit) and FM3 (per capita profit). FM4 (project cost-saving), FM5 (success of project claims) and FM6 (project payment default) during the project construction stage are used to measure the funds and cost management ability. Beyond that, FM7 (the financial management plan) is crucial to the success of the project.

Indicators for Safety Management

Project safety management is critical to the success of the project. In the database, safety management contains two aspects, which are safety management results and security awareness. SM1 (safety and security inspections passed rate), SM2 (safety accident occurrence), and SM3 (gracious construction award) are used to measure the safety management performance. In addition, construction enterprises should raise their security awareness in terms of SM4 (safety education training), SM5 (hazard identification) and SM6 (safety construction facilities).

Indicators for Schedule Management

According to the database, in previous historical projects, the project managers mainly controlled the project schedule from three aspects: Whether the project was completed on time, whether the sub-contract project duration was reached, and whether the schedule plan was practical and rational. The ScM1 (time saving rate: The difference between the planned and the actual schedule) and the ScM2 (unit-project competed rate) are used to measure the completion of the schedule. The sub-contracts schedule conditions are measured by ScM3 (completion rate of labor sub-contracts) and ScM4 (specialized sub-contracts). In the schedule management system, managers need to input the ScM5 (image progress plan record), and ScM6 (realistic schedule planning) determines the outcome to a large extent.

Indicators for Quality Management

Management situation of project quality is mainly evaluated from three aspects: The quality acceptance rate, the quality self-inspection situation, and the quality plan. Quality acceptance rate is divided into three categories: QM1 (the outcome), QM2 (sub-engineering), and QM3 (construction unit) qualification rate. The quality self-examination of the project is measured by QM4 (worker self-examination passed rate). Moreover, QM5 (quality accident occurrence) and QM6 (quality planning) are used as the evaluation indicators of the quality plan.

Indicators for Informatization Management

Because informatization is an important means to improve the management level, project managers should enhance the level of informatization from two aspects: The technology and the management. There are four indicators to measure the level of information technology which are the IM1 (information modelling), IM2 (information management), IM3 (software and hardware configuration), and IM4 (training of personnel of building information management). Furthermore, the management of informatization also should consider the IM5 (information platform), IM6 (dynamic tracking management) and IM7 (enterprise database establishment).

Indicators for Greenization Management

Greenization is the future direction of the construction industry, and the greenization management of construction projects is mainly from the view of green construction. The managers should consider three aspects: The greenness of the construction site, the greenness of the construction materials, the disposal of construction waste. The GM1 (satisfaction degree of residents around), GM2 (green site award) and GM3 (green inspection passed rate) are the key indicators to measure the greenness of the construction site. The greenness of the materials was measured by GM4 (green material usage) and GM5 (resource-saving situation). Otherwise, GM6 (the disposal of construction waste which contains noise, flying dust, effluent, and garbage) is also a big problem for project managers. Furthermore, GM7 (green construction plan) is an important guarantee.

However, the importance of each indicator may vary from one project to another. It is necessary to develop a methodology to quantify the importance of SCPM performance indicators. Therefore, this research proposes a new approach that combines an SPA and a combination weighting of iG1 to establish a performance evaluation of the SCPM performance model. Based on a questionnaire survey, this approach evaluates the extent to which the project participation sides value each performance indicator. Then the quantified extents are converted into an importance matrix through iG1. The proposed method is detailed in the next chapter.

3.2. *Assessment Model*

3.2.1. The iG1 Weight Method

Compared with the analytical hierarchical process (AHP), iG1 weight method has the following three advantages: (1) It is applicable to a situation where the judgment matrix is an inconsistent matrix. (2) Multiple experts have the rank of importance of different indicators. (3) The amount of calculation is small so as to reduce errors. In order to reduce the interference of subjective factors in application, it is usual to invite multiple experts to judge the indices by importance at the same time. In addition, the problem of expert group judgment is that the judgment matrix is usually inconsistent, and different opinions are more likely to arise when determining the relative importance ratio of indicators. Therefore, this research offers some improvements to the iG1 weight method. Finally, the rank of these factors by the individual degree of their influence on construction management converts the component values, and then the ratio of the score standard deviation of adjacent indicators determines the importance degree of the indicators, making the importance weight of the indicators more scientific and reasonable.

Firstly, the ranking of multiple experts is conducted in the second comprehensive ranking, the ranking of indicators is converted into scores R_{ij} (the first important score is m, the second important score is $m-1$, etc.), and then the score of each expert is summarized to obtain the comprehensive scores R_i^* of indicators.

$$R_i^* = \frac{1}{n}\sum_{j=1}^{n} R_{ij} (j = 1, 2, 3, \ldots, n) \tag{1}$$

Secondly, according to the score R_i^*, the higher the score, the higher the ranking, and vice versa. If two indicators score the same, the ranking is determined by calculating the standard deviation of different scores. The comprehensive ranking of indicators is R_k.

$$\sigma_i = \sqrt{\frac{1}{n}\sum_{j=1}^{n}(R_{ij} - R_i^*)^2} \tag{2}$$

If the ranking is based on the standard deviation of the score, the higher the score, the lower the ranking, and vice versa. If the scores of the two indicators are the same, the ranking is the same.

After the comprehensive ranking is determined, the ratio of x_{i-1} to x_i degree of importance is determined by the standard deviation of the indicator score.

$$x_i^* = \frac{1}{n}\left(x_{i,1} + x_{i,2} + \ldots + x_{i,j}\right) \tag{3}$$

$$s_i = \sqrt{\frac{1}{n}\left[\left(x_{i,1} - x_i^*\right)^2 + \ldots + \left(x_{i,j} - x_i^*\right)^2\right]} \tag{4}$$

$$r_i = \begin{cases} s_{i-1}/s_i, s_{i-1} \geq s_i \\ 1, s_{i-1} < s_i \end{cases} \tag{5}$$

$$\omega_m^* = \left(1 + \sum_{k=2}^{n}\prod_{i=k}^{n} r_i\right)^{-1} \xrightarrow{\text{recursion}} \omega_{m-1}^* = r_i\omega_m^* \tag{6}$$

where x_{ij} represents the score of the jth expert to the ith indicator in their dimension, s_i represents the standard deviation of the ith indicator, and x_i^* is the average score of all experts. The final combined weight is $w_m^*, w_{m-1}^*, \ldots, w_1^*$.

3.2.2. The SPA Evaluation Model

To reflect the realistic circumstance in a construction project, only the hierarchy of sustainable performance is not adequate to evaluate the project management comprehensively. To accommodate this, the set pair analysis model is introduced to calculate the nearness degree of sustainable performance by using the association degree function. It is not only able to judge the tightness of the relationship between the appraised object and the various hierarchies, but also to find out the transformation trends of adjacent hierarchies, thus, the accuracy of the evaluation results has been improved.

The SPA method, proposed by Zhao in 1989 [76], is a systematic methodology considering both certainties and uncertainties as an integrated certain-uncertain system and depicting the certainty and uncertainty systematically from three aspects: Identity, discrepancy, and contrary. Set pair refers to a couple that consists of two interrelated sets. The basic idea of SPA is to analyze the features of a set pair and set up a connection degree formula of these two sets including identity degree, discrepancy degree and contrary degree under certain circumstances. Based on the connection degree formula, a series of SPA-based researches have been conducted [77,78]. In the process of performance assessment for SCPM, the dimension can be considered as set A, and the evaluation elements of performance assessment can be considered as set B. And the evaluation elements of performance assessment can be considered as set B. Then two sets constitute a set pair $\mu = (A, B)$. Connection degree $\mu_{(A-B)}$ is used to analyze the mathematics property of set pair $\mu_{(A-B)} = a + bi + cj$. This is the three-member connection degree, the fundamental formula of SPA. Because there are usually more than three levels in practice, b_i is usually extended as: $b_i = b_1i_1 + b_2i_2 + \ldots + b_{r-2}i_{r-2}$. Then the connection could be expressed as:

$$\mu_{(A-B)} = a + b_1i_1 + b_2i_2 + \ldots + b_{r-2}i_{r-2} + cj.$$

where a, b and c are the identical degree, discrepancy degree and contrary, respectively. $b_1, b_2, \ldots, b_{r-2}$ are the partial quantities of discrepant coefficient and $a + b_1 + b_2 + \ldots + b_{r-2} + c = 1, i \in [-1, 1], j = -1$, $i_1, i_2, \ldots, i_{r-2}$ are the uncertainly coefficient of discrepancy degree, j is the coefficient of contrary degree, $r = 1, 2, \ldots, r$ and r is the number of members in A or B.

Where $t = 1, 2, \ldots, m$, and m is the number of indicators, at-coefficient that the tth member in A are in the same level with and the member in set B, $b_{t,1}, b_{t,2}, \ldots, b_{t,r-2}$-coefficient that the partial quantities of discrepant coefficient and $b_{t,i}$ denotes the two members from A and B, respectively, have a distance of i levels, $i_1, i_2, \ldots, i_{r-2}$-the discrepant marking coefficient, c_t is the coefficient means that the two members have the farthest distance of levels, and $a + b_1 + b_2 + \ldots + b_{r-2} + c = 1$. The formulas to calculate the identity–discrepant–contrary coefficients are as below (set four-elemental connection degree for example).

1. When the evaluation indicator is in grade I,

$$\mu_{I1} = \begin{cases} 1 & X_t \geq S_{1t} \\ 1 + \frac{2(X_t - S_{1t})}{S_{1t} - S_{2t}} & S_{2t} \leq X_t < S_{1t} \\ -1 & X_t < S_{2t} \end{cases} \quad (7)$$

2. When the evaluation indicator is in grade II,

$$\mu_{I2} = \begin{cases} 1 & S_{2t} \leq X_t < S_{1t} \\ 1 + \frac{2(X_t - S_{1t})}{S_{1t} - S_{0t}} & X_t > S_{1t} \\ 1 + \frac{2(X_t - S_{2t})}{S_{2t} - S_{3t}} & S_{3t} \leq X_t < S_{2t} \\ -1 & X_t < S_{3t} \end{cases} \quad (8)$$

Similarly, the connection degree calculation equation can obtain, when the evaluation indicator is in grade III, IV.

3. When the evaluation indicator is in grade IV,

$$\mu_{I4} = \begin{cases} 1 & S_{4t} \leq X_t < S_{3t} \\ 1 + \frac{2(X_t - S_{1t})}{S_{1t} - S_{2t}} & S_{3t} \leq X_t < S_{2t} \\ -1 & X_t > S_{2t} \end{cases} \quad (9)$$

S_{1t}, S_{2t}, S_{3t}, S_{4t} and are the corresponding thresholds of evaluation degree I, II, III, IV respectively.

From the third chapter, the weight of key performance indicators is $W = (\omega_1, \omega_2, \ldots, \omega_n)^T$, Therefore the connection degree of $\mu_{(A-B)}$ is defined as:$\mu_{(A-B)} = \sum_{t=1}^{m} (\omega_t - \mu_{It})$.

When $\mu_{(A-B)} = \max\{\mu_{st}\}$, the t indicator belongs to the level i, the association degree for the t indicator to the level i is μ, which must be greater than the association degree to other levels. Combined with the characteristics of the project and the key points of construction, it can make the corresponding recommendations for SCPM performance.

3.3. Determination of the Weights

As discussed in the previous section, the SCPM performance index system has been identified. The next step is to score these indicators against certain dimensions. As the weight for these indicators can be complex, iG1 is then applied to reduce the impact of subjective elements and to analyze the final weight of the sustainability dimension in the entire system.

4. Data Collection

4.1. Survey Population

As a systematic technique of data collection, the questionnaire survey method has been widely used to collect professional opinions on the issues influencing the adoption of various innovations in construction management research. This study identified and selected respondents to the questionnaire that were especially knowledgeable or experienced within the construction industry. The participants were developers, constructors, designers, surveyors, and others who were involved and knowledgeable in SCPM. These people altogether constituted the sustainable building development team. An internet survey was carried out to rank the SCPM performance indicators. Out of the 64 questionnaires received, nine were rejected because of incomplete responses. Again, some were not properly filled. In this study, the remaining 55, through the questionnaire consistency check (Cronbach alpha $\alpha = 0.823 > 0.7$), were used for analysis.

The survey questionnaire consisted of two parts, part 1 aimed at determining the basic information about interviewees and part 2 aimed at their views on ranking the performance indicators of SCPM.

The certified 55 questionnaires indicated the academic background of the respondents as follows: 72% were degree holders, 36% were postgraduates. Thus, it is evident that the respondents had either university or polytechnic education. On the unit of analysis, the response rates showed that developers constituted 20% of the respondents, constructors 45.5%, surveyors and valuers 16.4%, designers 12.7%, and others 5.4%. Regarding years of professional experience, 38.2% have one to three years, 18.2% have four to six years, 14.5% have seven to nine years, and 29.1% have 10 years or above of experience. As to the frequency of performance evaluation, the annual part makes up a share of 20%, the quarterly part accounts for 60.0%, the monthly part 9.1 %, and the part of only once 10.9%. As for project performance evaluation, about 32.7% of the informants deem it very important, 60% think it is important and 5.5% consider it general important while 1.8% feel it is unimportant.

4.2. Survey Questionnaire

The scoring system which is commonly used in assessments of performance management usually includes the following two procedures: Setting up scoring dimension and basing the scoring on those dimensions. The largest characteristic is that the questionnaire focuses on the importance degree of multiple indicators in their respective criteria instead of the rank of a single indicator. For example, the order of importance of the indicators in informatization management by the *i*th. It can be expressed as $IM_j = \left\{IM_{1j}, IM_{2j}, \ldots, IM_{ij}\right\}$.

4.3. Data Analysis

All total scores for the comprehensive SCPM performance indicators which get from questionnaire results, were imported into iG1 weight method to calculate the subordinate weighting. Figure 3 shows the importance of indicators for six dimensions.

From the results of the weight shown in Figure 3 (finance 0.206, schedule 0.206, quality 0.185, safety 0.134, informatization 0.134, and greenization 0.134), there is not much difference among the six sustainable dimensions. The results of the survey further verify that in addition to the traditional goals of finance, quality, schedule, and safety, greenization and informatization have become the current focus. The informatization and greenization management dimensions have the same weight as safety management, all of which are 0.134. It shows that the professionals have come to realize the importance of informatization and greenization in SCPM.

In the financial management dimension, the FM1 (profit ratio) and FM2 (unit profit) are the key indicators, and the importance of the remaining four is well-balanced. The weight of ScM1 (the time-saving rate) is the highest in the schedule management, while the other indicators' weights are close to each other. Within the quality management, QM1 (the final quality), QM2 (the sub-engineering quality), and QM5 (the quality accident occurrence) are the key indicators, the measures of variation is

0.141 in this dimension, which is a little high. The weight of SM1 (safety and security inspections passed rate) is the highest in safety management, the weights of remaining indicators are 0.150 around. As BIM has been promoted by a compulsory provision in China, the sum of the weights of the two indicators IM1 and IM2 is 0.218, which represent people's attention to BIM. With respect to the greenization management GM1 (the satisfaction degree of residents around) and GM2 (green site award) are the most important indicators. From the above weight index distribution, it can be concluded that the importance of each indicator is basically consistent with the current situation of the construction industry in China.

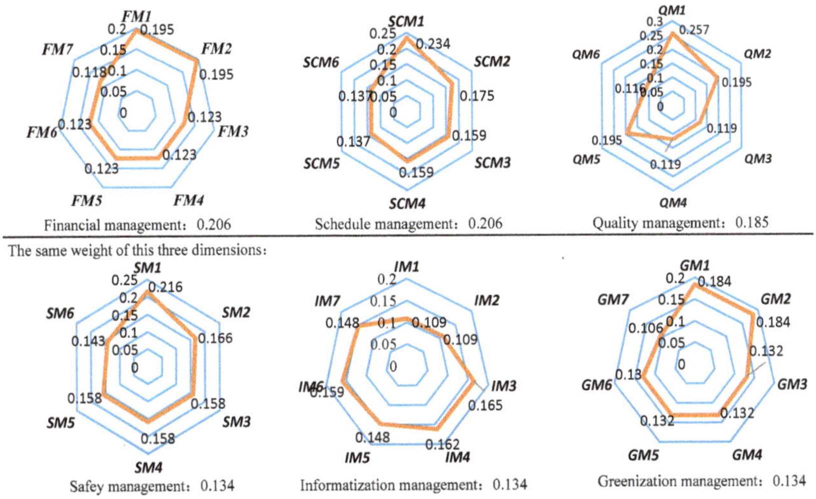

Figure 3. The weight of the index system.

5. Case Application

5.1. Project Background

In order to verify the effectiveness of the evaluation indicators and model for CPM performance, this research takes the electromechanical installation of the GZML7, which was constructed by Sichuan Huashi group corporation limited, as an example. The Zhongcun station covers a floor area of about 23,954 m^2 and is mainly a two-story underground building, with the length of 471.6 m and the width of 19.9 m. The other station is a three-story underground building, which is 156.8 m long and 25.9 m wide.

The composition of the electromechanical installation for GZML7 is composed of the following 10 aspects: Plumbing and fire extinguishing system, lighting and low voltage distribution system, ventilation and air condition system, intelligent building fire alarm system, building automation system, high-pressure water mist system, access control system, decoration engineering, relocating and rewiring engineering, and cooperation and coordination system.

5.2. The General Situation of SCPM

During the construction stage of the GZML7, the outcome of financial and schedule management was pretty good. Compared to the planning of construction, in reality, the project was completed ten days ahead of schedule, and only 69 million yuan was spent in reality. Besides, the outcome quality qualification rate is 100%, no safety accident occurred during the construction stage, safety construction facilities were fully equipped, and safety education training for workers was carried out well.

The GZML7 also won the 'Libing Cup of the civil engineering award and national quality engineering award' issued by the Ministry of Housing and Urban–Rural Development in 2017.

The greenization management of construction projects has achieved remarkable outcomes, reducing environmental pollution, controlling the noise and flying dust well, and having less impact on residents around. Moreover, the effluent and garbage have been reduced through the optimization of green construction planning.

In the field of informatization management for electromechanical installation, the arrangement of pipelines was complex. The collision inspection was carried out by using the BIM during the design stage, and the unreasonable areas were corrected. In the construction stage, the BIM technology was used to simulate the construction process for the complex construction nodes, which could improve the familiarity of workers' operation and avoid problems such as the installation of pipeline equipment and elevation errors.

5.3. *Performance Evaluation*

5.3.1. Connectivity and Evaluation Level Analysis

After obtaining the basic information, the project managers were invited to introduce the detail situation of GZML7 and established an evaluation team consisting of the managers in this construction enterprise. With a full understanding of the level standard, the actual value of the GZML7 project was calculated based on the specific data. According to the calculation formula in the last chapter, the research calculated the identity–discrepant–contrary coefficients of the evaluation indicators and determined the evaluation level. Only the association degree and evaluation level of the greenization and informatization dimensions are listed in Table 1.

Table 1. Association degree and the level of greenization and informatization management dimensions.

Indicator	Level Standards				Actual Value	Association Degree				Level
	Poor IV (0, S3)	Fair III (S3, S2)	Good II (S2, S1)	Excellent I (S1, -)		μ_{I4}	μ_{I3}	μ_{I2}	μ_{I1}	
GM1	(0,70)	(70,80)	(80,90)	(90,100)	92	−1	−1	0.6	1	I
GM2	(0,70)	(70,85)	(85,95)	(95,100)	91	−1	−0.2	1	0.2	II
GM3	(0,70)	(70,80)	(80,90)	(90,100)	65	1	0.85	−1	−1	IV
GM4	(0,0.1)	(0.1,0.2)	(0.2,0.3)	(0.3, -)	0.22	−1	0.6	1	−0.6	II
GM5	(0,0.1)	(0.1,0.2)	(0.2,0.3)	(0.3, -)	0.26	−1	−0.2	1	0.2	II
GM6	(0,0.7)	(0.7,0.8)	(0.8,0.9)	(0.9,1)	0.85	−1	0	1	0	II
GM7	(0,70)	(70,80)	(80,90)	(90,100)	77	−0.4	1	0.4	−1	III
comprehensive score/association degree					63.58	−0.67	0.05	0.60	−0.07	II
IM1	(0,70)	(70,80)	(80,90)	(90,100)	96	−1	−1	0.8	1	I
IM2	(0,70)	(70,80)	(80,90)	(90,100)	87	−1	−0.4	1	0.4	II
IM3	(0,70)	(70,80)	(80,90)	(90,100)	95	−1	−1	0	1	II
IM4	(0,70)	(70,80)	(80,90)	(90,100)	85	−1	0	1	0	II
IM5	(0,70)	(70,80)	(80,90)	(90,100)	88	−1	−0.6	1	0.6	II
IM6	(0,64)	(64,80)	(80,96)	(96,100)	97	−1	−1	0.5	1	I
IM7	(0,64)	(64,80)	(80,96)	(96,100)	85	−1	0.37	1	−0.37	II
comprehensive score/association degree					90.51	−1	−0.55	0.75	0.55	II

Table 1 shows that the comprehensive score of greenization and informatization management is 63.58 and 90.51, respectively. It is irrational to compare each dimension only based on comprehensive scores because of the differences in the scoring standard of them. Combining with the association degree, the highest score of greeniztion management is 0.60, which belongs to the level II of good, and the informatization management association degree is up to 0.75, which also reaches a good level. In the greenization dimension, the evaluation level of the secondary indicators indicates that the level of GM3 and GM7 is the lowest. This is consistent with the fact that the project has not been awarded as a green construction site and that the green construction planning is not good.

The project managers attached much importance to the application of information technology. The full use of the information management system, the complete configuration of software and hardware, the detailed application process of BIM technology in the construction stage, and the utilization of information technology to guide construction are the reasons for the project to achieve remarkable results. Therefore, the comprehensive evaluation of information indicators is better.

Similarly, the comprehensive evaluation levels of other dimensions can be calculated based on 13 valuation indicators and evaluation level standards.

Table 2 shows that the comprehensive evaluation level of the project is I. The company has rich management experience in the aspects of finance, safety, schedule, and quality. The performance evaluation of the informatization and greenization management of the construction project was lower in the historical cases. In the GZML7 project, the level of informatization and greenization management has reached the level II of good, which could give a strong reference value for other projects.

Table 2. The comprehensive evaluation levels of six performance management dimensions.

NO.	IM	GM	FM	SM	SCM	QM
level	II	II	I	I	I	I

5.3.2. Coordination Analysis

The coordination analysis of the dimension of informatization and greenization management based on the identity–discrepant–contrary coefficients are displayed in Table 3.

Table 3. The coordination analysis of greenization and informatization management.

Dimension	a	b1	b2	c
Greenization management	0.57	0.14	0.14	0.14
Informatization management	0.71	0.29	0	0

The value of a, b and c is the gauge of the coordination analysis, the greenization management system has good coordination, but the trend towards homogeneity and the improvement is weak. On the other hand, the association degree between the greenization dimension and level II is 0.60, the association with level III is 0.05, and the association degree with level I is 0.07. As discussed above, it shows that the ability of the greenization management system to convert from level II to level I is weak. In order to improve the situation where the trend of the system to excellent transformation is weak, construction enterprises should pay more attention to the indicators with the lowest score, such as green site award situation and whether the green construction planning is good or not. Similarly, the coordination analysis of informatization management can be obtained.

All in all, as a comprehensive evaluation method, SPA can reflect the overall state and make a quantitative evaluation, and can totally ensure the reliability and accuracy of the evaluation. In addition, all the single indicators are combined together organically by the calculation of the comprehensive association degree, making the evaluation of SCPM easier to understand. It plays a guiding role in the optimization and improvement of SCPM competence.

6. Conclusions

Performing the sustainable evaluation for the SCPM is a significant measure to increase environmental protection and informatization utilization, to enhance the competitive edge of construction companies to market changes. The process adoption began by putting forward a SCPM performance evaluation index system, based on existing norms and established research, to which the dimensions of greenization and informatization were added. The final index system contains the above two dimensions and the remaining four dimensions: Finance, safety, schedule, quality management. Next, the identified six dimensions of SCPM performance indicators were further analyzed by 55 expert questionnaires. The iG1 method was then applied to reduce the impact of subjective elements and to analyze the final weight of the dimensions in the entire system. Finally, based on the iG1 weight method and the SPA, the case of GZML7 was evaluated for SCPM performance. This SCPM evaluation model can not only provide a guideline for construction companies in China, but also serve as reference cases for other countries to carry out relevant research work.

In this research, the SCPM performance evaluation index system consisted of 39 secondary indicators from six dimensions. From the results of the weight (finance 0.206, schedule 0.206, quality 0.185, safety 0.134, informatization 0.134, and greenization 0.134), finance, schedule and quality management were still top three important in SCPM, but surprisingly, the greenization and informatization management were as significant as safety management. The importance of greenization and informatization gained popular recognition of construction professionals. Through a GZML7 case application by SPA, it shows that the evaluation level of the greenization management is II and the system converts from level II to level I was weak. Similarly, the evaluation level and coordination analysis of other dimensions could be obtained. Based on the results, the project managers could go back to find worse indicators and improve the management system in construction enterprises and successfully support sustainable development.

Flowing from this research, the following conclusions are obtained. Firstly, the establishment of evaluation index system based on the Huashi construction enterprise management database, which was valid and implementable. Secondly, this integrated assessment approach could reflect the coordination and development trend of the various dimensions of the system, and provided a theoretical basis and practical reference for the optimization of project management. Finally, this study offered a basis and useful reference to construction enterprises aiming to set up their own SCPM performance evaluation system, which is a guideline for project managers to improve the competitive edge of construction enterprises.

A limitation of this study, however, is that only representative indicators were employed, given the limited data available. Other indicators, such as human resources, procurement and communication, can also reflect the construction industry's level of sustainable development. These indicators will be analyzed in future research, as and when data become available. On the other hand, as the increasing level of informatization, we will further study the optimization model and methods for SCPM by data mining for selecting and optimizing the management databases.

Author Contributions: N.D. set up the framework and revised the paper; Y.F. wrote this paper; L.L. collected and analyzed the questionnaire data. F.X., Y.A. and I.M. revised the paper. All authors read and approved the final manuscript.

Funding: This research for this paper was funded by the Science and Technology Institute of Sichuan Province (No. 19H0469).

Conflicts of Interest: All authors declare that they have no conflict of interest.

References

1. Ahmad, M.; Zhao, Z.-Y.; Li, H. Revealing stylized empirical interactions among construction sector, urbanization, energy consumption, economic growth and CO_2 emissions in China. *Sci. Total Environ.* **2019**, *657*, 1085–1098. [CrossRef] [PubMed]
2. National Statistics Bureau. *China Statistical Yearbook—2018*; China Statistics: Beijing, China, 2018.
3. Edyta, P.; Michal, J.; Renata, K. Trends, costs, and benefits of green certification of office buildings: A Polish perspective. *Sustainability* **2019**, *11*, 2359.
4. Ma, L.; Wang, L.; Wu, K.-J.; Tseng, M.-L. Assessing co-benefit barriers among stakeholders in Chinese construction industry. *Resour. Conserv. Recycl.* **2018**, *137*, 102–112. [CrossRef]
5. Hwang, B.-G.; Zhao, X.; Toh, L.P. Risk management in small construction projects in Singapore: Status, barriers and impact. *J. Manag. Eng.* **2014**, *32*, 116–124. [CrossRef]
6. Ngowi, A.B. Creating Competitive Advantage by using environment-friendly building processes. *Build. Environ.* **2001**, *36*, 291–298. [CrossRef]
7. Ma, G.; Jia, J.; Ding, J.; Shang, S.; Jiang, S. Interpretive structural model based factor analysis of BIM adoption in Chinese construction organizations. *Sustainability* **2019**, *11*, 1982. [CrossRef]

8. He, L.; Zhang, L.; Zhong, Z.; Wang, D.; Wang, F. Green credit, renewable energy investment and green economy development: Empirical analysis based on 150 listed companies of China. *J. Clean. Prod.* **2019**, *208*, 363–372. [CrossRef]
9. Kelly, J.; Male, S.; Graham, D. *Value Management of Construction Projects*; Blackwell Science: Oxford, UK, 2008.
10. PMI. *A Guide to the Project Management Body of Knowledge*, 5th ed.; Project Management Institute, Pennsylvania Inc.: Delaware, PA, USA, 2013; pp. 12–24.
11. Brioso, X. Integrating ISO 21500 guidance on project management, Lean Construction and PMBOK. *Procedia Eng.* **2015**, *123*, 76–84. [CrossRef]
12. Varajão, J.; Colomo-Palacios, R.; Silva, H. ISO 21500: 2012 and PMBoK 5 processes in information systems project management. *Comp. Stand. Int.* **2017**, *50*, 216–222. [CrossRef]
13. Chan, A.C.; Scott, D.; Lam, E.M. Framework of success criteria for design/build projects. *J. Manag. Eng.* **2002**, *18*, 120–128. [CrossRef]
14. Zhao, D.; Mccoy, A.; Kleiner, B.; Mills, T.; Lingard, H. Stakeholder perceptions of risk in construction. *Saf. Sci.* **2016**, *82*, 111–119. [CrossRef] [PubMed]
15. Krzemiński, M. The scheduling of construction work under the assumption of brigade multitasking. *Procedia Eng.* **2017**, *208*, 63–68. [CrossRef]
16. Kong, L.; Li, H.; Luo, H.; Ding, L.; Zhang, X. Sustainable performance of just-in-time (JIT) management in time-dependent batch delivery scheduling of precast construction. *J. Clean. Prod.* **2018**, *193*, 684–701. [CrossRef]
17. Bassioni, H.A.; Price, A.D.F.; Hassan, T.M. Performance measurement in construction. *J. Manag. Eng.* **2004**, *20*, 42–50. [CrossRef]
18. Papke-Shields, K.E.; Beise, C.; Quan, J. Do project managers practice what they preach, and does it matter to project success? *J. Manag. Eng.* **2010**, *28*, 650–662. [CrossRef]
19. Ibarrondo-Dávila, M.P.; López-Alonso, M.; Rubio-Gámez, M.C. Managerial accounting for safety management. The case of a Spanish construction company. *Saf. Sci.* **2015**, *79*, 116–125. [CrossRef]
20. Didkovskaya, O.V.; Mamayeva, O.A.; Ilyina, M.V. Development of cost engineering system in construction. *Procedia Eng.* **2016**, *153*, 131–135. [CrossRef]
21. Tam, C.M.; Tong, T.K.L.; Chiu, G.C.W.; Fung, I.W.H. Non-structural fuzzy decision support system for evaluation of construction safety management system. *J. Manag. Eng.* **2002**, *20*, 303–313. [CrossRef]
22. Törner, M.; Pousette, A. Safety in construction—A comprehensive description of the characteristics of high safety standards in construction work, from the combined perspective of supervisors and experienced workers. *J. Saf. Res.* **2009**, *40*, 399–409. [CrossRef]
23. Li, H.; Lu, M.; Hsu, S.C.; Gray, M.; Huang, T. Proactive behavior-based safety management for construction safety improvement. *Saf. Sci.* **2015**, *75*, 107–117. [CrossRef]
24. Chen, L.J.; Luo, H. A BIM-based construction quality management model and its applications. *Autom. Const.* **2014**, *46*, 64–73. [CrossRef]
25. Bragadin, M.A.; Kähkönen, K. Safety, space and structure quality requirements in construction scheduling. *Proc. Econ. Financ.* **2015**, *21*, 407–414. [CrossRef]
26. Lukichev, S.; Romanovich, M. The quality management system as a key factor for sustainable development of the construction companies. *Procedia Eng.* **2016**, *165*, 1717–1721. [CrossRef]
27. United Nations. Working Arrangements for the 2016 Session of the Economic and Social Council, 24 July 2015–27 July 2016. Available online: http://www.un.org/ecosoc/en/sustainable-development (accessed on 8 May 2017).
28. Zemigala, M. Tendencies in research on sustainable development in management Sciences. *J. Clean. Prod.* **2019**, *218*, 796–809. [CrossRef]
29. WCED. *Our Common Future*; Oxford University Press: Oxford, NY, USA, 1987.
30. Chan, R.; Yao, S. Urbanization and sustainable metropolitan development in China: Patterns, problems and prospects. *Geojournal* **1999**, *49*, 269–277. [CrossRef]
31. Rose, D.C.; Sutherland, W.J.; Barnes, A.P.; Borthwick, F.; Ffoulkes, C.; Hall, C.; Moorby, J.M.; Nicholas-Davies, P.; Twining, S.; Dicks, L.V. Integrated farm management for sustainable agriculture: Lessons for knowledge exchange and policy. *Land Use Policy* **2019**, *81*, 834–842. [CrossRef]
32. Uehara, T.; Mineo, K. Regional sustainability assessment framework for integrated coastal zone management: Satoumi, ecosystem services approach, and inclusive wealth. *Ecol. Indic.* **2017**, *73*, 716–725. [CrossRef]

33. Baba, C.; Hack, J. Economic valuation of ecosystem services for the sustainable management of agropastoral dams. A case study of the Sakabansi dam, northern Benin. *Ecol. Indic.* **2019**, *107*, 105648. [CrossRef]
34. Kourula, A.; Pisani, N.; Kolk, A. Corporate sustainability and inclusive development: Highlights from international business and management research. *Curr. Opin. Environ. Sustain.* **2017**, *24*, 14–18. [CrossRef]
35. Fan, Y.; Chen, Y.; Xia, M.; Zhang, Y. The influence of social embeddedness on organizational legitimacy and the sustainability of the globalization of the sharing economic platform: Evidence from Uber China. *Resour. Conserv. Recycl.* **2019**, *151*, 104490. [CrossRef]
36. Liu, Y.; Qu, Y.; Lei, Z.; Jia, H. Understanding the evolution of sustainable consumption research. *Sustain. Dev.* **2017**, *25*, 414–430. [CrossRef]
37. Abdel-Shafy, H.I.; Mansour, M.S.M. Solid waste issue: Sources, composition, disposal, recycling, and valorization. *Egypt. J. Pet.* **2018**, *27*, 1275–1290. [CrossRef]
38. Amato, A.; Becci, A.; Birloaga, I.; De Michelis, I.; Ferella, F.; Innocenzi, V.; Ippolito, N.M.; Pillar Jimenez Gomez, C.; Vegliò, F.; Beolchini, F. Sustainability analysis of innovative technologies for the rare earth elements recovery. *Renew. Sustain. Energy Rev.* **2019**, *106*, 41–53. [CrossRef]
39. Joyram, H. A critical evaluation on the factors impacting the adoption of eco-block as a green construction material: From a Mauritian perspective. *J. Build. Eng.* **2019**, *25*, 100789. [CrossRef]
40. Ding, G.K.C. Sustainable construction—The role of environmental assessment tools. *J. Environ. Manag.* **2008**, *86*, 451–464. [CrossRef]
41. Whang, S.W.; Kim, S. Balanced sustainable implementation in the construction industry: The perspective of Korean contractors. *Energy Build.* **2015**, *96*, 76–85. [CrossRef]
42. Trotta, G. The determinants of energy efficient retrofit investments in the English residential sector. *Energy Policy* **2018**, *120*, 175–182. [CrossRef]
43. Bryde, D.; Broquetas, M.; Volm, J.M. The project benefits of building information modelling (BIM). *Int. J. Proj. Manag.* **2013**, *31*, 971–980. [CrossRef]
44. Cao, D.; Li, H.; Wang, G.; Luo, X.; Tan, D. Relationship network structure and organizational competitiveness: Evidence from BIM implementation practices in the construction industry. *J. Manag. Eng.* **2018**, *34*, 04018005. [CrossRef]
45. Lu, Y.; Wu, Z.; Chang, R.; Li, Y. Building information modeling (BIM) for green buildings: A critical review and future directions. *Autom. Constr.* **2017**, *83*, 134–148. [CrossRef]
46. Isikdag, U.; Underwood, J. Two design patterns for facilitating building information model-based synchronous collaboration. *Autom. Constr.* **2010**, *19*, 544–553. [CrossRef]
47. Chen, H.-M.; Hou, C.-C. Asynchronous online collaboration in BIM generation using hybrid client-server and P2P network. *Autom. Constr.* **2014**, *45*, 72–85. [CrossRef]
48. Silvius, G.; Tharp, J.; Silvius, G.; Tharp, J. *Sustainability Integration for Effective Project Management*; Business Science Reference: Hershey, PA, USA, 2013.
49. Xu, S.; Xu, D.; Liu, L. Construction of regional informatization ecological environment based on the entropy weight modified AHP hierarchy model. *Sustain. Comput. Inf.* **2019**, *22*, 26–31. [CrossRef]
50. Li, X.; Xu, J.; Zhang, Q. Research on construction schedule management based on BIM technology. *Procedia Eng.* **2017**, *174*, 657–667. [CrossRef]
51. Li, C.Z.; Zhong, R.Y.; Xue, F.; Xu, G.; Chen, K.; Huang, G.G.; Shen, G.Q. Integrating RFID and BIM technologies for mitigating risks and improving schedule performance of prefabricated house construction. *J. Clean. Prod.* **2017**, *165*, 1048–1062. [CrossRef]
52. Azar, E.; Nikolopoulou, C.; Papadopoulos, S. Integrating and optimizing metrics of sustainable building performance using human-focused agent-based modeling. *Appl. Energy* **2016**, *183*, 926–937. [CrossRef]
53. Oke, A.; Aghimien, D.; Aigbavboa, C.; Musenga, C. Drivers of sustainable construction practices in the zambian construction industry. *Energy Proc.* **2019**, *158*, 3246–3252. [CrossRef]
54. Zheng, M.; Cai, J. Study on construction industry's sustainable development based on factor and cluster analysis. *Sci. Technol. Manag. Res.* **2014**, *310*, 223–227. (In Chinese)
55. Banihashemi, S.; Hosseini, M.R.; Golizadeh, H.; Sankaran, S. Critical success factors (CSFs) for integration of sustainability into construction project management practices in developing countries. *Int. J. Proj. Manag.* **2017**, *35*, 1103–1119. [CrossRef]
56. Kiani, M.R.; Standing, C. Critical success factors of sustainable project management in construction: A fuzzy DEMATEL-ANP approach. *J. Clean. Prod.* **2018**, *194*, 751–765. [CrossRef]

57. Pham, K.; Kim, S.-Y. The effects of sustainable practices and managers' leadership competences on sustainability performance of construction firms. *Sustain. Prod. Consum.* **2019**, *20*, 1–14. [CrossRef]
58. Bamgbade, J.A.; Kamaruddeen, A.M.; Nawi, M.N.M. Malaysian construction firms' social sustainability via organizational innovativeness and government support: The mediating role of market culture. *J. Clean. Prod.* **2017**, *154*, 114–124. [CrossRef]
59. Tan, T.; Xiong, Z. Comparative study on the evaluation indicator system of project performance. *Sci. Technol. Manag. Res* **2014**, *23*, 81–90. (In Chinese)
60. Salem, D.; Bakr, A.; El Sayad, Z. Post-construction stages cost management: Sustainable design approach. *Alex. Eng. J.* **2018**, *57*, 3429–3435. [CrossRef]
61. Poshdar, M.; González, V.A.; Raftery, G.M.; Orozco, F.; Cabrera-Guerrero, G.G. A multi-objective probabilistic-based method to determine optimum allocation of time buffer in construction schedules. *Autom. Constr.* **2018**, *92*, 46–58. [CrossRef]
62. De Soto, B.G.; Rosarius, A.; Rieger, J.; Chen, Q.; Adey, B.T. Using a Tabu-search algorithm and 4D models to improve construction project schedules. *Procedia Eng.* **2017**, *196*, 698–705. [CrossRef]
63. Demirkesen, S.; Ozorhon, B. Impact of integration management on construction project management performance. *Int. J. Proj. Manag.* **2017**, *35*, 1639–1654. [CrossRef]
64. Wang, Y.; Wang, X. Evaluation study of large scale construction project safety risk based on gray-shaplely. *Archit. Technol.* **2017**, *48*, 289–292. (In Chinese)
65. Kim, J.; Koo, C.; Kim, C.-J.; Hong, T.; Park, H.S. Integrated CO2, cost, and schedule management system for building construction projects using the earned value management theory. *J. Clean. Prod.* **2015**, *103*, 275–285. [CrossRef]
66. Xia, N.; Zou, P.X.W.; Griffin, M.A.; Wang, X.; Zhong, R. Towards integrating construction risk management and stakeholder management: A systematic literature review and future research agendas. *Int. J. Proj. Manag.* **2018**, *36*, 701–715. [CrossRef]
67. Yiu, N.S.N.; Chan, D.W.M.; Shan, M.; Sze, N.N. Implementation of safety management system in managing construction projects: Benefits and obstacles. *Saf. Sci.* **2019**, *117*, 23–32. [CrossRef]
68. Niu, Y.; Lu, W.; Xue, F.; Liu, D.; Chen, K.; Fang, D.; Anumba, C. Towards the "third Wave": An SCO-enabled occupational health and safety management system for construction. *Saf. Sci.* **2019**, *111*, 213–223. [CrossRef]
69. Tang, N.; Hu, H.; Xu, F.; Zhu, F. Personalized safety instruction system for construction site based on internet technology. *Saf. Sci.* **2019**, *116*, 161–169. [CrossRef]
70. Krzemiński, M. Chosen criteria of construction schedule evaluation. *Procedia Eng.* **2016**, *153*, 345–348. [CrossRef]
71. Krzemiński, M. Optimization of the construction schedule for paving a parking area with concrete. *Procedia Eng.* **2016**, *153*, 349–354. [CrossRef]
72. Chin, L.S.; Hamid, A.R.A. The practice of time management on construction project. *Procedia Eng.* **2015**, *125*, 32–39. [CrossRef]
73. Ma, Z.; Cai, S.; Mao, N.; Yang, Q.; Feng, J.; Wang, P. Construction quality management based on a collaborative system using BIM and indoor positioning. *Autom. Constr.* **2018**, *92*, 35–45. [CrossRef]
74. Acıkara, T.; Aynur, K.; Ulubeyli, S. Evaluations of construction project participants' attitudes toward quality management in Turkey. *Procedia Eng.* **2017**, *196*, 203–210. [CrossRef]
75. Mazari, M.; Nazarian, S. Mechanistic approach for construction quality management of compacted geomaterials. *Transp. Geotech.* **2017**, *13*, 92–102. [CrossRef]
76. Zhao, K.; Xuan, A. Set pair theory-a new theory method of non-define and its applications. *Syst. Eng.* **1996**, *14*, 18–23. (In Chinese)
77. Jiang, Y.-L.; Xu, C.-F.; Yao, Y.; Zhao, K.-Q. Systems information of set pair analysis and its applications. In Proceedings of the Third International Conference on Machine Learning and Cybernetics, Shanghai, China, 26–29 August 2004; pp. 1717–1722.
78. Li, C.; Sun, L.; Jia, J.; Cai, Y.; Wang, X. Risk assessment of water pollution sources based on an integrated K-means clustering and set pair analysis method in the region of Shiyan, China. *Sci. Total Environ.* **2016**, *557*, 307–316. [CrossRef] [PubMed]

sustainability

Article

Developing A Semi-Markov Process Model for Bridge Deterioration Prediction in Shanghai

Yu Fang and Lijun Sun *

The Key Laboratory of Road and Traffic Engineering, Ministry of Education, Tongji University, Shanghai 201804, China; 1610049@tongji.edu.cn

* Correspondence: ljsun@tongji.edu.cn; Tel.: +86-021-69583810

Received: 15 September 2019; Accepted: 2 October 2019; Published: 7 October 2019

Abstract: The performance of urban bridges will deteriorate gradually throughout service life. Bridge deterioration prediction is essential for bridge management, especially for maintenance planning and decision-making. By considering the time-dependent reliability in the bridge deterioration process, a Weibull distribution based semi-Markov process model for urban bridge deterioration prediction was proposed in this paper. Historical inspection records stored in the Bridge Manage System (BMS) database in Shanghai since 2004 were investigated. The Weibull distribution was used to characterize the bridge deterioration behavior within each condition rating (CR), and the semi-Markov process was used to calculate the bridge transition probabilities between adjacent CRs. After that, the service life expectancy of urban bridges, the transition probabilities of the deck system and the substructure, and the future CR proportion change caused by deterioration was predicted. The prediction results indicate that the life expectancy of concrete beam bridges is about 77 years. The decay rate of the deck system is the fastest among three major parts, and the substructure has a much longer life expectancy. It suggests that the overall prediction accuracy of the semi-Markov model in network-level is better than the regression analysis method. Furthermore, the proportion of bridges in intact condition will gradually decrease in the next few decades, while the percentage of bridges in the qualified and bad state will increase rapidly. The prediction results show a good agreement with the actual deterioration trend of the urban bridges in Shanghai. In order to alleviate the pressure of bridge maintenance in the future, it is necessary to adopt a more targeted preventive maintenance strategy.

Keywords: bridge deterioration; prediction model; semi-Markov process; Weibull-distribution; condition rating

---

1. Introduction

With environmental actions and traffic loads, the performance of urban bridges, especially reinforced concrete (RC) bridges, will deteriorate gradually throughout the bridge service life (i.e., concrete damage and reinforcement corrosion may occur, with cross-section reduction leading to a decrease in the geometric dimensions and materials properties, and thereby to a serviceability degradation or structure failure) [1,2]. Bridge management systems (BMS) have been developing since the early 1990s to assist the management of bridges, which can maximize the network-level bridge performance and minimize the failure probability [3]. Based on the current state evaluation and future bridge condition prediction for the bridge performance, the principal objective of BMS is to develop an effective bridge maintenance, repair, and rehabilitation (MRR) strategy under a limited financial budget [4]. In Shanghai city, the BMS has been applied to the management of small or moderately sized urban bridges since 2004. Therefore, there were 2377 urban bridges across the city included in the BMS database by the end of 2016, in which reinforced concrete beam bridges and prestressed concrete beam bridges account for nearly 85% of the total urban bridge numbers. The construction time of bridges in Shanghai is shown as the following. (Figure 1).

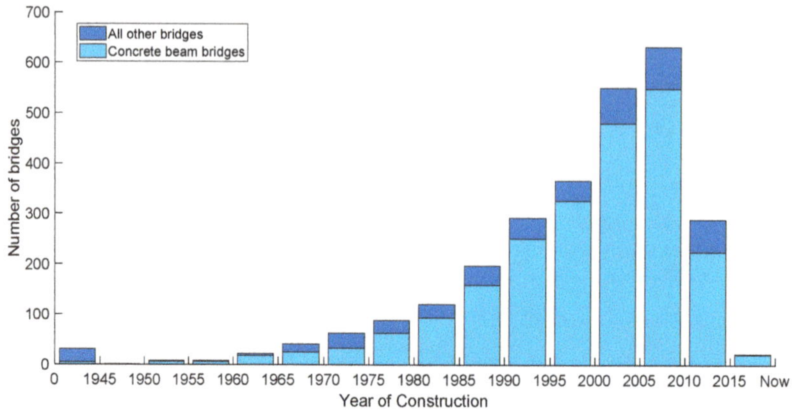

Figure 1. The distribution of bridge construction by year in Shanghai.

The current state assessment of urban bridges is the foundation of further bridge decay process forecasting and maintenance decision making. In China, the routine evaluation of urban bridges mainly depends on the annual visual inspection, and the bridge condition index (BCI) is used to assess the current state of categories II–V bridges (i.e., small or moderately sized bridges) [5]. Through a hierarchy-weight evaluation method based on bridge components, the whole bridge is comprised of the form "bridge-part-component." Therefore, the BCI score can reflect the overall technical condition of the whole bridge by comprehensively weighted damage degrees and positions of three major bridge parts (i.e., the deck system, the superstructure, and the substructure) which are made up of other more detailed components [6]. According to the Chinese Technical Code of Maintenance for City Bridge (CJJ 99-2017) [7], BCI scores from 0 to 100 correspond to the bridge condition ratings (CR) from A to E as shown in Table 1 [6]. It is generally believed that BCI scores evaluated through the annual inspection will affect the MRR strategy of urban bridges in the next year, as shown in Table 1. Where, merely routine maintenance, minor repair, or special inspection is needed for bridges with CR from A to C. As for bridges with CR D or E, medium repair, major repair, or rehabilitation are required in the next year.

Table 1. Chinese bridge condition ratings (CJJ 99-2017).

Rating	BCI Score	Condition	MRR Strategy
A	[90,100]	Intact	Routine maintenance
B	[80,89]	Good	Routine maintenance
C	[66,79]	Qualified	Minor repair
D	[50,65]	Bad	Routine maintenance
E	[0,49]	Dangerous	Minor repair

The reliability of bridge serviceability behavior will gradually decrease over time due to material degradation, increasing traffic load, and the external environment. Some of the typical problems of RC bridges, namely, concrete carbonization, surface delamination and spalling, crack propagation, steel rebar corrosion, cross-section reduction, and delayed debonding of the external reinforcement may occur [8,9]. The prediction results of the bridge decay process can provide a reliable basis for the further choice of maintenance measures and allocation of finance funds. Accordingly, deterioration forecasting models are essential for the management of urban bridges, especially for network-level or long-term bridge maintenance planning and decision-making. Nevertheless, the existing prediction model for bridge deterioration forecasting in the BMS of Shanghai city still uses the deterministic regression method, which has difficulty reflecting the uncertainty and randomness of the bridge deterioration process. Consequently, this paper proposes a Weibull distribution based semi-Markov process model for the deterioration prediction of urban bridges by considering the time-dependent reliability in the process of bridge deterioration. Consequently, the Weibull distribution was used to characterize the service-life behavior of bridge deterioration within each condition rating (CR) and the semi-Markov process was used to evaluate the transition probabilities of bridge deterioration process between adjacent CRs.

2. Summary of Basic Theory

The bridge deterioration process has often been modeled as the decay of performance over time, which means that the remaining service-life and the future performance of bridges can be forecasted by estimating the rate and pattern of bridge decay. According to different model assumptions, bridge deterioration prediction models can be divided into deterministic models and probabilistic models [10]. Deterministic models assume that the bridge decay trend is specific [11], and mostly estimate the bridge deterioration rate through the regression analysis method. For example, the existing deterioration model in the BMS of Shanghai uses a deterministic regression equation, as shown in Equation (1). Deterministic models have the advantages of simple modeling and convenient correcting [12], but it is quite challenging to reflect the uncertainty and randomness of the deterioration process. Besides, the requirement of high-quality historical data is usually hard to be satisfied, and the preprocessing of original data may also cause a risk of subjective judgment [13].

$$BCI = BCI_0[1 - \exp(-\frac{\alpha}{y})^{\beta}], \tag{1}$$

where BCI is the bridge condition index, $BCI_0 = 100$ is the initial BCI, y is the bridge age, α is the bridge life factor, and β is the curve shape factor.

In probabilistic models, bridge deterioration is seen as the result of stochastic processes, and bridge performance can be predicted by describing the probability distribution of the bridge condition state in the future. These models can reflect the uncertainty of bridge deterioration caused by external factors such as traffic loads, materials, environment, and maintenance [14]. Based on the concept of probabilistic cumulative damage, these stochastic models can be further subdivided into state-based or time-based models [15]. The state-based models such as Markov chains are modeled through a discrete-time transition probability matrix from one condition state to another, and the distribution of the future condition is not dependent on the past. Bridge management systems in the U.S like Pontis [16] and BRIDGIT [17] have adopted the Markov chain theory to calculate the bridge deterioration rates [18]. The state-based models have advantages in reflecting the uncertainties of the bridge deterioration process and can be easily applied to network-level bridge management.

Meanwhile, the time-based models such as Weibull distribution assume the duration time of a bridge maintains a condition rating as a random variable which can be modeled through the Weibull probability density functions [19]. Probabilistic models have the advantages of versatility and simplicity, which can clearly describe the service-life behaviors of bridge deterioration (i.e., failure rates, reliability, mean durations, and quantile statistics) [20]. The limitations of probabilistic models lie in need of many accurate observational data and empirical model assumptions [18].

2.1. Defects of Markov Chains

The Markov process describes that a system can be in one of several states, and each state can be passed to another state at a fixed probability over each time step. The Markov chain is a particular case of the Markov process, for which time and state parameters are both discrete. For a stochastic process $\{X(t), t \in T\}$, if the conditional probability can be expressed as Equation (2), then $\{X(t), t \in T\}$ is a Markov chain with discrete parameters [5,21,22].

$$P(X_{t+1} = i_{t+1}|X_0 = i_0, X_1 = i_1, \cdots, X_t = i_t) = P(X_{t+1} = i_{t+1}|X_t = i_t), \tag{2}$$

where i_t is the process state at the time t, and P is the conditional probability of a future event.

In Markov-chain based deterioration models, the performance of the bridge changes from one condition state (CR) to another according to a set of transition probabilities, as shown in Equation (3).

$$P = \left[\begin{pmatrix} p_{11} & \cdots & p_{1n} \\ \vdots & \ddots & \vdots \\ p_{n1} & \cdots & p_{nn} \end{pmatrix}\right] \left\{ \begin{array}{ll} p_{ij \geq 0} & i, j \in I \\ \sum\limits_{j \in I} p_{ij} = 0 & i \in I \end{array} \right., \tag{3}$$

where P is the transition probability matrix, p_{ij} is the transition probability from the state i to the state j, and n is the number of bridge condition states.

Therefore, the deterioration process $\{C(t), t \in T\}$ of bridges can be recognized as a stochastic process represented by the initial bridge condition vector $C(0)$ and the transition probability matrix (TPM), as shown in Equation (4).

$$C(t) = C(0) \times P^t, \tag{4}$$

where C is the condition rating vector, P is the transition probability matrix, and t is the time step.

However, the Markov chain based model has two underlying assumptions, namely, memoryless and homogeneous [9]. Memoryless means the future state of the deterioration process depends only on the current state and has nothing to do with the past, and homogeneous requires the probability of transition from one state to another remain constant throughout the time [23]. Therefore, this paper develops a semi-Markov process model for bridge deterioration prediction in Shanghai. In which, the time-dependent reliability theory was used to characterize the service-life behavior of bridge deterioration within each condition rating (CR) and the semi-Markov process was used to evaluate the transition probabilities of the bridge deterioration process between adjacent CRs. Thus, in the semi-Markov process model, the bridge future state is not only based on the current state but also associated with the history state, and the transition probability from one state to another also could change over time [24].

2.2. Time-Dependent Reliability

When the time-dependent reliability theory can be used to describe the bridge deterioration duration time within an individual CR, the underlying assumption is the existence of a probabilities relationship between the observed CR of discrete-state and the unobserved continuous deterioration process [4]. The CR of bridge performance is treated as a response variable subject to service time and other exponential variables. The bridge performance continues to decay with the increase of time [10]. Furthermore, once the unobserved continuous deterioration reaches or exceeds the threshold of boundaries between different CRs, the time $[0, t]$ will be recorded as the duration T_i of a particular $CR = i$.

As shown in Equation (5), the cumulative distribution function $F_i(t)$ of the duration T_i in $CR = i$ describes the probability that the bridge will transition out of $CR = i$ by the time t.

$$F_i(t) = prob(T_i \leq t) = \int_0^t f_i(x)dx, \tag{5}$$

where T_i is the duration of $CR = i$, and $f_i(t)$ is the probability distribution function of T_i.

Moreover, the survivor function $S(T_i)$ is the complement of $F_i(t)$ which means the probability that the bridge element would still be in $CR = i$ by the time t, and is defined as Equation (6).

$$S_i(t) = prob(T_i > t) = 1 - F_i(t), \tag{6}$$

where T_i is the duration of $CR = i$, and $F_i(t)$ is the cumulative distribution function of T_i.

Then the transition probability $p_{ij}(t,\Delta)$ out of $CR = i$ into a worse condition state $CR = j$ within the period Δ after the time t can be defined as Equation (7)

$$p_{ij}(t,\Delta) = prob(t < T_i < t+\Delta | T_i > t) = \frac{F_i(t+\Delta) - F_i(t)}{S_i(t)} = 1 - \frac{S_i(t+\Delta)}{S_i(t)}, \quad (7)$$

where T_i is the duration of $CR = i$, $F_i(t)$ is the cumulative distribution function of T_i, and $S_i(t)$ is the survivor function of T_i.

As the complement of $p_{ij}(t,\Delta)$, the transition probability $p_{ii}(t,\Delta)$ of remaining in the same state $CR = i$ can also be given as Equation (8).

$$P_{ii}(t,\Delta) = 1 - P_{ij}(t,\Delta) = \frac{S_i(t+\Delta)}{S_i(t)}, \quad (8)$$

where $S_i(t)$ is the survivor function of T_i.

Therefore, the hazard function $h_i(t)$ which describes the instantaneous rates of failure at the time t given that individual survives up till the time t can be calculated as Equation (9).

$$h_i(T) = \lim_{\Delta \to 0} \frac{p_{ij}(t,\Delta)}{\Delta} = \lim_{\Delta \to 0} \frac{F_i(t+\Delta) - F_i(t)}{\Delta S_i(t)} = \frac{f_i(t)}{S_i(t)} i, \quad (9)$$

where $p_{ij}(t,\Delta)$ is the transition probability from condition rating i to j within the period Δ, $f_i(t)$ is the probability distribution function of T_i, $F_i(t)$ is the cumulative distribution function of T_i, and $S_i(t)$ is the survivor function of T_i.

2.3. Semi-Markov Process

The transition process between different CRs of bridge deterioration can be described as a semi-Markov process Y_n with m feasible states. The associated random variables (X_n, TS_n) are the successive states and times of the n th transition, and the length of a sojourn time interval (TS_n, TS_{n+1}) is same as the random variable duration T_i of a specific condition state $X_n = i$ described in the time-dependent reliability theory. The distribution of duration T_i depends on both the state X_n being visited and the state X_{n+1} to be visited next [25].

For many states $i, j = 1, 2, \cdots, m$ and the time $t \geq 0$, the associated matrix of the transition probabilities Q_{ij}, formally called the semi-Markov kernel can be defined as shown in Equation (10).

$$\begin{aligned} Q_{ij}(t) &= P(X_{n+1} = j, TS_{n+1} - TS_n \leq t | (X_0, TS_0), (X_1, TS_1), \cdots (X_n, TS_n)) \\ &= P(X_{n+1} = j, TS_{n+1} - TS_n \leq t | X_n = i) \end{aligned} \quad (10)$$

where X_n is the successive state visited of the n th transition, and TS_n is the successive time of the n th transition.

Then the cumulative distribution function $F_i(T)$ is the same as described in Equation (5) which represents the duration T_i in the state i here, can be given as Equation (11).

$$F_i(T) = P(TS_{n-1} - TS_n \leq t | X_n = i) = \sum_{j=1}^{m} Q_{ij}(t), \quad (11)$$

where $Q_{ij}(t)$ is the semi-Markov kernel, and X_n is the successive state visited of the n th transition.

As well known, the eventual transition probability p^e_{ij} shown as Equation (12) represents the probability that the process can move from the state i to the state j neglecting the duration T_i in the state i.

$$p^e_{ij} = \lim_{t \to \infty} Q_{ij}(t) = P(X_{n+1} = j | X_n = i),\ i, j \in E, \quad (12)$$

where $Q_{ij}(t)$ is the semi-Markov kernel, and X_n is the successive state visited of the n th transition.

Additionally, the matrix $P = [p^e_{ij}]$ is called the transition probability of the embedded Markov chain, and the following conditions must be satisfied:

$$\begin{cases} p^e_{ij} \geq 0 \\ \sum\limits_{i,j=1}^{m} p^e_{ij} \geq 0 \end{cases}. \tag{13}$$

Then the conditional distribution function $G_{ij}(t)$ given both the current state i and next state j can be defined as shown in Equation (14).

$$G_{ij}(t) = P(TS_{n+1} - TS_n \leq t | X_n = i, X_{n+1} = j) = \begin{cases} \frac{Q_{ij}(t)}{p^e_{ij}} & p^e_{ij} \neq 0 \\ 1 & p^e_{ij} = 0 \end{cases}, \tag{14}$$

where $Q_{ij}(t)$ is the semi-Markov kernel, X_n is the successive state visited of the n th transition, and TS_n is the successive time of the n th transition.

Therefore, the semi-Markov process kernel Q_{ij} could be defined as:

$$Q_{ij}(t) = p^e_{ij} G_{ij}(t),\ if\ p_{ij} \neq 0, \tag{15}$$

where p^e_{ij} is the eventual transition probability, and $G_{ij}(t)$ is the conditional distribution function.

Therefore, the transition probability of the semi-Markov process can be defined in the following:

$$p_{ij}(t) = P(X_{n+1} = j | X_n = i), \tag{16}$$

where is the successive states visited of the n th transitions.

If we assume a semi-Markov process has spent initial time t_0 in the state i before the time $t = 0$ (i.e., the age of bridge when the observation starts), with a destination state j, and the system could go through an intermediate state l. Let τ be the time measured for the stay in the state i from the time $t = 0$ to the time of transition to the state k. Then the transition probabilities $\phi_{ij}(t)$ of the semi-Markov process can be calculated as following [26]:

$$\begin{aligned} p_{ij}(t) &= \delta_{ij}(1 - F_i(t)) + \sum_{l=1}^{m} \int_0^t p_{lj}(t-\tau) dQ_{il}(\tau) \\ &= \delta_{ij}(1 - F_i(t)) + \sum_{l=1}^{m} \int_0^t Q'_{il}(\tau) p_{lj}(t-\tau) d\tau \end{aligned} \tag{17}$$

where δ_{ij} is the Kronecker δ (i.e., if $i = j, \delta_{ij} = 1$ or if $i \neq j, \delta_{ij} = 0$), $F_i(t)$ is the cumulative distribution function, and $Q_{il}(\tau)$ is the semi-Markov process kernel.

Equation (17) is a complicated recursive equation, which can only be solved by Laplace transforms. However, let h be the step measure (i.e., $t = kh$), and using a simplified discretization algorithm. The approximate numerical solution of the transition probability can be calculated as following [27].

$$p_{ij}(kh) = \delta_{ij}(1 - F_i(kh)) + \sum_{l=1}^{m} (\sum_{\tau=0}^{k} w_{k\tau} p_{ij}(kh - \tau h) Q'_{il}(\tau h)), \tag{18}$$

where h is the step measure, δ_{ij} is the Kronecker delta (i.e., if $i = j, \delta_{ij} = 1$ or if $i \neq j, \delta_{ij} = 0$), $F_i(kh)$ is the cumulative distribution function, $w_{k\tau}$ is the weight related to the quadrature formula, and Q_{il} is the semi-Markov process kernel.

3. Deterioration Prediction Model

The BMS has been applied to the management of urban bridges in Shanghai since 2004, and most urban bridges across the city are inspected annually. The original data used in this study is derived from a total of 3185 urban bridge records and 13 years of historical inspection records stored in the BMS database. Besides, the number of inspection data records in the BMS database is as following (Table 2). It should be noted that not every bridge in the database was examined annually, owing to some reasons that include the water level is too high; the bridge is under repair; or the structure form does not apply to the BCI evaluation method.

Table 2. Data records in the Bridge Manage System (BMS) database of Shanghai (2004–2016).

Year	All Bridges		Concrete Beam Bridges		Valid Data Records
	Bridge Numbers	Inspection Records	Bridge Numbers	Inspection Records	
2004	1390	1091	1116	899	906
2005	1550	1357	1270	1152	830
2006	1601	1447	1335	1248	932
2007	1644	1486	1379	1283	1030
2008	1748	1608	1478	1391	1142
2009	1804	1590	1528	1373	1159
2010	1899	1686	1606	1463	1241
2011	1972	1727	1663	1495	1273
2012	1911	1756	1609	1500	1267
2013	1982	1801	1655	1547	1234
2014	2177	1992	1787	1668	1345
2015	2305	2027	1804	1673	1245
2016	2377	1799	1860	1545	1150
Total	3185	21,367	2468	18,237	14,754

Although there are a variety of different structural forms of the urban bridges, the concrete beam bridge accounted for most of the total number. The necessary information about these concrete beam bridges is also more accurate than others. Thus, the deterioration process of concrete beam bridges is focused on in this paper.

3.1. Data Preparation Process

It is essential to avoid the interference of the inspector's subjective deviation and the external intervention of MRR activities, which could guarantee that data is analyzed consistent with the deterioration phenomenon. Therefore, an active process of data preparation is crucial when developing a bridge deterioration prediction model. In this paper, the following main steps of data preparation were used before the development of the deterioration prediction model:

- Cleaning abnormal or null records among the inspection records and bridge ages;
- Filtering inspection records with CR non-monotonically decline or decline over two ratings in two adjacent years;
- Filling the missing records with the same CR value if the CR is the same for three adjacent consecutive years.
- Resetting the age of bridges with rating A which just upgraded from rating D–E to 0;
- Using the analysis-calibration method to set a reasonable data range threshold for bridge ages and CRs based on the bridge engineering experience. (For example, the design service life of urban bridges in China is generally 100 years. Bridges cannot decay from A to D within 20 years, thus determining the lower boundary. Furthermore, bridges cannot remain at A for more than 40 years, then the upper limit of the reasonable data range can be set.)

After the data processing process discussed above, the obtained number of valid data is as shown in the last column of Table 2. For example, in the year of 2016, there are 1150 accurate inspection records of concrete beam bridges meeting the requirement for subsequent analysis and modeling. Overall, the total number of valid data accumulated from 2004 to 2016 is 14,754. Additionally, each of the records contains five sub-records, including the bridge age and CRs of the whole bridge and three main components. For ease of calculation, the condition ratings stored in the BMS database are simplified from a qualitative rating [A, B, C, D, E] to an ordinal system [1, 2, 3, 4, 5].

3.2. *Weibull-Distribution Parameter Estimation*

Estimation of the reliability function by fitting the distribution parameters of service-life data is the main subject of survival analysis which has been widely used for machine reliability testing and life expectancy prediction. However, when using this method, the presence of censored observation should be taken into account [2,23]. Censored observations are incomplete, and the different influence of the complete, the right-censored, and the left-censored observation is considered as the following [21].

- Complete observation: if a bridge reaches a well-defined threshold value at a certain age, then we have a complete observation about the service lifetime of the bridge.
- Right-censored observation: if a bridge had not reached the threshold value at the time of bridge inspection, then we have a right-censored observation. The right-censored observation can tell us the service lifetime of the bridge goes beyond its present age.
- Left-censored observation: if a bridge had already surpassed the threshold value at the time of inspection, then we have a right-censored observation. The left-censored observation can tell us that the lifetime of the bridges is less than or equal to the present age.

In this paper, the data size of different kinds of observations is shown as in Table 3. Compared with the left-censored observation, the sample number of right-censored observation is relatively large and can be more easily used to estimate parameters. Therefore, the records of left-censored observations were eliminated, and the right-censored observation records were used in the estimation of the Weibull-distribution parameters together with the complete observation records after the tag processing.

Table 3. Data size of different observations.

CR	Component Type	Number of Different Observations		
		Complete	Right-Censored	Left-Censored
1	Whole bridge	1041	1023	51
	Deck system	1049	912	45
	Superstructure	1016	1112	53
	Substructure	783	1362	54
2	Whole bridge	631	463	36
	Deck system	945	385	35
	Superstructure	555	308	27
	Substructure	691	240	28
3	Whole bridge	234	141	18
	Deck system	832	243	54
	Superstructure	427	158	14
	Substructure	228	28	11
4	Whole bridge	149	16	42
	Deck system	549	68	32
	Superstructure	174	47	15
	Substructure	45	6	8

The Weibull distribution has been used in the analysis of time-dependent reliability and service life behavior due to its flexibility in fitting different types of service life data. More than that, with different values of the shape parameter β, the Weibull distribution can be related to a number of other probability distributions such as the normal distribution, the exponential distribution, and even the Rayleigh distribution. Once the Weibull parameters are obtained, the reliability function of Weibull distribution can be used to model a variety of bridge service-life behaviors within different CRs (including the survival function, the hazard function, the mean value, and the quantile statistics).

In this paper, the Weibull distribution method was used to model the observed duration T_i, which represents the time of the bridge staying at a particular $CR = i$. The Weibull distribution is mathematically defined by its pdf equation, and the two-parameter pdf expression of Weibull distribution is defined as follows.

$$f(T_i) = \frac{\beta_i}{\eta_i}\left(\frac{T_i}{\eta_i}\right)^{\beta_i - 1} e^{-\left(\frac{T_i}{\eta_i}\right)^{\beta_i}}, \tag{19}$$

where $T_i \geq 0$ is the duration, $\beta_i > 0$ is the shape parameter, and $\eta_i > 0$ is the scale parameter.

The shape parameter β_i and scale parameter η_i of Weibull distribution can be both obtained by fitting the observed durations of the whole bridge, the deck system, the superstructure, and the substructure at different CRs. The estimated results of the Weibull distribution parameters are as shown in Table 4.

Table 4. Estimated parameters of the Weibull-distribution at different condition ratings (CRs).

CR	Component Type	Estimated Weibull Distribution Parameters			
		Shape β_i	Scale η_i	Mean Value	Standard Deviation
1	Whole bridge	1.458	27.531	24.944	17.388
	Deck system	1.411	25.587	23.293	16.739
	Superstructure	1.421	29.027	26.395	18.838
	Substructure	1.429	37.049	33.663	23.907
2	Whole bridge	1.599	26.025	23.334	14.940
	Deck system	1.510	21.387	19.292	13.022
	Superstructure	1.633	23.881	21.373	13.423
	Substructure	1.484	21.868	19.768	13.555
3	Whole bridge	1.328	31.788	29.237	22.231
	Deck system	1.502	22.127	19.971	13.540
	Superstructure	1.402	22.571	20.567	14.865
	Substructure	1.616	20.832	18.661	11.833
4	Whole bridge	1.217	21.266	19.933	16.461
	Deck system	1.417	19.824	18.034	12.906
	Superstructure	1.315	29.686	27.356	20.996
	Substructure	1.597	23.985	21.508	13.784

Through the fitting results of Weibull parameters listed in the above table, we can find that all the shape parameters are never estimated as 1. In which, the Weibull distribution would be the same with exponential distribution as a special case. Therefore, the memoryless and homogeneous assumptions of the Markov chain theory seem invalid in the deterioration process of urban bridges. In fact, the Weibull distributions under the circumstance where $\beta_i > 1$ are known as wear-out failures, which means the failure rates of concrete beam bridges in Shanghai are increasing with time.

By using the mean value of durations at each CR, the service life expectancy of urban bridges can be predicted, as shown in Figure 2. The prediction result reveals the service life expectancy of concrete beam bridges in Shanghai is about 77 years, while the mean life of the deck system is only 62 years. The decay rate of the deck system is the fastest when compared with the other two parts. Meanwhile, the substructure has a much longer life expectancy, and that is also consistent with the actual bridge maintenance experience. As the bridge deck system directly bears the load effect, which makes components like the deck pavement, expansion joints, etc., more susceptible to damage. The substructure is generally less damaged, and the abutment foundation is relatively stable.

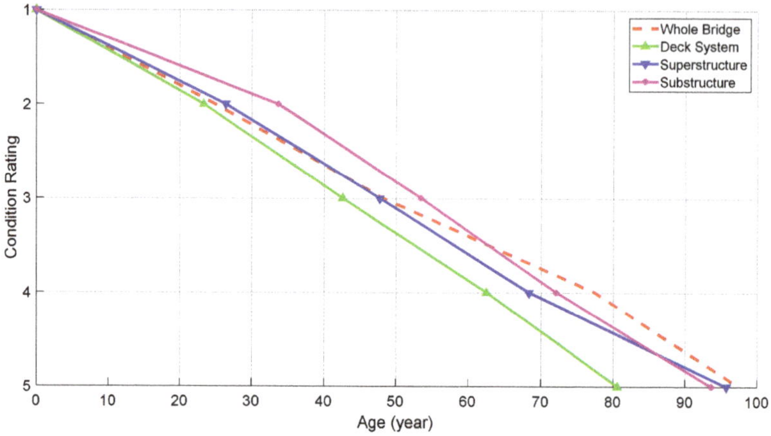

Figure 2. Deterioration prediction of urban bridges at different CRs.

3.3. Semi-Markov Transition Probability Evaluation

As mentioned above, urban bridges in Shanghai are inspected once a year, so the step measure h in Equation (18) could be set to 1. In addition, considering the simplest quadrature method (i.e., rectangle formula) [26], the transition probability $p_{ij}(t)$ can be obtained by the following equation after the necessary simplification [24].

$$\begin{cases} p_{ij}(t) = S_i(t), & i = j \\ p_{ij}(t) = \sum_{k} \sum_{x=1}^{t} f_{ik}(x) p_{kj}(t-x), & i \neq j \end{cases}, \tag{20}$$

where $S_i(t)$ is the cumulative distribution function, k is the intermediate state between i and j, and $f_{ik}(x)$ is the probability density function of duration T_i from $CR = i$ to $CR = k$.

In this paper, it was assumed that bridges can degrade no more than two CRs in the interval of two adjacent years, and $t = 0$ and $\Delta = t - x$ were set in Equation (7) [24]. Then the calculation equation of the semi-Markov transition probability can be rewritten as Equation (21).

$$\begin{aligned} p_{ij}(t) &= \sum_{k} \sum_{x=1}^{t} f_{ik}(x) \left[\frac{F_{kj}(t-x) - F_{kj}(0)}{1 - F_{kj}(0)} \right], \; i \neq j \\ &= \sum_{k} \sum_{x=1}^{t} f_{ik}(x) F_{kj}(t-x) \end{aligned} \tag{21}$$

where $f_{ik}(x)$ is the probability density function of duration T_i from $CR = i$ to $CR = k$, and $F_{kj}(t-x)$ is the cumulative density function of duration T_k from $CR = k$ to $CR = j$.

For example, the CR transition probabilities p_{ij} in different years of the deck system and the substructure starting at $CR = 1$ can be calculated through the semi-Markov method. As can be seen from Figure 3, transition probabilities of the deck system decay to a worse CR are much larger than the substructure at the same bridge age. The deck system has more than 50% probability that it may decay to $CR = 2$ and $CR = 3$ after the 19th year, while the substructure has more than half the chance to remain at $CR = 1$ till the 28th year. Additionally, the prediction result shows that the deck system has a better chance of remaining at $CR = 1$ in the first 24 years and then the deck system will have about 40% probability decay to $CR = 2$ between 24 and 34 years. Finally, the deck system will have more than 40% probability transit to $CR = 3$ after 34 years later. As for the substructure, transition probabilities of remaining at $CR = 1$ are more likely than decaying to $CR = 2$ or $CR = 3$ in the first 39 years. Then the substructure is more likely to decay to $CR = 3$.

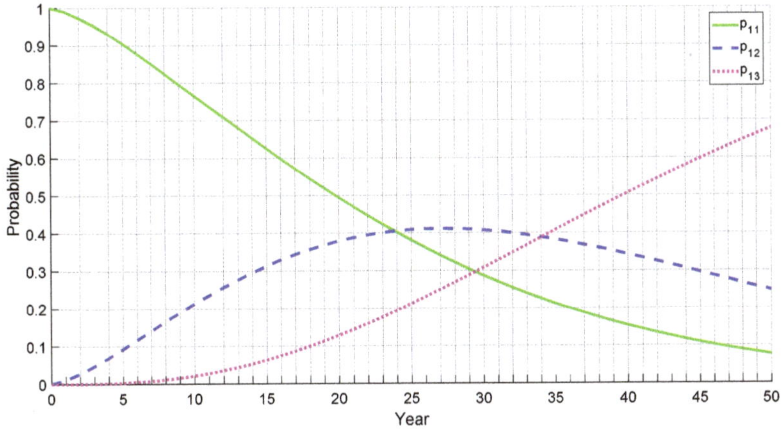

(**a**)

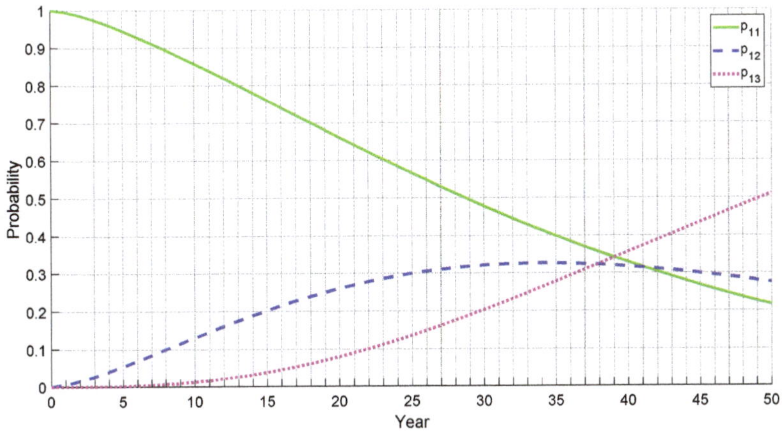

(**b**)

Figure 3. Semi-Markov transition probabilities change with bridge service time: (**a**) transition probability of the deck system; (**b**) transition probability of the substructure.

4. Discussion

The initial condition $CR(0)$ at the time $t = 0$ is needed in the form of a state vector, as shown in Equation (22).

$$CR(0) = \begin{bmatrix} p_1 & p_2 & p_3 & p_4 & p_5 \end{bmatrix}, \tag{22}$$

where p_i is the proportion of the whole bridge inventory in $CR = i$ at the time $t = 0$.

Then the predicted $CR(t)$ of the semi-Markov process would be a product of the initial condition vector and the transition probability matrix, as shown below.

$$CR(t) = CR(0) \times P_{ij}(t), \tag{23}$$

where $P_{ij}(t) = \begin{bmatrix} p_{11}(t) & p_{12}(t) & p_{13}(t) & 0 & 0 \\ 0 & p_{22}(t) & p_{23}(t) & p_{24}(t) & 0 \\ 0 & 0 & p_{33}(t) & p_{34}(t) & p_{35}(t) \\ 0 & 0 & 0 & p_{44}(t) & p_{45}(t) \\ 0 & 0 & 0 & 0 & 1 \end{bmatrix}$ is the transition probability matrix of the semi-Markov process at the time t.

If we assume the proportion of the whole bridge, the deck system, the superstructure, and the substructure with different CRs in 2016 as the initial proportion of the bridge inventory, then the future CR proportion change caused by performance deterioration can be predicted through the semi-Markov process model. The comparison between prediction results of the semi-Markov method and the regression analysis method currently used by Shanghai BMS in 2017 and 2018 are shown in Table 5. The predicted values of the semi-Markov method are closer to the actual values, while the traditional approach has a significant deviation. It suggests that the overall prediction accuracy of the semi-Markov model is better than the regression analysis method. The relative errors of the semi-Markov model for bridges at $CR = 1$ are nearly 6% while the traditional ones are more than 10%. Furthermore, for bridges at $CR = 2$, the relative errors are less than 6% of the proposed model; however, the traditional model has about 50% relative error. As for bridges at $CR = 3$, when the relative errors of the conventional method are about 300%, the relative error of the new approach is 5% in 2017 and 35% in 2018. For bridges with more deteriorated CRs, smaller data size and external interference may have negative impacts on deterioration prediction. However, compared to the large deviations of the regression analysis, the prediction accuracy of the semi-Markov model for bridges at $CR = 4$ and $CR = 5$ is also acceptable.

Table 5. Comparison of the regression analysis and semi-Markov model prediction results.

CR	Prediction Model	2017			2018		
		Actual Value	Expected Value	Relative Error	Actual Value	Expected Value	Relative Error
1	Regression analysis	52.40%	46.34%	11.56%	52.24%	38.47%	26.36%
	Semi-Markov		49.36%	5.81%		48.67%	6.83%
2	Regression analysis	39.61%	13.73%	65.34%	41.27%	20.84%	49.51%
	Semi-Markov		41.72%	5.32%		41.94%	1.63%
3	Regression analysis	7.15%	28.22%	294.73%	5.82%	28.99%	398.25%
	Semi-Markov		7.59%	6.18%		7.93%	36.32%
4	Regression analysis	0.80%	9.62%	1103.57%	0.59%	9.62%	1541.58%
	Semi-Markov		1.22%	52.25%		1.29%	119.29%
5	Regression analysis	0.04%	2.09%	4606.68%	0.08%	2.09%	2396.51%
	Semi-Markov		0.12%	165.75%		0.16%	95.55%

As shown in Figure 4, the Weibull distribution based semi-Markov process bridge deterioration model can be used to predict the overall deterioration trend of urban bridges at network-level. It is easy to find that the overall decay rate of the deck system is the fastest, the superstructure is the second, and the substructure is the slowest. As for the whole bridge, there will be about 51% of bridges remaining in $CR = 1$, 34% remaining in $CR = 2$, and 12% remaining in $CR = 3$ in the next decade. It follows that the proportion of bridges with $CR = 1$ will gradually decrease, while the proportion of bridges with $CR = 3$ and $CR = 4$ will increase rapidly within the next 50 years. The predicted results are in a good agreement with the actual degrade trend of the overall urban bridges in Shanghai. In order to reduce the bridge maintenance pressure in the future, it is necessary to adopt a targeted preventive maintenance strategy and to delay the decay rate of bridge performance.

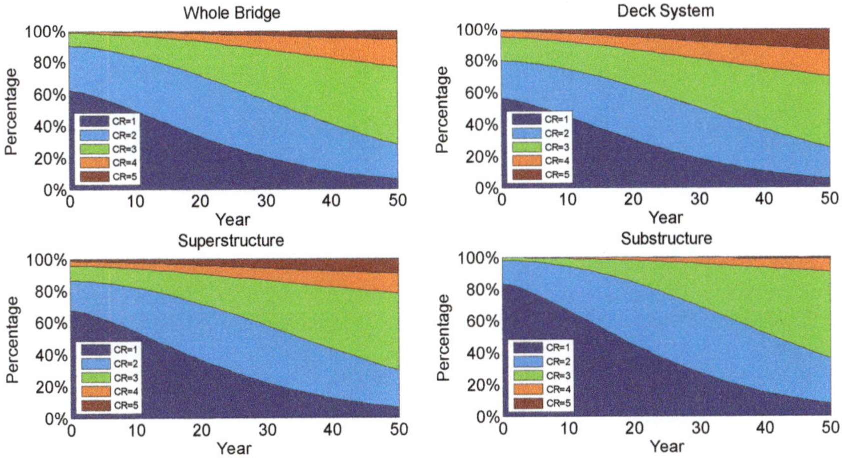

Figure 4. Prediction of the concrete beam bridge deterioration in Shanghai.

5. Conclusions

The performance of urban bridges will gradually deteriorate with the increase of service time. The existing bridge deterioration prediction model of the BMS in Shanghai uses the regression analysis method. This paper proposed a Weibull distribution based semi-Markov process prediction model by using the bridge inspection data in BMS since 2004. The Weibull distribution was used to characterize the bridge service-life behavior of duration time within each CR, and the semi-Markov process was used to evaluate the transition probabilities of bridge deterioration between different CRs. The service life expectancy of urban bridges, the transition probabilities of the deck system and the substructure, and the future CR proportion change caused by deterioration was predicted.

The prediction result reveals that the estimated shape parameters of Weibull distribution are all larger than 1, which indicates the failure rates of urban bridges in Shanghai are not homogeneous but increase with time. The service life expectancy of concrete beam bridges in Shanghai is about 77 years. Among the three major parts of urban bridges, the decay rate of the deck system is the fastest, and the substructure has a much longer life expectancy. It may be because the bridge deck system directly bears the load effect, which makes it more susceptible to damage. By comparison, the overall prediction accuracy of the semi-Markov model is better than the regression analysis method. It follows that the proportion of bridges $CR = 1$ with $CR = 2$ will gradually decrease, while the proportion of bridges with $CR = 3$ and $CR = 4$ will increase rapidly within the next 50 years. The prediction results show it is necessary to strengthen the maintenance of the deck system and adopt a preventive maintenance strategy for bridges now with less deteriorated CRs to reduce the future bridge maintenance pressure.

Compared with the existing regression analysis method, the Weibull distribution based semi-Markov process bridge deterioration model can describe the performance decay behavior of the bridge in more detail, and the prediction accuracy is relatively higher. Therefore, when the data size of historical bridge inspection records is big enough, this model can be well applied to the deterioration prediction of urban bridges both at project-level and network-level. At the same time, the proposed model can provide some basis for bridge maintenance decision-making and financial allocation optimization.

Author Contributions: Conceptualization, L.S.; methodology, Y.F.; software, Y.F.; validation, Y.F.; formal analysis, Y.F.; investigation, Y.F.; resources, L.S.; data curation, L.S.; writing—original draft preparation, Y.F.; writing—review and editing, Y.F. and L.S.; visualization, Y.F.; supervision, L.S.; project administration, L.S.; funding acquisition, L.S.

Funding: This research was funded by the National Key R&D Program of China, grant number No. 2018YFB1600100.

Acknowledgments: Thanks to the Shanghai Road Administration Bureau for kindly providing the bridge historical inspection records in the BMS database for this case study.

Conflicts of Interest: The authors declare no conflict of interest. The funders had no role in the design of the study; in the collection, analysis, or interpretation of data; in the writing of the manuscript; or in the decision to publish the results.

References

1. Foraboschi, P. Analytical model to predict the lifetime of concrete members externally reinforced with FRP. *Theor. Appl. Fract. Mech.* **2015**, *75*, 137–145. [CrossRef]
2. Branco, F.A.; Brito, J.D. *Handbook of Concrete Bridge Management*; American Society of Civil Engineers: Reston, VA, USA, 2004.
3. Morcous, G. Performance prediction of bridge deck systems using Markov chains. *J. Perform. Constr. Fac.* **2006**, *20*, 146–155. [CrossRef]
4. Mishalani, R.G.; Madanat, S.M. Computation of infrastructure transition probabilities using stochastic duration models. *J. Infrastruct. Syst.* **2002**, *8*, 139–148. [CrossRef]
5. Li, L.; Li, F.; Chen, Z.; Sun, L. Use of Markov Chain Model Based on Actual Repair Status to Predict Bridge Deterioration in Shanghai, China. *Transp. Res. Record* **2016**, *2550*, 106–114. [CrossRef]
6. Fang, Y.; Li, L.; Chen, Z.; Sun, L. Prediction Model of Concrete Girder Bridge Deterioration in Shanghai Using Weibull-Distribution Method. 2017. Available online: https://trid.trb.org/view/1438643 (accessed on 8 December 2016).
7. China, M.O.C.O. Technical Code of Maintenance for City Bridges. *S.* In CJJ 99-2003, China; 2017. Available online: http://www.mohurd.gov.cn/wjfb/201801/t20180104_234665.html (accessed on 31 July 2017).
8. Foraboschi, P. Shear strength computation of reinforced concrete beams strengthened with composite materials. *Compos.: Mech. Comput. Appl.: Int. J.* **2012**, *3*, 227–252. [CrossRef]
9. Foraboschi, P. Structural layout that takes full advantage of the capabilities and opportunities afforded by two-way RC floors, coupled with the selection of the best technique, to avoid serviceability failures. *Eng. Fail. Anal.* **2016**, *70*, 387–418. [CrossRef]
10. Agrawal, A.K.; Kawaguchi, A.; Chen, Z. Deterioration rates of typical bridge elements in New York. *J. Bridge. Eng.* **2010**, *15*, 419–429. [CrossRef]
11. Lu, P.; Pei, S.; Tolliver, D.; Jin, Z. Data-based Evaluation of Regression Models for Bridge Component Deterioration. 2015. Available online: https://trid.trb.org/view/1337198 (accessed on 30 December 2014).
12. Chen, Z. Research on Technology Structure of Transportation Infrastructure Management System. Ph.D. Thesis, Tongji University, Shanghai, China, July 2005.
13. Zambon, I.; Vidovic, A.; Strauss, A.; Matos, J.; Amado, J. Comparison of stochastic prediction models based on visual inspections of bridge decks. *J. Civ. Eng. Manag.* **2017**, *23*, 553–561. [CrossRef]
14. Su, D.; Nassif, H.; Hwang, E. Probabilistic Approach for Forecasting Long Term Performance of Girder Bridges. 2015. Available online: https://trid.trb.org/view/1339284 (accessed on 30 December 2014).
15. Madanat, S.; Mishalani, R.; Ibrahim, W.H.W. Estimation of infrastructure transition probabilities from condition rating data. *J. Infrastruct. Syst.* **1995**, *1*, 120–125. [CrossRef]
16. Golabi, K.; Shepard, R. Pontis: A system for maintenance optimization and improvement of US bridge networks. *Interfaces* **1997**, *27*, 71–88. [CrossRef]
17. Hawk, H.; Small, E.P. The BRIDGIT bridge management system. *Struct. Eng. Int.* **1998**, *8*, 309–314. [CrossRef]
18. Reardon, M.F.; Chase, S.B. Migration of Element-Level Inspection Data for Bridge Management System. 2016. Available online: https://trid.trb.org/view/1392750 (accessed on 1 December 2016).
19. Ng, S.K.; Moses, F. Prediction of bridge service life using time-dependent reliability analysis. *Bridge Manag.* **1996**, *3*, 26–32.
20. Medjoudj, R.; Aissani, D.; Boubakeur, A.; Haim, K.D. Interruption modelling in electrical power distribution systems using the Weibull—Markov model. *Proc. Inst. Mech. Eng. Part O: J. Risk Reliab.* **2009**, *223*, 145–157. [CrossRef]
21. Madanat, S.M.; Karlaftis, M.G.; McCarthy, P.S. Probabilistic infrastructure deterioration models with panel data. *J. Infrastruct. Syst.* **1997**, *3*, 4–9. [CrossRef]

22. Micevski, T.; Kuczera, G.; Coombes, P. Markov model for storm water pipe deterioration. *J. Infrastruct. Syst.* **2002**, *8*, 49–56. [CrossRef]
23. Wellalage, N.W.; Zhang, T.; Dwight, R.; El-Akruti, K. Bridge deterioration modeling by Markov Chain Monte Carlo (MCMC) simulation method. In *Engineering Asset Management-Systems, Professional Practices and Certification*; Springer: Cham, Switzerland, 2015; pp. 545–556.
24. Sobanjo, J.O. State transition probabilities in bridge deterioration based on Weibull sojourn times. *Struct. Infrastruct. E* **2011**, *7*, 747–764. [CrossRef]
25. Ng, S.; Moses, F. Bridge deterioration modeling using semi-Markov theory. *A. A. Balkema Uitgevers B. V, Struct. Saf. Reliab.* **1998**, *1*, 113–120.
26. Corradi, G.; Janssen, J.; Manca, R. Numerical treatment of homogeneous semi-Markov processes in transient case—A straightforward approach. *Methodol. Comput. Appl.* **2004**, *6*, 233–246. [CrossRef]
27. Kallen, M.J.; Van Noortwijk, J.M. Statistical Inference for Markov Deterioration Models of Bridge Conditions in the Netherlands. 2006. Available online: http://citeseerx.ist.psu.edu/viewdoc/download?doi=10.1.1.74.5105&rep=rep1&type=pdf (accessed on 2 March 2015).

sustainability

MDPI

Article

Implementing LEED v4 BD+C Projects in Vietnam: Contributions and Challenges for General Contractor

Duy Hoang Pham [1], Joosung Lee [2,*] and Yonghan Ahn [3,*]

1 Department of Architectural Engineering, Hanyang University, 55 Hanyangdaehak-ro, Sangrok-gu, Ansan-si, Gyeonggi-do 15588, Korea; phamduyhoang@hangyang.ac.kr

2 Innovative Durable Building and Infrastructure Research Center, Hanyang University, 55 Hanyangdaehak-ro, Sangrok-gu, Ansan-si, Gyeonggi-do 15588, Korea

3 School of Architecture and Architectural Engineering, Hanyang University, 55 Hanyangdaehak-ro, Sangrok-gu, Ansan-si, Gyeonggi-do 15588, Korea

* Correspondence: js4ever@hanyang.ac.kr (J.L.) yhahn@hanyang.ac.kr (Y.A.)

Received: 13 August 2019; Accepted: 29 September 2019; Published: 1 October 2019

Abstract: Sustainable construction addresses both rising housing demand and the need to reduce energy consumption, but is not yet widespread in Vietnam, although the number of Green Building Certified projects has increased significantly since 2015, especially those with LEED certification. Certification adds value to the project but also incurs additional costs and implementation challenges for general contractors (GCs) and other stakeholders. The growing popularity of LEED buildings requires GCs to understand their role in fulfilling the LEED requirements. We therefore conducted a thorough review of the previous research on LEED v4 guidelines and their local equivalents to determine what is expected of GCs working on LEED v4 BD + C projects in Vietnam. A survey of 72 experts, engineers, and architects working in the construction sector identified the LEED tasks where Vietnamese GCs lack experience and suggested solutions to address this shortfall were developed. In particular, Vietnamese GCs lacked experience in implementing their sustainable roles. These results provide a useful foundation for Vietnamese GCs seeking to expand the scope of their LEED work and identified necessary training. Our findings will also guide future research to help GCs in Vietnam adapt to LEED's sustainability requirements and support sustainable construction in Vietnam.

Keywords: sustainable roles; LEED; contractors; Vietnam

---

1. Introduction

Sustainable development is vital if humanity is to meet the needs of the current world population without adversely affecting the lives of the next generation [1,2]. Over the last decade, many countries have begun to be severely affected by the effects of climate change, global warming, pollution, and the depletion of natural resources, all of which are the direct consequences of the industrial revolution [2–4]. Our growing awareness of these problems has made sustainable development a priority as it represents an important way to mitigate some of the most serious environmental impacts of industrialization [5,6]. The construction industry is a significant consumer of energy and raw materials and is thus a major contributor to global warming [7–11]. Several previous studies have indicated that globally, the construction industry consumes ~50% of the energy produced, nearly 40% of the raw materials, and 16% of the water, and is responsible for 35 % of the CO_2 produced [12–19]. It also contributes to problems such as air pollution, noise pollution, and waste pollution, among others [1,18,19]. To address these issues, sustainable construction practices and green building solutions are being implemented in countries around the world, optimizing resource use, reducing energy consumption, and minimizing the impact of buildings on the surrounding environment [20,21]. Sustainable construction has become a major global trend in the construction industry [22,23] because of its benefits and its ultimate goals of preserving our quality of life by minimizing negative environmental impacts and protecting valuable resources that will be needed by future generations. [24–26].

Green building certification systems promote sustainable construction activities by providing guidance and rewarding efforts to deliver sustainable, energy-efficient buildings [27]. Many countries have developed certification systems, including the United States (LEED; Green Building Council), the United Kingdom (BREEAM; Building Research Establishment), and France (HQE; Association pour la Haute Qualité Environnementale) [28–30]. Among these, LEED, which was first introduced in 1993, aims to minimize the adverse environmental impacts of construction activities and ensure energy/water efficiency. LEED has played a leading role in popularizing green building certification systems worldwide [6,8,31–38], and a majority of existing green building projects are now implemented based on the guidance provided by the LEED rating system or its local equivalent [39].

As in many other developing countries, Vietnam's rapid economic growth is creating considerable pressure on housing demand and raw resources, as well as environmental pollution [40,41]. Vietnam's long coastline and low-lying and densely populated delta regions make it one of the five nations in the world that are most severely affected by rising sea levels and climate change [42–44]. As the construction industry made up ~34.28 % of Vietnam's GDP in 2018 [45], the government considers promoting sustainable construction to be a potential solution that not only addresses issues related to housing demand and energy consumption but also minimizes the nation's environmental vulnerability [46]. Nguyen et al. found that the low cost of electricity in Vietnam is one of the worst drivers of investment decisions that ignore energy efficiency because of the long payback time [41]. In an attempt to address this issue, the government increased electricity prices by 8.36% in 2018, significantly increasing operating costs for buildings. It has also implemented a policy of purchasing electricity (paying US$0.0935/kWh) from owners who install renewable energy systems [47]. Due to these incentives and the better payback compared to previous green investment opportunities [6], the number of developers interested in green building projects has been increasing [41], as shown by the significant increase in the number of green certifications issued in recent years. As of December 2018, there were 174 LEED projects [48], 34 EDGE projects [49], and 51 LOTUS [50] projects in Vietnam, along with a few other certifications. These figures indicate that LEED certification is by far the most popular rating system in Vietnam.

Green building projects are increasing in popularity because of the many advantages they bring for the buildings' occupants [20]. Although LEED and other rating systems provide a way to measure a green building's quality, they also support sustainability and bring other benefits to the building users [41] due to their more stringent construction requirements and well-considered building designs. LEED's Integrative Project credits (IPp1) also require substantial input from the architects and engineers; the project manager; and all the contractors, construction consultants, and building operation experts involved from the early-project phase onward [51–53]. Currently, green building projects in Vietnam face many difficulties because of the participants' lack of relevant experience and the limited technical support available [54], which creates risks simply due to the lack of information, the need to rework sections and the increased costs incurred for construction participants such as GCs, architects and engineers, subcontractors, and the other partners involved in implementing green building projects [20,55,56]. This is especially true for GCs, who are directly responsible for ensuring the sustainability of the project [57]. With a solid understanding of their roles related to individual sustainability construction concepts, contractors will be able to achieve a better performance in delivering their sustainable project [57]. Thus, GCs need and want to understand their precise roles in implementing a LEED project, which will enable them to identify cost-efficient methods to satisfy the sustainable needs of both the building developer and the ultimate users [58]. To meet the requirements of LEED certification with a minimal increase in the need for reworking, and the additional costs this incurs, the GCs must gain more knowledge in order to fully understand what is expected of them when undertaking a LEED project [59]. This study, therefore, focuses on the sustainable roles of GCs working on a LEED v4 BD + C project by reviewing the updated "LEED v4 BD+C Technical Guidelines", previous papers examining this topic and local conditions and regulations. In order to identify the risks that arise due to inexperience and the associated need to carry out extensive remediation work, a questionnaire survey was utilized to determine the level of experience of the GCs in Vietnam who are implementing sustainable green building projects.

2. Literature Review

2.1. Previous Studies on LEED and GCs

GCs play a vital role in the success and costefficiency of any construction project; this is especially true for a sustainable building or a successful LEED project [53,57,59]. Many studies in the field of construction management and sustainable building have therefore focused on the roles of GCs and their contribution to the success of LEED projects. These studies have generally highlighted the need for greater awareness of the importance of GCs' roles in accomplishing sustainable building projects on time and within budget.

A number of studies have pointed out the GCs' roles and responsibilities and identified their sustainable site roles. Mollaoglu-Korkmaz et al. (2013) analyzed the impact of each level of integration achieved on success in meeting the sustainable goals of the project set by the owner and pointed out, in particular, the potential role of the constructor in the earlier phases of the project [60]. The U.S. Green Building Council identified 17 credits that are significant for GCs and their responsibility for implementing and documenting LEED requirements [61]. Glavinich reviewed several case studies to evaluate the role of GCs in the design process, subcontractor management, material management, LEED file collection, and the commissioning process (Cx) [62]. In his review of eight LEED projects in the US, Klehm pointed out that GCs perform three management functions in LEED project, namely, subcontractor management, materials management, and activity management [63]. Syal et al. examined the impacts of LEED credits on the GCs' function and the sustainable site practices required for green building projects [64] and analyzed the roles of GCs working on LEED NC projects and its impact on their construction management practices [65]. Frattari et al. studied the sustainable roles of the GCs in two LEED project in Italy [66], demonstrating the international significance of the LEED certification system.

Several studies focus on implementation guides to help GCs achieve their sustainability goals. Kibert (2012) described GBs systems and implementation [25]. Ahn et al. developed a process model that includes identifying the sustainable goals, implementation process, and roles of participating teams, including GCs, throughout the construction phase [67]. Bayraktar et al. (2009) provide a set of LEED implementation guidelines for GCs and others to assist them with the certification process [68]. Schaufelberger and Cloud also provided a four-step guide for GCs [69]. Son et al. surveyed GCs in Korea and the U.S. to determine their level of awareness of sustainable construction measures during the actual construction phase and developed a tool to assess and monitor the effectiveness of construction waste management, listing 17 factors that affect the performance of contractors' waste management plans [16].

Other researchers have focused on analyzing the characteristics of GCs' work on LEED projects. Opoku et al. examined the role of integrative team coordination and its influence on achieving LEED certification [70], Mollaoglu-Korkmaz et al. looked at the relationships and influence of project delivery attribute in their study of 12 green office buildings in the US [60], and Uğur and Leblebici showed the critical role played by GCs for cost estimation and the analysis of construction cost-benefit and payback periods [31]. Research by Robichaud and Anantatmula highlighted the difference it makes when GCs have LEED experience and participate early in the project [53], whereas Pulaski and Horman proposed a "continuous value enhancement process" to improve the effectiveness of management practices in sustainable construction projects [71,72]. Finally, the U.S. Environmental Protection Agency issued a set of guidelines to assist with the selection of appropriate recycled content products for construction [73].

In summary, there exists a significant need to define the role of GCs and stress the advantages gained by their involvement in the early project phases [74–76]. Previous studies have clarified the duties of GCs and other partners working on LEED projects and provided guides for improving the effectiveness of GCs. However, most have focused on GCs and LEED projects in well-developed countries [8]. There are significant differences in both the GCs' qualifications and local technical standards between Vietnam and more well-developed LEED markets [41]. Therefore, the roles of GCs in implementing site sustainability and better support systems to help them perform those roles is needed if we are to improve the competitiveness of Vietnamese GCs and enable them to participate effectively in the trend towards sustainable construction.

2.2. *Identification of GCs Sustainable Roles in LEED v4 Projects*

Although GCs are not the most significant contributors in terms of the additional work caused by LEED implementation, their involvement directly impacts the effectiveness and success of the project [53,57,59]. The GC or construction manager should thus be involved as early as possible in a LEED project [74–77]. Unlike traditional projects, GCs are involved in all three stages, namely, the pre-construction, construction, and closeout phases [67,77,78]. The literature examined for this study identified 18 potential roles for GCs in LEED v4 projects and these are summarized in Table 1.

(A) Pre-Construction Phase: An integrated design (ID) is a prerequisite to satisfy the requirements of LEED v4 (healthcare project only) [78]. An integrative design team should be formed to facilitate cooperation between the many different stakeholders involved, including GCs and other participants [60,79,80]. The earlier the GC begins to work with the project team, the better for the project as this will lower costs and shorten the construction schedule by setting suitable project goals at every stage [16,53]. GCs provide valuable practical advice and support the ID team's efforts to solve problems, such as procuring sustainable materials, recycling materials, creating practical designs that are appropriate for the region, developing realistic materials delivery schedules, and minimizing the pollution created by construction processes [16]. GCs are thus better able to fulfill the intentions of the design team if they have a good understanding of sustainability and its benefits, as this allows them to develop more accurate cost estimates and assess the reliability of selected products [16,53,81]. In summary, GCs should be active participants in the integrative team to enable the team to evaluate options and select LEED project goals that can be achieved at a reasonable cost and are feasible from a construction perspective.

(B) Construction Phase: The GC is responsible for the majority of the construction credits as well as several credits related to site and energy that require the involvement of other team members [78]. According to the LEED v4 guidelines, the contractor is responsible for creating an erosion and sedimentation control (ESC) plan, general LEED requirements, LEED product requirements, waste management and disposal, and indoor air quality [66]. The contractor plays an essential role in selecting the appropriate construction method and assessing constructability to meet the LEED goals [64]. The GC must also carefully review and approve all materials before they are installed, as well as being primarily responsible for collecting information from the subcontractors and submitting progress reports [77]. The GCs' on-site supervisor will oversee waste management implementation [77]. According to LEED BC+D v4 (2014), GCs play an essential role in construction activity pollution prevention (CAPP) and construction and demolition waste management planning (CDWM) [78]. The CAPP plan must meet the requirements laid down in the erosion and sedimentation section of the 2012 U.S. Environmental Protection Agency (EPA) Construction General Permit (CGP) regulations or their local equivalent [78]. As these LEED requirements are more rigorous than the local equivalent in Vietnam [82–84], GCs need to understand the rules and be prepared to comply with the more stringent LEED requirements. Regarding fundamental commissioning and verification criteria, GCs only take the lead role for projects where the ground floor area (GFA) is below 1860 m^2 [78]. For larger projects, a commissioning authority (CxA) will assume the primary responsibility with support from the GC. It is essential for GCs to collect all the material data sheets and take extensive job site photographs. Many credits require documentation to be prepared continuously for submission as part of the LEED certification process. GCs must coordinate with the LEED consultant when collecting data on materials and the installation process for mid-project audits to review compliance with the LEED credit requirements [66]. They must also periodically inspect the indoor air quality control measures implemented, identify any issues that need to be corrected, and record their observations on a checklist [77]. At the end of construction phase, the flush-out and air testing are conducted by GC or a special contractor.

(C) Closing Phase: The GCs' most crucial task in this phase of the project is to gather the necessary documents and provide these to the LEED consultant. These include a list of LEED related materials, the quantity of each utilized, purchase reports, datasheets from manufacturers, environmental certificates, and so on (Frattari, 2012; Schaufelberger & Cloud, 2009). GCs that lack previous LEED experience are likely to need assistance from LEED consultants when developing their LEED credit template and LEED site checklist, as well as when preparing documents for submission [66]. GCs may also be responsible for performing an additional building flush-out test before the building's occupant's move in [78], but this is optional, quite costly, and uncommon in Vietnam.

Table 1. List of GCs' roles in LEED projects identified in the literature.

ID [1]	LEED v4 Credits	GC's Sustainable Roles/Function
IPp1	Integrative project team member	IP1: Integrative Process worksheet of GC [66,67,69,70,77,78,84–90] **(A)**.
SSp1 *	Construction activity pollution prevention	SS1: Erosion and sedimentation control (ESC) plan [25,64,66,78,87,89,90] **(A)**. SS2: ESC Implementing weekly report [25,66,69,70,77,78,82–84,89–92] **(B)**. SS3: Report of compliance with EPA CGP [25,64,66,69,77,78,89,90] **(B,C)**.
SSc2	Construction activity management for greenfield protection	Following the strategies listed in SSp1 "Construction Activity Pollution Prevention" [17,78,90] such as preventing construction damage to green fields, soil poisoning, soil compaction and so on.
WE	All water efficiency credits	WE1: Report of the irrigation systems purchase [66,77,78] **(C)**. WE2: Report of the water use equipment purchase (it can be replaced by on-site testing) [64,66,77,78,90] **(C)**. WE3: Report of the water consumption monitoring system purchase [64,66,78] **(C)**.
EAp1 *	Fundamental commissioning (GFA < 1860 m^2—major role)	EA1: Fundamental commissioning (Cx) & verification plan [64,69,78,89,90] **(B)**. EA2: Fundamental Cx & verification implementing report [64,78] **(B,C)**.
	(GFA > 1860 m^2—supporting role)	EA3: Supporting fundamental Cx contractors by providing the necessary documentation [64,66,77,78] **(B,C)**.
EAc1	Enhanced commissioning. v4	Not a GC roles. GCs only provide the necessary documentation (Envelope material) [64,66,77,78,90].
EQc4	Indoor air quality assessment v4	Made by the "Flush-Out" or "Air testing" contractor, GCs support for the on-site preparation **(C)**.
EAc6	Enhanced refrigerant management	EA4: Report of the fundamental refrigerant management plan **(C)** [66,78].
MRp2	CDWM Planning	MR1: Construction and demolition waste management plan **(B)** [64,66,69,78,90].
MRc2	Environmental product declaration v4	AD1: Report of compliance with LEED's material purchase requirements **(B,C)** [25,59,64,66,69,77,78,89,90].
MRc3	Sourcing of raw materials v4	
MRc4	Material ingredients v4	
EQc2	Low-emitting materials v4	
MRc5 *	Construction & demolition waste management (CDWM)	MR2: Report of CDWM Plan implementing [64,66,69,77,78,89,90,92] **(B)**. MR3: Report of implementing the construction and demolition waste management result [64,66,69,77,78,89–91] **(C)**.
EQc3 *	Construction indoor air quality management	EQ1: Report of Implementing the construction indoor air quality management plan [25,64,66,69,77,78,90] **(B)**. EQ2: Environmental tobacco smoke control policy [66,78] **(B,C)**.

[1]: The IDs used for the credits are the same as those used on the leeduser website; * indicates the LEED V4 credits where GCs play a significant role; v4 these credits differ markedly from those used in LEED V3. **(A)** The main phase is pre-construction phase; **(B)** the main phase is construction phase; **(C)**: the main phase is closing phase.

3. Research Methods

3.1. Questionnaire Design

Questionnaire surveys are used extensively to collect data on professional opinion in the construction management field and green building research [92]. A two-part survey questionnaire was therefore created for this study to gather expert perspectives on the current experience of Vietnamese GCs implementing LEED projects. The first part of the questionnaire gathers background information on the responders, including the field they work in, their years of experience, and their green building experience. Respondents working in non-related fields or with less than one year of experience were eliminated. The second questionnaire part was developed based on the "LEED Reference Guide for Building Design and Construction", 2013 Edition, and the findings of previous researchers presented above in the literature review. A pilot questionnaire was tested to ensure its comprehensibility and suitability for the intended purpose by administering it to a green building expert with nearly ten years of experience in Vietnam. Four of the 22 GC roles were removed based on the expert's feedback, as they are seldom applicable to LEED projects in Vietnam. The final questionnaire consisted of 18 questions examining the topics presented in Table 1 and respondents were asked to indicate their experience of the GCs implementing the various LEED roles using a 5-point Likert scale ranging from 1 (strongly disagree) to 5 (strongly agree). Respondents were also asked to assess the LEED experience levels of contractors that they had worked with, or coordinated in previous building projects. The 5-point Likert scale is widely used in green building studies because it is unambiguous and easily interpreted [93].

3.2. Data Collection

The target population for this research was the engineers and architects working in the construction field in Vietnam who had extensive work experience and understanding of Vietnamese GCs. As sampling this entire population would be costly to implement, the nonprobability sampling technique was deemed suitable for selecting the sample based on the purpose of the study. Following the study of B.G. Hwang (2017) [92], Equation (1) was utilized to determine the minimum sample size required for significance. To maximize the number of samples collected, a relatively high standard deviation of 4 was selected and a minimum error factor of 1; the minimum sample size of the survey was thus 62.

$$n = \left(\frac{Zs}{E}\right)^2 = \left(\frac{1.96 \times 4}{1}\right)^2 = 61.5 \quad (1)$$

- n: minimum sample size.
- S: sample standard deviation
- E: error factor,
- Z = 1.96, equal to 95 percent of the confidence

The survey email was sent to potential participants on a list of surveyors' email collected by researchers in the construction department of Ho Chi Minh City University of Technology (HCMUT). This list includes experts working in the construction sector in the southern region in Vietnam, as was originally collected and used for previous studies in the construction management field by graduate students attending HCMUT. The southern region is a useful location for green building studies since the most recent data (January 2019) indicates 64 of the country's 88 certified green building projects are located in this area. Two versions of the survey questionnaire based on the topics in Table 1, in English and Vietnamese, were created using the Google survey. A link to the questionnaire was included in an email describing the objectives of the study and inviting recipients to participate in the survey. Emails were sent to 542 people on the HCMUT email list and posted on several construction and architecture forums in Vietnam. Recipients were also encouraged to share the survey with any of their colleagues who are interested in green building projects. A total of 72 respondents completed the survey, ~13.3% of the emails sent out. This result is lower than the usual response rate of approximately 20 to

30% [92], suggesting that many Vietnamese construction experts still have little interest in sustainable construction. The sample size is larger than the minimum number of samples required based on the calculation in Equation (1).

The survey respondents had diverse backgrounds and were all active in the construction sector. As the data presented in Figure 1 demonstrate, the majority had 1–5 years of experience, with only a few having more than ten years of experience. Given the growth in green building projects in recent years, it is not surprising that the most experienced respondents were engineers. There were two main groups of experts: general contractor's staff and non-general constructor's staff. Half of the respondents were green building experts working for GCs, and the remaining half were evenly split between the other groups. This diversity was useful, introducing more perspectives on the characteristics needed for GCs working on LEED projects.

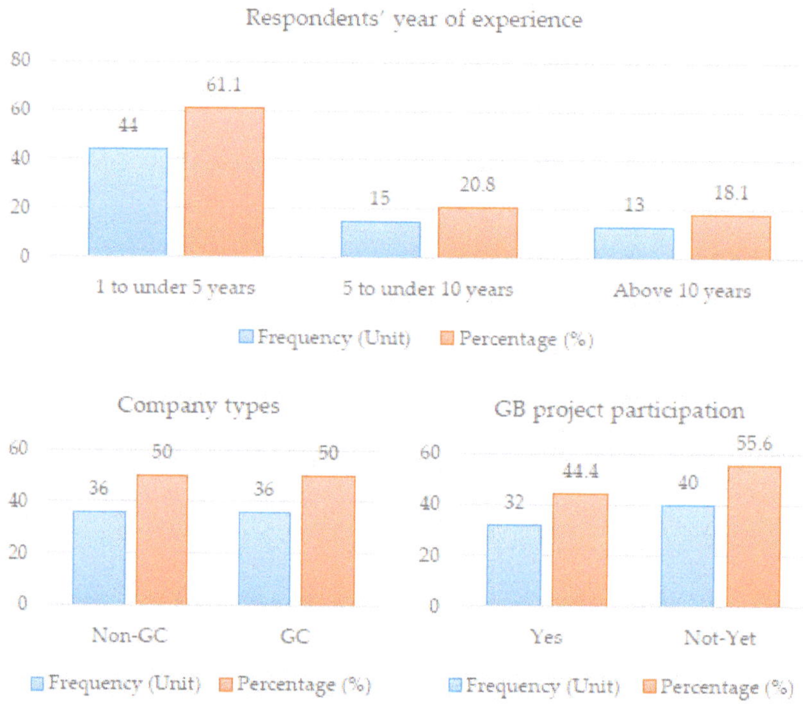

Figure 1. The respondents' professions.

3.3. Statistical Analysis

The data collected were statistically analyzed using the five-step data analysis framework, which is following the studies of A.Darko et al (2017), B.G. Hwang (2017), and S. Pushkar (2018) [1,92,94]. The data were analyzed by using the SPSS 20.0 software package.

1. Cronbach's alpha was used to evaluate the reliability of data for further analysis.

2. After that, the mean value ranking technique was applied to identify and prioritize the roles where Vietnamese GCs are most inexperienced.

3. The Shapiro–Wilk test was then utilized to test the normality of the data collected with the null hypothesis that each sample data came from a normally distributed population. At a 95% level of confidence, the null hypothesis of this test is that the collected data is normally distributed. If the p-value is less than the 0.05, the null hypothesis is rejected and there is evidence that the data tested are not normally distributed.

4. Identifying the roles where GCs lacked experience-based on the significance of the mean values via a t-test to determine whether the difference of the mean of each GCs' LEED role was close to the mean value on the Likert-scale. At a 95% level of confidence (*p*-value: 0.05), and against a test value of 3.00 (neutral on the rating scale), the one-sample t-test was therefore conducted for significant differences in the mean values for the experience levels of the Vietnamese GCs implementing LEED projects. The null hypothesis of the t-test in is H0, and the mean value is not statistically significant. The null hypothesis H0 would be accepted if the *p*-value of the GCs' role task is more than 0.05, and the alternative hypothesis H1 "the mean value is statistically significant" would be rejected.

5. Following the studies of A. Darko at al. (2017), the ANOVA test (parametric tests) or Kruskal–Wallis test (nonparametric tests) are applied to identify any statistically significant differences [1]. However, in this study, the Wilcoxon–Mann–Whitney (WMW) and Cliff's effect size nonparametric tests were replaced to compare the significance of the mean values of two or more groups because of homoscedasticity uncertainty of the data. The data are presented as the median ± interquartile range (IQR: 25th to 75th percentile). According to suggested sampling structures of S. Pushkar (2018) study, "Cliff's δ was used to measure the substantive significance (effect size) between two unpaired groups", and is calculated as $\delta = \#(x1 > x2) - \#(x1 < x2)/(n1n2)$ [94]. As a result, the effect size is considered "negligible" if $|\delta| < 0.147$, "small" if $0.147 \leq |\delta| < 0.33$, "medium" if $0.33 \leq |\delta| < 0.474$, or "large" if $|\delta| \geq 0.474$ [94]. Where x1 and x2 are scored within each ground and n1 and n2 are the sizes of the sample groups, and the cardinality symbol # indicates counting. Furthermore, an approximate WMW was conducted while the sample sizes were $n1 = n2 \geq 9$ to determine statistical difference (*p*-value) between two unpaired groups.

4. Results and Discussion

4.1. Data Analyzing

In this study, the relative importance of the GCs' experience of various LEED roles was evaluated using a 5-point Likert scale. The Cronbach's alpha coefficient of the data packet was found to be 0.918, which is above the high-reliability threshold of 0.7 [95], thus indicating the data has high reliability. In the second step, the full results of the survey, as well as the relevant statistical test results, are shown in Table 2. From the results of the Shapiro–Wilk test, no *p*-value values of the entire GCs' roles were above a confidence level of 0.05, indicating that the data collected from the survey was not a normal distribution.

Based on the results presented in Table 2, 89% of the survey results had mean values below 3 (neutral level). The respondents' overall rating for the GCs' experience in implementing LEED roles was only 2.66, indicating that respondents do not consider that GCs have adequate experience for in implementing their LEED roles. This shows a serious lack of experience and indicates that GCs need to be provided with more training if they are to fulfill their responsibility for implementing these LEED roles, which concern fundamental commissioning, refrigerant management, material documentation, managing indoor air quality during construction, and environment product declarations. Only two GC LEED tasks had a mean value greater than 3.0, namely IP1 (Integrative design support; 3.39) and EAQ2 (Smoking control policy; 3.42). This suggests that GCs in Vietnam have approached and supported investors in the process of project formulation, possibly because the GCs who win these contracts have worked with investors and thus have a greater understanding of the projects and their aims. Also, most reputable contractors have implemented a smoking control policy on sites since the Vietnamese government issued a smoking ban for workplaces in June 2018 [96].

Table 2. Mean value ranking, t-test and Shapiro–Wilk test result in the GC LEED roles.

All (n = 72)											
Code	**Mean**	**SD**	**Rank**	***p*-value (t-test)**	**Sig. (SW)**	**Code**	**Mean**	**SD**	**Rank**	***p*-value (t-test)**	**Sig. (SW)**
EQ2	3.42	1.3	1	0.006 *	0.00 [a]	SS1	2.58	1	10	0.000 *	0.00 [a]
IP1	3.39	1	2	0.002 *	0.00 [a]	SS3	2.54	1	11	0.000 *	0.00 [a]
WE2	2.89	1	3	0.369	0.00 [a]	EA3	2.53	1	12	0.000 *	0.00 [a]
MR1	2.85	1	4	0.206	0.00 [a]	EA1	2.42	1	13	0.000 *	0.00 [a]
WE1	2.83	1	5	0.165	0.00 [a]	EA4	2.33	1	14	0.000 *	0.00 [a]
MR2	2.82	1.1	6	0.16	0.00 [a]	EA2	2.32	1.1	15	0.000 *	0.00 [a]
WE3	2.76	0.9	7	0.037 *	0.00 [a]	EQ1	2.32	1	16	0.000 *	0.00 [a]
MR3	2.67	1	8	0.005 *	0.00 [a]	AD1	2.31	1.2	17	0.000 *	0.00 [a]
SS2	2.6	0.9	9	0.001 *	0.00 [a]						

Notes: * t-test results with *p*-values below 0.05, suggesting a significant difference in the mean value. [a]: Shapiro–Wilk test results with a *p*-value less than 0.05, indicating the data is not a normal distribution.

This study used statistical methods including mean values, ranking techniques, and one-sample t-tests to identify any statistical significant roles that GCs are lacking experience. Although the data from the survey is not a normal distribution, according to the central limit principle the size of samples is more significant than 30, and therefore can be accepted as a normal distribution, so the t-test result can be considered reasonable [97]. The results of the t-test, shown in Table 2, suggested that Vietnamese GCs are lacked experience in most of their roles. The one-sample t-test results and the mean-value present 12/17 GCs roles significant lacked experience when carrying out these critical LEED roles. It included credits, which GCs have played a significant contribution in the success of LEED projects, such as "Construction activity pollution prevention", "material criteria", "Construction indoor air quality management", and "Construction & demolition waste management". Finally, an approximate WMW test was conducted. However, if experimental units are pooled from the different sampling frames, the problem of sacrificial pseudo-replication can occur [94]. Thus, the sample from each compared group was randomly selected. To perform an approximate WMW test, minimum sample size was $n1 = n2 \geq 9$. The number of sample size with the total number of respondents and number of randomly selected respondents' data which is presented in Table 3.

Table 3. The number of sample size and the number of randomly selected.

Company Types	**Sample Size (n)**	
	Yes	**Not-yet**
Non-GC	18 (18)	18 (18)
GC	14 (14)	22 (14)

There are considerable differences in opinion among expert groups that can lead to errors in mean-value, particularly those who had (or not yet) experience in LEED project or the GC/non-GC. Therefore, the group of respondents who have (or never) participated in the LEED project in the group from was statistically evaluated individual working field (GC/non-GC) sampling frame analysis (by an approximate WMW test). Data analysis results using the WMW test and Cliff's delta by grouped experienced/non-experience and GC/non-GC groups are presented in Table 4. According to Tables 2 and 4, the author divides the roles of GCs through the survey into four main groups: (i) experienced and no statistical difference, (ii) experienced and statistically different, (iii) inexperienced and no different Statistical differences, and (iv) non-experienced and statistical differences. As the result of GC respondents, group (i) includes EQ2, group (ii) includes IP1, WE1, WE2, MR1, and MR2; group (iii) includes SS1, SS2, and EA3; and group (iv) includes SS3, WE3, EA1,

EA2, EA4, MR3, and EQ1. For Non-GC respondents, group (i) includes MR1; group (ii) includes IP1, WE1, WE2, MR2, and QE2; group (iii) includes SS1, SS2, EA2: EA4, and MR3; and group (iv) includes SS3, WE3, EA1, EA3, and EQ1. A summary of Group (iv) for the GC/Non-GC roles WE3, EA1, EA2, EA3, EA4, MR3, and EQ1 is linked to lack of experience, but there are statistical differences between groups of people answer each participant/have never participated in the green building project. Respondents who participated in green building projects have a larger mean-value than the other groups. It demonstrates that the roles in this group can be performed better than the overall respondents expected. Summary of the group (iii) of GC and Non-GC, the roles SS1; SS2; EA2: EA3; EA4; MR3 are agreed by both groups of respondents that GC lacks experience in implementing that Vietnamese contractors need to give priority to the training of their staff.

Table 4. Cliff's δ and the Wilcoxon–Mann–Whitney (WMW) test result.

ID	Position	Sample Size n1 = n2	GB Experience						Cliff's Delta	*p*-Value	NFSAs
			Yes			No					
IP1	Non-GC	18	3.78	±	0.79	3.11	±	1.15	0.353	0.071	Suspended
	GC	14	3.71	±	0.88	3.00	±	1.00	0.418	0.049	Positive
SS1	Non-GC	18	2.72	±	0.73	2.56	±	1.07	0.114	0.592	Negative
	GC	14	2.57	±	0.90	2.36	±	0.97	0.184	0.381	Negative
SS2	Non-GC	18	2.83	±	0.69	2.50	±	1.12	0.199	0.295	Negative
	GC	14	2.43	±	0.98	2.57	±	0.98	-0.036	0.866	Negative
SS3	Non-GC	18	2.72	±	0.73	2.28	±	1.15	0.301	0.140	Suspended
	GC	14	2.86	±	1.12	2.29	±	0.80	0.306	0.150	Suspended
WE1	Non-GC	18	3.22	±	0.92	2.39	±	0.89	0.461	0.011	Positive
	GC	14	3.14	±	0.99	2.36	±	0.81	0.408	0.054	Suspended
WE2	Non-GC	18	3.28	±	0.93	2.44	±	0.90	0.464	0.010	Positive
	GC	14	3.36	±	1.04	2.36	±	0.89	0.515	0.016	Positive
WE3	Non-GC	18	3.11	±	0.87	2.33	±	0.82	0.454	0.012	Positive
	GC	14	3.21	±	0.86	2.36	±	0.89	0.505	0.016	Positive
EA1	Non-GC	18	2.67	±	1.05	2.11	±	1.10	0.261	0.111	Suspended
	GC	14	2.79	±	1.01	2.21	±	0.56	0.301	0.137	Suspended
EA2	Non-GC	18	2.44	±	1.12	2.06	±	1.08	0.180	0.263	Negative
	GC	14	2.79	±	1.01	2.21	±	0.67	0.296	0.155	Suspended
EA3	Non-GC	18	2.94	±	1.03	2.22	±	1.13	0.366	0.046	Positive
	GC	14	2.71	±	0.80	2.36	±	0.61	0.260	0.201	Negative
EA4	Non-GC	18	2.67	±	1.15	2.61	±	1.06	0.059	0.794	Negative
	GC	14	3.00	±	0.76	2.14	±	0.74	0.551	0.008	Positive
MR1	Non-GC	18	2.72	±	0.93	2.50	±	0.96	0.088	0.712	Negative
	GC	14	3.14	±	0.91	2.43	±	0.90	0.418	0.047	Positive
MR2	Non-GC	18	3.00	±	0.94	2.50	±	1.17	0.261	0.185	Suspended
	GC	14	3.36	±	0.72	2.36	±	0.89	0.577	0.007	Positive
MR3	Non-GC	18	2.89	±	0.81	2.44	±	1.26	0.242	0.229	Negative
	GC	14	3.36	±	0.89	2.50	±	1.05	0.429	0.045	Positive
EQ1	Non-GC	18	2.67	±	1.00	2.22	±	1.18	0.245	0.178	Suspended
	GC	14	2.57	±	0.73	2.00	±	0.76	0.429	0.036	Positive
QE2	Non-GC	18	3.44	±	1.07	2.67	±	1.37	0.337	0.113	Suspended
	GC	14	3.86	±	0.99	3.36	±	1.23	0.219	0.307	Negative

4.2. Differences with Developed Countries

In general, the role of GCs in LEED projects in Vietnam is quite similar to projects in other countries. For most criteria, like construction activity pollution prevention, commissioning activities are compliant with equivalent standards in the US. However, as a result of inexperience, the Vietnamese GC's role is limited and replaced by LEED consultants or foreign contractors, especially jobs involving Cx and flush out or air testing. Compared to well-developed countries such as the US, contractors in Vietnam lacked experience in the majority of LEED's work. Especially the group roles (iii) such as SS1, SS2, EA2: EA3, EA4, and MR3. The group of sustainable site credits (SS1 and SS2), which comply with U.S. EPA requirements, creates a new challenge for Vietnamese general contractors. The Cx process and flush-out or air testing (EA2–EA4) are not yet common in Vietnam and may not be mentioned in the bidding process. Construction waste management implementation (MR3) is still a big problem for Vietnamese contractors, as the government has regulations on solid waste management in construction, but it has not yet been applied in real conditions [96]. Furthermore, the construction materials in Vietnamese market are also often lack of necessary green certificates for LEED evident, and it also creates confusion for the GC in the process of purchasing materials. Besides, language differences are also a barrier for contractors in Vietnam because it generates extra documents and consumes human resources as well as time. Vietnamese general contractor also lacks experience in collecting LEED records of materials and equipment. Therefore, hiring LEED specialists and conducting training programs is necessary for contractors to improve the effectiveness of their first LEED jobs. The results of the study also indicate that contractors need to prioritize training and pay more attention to the work in the group (iii), and group (iv), especially the jobs in which they play a major role.

5. Conclusions

In recent years, the demand for sustainable construction to offset the shortage of resources and minimize the environmental impacts of fast-growing industries is evidenced by the increase in the number of green-certified buildings. If they are to remain competitive and thrive in this new market, the study helped the Vietnamese GCs to develop a better understanding of their sustainable roles and learn how to adapt cost-effectively and engage in more sustainable green building practices. The following was identified.

1. This review found that Vietnamese CGs are expected to play a significant role in four types of LEED v4 credits, namely, the implementation of the Erosion and Sedimentation Control Plan, Construction & Demolition Waste Management, Construction Indoor Air Quality Management, and Fundamental Commissioning. In particular, Cx and flush out or air testing can be done by other (foreign) contractors/experts. However, it is necessary to define the role of GC in the process of bidding and contract making.
2. The Vietnamese GCs lack experience in implementing LEED credits, which included SSp1 "Construction activity pollution prevention" (SS1–SS3), EAp1"Fundamental Commissioning" (EA1–EA3), "Construction & demolition waste management" (MR3), and "Construction indoor air quality management". Moreover, Vietnamese GC also lack experience in selecting and documenting the purchasing of LEED materials.
3. Currently, Vietnamese regulations and standards do not have such strict requirements compare to LEED requirements. Therefore, works such as ESC and Cx should be planned according to the requirements or standards which are mentioned by LEED.
4. The results of the study showed that the Vietnamese general contractors were lacked experienced in most LEED roles. In which priority is given to training programs on LEED for the following jobs (iii); SS1, SS2, EA2: EA4, and MR3, which included "Erosion and sedimentation control" (SS1 and SS2), "Fundamental commission for project less than 1860m2" (EA2), "Enhanced refrigerant management" (EA4), and "Construction and demolition waste management" (MR3).

Given the limited empirical studies on GCs' experience in implementing LEED roles, especially in Vietnam, the practical results of this study make a substantial contribution to the green building body of knowledge and are expected to encourage sustainable construction in Vietnam. Furthermore, the findings of this study will serve as a valuable reference to assist students and project managers, developing a practical construction implementation to achieve more sustainable building developments. These findings highlight the importance of involving the GC integrated design process to provide a more efficient building project and sustainable construction. Providing the GC with an outline of the scope of work, and the advantages and disadvantages inherent in a LEED project, will improve the GC's performance and support the developer's decision to invest in a green building project, thus contributing to sustainable construction in Vietnam.

As with most research, this study suffers from a number of limitations that remain to be addressed in future research. Regarding the population of the study, most of the experts participating in the survey come from the southern part of Vietnam, where most of the LEED-certified projects are concentrated. However, a survey dataset with a larger population and nationwide distribution may be more reliable. Although the goal is to promote the construction of sustainable green building practices and sustainable construction in Vietnam, this research focuses solely on GCs' roles in LEED projects, building a foundation for future studies on the impact of LEED certification on GCs and the optimization of LEED-related tasks for Vietnamese GCs. The GC roles identified will provide a useful starting point for efforts to reduce the effects of LEED's incremental work on GCs and optimize these tasks to introduce greater cost efficiency for projects and GCs. However, these issues were outside the scope of this research to investigate the GCs' roles in implementing LEED project, which identified their lack of experience related to many of their expected roles. Future studies should extend the original research topics to examine the roles of the building developers and/or the project management teams in implementing popular green building rating systems such as LEED in Vietnam.

Author Contributions: Supervision, J.L.; Writing—original draft, D.P.; Writing—review & editing, Y.A.

Funding: This research received no external funding.

Acknowledgments: This work was supported by the Korean Institute of Energy Technology Evaluation and Planning (**KETEP**) and the Ministry of Trade, Industry & Energy (MOTIE) of the Republic of Korea (No. 20172010000370).

Conflicts of Interest: The authors declare no conflict of interest.

References

1. Darko, A.; Chan, A.P.; Owusu-Manu, D.G.; Ameyaw, E.E. Drivers for implementing green building technologies: An international survey of experts. *J. Clean. Prod.* **2017**, *145*, 386–394. [CrossRef]
2. U.S. Green Building Council; Sustainable Buildings Industry Council. What is a Green Building? *San Marcos* **2016**, *888*, 336–7553.
3. Wu, P.; Xia, B.; Pienaar, J.; Zhao, X. The past, present, and future of carbon labeling for construction materials—A review. *Build. Environ.* **2014**, *77*, 160–168. [CrossRef]
4. Zhao, X.; Hwang, B.G.; Gao, Y. A fuzzy synthetic evaluation approach for risk assessment: A case of Singapore's green projects. *J. Clean. Prod.* **2016**, *115*, 203–213. [CrossRef]
5. Hwang, B.G.; Zhao, X.; Tan, L.L.G. Green building projects: Schedule performance, influential factors, and solutions. *Eng. Constr. Archit. Manag.* **2015**, *22*, 327–346. [CrossRef]
6. Mangialardo, A.; Micelli, E.; Saccani, F. Does sustainability affect real estate market values? Empirical evidence from the office buildings Market in Milan (Italy). *Sustainability* **2019**, *11*, 12. [CrossRef]
7. Micelli, E.; Mangialardo, A. Recycling the city new perspective on the real-estate market and construction industry. In *Smart and Sustainable Planning for Cities and Regions*; Bisello, A., Vettorato, D., Stephens, R., Elisei, P., Eds.; SSPCR 2015. Green Energy and Technology; Springer: Basel, Switzerland, 2015; pp. 115–125.
8. Amiri, A.; Ottelin, J.; Sorvari, J. Are LEED-certified buildings energy-efficient in practice? *Sustainability* **2019**, *11*, 1672. [CrossRef]
9. Nilforooshan, R.; Adamo-Villani, N.; Dib, H. A study of the effects of computer animation on college students' learning of Leadership in Energy and Environmental Design-LEED. *ICST Trans. eEduc. eLearn.* **2013**, *1*, e3. [CrossRef]
10. Gurgun, A.; Arditi, D. Assessment of energy credits in LEED-certified buildings based on certification levels and project ownership. *Buildings* **2018**, *8*, 29. [CrossRef]
11. Ferreira, J.; Pinheiro, M.D.; de Brito, J. Portuguese sustainable construction assessment tools benchmarked with BREEAM and LEED: An energy analysis. *Energy Build.* **2014**, *69*, 451–463. [CrossRef]

12. Mao, C.; Shen, Q.; Shen, L.; Tang, L. Comparative study of greenhouse gas emissions between off-site prefabrication and conventional construction methods: Two case studies of residential projects. *Energy Build.* **2013**, *66*, 165–176. [CrossRef]
13. Yang, R.J.; Zou, P.X. Stakeholder-associated risks and their interactions in complex green building projects: A social network model. *Build. Environ.* **2014**, *73*, 208–222. [CrossRef]
14. Yang, R.J.; Zou, P.X.; Wang, J. Modelling stakeholder-associated risk networks in green building projects. *Int. J. Proj. Manag.* **2016**, *34*, 66–81. [CrossRef]
15. Yuan, H. SWOT analysis of successful construction waste management. *J. Clean. Prod.* **2013**, *39*, 1–8. [CrossRef]
16. Son, H.; Kim, C.; Chong, W.K.; Chou, J.S. Implementing sustainable development in the construction industry: Constructors' perspectives in the US and Korea. *Sustain. Dev.* **2011**, *19*, 337–347. [CrossRef]
17. Sev, A. How can the construction industry contribute to sustainable development? A conceptual framework. *Sustain. Dev.* **2009**, *17*, 161–173. [CrossRef]
18. Berardi, U. Clarifying the new interpretations of the concept of sustainable building. *Sustain. Cities Soc.* **2013**, *8*, 72–78. [CrossRef]
19. Shen, L.Y.; Tam, V.W. Implementation of environmental management in the Hong Kong construction industry. *Int. J. Proj. Manag.* **2002**, *20*, 535–543. [CrossRef]
20. Shen, W.; Tang, W.; Siripanan, A.; Lei, Z.; Duffield, C.; Hui, F. Understanding the green technical capabilities and barriers to green buildings in developing countries: A case study of Thailand. *Sustainability* **2018**, *10*, 3585. [CrossRef]
21. Zuo, J.; Zhao, Z.Y. Green building research–current status and future agenda: A review. *Renew. Sustain. Energy Rev.* **2014**, *30*, 271–281. [CrossRef]
22. Ahn, Y.H.; Pearce, A.R.; Wang, Y.; Wang, G. Drivers and barriers of sustainable design and construction: The perception of green building experience. *Int. J. Sustain. Build. Technol. Urban Dev.* **2013**, *4*, 35–45. [CrossRef]
23. Serpell, A.; Kort, J.; Vera, S. Awareness, actions, drivers and barriers of sustainable construction in Chile. *Technol. Econ. Dev. Econ.* **2013**, *19*, 272–288. [CrossRef]
24. OECD. Environmentally Sustainable Buildings, Challenges and Policies. 2003. Available online: http://www.oecd.org/env/consumption-innovation/2715115.pdf (accessed on 22 June 2019).
25. Kibert, C.J. *Sustainable Construction: Green Building Design and Delivery*, 3rd ed.; John Wiley & Sons: Hoboken, NY, USA, 2012.
26. Pham, H.; Kim, S.Y.; Luu, T.V. Managerial perceptions on barriers to sustainable construction in developing countries: Vietnam case. *Environ. Dev. Sustain.* **2019**, 1–25. [CrossRef]
27. Green Building Standards and Certification Systems. Available online: https://www.wbdg.org/resources/green-building-standards-and-certification-systems (accessed on 6 May 2019).
28. Griffiths, K.; Boyle, C.; Henning, T. Beyond the certification badge—How infrastructure sustainability rating tools impact on individual, organizational, and industry practice. *Sustainability* **2018**, *10*, 1038. [CrossRef]
29. Shin, M.; Kim, H.; Gu, D.; Kim, H. LEED, its efficacy and fallacy in a regional context—An urban heat island case in California. *Sustainability* **2017**, *9*, 1674. [CrossRef]
30. Potrč Obrecht, T.; Kunič, R.; Jordan, S.; Dovjak, M. Comparison of health and well-being aspects in building certification schemes. *Sustainability* **2019**, *11*, 2616. [CrossRef]
31. Uğur, L.O.; Leblebici, N. An examination of the LEED green building certification system in terms of construction costs. *Renew. Sustain. Energy Rev.* **2018**, *81*, 1476–1483. [CrossRef]
32. Pearce, D. Is the construction sector sustainable? Definitions and reflections. *Build. Res. Inf.* **2006**, *34*, 201–207. [CrossRef]
33. Jeong, J.; Hong, T.; Ji, C.; Kim, J.; Lee, M.; Jeong, K. Development of an evaluation process for green and non-green buildings focused on energy performance of G-SEED and LEED. *Build. Environ.* **2016**, *105*, 172–184. [CrossRef]
34. Donghwan, G.; Yong, K.H.; Hyoungsub, K. LEED, its efficacy in regional context: Finding a relationship between regional measurements and urban temperature. *Energy Build.* **2015**, *86*, 687–691. [CrossRef]
35. Zhao, J.; Lam, K.P.; Biswas, T.; Wang, H. An online platform to automate LEED energy performance evaluation and submission process. *Constr. Innov.* **2015**, *15*, 313–332. [CrossRef]

36. Reichardt, A. Operating expenses and the rent premium of energy star and LEED certified buildings in the central and eastern US. *J. Real Estate Financ. Econ.* **2014**, *49*, 413–433. [CrossRef]
37. Tsai, W.H.; Lin, S.J.; Lee, Y.F.; Chang, Y.C.; Hsu, J.L. Construction method selection for green building projects to improve environmental sustainability by using an MCDM approach. *J. Environ. Plan. Manag.* **2013**, *56*, 1487–1510. [CrossRef]
38. Wu, P.; Song, Y.; Shou, W.; Chi, H.; Chong, H.Y.; Sutrisna, M. A comprehensive analysis of the credits obtained by LEED 2009 certified green buildings. *Renew. Sustain. Energy Rev.* **2017**, *68*, 370–379. [CrossRef]
39. Xia, B.; Chen, Q.; Xu, Y.; Li, M.; Jin, X. Design-build contractor selection for public sustainable buildings. *J. Manag. Eng.* **2014**, *31*, 04014070. [CrossRef]
40. Massoud, M.A.; Tarhini, A.; Nasr, J.A. Decentralized approaches to wastewater treatment and management: Applicability in developing countries. *J. Environ. Manag.* **2009**, *90*, 652–659. [CrossRef]
41. Nguyen, H.T.; Skitmore, M.; Gray, M.; Zhang, X.; Olanipekun, A.O. Will green building development take off? An exploratory study of barriers to green building in Vietnam. *Resour. Conserv. Recycl.* **2017**, *127*, 8–20. [CrossRef]
42. Carew-Reid, J. *Rapid Assessment of the Extent and Impact of Sea Level Rise in Viet Nam*; International Centre for Environment Management (ICEM): Brisbane, Australia, 2008; Volume 82.
43. Adger, W.N. Social vulnerability to climate change and extremes in coastal Vietnam. *World Dev.* **1999**, *27*, 249–269. [CrossRef]
44. Mackay, A. Climate change 2007: Impacts, adaptation and vulnerability. Contribution of Working Group II to the fourth assessment report of the Intergovernmental Panel on Climate Change. *J. Environ. Qual.* **2008**, *37*, 2407. [CrossRef]
45. Overview of Vietnam Economy and Society in 2018. Available online: https://www.gso.gov.vn/default.aspx?tabid=382&idmid=2&ItemID=19041 (accessed on 1 June 2019).
46. Nguyen, H.T.; Gray, M. A review on green building in Vietnam. *Procedia Eng.* **2016**, *142*, 314–321. [CrossRef]
47. Vietnam Energy Industry, Media Climate Net, Update Report 2019. Available online: https://moit.gov.vn/tin-chi-tiet/-/chi-tiet/chinh-phu-quy-%C4%91inh-gia-ban-%C4%91ien-mat-troi-tai-viet-nam-la-9-35-uscents-kwh-2650-136.html (accessed on 12 June 2019).
48. LEED Projects. Available online: https://www.usgbc.org/projects/list?keys=vietnam (accessed on 12 June 2019).
49. EDGE Projects. Available online: https://www.edgebuildings.com/projects/ (accessed on 12 June 2019).
50. LOTUS Projects. Available online: https://vgbc.vn/en/lotus-projects/ (accessed on 12 June 2019).
51. Rohracher, H. Managing the technological transition to sustainable construction of buildings: A socio-technical perspective. *Technol. Anal. Strateg. Manag.* **2001**, *13*, 137–150. [CrossRef]
52. Pulaski, M.H.; Horman, M.J.; Riley, D.R. Constructability practices to manage sustainable building knowledge. *J. Archit. Eng.* **2006**, *12*, 83–92. [CrossRef]
53. Robichaud, L.B.; Anantatmula, V.S. Greening project management practices for sustainable construction. *J. Manag. Eng.* **2010**, *27*, 48–57. [CrossRef]
54. Nguyen, H.D.; Nguyen, L.D.; Chih, Y.Y.; Le-Hoai, L. Influence of participants' characteristics on sustainable building practices in emerging economies: Empirical case study. *J. Constr. Eng. Manag.* **2017**, *143*, 05017014. [CrossRef]
55. Chan, A.P.; Scott, D.; Chan, A.P. Factors affecting the success of a construction project. *J. Constr. Eng. Manag.* **2004**, *130*, 153–155. [CrossRef]
56. Inayat, A.; Melhem, H.; Esmaeily, A. Critical success factors in an agency construction management environment. *J. Constr. Eng. Manag.* **2014**, *141*, 06014010. [CrossRef]
57. Pulaski, M.H.; Horman, M.; Riley, D.; Dahl, P.; Hickey, A.; Lapinski, A.; Holland, N. *Field Guide for Sustainable Construction*; Pentagon Renovation and Construction Program Office Safety Sustainability and Environment IPT and Washington Headquarters Services Defense Facilities Directorate: Washington, DC, USA, 2004.
58. Shen, L.Y.; Lu, W.S.; Hong, Y.; Wu, D.H. A computer-based scoring method for measuring the environmental performance of construction activities. *Autom. Constr.* **2005**, *14*, 297–309. [CrossRef]
59. Bayraktar, M.E.; Owens, C.R.; Zhu, Y. State-of-practice of LEED in the United States: A contractor's perspective. *Int. J. Constr. Manag.* **2011**, *11*, 1–17. [CrossRef]
60. Mollaoglu-Korkmaz, S.; Swarup, L.; Riley, D. Delivering sustainable, high-performance buildings: Influence of project delivery methods on integration and project outcomes. *J. Manag. Eng.* **2011**, *29*, 71–78. [CrossRef]

61. U.S. Green Building Council. Green Building Facts. US Green Building Council, 2008. Available online: https://www.usgbc.org/ShowFile.aspx. (accessed on 12 June 2019).
62. Glavinich, T.E. *Contractor's Guide to Green Building Construction: Management, Project Delivery, Documentation, and Risk Reduction*; John Wiley & Sons: Hoboken, NJ, USA, 2008.
63. Klehm, C. Value Added General Contracting for LEED™ Rated Projects. Available online: https://www.usgbc.org/drupal/legacy/usgbc/docs/Archive/MediaArchive/608_Klehm_AB768.pdf (accessed on 9 August 2019).
64. Syal, M.M.; Mago, S.; Moody, D. Impact of LEED-NC credits on contractors. *J. Archit. Eng.* **2007**, *13*, 174–179. [CrossRef]
65. Syal, M.; Li, Q.; Abdulrahman, K.; Mago, S. LEED® requirements and construction project management. *Int. J. Constr. Proj. Manag.* **2011**, *3*, 257.
66. Frattari, A.; Dalprà, M.; Salvaterra, G. The role of the general contractor in sustainable green buildings: The case study of two buildings in the LEED certification in Italy. *Int. J. Hous. Sci. Its Appl.* **2012**, *36*, 139.
67. Ahn, Y.H.; Jung, C.W.; Suh, M.; Jeon, M.H. Integrated construction process for green building. *Procedia Eng.* **2016**, *145*, 670–676. [CrossRef]
68. Bayraktar, M.E.; Owens, C.R. LEED implementation guide for construction practitioners. *J. Archit. Eng.* **2009**, *16*, 85–93. [CrossRef]
69. Schaufelberger, J.; Cloud, J. LEED certification: A constructor's perspective. In Proceedings of the Construction Research Congress 2009: Building a Sustainable Future, Seattle, WA, USA, 5–7 April 2009; Ariaratnam, S.T., Rojas, E.M., Eds.; American Society of Civil Engineers: Reston, VA, USA, 2009; pp. 598–607. [CrossRef]
70. Opoku, A.; Ahmed, V.; Cruickshank, H.; Senaratne, S.; Hewamanage, P.R. The role of team leadership in achieving LEED certification in a green building project. *Built Environ. Proj. Asset Manag.* **2015**, *5*, 170–183.
71. Korkmaz, S.; Riley, D.; Horman, M. Piloting evaluation metrics for sustainable high-performance building project delivery. *J. Constr. Eng. Manag.* **2010**, *136*, 877–885. [CrossRef]
72. Pulaski, M.; Horman, M. The continuous value enhancement process to efficiently achieve sustainable project objectives. In Proceedings of the Construction Research Congress 2005: Broadening Perspectives, San Diego, CA, USA, 5–7 April 2005; Tommelein, I.R., Ed.; American Society of Civil Engineers: Reston, VA, USA, 2005; pp. 1–10. [CrossRef]
73. U.S. Environmental Protection Agency. EPA Comprehensive Procurement Guideline Program. Available online: https://www.epa.gov/sites/production/files/2016-02/documents/cpg-fs.pdf (accessed on 12 June 2019).
74. Samaras, C. Sustainable development and the construction industry: Status and implications. *Carnegie Mellon Univ. Retrieved Oct.* **2004**, *26*, 2007.
75. Riley, D.; Pexton, K.; Drilling, J. Procurement of sustainable construction services in the United States: The contractor's role in green buildings. *Ind. Environ.* **2003**, *26*, 66–69.
76. Mogge, J.W., Jr. Breaking Through the First Cost Barriers to Sustainable Planning, Design and Construction. Ph.D. Thesis, Georgia Institute of Technology, Atlanta, GA, USA, 2004.
77. Yellamraju, V. *LEED-New Construction Project Management (GreenSource)*; McGraw Hill Professional: New York, NY, USA, 2010.
78. U.S. Green Building Council. *Reference Guide for Building Design and Construction v4*; USGBC: Washington, DC, USA, 2014.
79. Klotz, L.; Horman, M.; Riley, D.; Bechtel, J. Process transparency for sustainable building delivery. *Int. J. Sustain. Eng.* **2009**, *2*, 298–307. [CrossRef]
80. Riley, D.; Pexton, K.; Drilling, J. *Defining the Role of Contractors on Green Building Projects*; International Council for Research and Innovation in Building and Construction: Deft, Switerland, 2003.
81. Alawneh, R.; Ghazali, F.E.M.; Ali, H.; Asif, M. Assessing the contribution of water and energy efficiency in green buildings to achieve United Nations Sustainable Development Goals in Jordan. *Build. Environ.* **2018**, *146*, 119–132. [CrossRef]
82. Ministry of Natural Resources and Environment of Vietnam. Circular No. 27/2015/TT-BTNMT on Strategic Environmental Assessment, Environmental Impact Assessment, and Environmental Protection Plans. Available online: https://www.ecolex.org/details/legislation/circular-no-272015tt-btnmt-on-strategic-environmental-assessment-environmental-impact-assessment-and-environmental-protection-plans-lex-faoc168538/ (accessed on 12 June 2019).

83. The Government of Vietnam. Decree No. 18/2015/ND-CP Prescribing Environmental Protection Master Plan, Strategic Environmental Assessment, Environmental Impact Assessment and Environmental Protection Plan. Available online: https://www.ecolex.org/details/legislation/decree-no-182015nd-cp-prescribing-environmental-protection-master-plan-strategic-environmental-assessment-environmental-impact-assessment-and-environmental-protection-plan-lex-faoc168510/ (accessed on 12 June 2019).
84. The Ministry of Construction of Vietnam. Circular No. 02/2018/TT-BXD on Environmental Protection in Construction and Construction and Reporting on Environmental Protection in the Construction Industry. Available online: http://www.moc.gov.vn/home/-/legal/2pBh/vi_VN/18/428701/37 (accessed on 12 June 2019).
85. Watson, D. *Environmental Design Charrette Workbook*; AIA Committee on the Environment: Washington, DC, USA, 1996.
86. Gibson, G.E., Jr.; Gebken, R.J., II. Planning charrettes using the project definition rating index. In *Architectural Engineering 2003: Building Integration Solutions*; Liu, M.S., Parfitt, K.M., Eds.; American Society of Civil Engineers: Reston, VA, USA, 2003; pp. 1–7. [CrossRef]
87. Gibson, G.E., Jr.; Whittington, D.A. Charrettes as a method for engaging industry in best practices research. *J. Constr. Eng. Manag.* **2009**, *136*, 66–75. [CrossRef]
88. Knox, M.W.; Clevenger, C.M.; Dunbar, B.H.; Leigh, K.E. Impact of charrettes and their characteristics on achieved LEED certification. *J. Archit. Eng.* **2013**, *20*, 04013012. [CrossRef]
89. Loppnow, S.W. Current Industry Perceptions of the Role of the Contractor in the LEED Certification Process. Ph.D. Thesis, Colorado State University, Fort Collins, CO, USA, 2011.
90. Gürgün, A.P.; Polat, G.; Bayhan, H.G. Contractor Perspective on Material Risks in Green Building Projects. In Proceedings of the 12th International Congress on Advances in Civil Engineering, Istanbul, Turkey, 21–23 September 2016; Özturan, T., Luş, H., Eds.; ACE: Istanbul, Turkey, 2016; pp. 21–23.
91. Arif, M.; Syal, M.; Li, Q.; Turner, N. Constructors and innovation credits in green building projects. *Constr. Innov.* **2013**, *13*, 320–338.
92. Hwang, B.G.; Shan, M.; Xie, S.; Chi, S. Investigating residents' perceptions of green retrofit program in mature residential estates: The case of Singapore. *Habitat Int.* **2017**, *63*, 103–112. [CrossRef]
93. Chan, A.P.C.; Darko, A.; Ameyaw, E.E. Strategies for promoting green building technologies adoption in the construction industry—An international study. *Sustainability* **2017**, *9*, 969. [CrossRef]
94. Pushkar, S. Sacrificial Pseudoreplication in LEED Cross-Certification Strategy Assessment: Sampling Structures. *Sustainability* **2018**, *10*, 1353. [CrossRef]
95. Norusis, M. *IBM SPSS Statistics 19 Guide to Data Analysis: International Edition*; Pearson: London, UK, 2012.
96. Office of the National Assembly, Consolidated text 08/VBHN-VPQH on Prevention and Control of Tobacco Harm Law. Available online: https://vanbanluat.com/y-te/van-ban-hop-nhat-08-vbhn-vpqh-van-phong-quoc-hoi-28a30.html#noidung (accessed on 12 June 2019).
97. Odiase, J.I.; Ogbonmwan, S.M. JMASM20: Exact permutation critical values for the Kruskal-Wallis one-way ANOVA. *J. Mod. Appl. Stat. Methods* **2005**, *4*, 28. [CrossRef]

sustainability

MDPI

Article

Smartphone-Based Data Collection System for Repetitive Concrete Temperature Monitoring in High-Rise Building Construction

Hyunsu Lim [1] and Taehoon Kim [2,*]

1 Department of Architecture, Soonchunhyang University, Asan 31538, Korea; hslim@sch.ac.kr
2 School of Architecture, Chosun University, Gwangju 61452, Korea
* Correspondence: thoonkim@chosun.ac.kr; Tel.: +82-62-230-7145

Received: 6 August 2019; Accepted: 18 September 2019; Published: 23 September 2019

Abstract: The systematic collection and management of on-site information in high-rise building construction are important factors in construction management. Recently, wireless sensor network (WSN) technology has been utilized to manage the various tasks involved in high-rise construction efficiently and in a timely manner. However, because of the repeated installation of sensors and repeaters along with the construction progress, the existing WSN technology is ineffective when applied to the temperature management of concrete in structural work. Here, we propose a new data collection method in which a worker uses a smartphone to repeatedly monitor concrete temperature. In field implementation, the proposed system enables concrete temperature management without a transmission gap for monitoring in 60-min intervals with smartphones provided to 20% of the structural workers. Next, a case study was performed on a high-rise building construction site to analyze the effectiveness of the proposed system in terms of cost savings by avoiding schedule delay. The results of the case study show that the proposed system can reduce the additional work costs resulting from delays in concrete curing and save up to $18,907 in labor costs. In addition, this system can reduce the temperature management time of the quality manager and enable more efficient management. It is also expected that this system will contribute to on-site waste management by reducing the number of embedded sensors.

Keywords: data collection system; worker's smartphone; concrete temperature monitoring; high-rise building construction

1. Introduction

Accurate and timely management of on-site information in high-rise building construction is a crucial element for the success of a project [1]. In high-rise construction, a large amount of real-time information is available because various tasks are performed simultaneously [2]. If on-site information for timely management is missing or delayed in transfer, subsequent processes may in turn be delayed or rework may be required [3]. As a result, increasing efforts are being made to collect information in real time for on-time action in high-rise construction.

Various wireless sensor networks (WSNs) have been applied for timely and efficient construction management in high-rise building sites. WSNs collect real-time information using the sensors installed on the site and transfer the information to the manager's device wirelessly through a router. It is possible to reduce the information collection task of the manager and to share accurate information by collecting real-time information automatically. In the construction field, WSNs are mainly applied to progress monitoring, material tracking, and safety monitoring, in which real-time information is required [4].

However, existing WSN-based data acquisition methods are not efficient for use in repetitive tasks such as concrete temperature monitoring in high-rise building construction. In structural work, the temperature of concrete is an important element to estimate the strength of concrete in the corresponding time [5]. Because the construction process of the next floor can begin when the strength of concrete is met, managing the temperature for strength development is very important in the high-rise construction schedule, which has generally a short floor cycle [6]. Furthermore, it is important to measure the concrete temperature in real time because it varies considerably depending on the outside weather. Consequently, sensors and routers must be installed on every floor to apply a WSN to structural work. Repeated installation of hardware on every floor requires additional work cost and time and may negatively affect construction waste management due to the number of embedded sensors required [7]. In addition, obstacles such as the steel materials of formwork may lower transfer rates from sensors to routers. Maintenance of the devices installed on a site is also difficult because of the rough environment, for example, with dust at the construction site or changes in outside weather.

Data collection by workers using smartphones can be an efficient solution for repetitive concrete temperature monitoring in high-rise construction. Because people commonly use smartphones every day, data collection by workers using the ubiquitous sensors embedded in smartphones has been attempted at construction sites [8]. Smartphones can replace the sensors or routers of the existing WSNs because they have various sensors and communication environments [9]. If smartphones are used instead of routers in high-rise construction, data can be transferred without router installation on every floor and the cost of devices can be saved. In addition, because workers continuously move in the work area, data transmission failure because of obstacles can be minimized.

This study proposes a new data collection method using smartphones to improve the efficiency of repetitive data collection such as monitoring concrete temperature in high-rise frame construction. In this study, the concrete temperature monitoring systems using existing WSNs are reviewed. Then, the architecture of a new monitoring system with workers using smartphones is outlined, and major technologies are demonstrated. Next, the system is implemented through a field test on an actual high-rise building construction site. Based on the collected data, the feasibility of the system is discussed by analyzing the transfer rate. Lastly, the effectiveness of the proposed system is analyzed in terms of cost savings by preventing schedule delay through a case study. It is expected that the results of this study will enable efficient data collection and monitoring in high-rise frame construction and contribute to preventing construction delay caused by the concrete temperature and securing excellent quality.

2. Literature Review

2.1. Monitoring System for Concrete Temperature

In general, existing concrete temperature monitoring systems using WSNs consist of sensor nodes, a repeater, a router, and a server (Figure 1) [10]. Sensor nodes refer to sensors and devices with a communication function to transfer sensor data. The information measured by a sensor is transferred to a nearby repeater or router using a local area network (LAN). The repeater also transfers the information from a sensor node to a router if the transfer distance of a sensor node is long. The router transfers the information to a server using a wide area network (WAN). The LAN of existing monitoring systems uses Zigbee, Bluetooth, and radio frequency identification (RFID), while WAN performs wireless transfer to servers using wideband code division multiple access (W-CDMA). Existing monitoring systems have been applied to mass concrete with high hydration heat or cold weather concrete to monitor the concrete temperature.

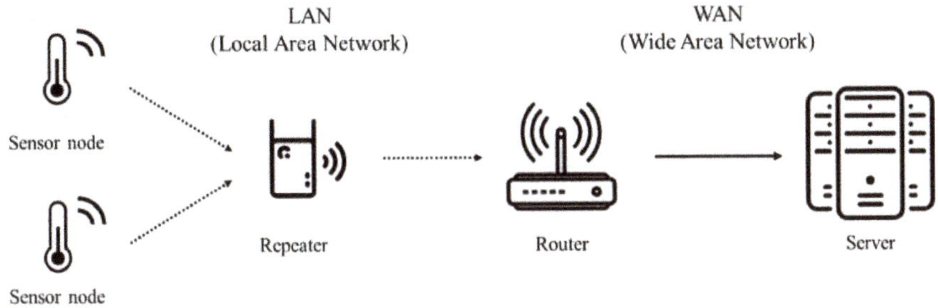

Figure 1. Existing concrete temperature monitoring process using a wireless sensor network (WSN).

Because existing concrete monitoring systems were developed to be used only during a required period, they are not suitable for repeated structural work in building construction. For repeated use, sensor nodes and routers must be continually moved to the required area. Such additional processes increase costs and construction time, and construction waste also increases because of the required number of embedded sensors. Furthermore, because formwork materials and structures may interfere with transmission, more repeaters may be required depending on the building type. Thus, when applying a WSN-based monitoring system to structural work in high-rise construction, a method capable of minimizing the reinstallation of devices and improving transfer rates is required.

Most studies on using WSNs for concrete curing temperature monitoring are targeted at mass concrete and focused on improving sensor accuracy and economic efficiency. Lee et al. [11] developed a ubiquitous sensor network (USN)-based mass concrete curing management system and applied it to foundation concrete. Embedded sensors and a radio frequency (RF) modem were used, and the RF modem was connected to a server by cable. Because the system was installed in an open space for mass concrete, obstacles against transmission and system reuse were not considered. Norris et al. [12] proposed a system capable of measuring the concrete temperature and humidity using nanotechnology/microelectromechanical devices. Accuracy was improved compared with existing temperature sensors, but wireless communication was not considered because data were transferred to a server by cable. Chang and Hung [13] suggested a new temperature measurement technology considering a large-scale construction site environment. They studied the communication distance in an environment with external obstacles using an RF integrated circuit and temperature/humidity sensors. However, their experiment used fixed obstacles in an open area and thus had limits for application in a site environment where actual obstacles change frequently. Lee and Park [14] developed a reusable sensor node that can be attached to formwork. The node is reusable because the temperature sensor is embedded in concrete and the node can be separated from the formwork. However, in this case, cost and time increase because of the need to embed probe-type sensors and install sensor nodes on every floor. Barroca et al. [5] developed a durable and economical system for temperature and humidity monitoring using WSN. The temperature–humidity sensor embedded in concrete was connected to a communication module and monitored through an external device. However, reuse of the sensor was difficult because it was embedded and LAN has limited transfer distance. Kim et al. [15] proposed a new temperature measurement method using the surface acoustic wave. Liu et al. [16] improved the transfer distance of the embedded module using the embedded RFID sensor tag and outlined an economical system. However, again, those two studies had limits with respect to reusability because embedded sensor nodes were used. Cabezas et al. (2018) [17] developed and field-tested a concrete curing monitoring system using a compact embedded wireless sensor. That study focuses on determining how to withstand the harsh environments where embedded sensor nodes are installed and how to save energy in sensor node communications. Jianshu et al. [18] proposed a framework of cracking control for a mass concrete structure by taking advantage of Distributed Temperature Sensing (DTS) system and showed higher efficiency in temperature data acquisition than conventional electric converters. Angat et al. [19] developed an integrated system by analyzing combined measurements of RFID and optical fibers. That study focused on the measurement method for improving the measurement performance of a concrete curing monitoring.

Previous studies conducted experiments in laboratory environments and thus focused on sensor accuracy, economic efficiency, and transfer rate of information. Thus, research on a new data collection method is required for operation inside building construction that requires many repeated installations and has many obstacles in the structural construction of a high-rise building.

2.2. Data Collection Systems Using Smartphones

To address the drawbacks involved in sensor installation and maintenance of existing USN technology at construction sites, data collection methods using smartphones with built-in sensors have been utilized [5]. Because smartphones have various sensors and communication modules capable of accessing both WAN and LAN and are easy to carry, they have replaced most devices previously used at construction sites.

Previous studies on smartphones focused on the manager's smartphone and improved the efficiency of on-site information management. Dong et al. [20] improved the information management efficiency between the site and the office using mobile phones for construction defect management. Lee and Kim [21] suggested measures to streamline the lifted material management process using smartphones. Kim et al. [1] developed an on-site management system capable of performing the functions of site monitoring, task management, and real-time information sharing using mobile computing technology. Park et al. [22] proposed proactive construction defect management using mobile devices and augmented reality.

As smartphone penetration continuously increases [23], studies that use workers' smartphones have been conducted. Cobili et al. [24] suggested a mobile system that can confirm the activities and on-site presence of workers. Ahn et al. [25] suggested a system for checking the attendance of workers at a construction company using the smartphone's Global Positioning System(GPS). Dominics et al. [26] analyzed the usability of workers' smartphones to identify their work position information. Dzeng et al. [27] presented an algorithm that can detect falls or signs of falls of workers using the acceleration sensor of workers' smartphones. Akhavian et al. [28] suggested an application capable of recognizing the activities of workers for each process by classifying the activities for each work process and utilizing the smartphone's gyroscope and acceleration sensor. Yan et al. [29] developed a real-time motion warning system to help prevent injuries that can recognize the incorrect movements of workers using smartphones attached to helmets and vests. Zhang et al. [30] proposed a decision tree algorithm that can more accurately recognize a worker's activity by using the accelerometers and gyroscope sensors of a smartphone. Zhang et al. [31] explored potential applications of the worker's smartphone as a data acquisition tool to detect near-miss falls on the basis of an artificial neural network.

Thus, existing studies on data acquisition systems using smartphones at construction sites have evolved from a focus on real-time site information using the manager's smartphone to those using workers' smartphones. This trend means that workers' smartphones can be expanded to individual monitoring systems that include sensors and network functions, thereby measuring a variety of on-site data. However, previous studies only covered information that could be measured by the built-in sensor of a worker's smartphone. To better manage on-site information, the worker's smartphone can be used as a router for transmitting information. In this way, the smartphone can be utilized as individual communication nodes that construct a network on site. Here, we present a new method to monitor concrete temperature information by using workers' smartphones as a router.

Although Lim et al. [32] conducted a preliminary analysis of a new data acquisition method using smartphones, they only analyzed the economic aspect of the concept model for the proposed method. The present study proposes the specific method and technical details of a fully developed system and verifies its practical aspects through a field experiment.

3. A Smartphone-Based Data Collection System (S-DCS)

3.1. Research Method

The research method of this study is shown in Figure 2. First, the framework of S-DCS was drawn up, and then each elementary technology was developed to produce a prototype. Next, to test the feasibility of the S-DCS, a field test was performed to analyze the data rate for each variable. Lastly, to verify the effectiveness of S-DCS, the cost effectiveness of this system was analyzed through a case study of a high-rise construction site.

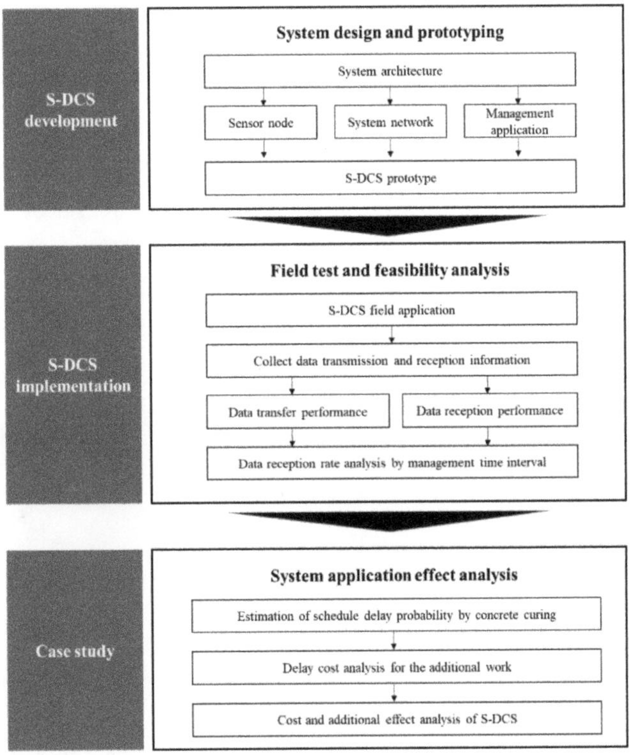

Figure 2. Research process. S-DCS: Smartphone-Based Data Collection System.

In the system development part, S-DCS was designed based on the requirements for repetitive concrete temperature monitoring by dividing by sensor node, system network, and management application. Next, we prototyped the sensor node and application for the field test of the system.

In the S-DCS implementation part, feasibility was tested by applying S-DCS to a 42-story high-rise building construction. The sensor node was installed in the slab formwork, and the workers were given a smartphone to measure the concrete temperature in the construction of one floor. During the construction period, a video was taken to identify the location of the smartphone held by the worker at the time the data were received. The time and number of times sent from the sensor node were checked for the log data of each sensor node, and the received time and number of smartphones were checked through the application of the smartphone. The data transfer performance and data reception performance of smartphones are analyzed by the distance between the sensor node and the smartphone, transmission time of the temperature data, reception time, and number of times transmitted. In addition, based on the collected data, we analyzed the feasibility of this system through data reception rate analysis by management time interval.

In the case study part, the effectiveness of S-DCS was analyzed in a high-rise building site consisting of three buildings: 77 floors, 66 floors, and 42 floors. The probability of schedule delay because of concrete curing at the site was calculated, and the delay cost for the additional work was analyzed. In addition, cost and additional effects were analyzed when delays did not occur through monitoring using S-DCS.

3.2. System Architecture

An S-DCS is designed so that data are transferred using workers' smartphones instead of routers attached to formwork for repeated use on every floor in high-rise construction. Figure 3 shows the conceptual framework of an S-DCS. The sensor node installed in the formwork records the temperature of the curing concrete and transfers the data to the smartphone of a nearby worker through LAN at regular intervals. The worker's smartphone transfers the data to a server using W-CDMA, the data communication network of smartphones. The site manager can access the server and monitor the temperature of the structure.

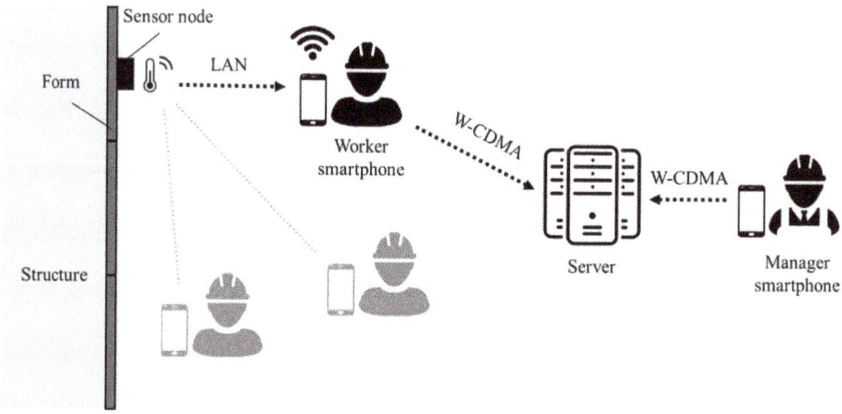

Figure 3. Architecture of an S-DCS. LAN: local area network; W-CDMA: wideband code division multiple access.

Because the sensor node of the S-DCS is attached to the formwork, repeated installation of the sensor node is not required. In addition, there is no need for additional routers or installation work because the smartphone functions as a router. Furthermore, the continuous movements of a worker's smartphone in the workspace can minimize the transfer gaps caused by obstacles such as walls and formwork materials.

3.3. Sensor Node

The sensor node in an S-DCS is attached to the bottom of the slab form and the probe thermistor penetrates the formwork and measures the temperature inside the concrete (Figure 4). Because the formwork must be installed with the sensor node attached, the sensor is attached to the formwork using four bolts. The vertical probe thermistor attached to the node penetrates the formwork so that the sensor can remain vertical during concrete pouring and can be easily separated from the formwork during disassembly. In addition, a protective cap covers the probe thermistor so that it is not in direct contact with the concrete during curing and be easily separated. Two types of probe lengths were designed to measure the temperatures of the center and the surface of the slab concrete. The slab thickness of the test building to which the sensor node prototype was to be applied was

250 mm, the probe length for the center was 120 mm, and the probe length for the surface was 20 mm. The measurement range of the temperature sensor was set between −40 °C and 110 °C, and the experiment confirmed that the error range was within ±1 °C [33].

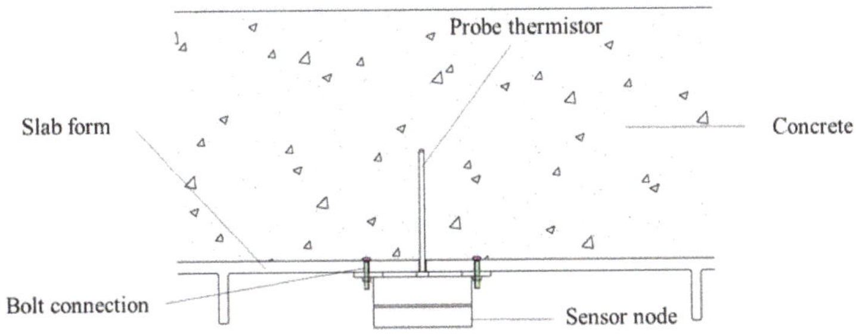

Figure 4. Sensor node installation.

The sensor node activates the temperature sensor at 5-min intervals. Because continuous temperature measurement and transfer can rapidly consume power and may cause additional work because of frequent charging, the sensor node is only activated every 5 min, and with a 700-mAh battery, it can be used for up to 30 days without charging. In addition, the sensor node has Light-Emitting Diodes (LED) that can display the status of the sensor node. The LEDs display the operation, error, and charge of the sensor node to allow the worker to identify the status at the site (Figure 5). The remaining battery capacity, activation status, and transfer error of the sensor node along with the temperature information are transferred to the server so that the manager can identify the status of each sensor node in the office.

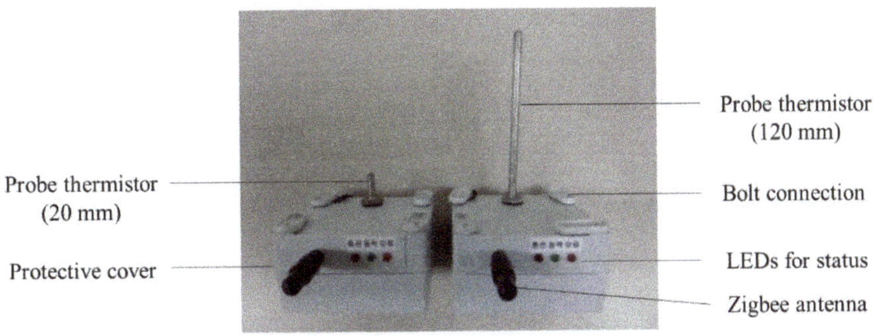

Figure 5. Sensor node configuration. LEDs:

A cover protects the inside of the sensor node against dust and water on site. The cover and fixing pin are made of reinforced plastic to prevent corrosion caused by water entering through a hole in the formwork. In addition, a 3-mm-thick rubber cover was installed on the inside edge of the sensor node to prevent dust and water from entering, and dual rubber rings were installed where the probe thermistor is mounted on the formwork to prevent foreign matter from entering.

3.4. System Network

The local network of an S-DCS uses the Zigbee method. The local networks that can be used on smartphones include Wireless Fidelity (WiFi), Bluetooth, Zigbee, and Near-Field Communication (NFC). WiFi consumes much power and is, therefore, not suitable when continuously used because of increased maintenance work. NFC transfers data through a tag within 10 cm and is not suitable because the measurement work of the worker increases. Bluetooth and Zigbee have suitable transfer speed and power consumption for high-rise construction and are economical. Because routers move in an S-DCS, information must be simultaneously transferred to as many smartphones as possible to improve the continuity of transferred information. Zigbee can access several smartphones simultaneously, and thus, is more suitable for an S-DCS than Bluetooth, which has a limited number of simultaneous connections.

Because smartphones simultaneously and continuously move and receive data, an algorithm for missing and duplicated data was established. Figure 6a shows the data processing algorithm of the sensor node. To reduce power consumption, the sensor node selects a power supply that is activated every 5 min. When the sensor node is activated, the temperature data are collected and stored in the internal memory. Then, the sensor node checks the smartphone within the communication radius. If there is no smartphone, the sensor node is deactivated and reactivated after 5 min. If smartphones are identified, the first received smartphone is selected and the temperature data in the internal memory are transferred. When data reception confirmation is received from the smartphone, the stored data are deleted to secure memory capacity and the sensor node is deactivated again.

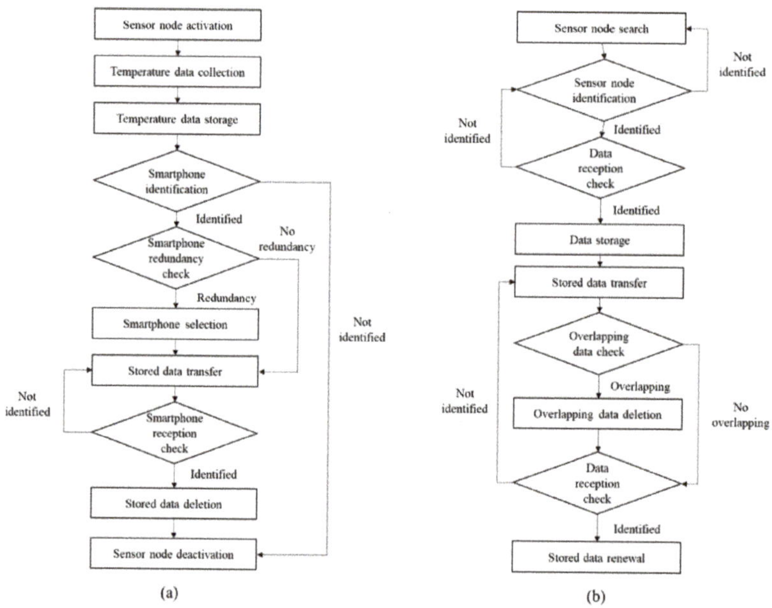

Figure 6. Data processing algorithm: (**a**) sensor node; (**b**) smartphone.

Figure 6b shows the data processing algorithm of the smartphone, and the smartphone searches for nearby sensor nodes using a Zigbee network. If a nearby sensor node is identified, data are received. If multiple sensor nodes are recognized, the data are classified and stored by each sensor node. When the data are received, the smartphone confirms the reception and stores the data. The stored data are transferred to a database through the W-CDMA network, and the transferred data are checked again for overlapping data. Any overlapping data are deleted, and the database is renewed.

3.5. Management Application

The workers' smartphones require an application to transfer and manage data. Figure 7a illustrates the interface of a function that shows the data collection status. There is a function window for data collection and stop, and a data reception and storage list is displayed in the center of the screen. The list shows the received sequence number, sensor node serial number, reception date and time, and number of data points. The monitoring application can check the device list and work area. Figure 7b shows details of the device information when the device icon is selected. The application displays the remaining battery capacity, measuring start and end time, installation location, and measured temperature graph. The sensor node information is also displayed on the floor plan when the drawing tab is selected (Figure 7c).

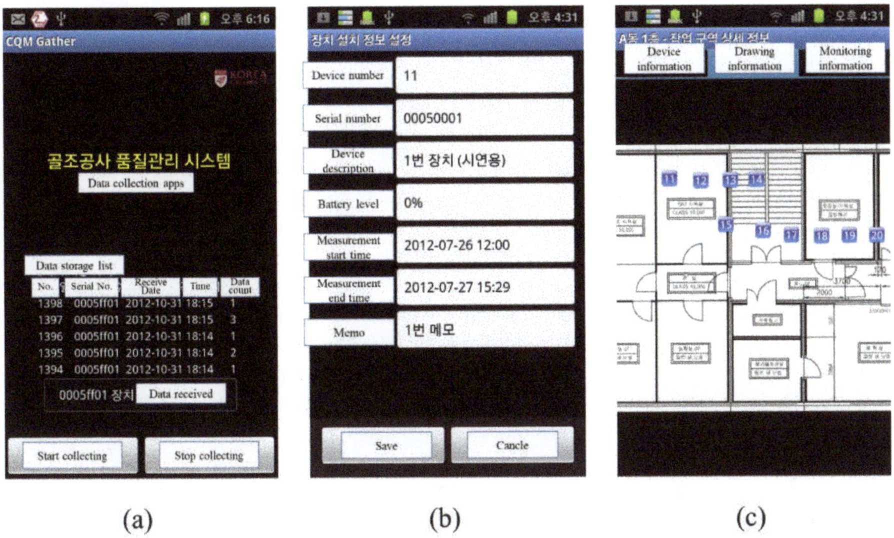

Figure 7. Application for the smartphone: (**a**) function; (**b**) node information; (**c**) floor plan.

The interface of an application for the server was designed based on the smartphone application. The floor plan of the work area can be loaded, and the sensor node and status information can be displayed at corresponding locations for management. When a sensor node is selected, the screen is converted to a monitoring screen composed of the sensor node information, and the temperature data over time are displayed (Figure 8a). For monitoring, the strength estimation graph of a structure is displayed based on the accumulated temperature. When each structure reaches the formwork-removal criterion strength, the manager is notified through the alarm window (Figure 8b).

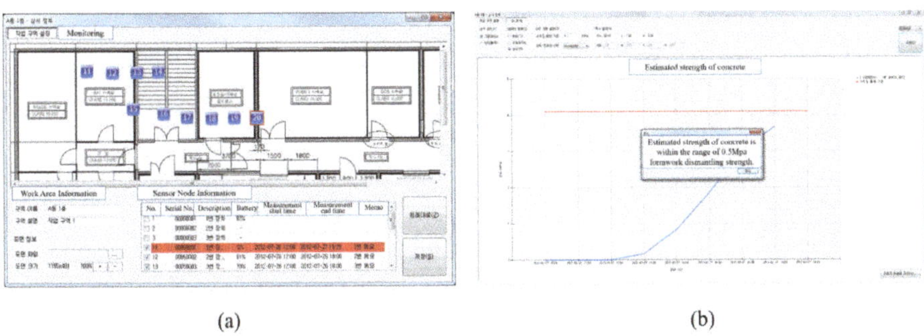

Figure 8. Application for the server: (**a**) node information; (**b**) strength estimation graph.

4. S-DCS Implementation

4.1. Outline of the Field Experiment

To test the feasibility of the S-DCS on an actual construction site, a field experiment was conducted by applying the S-DCS to structural work during construction of a tall residential building. The site was for a building with five underground floors and 42 ground floors, and it had a flat plate/slab structure. Panelized aluminum forms were used as the slab and wall formwork. One floor was divided into two work zones, and it took four working days to complete construction for each zone. The field experiment was conducted during the construction of the 31st floor of zone A, considering that the same floor construction was repeated on every floor.

Sensor nodes were installed on the slab form of a total of five areas; each area had a radius of 10 m, considering the building type and data transfer rate (Figure 9). Sensor nodes of 20 mm and 120 mm were installed side by side on a 250-mm-thick slab to examine the temperature difference according to the concrete penetration height. Two sensor nodes were supposed to be installed in each of the five areas, but one sensor node with a 20-mm probe in area E was broken while being installed; thus, a total of nine sensor nodes were installed. Because there was only one sensor node in the E area, the absolute number of transmitted and received data items may be smaller than in other areas. However, the analysis based on area compares the ratio of received data by area. Therefore, the difference in the number of data items because of the difference in the number of sensor nodes does not significantly affect this field test feasibility analysis. Thus, field tests were conducted with nine sensor nodes, and the nodes in area A were sequentially assigned a number, as shown in Figure 9.

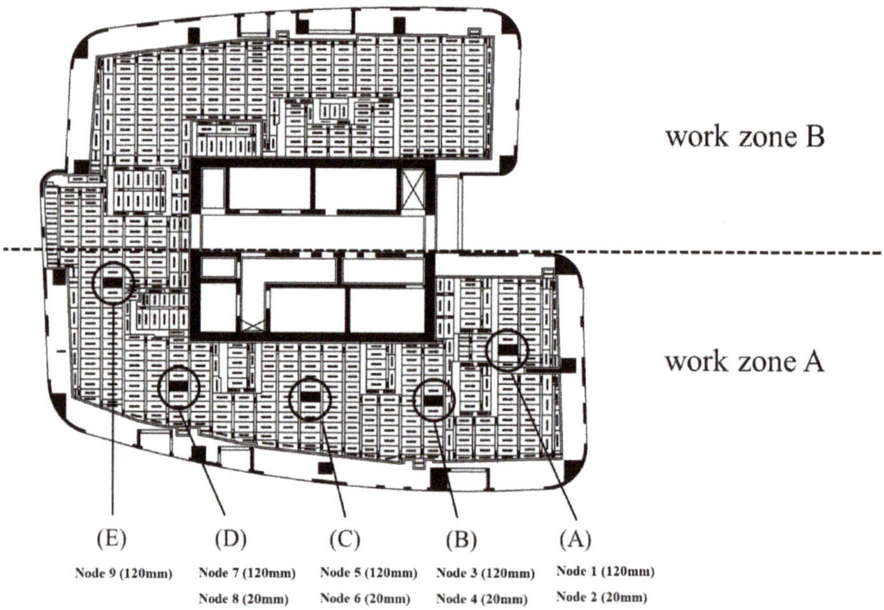

Figure 9. Sensor node installation layout.

Table 1 shows the number of workers to be provided with a smartphone during the test period. On day 1, smartphones were provided to three concrete workers and one formwork inspector in zone A and two slab formwork workers in zone B. On day 2, smartphones were provided to three workers for dismantling the wall formwork on the 30th floor of zone A and three wall rebar workers on the 31st floor. On day 3, smartphones were provided to three workers for installing the wall formwork on the

31st floor of zone A and three wall reinforcement workers. On day 4, smartphones were provided to two workers for slab form dismantling on the 30th floor of zone A, to two workers for installing wall formwork on the 31st floor, and to two workers pouring concrete on the 31st floor of zone B. The total number of workers for the four days was 135 and smartphones were provided to 24 workers (18%).

Table 1. Number of smartphones provided to workers.

Work Zone	Worker Type	Number of Smartphones Provided Per Day			
		Day 1	Day 2	Day 3	Day 4
A	Formwork worker	1	3	3	4
	Rebar worker		3	3	
	Concrete worker	3			
B	Formwork worker	2			
	Concrete worker				2
Total number of smartphones		6	6	6	6
Total number of structural workers		27	38	43	27
Provision ratio		22%	16%	14%	22%

4.2. Data Transfer Performance

Table 2 provides an overview of the temperature data transferred to the server. A total of 7538 temperature data points were transferred from nine sensor nodes to the server. The number of planned data calculated from the measurement time was 7591, which differs from the number of data points actually measured, and was caused by a delay in data transfer time and missing data. Analysis of the time intervals of the transferred data indicates that differences of a few seconds from the set time unit were observed; these accumulated and resulted in the difference from the planned number of data points, which was set as the data difference. Data that had a transferred interval of over 5 min were regarded as missing data; 46 such data points occurred because of the transfer time delay, and seven data points were missing because of transfer failures.

Table 2. Number of data transmitted per sensor node.

Node	Measuring Time (min)	Number of Data (EA)				Sensor Node Performance Ratio (%)	Data Transfer Rate (%)
		Planned Data Points	Measured Data Points	Transfer Time Delay Data Points	Transfer Failure Data Points		
node 1	4122	824	818	6	0	99.27%	100.00%
node 2	4247	849	842	6	1	99.18%	99.88%
node 3	4248	849	845	1	3	99.53%	99.65%
node 4	4249	849	843	5	1	99.29%	99.88%
node 5	4243	848	843	5	0	99.41%	100.00%
node 6	4124	824	818	6	0	99.27%	100.00%
node 7	4248	849	843	6	0	99.29%	100.00%
node 8	4246	849	843	5	1	99.29%	99.88%
node 9	4250	850	843	6	1	99.18%	99.88%
Total	37,977	7591	7538	46	7	-	-
Average	4219.67	843.44	837.56	5.11	0.78	99.30%	99.91%

The performance ratio of a sensor node can be expressed by the ratio of the number of measured data to planned data. As an index to determine the accurate frequency of when a sensor node performed data transfer, the ratio of the number of measured data to planned data excluding missing data was used. As noted, the difference between the number of planned and measured data was 46, that is, 5.11 per sensor node. The performance ratios of the sensor nodes ranged from 99.18% to 99.53%

(average, 99.30%). The results indicate that continuous monitoring is possible because the sensor nodes perform the measurement and transfer every 5 min with >99% accuracy. The performance ratio could be further improved by setting a shorter transfer time interval.

The data transfer rate between a smartphone and a server can be measured by considering the measured data and missing data. Because a sensor node sends all previous records even if transfer to a smartphone is missing, missing data occur during the transfer from a smartphone to a server. The data transfer rate is represented by the ratio of the number of measured data to that of data before the missing data are removed, excluding the data difference caused by the time difference. The data transfer rate ranged from 99.65% to 100% (average, 99.91%), indicating that monitoring is feasible almost without data transfer gaps because the server transfer rate was >99%.

4.3. Data Reception Performance of Smartphones

4.3.1. Reception Ratio by Distance

The data reception ratio of the smartphones according to various variables was analyzed to examine their performance as routers at the construction site. Only the data during working hours when the workers actually carried the smartphones were analyzed, that is, 1394 data points measured for a total of 1290 min over four days. The data reception time and transfer from sensor nodes were recorded through the log files of the smartphones, and distances to the sensor nodes were measured from the worker's location at the time of reception using recorded videos.

Table 3 summarizes the number of received data from each sensor node at a 5-m interval to identify the reception rate by distance. The distances at which most data were received were 5–10 m, accounting for 33.43% of all data. The reception rate, at 15–20 m, was the second highest at 24.46%. At shorter distances of 0–5 m, a reception rate of just 9.9% was observed. Thus, the reception ratio was higher at distances of 5–10 m and beyond than when workers were nearest the sensor node antenna. Most of the reception data were attributed to workers' smartphones received in zone B. This is likely because when area B workers approached sensor nodes when there were no workers with smartphones in zone A, reception occurred at 15–20 m. Because the number of data points by distance did not differ greatly between 5 m and 20 m, where almost 80% of data were concentrated, the results indicate that installing sensor nodes at a radius of 5–20 m can improve data reception. The maximum measurement distance was 28.34 m, indicating that measurement was possible throughout the entire work zone.

Table 3. Number of received data by distance.

Node	Number of Data Points					
	0–5 m	5–10 m	10–15 m	15–20 m	20–25 m	25–30 m
node 1	20	28	19	49	7	2
node 2	13	29	34	30	21	0
node 3	25	47	11	33	13	4
node 4	11	44	40	66	17	3
node 5	11	101	23	20	27	4
node 6	11	19	29	47	23	2
node 7	18	31	19	41	15	8
node 8	22	109	26	26	8	1
node 9	7	58	70	29	19	4
Total	138	466	271	341	150	28
Rate	9.9%	33.43%	19.44%	24.46%	10.76%	2.01%

4.3.2. Reception Ratio by Work Type

The smartphone reception rate by work date was analyzed to identify the reception rate by work type. The reception ratio of the smartphone is calculated by the ratio of the number of data points received from the smartphone and the number of those transmitted from the sensor node every 5 min. In other words, the reception ratio of the smartphone is an index that measures how much transmission gap is generated in real-time monitoring. Table 4 shows the smartphone reception rate by work date. The reception rate was lowest at 51.28% during the pouring work on the day 1 and the highest at 81.26% on day 2. The rates were 60.13% and 61.88% on days 3 and 4, respectively, giving an average of 62.85% over the four days. The reception rate of day 2 showed the largest difference of ~20% from the average due to work being stopped when it rained during the morning of day 2. As the workers were waiting without moving within the radius of the sensor nodes, the data reception rate increased. The reception rate of day 3, when the same work type was performed as that of day 2, was 60.31% and that of day 4 with a different work type was 61.88%, representing values similar to the average. These results indicate that the reception rate is affected more by the behavior characteristics of workers than by the work type.

Table 4. Smartphone reception rate by work type.

Work Date (Month/Day)	Work Type	Number of Received Data Points	Number of Transmitted Data Points	Reception Ratio (%)
day 1	Pouring concrete	221	431	51.28%
day 2	Installing wall rebar Dismantling wall form	347	427	81.26%
day 3	Installing wall rebar Dismantling and lifting wall form	537	893	60.13%
day 4	Installing wall form Dismantling slab form	289	467	61.88%
Total		1394	2218	62.85%

4.3.3. Reception Ratio by Work Area

The reception rate by work area was analyzed to identify the reception rate according to the installation area of sensor nodes. Table 5 shows the reception ratio of smartphones according to the sensor node installation area. The rate was lowest at 52.89% in (A) and highest at 75.61% in (E), as shown in Figure 9. The reception rate increased as the area moved from (A) to (E), and this was attributed to the characteristics of the work area.

Table 5. Smartphone reception rate by work area.

Construction Area		Number of Received Data Points	Number of Temperature Data Points	Reception Rate (%)
Location	Node			
(A)	nodes 1, 6	256	484	52.89%
(B)	nodes 2, 7	259	482	53.73%
(C)	nodes 3, 8	325	499	65.13%
(D)	nodes 4, 9	368	507	72.58%
(E)	node 5	186	246	75.61%
Total		1394	2218	62.85%

In zone A, the work quantities of form, rebar, and concrete are not significantly different between the left and right sides. When the work quantity is similar, the workers' movements can affect the exposure frequency of the sensor nodes. The form workers move frequently between the upper and lower floors to lift the dismantling formwork and share tools. Workers move mostly along the left path because the right path is blocked by the concrete placing boom (CPB) to go downstairs. By moving along the left path, workers are more exposed to the sensor nodes on the left side. The rebar workers move rebar from the stacked area to the work location. Given that the rebar was mostly stacked on the right side on the floor by the inside left wall, the rebar workers continuously moved to the left side of the workspace and thus the reception rate on the left side was higher.

The results show that the reception rate of smartphones—a key element for the continuity of monitoring—is affected by the distance between the sensor nodes, the behavior of the workers, and the characteristics of the work area. The distance between the sensor nodes must be within a radius of 20 m for a high reception rate, but it can be extended to 30 m under unavoidable conditions because of site characteristics. The results show that the work type does not significantly affect the reception rate if the numbers or proportions of the provided smartphones are identical. It is possible to improve the reception ratio by providing smartphones first of all to workers who move little during their work because the rate is affected by the movement frequency of workers. The reception rate is also affected by the movement path of workers according to the work area. Therefore, the reception rate of smartphones can be increased by positioning more sensor nodes at locations with low movement frequency after analyzing the movement paths of workers according to the building type, work quantity, location of stairs, and material storage location of the project.

4.4. Data Reception Rate Analysis by Management Time Interval

Because different management information requires different management time intervals, it is important to verify if data are received without data gaps within the management time for practical application of the system. In general, the maturity of concrete is measured by a time unit. Therefore, the reception rate of data was analyzed with the management time interval set to 5, 10, 30, and 60 min. Because two sensor nodes were installed in each area, the data of the two sensor nodes were combined for analysis of the reception rate.

Table 6 shows the data reception rate for each area during each time interval. For monitoring at a 5-min interval, the transfer rate was 62.85%, which was the same as the smartphone reception rate. The transfer rate increased to 80.36% for the 10-min unit and 91.67% for the 30-min unit. As the management unit time increased, data transfer gaps decreased. In particular, for the 60-min unit, the transfer rate was 100% for all areas. The maximum transfer gap occurred in area (D), where reception of the smartphones was not examined because there was no movement of the workers for 45 min.

Table 6. Data reception rate by management time interval.

Construction Area	Management Time Interval			
	5 min	10 min	30 min	60 min
(A)	52.89%	79.73%	91.67%	100.00%
(B)	53.73%	81.82%	96.00%	100.00%
(C)	65.13%	80.65%	90.00%	100.00%
(D)	72.58%	80.28%	91.30%	100.00%
(E)	75.61%	78.85%	87.50%	100.00%
Average	62.85%	80.36%	91.67%	100.00%

The results indicate that the S-DCS is optimal for concrete temperature monitoring at the 60-min interval with six smartphones at the site. Because the smartphones used as routers were provided to ~20% of the structural workers, it would be possible to reduce the transfer gap time and management time unit if the application were carried by more workers. Furthermore, a higher reception rate can be achieved when there is much work quantity and many workers. Considering the characteristics of high-rise sites and curing management, the S-DCS shows sufficient feasibility in terms of continuity of data transfer.

5. Case Study

5.1. Case Description

To meet the objective of S-DCS development, a high-rise building site with more than 50 floors with many repetitive floors and short construction period was selected as the case site. The case site is a high-rise building located in Busan consisting of three buildings of 77 floors, 66 floors, and 42 floors. The time for concrete curing is very short because construction proceeds at a rate of one floor every four days. Therefore, if the curing temperature is not properly controlled, the concrete strength cannot be met and the construction is delayed. To analyze the effect of the proposed system, a high-rise building under construction in winter was selected.

The time required for dismantling the slab form after pouring was 28 h. The site was constructed with a 4-day cycle for a typical floor, and heat curing and concrete temperature management were performed because of the lack of slab-curing time during the day when the average temperature was below 10 °C. During the heating curing period, floors 29–49 of building A were built, floors 27–49 of building B, and floors 18–31 of building C; concrete was poured 174 times. Table 7 shows the slab formwork area and the number of workers: 22 in building A, 19 in building B, and 21 in building C.

Table 7. Slab area and form workers by building.

	Building A	Building B	Building C
Slab area (m^2)	636	566	602
Number of workers	22	19	21

5.2. Delay Cost Analysis

In the existing curing temperature management process, the concrete may not reach the designed compressive strength because of various environmental factors. Management cannot immediately respond to variable factors that arise during the night by installing a temperature measuring device and checking the next day. Consequently, scheduling delay occurs because of insufficient management, damage to heating facilities, and sudden changes in temperature during the night. From interviews with the quality managers of three high-rise building construction projects over 50 floors, schedule delays caused by curing occurred in just 5–10% of instances of pouring concrete.

If the curing strength is not satisfactory at the time of dismantling the slab formwork, the schedule for slab formwork will be delayed, and additional work for 2 h will be required to install the slab formwork the next day to proceed as planned. The additional work cost can be estimated as the labor cost of overtime work for 2 h, and calculated by multiplying the number of slab form workers by overtime wages times the number of delays [34].

Table 8 shows the additional costs for different numbers of delays. Because the concrete is poured 174 times, delays could occur between nine and 17 times. The minimum cost of delays (9× in building B) is $8645 and the maximum (17× times in building A) is $18,907.

Table 8. Additional work costs based on the number of delays.

Number of Delays	Additional Work Costs ($)		
	Building A	Building B	Building C
9	10,010	8645	9555
10	11,122	9605	10,616
11	12,234	10,566	11,678
12	13,346	11,526	12,740
13	14,459	12,487	13,801
14	15,571	13,447	14,863
15	16,683	14,408	15,925
16	17,795	15,369	16,986
17	18,907	16,329	18,048

The S-DCS can prevent delays through immediate response by monitoring in real time the strength degradation of the concrete due to various environmental factors. Applying the S-DCS to our case site will save additional work costs of $8645 to $18,907. In addition to cost-effectiveness, real-time management also helps quality managers to reduce the time needed to install devices and manage concrete temperature, ensuring that quality managers use their time more efficiently. The number of temperature sensors embedded in concrete for temperature management can also be reduced, contributing significantly to waste reduction in the field.

6. Conclusions

In this study, a new data acquisition system was proposed for concrete temperature monitoring in high-rise construction. A network system suitable for repeated processes in high-rise construction was designed using smartphones provided to workers. The feasibility of the S-DCS was verified by applying the proposed system to a high-rise construction site. The results showed that the S-DCS can be used to monitor the 60-min management time unit without data gaps regardless of the movements of workers/smartphones.

Our study outlines the practical application of a smartphone-based data acquisition method and the variables to be considered according to the construction site. To perform real-time monitoring without transmission gaps when using this method, the following aspects should be considered. (1) If the distance of the sensor nodes exceeds 20 m, the reception ratio is lower. (2) Because the reception ratio is influenced by the behavioral characteristics of workers, it is possible to increase the reception ratio by equipping with smartphones the workers who remain for a long time in a specific area, which is the case with, for example, rebar workers. (3) It is possible to increase the reception ratio by installing more nodes in an area where there are few moving paths of workers, by considering the layout plan of, for example, material storage, the location of stairs, and floor shape. (4) The S-DCS can be practically applied in 60-min units without transmission gaps in management, and equipping >20% of structural workers with smartphones can shorten the management time interval.

The proposed system can prevent schedule delays resulting from improper concrete curing, and utilize managers' time more efficiently by omitting inconvenient tasks in the monitoring process. This system can also improve the sustainability of the construction field by avoiding material waste through the reuse of sensors and improving the structure quality through concrete monitoring. The proposed method can be applied to other data collection tasks in repetitive high-rise building construction.

The proposed system can also be extended to the monitoring of various structures through data collection methods using smartphones. For example, Chiara Bedon [35] used micro-electromechanical systems sensors to measure the information of the structure of a glass suspension footbridge and to perform diagnostic analysis. If an information collection method using smartphones is applied to the glass suspension footbridge, it is possible to efficiently monitor the information such as temperature, humidity, and vibration through the smartphone of the maintenance manager or safety manager, and the structural safety can be predicted by accumulating data. In addition, it is possible to practically apply the data collection method of this study to various structural health monitoring systems using wireless sensors.

Additional research is required on a more efficient network construction method for non-working hours when workers are not present.

Author Contributions: In this paper, H.L. developed the research ideas and completed the writing; H.L. and T.K. collected and analyzed data; T.K. designed and organized the overall research flow.

Funding: This research was supported by the Soonchunhyang University Research Fund (No. 20181000) and by research fund from Chosun University, 2016.

Conflicts of Interest: The authors declare no conflict of interest.

References

1. Kim, C.; Park, T.; Lim, H.; Kim, H. On-site construction management using mobile computing technology. *Autom. Constr.* **2013**, *35*, 415–423. [CrossRef]
2. Quesada-Olmo, N.; Jimenez-Martinez, M.J.; Farjas-Abadia, M. Real-time high-rise building monitoring system using global navigation satellite system technology. *Measurement* **2018**, *123*, 115–124. [CrossRef]
3. Louis, J.; Dunston b, P.S. Integrating IoT into operational workflows for real-time and automated decision-making in repetitive construction operations. *Autom. Constr.* **2018**, *94*, 317–327. [CrossRef]
4. Soman, K.; Rapheal, B.; Varghese, K. A system identification methodology to monitor construction activities using structural responses. *Autom. Constr.* **2017**, *75*, 79–90. [CrossRef]
5. Barroca, N.; Borges, M.; Velez, J.; Monteiro, F.; Górski, M.; Castro-Gomes, J. Wireless sensor networks for temperature and humidity monitoring within concrete structures. *Constr. Build. Mater.* **2013**, *40*, 1156–1166. [CrossRef]
6. Kim, H.R.; Kim, H.G.; Jeon, J.H.; Kim, I.S. Curing Temperature Management of In-Placed Concrete at Early-Ages by Application of the Wireless Sensor Network. *J. Korea Concr. Inst.* **2009**, *21*, 50–57.
7. Lee, S.; Bea, K.; Lee, D. Application for Measurement of Curing Temperature of Concrete in a Construction Site using a Wireless Sensor Network. *Korea Inst. Build. Constr.* **2011**, *11*, 283–291. [CrossRef]
8. Akhavian, R.; Behzadanb, H. Smartphone-based construction workers' activity recognition and classification. *Autom. Constr.* **2016**, *71*, 198–209. [CrossRef]
9. Kim, S.; Kim, T.; Ok, H. A Study on Developing the System for Supporting Mobile-based Work Process in Construction Site. *Smart Media J.* **2017**, *6*, 50–57.
10. Kim, K. A Study on the Implementation of USN Technologies for Safety Management Monitoring of Architectural Construction Sites. *Korea Inst. Build. Constr.* **2009**, *9*, 103–109.
11. Lee, U.K.; Kang, K.I.; Kim, G.H. Mass concrete curing management based on ubiquitous computing. *Comput. Aided Civ. Infrastruct. Eng.* **2005**, *21*, 148–155. [CrossRef]
12. Norris, A.; Saafi, M.; Romin, P. Temperature and moisture monitoring in concrete structures using embedded nanotechnology/microelectromechanical systems (MEMS) sensors. *Constr. Build. Mater.* **2008**, *22*, 111–120. [CrossRef]
13. Chang, C.; Hung, S. Implementing RFIC and sensor technology to measure temperature and humidity inside concrete structures. *Constr. Build. Mater.* **2012**, *26*, 628–637. [CrossRef]
14. Lee, S.; Park, S. Development of integrated wireless sensor network device with mold for measurement of concrete temperature. *J. Korea Inst. Struct. Maint. Insp.* **2012**, *16*, 129–136. [CrossRef]
15. Kim, J.; Luis, R.; Smith, S.; Figueroa, A.; Maloch, C.; Boo, H. Concrete temperature monitoring using passive wireless surface acoustic wave sensor system. *Sens. Actuators A Phys.* **2015**, *224*, 131–139. [CrossRef]
16. Liu, Y.; Deng, F.; He, Y.; Li, B.; Liang, Z.; Zhou, S. Novel concrete temperature monitoring method based on an embedded passive RFID sensor tag. *Sensors* **2017**, *17*, 1463. [CrossRef]
17. Cabezas, J.; Sanchez-Rodriguez, T.; Gomez-Galan, J.; Cifuentes, H.; Carvajal, R. Compact embedded wireless sensor-based monitoring of concrete curing. *Sensors* **2018**, *18*, 876. [CrossRef]

18. Ouyang, J.; Chen, X.; Huangfu, Z.; Lu, C.; Huang, D.; Li, Y. Application of distributed temperature sensing for cracking control of mass concrete. *Constr. Build. Mater.* **2019**, *197*, 778–791. [CrossRef]
19. Bhatia, A.; Lopez, R.; Moein, T.; Chon, J.; Moon, S. Integrated System for Concrete Curing Monitoring: RFID and Optical Fiber Technologies. *J. Mater. Civ. Eng.* **2019**, *31*, 06018028. [CrossRef]
20. Dong, A.; Maher, M.; Kim, M.; Gu, N.; Wang, X. Construction defect management using a telematic digital workbench. *Autom. Constr.* **2009**, *18*, 814–824. [CrossRef]
21. Lee, J.; Kim, Y. The effective process of tower crane lifting & material management using smart-phone. *J. Archit. Inst. Korea Struct. Constr.* **2012**, *28*, 141–150.
22. Park, C.; Lee, D.; Kwon, O.; Wang, X. A framework for proactive construction defect management using BIM, augmented reality and ontology-based data collection template. *Autom. Constr.* **2013**, *33*, 61–71. [CrossRef]
23. Korea Development Institute. *2016 Mobile Trend Forecast*; Digieco: Seoul, Korea, 2016.
24. Covili, J.; Ochoa, S. A lightweight and distributed middleware to provide presence awareness in mobile ubiquitous systems. *Sci. Comput. Program* **2013**, *78*, 2009–2025. [CrossRef]
25. Ahn, C.; Yoon, S.; Chin, S. Diligence and Indolence Management System for Specialty Contractor on Construction Site -Using GPS of Smart Phone. *Korean J. Constr. Eng. Manag.* **2012**, *13*, 56–66. [CrossRef]
26. Dominicis, C.M.; Depari, A.; Flammini, A.; Rinaldi, S.; Sisinni, E. Smartphone based localization solution for construction site management. In Proceedings of the 2013 IEEE Sensors Applications Symposium, Galveston, TX, USA, 19–21 February 2013; pp. 83–88. [CrossRef]
27. Dzeng, R.; Fang, Y.; Chen, I. A feasibility study of using smartphone built-in accelerometers to detect fall portents. *Autom. Constr.* **2014**, *38*, 74–86. [CrossRef]
28. Ranza, J.; Aparicioa, S.; Fuenteb, J.V.; Anayaa, J.J.; Hernándeza, M.G. Monitoring of the Curing Process in Precast Concrete Slabs: An Experimental Study. *Constr. Build. Mater.* **2016**, *122*, 406–416. [CrossRef]
29. Yan, X.; Li, H.; Li, A.; Zhang, H. Wearable IMU-based real-time motion warning system for construction workers' musculoskeletal disorders prevention. *Autom. Constr.* **2017**, *74*, 2–11. [CrossRef]
30. Zhang, M.; Chen, S.; Zhao, X.; Yang, Z. Research on construction workers' activity recognition based on smartphone. *Sensors.* **2018**, *18*, 2667. [CrossRef]
31. Zhang, M.; Cao, T.; Zhao, X. Using Smartphones to Detect and Identify Construction Workers' Near-Miss Falls Based on ANN. *J. Constr. Eng. Manag.* **2018**, *145*, 04018120. [CrossRef]
32. Lim, H.; Lee, J.; Kim, T.; Cho, K.; Cho, H. Economic analysis of USN-based data acquisition systems in tall building construction. *Sustainability* **2017**, *9*, 1360. [CrossRef]
33. Jo, A.; Kim, T.; Cho, H.; Kang, K.I. Data acquisition method using a smartphone on construction site. *ICCEPM* **2013**, *5*, 245–248.
34. Construction Association of Korea. *Construction Industry Wage Survey Report*; CAK: Seoul, Korea, 2018.
35. Bedon, C. Diagnostic analysis and dynamic identification of a glass suspension footbridge via on-site vibration experiments and FE numerical modelling. *Compos. Struct.* **2019**, *216*, 366–378. [CrossRef]

sustainability

MDPI

Article

Identifying Significant Risks and Analyzing Risk Relationship for Construction PPP Projects in China Using Integrated FISM-MICMAC Approach

Xiaoyan Jiang [1,*], Kun Lu [1], Bo Xia [2], Yong Liu [3] and Caiyun Cui [4]

1 School of Civil Engineering, Hefei University of Technology, Hefei 230009, Anhui, China; lukun@mail.hfut.edu.cn

2 School of Civil Engineering and Built Environment, Queensland University of Technology, Gardens Point, Brisbane, QLD4001, Australia; paul.xia@qut.edu.au

3 School of Civil Engineering and Architecture, Zhejiang Sci-Tech University, Hangzhou 310018, China; jhly1007@163.com

4 North China Institute of Science and Technology, Langfang 065201, China; cuicaiyun@163.com

* Correspondence: jiangxiaoyan@hfut.edu.cn

Received: 30 August 2019; Accepted: 17 September 2019; Published: 23 September 2019

Abstract: To meet the growing demand for public facilities and services, many developing countries, including China, have adopted the concept of public–private partnership (PPP). However, there are many risks in PPP projects. Furthermore, these risks affect each other, which may lead to project failure. However, the existing research on the PPP risk relationship has not gone into sufficient detail. Therefore, in order to fill this literature gap, this study proposes a procedural method to analyze the correlation between PPP risks. Firstly, this study, identifies the risks of construction PPP projects in China by combining the literature review with a case study and interviews. Then, fuzzy interpretative structural modeling (FISM) is used to reflect the relationships between these risks and reveal the failure mechanisms of PPP projects. In addition, based on matrix impact cross-reference multiplication applied to a classification (MICMAC) analysis, the risk is divided into four clusters, according to the driving and dependence power, to show the relationship level of the risk. Finally, the paper compares and discusses the research results with other studies and puts forward some suggestions on PPP risks. The FISM-MICMAC method adopted in this study considers the fuzzy of the PPP risk relationship and improves upon previous studies. In addition, the method of FISM-MICMAC can provide a new risk assessment tool for risk management strategies in the field of construction engineering and management.

Keywords: Public–private partnership (PPP); risk identification; risk relationship; triangular fuzzy number; ISM; MICMAC

1. Introduction

A PPP (public–private partnership) refers to a partnership between the government and private investors to provide public infrastructure projects, public goods, and services [1,2]. PPP originated in the UK [3]. As it can deliver high-quality results within the concession period and budget [4], PPP has attracted extensive attention from the public sector and has been adopted in a number of countries [5–7].

However, PPP projects are characterized by large scale investments, long contract concession periods, and complex technologies, which give rise to many potential risk factors in the implementation process, which could also lead to the failure of PPP projects [8–10]. According to the World Bank, 279 PPP projects have been "Cancelled" since 1990 [5].

In the wake of the 2007–2008 global financial crisis, the Chinese government has become increasingly interested in PPP as a way for local governments to transfer their debts [10]. However, compared with developed countries, the growth of PPP in China is still in its infancy [11]. Furthermore, some PPP projects cannot be successful in China [12,13]. As a result, on 16 November 2017, China's Ministry of Finance announced the "Notice of standardizing project library in PPP integrated information platform" (hereafter referred to as "No. 92") [14], designed to correct the current problem of deviation and variation in the course of PPP project implementation, further improve the quality of the project library storage, as well as project effectiveness and information given to the public, and improved social supervision. Since the release of No. 92, from 16 November 2017 to 16 August 2019, the number of "withdrawal projects" reduced by the PPP project library of the Ministry of Finance has reached 6955, and there are 12,561 existing PPP projects in the PPP project library [15].

PPP project failures and cost overruns are not rare, and they pose a threat to sustainable development [16]. The failure of a PPP is largely due to PPP risks [10,17]. There are many risks in PPP projects which, furthermore, are correlated [18,19]. According to accident causal chain theory, accidents are usually caused by a chain of events in the system and the sequential occurrence of causality [20,21]. The failure of PPP projects is usually not caused by a single risk factor, but by a single risk inducing other risks, forming a chain of PPP project risks and, finally, leading to the failure of the PPP [18,19].

Existing studies have mainly focused on the key success factors of PPP [10], risk factors [18,22], risk allocation [17,23], and risk assessment [11,24], etc. However, no in-depth studies on the relationships between PPP risks have been carried out. It is crucial to know what relationships exist between PPP risks, and which risks play a leading role in triggering project failure, in order to help PPP managers to better understand the impact of the relationship between risks on project failure, and to distinguish the key points in future risk management and control.

This study identifies the key risk factors of PPP projects by literature survey, case study, and interviews. Then, fuzzy interpretative structural modeling (FISM), which combines fuzzy theory and interpretative structural modeling (ISM), is used to reflect the relationships between these risks and reveal the failure mechanisms of PPP projects. In addition, key risk factors are classified with the MICMAC (matrix impact cross-reference multiplication applied to a classification) method, and the driving effect of each risk factor on the failure of PPP project is analyzed. Finally, the similarities and differences between this study and other studies are analyzed, and suggestions for managing key risk factors of PPP projects are given.

2. Literature Review

2.1. General Review of PPP Risks

The impact of risks on the completion of a PPP project is usually significant [11]. In PPP projects, various risks exist, not only due to the complexity of the financial and organizational structures of the project, but also due to the large amount of investment, long operation period, sophisticated technical know-how of the project, political impact, and government involvement [25].

The existing studies on PPP risks have mainly been focused on risk factors, risk allocation, risk assessment, and risk identification, etc. Only a limited number of studies have investigated the relationships of PPP risks. Some researchers have sought to identify the risk factors associated with PPPs in specific projects or in specific countries [26,27], and have generally categorized them in terms of being equally shared by both parties or mostly allocated either to the public or private partners [28]. In addition, many scholars have also studied proper allocation of risk factors between the project participants [1,29,30]. Jin combined fuzzy logic with artificial neural network techniques to design a neuro-fuzzy model to help the effective allocation of risks in PPP projects [31]. Ameyaw and Chan's approach based on fuzzy set theory outlined the risk allocation principle, explained the fuzziness inherent in the human cognitive process, and made a case study of risk allocation in a PPP water supply project in Ghana [32]. According to the findings of Jayasuriya et al., there has often been disagreement between PPP regulators (public partner) and operators (private partner) about the preferred risk allocation [28]. For PPP risk assessment, efforts have also been made to develop models to evaluate PPP risk values [11,24,33,34]. Li and Zou applied a fuzzy analytic hierarchy process (FAHP) to the risk assessment of PPP projects [24]. Mazher et al. proposed a PPP infrastructure project risk assessment method based on a fuzzy measure and a non-additive fuzzy integral [35]. Thomas et al. proposed a risk probability and impact assessment framework for build–operate–transfer (BOT) roads, based on a fuzzy-fault tree and the Delphi method [36]. Bai et al., based on the methods of fuzzy comprehensive evaluation model and failure mode, conducted an effects and criticality analysis for evaluating the sustainability risk level of PPP projects [37].

PPP risks identification and relationships among PPP risks will be described in the following subsections.

2.2. Identification Methods of PPP Risks

As mentioned, some scholars have systematically reviewed the risk factors. PPP risks are specific for different types of projects. For sponge city projects [38], water projects [29], highway projects [39], waste-to-energy projects [26], marine projects [40], and other project types, the inherent PPP risks are different. In addition, for different countries, such as Indonesia [41], China [26], the United States [42], Iran [9], and Malaysia [43], the PPP risks are also different.

Due to the specificity of PPP project risks, specific risks should be extracted for different types of PPP projects in different countries. Some scholars and risk managers have attempted to introduce a variety of different risk identification methods into PPP projects, among which the three most frequently used risk identification methods are listed in Table 1.

Table 1. Common methods for identifying public–private partnership (PPP) risks.

Method	Advantage	Disadvantage	Reference
Interview method	Experts have rich experience and knowledge, which is conducive to the rapid identification of risks.	Subject to the subjective influence of experts; experts require higher ability.	Xu et al. [30] Ke et al. [17] Chan et al. [44] Ke et al. [45]
Literature survey	Risk identification is easy to operate and is low cost, which is conducive to reducing the interference of subjective factors.	It is necessary to consider the influence of different industrial categories, regions, and other factors on risk identification.	Osei-Kyei et al. [10] Xu et al. [11] Ameyaw et al. [46] Ameyaw et al. [47]
Case study	High credibility with the actual situation.	Case collection and analysis requires much time and money.	Xu et al. [48] Ameyaw et al. [46] Ameyaw et al. [47]

2.3. Relationship among PPP Risks

From the above analysis, it is evident that there are relationships between PPP risks that lead to the mutual influence of risks [18,19]. In analyzing risks at different project stages, it is important to consider their interrelationships, because an understanding of the relationships between the risks can facilitate a more holistic risk identification and assessment. The existence of risk at an earlier stage may contribute to increased risk manifesting in later stages, or it may not. Sometimes a less important risk can affect a more significant risk, both in terms of likelihood and severity [49].

For example, Valipour, A. et al. identify PPP shared risks using an approach in the form of a hybrid Fuzzy method and Cybernetic Analytic Network Process (CANP) model [50]. Wang et al. proposed a risk model, named the Alien Eye's Risk Model, to show the hierarchical levels of the risks and the influential relationship among the risks in a risk influence matrix [51]. Aladağ and Işık used FAHP to determine the priority factors of design and construction risks in build–operate–transfer (BOT) type mega transportation projects [52]. Furthermore, a fuzzy analytic network process (FANP) method was applied for overcoming the problems of interdependencies and feedback among different risk-ranking alternatives in freeway PPP projects [53]. Multiple-regression analysis was also proposed to estimate and quantify risk interrelationship [54].

2.4. Research Gap and Contribution

These studies have used techniques, such as fuzzy synthetic evaluation [11], fuzzy analytic hierarchy processes [24,33], 2-dimensional linguistic information [34], ISM-MICMAC method [19], hybrid fuzzy cybernetic analytic network process model [50] and multiple-regression analysis [54] etc., to analyze PPP risks and their relationships. However, studies on these relationships and influences based on failed PPP cases in China is lacking. Moreover, it is common to utilize a single approach, but less common to utilize a fuzzy-ISM integrated MICMAC method comprehensively in PPP risks relationship analysis.

The main contribution of the study is to present the application of the fuzzy-ISM technique integrated MICMAC analysis on the relationship among significant PPP risks, which affect the implementation of PPP projects in China, and also to analyze which risks play a leading role in triggering project failure. The 28 cases in this study also make an important contribution to PPP case studies, because failed PPP projects in China can be used as important references for other countries. In addition, this study makes a contribution to relevant literature on common methods for identifying PPP risks and to our understanding of failed PPP projects by summarizing some abnormal phenomena, such as government buyback, project predicament or no operations, severe losses, management right transfers, failure to realize value for money (VFM), and contract cancellations.

3. Research Methods

This study draws on literature, case studies and expert interviews to collect data, and identifies the significant PPP risks in China. Then, the fuzzy ISM (FISM) approach is adopted to clarify the interaction relationships among the PPP risks and to establish a hierarchical structure of these risks. Finally, the MICMAC integrated with fuzzy-ISM approach is employed to determine the significant risks from the perspective of their interaction status. There are six analysis steps in this study, as shown in Figure 1. The details of these six steps will be described in the following subsections.

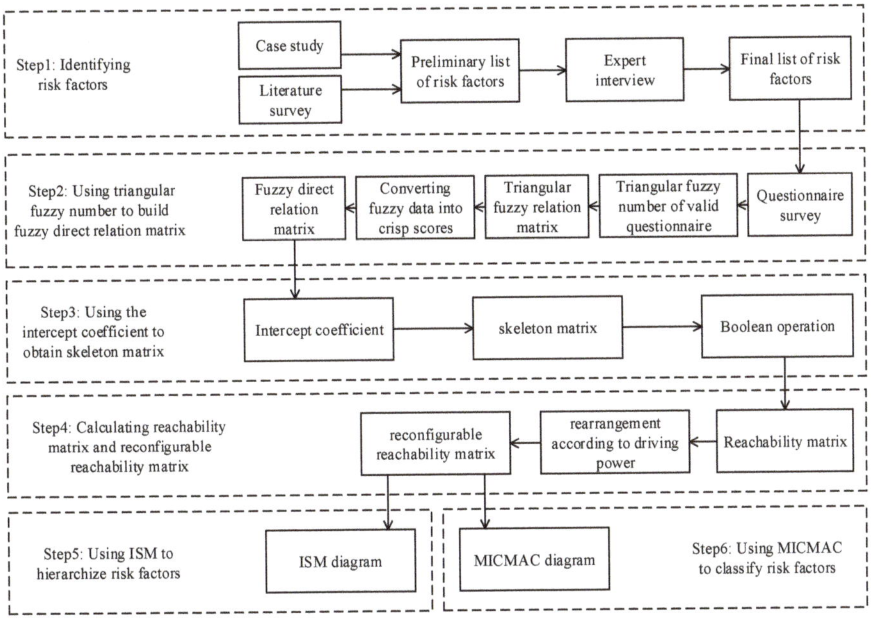

Figure 1. Research framework of the study.

3.1. *Step 1: Collecting Data and Identifying Risk Factors*

The preliminary factor list is extracted by case analysis and literature review, following which, the factor list is modified and simplified by expert interview.

The 28 failed PPP projects are from the World Bank website [55] and China Public–Private Partnerships Center [56]. In order to comply with Chinese national conditions, all of these cases and the literature were based on the actual situation of China. But we cannot find a unified definition on "failed project". Based on prior studies [57,58], this study deems a PPP "failure" as when the following phenomena occur: government buyback, project predicament or no operations, severe losses, management right transfers, failure to realize VFM, and contract cancellations. The cases were chosen as typical examples of the failed PPP projects, with their unique characteristics and failure causes were an excellent match to the critical PPP risk identification. Details of the 28 cases can be found in Appendix A. We have adopted a broadly similar format where possible, in each case drawing out the key risk factors relevant to failure or problem description.

3.2. *Step 2: Using Triangular Fuzzy Number to Build Fuzzy Direct Relation Matrix*

This study utilizes triangular fuzzy number to build fuzzy direct relation matrix. It has been found that fuzzy research, as applied in the construction management discipline, can be divided into two broad fields—fuzzy sets and hybrid fuzzy techniques [59]. The triangular fuzzy number method belongs to the former category [60].

Since the publication of the seminal work "Fuzzy sets" by Zadeh [61], defuzzification by using the fuzzy approach has led to many successful practical applications. For example, Paek applied a fuzzy set approach to price to analyze construction risk [62]. Zhang and Zou used a fuzzy analytic hierarchy process (FAHP) for the appraisal of the risk environment pertaining to joint ventures, in order to support the rational decision-making of project stakeholders [63]. Abdelgawad and Fayek used Fuzzy fault-tree analysis to quantitatively assess risk events in the construction industry [64].

Zadeh proposed the concept of fuzzy sets in order to solve problems described as semi-structured or ill-structured, which allows us to process and transform imprecise information effectively and flexibly [57]. Because of the uncertainty of objective things and the fuzziness of human thought, fuzzy decision-making based on fuzzy sets theory has become a basic method for decision-making, meanwhile triangular fuzzy numbers have been extensively applied in fuzzy control and fuzzy decision-making [65–67].

In this study, the fuzzy triangle number of the relationship between various factors is obtained by expert scoring. A triangular fuzzy number is usually represented by three letters: l, m, and r. These three parameters, respectively, represent the minimum possible value, the median value, and the maximum possible value, (i.e., $l \le m \le r$) [49]. Referring to the value table of language operators and triangular fuzzy numbers given by Li [68] (as Table 2), we let experts judge the strength of the relationship between the two risk factors.

Table 2. Triangular fuzzy number corresponding to language operator.

Language Operator	Triangular Fuzzy Number
Very low impact (VL)	(0, 0, 0.25)
Low impact (L)	(0, 0.25, 0.5)
medium impact (M)	(0.25, 0.5, 0.75)
High impact (H)	(0.5, 0.75, 1)
Very high impact (VH)	(0.75, 1, 1)

According to the results of the questionnaire, a triangular fuzzy relation matrix is established to judge the strength of the relationship between failure risk factors. In triangular fuzzy relation matrix $\widetilde{D}^k$, the triangular fuzzy number $\widetilde{d}^k_{ij} = \left(l^k_{ij}, m^k_{ij}, r^k_{ij}\right)$ is used to represent the judgment result of the influence degree of the risk factor R_i on R_j, given by the expert (as follows).

$$\widetilde{D}^k = \begin{bmatrix} 0 & \widetilde{d}^k_{12} & \cdots & \widetilde{d}^k_{1n} \\ \widetilde{d}^k_{21} & 0 & \cdots & \widetilde{d}^k_{2n} \\ \vdots & \vdots & \vdots & \vdots \\ \widetilde{d}^k_{n1} & \widetilde{d}^k_{n2} & \cdots & 0 \end{bmatrix} \tag{1}$$

Then, the fuzzy direct relation matrix D. of failure risk factors can be obtained by converting fuzzy data into crisp scores (CFCS) method [69]. The specific steps are shown below:

1. Standardize the triangular fuzzy number

$$\begin{aligned} a^k_{ij} &= \left(l^k_{ij} - \min l^k_{ij}\right)/\Delta^{max}_{min} \\ b^k_{ij} &= \left(m^k_{ij} - \min m^k_{ij}\right)/\Delta^{max}_{min} \\ c^k_{ij} &= \left(r^k_{ij} - \min r^k_{ij}\right)/\Delta^{max}_{min} \\ \Delta^{max}_{min} &= \max r^k_{ij} - \min l^k_{ij} \end{aligned} \tag{2}$$

2. Calculate the left and right limits of the standardized values u^k_{ij} and v^k_{ij}

$$\begin{aligned} u^k_{ij} &= b^k_{ij}/\left(1 + b^k_{ij} - a^k_{ij}\right) \\ v^k_{ij} &= c^k_{ij}/\left(1 + c^k_{ij} - b^k_{ij}\right) \end{aligned} \tag{3}$$

3. Calculate the total value of the standardized values

$$w^k_{ij} = \frac{u^k_{ij}\left(1 - u^k_{ij}\right) + \left(v^k_{ij}\right)^2}{1 - u^k_{ij} + v^k_{ij}} \tag{4}$$

4. Calculate the exact value of expert triangle fuzzy judgment d^k_{ij}

$$d_{ij}^{k} = min\ l_{ij}^{k} + w_{ij}^{k} \cdot \Delta_{min}^{max} \tag{5}$$

5. Calculate the standardized accurate value d_{ij}, as evaluated by experts

$$d_{ij} = \frac{1}{p} \cdot \sum_{k=1}^{p} d_{ij}^{k} \tag{6}$$

6. Determine the fuzzy direct relation matrix D of key risk factors

$$D = \begin{bmatrix} 0 & d_{12} & \cdots & d_{1n} \\ d_{21} & 0 & \cdots & d_{2n} \\ \vdots & \vdots & \ddots & \vdots \\ d_{n1} & d_{n2} & \cdots & 0 \end{bmatrix} \tag{7}$$

3.3. *Step 3: Using the Intercept Coefficient to Obtain the Skeleton Matrix*

By using appropriate intercept coefficients α, the fuzzy direct relation matrix is transformed into a skeleton matrix S, which can be directly used for Boolean operation. When the elements in the fuzzy direct relation matrix are greater than or equal to α, the corresponding position of the elements in the skeleton matrix is set to 1; otherwise it is set to 0.

$$s_{ij} = \begin{cases} 1\left(d_{ij} \geq \alpha\right) \\ 0\left(d_{ij} < \alpha\right) \end{cases} \tag{8}$$

3.4. *Step 4: Calculating Reachability Matrix and Reconfigurable Reachability Matrix*

ISM is an effective model that can clearly define inter-relationships between multiple elements of a problem [70]. In order to study the correlation between evaluation objects, Mandal and Deshmukh introduced interpretative structural modeling (ISM) [71]. ISM has become an analytical tool widely used by scholars to study the interdependence and interaction of factors, and is widely used in policy analysis, supply chains and in other fields [72,73]. For example, Tseng et al. used ISM to construct a systematic model of interconnected natural disaster risks [74]. Yanmei et al. identified power overload risk factors and established ISM to analyze the relationships between these factors [75]. Li et al. also used ISM to assess the risks of India's thermal power plants [76].

The advantage of ISM is that it requires fewer questionnaires than other methods, such as structural equation modeling and the Delphi technique, and is able to extract a clear structural view from unstructured models [77,78]. However, in ISM, only the existence of an influencing relationship between two factors is investigated; if there is a relationship between factors, it is denoted by "1", otherwise, it is denoted by "0" [78]. However, the association between the factors is fuzzy and cannot be easily divided into "related" or "unrelated" [79]. To solve this problem, many scholars have applied fuzzy set theory to ISM, in order to consider the degree of influence between factors, in an approach called fuzzy interpretative structural modeling (FISM) [78,80,81].

As described in the previous section, this study uses the triangular fuzzy number method integrated with ISM in this paper. The process is as follows.

First, the Boolean operation of the skeleton matrix and the identity matrix is performed, where the operation rules are as follows:

$$0 + 0 = 0,\ 0 + 1 = 1,\ 1 + 1 = 1,\ 0 * 0 = 0,\ 0 * 1 = 0,\ 1 * 0 = 0,\ \text{and}\ 1 * 1 = 1$$

Meanwhile, it must follow the shift law characteristic of matrix operations; that is, when a directly reaches b through a path of length 1 and b directly reaches c through a path of length 1, then a must reach c through a path of length 2.

Then, the reachability matrix R can be obtained through a Boolean operation of the skeleton matrix S. When all products are equal, the reachability matrix R can be obtained.

$$(S+I) \neq (S+I)^2 \neq \cdots \neq (S+I)^{r-1} \neq (S+I)^k = (S+I)^{k+1} = R \quad (9)$$

Lastly, based on the reachability matrix, the risk factors of failure are rearranged according to the value of driving power and the reconstituted reachability matrix R^* is obtained.

3.5. Step 5: Using ISM to Hierarchize Risk Factors

According to the interaction relation of failure risk factors, the hierarchical results can be obtained by referring to the values of elements in the reconstructed reachability matrix. Connecting all the factors with arrow lines, the ISM diagram of the interaction of failure risk factors will be achieved.

3.6. Step 6: Using MICMAC to Classify Risk Factors

After ISM, MICMAC (matrix impact cross-reference multiplication applied to a classification) is often used as a complementary method to analyze the driver power and dependence power of the risks [82]. The driver power of a risk means the total number of risks it can influence, whereas the dependence power of a risk means the total number of risks which can influence it [83]. According to the driver and dependence powers of the factors, they can be divided into four groups: autonomous, dependent, linkage, and independent. Issues having weak driving and dependence powers are in the autonomous group, whereas issues having strong dependent and driving powers are in the linkage group [78]. Issues which have strong driving and weak dependence powers are known as independent issues, whereas issues which have strong dependence power and weak driving power are known as dependent issues [78].

Many scholars have applied ISM-MICMAC to PPP risk analysis. Iyer et al. analyzed the hierarchical structure of PPP risks of Indian roads for the first time by combining ISM and MICMAC [19]. Han et al. used ISM-MICMAC to analyze PPP risks in brownfield remediation projects in China [83]. Li and Wang used a fuzzy analytic network process and ISM-MICMAC for PPP risk assessment [18].

Although these analyses based on ISM-MICMAC have achieved some beneficial results, they have all ignored the notion that the impact degree of PPP risks is vague and cannot be judged simply by the impact or no impact. Based on this, FISM is adopted in this paper to consider the fuzziness of the relationships between PPP risks, replacing the traditional "0 or 1" judgment with a fuzzy number, in order to reduce the subjectivity of expert judgment.

In this study, the categories of risk factors are determined by FISM integrated MICMAC method. Based on the reconstituted reachability matrix R^*, each factor is plotted as a point in the four quadrants of the rectangular coordinate system, according to the driving power and dependency power of each failure risk factor, and the key risk factors are classified.

4. Results and Findings

By following Step 1 in Section 3, a literature review and case study analysis was conducted to elicit the preliminary list of risk factors. A total of 29 preliminary risk factors were extracted to the preliminary risk list.

To mitigate the deficiency of the literature review and case study, we invited five experts for two rounds of interviews to validate the preliminary list. We chose experts who had sufficient time available for being interviewed and for summarizing these risk factors. Other experts that were considered, but who were too busy, were excluded from the risk identification stage. These experts included one lawyer involved in PPP consulting, one manager of a PPP project company, two managers of a PPP consulting company and one PPP research scholar, all of whom have participated in the management of PPP projects and have a rich knowledge of PPP theory and practice.

After the first round of interviews, based on the opinions of various experts and further confirmation in the second round of interviews, the risk factors that led to the failure of PPP projects were summarized into 20 risks (see Table 3). The literature and case sources of these 20 risks can be seen in Tables 4 and 5.

These 20 risks are: government decision-making approval risk (R1); policy and regulatory change risk (R2); government credit risk (R3); government regulatory risk (R4); planning and design risk (R5); bidding risk (R6); contract risk (R7); facility matching risk (R8); financing risk (R9); economic risk (R10); project change risk (R11); construction risk (R12); project income risk (R13); parallel project competitive risk (R14); operation and maintenance risk (R15); force majeure risk (R16); public opposition risk (R17); organizational coordination risk (R18); environmental risk (R19); and project company violates laws and regulations (R20).

In this study, in addition to the five original experts who participated in expert interviews, another 10 experts were invited to conduct a questionnaire survey on the relationship between PPP risk factors by means of e-mail, WeChat and on-site survey (see Appendix B). All 15 experts have rich knowledge of PPP theory and practice, and their background information is shown in the Table 6. Through comparative analysis of questionnaire data, the high impact of other risks on the force majeure risk of R16 was taken as an invalid judgment standard to eliminate the invalid questionnaires. In total, there were 9 valid questionnaires.

Table 3. Preliminary risk list and final risk list.

Preliminary Risk List (29 Risks)	Final Risk List (20 Risks)
Lengthy government decision-making and approval Lack of PPP co-operation experience	Government decision-making approval risk (R1)
Policy and regulatory change risk	Policy and regulatory change risk (R2)
Government official corruption Government intervention	Government credit risk (R3)
Government regulatory risk	Government regulatory risk (R4)
Planning and design risk	Planning and design risk (R5)
Bidding risk	Bidding risk (R6)
Imperfect price adjustment mechanism Unreasonable profit distribution Imperfect dispute settlement mechanism Incomplete contract risk	Contract risk (R7)
Facility matching risk	Facility matching risk (R8)
Financing risk	Financing risk (R9)
Economic risk	Economic risk (R10)
Project change risk	Project change risk (R11)
Construction cost overruns Construction quality risk Construction schedule risk Completion risks	Construction risk (R12)
Changes in market demand Insufficient expense payment	Project income risk (R13)
Parallel project competitive risk	Parallel project competitive risk (R14)
Operation and maintenance risk	Operation and maintenance risk (R15)
Force majeure risk	Force majeure risk (R16)
Public opposition risk	Public opposition risk (R17)
Organizational coordination risk	Organizational coordination risk (R18)
Environmental risk	Environmental risk (R19)
Project company violates laws and regulations	Project company violates laws and regulations (R20)

Table 4. Literature statistics for PPP risk factors.

Risk Factors	1 [38]	2 [83]	3 [84]	4 [18]	5 [39]	6 [85]	7 [86]	8 [34]	9 [29]	10 [26]	11 [12]	12 [11]	13 [17]	14 [87]	15 [27]	16 [44]
Government decision-making approval risk (R1)	O	O	O	O	O	O	O	O	O	O	O	O	O	O	O	O
Policy and regulatory change risk (R2)	O	O	O	O	O	O	O	O	O	O	O	O	O	O	O	O
Government credit risk (R3)	O	O	O	O	O	O	O	O	O	O	O	O	O	O	O	O
Government regulatory risk (R4)	O	O	O	O	O	O	O	O	O	O	O	O	O	O	O	O
Planning and design risk (R5)	O	O	O	O	O	O	O	O	O	O	O	O	O	O	O	O
Bidding risk (R6)		O	O	O		O				O		O	O	O		O
Contract risk (R7)		O	O	O	O	O	O	O	O	O	O	O	O	O	O	O
Facility matching risk (R8)	O	O			O				O			O	O	O	O	O
Financing risk (R9)	O	O	O	O	O	O	O	O	O	O	O	O	O	O		O
Economic risk (R10)	O	O	O	O	O	O	O	O	O	O	O	O	O	O		O
Project change risk (R11)		O	O	O	O	O	O	O	O	O		O	O	O	O	O
Construction risk (R12)	O	O	O	O	O	O	O	O	O	O	O	O	O	O	O	O
Project income risk (R13)	O	O	O	O	O	O	O	O	O	O	O	O	O	O	O	O
Parallel project competitive risk (R14)				O	O	O	O		O			O	O	O	O	
Operation and maintenance risk (R15)	O	O	O	O	O	O	O	O	O	O	O	O	O	O	O	O
Force majeure risk (R16)	O	O	O	O	O		O	O	O	O	O	O	O	O	O	O
Public opposition risk (R17)		O	O	O	O		O	O	O	O		O	O	O	O	O
Organizational coordination risk (R18)		O	O	O	O	O	O	O	O	O		O	O	O		O
Environmental risk (R19)		O	O	O	O		O	O	O	O	O	O	O	O	O	O
Project company violates laws and regulations (R20)		O	O	O	O	O			O	O						

Table 5. Case statistics for PPP risk factors.

Risk Factors \ Cases	1	2	3	4	5	6	7	8	9	10	11	12	13	14	15	16	17	18	19	20	21	22	23	24	25	26	27	28
R1		√	√	√	√	√	√	√														√						
R2	√				√			√						√				√						√			√	
R3	√				√	√	√		√	√		√		√	√	√	√	√		√	√		√			√		√
R4	√	√								√	√	√				√				√							√	
R5								√				√	√				√		√	√	√	√				√		√
R6											√													√	√		√	√
R7		√	√		√	√	√	√	√	√			√	√				√					√			√		
R8			√																√				√					
R9				√																				√			√	
R10													√															
R11				√											√						√							√
R12				√					√				√		√	√	√							√	√			√
R13	√		√	√		√	√	√			√	√	√	√	√	√		√	√	√	√		√		√			
R14													√	√	√	√										√		
R15		√								√															√			
R16																	√											
R17		√		√						√							√					√			√	√		
R18											√						√							√	√		√	
R19	√	√								√		√								√		√	√					
R20	√	√								√	√										√							

Table 6. Background information of 15 experts.

No.	Education Profile	Work Experience	Number of PPP Participated	Company
1	PhD	6 to 10 years	Three	University
2	PhD	Under 5 years	One	University
3	Master	6 to 10 years	Two	Government
4	Master	11 to 15 years	More than 4	Consulting
5	Master	6 to 10 years	Three	Consulting
6	Master	Under 5 years	Two	Consulting
7	Bachelor	11 to 15 years	Three	Consulting
8	Master	6 to 10 years	One	Private
9	Master	6 to 10 years	One	Private
10	Master	11 to 15 years	Two	Private
11	Master	6 to 10 years	One	Private
12	Master	6 to 10 years	Two	Private
13	Bachelor	More than 16 years	Two	Private
14	Bachelor	More than 16 years	Two	Private
15	Bachelor	11 to 15 years	Two	Private

The triangular fuzzy number of 9 valid questionnaires were filled in the triangular fuzzy relation matrix. Then, we converted triangular fuzzy relation matrix to fuzzy direct relation matrix through converting fuzzy data into crisp scores (CFCS) method. The final fuzzy direct relation matrix is shown in Table 7.

Table 7. Fuzzy direct relation matrix of the PPP key risk factors.

Risk	R1	R2	R3	R4	R5	R6	R7	R8	R9	R10	R11	R12	R13	R14	R15	R16	R17	R18	R19	R20
R1	0	0.20	0.51	0.35	0.51	0.45	0.34	0.36	0.28	0.15	0.35	0.36	0.35	0.49	0.17	0.00	0.39	0.38	0.28	0.13
R2	0.38	0	0.54	0.40	0.15	0.23	0.54	0.12	0.43	0.41	0.63	0.38	0.35	0.24	0.33	0.00	0.19	0.14	0.46	0.20
R3	0.40	0.25	0	0.38	0.24	0.30	0.49	0.37	0.39	0.31	0.45	0.38	0.45	0.65	0.33	0.08	0.25	0.35	0.15	0.20
R4	0.29	0.05	0.40	0	0.33	0.53	0.31	0.41	0.35	0.28	0.30	0.60	0.35	0.31	0.33	0.02	0.46	0.37	0.52	0.49
R5	0.18	0.04	0.10	0.04	0	0.28	0.35	0.25	0.32	0.18	0.74	0.38	0.57	0.31	0.38	0.00	0.51	0.09	0.45	0.02
R6	0.10	0.15	0.19	0.23	0.17	0	0.43	0.23	0.31	0.22	0.37	0.45	0.60	0.07	0.57	0.02	0.15	0.32	0.19	0.45
R7	0.12	0.15	0.30	0.35	0.15	0.07	0	0.27	0.19	0.28	0.43	0.46	0.45	0.18	0.40	0.12	0.19	0.40	0.25	0.38
R8	0.17	0.05	0.17	0.07	0.19	0.15	0.33	0	0.24	0.12	0.33	0.60	0.38	0.08	0.49	0.00	0.09	0.10	0.15	0.22
R9	0.10	0.10	0.12	0.07	0.10	0.12	0.38	0.10	0	0.33	0.22	0.50	0.53	0.04	0.47	0.00	0.07	0.15	0.12	0.25
R10	0.25	0.32	0.27	0.14	0.04	0.17	0.31	0.18	0.59	0	0.16	0.35	0.53	0.10	0.35	0.00	0.19	0.12	0.12	0.41
R11	0.15	0.12	0.38	0.28	0.25	0.09	0.41	0.30	0.39	0.02	0	0.71	0.60	0.07	0.52	0.02	0.11	0.27	0.24	0.27
R12	0.04	0.07	0.27	0.25	0.16	0.07	0.43	0.15	0.33	0.02	0.45	0	0.65	0.07	0.35	0.13	0.35	0.20	0.35	0.33
R13	0.16	0.15	0.44	0.22	0.10	0.10	0.44	0.12	0.60	0.10	0.43	0.32	0	0.27	0.46	0.00	0.27	0.17	0.10	0.36
R14	0.12	0.12	0.57	0.20	0.15	0.15	0.52	0.12	0.31	0.07	0.36	0.26	0.52	0	0.24	0.00	0.14	0.17	0.04	0.41
R15	0.07	0.12	0.30	0.23	0.17	0.07	0.38	0.17	0.22	0.10	0.28	0.04	0.80	0.10	0	0.00	0.37	0.22	0.46	0.32
R16	0.12	0.15	0.33	0.22	0.10	0.10	0.33	0.22	0.38	0.24	0.52	0.43	0.47	0.04	0.55	0	0.12	0.17	0.29	0.22
R17	0.51	0.22	0.46	0.33	0.27	0.09	0.22	0.19	0.33	0.15	0.56	0.45	0.33	0.17	0.49	0.07	0	0.17	0.20	0.17
R18	0.33	0.09	0.32	0.43	0.33	0.32	0.46	0.23	0.35	0.04	0.33	0.57	0.49	0.17	0.52	0.02	0.25	0	0.15	0.41
R19	0.22	0.32	0.33	0.33	0.30	0.10	0.35	0.20	0.24	0.04	0.51	0.51	0.63	0.07	0.54	0.07	0.77	0.12	0	0.28
R20	0.04	0.11	0.35	0.28	0.18	0.23	0.55	0.17	0.38	0.04	0.41	0.54	0.57	0.12	0.48	0.00	0.60	0.30	0.60	0

After obtaining the fuzzy direct relation matrix, the intercept coefficient α was adopted to convert the fuzzy direct relation matrix into a skeleton matrix, which was directly used for Boolean operation, as shown in Table 8.

Table 8. Skeleton matrix of the PPP key risk factors.

Risk	R1	R2	R3	R4	R5	R6	R7	R8	R9	R10	R11	R12	R13	R14	R15	R16	R17	R18	R19	R20
R1	1	0	1	0	1	0	0	0	0	0	0	0	0	0	0	0	0	0	0	0
R2	0	1	1	0	0	0	1	0	0	0	1	0	0	0	0	0	0	0	0	0
R3	0	0	1	0	0	0	0	0	0	0	0	0	0	1	0	0	0	0	0	0
R4	0	0	0	1	0	1	0	0	0	0	0	1	0	0	0	0	0	0	1	0
R5	0	0	0	0	1	0	0	0	0	0	1	0	1	0	0	0	1	0	0	0
R6	0	0	0	0	0	1	0	0	0	0	0	0	1	0	1	0	0	0	0	0
R7	0	0	0	0	0	0	1	0	0	0	0	0	0	0	0	0	0	0	0	0
R8	0	0	0	0	0	0	0	1	0	0	0	1	0	0	0	0	0	0	0	0
R9	0	0	0	0	0	0	0	0	1	0	0	1	1	0	0	0	0	0	0	0
R10	0	0	0	0	0	0	0	0	1	1	0	0	1	0	0	0	0	0	0	0
R11	0	0	0	0	0	0	0	0	0	0	1	1	1	0	1	0	0	0	0	0
R12	0	0	0	0	0	0	0	0	0	0	0	1	1	0	0	0	0	0	0	0
R13	0	0	0	0	0	0	0	0	1	0	0	0	1	0	0	0	0	0	0	0
R14	0	0	1	0	0	0	1	0	0	0	0	0	1	1	0	0	0	0	0	0
R15	0	0	0	0	0	0	0	0	0	0	0	0	1	0	1	0	0	0	0	0
R16	0	0	0	0	0	0	0	0	0	0	1	0	0	0	1	1	0	0	0	0
R17	1	0	0	0	0	0	0	0	0	0	1	0	0	0	0	0	1	0	0	0
R18	0	0	0	0	0	0	0	0	0	0	0	1	0	0	1	0	0	1	0	0
R19	0	0	0	0	0	0	0	0	0	0	1	1	1	0	1	0	1	0	1	0
R20	0	0	0	0	0	0	1	0	0	0	0	1	1	0	0	0	1	0	1	1

Reachability matrix was obtained after Boolean operation of the skeleton matrix, and on the basis of the reachability matrix, all failure risk factors were rearranged according to the value of driving power to get the reconfigurable reachability matrix (see in Table 9). In order to reduce the amount of calculation, we used Python programming in the operation process.

Table 9. Reconfigured reachability matrix of the PPP key risk factors.

Risk	R7	R9	R12	R13	R8	R10	R15	R6	R11	R18	R3	R14	R16	R2	R1	R5	R17	R19	R20	R4	Dr
R7	1	0	0	0	0	0	0	0	0	0	0	0	0	0	0	0	0	0	0	0	1
R9	0	1	1	1	0	0	0	0	0	0	0	0	0	0	0	0	0	0	0	0	3
R12	0	1	1	1	0	0	0	0	0	0	0	0	0	0	0	0	0	0	0	0	3
R13	0	1	1	1	0	0	0	0	0	0	0	0	0	0	0	0	0	0	0	0	3
R8	0	1	1	1	1	0	0	0	0	0	0	0	0	0	0	0	0	0	0	0	4
R10	0	1	1	1	0	1	0	0	0	0	0	0	0	0	0	0	0	0	0	0	4
R15	0	1	1	1	0	0	1	0	0	0	0	0	0	0	0	0	0	0	0	0	4
R6	0	1	1	1	0	0	1	1	0	0	0	0	0	0	0	0	0	0	0	0	5
R11	0	1	1	1	0	0	1	0	1	0	0	0	0	0	0	0	0	0	0	0	5
R18	0	1	1	1	0	0	1	0	0	1	0	0	0	0	0	0	0	0	0	0	5
R3	1	1	1	1	0	0	0	0	0	0	1	1	0	0	0	0	0	0	0	0	6
R14	1	1	1	1	0	0	0	0	0	0	1	1	0	0	0	0	0	0	0	0	6
R16	0	1	1	1	0	0	1	0	1	0	0	0	1	0	0	0	0	0	0	0	6
R2	1	1	1	1	0	0	1	0	1	0	1	1	0	1	0	0	0	0	0	0	9
R1	1	1	1	1	0	0	1	0	1	0	1	1	0	0	1	1	1	0	0	0	11
R5	1	1	1	1	0	0	1	0	1	0	1	1	0	0	1	1	1	0	0	0	11
R17	1	1	1	1	0	0	1	0	1	0	1	1	0	0	1	1	1	0	0	0	11
R19	1	1	1	1	0	0	1	0	1	0	1	1	0	0	1	1	1	1	0	0	12
R20	1	1	1	1	0	0	1	0	1	0	1	1	0	0	1	1	1	1	1	0	13
R4	1	1	1	1	0	0	1	1	1	0	1	1	0	0	1	1	1	1	0	1	14
De	10	19	19	19	1	1	12	2	9	1	9	9	1	1	6	6	6	3	1	1	

Risk factors were hierarchized by using the ISM diagram. In the ISM diagram, the classification of a risk factor hierarchy can be determined by the driving power of a risk factor; that is, the risk factor with smaller driving power sits on the higher level in the risk system of ISM diagram, and the risk factor with higher driving power sits on the bottom position in the ISM diagram. We obtained the hierarchical partition results and ISM diagram according to the driving power of each risk in reconfigured reachability matrix. According to the positions of the risk factors in the ISM diagram, they were divided into three areas: upper, middle, and bottom, as shown in Figure 2. In Figure 2, we can see that:

1. The upper risk factors included R7 (contract risk), R9 (financing risk), R12 (construction risk), and R13 (project income risk), and there was a strong connection between R9, R12, and R13. These risks directly led to the failure of PPP projects.
2. The middle risk factors included R3 (government credit risk), R11 (project change risk), R14 (parallel project competitive risk), R15 (operation and maintenance risk), R6 (bidding risk), R2 (policies and regulations change risk), R8 (Facility matching risk), R10 (economic risk), R16 (force majeure risk), and R18 (organizational coordination risk).
3. The bottom risk factors included R4 (government regulatory risk), R20 (project company violates laws and regulations), R19 (environmental risk), R17 (public opposition risk), R1 (government decision-making approval risk), and R5 (planning and design risk). It can be considered that these risk factors are the most critical and fundamental in the risk system.

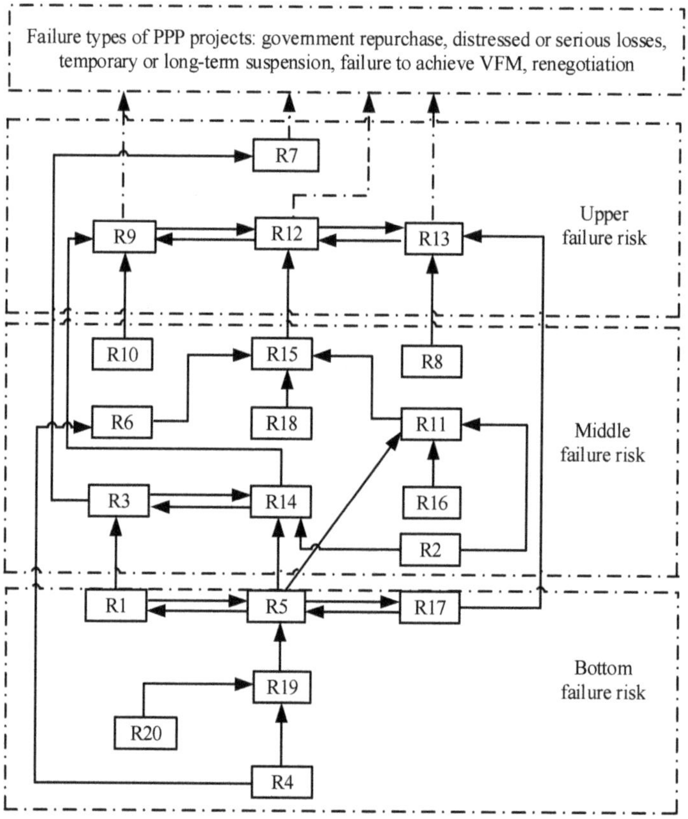

Figure 2. Interpretative structural modeling (ISM) diagram of the PPP key risk factors.

After obtaining the ISM diagram that represents the mutual influence of key risk factors in PPP, it is necessary to refer to the MICMAC method to determine the categories of risk factors, so as to provide corresponding prevention suggestions for different types of risk in a more targeted way.

Based on the reconfigurable reachability matrix, each factor was plotted as a point in the conventional X-Y coordinate system according to the values of driving power and dependence power. Thus, the 20 key risk factors were divided into four categories and distributed in four quadrants, as shown in Figure 3.

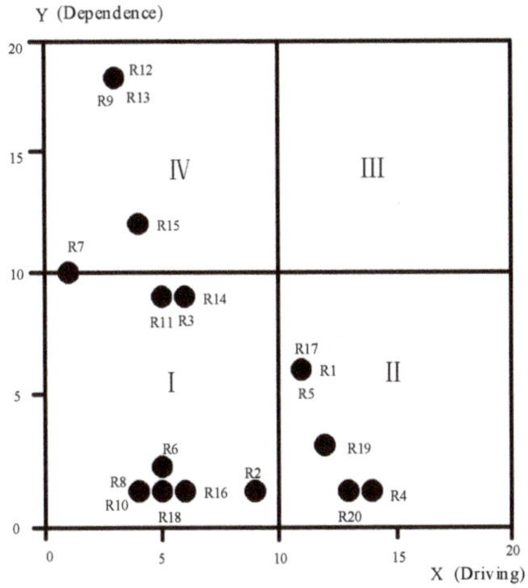

Figure 3. The driving–dependence power diagram of the PPP key risk factors.

Each quadrant contains risk factors as follows: Quadrant I (Autonomous risks) includes: R2 (policies and regulations change risk), R3 (government credit risk), R6 (bidding risk), R8 (facility matching risk), R10 (economic risk), R11 (project change risk), R14 (parallel project competitive risk), R16 (force majeure risk), R18 (organizational coordination risk). The driving power and dependence power of the risk factors are relatively low, which means they are not easily influenced by other risk factors, and also do not easy to lead to the occurrence of other risk factors;

Quadrant II (Dependent risks) includes: R1 (government decision-making approval risk), R5 (planning and design risk), R4 (government regulatory risk), R17 (public opposition risk), R19 (environmental risk), R20 (project company violates laws and regulations). They have a powerful driving power and weak dependence power, which are key risk factors and basic conditions in the system;

Quadrant III (Linkage risks): there is no any risk. It means that there is no risk that is susceptible to, and affects other, risks

Quadrant IV (Independent risks) includes: R7 (contract risk), R9 (financing risk), R12 (construction risk), R13 (project income risk), and R15 (operation and maintenance risk). They all have stronger dependence, meaning that the occurrence of these risk factors is largely due to the change or accumulation of other risk factors.

5. Analysis and Discussion

5.1. Discussion of Method

In this study, the FISM is adopted to understand the magnitude of the relationships between the risks [80]. This takes the fuzzy of relationships between PPP risks into account, which has a greater advantage than the traditional ISM method. It also helps in identifying the magnitude of cascading relationships between lower-level issues to higher-level issues [78]. In addition, the method of FISM-MICMAC can provide a new risk assessment tool for risk management strategies in the field of construction engineering and management.

FISM combines fuzzy theory and uses fuzzy language conversion to convert the subjective risks that are difficult to quantify, which is caused by too many uncertain factors in PPP, so as to realize the process of risk evaluation from subjective to objective, and to provide a more scientific theoretical basis for risk response. Compared with traditional ISM, FISM has shown greater advantages [78,80,81,88].

In this study, the reachability matrix is obtained by using the intercept coefficient to fuzzy direct relation matrix. This method shows that risk managers can obtain different reachability matrices through different intercept coefficients, according to their own requirements of the risk correlation degree.

5.2. Risk Structure Analysis

(1) The bottom risk factors of Figure 2 can be classified into two categories. One is likely to occur at any stage in the life cycle of a PPP project, including R4 (government regulatory risk), R20 (project company violates laws and regulations), R19 (environmental risk), and R17 (public opposition risk); whereas R1 (government decision-making approval risk) and R5 (planning and design risk) are pre-project risks. Once these occur, they may cause the formation of a risk chain. Therefore, in the early stages of a PPP project, attention should be paid to the government's decision-making and the planning and design of the PPP project. In the life cycle of PPP projects, there should be focus on government regulation, delinquency of private, environmental pollution and public opposition.

An empirical study in 2011 showed that the three most important risk factors for Chinese PPP projects are government intervention, government corruption, and poor public decision-making processes [89], which is consistent with our research. In addition, we also need to make a special description of R19 (environmental risk), as China imposed an environmental tax in January 2011 [90]. With increasingly strict environmental protection policies, the price of raw materials in the construction industry has been generally rising, which leads to the rising cost of many construction projects which, finally, leads to the failure of projects. From the above analysis, we can see that the fundamental risks affecting PPP projects are not invariable, but rather, are closely related to the actual situation.

(2) From Figures 2 and 3, we found that the R7 (contract risk), R9 (financing risk), R12 (construction risk), R13 (project income risk), and R15 (operation and maintenance risk) have higher dependency values (R15 is in the middle of Figure 2, but in the Quadrant IV of Figure 3), indicating that they are easily affected by other risks. The occurrence of these risks is often caused by the accumulation of other risks and, when these risks occur, they can easily and directly cause the failure of the project, in agreement with the study of Li and Wang [18]. We can prevent such risk factors by controlling the occurrence of the other risks that have an impact on them. In addition, it should be realized that, as these risk factors are easily influenced by other risk factors, their own performance can reflect whether the risks affecting them have been well-controlled or not; that is to say, they can also serve as an indicator of the effect of risk management.

(3) In addition, it can be seen, from Figure 3, that no risk factors are located in Quadrant III, which reflects that there is no PPP risk that can be affected by many factors and can affect many factors, which is consistent with the research of Iyer et al. [19]. This is demonstrates that the key risk factors usually do not lead to project failure through a singular role, but rather, increase the impact on the project through a correlation between risks, cumulatively contributing to the failure of a PPP project.

5.3. Comparison with Other Studies

By comparing Iyer's study on the Indian highway PPP [19], Han's study on the Chinese brownfield remediation projects PPP [79], and Li's ISM study on Chinese PPP [18], we find that:

1. In Iyer's study on PPP of Indian highways [19], environmental risk was also at the bottom of the risk level, which could easily induce permit risk and direct political risk, eventually triggering schedule risk and cost overrun risk, which is consistent with our study and indicates that there are similarities between India and China, in this respect.
2. Comparing the PPP risk studies of Han [83] and Li [18], based on Chinese national conditions, Han believed the bottom of the risk was an immature legal system, inadequate experience, inadequate data and research, lack of a standard PPP contract template, private inability, and technique risk. Meanwhile, in Li's research, legal and policy risk, organization and coordination risk, rights, responsibility, and risk allocation between the co-operative parties were considered to be at the bottom. Although they each had different opinions of risk, their research was consistent:

a government with weak supervision and ability, a private party with inability and irregularities, and the lack of organization and coordination between them, are projected at the bottom of the risk. These risks can be controlled, and the managers and decision-makers must work to control these risks.

3. Previous studies and this study, have found that both contract risk and financing risk can induce the failure of PPP projects, as triggered by construction risk (including schedule risk and cost overrun risk).
4. As for the MICMAC analysis, each study differed. In our study, there was no risk found in Quadrant III. In the study of Iyer [19], there was no risk in Quadrant III and IV; in the study of Li [14], there was no risk in Quadrant I; meanwhile, in the study of Han [83], there was risk in all quadrants. Therefore, in the present study and that of Iyer, the risks were more independent of each other. In contrast, the degree of risk interrelation in the studies of of Han and Li was higher. This difference was caused by one the of factors influencing the existence of ambiguity, which further proves the advantages of FISM-MICMAC method. For a different cutting coefficient, we could control the threshold value of correlation degree between various factors.

5.4. Suggestions to the Government

1. China's current PPP related policies and regulations are not perfect [91,92]. The government should summarize the many problems existing in current PPP projects, and formulate special PPP laws and regulations, as soon as possible, to protect the rights and interests of both parties, reduce the negative effects caused by changes in policies and regulations, and enhance the investment confidence of the private parties involved.
2. The choice of private sector is very important. A private sector with strong financing, technical ability, and high reputation will provide higher advantages in a PPP project, such as by realizing project appreciation, reducing financing risk, construction risk, operation and maintenance risk, and the risk of the project company violating laws and regulations. In No. 27 case in Appendix A, the wrong choice of private party leads to improper subcontracting of PPP projects, which ultimately leads to the failure of the PPP projects.
3. In addition, the government should regard PPP as a governance scheme, rather than a pragmatic economic tool [92]. Governmental departments should strengthen the PPP-relevant theoretical knowledge and practical experience, learn from similar projects, pay attention to the feasibility in early stages of the project, establish and perfect project decision-making mechanisms, and conduct in-depth research on the project for scientific planning and design. In addition, governmental departments should strengthen supervision and management, perfect their regulation systems, and reduce the PPP project risk that exists because of inadequate supervision.
4. The government should realize that the relationship with stakeholders in a PPP project is key to the sustainable development of the project [93,94]. The government should attach importance to the relationship with private parties, avoid the risk of governmental decision-making approval as much as possible, reduce late-project changes, avoid competition of similar projects caused by unreasonable commitments in the early stage, and gradually improve government credit.
5. In terms of PPP project contract, the rights and obligations of the public and private should be reasonably distributed, the contents (such as the distribution of benefits and risks) of dispute resolutions and renegotiation mechanisms should be clarified, and the behaviors of both parties should be regulated [1,29,30]. According to the actual situation of the project, a reasonable risk sharing mechanism should be determined for planning and design risks, financial risks, environmental risks, change risks, force majeure risk, and public opposition risks.
6. In the past, government departments have paid too much attention to economic evaluation while ignoring social evaluation when evaluating investment projects, leading to higher rates of public opposition [95]. In the implementation of PPP projects in the future, the period for soliciting public opinion should be extended, and public information should be kept open throughout the

period for soliciting public opinion, so as to improve public participation, give full play to the role of public supervision, and ensure the normal operations of the project.

5.5. Limitations of this Study

This study, has some limitations:

1. The cases in this paper were consulted on the Internet, without field verification, meaning that certain risks may be missed. However, the combination method of case study, literature survey and expert interview makes up some deficiencies. Moreover, with the development of PPP projects, there will be increasingly similar projects. Future work will further supplement the cases.
2. Due to the limited number of PPP experts and their limited time availability, in addition to the means of on-site, we also used WeChat, e-mail and other methods in questionnaire survey, which may could have affected the quality of the questionnaire. It takes much time and effort to complete 380 triangular fuzzy numbers for the 20 risk correlations in one questionnaire. Some experts did not have enough patience and/or time to fill out the questionnaire, so the data through WeChat and e-mail are often chaotic. However, by taking the high impact of other risks on the R16 (force majeure risk) as an invalid judgment standard, the validity of the questionnaire could be guaranteed.
3. The risks in this study are mainly about PPP risks in China, and the relationships between analyzed risks are mainly applicable to the current situation in China. For example, the environmental risk (R19) is at the bottom of the risk hierarchy, which is related to China's sudden environmental policies. This may be not in line with other countries. However, the research methods and framework proposed in this study, can be applied to other studies on the relationship between PPP risks in different countries and regions.
4. A total of 20 items are identified as PPP project failure risk factors, which is a broadly representative result, suitable for the key risk factors identified in most PPP projects. However, in a specific project, depending on the specific situation of a PPP project, it may be appropriate to supplement or adjust the key risk factors. For example, in a PPP project for waste incineration, there is often public opposition, while in PPP projects for sponge cities or wetland parks, there is usually no public opposition. In these circumstances, public opposition risk (R17) should be adapted according to the characteristics of the PPP projects.
5. Although fuzzy theory reduces the subjectivity of expert judgment, it also increases the workload of data processing, requiring more resources to be invested into modeling and analysis. Thus, the method in this paper would benefit from programs to facilitate its application.
6. The triangular fuzzy number given by experts will have significant impact on the results. If there are disputes among experts about the risk relationship (the triangular fuzzy number of different experts varies greatly), it will be difficult to calculate the results. In addition, the selection of the intercept coefficient requires PPP experts with abundant experience and professional insight.

6. Conclusions and Future Research

There is a close interaction between PPP project risks, which has been neglected in relevant literature. Furthermore, the existing research on the interaction of PPP risks lacked consideration of complexity and ambiguity among risks. Based on this, this study has adopted the FISM-MICMAC method to discuss the risk factors and their hierarchical relationship of PPP projects. In addition, the method of FISM-MICMAC can provide a new risk assessment tool for risk management strategies in the field of construction management.

This study started by identifying key risk factors that lead to the failure of PPP projects, using a combination of various risk identification methods, such as case study, literature survey, and expert interviews, in order to identify 20 key risk factors that have an important impact on the success or failure of PPP projects. Then, FISM was used to discuss the PPP project risk factors and their hierarchical relationship, where the hierarchical structure chart of the key risk factors of PPP projects was constructed. Moreover, MICMAC was used to classify the driving power and dependence power of the risk factors, so as to solve the complex problems of the interaction and feedback relationships between key risk factors in PPP projects. The final sequence of factors, which have the obvious characteristics of network, was obtained as a result of an analysis of the dependencies between risk factors.

On the basis of this study, further research can be carried out in the future: (1) Although we have ascertained relationships of influence between risks, the degree of influence, impact speed, and impact mechanisms remain to be further studied; (2) The coupling effect between risk factors needs to be further analyzed, and the overall risk of the project under the coupling effect needs to be evaluated; (3) In future studies, risk factors for the failure of PPP projects can be combined with liability sharing to establish a liability sharing model under the interaction of risks, which can provide a decision-making basis for clearly defining responsibility and compensation for failure.

Author Contributions: Conceptualization, X.J.; methodology, X.J. and K.L.; software, K.L.; formal analysis, X.J. and B.X.; investigation, Y.L. and C.C.; data collection, K.L. and Y.L.; writing—original draft preparation, X.J. and K.L.; writing—review and editing, B.X.; funding acquisition, X.J. and Y.L.; supervision, X.J.

Funding: The research was supported by the National Natural Science Foundation of China (71672180), the Soft Science Research Plan of Department of Housing and Urban–Rural Development of Anhui Province in China (No. JS2016AHST0011), Innovation Research Plan of Anhui Construction Engineering Group in China (No. W2018JSZX0002).

Conflicts of Interest: There is no conflict of interest.

Appendix A Case Description and Risk Factor Extraction for 28 Failed PPP Projects

Table A1. Project overview, problem description, and risk extraction of 28 failed PPP projects.

No.	Project	Project Overview and Results	Failure or Problem Description	Risk Factor Extraction
(1)	Changchun Huilu sewage treatment plant	Signed the contract in 2000 with an investment scale of $32 million dollars, and defaulted on the water bill in 2002. Government made buyback in 2005.	The government abolished franchise schemes and stopped paying water bills. It was difficult to maintain the normal operations of the project. The sewage plant has failed in the appeal. The project experienced two years of legal disputes.	Policy and regulatory change risk; Government credit risk; Project income risk; Government regulatory risk; Project company violates laws and regulations; Environment risk.
(2)	Qingdao Veolia sewage treatment plant	Signed in 2003, with an investment scale of $42.8 million. After renegotiation, $32.7 million of reconstruction in 2015.	The government had limited understanding of PPP. After the contract was signed, the government thought the price was unfair and unilaterally asked for renegotiation. Lack of effective supervision and pollution incidents led to public protests.	Government credit risk; Lack of PPP co-operation experience; Operation and maintenance risk; Government regulatory risk; Project company violates laws and regulations; Environmental risks; Imperfect price adjustment mechanism; Public opposition risk.
(3)	Wuhan Townsend sewage treatment plant	Construction in 2001; $12.8 million investment; Government made buyback in 2004.	The surrounding sewage network was not equipped, there was no water source for operation, and the problem of a sewage treatment fee was not solved. The project is idle after completion.	Government decision-making approval risks; Changes in market demand; Imperfect price adjustment mechanism; Facility matching risk.
(4)	Beijing No.10 water plant	The project was initiated in 1998 with an investment scale of $326.6 million and started in 2012. It was not officially operational in 2015.	Long-term failure to sign a contract after winning the bid. Market supply and demand upside down, the risk was mainly borne by investors. Banks did not lend. Difficulties in land allocation and rising construction costs. Network changes. Water prices raised public opposition.	Lengthy government decision-making and approval; Planning and design risk; Changes in market demand; Construction cost overrun; Public opposition risk; Financing risk.
(5)	Shanghai Dayang water plant	Signed in 1996 with an investment of $70 million. Government made buyback in 2004.	Delay in project approval. Water price below cost was difficult to increase. Government fixed rate of return, conflict with policy, renegotiation failed.	Lengthy government decision-making and approval; Imperfect price adjustment mechanism; Unreasonable profit distribution; Government credit risk; Policy and regulatory change risk.
(6)	Lianjiang zhongfa water plant	Signed in 1997, invested $16.69 million. It sat idle and sued in 1999, and government made buyback in 2009.	Water quantity and price agreement could not be implemented. The price of water was too high for the public to afford, and the government could not continue to pay for it. Chinese and French water companies suffered losses.	Lack of PPP co-operation experience; Government official corruption; Government credit risk; Imperfect price adjustment mechanism; Project income risk.
(7)	Shenyang No.9 water plant	Equity transfer negotiated in 1996, and contract changed in 2000. Government made two buybacks in 2001 and in 2006.	The sewage treatment fee was higher than the market price, and Shenyang water supply company lost money. Contract negotiation and modification.	Lack of PPP co-operation experience; Government official corruption; Imperfect price adjustment mechanism; Unreasonable profit distribution.

Table A1. *Cont.*

No.	Project	Project Overview and Results	Failure or Problem Description	Risk Factor Extraction
(8)	Chengdu Veolia No.6 water plant	Built in 2000 with an investment of $106.5 million. It was put into operation in February 2002. The project suffered heavy losses.	Some of the projects changed for legal reasons were BT projects. The economic situation was not reasonably predicted and the water supply market was seriously surplus. The water company suffered heavy losses.	Planning and design risk; Lack of PPP co-operation experience; Unreasonable profit distribution; Changes in market demand; Policy and regulatory change risk.
(9)	Quanzhou Donghai municipal sewage treatment plant	In Phase I, investment was 6.66 million yuan, bidding price was $0.12 dollar/m3. Project completed in 2011 and did not implement VFM.	During the construction period, the government required speeding up of the construction, with construction cost overruns. According to the contract, the adjustment period of water price was five years, which led to great market risks. Suspension of franchise period and adjustment ambiguity were controversial.	Government intervention; Construction cost overrun; Imperfect price adjustment mechanism; Imperfect dispute settlement mechanism.
(10)	Lanzhou Veolia water	In January 2007, the contract was signed, and the private contributed $242.8 million. In 2014, the tap water was polluted. The project did not implement VFM.	The government took most of the premium to the state-owned assets supervision and administration commission, which is unfavorable to the operation of a joint venture. Public opposition to water price increases. The separation of pipe network water plant ownership, unfavorable maintenance of equipment, resulting in water pollution incidents	Operation and maintenance risk; Government credit risk; Government regulatory risk; Environmental risks; Public opposition risk; Project company violates laws and regulations; Imperfect price adjustment mechanism.
(11)	Shanghai Huqingping (Huyu) expressway	Won the bid in 2000, completed the whole line in 2005. In July 2008, the contract was suspended ahead of schedule and government withdrew the operation.	The selection process of social capital was not transparent, the project company had poor management, and there were many violations. The project capital and financial supervision were not in place, the project payment is in arrears, and the actual income is low.	Government regulatory risk; Construction risk; Operation and maintenance risks; Bidding risks; Changes in market demand; Project company violates laws and regulations.
(12)	Jinzhou municipal sewage treatment plant	Constructed in 2001 with an investment of $2.56 million and put into operation in January 2004. Intermittent stop after operation.	Sewage discharge was not up to standard, and the government did not pay sewage treatment fees. In 2005, water pollution caused by intermittent shutdown. Design process defected. The inlet water quality exceeded the standard. After many consultations, consensus was reached for renovation.	Environment risk; Government credit risk; Planning and design risk; Government regulatory risk; Insufficient expense payment.
(13)	Hangzhou Bay sea-crossing bridge	Started in 2003 and opened in 2008 with a total investment of about $1675.5 million yuan. It lost money in 2013 and government made buyback.	Private capital transfer and withdrawal, the government to buy back 80% of the equity. During the construction period, the government repeatedly proposed to change the rate of return on investment. Suffer from interest rate increase inflation, serious losses, unable to recover the principal on schedule.	Planning and design risk; Construction cost overrun; Change in market demand; Parallel project competitive risk; Contract risk; Economic risk.
(14)	Quanzhou Wiliongtong bridge	Project started in May 1995 with a total investment of $35.5 million and was put into operation in 1997. Government made buyback in 2016.	Quanzhou government to build a number of free competitive Bridges to disperse traffic. Defects of legal contract, unclear division of responsibilities and rights. The contract stipulated that the concession period exceed the state regulation, and fee increase was not permitted	Policy and regulatory change risk; Parallel project competitive risk; Government credit risk; Insufficient expense payment; Incomplete contract risk.

Table A1. *Cont.*

No.	Project	Project Overview and Results	Failure or Problem Description	Risk Factor Extraction
(15)	Changtan west expressway	Started construction in 2004 with an investment of $122.1 million and opened to traffic in 2007. The project suffered heavy losses.	The government defaulted, the compensation for land expropriation and demolition failed, and the toll station was repeatedly changed during construction. A number of competitive toll highways existed.	Government intervention; Parallel project competitive risk; Insufficient expense payment; Planning and design risk; Completion risk.
(16)	Wuhan Yangtze river three bridges	Construction started in 1997 with an investment of $156.2 million and opened to traffic in 2000. The project suffered heavy losses.	The government failed to provide a fixed return and non-competitive guarantee, as agreed, to build several Bridges. Long operation period for maintenance, overload was a serious issue.	Government credit risk; Parallel project competitive risk; Project income risk; Government regulatory risk; Construction quality risk.
(17)	Xianyang Weihe river No.3 bridge	Started in 2003 and was opened to traffic in 2004, with a total investment of about $17.0 million. Government made buyback in 2011.	Affected by SARS in the early stages of project construction, labor shortage. The traffic volume forecast was insufficient, the construction scale was insufficient. In the operation stage, the service quality was not up to standard, and traffic jams on the bridge caused public opposition.	Government credit risk; Force majeure risk; Parallel project competitive risk; Public opposition risk; Planning and design risk; Organizational coordination risk.
(18)	Shandong Zhonghua power generation project	Initiated in 1997 and completed in 2004. Project collaboration is in trouble.	Reform of electric power system in the operation period. The government promised that the minimum purchase power was insufficient, so could not co-operate. In 2002, fees were reduced and revenues reduced.	Lack of PPP co-operation experience; Government credit risk; Changes in market demand; Contract risk; Policy and regulatory change risk.
(19)	Nanjing Yangtze river three bridges	Started construction in 2002 and opened to traffic in 2005 with a total investment of $438.7 million. After opening to traffic, the project suffered serious losses.	The location was remote, the road network on both sides was not perfect, the traffic flow was seriously lower than expected, and the financial cost was high, resulting in a project loss. In 2009, the government negotiated an extension of the franchise to 30 years.	Planning and design risk; project income risk; Facility matching risk.
(20)	Tianjin Shuanggang waste incineration power plant	Started construction in 2003 and went into operation in 2005 with a total investment of $76.7 million. Project is in trouble.	Waste incineration pollution, poor government supervision, public opposition. The amount of subsidy promised by the government was not clear, and the project was not profitable enough	Planning and design risk; Environmental risk; Government credit risk; insufficient expense payment; Government regulatory risk.
(21)	Shandong Heze waste-to-power plant	In 2013, a co-operation agreement was signed, with an initial investment of $49.7 million. The project suffered serious losses.	The technology was not mature and the argument was not sufficient. Accumulated losses of more than $2.1 million in four years of operation. Failure to implement government subsidy schemes. Mix-in coal amount exceeded national regulation, became small thermal power plant.	Planning and design risk; Policy and regulatory change risks; Government credit risk; Project changes risk; Project income risk; Project company violates laws and regulations.
(22)	Beijing Liulitun waste-to-power plant	Approved in 2005 and planned to start construction in 2007. In 2007, the state environmental protection administration ordered a moratorium, and the project was halted in 2011.	Project planning was lagging behind, the site was close to a diversion canal and residential air outlet, and the odor affected the surrounding environment, strongly opposed by residents. The state environmental protection administration ordered a moratorium on construction, and the project has been suspended for a long time.	Government decision-making approval risk; Planning and design risk; Public opposition risk; Environmental risk.

Table A1. *Cont.*

No.	Project	Project Overview and Results	Failure or Problem Description	Risk Factor Extraction
(23)	Shenzhen Nanshan waste-to-power plant	Started in 2002 and put into operation in December 2003, with a total investment of $51.4 million. The project did not implement VFM.	The voltage of the power grid was unstable, the flue gas treatment system could not work normally, and toxic gas was discharged directly. No agreement was been reached on garbage disposal fee and electricity price after production. Garbage disposal fee and Internet electricity fee payment delayed.	Environmental risk; Government credit risk; Facility matching risk; Insufficient expense payment; Contract risk.
(24)	Chaojie highway	In 2002, the total investment was $174.6 billion. Construction had not started for 8 years, and resumed in 2010. Shenzhen Taibang company took over the project.	The introduction of new laws and regulations made it more difficult to raise the cost of land acquisition. Social capital financing difficulties, request to adjust the budget, and the government conflict.	Policy and regulatory change risk; Organizational coordination risk; Financing risk; Bidding risk; Construction cost overrun.
(25)	Beijing bird's nest stadium	Started in 2003 and completed in 2008, with a total investment of $445.7 million. In 2009, it was taken over by State-owned assets supervision and administration commission. The project did not implement VFM.	Social capital side had serious lack of experience in the operations and management of stadiums, gymnasiums, and sports industry resources. After one year of operation, the joint-stock reform agreement was re-signed, and the project company converted the 30-year franchise into equity.	Public opposition risk; Contract risk; project income risk; Bidding risks; Organizational coordination risk; Construction cost overrun; Operation and maintenance risk.
(26)	Shenzhen Wutong mountain tunnel	In 1997, 50% of the shares were transferred to the Hong Kong Dajia company. In 2004, the government repurchased the shares, but the negotiation was unsuccessful.	More traffic than expected, public opposition to congestion. Lack of perfect contract terms, the repo two sides bid a large gap, negotiations reached a deadlock. The government built new roads around it for free, and social capital was forced to abandon franchises.	Planning and design risk; Public opposition risk; Government credit risk; Incomplete contract risk; Parallel project competitive risk.
(27)	Chenganyu highway	Started in 2009, and the planned construction period was three years. In 2016 the government cancelled the concession agreement.	Illegal subcontracting of the project resulted in economic contract disputes, resulting in a disturbance of visiting events, project progress lagging behind, and a state of repeated suspension and resumption of work. The relevant national loan policy adjustment, the project company's own capital turnover was difficult, loan difficulty, capital chain interruption.	Bidding risks; Government regulatory risk; Policy and regulatory change risk; Financing risk; Organizational coordination risk; Project company violates laws and regulations.
(28)	Changan highway	Started in 2010, with a planned period of three years, to stop at the beginning of 2013. In 2015, the government took over the project.	The government has repeatedly proposed changes to increase the construction cost of the project, resulting in fund turnover difficulties and forced large-scale shutdown of the project. Invitation to tender, social capital review was not in place, and the cost estimate was inaccurate.	Planning and design risk; Project change risk; Bidding risk; Government credit risk; construction cost overruns; Construction schedule risk

Table A2. 28 failed projects and their links to websites.

No.	Project	Links to Websites (Accessed on 19 September 2019)
(1)	Changchun Huilu sewage treatment plant	http://huanbao.bjx.com.cn/news/20150305/594592.shtml
(2)	Qingdao Veolia sewage treatment plant	http://www.h2o-china.com/news/243917.html
(3)	Wuhan Townsend sewage treatment plant	http://www.h2o-china.com/news/243917.html
(4)	Beijing No.10 water plant	http://www.xcar.com.cn/bbs/viewthread.php?tid=26304486
(5)	Shanghai Dayang water plant	http://blog.sina.com.cn/s/blog_99ca571e0102vzcd.html
(6)	Lianjiang zhongfa water plant	http://mini.eastday.com/mobile/180326172138091.html#
(7)	Shenyang No.9 water plant	http://mini.eastday.com/mobile/180326172138091.html#
(8)	Chengdu Veolia No.6 water plant	http: //www.cspea.org.cn/article/hydt/hydt/201708/20170800006038.shtml
(9)	Quanzhou Donghai municipal sewage treatment plant	https://max.book118.com/html/2017/1126/141657699.shtm
(10)	Lanzhou Veolia water	http://www.tanpaifang.com/ppp/201612/0757836_4.html
(11)	Shanghai Huqingping (Huyu) expressway	https://www.zjtcn.com/baike/hygsgl
(12)	Jinzhou municipal sewage treatment plant	https://www.tianyancha.com/company/2333146105
(13)	Hangzhou Bay Sea-Crossing Bridge	http://www.360doc.com/content/17/0210/11/366082_627991592.shtml
(14)	Quanzhou Wiliongtong bridge	http://opinion.caixin.com/2014-09-25/100732936.html
(15)	Changtan west expressway	https://www.sohu.com/a/217236864_99929980
(16)	Wuhan Yangtze river three Bridges	https://www.sohu.com/a/205755812_100012673
(17)	Xianyang Weihe river No.3 bridge	https://www.docin.com/p-1434785098.html
(18)	Shandong Zhonghua power generation project	https://www.zhihu.com/question/28741657?sort=created
(19)	Nanjing Yangtze river three Bridges	http://www.360doc.com/content/16/0722/13/366082_577533747.shtml
(20)	Tianjin Shuanggang waste incineration power plant	http://mini.eastday.com/mobile/180326172138091.html#
(21)	Shandong Heze waste-to-power plant	http://huanbao.bjx.com.cn/news/20171227/870276.shtml
(22)	Beijing Liulitun waste-to-power plant	https://news.qq.com/a/20110120/000136.htm
(23)	Shenzhen Nanshan waste-to-power plant	http://www.sohu.com/a/214527318_100053329
(24)	Chaojie highway	http://www.gongdaogc.com/show.asp?id=40917
(25)	Beijing bird's nest stadium	https://www.sohu.com/a/142496023_798724
(26)	Shenzhen Wutong mountain tunnel	https://www.sohu.com/a/133919852_739751
(27)	Chenganyu highway	http://www.sohu.com/a/19689655_116913
(28)	Changan highway	http://bbs.tianya.cn/post-293-152478-1.shtml

B. Questionnaire on the Risk Relationship of PPP Projects

Dear experts,

Hello! Research Group of Dr. Jiang, School of Civil Engineering, Hefei University of Technology, is carrying out research on " the risk relationship of PPP projects ". Our research group sincerely invites you to give your opinion on this survey based on your past experience and relevant knowledge in PPP projects. Your opinion is very valuable and will play an important role in our research. The data and information collected in this questionnaire will only be used for academic research and will not negatively affect your daily work and life. Thank you for your understanding and support!

Jiang Group

Table 1 shows the list of key risk factors of PPP and their meaning explanation. This survey gives a list of risk factors according to the past practice in the 28 failure PPP project cases, related literature survey, and expert interview. After familiarizing the risk factors in Table 1, please complete Table 3 according to the reminders in Table 2.

Table 3. Risk factors and their meaning explanation for PPP projects.

Risk Factor	Meaning Explanation
Government decision-making approval risk (R1)	Government decision-making procedures are irregular. The government has limited understanding of PPP. Project decisions go wrong. The approval process is lengthy 政府决策程序不规范。政府对PPP的理解有限。项目决策存在错误。审批过程很冗长。
Policy and regulatory change risk (R2)	Existing PPP-related policies and regulations are unsound, including low legislative level, poor operability, and conflict. 现有的PPP相关政策法规不健全、立法层级低、可操作性差、相互冲突;
Government credit risk (R3)	The government does not fulfill the responsibilities and obligations stipulated in the contract. The government excessively interferes in the construction or operation of the project. There is official corruption in government. 政府不履行合同规定的责任和义务。政府过分干预项目的建设或运行。政府中有官员腐败。
Government regulatory risk (R4)	The government does not have adequate supervision over the project, and the supervision system is not perfect. 政府对项目监管不到位, 监管体系不完善。
Planning and design risk (R5)	The project lacks detailed argumentation and investigation. The planning and design scheme (including project scale, technology, equipment standard, etc.) is unreasonable and unfeasible. It does not meet the requirements of practical application and future development. 该项目缺乏详细的论证和调查。规划设计方案(包括项目规模、技术、设备标准等) 不合理、不可行。不符合实际应用和未来发展的要求
Bidding risk (R6)	These are vicious competition or insufficient competition among private capital, unreasonable bidding, tendering procedures, lax qualification examination conditions and insufficient supervision, etc. The government choose the incompetent private capital. 社会资本方之间存在恶性竞争或竞争不足、招标投标程序不合理、资格审查条件不严格、监督力度不够等。政府选择了不称职的私人资本方。
Contract risk (R7)	There are problems with the contract design, including: contract ambiguity, contract document conflict, contract storage not in place, imperfect price adjustment mechanism, unreasonable profit distribution, imperfect dispute settlement mechanism, and incomplete contract. Risk allocation and income distribution are not reasonable. The scope of rights and responsibilities is not clear. 合同设计存在问题，包括:合同歧义、合同文件冲突、合同保管不到位、价格调节机制不完善、利润分配不合理、纠纷解决机制不完善、合同不完整等。风险配置和收益分配不合理。权利和责任范围不明确。
Facility matching risk (R8)	The project is idle due to the lack of supporting infrastructure and insufficient supply. 相关的基础设施配套不到位、供应不足等致使项目闲置,
Financing risk (R9)	The financial market is not sound, financing channels are not smooth, financing structure is not reasonable with high debt ratio. It is difficult to raise funds for the project. 融资市场不健全，融资渠道不畅通，融资结构不合理，负债率高。为这个项目筹集资金是困难的

Table 3. *Cont.*

Risk Factor	Meaning Explanation
Economic risk (R10)	It refers to the uncertainty of expected cash flow of PPP projects caused by changes in macroeconomic factors and fragility of the financial system, as well as the possibility of expected income loss of PPP projects caused by changes in exchange rate, interest rate, commodity price and inflation. 指宏观经济因素的变动以及金融体系内在的不稳定性和脆弱性而带来的PPP项目预期现金流的不确定性，以及由于汇率、利率和商品价格变动以及通货膨胀带来的PPP项目预期收益损失的可能性变动
Project change risk (R11)	The project changes, including the change of the construction object, the change of the operation scope, the change of the contract 项目变更，包括建设对象的变更、运营范围的变更、合同的变更
Construction risk (R12)	Project construction management level is poor. Construction cost overruns. Project delays. The quality is not up to standard. Project construction cannot be completed. 项目建设管理水平较差。建设成本超支。项目延迟。质量不符合标准。项目建设不能完工。
Project income risk (R13)	Due to the change of market demand or insufficient expense payment, project investors cannot recover the investment cost or achieve the predetermined income level 市场需要的变化或者支付能力不足，导致项目投资者不能收回投资成本或不能达到预定收益水平
Parallel project competitive risk (R14)	The government or other investors build or rebuild similar projects, which form substantial commercial competition with the original PPP projects. 政府或其他投资人新建或改建相似的项目，与原有项目形成实质性商业竞争
Operation and maintenance risk (R15)	The project company's operation efficiency is low. The price of raw materials, fuels and power increases. The equipment is of poor quality and often maintained. The operation cost and maintenance cost increase. 项目公司运营效率低。原材料、燃料和电力价格上涨。设备质量差，经常维修。运营和维护成本增加。
Force majeure risk (R16)	Force majeure means that both parties have no control over the event or situation that cannot be avoided or overcome before the signing of the contract，such as war, embargo, terrorism, and other social force majeure; floods, typhoons, earthquakes, fires, epidemics, and other natural disasters. 不可抗力是指双方合同签订前无法避免或克服的事件或情况，如战争、禁运、恐怖主义等社会不可抗力;洪水、台风、地震、火灾、疫病等自然灾害
Public opposition risk (R17)	Due to the lack of protection of public interests, the public is dissatisfied with the project and even opposes it. In order to protect the public interest and social stability, the government usually negotiates with private capital. 由于社会公共利益得不到保护或受损，引起公众对项目产生不满甚至反对项目建设，而政府为维护公众利益和社会稳定，通常会与私人资本谈判。
Organizational coordination risk (R18)	Coordination between government departments is difficult. The organization and coordination ability of the project company is insufficient. Communication costs among participants are increase, and conflicts arise among participants. 政府部门之间协调困难。项目公司组织协调能力不足。参与者之间的沟通成本增加，参与者之间存在冲突。
Environmental risk (R19)	During the implementation of the project, the improvement of environmental protection standards may lead to the increase of investment in new equipment. The project company failed to take effective control measures and emergency measures due to the environmental damage. 在项目实施过程中，环保标准的提高可能会导致新设备投资的增加。项目公司未能采取有效的控制措施和应急措施导致环境破坏。
Project company violates laws and regulations (R20)	During the implementation of the project, the project company or private capital violates laws and regulations, such as using projects for fraud or modifying the project data maliciously。 在项目实施过程中，项目公司或私人资本方违反法律法规，比如，利用项目进行欺诈或者恶意修改项目数据

According to the scoring in Table 2, please judge the mutual relationship of two factors, and fill the evaluation results into the corresponding positions in Table 3.

Table 4. Language operator, triangular fuzzy number, and scoring.

Language Operator.	Triangular Fuzzy Number ($\tilde{d}_{ij}^{k}$)	Scoring
Very low impact (VL)	(0, 0, 0.25)	1
Low impact (L)	(0, 0.25, 0.5)	2
medium impact (M)	(0.25, 0.5, 0.75)	3
High impact (H)	(0.5, 0.75, 1)	4
Very high impact (VH)	(0.75, 1, 1)	5

Example: when you consider the impact of the risk factor R1 on R2 to be very low impact, you should fill the "1" in the second row and third column of Table 3.

Table 5. Mutual relationship intensity between two risk factors.

	R1	R2	R3	R4	R5	R6	R7	R8	R9	R10	R11	R12	R13	R14	R15	R16	R17	R18	R19	R20
R1	0																			
R2		0																		
R3			0																	
R4				0																
R5					0															
R6						0														
R7							0													
R8								0												
R9									0											
R10										0										
R11											0									
R12												0								
R13													0							
R14														0						
R15															0					
R16																0				
R17																	0			
R18																		0		
R19																			0	
R20																				0

Note: the risk factors affecting are column and the risk factors bearing the influence are row; there is an asymmetric relationship between risk factors. Please complete this questionnaire as soon as possible. Thank you again for your patience.

References

1. Ma, G.F.; Du, Q.J.; Wang, K.D. A Concession Period and Price Determination Model for PPP Projects: Based on Real Options and Risk Allocation. *Sustainability* **2018**, *10*, 21.
2. Hodge, G.A.; Greve, C. On public–private partnership performance: A contemporary review. *Public Work. Manag. Policy* **2017**, *22*, 55–78. [CrossRef]
3. Delorme, C.D.; Thompson, H.G.; Warren, R.S. Public infrastructure and private productivity: A stochastic-frontier approach. *J. Macroecon.* **1999**, *21*, 563–576. [CrossRef]
4. Yuan, J.; Wang, C.; Skibniewski, M.J.; Li, Q. Developing Key Performance Indicators for Public-Private Partnership Projects: Questionnaire Survey and Analysis. *J. Manag. Eng.* **2012**, *28*, 252–264. [CrossRef]
5. BANK, W. Private Participation in Infrastructure Database. Available online: http://ppi.worldbank.org/customquery (accessed on 19 September 2019).
6. Węgrzyn, J. The perception of critical success factors for PPP projects in different stakeholders groups. *Entrep. Bus. Econ. Rev.* **2016**, *4*, 81–92. [CrossRef]

7. Wojewnik-Filipkowska, A.; Trojanowski, D. Principles of public-private partnership financing–Polish experience. *J. Prop. Invest. Financ.* **2013**, *31*, 329–344. [CrossRef]
8. Tang, L.; Shen, Q.; Cheng, E.W.L. A review of studies on Public–Private Partnership projects in the construction industry. *Int. J. Proj. Manag.* **2010**, *28*, 683–694. [CrossRef]
9. Heravi, G.; Hajihosseini, Z. Risk Allocation in Public-Private Partnership Infrastructure Projects in Developing Countries: Case Study of the Tehran-Chalus Toll Road. *J. Infrastruct. Syst.* **2012**, *18*, 210–217. [CrossRef]
10. Osei-Kyei, R.; Chan, A.P.C. Review of studies on the critical success factors for public-private partnership (PPP) projects from 1990 to 2013. *Int. J. Proj. Manag.* **2015**, *33*, 1335–1346. [CrossRef]
11. Xu, Y.; Yeung, J.F.Y.; Chan, A.P.C.; Chan, Q.W.M.; Wang, S.Q.; Ke, Y. Developing a risk assessment model for PPP projects in China—A fuzzy synthetic evaluation approach. *Autom. Constr.* **2010**, *19*, 929–943. [CrossRef]
12. Shen, L.Y.; Platten, A.; Deng, X.P. Role of public private partnerships to manage risks in public sector projects in Hong Kong. *Int. J. Proj. Manag.* **2006**, *24*, 587–594. [CrossRef]
13. Chen, C.; Doloi, H. BOT application in China: Driving and impeding factors. *Int. J. Proj. Manag.* **2008**, *26*, 388–398. [CrossRef]
14. Ministry of Finance. *Notice of Standardizing Project Library in PPP Integrated Information Platform "No. 92"*; Ministry of Finance, Ed.; Ministry of Finance: Beijing, China, 2017.
15. Bridata. Data Custom Download. Available online: http://bridata.com/t#/tools (accessed on 19 September 2019).
16. Liang, Y.; Wang, H. Sustainable Performance Measurements for Public–Private Partnership Projects: Empirical Evidence from China. *Sustainability* **2019**, *11*, 3653. [CrossRef]
17. Ke, Y.J.; Wang, S.Q.; Chan, A.P.C.; Lam, P.T.I. Preferred risk allocation in China's public-private partnership (PPP) projects. *Int. J. Proj. Manag.* **2010**, *28*, 482–492. [CrossRef]
18. Li, Y.; Wang, X.Y. Using fuzzy analytic network process and ISM methods for risk assessment of public-private partnership: A China perspective. *J. Civ. Eng. Manag.* **2019**, *25*, 168–183. [CrossRef]
19. Iyer, K.C.; Sagheer, M. Hierarchical Structuring of PPP Risks Using Interpretative Structural Modeling. *J. Constr. Eng. Manag.* **2010**, *136*, 151–159. [CrossRef]
20. Heinrich, H.W. *Industrial Accident Prevention*; McGraw-Hill: New York, NY, USA, 1979.
21. Bird, F.E. *Management Guide to Loss Control*; International Loss Control Institute: Atlanta, GA, USA, 1974.
22. Choi, J.H.; Chung, J.; Lee, D.J. Risk perception analysis: Participation in China's water PPP market. *Int. J. Proj. Manag.* **2010**, *28*, 580–592. [CrossRef]
23. Ke, Y.; Wang, S.; Chan, A.P.C. Risk Allocation in Public-Private Partnership Infrastructure Projects: Comparative Study. *J. Infrastruct. Syst.* **2010**, *16*, 343–351. [CrossRef]
24. Li, J.; Zou, P.X.W. Fuzzy AHP-based risk assessment methodology for PPP projects. *J. Constr. Eng. Manag.* **2011**, *137*, 1205–1209. [CrossRef]
25. Zhang, S.; Chan, A.P.C.; Feng, Y.; Duan, H.; Ke, Y. Critical review on PPP Research—A search from the Chinese and International Journals. *Int. J. Proj. Manag.* **2016**, *34*, 597–612. [CrossRef]
26. Liu, Y. Identification of Risk Factors Affecting PPP Waste-to-Energy Incineration Projects in China: A Multiple Case Study. *Adv. Civ. Eng.* **2018**, *16*, 4983523. [CrossRef]
27. Song, J.; Sun, C.; Xia, B.; Liu, S.; Skitmore, M. Risk identification for PPP waste-to-energy incineration projects in China. *Energy Policy* **2013**, *61*, 953–962. [CrossRef]
28. Jayasuriya, S. Challenges in public private partnerships in construction industry. *Built Environ. Proj. Asset Manag.* **2019**, *9*, 172–185. [CrossRef]
29. Shrestha, A.; Chan, T.K.; Aibinu, A.A.; Martek, L.; Chen, C.; Asce, A.M. Risk Allocation Inefficiencies in Chinese PPP Water Projects. *J. Constr. Eng. Manag.* **2018**, *144*, 04018013. [CrossRef]
30. Xu, Y.; Chan, A.P.C.; Yeung, J.F.Y. Developing a Fuzzy Risk Allocation Model for PPP Projects in China. *J. Constr. Eng. Manag.* **2010**, *136*, 894–903. [CrossRef]
31. Jin, X.H. Model for efficient risk allocation in privately financed public infrastructure projects using neuro-fuzzy techniques. *J. Constr. Eng. Manag.* **2011**, *137*, 1003–1014. [CrossRef]
32. Ameyaw, E.E.; Chan, A.P.C. Risk allocation in public-private partnership water supply projects in Ghana. *Constr. Manag. Econ.* **2015**, *33*, 187–208. [CrossRef]
33. Zou, P.X.W.; Li, J. Risk identification and assessment in subway projects: Case study of Nanjing Subway Line 2. *Constr. Manag. Econ.* **2010**, *28*, 1219–1238. [CrossRef]

34. Wu, Y.; Xu, C.; Li, L.; Wang, Y.; Chen, K.; Xu, R. A risk assessment framework of PPP waste-to-energy incineration projects in China under 2-dimension linguistic environment. *J. Clean. Prod.* **2018**, *183*, 602–617. [CrossRef]
35. Mazher, K.M.; Chan, A.P.C.; Zahoor, H.; Khan, M.I. Fuzzy Integral-Based Risk-Assessment Approach for Public-Private Partnership Infrastructure Projects. *J. Constr. Eng. Manag.* **2018**, *144*, 04018111. [CrossRef]
36. Thomas, A.V.; Kalidindi, S.N.; Ganesh, L.S. Modelling and assessment of critical risks in BOT road projects. *Constr. Manag. Econ.* **2006**, *24*, 407–424. [CrossRef]
37. Bai, L.B.; Li, Y.; Du, Q.; Xu, Y. A Fuzzy Comprehensive Evaluation Model for Sustainability Risk Evaluation of PPP Projects. *Sustainability* **2017**, *9*, 1890. [CrossRef]
38. Zhang, L.; Sun, X.J.; Xue, H. Identifying critical risks in Sponge City PPP projects using DEMATEL method: A case study of China. *J. Clean. Prod.* **2019**, *226*, 949–958. [CrossRef]
39. Zheng, C.J.; Yuan, J.; Li, L.; Skibniewski, M.J. Process-Based Identification of Critical Factors for Residual Value Risk in China's Highway PPP Projects. *Adv. Civ. Eng.* **2019**, *21*, 5958904. [CrossRef]
40. Liu, Y.; Hao, Y.; Lu, Y.L. Improved Design of Risk Assessment Model for PPP Project under the Development of Marine Architecture. *J. Coast. Res.* **2018**, *83*, 74–80. [CrossRef]
41. Wibowo, A.; Kochendoerfer, B. Selecting BOT/PPP Infrastructure Projects for Government Guarantee Portfolio under Conditions of Budget and Risk in the Indonesian Context. *J. Constr. Eng. Manag.* **2011**, *137*, 512–522. [CrossRef]
42. Nguyen, D.A.; Garvin, M.J.; Gonzalez, E.E. Risk Allocation in US Public-Private Partnership Highway Project Contracts. *J. Constr. Eng. Manag.* **2018**, *144*, 04018017. [CrossRef]
43. Alireza, V.; Mohammadreza, Y.; Zin, R.M.; Yahaya, N. An enhanced multi-objective optimization approach for risk allocation in public-private partnership projects: A case study of Malaysia. *Can. J. Civ. Eng.* **2014**, *41*, 164–177. [CrossRef]
44. Chan, A.P.C.; Lam, P.T.I.; Wen, Y.; Ameyaw, E.E.; Wang, S. Cross-sectional analysis of critical risk factors for PPP water projects in China. *J. Infrastruct. Syst.* **2015**, *21*, 04014031. [CrossRef]
45. Ke, Y.; Wang, S.; Chan, A.P.C.; Cheung, E. Understanding the risks in China's PPP projects: Ranking of their probability and consequence. *Eng. Constr. Archit. Manag.* **2011**, *18*, 481–496. [CrossRef]
46. Effah Ameyaw, E.; Chan, A.P.C. Identifying public-private partnership (PPP) risks in managing water supply projects in Ghana. *J. Facil. Manag.* **2013**, *11*, 152–182. [CrossRef]
47. Ameyaw, E.E.; Chan, A.P. Risk ranking and analysis in PPP water supply infrastructure projects. *Facilities* **2015**, *33*, 428–453. [CrossRef]
48. Xu, Y. Critical risk factors affecting the implementation of PPP waste-to-energy projects in China. *Appl. Energy* **2015**, *158*, 403–411. [CrossRef]
49. Shrestha, A. Risks in PPP water projects in China: Perspective of local governments. *J. Constr. Eng. Manag.* **2017**, *143*, 05017006. [CrossRef]
50. Valipour, A.; Yahaya, N.; Md Noor, N.; Mardani, A.; Antuchevičienė, J. A new hybrid fuzzy cybernetic analytic network process model to identify shared risks in PPP projects. *Int. J. Strateg. Prop. Manag.* **2016**, *20*, 409–426. [CrossRef]
51. Wang, S.Q.; Dulaimi, M.F.Y.; Aguria, M. Risk management framework for construction projects in developing countries. *Constr. Manag. Econ.* **2004**, *22*, 237–252. [CrossRef]
52. Aladağ, H.; Işık, Z. Design and construction risks in BOT type mega transportation projects. *Eng. Constr. Archit. Manag.* **2019**.
53. Valipour, A. A Fuzzy Analytic Network Process Method for Risk Prioritization In Freeway Ppp Projects: An Iranian Case Study. *J. Civ. Eng. Manag.* **2015**, *21*, 933–947. [CrossRef]
54. Jang, G.W. *Bids-Evaluation Decision Model Development and Application for PPP Transport Projects: A Project Risks Modeling Framework;* Colorado State University: Denver, CO, USA, 2010.
55. World Bank Group. Available online: https://ppi.worldbank.org/en/ppi (accessed on 29 August 2019).
56. China Public-Private Partnerships Center. Available online: http://www.cpppc.org/ (accessed on 29 August 2019).
57. Pinto, J.K.; Mantel, S.J. The causes of project failure. *IEEE Trans. Eng. Manag.* **1990**, *37*, 269–276. [CrossRef]
58. Jamali, D. Success and failure mechanisms of public private partnerships (PPPs) in developing countries: Insights from the Lebanese context. *Int. J. Public Sector Manag.* **2004**, *17*, 414–430. [CrossRef]

59. Chan, A.P.C.; Chan, D.W.M.; Yeung, J.F.Y. Overview of the Application of "Fuzzy Techniques" in Construction Management Research. *J. Constr. Eng. Manag.* **2009**, *135*, 1241–1252. [CrossRef]
60. De, P.K.; Rawat, A. A Fuzzy Inventory Model Without Shortages Using Triangular Fuzzy Number. *Fuzzy Inf. Eng.* **2019**, *3*, 59–68. [CrossRef]
61. Zadeh, L.A. Fuzzy sets. *Inf. Control* **1965**, *8*, 338–353. [CrossRef]
62. Paek, J.H.; Lee, Y.W.; Ock, J.H. Pricing construction risk: Fuzzy set application. *J. Constr. Eng. Manag.* **1993**, *119*, 743–756. [CrossRef]
63. Zhang, G.; Zou, P.X.W. Fuzzy analytical hierarchy process risk assessment approach for joint venture construction projects in China. *J. Constr. Eng. Manag.* **2007**, *133*, 771–779. [CrossRef]
64. Abdelgawad, M.; Fayek, A.R. Fuzzy reliability analyzer: Quantitative assessment of risk events in the construction industry using fuzzy fault-tree analysis. *J. Constr. Eng. Manag.* **2011**, *137*, 294–302. [CrossRef]
65. Fahmi, A. Expected Values of Aggregation Operators on Cubic Triangular Fuzzy Number and Its Application to Multi-Criteria Decision Making Problems. *Eng. Math.* **2018**, *2*, 1–11. [CrossRef]
66. Wang, J. A synthetic method for knowledge management performance evaluation based on triangular fuzzy number and group support systems. *Appl. Soft Comput.* **2016**, *39*, 11–20. [CrossRef]
67. Chakraborty, D.; Jana, D.K.; Roy, T.K. A new approach to solve fully fuzzy transportation problem using triangular fuzzy number. *Int. J. Oper. Res.* **2016**, *26*, 153–179. [CrossRef]
68. Li, R.J. Fuzzy method in group decision making. *Comput. Math. Appl.* **1999**, *38*, 91–101. [CrossRef]
69. Wu, W.W.; Lee, Y.T. Developing global managers' competencies using the fuzzy DEMATEL method. *Expert Syst. Appl.* **2007**, *32*, 499–507. [CrossRef]
70. Govindan, K. Analysis of third party reverse logistics provider using interpretive structural modeling. *Int. J. Prod. Econ.* **2012**, *140*, 204–211. [CrossRef]
71. Mandal, A.; Deshmukh, S.G. Vendor Selection Using Interpretive Structural Modelling (ISM). *Int. J. Oper. Prod. Manag.* **1994**, *14*, 52–59. [CrossRef]
72. Govindan, K.; Shankar, K.M.; Kannan, D. Application of fuzzy analytic network process for barrier evaluation in automotive parts remanufacturing towards cleaner production—A study in an Indian scenario. *J. Clean. Prod.* **2016**, *114*, 199–213. [CrossRef]
73. Luthra, S. Adoption of smart grid technologies: An analysis of interactions among barriers. *Renew. Sustain. Energy Rev.* **2014**, *33*, 554–565. [CrossRef]
74. Tseng, C.P.; Chen, C.W.; Tu, Y.P. A new viewpoint on risk control decision models for natural disasters. *Nat. Hazards* **2011**, *59*, 1715–1733. [CrossRef]
75. Yanmei, L.; Yanmei, L.; Zi, C. Analysis of load factors based on interpretive structural model. *J. Comput.* **2012**, *7*, 1704–1711.
76. Li, Y. Risks assessment in thermal power plants using ISM methodology. *Ann. Oper. Res.* **2019**, *279*, 89–113. [CrossRef]
77. Yadav, D.K.; Barve, A. Analysis of critical success factors of humanitarian supply chain: An application of Interpretive Structural Modeling. *Int. J. Disaster Risk Reduct.* **2015**, *12*, 213–225. [CrossRef]
78. Chakraborty, K.; Mondal, S.; Mukherjee, K. Critical analysis of enablers and barriers in extension of useful life of automotive products through remanufacturing. *J. Clean. Prod.* **2019**, *227*, 1117–1135. [CrossRef]
79. Tseng, M.L.; Lim, M.K.; Wu, K.J. Improving the benefits and costs on sustainable supply chain finance under uncertainty. *Int. J. Prod. Econ.* **2019**, *218*, 308–321. [CrossRef]
80. Chaudhuri, A. Risk propagation and its impact on performance in food processing supply chain: A fuzzy interpretive structural modeling based approach. *J. Model. Manag.* **2016**, *11*, 660–693. [CrossRef]
81. Ragade, R.K. Fuzzy interpretive structural modeling. *J. Cybern.* **1976**, *6*, 189–211. [CrossRef]
82. Bhosale, V.A.; Kant, R. An integrated ISM fuzzy MICMAC approach for modelling the supply chain knowledge flow enablers. *Int. J. Prod. Res.* **2016**, *54*, 7374–7399. [CrossRef]
83. Han, Q. Public private partnership in brownfield remediation projects in China: Identification and structure analysis of risks. *Land Use Policy* **2019**, *84*, 87–104. [CrossRef]
84. Zhang, L. Risk identification and analysis for PPP projects of electric vehicle charging infrastructure based on 2-tuple and the DEMATEL model. *World Electr. Veh. J.* **2019**, *10*, 4. [CrossRef]
85. Wu, Y. Risk management of public-private partnership charging infrastructure projects in China based on a three-dimension framework. *Energy* **2018**, *165*, 1089–1101. [CrossRef]

86. Li, Y.; Wang, X.Y. Risk assessment for public-private partnership projects: Using a fuzzy analytic hierarchical process method and expert opinion in China. *J. Risk Res.* **2018**, *21*, 952–973. [CrossRef]
87. Ke, Y.J.; Wang, S.Q.; Chan, A.P.C. Risk Misallocation in Public-Private Partnership Projects in China. *Int. Public Manag. J.* **2013**, *16*, 438–460. [CrossRef]
88. Wang, W. Assessing contributory factors in potential systemic accidents using AcciMap and integrated fuzzy ISM—MICMAC approach. *Int. J. Ind. Ergon.* **2018**, *68*, 311–326. [CrossRef]
89. Chan, A.P.C. Empirical Study of Risk Assessment and Allocation of Public-Private Partnership Projects in China. *J. Manag. Eng.* **2011**, *27*, 136–148. [CrossRef]
90. The National People's Congress of the People's Republic of China. Environmental protection tax law of the People's Republic of China. 2018. Available online: http://www.npc.gov.cn/npc/c12435/201811/5e7d3cfb3afa4ef79428c0ff7a29fdd7.shtml (accessed on 29 August 2019).
91. Zhang, S. PPP application in infrastructure development in China: Institutional analysis and implications. *Int. J. Proj. Manag.* **2015**, *33*, 497–509. [CrossRef]
92. Chen, C.; Li, D.; Man, C.X. Toward Sustainable Development? A Bibliometric Analysis of PPP-Related Policies in China between 1980 and 2017. *Sustainability* **2019**, *11*, 142. [CrossRef]
93. Wojewnik-Filipkowska, A.; Wegrzyn, J. Understanding of Public-Private Partnership Stakeholders as a Condition of Sustainable Development. *Sustainability* **2019**, *11*, 1194. [CrossRef]
94. Li, L. Enhanced Cooperation among Stakeholders in PPP Mega-Infrastructure Projects: A China Study. *Sustainability* **2018**, *10*, 2791. [CrossRef]
95. Henjewele, C.; Fewings, P.; Rwelamila, P.D. De-marginalising the public in PPP projects through multi-stakeholders management. *J. Financ. Manag. Prop. Constr.* **2013**, *18*, 210–231. [CrossRef]

sustainability

MDPI

Article

Development of a Bridge Management System Based on the Building Information Modeling Technology

Chunfeng Wan [1,*], Zhenwei Zhou [1], Siyuan Li [1], Youliang Ding [1], Zhao Xu [1], Zegang Yang [2], Yefei Xia [2] and Fangzhou Yin [1]

1 Key Laboratory of concrete and pre-stressed concrete structure of Ministry of Education, Southeast University, Nanjing 210096, Jiangsu, China
2 Huatong Engineering Co, Ltd, Nanjing 210000, Jiangsu, China
* Correspondence: wan@seu.edu.cn

Received: 20 July 2019; Accepted: 20 August 2019; Published: 23 August 2019

Abstract: With the development of the Chinese transportation industry, the number of bridges has increased significantly, but this results in high pressure of structural maintenance and management. Bridge management system (BMS) is critical for efficient maintenance and ensured safety of bridge structures during long-term operation. Building information modeling (BIM) is an emerging technology with powerful visualization and informatization capability, making it an ideal tool for developing modern management systems. This paper introduces the development of a bridge management system based on the BIM technology. Industry Foundation Classes (IFC) and International Framework for Dictionaries (IFD) standards are studied and extended, and coding rules are proposed for the Chinese bridge industry. Also, a standard structural modeling method is proposed to fast build the bridge BIM model. Web-BIM oriented bridge management is proposed, and portable devices are introduced into the system. Collaborative management is realized for different users. The BIM-based maintenance management system is designed. Finally, a practical BIM-based BMS is established for a long-span cable-stayed bridge in China. This system integrates the BIM with the geographic information system (GIS) and contains information management, inspection management, technical condition evaluation, and enables users to cooperate with each other. Such a BMS could help to improve the management efficiency and ensure its normal operation, providing a useful platform for the maintenance of massive bridges in China.

Keywords: BIM; information model; bridge; maintenance; management system

---

1. Introduction

In recent years, with the rapid development of the economy and the advancement of urbanization, China's bridge construction has reached its crescendo. Bridges play an important role in the transportation system by supporting economic and social development and are therefore regarded as one of the most important and indispensable infrastructures. According to the statistics, there are currently more than 800,000 road bridges in China, ranking first in the world [1], which gives severe pressure to their maintenance. However, at the present stage in China, people still mainly focus on construction but not maintenance. During the long-time operation of the bridges, bridge structures suffer different levels of damages caused by loads and environmental effects, which then reduces the structures' reliability and shortens their expected service life-time. For example, severe rust and breakage of the boom can result in bridge collapse, such as the destruction of the Yi Bin south gate bridge in China. There are also many similar cases of bridge defect or even destruction accidents, and the main cause is the lack of maintenance [2,3].

Bridge management system (BMS) is an integrated computer system to provide decision support throughout design, construction, operation, and maintenance phases. BMS can improve the efficiency of management and decrease redundant costs in dealing with infrastructure management issues [4]. Many countries invested a great deal of resources and efforts to develop efficient BMSs, including Finland [5], Denmark [6], Germany [7], and Japan [8]. In China, Chinese Bridge Management System (CMBS) was developed by the Ministry of Transport of China (MOT) in 1993 and has been widely used in China after continuous improvement. The main functions of CBMS are to master the basic data of highway bridges, supervise the regular inspection, and provide technical support for maintenance [9]. However, there are still many problems during the application of CBMS. Firstly, the information stored in the system is fragmented and isolated. Secondly, it is hard to realize visualization when the maintenance information is separated from the bridge model and makes it difficult to handle for users. Additionally, bridge engineers and managers find it difficult to collaborate with each other through the system.

In recent years, building information modeling (BIM) has become the most flourishing technology in the building industry, and it has been extended to infrastructure engineering [10]. This technology is a new approach and can be used for design, construction, and facilities management of structures, wherein a digital representation of the building process is used to facilitate the exchange and the interoperability of information [11]. The application of BIM brings about cost reduction, quality control, and efficiency improvement throughout the life cycle of the project. A survey by Stanford University in the United States pointed out that BIM can eliminate 40% of extra changes, reduce the contract price by 10% by discovering and resolving conflicts, and shorten the project duration by 7% [12]. With these advantages, BIM has also been adopted by commercial software, such as Autodesk Revit, ArchiCAD, and Allplan [13]. Furthermore, BIM has been successfully implemented in many bridges' designs [14,15] and constructions [16–18]. However, the application in the maintenance phase started relatively late. McGuire and Atadero [19] utilized the BIM to manage the inspection and evaluate information. Abudayyeh and Al-Battaineh [20] adopted the as-built bridge information model for maintenance and management. Some scholars [21–23] also combined the BIM with the traditional management system to improve maintenance efficiency.

In China, however, the lifecycle bridge management system based on the BIM technology is still under development. This paper introduces the development of a BIM-based BMS, which integrates the geographic information system (GIS), develops the necessary Industry Foundation Classes (IFC) and International Framework for Dictionaries (IFD) standards, which are still in blank in China, implements the Web-BIM oriented management, and realizes the collaborative work among all related users. This BIM-based bridge management system is supposed to improve the efficiency of maintenance management in order to satisfy the keen demand of maintenance management from substantive bridges in China.

2. System Goal and Requirement

With the huge inventory of Chinese bridges, maintenance of these bridges becomes especially important and challenging. An informationalized management system is helpful and necessary. Many BMS have been developed, and fairly good practical results have been achieved, such as GIS-based BMS [24], visual inspection data-based BMS [25], and consecutive condition assessments-based BMS [26]. In this paper, we introduce a BIM-based BMS. The main target is to increase the management efficiency, make it visualized and easy to use, and provide a collaborative management platform for all people involved in the projects. To realize it, the bridge management system should:

(1) Integrate the BIM with the GIS. As a node of the transportation network, geographic information is crucial to the bridge. Moreover, our management is not only focused on a single bridge but the bridges at a certain transportation line or a certain area. Information of the group of the bridges helps to grasp the overall condition of the transportation and gives us guidance in maintenance.

(2) Realize the Web operable BIM system. Confining the users to computers is no longer welcome. The system should be operable at the Web, reducing the dependency on heavy BIM software, and, most importantly, making the portable device based operation possible.

(3) Have a unified database. The database of the system should support different kinds of data. Information should have the same coding rule so that all data can be searched, retrieved, and analyzed. Interaction between systems can also be implemented.

(4) Realize the function of structural inspection and monitoring. Condition of the bridges reveals the safety of the bridges. Structural inspection, health monitoring, as well as the performance evaluation are therefore the most important in the management system.

(5) Realize the collaborative management. By collaborating, all people with different jobs work together; collaborative management is a key issue to increase the efficiency of the management.

(6) Achieve the multi-scale visualization during the management. Visualization is a significant advantage of the BIM-based management system. It makes the system easy to understand and easy to grasp for people.

As a popular informationalized method, the term BIM itself has several definitions, such as a product, a collaborative process, and a facility life cycle management requirement [27]. It can also be regarded as a combination of the concepts of building information modeling and building information management. Three-dimensional representation, information integration, and digital models are the three significant features of BIM. The collaborative management system utilizes the advantages of BIM to provide better service for bridge maintenance, which makes up for deficiencies of the traditional management system. Considering the merit of the BIM technology, a BIM-based bridge management system is developed in this paper to improve the maintenance efficiency and release the management pressure of large numbers of existing bridges.

3. Key Points of the Proposed Bridge Management System

3.1. BIM Standard Extension

As mentioned earlier, some relevant standards are still incomplete or blank in the bridge industry in China. The lack of uniform BIM standards has become one of the main obstacles to the application and the development of BIM [28]. Compared with other industries, civil engineering projects usually include many stages—planning, design, and construction—and involve many participants. Different types of BIM software are adopted in the process. Without a unified BIM standard, it is hard to achieve information sharing among different stages and software, which leads to a large number of repetitive works. It is necessary to establish information storage and exchange standard. Otherwise, the values of BIM are not realized fully.

3.1.1. IFC Standard Extension

The IFC standard is an open international standard for the expression and the exchange of building product data, supporting data exchange, and sharing throughout the life of a building project. The architecture of the IFC standard consists of four levels; from the bottom to the top, these are resource layer, core layer, interoperability layer, and domain layer.

At present, the IFC standard is incomplete, and it is hard to cover the whole life-cycle information of a bridge [29]. Thus, it is necessary to make an extension by using EXPRESS language based on the existing IFC framework. The bridge member, such as beam and abutment, can directly refer to IfcBuildingElement in the existing IFC framework. However, structural defect, inspection, and evaluation information should be defined by extending the attribute set. Table 1 defines the attribute set of bridge defect information and reinforcement information. Tables 2 and 3 list the specific attribute name and the type of the attribute set. Table 4 lists the attribute definition of the concrete crack. Finally, the bridge maintenance information is attached to the bridge members through IFC relationship entity. In this way, the IFC framework targeting at bridge maintenance is built completely, which provides a uniform standard for information exchange and delivery. Bridge information models can be imported and exported by mainstream BIM software such as Autodesk, Bently, Dassault, etc. Since the unstructured data cannot be directly expressed in the existing IFC standard, it is necessary to extend to encoding the attribute information. Parameter and attribute information sets can be added to unstructured data, including disease information, assessment information, files, pictures, and videos. As shown in Figure 1, the attribute information set combined with the structured information through data sources is compressed together and integrated with IFC parsing. Thus, the entity model of the IFC extension for bridge disease and evaluation is completed, and the structural defect extension rules are formed.

Table 1. Attribute set definition of maintenance information.

Attribute Set Name	Applicable Entity	Applicable Type	Definition
Pset-diseasetype	IfcBuildingElement	UserDefined	Disease type
Pset-reinforcetype	IfcBuildingElement	UserDefined	Reinforcement method

Table 2. Attribute definition of structural defect information.

Attribute Name	Attribute Type	Data Type	Definition
Crack	IfcPropertySingleValue	IfcBoolean	Concrete crack
HungrySpots	IfcPropertySingleValue	IfcBoolean	Voids and pits of concrete
Spalling	IfcPropertySingleValue	IfcBoolean	Concrete spalling
Corrosion	IfcPropertySingleValue	IfcBoolean	Corrosion of steel bar
Deformation	IfcPropertySingleValue	IfcBoolean	Excessive deformation

Table 3. Attribute definition of reinforcement information.

Attribute Name	Attribute Type	Data Type	Definition
ElargeSection	IfcPropertySingleValue	IfcBoolean	Elarging section
BondFRP	IfcPropertySingleValue	IfcBoolean	Bonding FRP
BondSteel	IfcPropertySingleValue	IfcBoolean	Bonding steel plate
ExternalPrestress	IfcPropertySingleValue	IfcBoolean	External prestressing

Table 4. Attribute definition of crack.

Attribute Name	Attribute Type	Data Type	Definition
CrackDirect	IfcPropertySingleValue	Ifclabel	Direction of crack
CrackLength	IfcPropertySingleValue	Ifcreal	Length of crack
CrackWidth	IfcPropertySingleValue	Ifcreal	Width of crack

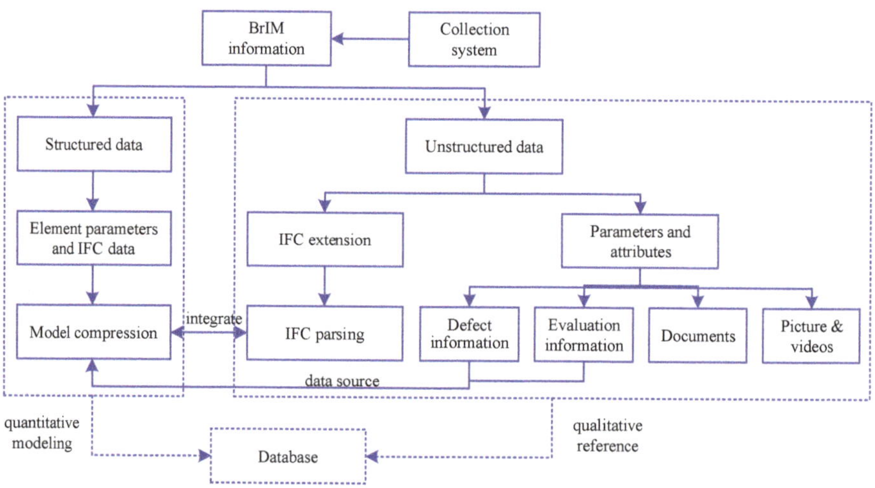

Figure 1. Expression of attribute information using Industry Foundation Classes (IFC) extension.

3.1.2. IFD Standard Extension

The descriptions of information are usually distinguished in different languages and under different situations, which makes information uncertain and inaccurate. The IFD defines a globally unique identifier for each conception. In China, IFD standards have been established in the field of architecture and railway industry [30]. However, the IFD standards in the highway field are still blank in China. This paper proposes the IFD standard for Chinese high way industry and makes a supplement for the highway bridge maintenance.

International Standard Organization (ISO) provides the information classification framework (ISO 12006-2). This framework [31] divides the building information into three categories: construction resources, construction processes, and construction results. Construction results are produced by applying construction resources to the construction process, as Figure 2 shows. Among them, the primary construction resources include products, tools of production, roles of bridge engineers, roles of bridge engineering organizations, information, materials, attributes, etc. Meanwhile, the construction process includes the behavior of bridge engineering, the project phase of bridge engineering, the professional field, etc. The construction results consist of the structures classified by function, morphology, and so on. The classification and the encoding of the bridge maintenance information can refer to or extend from the existing IFD standards. For the cases where the current standards cannot satisfy the needs, information needs to be classified and encoded based on the ISO 12006-2 framework.

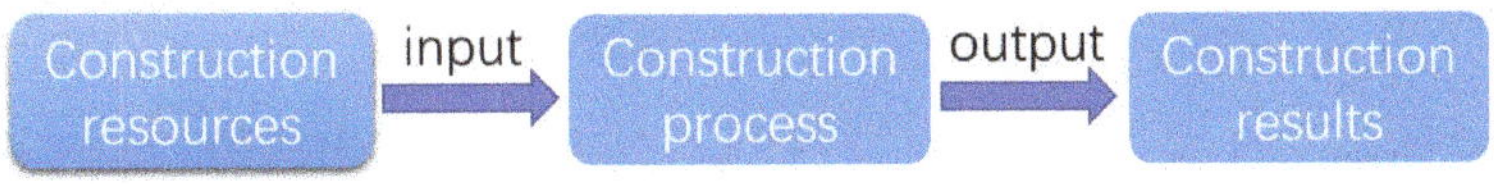

Figure 2. Classification framework.

The numerical method is applied to encode the information. The coding consists of two parts: table name and coding of different levels (each two numbers represent one level), which are separated by a dividing line. A part of coding is to directly quote IFD standards in the field of construction, while another part is to extend the building standards. Coding should independently compile the information. Among them, codes with initial numbers 71, 72, 41, 74, and 78 indicate bridge type, bridge component, attribute, bridge behavior, and bridge characteristics, respectively. The coding principles of the bridge type and the bridge component are shown in Figure 3. Because of the limitation of space, Table 5 only lists a part of the engineering behavior information coding in the bridge maintenance phase. Table 6 gives a part of the structural defect, evaluation, and reinforcement information coding of the bridge. Defect levels, associated maintenance, and repair priorities are defined as well.

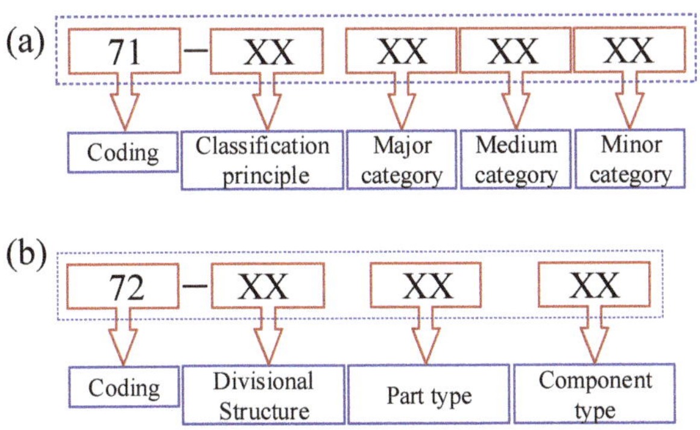

Figure 3. Coding principles using International Framework for Dictionaries (IFD) standard extension: (**a**) bridge type, (**b**) bridge component.

Table 5. Engineering behavior information.

Coding	First Level	Second Level	Third Level	Forth Level
74-10 00 00	Investment			
74-20 00 00	Design			
74-30 00 00	Construction			
74-40 00 00	Maintenance			
...	...	...	...	...
74-40 10 00		Inspection		
74-40 10 10			Daily Inspection	
74-40 10 10 10				Day patrol
74-40 10 10 20				Night patrol

Table 6. Bridge maintenance information.

Coding	First Level	Second Level	Third Level
78-10 00 00	Defect		
78-10 01 00		Crack	
78-10 01 01			Vertical crack
78-10 01 02			Horizontal crack
78-10 01 03			Longitudinal crack
...	...	...	...
78-10 01 04			Oblique crack
78-10 01 05			Bursting crack
78-10 01 06			Circumferential crack
78-10 01 07			Reticulate crack

3.2. *Structural Modeling Method*

In the bridge life-cycle planning, the information model is expected to be established during the design and the construction phases and then delivered to the maintenance phase. However, at present, most existing bridges did not apply the BIM technology in design and construction. Besides, some as-built information models make it hard to satisfy the needs of bridge maintenance due to the restriction of software. As a result, a bridge information model for maintenance needs to be rebuilt in most cases. This part introduces a structural modeling method by using Autodesk Inventor, taking the Grand Canal Bridge modeling process as an example. The background of Grand Canal Bridge is introduced later in Section 5.

Autodesk Inventor is modeling software that has strong curve design capability, powerful assembly, and information integration functions, which gives it the ability to build the model of the Grand Canal Bridge. Since the workload of modeling is large, it is necessary to unify the basic information such as coordinate system, unit, and storage path before building the model. The specific modeling steps are as follows:

(1) Establish component models. The Grand Canal Bridge is divided by structural components such as bridge decks, abutments, pile foundations, etc., and each component is modeled separately by creating a "part" in the Autodesk Inventor. However, to build all the components one by one is inefficient. The software provides a parametric method, which can be used for building different types of components effectively.

(2) Assembly of components. The separate components need to be assembled to form the bridge model. For components with geometric constraints, users can assemble them directly. For component types without direct constraint relationships, the reference surfaces and the reference points need to be created to help determine the location of the components. The assembly details can be seen in Figure 4.

(3) Level adjustment. The requirements of the model for maintenance are somewhat different from design and construction. The model should have a hierarchical characteristic in the maintenance phase because the structural safety assessment is usually carried out upwards from the bottom layer to the top layer. Engineers could adopt the degrading function of the Autodesk Inventor to adjust

the hierarchical relationship among the components. The final model of the Beijing-Hangzhou Grand Canal Bridge is shown in Figure 5.

(4) Information integration. After completing the model building, the information of design and construction need to be attached to the bridge model. Autodesk Inventor provides common information addition. If the types of information are not provided, users also can make a customized expansion. The Bill of Material (BOM) table can be used for adding information in bulk.

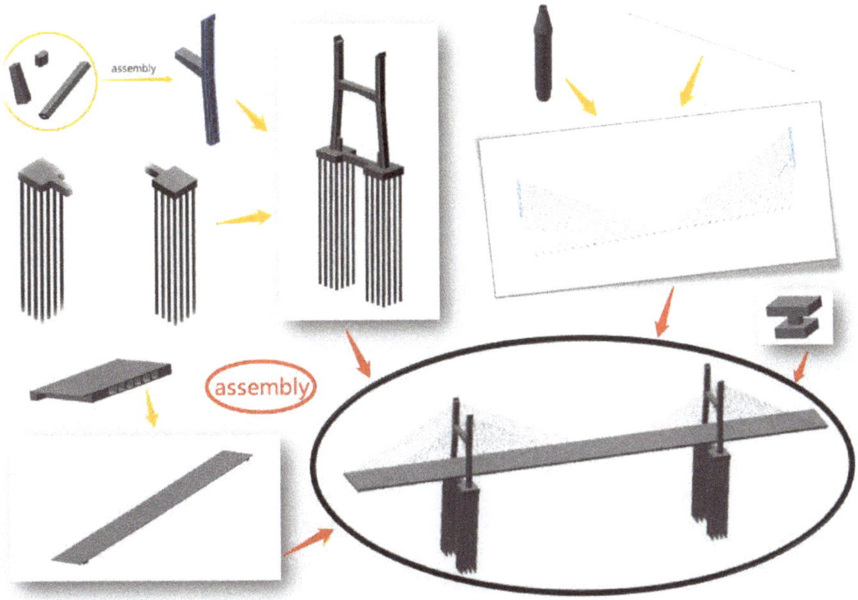

Figure 4. Modeling process.

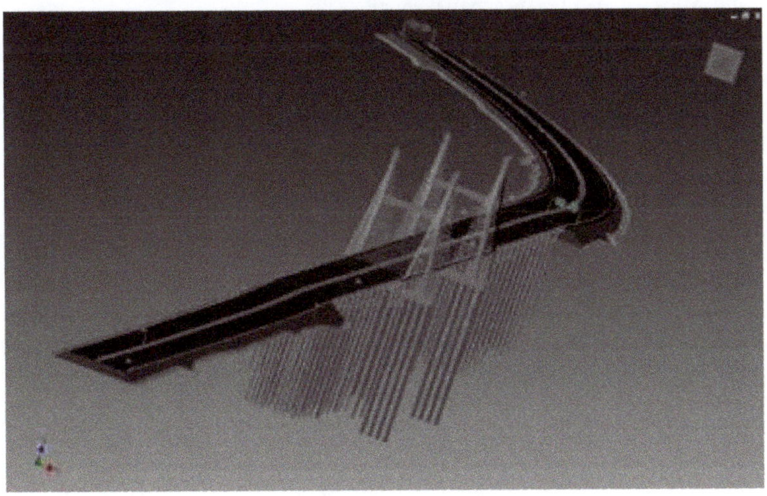

Figure 5. The Grand Canal Bridge model.

3.3. *Web-BIM Oriented Management*

Traditional bridge management systems are usually based on computers. Users need to install all the software at computers and finish all the work in front of the computers. However, from another point of view, users are confined to those computers. Nowadays, portable devices, such as pads and cell phones, give significant convenience and flexibility to their management works. However, portable devices do not have powerful enough central processing units (CPU) or sufficient memory and cannot support heavy BIM software such as the Revit. Also, expensive BIM software needs to be installed at each user's computer. Data such as inspection data and monitoring data usually need to be copied into some memory devices and then inputted into the system. All of these aspects make the management system expensive and inefficient. Web-BIM techniques have been developed and applied in the building management system but seldom in the bridge management system. Here, we propose the Web-BIM based management system with the lightweight techniques, thus the system can be operated at Web and could adapt to different users.

To develop the Web-BIM system, the structural model is required to be built using BIM software such as Revit, and then the model is exported and transferred to a certain file with OBJ format. Such OBJ files can then be uploaded to the server and used as a calling file to the webpage. By running the JavaScript software and calling the WebGL program, the structural model can then be transferred to a 3D web file and can be displayed on internet browsers.

For a bridge structure, especially the large-scale bridges, the converted structural model is often very huge and can hardly be displayed and operated smoothly. The lightweight model technique is necessary to reduce the data volume of the model. To realize that, a three-step lightweight approach is implemented as follows:

(1) Establish a hybrid space index. Separate the model into different scales and build the index. Trim the model to show only the information within the visual field.

(2) Establish the Levels of Detail (LOD) to the model data. Erase the information related to the unnoticeable details.

(3) Reduce the rendering batch numbers. Incorporate the objects with the same texture and render them together so that rendering efficiency can be significantly increased.

4. Design of BIM-based maintenance management system

The development of the bridge management system requires the cooperation of different specialties. The system takes management as the basic requirement, information technology as the means, and bridge professional knowledge as the basis. This part introduces the detailed process of the system design.

4.1. *Computer Network Design*

As Figure 6 shows, C/S (client/server) models and B/S (browser/server) models are currently mainstream network structures. The C/S model belongs to a distributed application structure that can partition tasks or workloads between the providers of a resource or service, called servers, and service requesters are called clients. Often, there is communication between clients and servers over a computer network on separate hardware, but both the client and the server may reside in the same system. The device utilization is highly efficient, and the security can be ensured. However, this structure is less scalable and is generally limited to local networks. The cost for the system's maintenance and upgrade is high. The B/S model developed along with the rise of the Internet. This model centralizes the core part of the system's functional implementation to the server, which greatly simplifies the client computer load and reduces the cost of system maintenance and up-gradation. Users only need a web browser to get access to the system without installing a client. B/S model's security is relatively poor, which is why some measures such as permission control are needed.

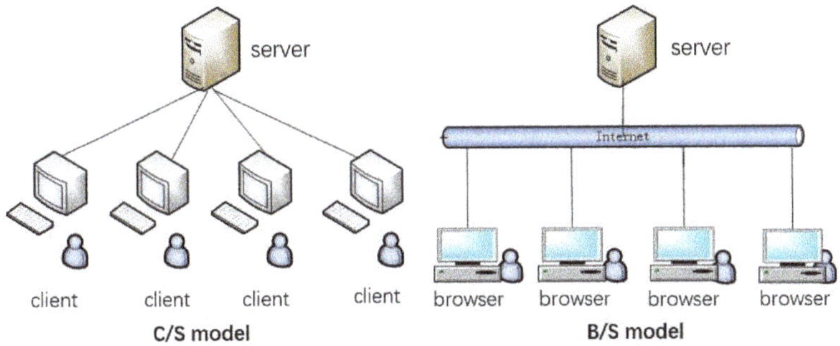

Figure 6. Computer network.

In the bridge maintenance project, a large number of stakeholders are involved. The management system usually has many users, including investors, managers, and engineers. They are likely to get access to the system from different locations and with different access methods [such as a local area network (LAN), a wide area network (WAN), and Internet/Intranet]. Taking the convenience and the cost into consideration, the B/S structure is in accordance with the Web-BIM oriented management and therefore more applicable to the BIM-based management system.

4.2. System Framework

The BIM-based system mainly includes four parts: data collection system, data center, model layer, and evaluation system. The framework of the system is shown in Figure 7, which displays the connections and the relationship between the different modules of the system.

The data center plays a significant role in the system, which consists of four layers: cloud resource layer, access layer, storage layer, and application layer. The cloud resource layer provides computing services by renting existing mature cloud computing resources for implementing various functions of the data center. The access layer provides a unified interface for the import of different types of data. The storage layer provides centralized storage of all bridge management-related data. The application layer includes the implementation of various user-oriented functions such as data query, data management, and database management tools.

The data center obtains static and dynamic data of bridges from scattered data resources through the data acquisition system of bridge maintenance. The dynamic data collection system includes two parts: the front-end collection program and the background management program. The front-end collection program is mainly the mobile APP software, which is responsible for real-time collection, input of maintenance data, and uploading information to the background management program. During the process, the background management program is responsible for managing the collected data and generating reports.

The bridge automatic analysis and evaluation system relies on the data center to obtain various types of data required for bridge condition assessment, bridge data statistics, and in-depth analysis. Through the calculation and the analysis of relevant models and algorithms, output evaluation, statistics, and analysis results are obtained and submitted to the data center. Also, the evaluation results would be presented in the bridge model.

The model layer is one of the important distinguishing points of the BIM-based bridge maintenance management system compared with the traditional systems. Because the system is based on the three-dimensional model for operations, it can make many maintenance services more intuitive and vivid. Considering that the size of the model is huge and takes up too much memory, the model is then transformed into a lightweight model to better display in the portable terminals. The model layer correlates the information such as inspection, evaluation, and maintenance reinforcement in the database with the component of the bridge, thus different information can be displayed through the model. Therefore, the model layer part effectively connects the data layer part and the functional layer part together and plays a central role in the whole system framework.

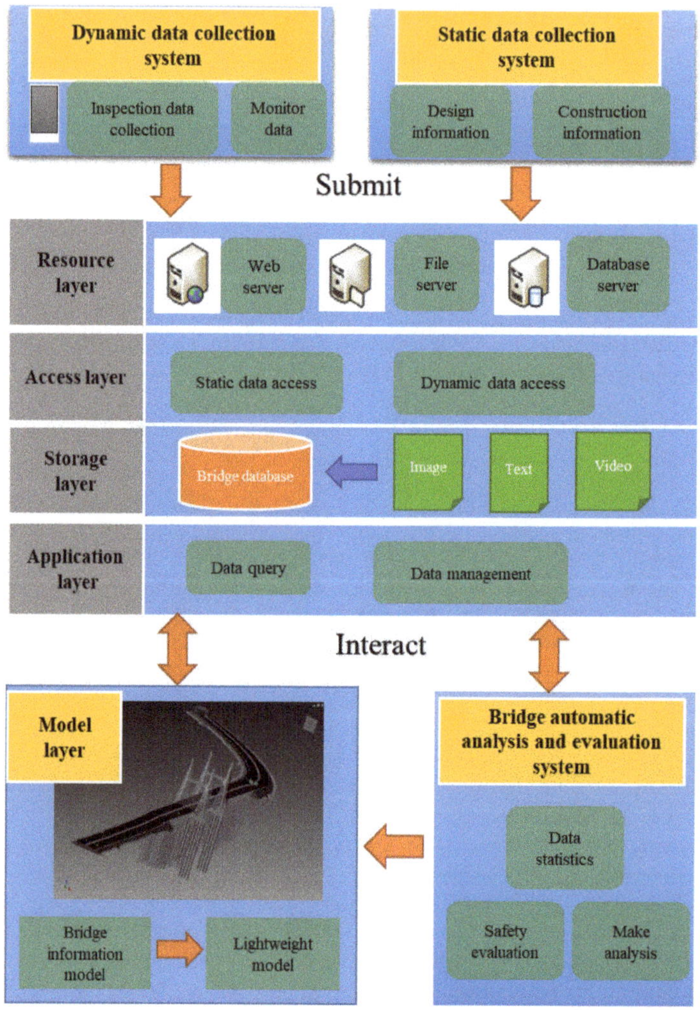

Figure 7. System framework.

5. The application on the Grand Canal Bridge

5.1. Project Overview

The Beijing–Hangzhou Grand Canal is the longest artificial river in the world, starting at Beijing, passing through Tianjin and provinces of Hebei, Shandong, Jiangsu, and Zhejiang, and finally arriving in the city of Hangzhou. The total length of the Grand Canal reaches 1776 kilometers. The Beijing-Hangzhou Grand Canal Bridge is located at the south of Yangzhou, Ning yang Expressway. As is shown in Figure 8, the main bridge across the Grand Canal is a double-tower semi-floating system concrete cable-stayed bridge, the length of which is 464 meters (108 m + 248 m + 108 m). On each side of the main bridge is 11-span (6 × 30 m + 5 × 30 m) prefabricated, prestressed concrete approach bridge.

(a) (b)

Figure 8. The Grand Canal Bridge: (**a**) bridge deck; (**b**) whole bridge.

5.2. Implementation in Grand Canal Bridge

Beijing-Hangzhou Grand Canal Bridge is classified in the second level according to its present structural condition. Many shrinkage cracks have appeared on the cable-stayed bridge tower and the main beams of the bridge. Some protective covers of the cable anchorages have deformed severely. Moreover, there are a lot of longitudinal cracks on the box girder of the approach bridge and transverse cracks on the wet joints. Some of the laminated rubber bearings have been damaged.

In order to realize the effective management and maintenance of the Beijing-Hangzhou Grand Canal Bridge, the BIM-based management system was built. Firstly, the bridge model was established using Inventor, and a special lightweight processing that could significantly reduce the data volume was implemented to form the model layer. Then, the lightweight model was connected to the GIS system, which showed the geographic information of the bridge, as Figure 9 shows. Also, it is associated with the bridge maintenance information database. The database is an important foundation for further evaluation and decision-making, which are the core parts of the function layer.

Figure 9. System homepage.

With the Web-BIM oriented approach, the stakeholders of the bridge can get access to the system through the Internet. The homepage of the system includes three parts: function modules, bridge model, and static information, as Figure 9 shows.

5.2.1. Static information module

Static information management in this BMS platform was established and is shown in Figure 10. This system accumulates the scattered information in the design and the construction phases to formulate the "bridge life card". The results indicate that users could choose the members in the bridge model to query corresponding static information such as design drawings, design parameters, material usage, etc. Moreover, the model tree is one type of dendrogram based on the hierarchy characteristic of the model, which is provided for accurate localization of members in the model. As Figure 11 shows, users can unfold the dendrogram to click the pile foundation of thirteenth span, which was highlighted in the model automatically.

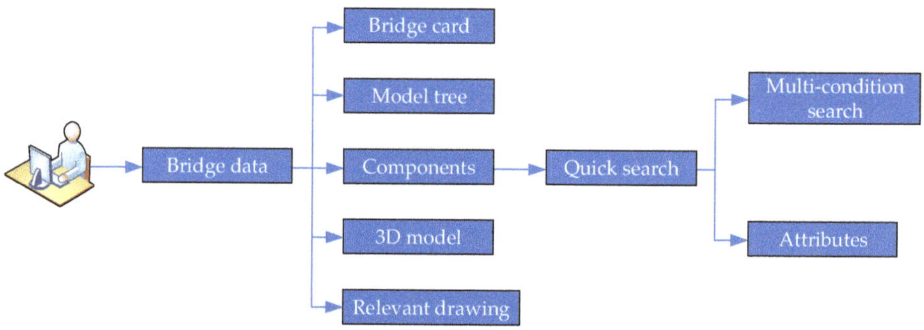

Figure 10. Visualization management of static information.

Figure 11. Pile foundation of thirteenth span in the model tree.

5.2.2. Inspection and evaluation module

Inspection and evaluation have great significance for BMS. Based on this, we firstly designed an inspection workflow to show how the bridge managers create and manage the inspection projects on the website. As Figure 12 shows, the inspectors could log into the system through a mobile application at the spots, collect all the structural defect information of the bridge, and then submit to the system. Large bridges have high numbers of members; inspectors can localize the member quickly by choosing it on the lightweight bridge model. In order to improve the accuracy of defect information description, the system provides a standardized template according to the statistics and the classification of common defects. After inspectors submit the defect information, managers can check it in the system, as Figure 13 shows.

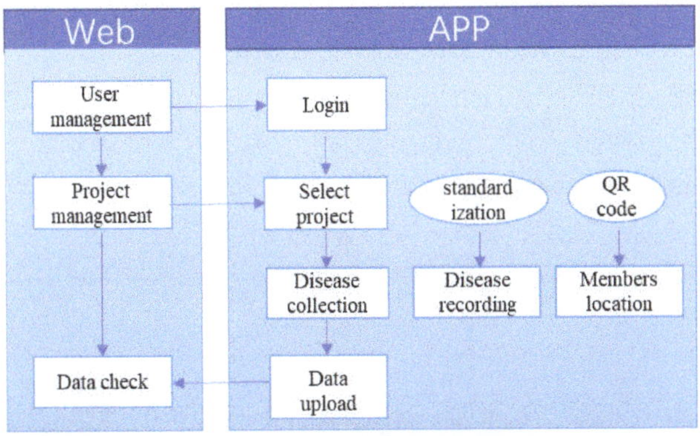

Figure 12. Inspection workflow.

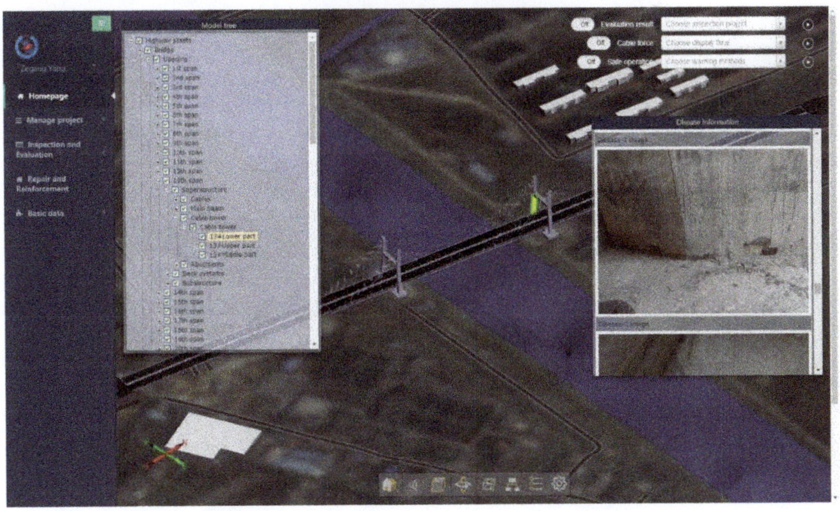

Figure 13. Structural defect information in inspection module.

Based on the defect information collected by the mobile terminal, the evaluation system can make automatic analysis and calculate the scores of the members according to the bridge maintenance standard. Then, the bridge information model displays different colors on the members to intuitively describe the technical condition of different levels of the whole bridge. Figure 14 shows the evaluation results of the Grand Canal Bridge. The members in green color stayed in a safe state, and the members in yellow suffered light damage. The red color means the members have been damaged severely, and corresponding measurement should be considered.

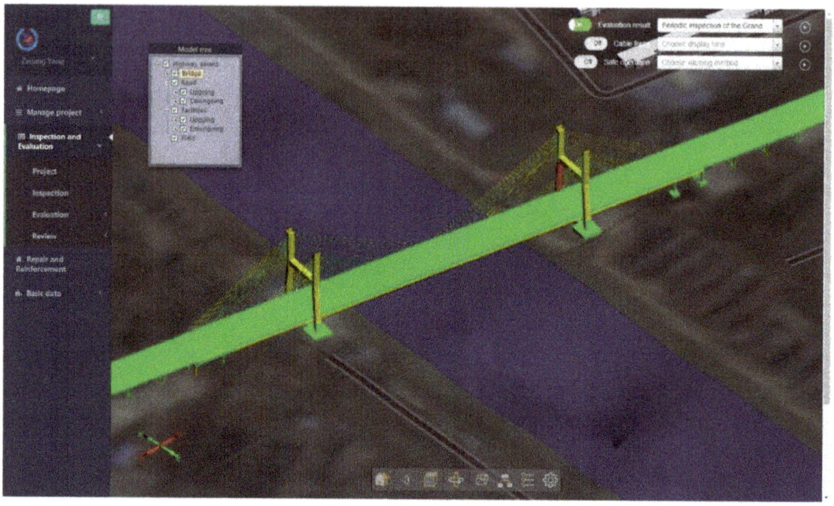

Figure 14. Evaluation module.

Repair and reinforcement information of the bridge is stored in the system and can be searched by users. The defects of the bridge, such as cracking, usually develop as the time grows. When the defect develops to a certain extent and threatens the structure safety, the system automatically sends text messages to bridge managers to remind them to create a repair and reinforcement project. Besides, the system regularly generates statistics, graphs, and tables according to inspection and monitoring information. Based on these functions, the BIM-based system helps to make auxiliary maintenance decisions. As shown in Figure 15, users can view the relevant information on the repair plan in this module. After this reinforcement and repair, one can determine whether the maintenance is qualified or not according to the actual situation. The maintenance can be displayed through the model, and the repair results can be check by the managers, as shown in Figure 16. Based on this, the detailed information of the location and the effects of this reinforcement can be obtained.

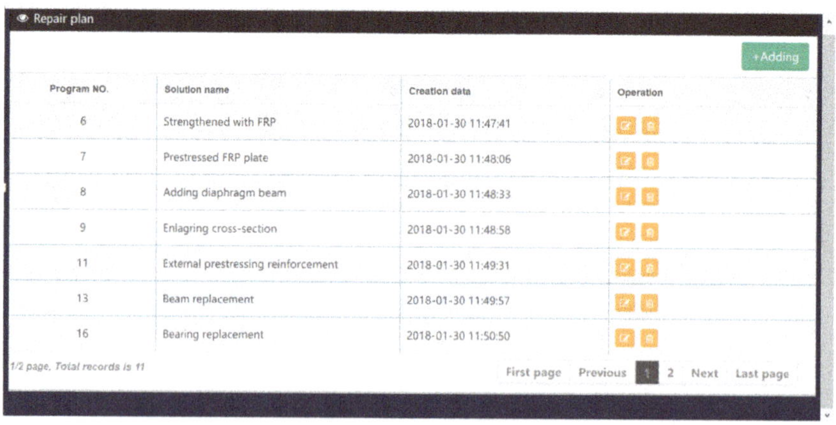

Repair plan

+Adding

Program NO.	Solution name	Creation data	Operation
6	Strengthened with FRP	2018-01-30 11:47:41	
7	Prestressed FRP plate	2018-01-30 11:48:06	
8	Adding diaphragm beam	2018-01-30 11:48:33	
9	Enlagring cross-section	2018-01-30 11:48:58	
11	External prestressing reinforcement	2018-01-30 11:49:31	
13	Beam replacement	2018-01-30 11:49:57	
16	Bearing replacement	2018-01-30 11:50:50	

1/2 page, Total records is 11 — First page Previous 1 2 Next Last page

Figure 15. The repair plan of bridge components.

Figure 16. The information of component maintenance in repair and reinforcement module.

5.3. Results and system benefits

After this maintenance system was built, it was applied to the company in charge of the maintenance of the Grand Canal Bridge. Feedback was quite positive. Web-BIM oriented management mode allows all users easy access to the system. Portable devices have also been applied in the bridge inspection procedure. By scanning the tag at a component, the component can be identified, and related structural defects can be directly reported by text description, picture, or report via the portable device on site. The established BMS also provides a good platform upon which all users can collaboratively work together. Owners, engineers, economists, managers, and other related staff can share their data, release/receive tasks, track the progress, and even communicate with each other. Multi-scale visualization of the bridge model helps to understand the structural condition easily. Structural analysis and diagnosis algorithms can be embedded in the system and make the system smart. All of these make the maintenance activity very convenient and efficient. It will be helpful for the maintenance of many bridges in China, and its application is promising.

6. Conclusions

The number of bridges has increased significantly within the last three decades in China. The large number of the bridges, however, brings great pressure to the owners, the industry, and even the government for their maintenance. An effective and efficient bridge management system has become indispensable to meet the massive demand of bridge maintenance.

In this paper, a bridge management system based on the BIM technology was proposed, developed, and introduced. Necessary IFC and IFD standards were proposed as supplements according to the actual needs of bridge maintenance, filling the blank in BIM standards of the bridge industry of China. Web-BIM oriented management was proposed, and portable devices were introduced to the system as well as the maintenance activity. The bridge modeling method was developed to build the proper information model for bridges. Lightweight methods were also introduced to reduce the volume of the model and make the system run smoothly. The system framework and functions were designed. A BMS for a real long-span cable-stayed bridge was also built and established, which incorporates the GIS and satisfies the main maintenance functions, such as information management, bridge inspection, condition evaluation, repair and reinforcement, multi-scale visualization, and collaborative management.

The proposed BIM-based bridge management system could improve management efficiency and satisfy the need for maintenance of substantive bridges in China. The future of such a management system is promising. However, at present, the BIM model cannot be connected to the finite element model (FEM), which strongly limits the structural analysis ability. Also, data in the system are not mined deep enough. Integration of the BIM model and the FEM model as well as the deep mining of data using big data technology will make the system much more powerful and useful and could provide a better platform for efficient bridge management.

Author Contributions: Conceptualization, C.W. and Y.D.; methodology, C.W. and Y.D.; software, Z.X., Z.Y. and Y.X.; formal analysis, Z.Z. and S.L.; investigation, Z.Z. and S.L.; data collection, Z.X. and F.Y.; writing—original draft preparation, Z.Z. and S.L.; writing—review and editing, C.W. and Z.Z.; funding acquisition, C.W. and Y.D.; supervision, C.W.

Funding: The research was supported by the National Key Research and Development Program of China (2017YFC0806001), the National Natural Science Foundation of China (No. 51578140), the Postgraduate Research & Practice Innovation Program of Jiangsu Province (No. SJCX18_0027).

Conflicts of Interest: There is no conflict of interest.

References

1. Zhang, L.; Zhou, G.; Han, Y.; Lin, H.; Wu, Y. Application of Internet of Things Technology and Convolutional Neural Network Model in Bridge Crack Detection. *IEEE Access* **2018**, *6*, 39442–39451. [CrossRef]
2. Qi, J.; Issa, R.R.A.; Olbina, S.; Hinze, J. Use of Building Information Modeling in Design to Prevent Construction Worker Falls. *J. Comput. Civ. Eng.* **2014**, *28*, A4014008. [CrossRef]
3. Gibb, A.; Haslam, R.; Gyi, D.; Hide, S.; Duff, R. *What Causes Accidents?* Institution of Civil Engineers: London, UK, 2006; Volume 159, pp. 46–50.
4. Yin, Z.H.; Li, Y.F.; Guo, J.; Li, Y. Integration Research and Design of the Bridge Maintenance Management System. *Procedia Eng.* **2011**, *15*, 5429–5434. [CrossRef]
5. Söderqvist, M.-K.; Veijola, M. The Finnish Bridge Management System. *Struct. Eng. Int.* **1998**, *8*, 315–319. [CrossRef]
6. Lauridsen, J.; Bjerrum, J.; Andersen, N.H.; Lassen, B. Creating a Bridge Management System. *Struct. Eng. Int.* **1998**, *8*, 216–220. [CrossRef]
7. Haardt, P. Development of a Bridge Management System for the German Highway Network. In Proceedings of the First International Conference on Bridge Maintenance, Safety and Management. IABMAS, Barcelona, Spain, 14–17 July 2002.
8. Miyamoto, A.; Kawamura, K.; Nakamura, H. Bridge Management System and Maintenance Optimization for Existing Bridges. *Comput. Civ. Infrastruct. Eng.* **2000**, *15*, 45–55. [CrossRef]
9. Dai, K.; Smith, B.H.; Chen, S.E.; Sun, L. Comparative study of bridge management programmes and practices in the USA and China. *Struct. Infrastruct. Eng.* **2014**, *10*, 577–588. [CrossRef]
10. Costin, A.; Adibfar, A.; Hu, H.; Chen, S.S. Building Information Modeling (BIM) for transportation Infrastructure—Literature review, applications, challenges, and recommendations. *Autom. Constr.* **2018**, *94*, 257–281. [CrossRef]
11. Eastman, C.; Teicholz, P.; Sacks, R.; Liston, K. *BIM Handbook: A Guide to Building Information Modeling For Owners, Managers, Designers, Engineers and Contractors*; Wiley: Hoboken, NJ, USA, 2011.
12. Gao, J.; Fischer, M. *Framework and Case Studies Comparing Implementations and Impacts of 3D/4D Modeling across Projects*; Stanford University: Stanford, CA, USA, 2008.
13. Martínez-Aires, M.D.; López-Alonso, M.; Martínez-Rojas, M. Building information modeling and safety management: A systematic review. *Saf. Sci.* **2018**, *101*, 11–18. [CrossRef]
14. Shin, H.; Lee, H.; Oh, S.; Chen, J. Analysis and Design of Reinforced Concrete Bridge Column Based on BIM. *Procedia Eng.* **2011**, *14*, 2160–2163. [CrossRef]
15. Liu, W.; Guo, H.; Li, H.; Li, Y. Retracted: Using BIM to Improve the Design and Construction of Bridge Projects: A Case Study of a Long-Span Steel-Box Arch Bridge Project. *Int. J. Adv. Robot. Syst.* **2014**, *11*, 125. [CrossRef]
16. Lee, K.M.; Lee, Y.B.; Shim, C.S.; Park, K.L.; Shim, C. Bridge information models for construction of a concrete box-girder bridge. *Struct. Infrastruct. Eng.* **2012**, *8*, 687–703. [CrossRef]
17. Shim, C.; Yun, N.; Song, H.; Shim, C. Application of 3D Bridge Information Modeling to Design and Construction of Bridges. *Procedia Eng.* **2011**, *14*, 95–99. [CrossRef]

18. Fanning, B.; Clevenger, C.M.; Ozbek, M.E.; Mahmoud, H. Implementing BIM on infrastructure: Comparison of two bridge construction projects. *Pract. Period. Struct. Des. Constr.* **2014**, *20*, 04014044. [CrossRef]
19. McGuire, B.; Atadero, R.; Clevenger, C.; Ozbek, M. Bridge Information Modeling for Inspection and Evaluation. *J. Bridg. Eng.* **2016**, *21*, 04015076. [CrossRef]
20. Abudayyeh, O.; Al-Battaineh, H.T. As-Built Information Model for Bridge Maintenance. *J. Comput. Civ. Eng.* **2003**, *17*, 105–112. [CrossRef]
21. Marzouk, M.M.; Hisham, M. Bridge Information Modeling in Sustainable bridge Management. In Proceedings of the Integrating Sustainability Practices in the Construction Industry, Kansas City, MI, USA, 23–25 March 2011; pp. 457–466.
22. Okasha, N.M.; Frangopol, D.M. Computational platform for the integrated life-cycle management of highway bridges. *Eng. Struct.* **2011**, *33*, 2145–2153. [CrossRef]
23. Shim, C.S.; Kang, H.R.; Dang, N.S.; Lee, D.K. Development of BIM-based bridge maintenance system for cable-stayed bridges. *Smart Struct. Syst.* **2017**, *20*, 697–708.
24. She, T.H.; Sarshar, M. A Geographic Information System (GIS)-Based Bridge Management System. *Comput. Civ. Infrastruct. Eng.* **1999**, *14*, 417–427. [CrossRef]
25. Estes, A.C.; Frangopol, D.M. Updating Bridge Reliability Based on Bridge Management Systems Visual Inspection Results. *J. Bridg. Eng.* **2003**, *8*, 374–382. [CrossRef]
26. Roelfstra, G.; Hajdin, R.; Adey, B.; Brühwiler, E. Condition Evolution in Bridge Management Systems and Corrosion-Induced Deterioration. *J. Bridg. Eng.* **2004**, *9*, 268–277. [CrossRef]
27. NIBS. *United States National Building Information Modeling Standard Version 1—Part 1: Overview, Principles, and Methodologies, Final Report*; National Institutes of Building Sciences: Washington, DC, USA, 2007.
28. Ma, Z.L.; Wei, Z.H.; Wu, S.; Zhe, L. Application and extension of the IFC standard in construction cost estimating for tendering in China. *Autom. Constr.* **2011**, *20*, 196–204.
29. Laakso, M.; Kiviniemi, A.O. The IFC standard: A review of history, development, and standardization, information technology. *ITcon* **2012**, *17*, 134–161.
30. Li, Z.; Man, Q. Research Front and Review of BIM Based on China Knowledge Resource Integrated Database. In Proceedings of the International Conference on Construction and Real Estate Management, Kunming, China, 27–28 September 2014; pp. 911–921.
31. ISO 120096-2. Building Construction-Organization of Information about Construction Works-Part2: Framework for Classification of Information. 2015. Available online: https://www.sis.se/api/document/preview/899471/ (accessed on 17 August 2019).

sustainability

MDPI

Article

Promoting Owners' BIM Adoption Behaviors to Achieve Sustainable Project Management

Hongping Yuan [1], Yu Yang [2,*] and Xiaolong Xue [1]

1 School of Management, Guangzhou University, Guangzhou 510006, China
2 School of Economics and Management, Southwest Jiaotong University, Chengdu 610031, China
* Correspondence: yangyu2017@my.swjtu.edu.cn

Received: 20 June 2019; Accepted: 12 July 2019; Published: 18 July 2019

Abstract: Although building information modeling (BIM) has a promising future in the architecture, engineering and construction industry, its wider adoption and implementation is desired. Grounded with a technology-organization-environment (TOE) framework and the theory of technology acceptance model (TAM), this study extracted "social influence", "organizational support", "BIM technical features", and "government BIM policies" as four key external antecedents—in reference to the particular BIM practices in China—and proposed a model to predict project owners' BIM adoption behaviors. To test the proposed model, structural equation modeling (SEM) analysis was applied for configuration analyses on a sample of 188 project owners from the Chinese construction industry. Results show that BIM technical features, and government BIM policies have positive effects on perceived usefulness, but social influence and organizational support have no significant influence on perceived usefulness. Furthermore, both social influence and BIM technical features have positive effects on perceived ease of use, while organizational support and government BIM policies have no significant influence on perceived ease of use. Attitude plays a significant intermediary role among perceived usefulness, perceived ease of use and behavior intention. Additionally, attitude significantly affects behavior intention, and behavior intention can also affect BIM adoption behavior. This study is the first attempt to investigate project owners' behaviors toward BIM adoption and the findings are expected to provide a better understanding of the essential elements of project owners' BIM adoption behaviors and guide industry practitioners in developing proper strategies to achieve more effective BIM implementation.

Keywords: building information modeling; project owner; attitude; behavior; technology acceptance model

1. Introduction

In the last decade, with the rise of information technologies (ITs), a paradigm shift of industrial informatization has translated into a critical national strategy [1–3]. As a pillar of the domestic economy, the architecture, construction and engineering (AEC) industry in China is on the cusp of transition from an extensive and high-consumption pattern to a new one driven by high efficiency, sustainability and informatization. According to Eastman et al. [4], BIM is "a new approach to design, construction, and facilities management, in which a digital representation of the building process (is used) to facilitate the exchange and interoperability of information in digital format". Therefore, BIM is an innovative paradigm of building information digitalization resorting on certain specific technologies or software which integrates cash flows, information flows, logistics throughout the project lifecycle and reduces the information asymmetry, unforeseen changes and re-doings effectively, turning the utopia of construction visualization to reality. In the meantime, existing literature suggests the integration of environment and economic assessment for the promotion of sustainable construction is considerably important [5,6], and it happens that BIM is an ideal tool which can integrate the assessment of sustainable construction as well as resource management efficiently, such as benefit-cost analysis of economically sustainable design, energy-consumption analysis for a sustainable built environment assessment, architectural information sharing for sustainable facilities management and stakeholder relationship management. Therefore, a wide adoption and application of BIM is bound to strike and even overturn the traditional development patterns of the Chinese AEC industry, embedding these sustainable assessments throughout project lifecycle and thus contributing to sustainable project management.

As such, BIM adoption has become one of the central topics among AEC studies over the past decade. Previous studies have attempted to determine the major factors motivating BIM adoption among project stakeholders, with the aid of various means including questionnaire surveys, interviews, and case studies. For instance, Cao et al. [7] found that the motivation of design units and general contractors in BIM adoption is closely linked to the characteristics of organizational nature and project scale. Meanwhile, evidence from the comparative case study in China and Australia indicated that BIM adoption strategies vary in building construction and infrastructure engineering industries [8]. There are also studies that have compiled and sorted a collection of factors that influence BIM adoption, such as effective leadership and organizational support [9–12], sufficient BIM human resources [13,14], and the availability of information and technology [11,15]. Similarly, the large amount of capital required for BIM adoption and application is also a Gordian knot to be unhitched by potential BIM participants [13,16,17]. Moreover, the lack of universal standards of BIM implementation [18] and the indistinct legal bounds of a series of work outcomes related to BIM (such as the BIM model) make the environment of BIM implementation immature, and this restricts the adoption and application of BIM [14].

The above studies are significant in promoting a wider adoption and application of BIM through addressing major barriers. However, unfortunately, the majority of current literature has ignored a vital stakeholder in AEC projects—the project owner, (e.g., government, real estate developers), who takes overall responsibility for project investment, initiation, construction or even the operation and management of facilities. Holding preponderant advantages in project planning and controlling the entire project lifecycle, the project owner could enhance project performance by requiring and driving other stakeholders (such as architects, general contractors and so on) to get involved in BIM adoption and implementation. Some implications from prior literature have proven the important role of the project owner in driving BIM adoption. For example, Ling et al. [19] pointed out that the superiority of project owners can effectively promote the application of innovative technologies. In a recent study, Cao et al. [7] also found that project owners' support for BIM application can facilitate better stakeholder cooperation and get more stakeholders engaged in BIM implementation. However, these studies are mainly conducted based on qualitative analysis or mostly focus on identifying factors hindering BIM acceptance and adoption and, thus, fail to reveal the mechanism that drives BIM adoption behavior. In addition, very limited studies have attempted to investigate project owners' BIM adoption behaviors, though some of them have suggested that project owners are critical in promoting BIM acceptance and adoption.

Given the research gap, we believe that understanding why the project owner adopts BIM is an important step in increasing the use of BIM within projects and potentially improving BIM adoption efficiency. The research questions we are attempting to answer are "What factors influence project owners' acceptance of BIM? And how do these factors result in project owners' final BIM adoption behaviors?" To address these research questions, we have developed and tested a model integrating the TOE framework and the theory of the technology acceptance model (TAM) to explain project owners' BIM adoption behaviors.

We also believe that the lack of a theoretical foundation for this stream of research has limited the contributions of previous research and prevented project stakeholders from understanding what makes effective BIM adoption and implementation. It is necessary to understand the effects of these factors hindering BIM adoption and determine the critical path impact on this behavior to develop effective

BIM adoption measures and design practical strategies that can lead to wider BIM application. The present study helps BIM researchers describe how project owners' adoption behavior can be driven and pilot project owners make informed decisions as to what strategies they can use to promote BIM application in their projects and organizations.

The rest of this paper is organized as follows. The next section briefly reviews research on BIM adoption factors, the theory of the TAM and the technology–organization–environment framework. Then, we present research hypotheses and the research model, followed by an introduction of the research method including the instrument development and validation process. After that, we present data analyses and results, and suggest the implications for research and practice as well as limitations of the current research. Finally, we conclude this paper with a brief summary in the conclusions.

2. Literature Review

2.1. *Influencing Factors of BIM Adoption*

BIM has been recognized as a pivotal information technology in the AEC sectors due to its strength of integrating the continuous flows of funds, information and logistics throughout the project lifecycle [4]. Admittedly, the adoption and application of BIM would inevitably make a profound impact on driving the development of informatization in the AEC industry. Thus, BIM has attracted a lot more attention from AEC researchers in recent years. For instance, Gu et al. [20] posited that although BIM develops with promising prospects, both technical and non-technical factors hamper its diffusion. Based on BIM implementation practices in China, Cao et al. [7] found that the BIM competitiveness of construction firms is closely related to the social network structure in which they are located. An empirical study by Son et al. [21] showed that top management support, subjective norms, and technical compatibility are most important in affecting designers' BIM adoption. In addition, qualified employees, efficient leadership, the availability of information, and the complexity of the project itself are also fundamental factors for successful BIM implementation [11,13]. Based on investigations on potential BIM adopters in the UK, Howard et al. [22] suggested that performance expectations do not directly affect the adoption bias of BIM potential adopters, but improving the strategic policy and the incentive mechanism would be a great help for accelerating BIM diffusion. Liu et al. [23] pay more attention to factors of BIM cooperation from the individual, technological and organizational dimensions among design and construction firms. For BIM users, more emphasis was placed on the information quality needed when implementing BIM and related exoteric services, because these factors will directly affect their satisfaction with BIM [24]. Additionally, based on institutional theory, Cao et al. [25] found that the homomorphism from both mandatory and imitation systems would significantly affect the application of BIM at the project level, and the support from project owners would be conducive to the acceptance of BIM in certain circumstances. By conducting a case study based on the theory of innovation diffusion, Gledson et al. [26] identified the inter-organizational factors driving BIM diffusion at the project level and provided exhaustive schemes to address the individual, managerial, environmental, and technological challenges experienced by construction firms in the process of BIM diffusion.

2.2. *Technology Acceptance Model*

Currently, the technology acceptance model proposed by Davis [27] (see Figure 1) has become a classical and parsimonious model, which has been widely used to explain the behavior of information technology adoption or acceptance. In line with the TAM, perceived usefulness and perceived ease of use are regarded as two essentials to explain the use of a technology [27]. According to the theoretical framework of the TAM, an individual's information technology adoption behavior is determined by his/her behavioral intention. Moreover, attitude and perceived usefulness influence his/her behavioral intention of using a technology, which would necessarily in turn affect the actual system usage. As a key construct, attitude is influenced by both perceived usefulness and perceived ease of use. Davis [28] also found that, to some extent, an individual's perceived ease of use of a particular technology will affect his/her perceived usefulness of this technology. In addition, external variables (such as technical features, user intervention, etc.) can indirectly affect user behavior through perceived usefulness and perceived ease of use [28]. Because there is no specific and strict constraint to external variables, the TAM is powerful in explaining user behavior of an information system with high parsimony and conciseness. As such, the TAM has been widely applied to predict various types of technology acceptance behaviors, including Smart Grid [29], virtual reality [30], LNG [31], and transportation [32]. In the context of BIM adoption, the research of Lee et al. [10] was one of the few existing studies that applied the TAM to investigate the BIM adoption of designers, contractors and engineers. However, it still did not approach the crucial role that project owners play in BIM adoption processes.

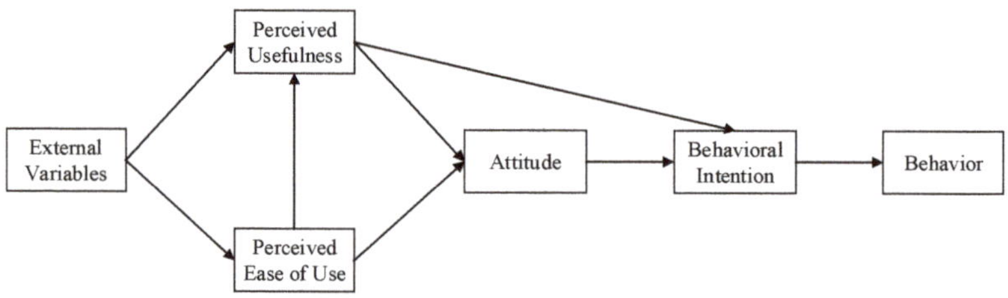

Figure 1. Classical technology acceptance model.

2.3. *Technology–Organization–Environment Framework (TOE)*

The TAM, however, has some limitations when extended beyond the workplace because its fundamental constructs do not fully reflect a variety of the user task environments and constraints. Furthermore, Legris et al. [33] also suggested that the TAM is a useful model but needs to be integrated into a broader model that includes variables related to both human and social factors. To take these limitations of the TAM theory into account, in this present study, we incorporated the technology–organization–environment framework (TOE) (see Figure 2) as the theoretical foundation to clustering various impacts on project owners' BIM adoption. This framework describes factors that influence technology adoption and its likelihood, and the process by which a firm adopts and implements technological innovations is jointly influenced by the technological, organizational, and environmental contexts.

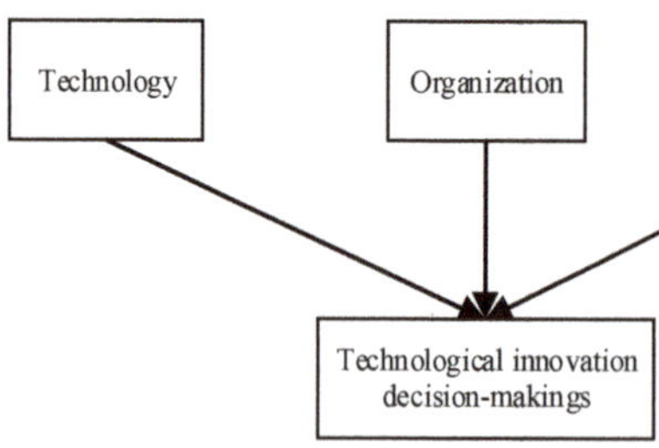

Figure 2. Technology–organization–environment framework.

Since its origination, the TOE framework has aroused increasing attention and been applied in the elements and factors studies of technology innovation among research fields such as tourism, manufacturing (3D printing, RFID, etc.), business (electronic data interchange, customer relationship management, etc.), and project management [34–39]. As indicated by previous studies, the TOE framework is supported by an abundance of empirical results, so it offers a solid foundation to unravel the conundrum behind project owners' BIM adoption decisions.

3. Theory and Hypotheses

In this study, we extend the classical TAM by identifying "social influence", "organizational support", "BIM features", and "government BIM policies" based on the TOE framework as external variables and propose a conceptual model to predict project owners' BIM adoption behavior (see Figure 3). The model is explained with detailed research hypotheses in the following section.

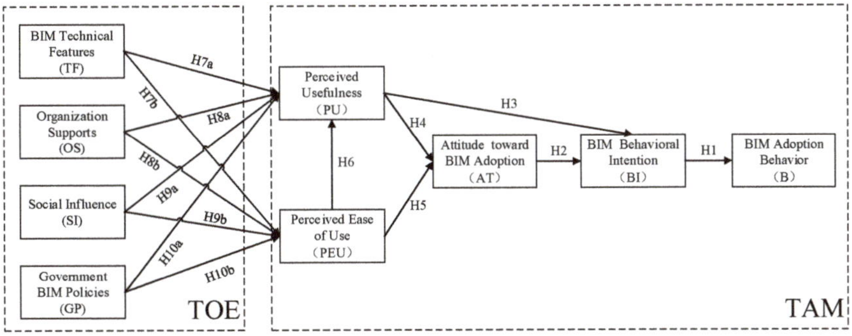

Figure 3. A technology–organization–environment (TOE)- and technology acceptance model (TAM)-based model of project owners' building information modeling (BIM) adoption behavior.

3.1. BIM Behavioral Intention and BIM Adoption Behavior

Behavioral intention is defined as "an indication of an individual's readiness to perform a given behavior". Therefore, behavioral intention is assumed to be an immediate antecedent of behavior [40]. Based on findings from case studies, Arayici et al. [41] found that the rapid promotion of BIM in the UK is benefitted precisely from the user's adoption decision, which leads to wide BIM adoption. Accordingly, we propose the following hypothesis:

H1. *Behavioral intention will have a positive influence on BIM adoption behavior.*

3.2. Attitude toward BIM and BIM Behavioral Intention

Attitude refers to one's subjective positive or negative judgment of a technology. Previous studies have shown that attitude has a certain influence on behavioral intention. Through investigating American consumers' behavior of car buying, Etter [42] found that attitudes can significantly affect purchase intentions. In addition, in a Taiwanese study focusing on online shopping, Wu [43] also found that attitudes directly influence purchasing decisions. Hereby, we propose the following hypothesis:

H2. *Attitude toward BIM adoption will positively affect the BIM behavioral intention.*

3.3. Perceived Usefulness and BIM Behavioral Intention

Perceived usefulness refers to the degree to which a person believes that using a particular technology would enhance his/her job performance [27]. Previous studies have proved that perceived usefulness has a direct effect on users' behavioral intention to use a technology [44,45]. Studies on BIM adoption in South Korea indicate that the perceived usefulness of BIM will significantly affect the behavioral intention of all parties involved in the construction industry [21,46]. Based on the comparative study of the acceptance of BIM in South Korea and the United States, Lee et al. [47] found that perceived usefulness had a significant impact on BIM behavioral intention at both the individual level and organizational level. Accordingly, we propose the following hypothesis:

H3. *Perceived usefulness will positively affect BIM behavioral intention.*

3.4. Perceived Usefulness, Perceived Ease of Use and Attitude toward BIM Adoption

In line with the TAM, perceived usefulness and perceived ease of use are essential variables framing an individual's technology acceptance behavior. Studies on the adoption of information technology have shown that perceived usefulness and perceived ease of use can have a significant impact on users' attitudes toward BIM Adoption [27,48]. In addition, in the context of BIM adoption, perceived ease of use also shows a positive effect on perceived usefulness [24]. We, thus, propose the following hypotheses:

H4. *Perceived usefulness of BIM will positively affect attitude toward BIM adoption.*

H5. *Perceived ease of use will positively affect attitude toward BIM adoption.*

H6. *Perceived ease of use has a positive effect on perceived usefulness of BIM.*

3.5. Social Influence, Perceived Usefulness and Perceived Ease of Use

Social influence refers to the degree of an individual's perception that most people who are important to him think he should or should not perform the behavior in question [49]. The rationale for a direct effect of social influence on perceived usefulness and perceived ease of use is that people may choose to accept the same perspective, even if they are not so favorable toward the opinions. If they believe one or more important referents (such as superiors, peers, partners, etc.), they are sufficiently motivated to comply with the referents (especially superiors) based on affiliation and/or respect [50]. In an early study on bandwagon innovation diffusion, Rosenkopf and Abrahamson [51] pointed out that bandwagons have a positive feedback loop in which information generated by more adoptions creates a stronger bandwagon pressure, and a stronger bandwagon pressure prompts more adoptions. Therefore, a successful BIM application by competitors and partners will, to some extent, affect project owners' perception of the usefulness and usability of this innovative technology, which, in turn, affects BIM adoption as a whole [11,25]. Accordingly, this study proposes the following hypotheses:

H7a. *Social influence has a positive influence on project owners' perceived usefulness of BIM.*

H7b. *Social influence has a positive influence on project owners' perceived ease of use of BIM.*

3.6. Organizational Support, Perceived Usefulness and Perceived Ease of Use

Organizational support refers to an individual's perception on the degree of policy, resources and other kinds of support provided by the organization for the use of technology. Herein, it refers to the support provided by the project organization to the project owner to adopt BIM. Organizational support carries great weight in motivating employees' potential, allocating resources and enhancing work performance [52–54]. It is easily understandable that sufficient organizational support will exert an incentive effect on employees and improve their sense of organizational backup. Gaining strong support from their organizations, employees will have a sense of being trusted which fulfills their expectation, making them more dedicated to their job, and more likely to demonstrate that they can achieve the organizational goals. However, if employees lose the necessary support (such as information, resources, equipment or training, etc.), their work procedures and work quality will be adversely affected, leading to employees' becoming upset and eventually frustrated [55,56].

In the information technology field, Lin et al. [57] found that organizational high-level support can improve employees' perceived usefulness and perceived ease of use of information technology. Recently, Song et al. [12] implied that, as a new project management technology, successful BIM adoption cannot be realized without the superincumbent financial and policy support in the early stage of software and hardware procurement and personnel training. Furthermore, there is some evidence showing that due to the fact that BIM is often not launched or advocated by the leader or decision-makers in organizations, it often fails to allocate sufficient human, material and financial resources to support BIM adoption [58]. Based on the above, we propose two hypotheses:

H8a. *Organizational support to adopt BIM will positively impact project owners' perceived usefulness of BIM.*

H8b. *Organizational support to adopt BIM will positively impact project owners' perceived ease of use of BIM.*

3.7. BIM Technical Features, Perceived Usefulness and Perceived Ease of Use

BIM technical features normally reflect the fitness, ease, compatibility and interoperability of BIM application. In line with a theory of innovation diffusion, Rogers [59] pointed out that the application of an innovation technology needs to be consistent with the existing value, demand and the experience of potential adopters. In particular, when it comes to introducing or adopting a new technology, firms will compare it with the existing technology, and consider the relevant advantages and characteristics of the two technologies in various aspects. Despite the huge potential value, if BIM is ineffective at interoperating or fitting current work procedures, it will not likely be accepted and adopted by project owners within a short period, as these project owners would be greatly concerned about the risk of abundant inputs (such as financial investments, human resources, etc.). Kim et al. [46] identified major obstacles to BIM adoption, including the actual software operation, the complexity of BIM workflows and the gap between the actual expectations of organizations. Evidence from previous studies clearly indicates that the lack of compatibility between different BIM software hinders the successful application of BIM in the construction industry [13,60]. For example, the compatibility of BIM will significantly affect users' perceived ease of use [21]. Accordingly, this study proposes the following hypotheses:

H9a. *BIM technical features will have a positive effect on project owners' perceived usefulness of BIM.*

H9b. *BIM technical features will have a positive effect on project owners' perceived ease of use of BIM.*

3.8. Government BIM Policies, Perceived Usefulness and Perceived Ease of Use

Government BIM policies generally refer to related policies issued by the government to promote BIM adoption. As shown in the Report of Business Value of BIM in China [58], the respondents who are project owners asserted that lacking first-hand experience deters them from joining BIM adoption and application practices. Even after the project owners adopted BIM, they usually need to emulate the existing projects which successfully applied BIM to guide their actual BIM applications and implementation. Thus, if the government could launch BIM pilot projects in batches and develop guidance for BIM application, this would significantly reduce the difficulty of BIM application, which would be likely to attract more project owners to adopt BIM in the first place. Some countries' experience has proven that appropriate financial subsidies would merit BIM adoption and application (such as in Singapore). For example, Succar [61] and Eadie et al. [62] indicated that government policies are among the primary factors influencing BIM adoption. Therefore, we propose the following hypotheses:

H10a. *Government BIM policies will have a positive effect on project owners' perceived usefulness of BIM.*

H10b. *Government BIM policies will have a positive effect on project owners' perceived ease of use of BIM.*

4. Research Method

4.1. Measurements and Pilot Survey

A structured questionnaire with two sections was designed and used for data collection. The first section covered demographic information of respondents including gender, age, education background, position, work experience and BIM experience. The second section included 24 measurement items (see Table 1 for details) which were designed to elicit project owners' assessments of BIM adoption on a five-point Likert-type scale, with 1–5 indicating "strongly disagree", "disagree", "generally", "agree" and "strongly agree", respectively. All of these measurements were adopted from existing studies and reworded to render the items relevant to BIM adoption for project owners in China. Specifically, perceived usefulness (PU), perceived ease of use (PEOU), attitude (AT), and behavioral intention (BI) were developed based on the measures previously validated by Davis [27,28,46,63,64] and Xu et al. [18], and were reworded in accordance with the context of BIM adoption among Chinese project owners. Social influence (SI) was adopted from Kim et al. [46], Venkatesh and Davis [63], and contains the two dimensions of authoritative influences by intra-firm and inter-firm individuals and associations. Organizational support (OS) was operationalized to reflect the different impacts that the organization's resources exert on BIM adoption. Similar items had previously been validated by Xu et al. [18] and Cao et al. [25]. The construct of BIM technical features (TF) is derived from the research of Xu et al. [18], Kim et al. [46] and Song et al. [12] on the basis of BIM adoption practice in mainland China, with three items indicating the degree of interoperability and compatibility and fitness of BIM in project owners' daily tasks. Referring to the study of Song et al. [12], we replenished and extended the connotations of the vital construct, "Government BIM Policies"(GP), with distinctive Chinese characteristics. Therewith, three BIM experts who have eminent experience in BIM research and application were invited to participate in the pilot survey. Shortly after the pilot survey was conducted, the experts were asked to provide advice regarding the refinement of items and their personal understanding of BIM adoption. According to the experts' advice, items with ambiguity were refined, and the items with tautology were eliminated. After that, we sent the modified measurement items to these experts again and asked them to review whether the amendments strictly complied with their intentions to ensure the applicability of measurement items. The ultimate measurement items of these constructs are provided in Table 1.

Table 1. Measurement items.

Variables		Items	Sources
Social Influence (SI)	SI1	My colleague suggests that I should use BIM at work.	[44,63]
	SI2	Top management thinks that I should use BIM at work.	
	SI3	Cooperative partners think that I should use BIM at work.	
Organization Support (OS)	OS1	My organization provides good BIM training for BIM use.	[16,23]
	OS2	My organization allocates sufficient funds for BIM facilities.	
	OS3	There are enough professionals in my organization to support BIM use.	
BIM Technical Features (TF)	TF1	BIM fits my daily tasks.	[16,44,56]
	TF2	BIM is equal to my work demand.	
	TF3	BIM is of interoperability with other technology platforms.	
Government BIM Policies (GP)	GP1	It is in favor of reducing the cost of BIM use if government exerts the subsidy policy for our BIM implementation.	[56]
	GP2	It will provide useful guidance for BIM use if government can launch a BIM pilot program.	
	GP3	It will promote BIM use if government streamlines the approval procedures of BIM projects.	
Perceived Usefulness (PU)	PU1	Using BIM will reduce the time of finishing tasks.	[25,26]
	PU2	Using BIM will enhance my job performance.	
	PU3	It would provide more chance to get promoted or raises if I can use BIM.	
Perceived Ease of Use (PEU)	PEU1	It is easy to learn and on top of BIM.	[25,26]
	PEU2	I can easily and skillfully use BIM to handle work tasks	
	PEU3	Overall, I think BIM is easy to use.	
Attitude toward BIM (AT)	AT1	I do not resist using BIM in my work.	[26,65]
	AT2	I like using BIM in my work.	
Behavioral Intention (BI)	BI1	I would like to use BIM in my work.	[44,63]
	BI2	I expect that my frequency of BIM will increase in the future.	
BIM Behavior (B)	B1	I will use BIM at work.	[65]
	B2	I will recommend BIM to others (colleagues, friends, etc.)	

4.2. Sampling and Data Collection

In this study, we targeted project owners in mainland China involved in BIM adoption as qualified respondents for data collection. A total of 300 questionnaires were distributed by means of face-to-face interviews (number: 200) and an online survey platform (number: 100). In the face-to-face part, before the formal survey, we conducted an interview with each interviewee to ensure that he/she had first-hand experience of BIM practice. As for the online channel, targeted delivery was the only step taken to send the questionnaire to the preselected project owners involved in BIM adoption. In addition, to obtain sufficient samples, a snowball sampling method was utilized to increase the sample size as we invited the surveyed respondents to share more information regarding knowledgeable participants in other BIM projects or organizations. Any questionnaire with incomplete information or missing values was excluded. Finally, 188 valid questionnaires were received (156 (83%) from the face-to-face interviews and 32 (17%) from the online platform). The valid response rate of face-to-face interviews was 78%, and that of the online survey was 32%. Among these respondents, 64.9% were male, and the remaining respondents were female. All of them were practitioners undertaking tasks directly related to BIM practice in the client departments. The demographics of the respondents under investigation are presented in Table 2.

Table 2. Demographics of the respondents (N = 188).

Variables	Category	Frequency	Percentage (%)
Gender	Male	122	64.9
	Female	66	35.1
Age	22~25	117	62.2
	26~35	63	33.5
	36~45	5	2.7
	Above 45	3	1.6
Education	Associate Degree and below	26	13.9
	Bachelor's Degree	124	66.0
	Master's Degree and above	38	20.1
Position	BIM operation specialist	133	70.7
	BIM engineer	40	21.3
	BIM program manager	9	4.8
	Executive BIM manager	6	3.2
Work experience	0~3 years	133	70.7
	3~5 years	22	11.7
	5~10 years	19	10.1
	Above 10 years	14	7.4
BIM experience	0~3 years	174	92.6
	3~5 years	9	4.8
	5~10 years	3	1.6
	Above 10 years	2	1.1

5. Data Analyses and Results

In this study, confirmatory and discriminant factor analyses of the measurement model were first conducted in order to assess the reliability and validity of the proposed constructs. Afterwards, the maximum likelihood estimate (MLE) method of the structural equation model (SEM) was employed to validate the hypotheses and the fitness of the proposed model.

5.1. Measurement Validation

In general, reliability and validity were the two most common indicators used to evaluate the measurement model. The reliability of the measurement for each construct can be assessed on the basis of Cronbach's α coefficient. Previous studies suggested that a Cronbach's α greater than 0.7 indicates acceptable reliability [66,67]. All of the Cronbach's α coefficient values in the present study are more than the threshold of 0.7 (see Table 3 for detailed values), which indicates good reliability.

With regard to validity, convergent and discriminant validity should both be taken into account. On one hand, convergent validity is usually assessed by three indices: composite reliability (CR), average variance extracted (AVE), and standardized factor loadings. For the composite reliability (CR), values of 0.7 or higher are recommended, according to Nunnally et al. [66] and Nunnally and Bemstein [68]. The CR values in this study range from 0.811 to 0.933, which satisfy the recommended value of 0.7 (see Table 3). In addition, as one of the indices to access the convergent validity, the AVE is often used by examining the construct relative to the amount of variance attributed to the measurement error [69]. With regard to Segars [70], the AVE value for each construct which exceeds the threshold of 0.5 is acceptable. In our study, the AVE values all meet the acceptable requirement (which all range from 0.633 to 0.846). Moreover, values of all standardized factor loadings in this study are above the threshold of 0.7. Therefore, all of the indices are satisfied at acceptable levels, demonstrating the convergent validity of the measure model.

Table 3. Convergent validity of the measurement model.

Variables	Item	Standardized Factor Loadings	*p*	Cronbach's α	KMO	Composite Reliability	Average Variance Extracted
Social Influence (SI)	SI1 SI2 SI3	0.843 0.898 0.716	*** *** ***	0.855	0.701	0.862	0.677
Organization Support (OS)	OS1 OS2 OS3	0.870 0.903 0.843	*** *** ***	0.903	0.744	0.905	0.761
BIM Technical Features (TF)	TF1 TF2 TF3	0.828 0.827 0.772	*** *** ***	0.850	0.725	0.851	0.655
Government BIM Policies (GP)	GP1 GP2 GP3	0.843 0.938 0.936	*** *** ***	0.930	0.747	0.933	0.822
Perceived Usefulness (PU)	PU1 PU2 PU3	0.797 0.903 0.670	*** *** ***	0.825	0.680	0.836	0.633
Perceived Ease of Use (PEU)	PEU1 PEU2 PEU3	0.737 0.887 0.901	*** *** ***	0.879	0.712	0.881	0.714
Attitude (AT)	AT1 AT2	0.827 0.893	*** ***	0.850	0.500	0.851	0.741
BIM-Behavioral Intention (BI)	BI1 BI2	0.912 0.927	*** ***	0.920	0.500	0.916	0.846
BIM Behavior (B)	B1 B2	0.820 0.832	*** ***	0.811	0.500	0.811	0.682

Note: *** $p < 0.001$.

On the other hand, discriminant validity is mainly used to demonstrate the non-correlation between one given construct and the others which ought not be correlated with the given one [69]. Normally, the discriminant validity of one item is judged based on whether it can be easily determined as good or bad by comparing the square root of the AVE for the given construct with the correlations between that construct and all others. If the square roots of the AVE of one given construct are greater than all correlation coefficients of other constructs, it implies that the given construct is more likely to be strongly correlated with its own indicators than the other constructs in the model. In this study, the square roots of all the average variances extracted (the diagonal elements) are greater than the values of the off-diagonal correlation coefficients in the corresponding columns in Table 4, which confirms good discriminant validity as a whole.

Table 4. Correlation matrix and the square of average variance extracted.

	GP	TC	OS	SI	PEU	PU	AT	BI	B
GP	0.907								
TC	0.573	0.809							
OS	0.581	0.570	0.872						
SI	0.626	0.465	0.659	0.823					
PEU	0.375	0.377	0.462	0.452	0.845				
PU	0.726	0.608	0.524	0.594	0.486	0.796			
AT	0.547	0.480	0.461	0.497	0.507	0.675	0.861		
BI	0.486	0.423	0.401	0.435	0.433	0.598	0.729	0.920	
B	0.344	0.300	0.284	0.308	0.307	0.424	0.516	0.680	0.826

Note: The diagonal numbers underlined represent the square of average variance extracted.

5.2. Hypotheses Testing

With the aid of AMOS 21.0, the maximum likelihood estimate (MLE) method in the structural equation model (SEM) was employed to validate the hypotheses and the fitness between the proposed model and the collected data.

The fitness of the proposed model is revealed by the indices of the ratio of the Chi-square model and degrees of freedom (χ^2/df), goodness-of-fit (GIF), root mean square error approximation (RMSEA), normed fit index (NFI), comparative fit index (CFI), incremental fit index (IFI), and Tacker–Lewis index (TLI). The recommend criteria of a goodness-of-fit and the values of these indices derived from this study are shown in Table 5. Despite the GFI and NFI being slightly lower than the recommended acceptable value of 0.90, they are close enough to suggest that the model fits the data reasonably well.

Table 5. Evaluation of overall fitness of the conceptual model.

Fitness Index	Recommended Value	Value
χ^2/df	<3	2.167
GIF	≥0.9	0.826
RMSEA	<0.08	0.079
CFI	≥0.9	0.925
NFI	≥0.9	0.871
IFI	≥0.9	0.926
TLI	≥0.9	0.911

Then, a path analysis is carried out to test the hypotheses. As the results in Table 6 show, nine of fourteen hypotheses are supported. Similar to findings in some previous studies [22,40,63,71], behavioral intention (BI) has a significant positive impact on behavior ($\beta = 0.698$, $t = 10.581$, $p < 0.001$), supporting H1. Furthermore, attitude has a significant positive impact on behavioral intention ($\beta = 0.886$, $t = 5.854$, $p < 0.001$), which means that H2 is supported. However, an unexpected outcome is that PU has no significant impact on BI; thus, H3 is not supported. Furthermore, both PU and PEOU have significantly positive impacts on attitude ($\beta = 0.476$, $t = 6.060$, $p < 0.001$; $\beta = 0.404$, $t = 4.391$, $p < 0.001$), and therefore H4 and H5 are supported. Meanwhile, H6 is also supported given that PEOU has a significantly positive influence on PU ($\beta = 0.282$, $t = 3.388$, $p < 0.001$). Social influence (SI), on the one hand, is found to have a significantly positive impact on PEOU ($\beta = 0.134$, $t = 2.901$, $p < 0.01$), supporting H7b, but on the other hand, SI has no significant influence on PU ($\beta = 0.134$, $t = 1.735$). OS has no significant impact on either PU or PEOU ($\beta = -0.123$, $t = -1.779$; $\beta = 0.134$, $t = 1.765$). In addition, the results show that TF has significant impacts on both PU and PEOU ($\beta = 0.489$, $t = 4.586$, $p < 0.001$; $\beta = 0.286$, $t = 2.565$, $p < 0.01$). Therefore, both H9a and H9b are supported by the empirical results. GP has a significant impact on PU ($\beta = 0.291$, $t = 4.309$, $p < 0.001$), while it has no significant impact on PEOU ($\beta = -0.012$, $t = -0.155$). Thus, H10a is supported but H10b is not.

Table 6. Results of the tested hypotheses.

Hypothesis	Relationship	β	Standardized Error	Critical Ratio (t-Value)	p	Results
H1	B←BI	0.698	0.066	10.581	***	Supported
H2	BI←AT	0.886	0.151	5.854	***	Supported
H3	BI←PU	0.054	0.122	0.439	0.661	Not supported
H4	AT←PU	0.476	0.079	6.060	***	Supported
H5	AT←PEOU	0.404	0.092	4.391	***	Supported
H6	PU←PEOU	0.282	0.083	3.388	***	Supported
H7a	PU←SI	0.130	0.075	1.735	0.083	Not supported
H7b	PEOU←SI	0.240	0.083	2.901	**	Supported
H8a	PU←OS	−0.123	0.069	−1.779	0.075	Not supported
H8b	PEOU←OS	0.134	0.076	1.765	0.078	Not supported
H9a	PU←TF	0.489	0.107	4.586	***	Supported
H9b	PEOU←TF	0.286	0.111	2.565	**	Supported
H10a	PU←GP	0.291	0.067	4.309	***	Supported
H10b	PEOU←GP	−0.012	0.075	−0.155	0.877	Not supported

Note: *** $p < 0.001$; ** $p < 0.01$; * $p < 0.05$.

6. Discussion and Implications

6.1. Discussion

The above results reveal that most of the proposed research hypotheses are well supported. As revealed by the results, project owners' behavior intention has a significantly positive impact on their BIM adoption behaviors, and their attitude will also positively influence their behavioral intention toward BIM adoption, which is consistent with the findings of Davis et al. [27], Ajzen [40], Yuan et al. [72] and Liu et al. [73]. However, we also found that the impact of perceived usefulness (PU) on behavioral intention is insignificant, which is contrary to the prediction of the classical TAM. Differing from previous TAM-based studies confirming that perceived usefulness (PU) has a significantly positive influence on attitude (AT), this study shows that attitude mediates between perceived usefulness and perceived ease of use. In BIM adoption cases for project owners, the impact of perceived usefulness (PU) on behavioral intention (BI) can be formed only across the "bridge" of attitude (AT). In addition, attitude is composed of three elements: inner feelings, emotions, and intentions, and these three elements are intersected with each other [74]. Therefore, perceived usefulness alone, without strong subjective inner feelings or desires, cannot transfer this perceived usefulness into a powerful driven force influencing behavioral intention, which echoes the dilemma that project owners are unwilling to step into actual BIM adoption although they have perceived the usefulness and great potential of BIM [58,75]. Both perceived usefulness (PU) and perceived ease of use (PEOU) positively influence the attitude. Specifically, the stronger the project owners' perceived usefulness and perceived ease-of-use of BIM, the more positive their attitude toward BIM adoption will be.

Additionally, social influence has no significant effect on perceived usefulness, while its impact on the perceived ease of use is significant. The explanation lies in that the impact of social influence on perceived usefulness will diminish as users' personal experience and cognition deepen over time [63]. Therefore, before the actual adoption, project owners' knowledge and beliefs about BIM are "vague and bandwagen", and they must therefore rely more on the opinions of others (such as top management and partners) as a basis for their intentions [76]. After implementation, when more information and details about BIM's strengths and weaknesses become clearer through direct experience, the social influence weakens [77]. On the other hand, social influence has a significant impact on perceived ease of use, indicating that when users' perceived ease of use of information technology is consistent with the external world, unlike perceived usefulness, their belief that the information technology (i.e., BIM) is easy to use will be further strengthened.

This study also presents an unexpected but interesting finding that organizational support has no significant influence on either perceived usefulness or perceived ease of use, which is in contrast to many previous research findings claiming that organizational support is a critical successful factor for BIM adoption and implementation [11,13,18]. These results could largely be explained by the differences in the demand of resources input and the actual effort exerted by top management to promote BIM adoption and implementation. Furthermore, the statements of the majority of survey respondents provide easily understandable reasons for why, in their organizations, management's support for BIM is inadequate to support comprehensive BIM application, which leads to a lack of necessity and incentive to use BIM. This parallels previous findings that employees' perception of organizational support influences whether employees will improve their organizational commitment and support organizational goals [55,78]. In turn, this finding indicates that organizational support will directly influence staff's perception and feelings regarding whether BIM is useful and easy to use, which consequently affects their ultimate adoption behavior.

Furthermore, it can be seen from Table 4 that BIM technical features (such as interoperability, compatibility, etc.) are the most important factors for determining project owners' decision on whether to adopt BIM. The result is consistent with the findings of previous studies [13], which state that the fitness and interoperability of BIM to the current tasks are critical factors influencing the owners' perceptions of the usefulness and ease of use of BIM and will ultimately constitute a BIM adoption behavior. Therefore, the matching degree of BIM's own task technology and how to improve users' perceived usefulness must be considered.

Besides, government BIM policies have a significantly positive effect on perceived usefulness. This is consistent with previous studies which revealed that the existence of government-led initiatives to promote BIM implementation within the industry is one of the critical success factors for extensive BIM adoption and diffusion [62,79]. On the one hand, government policies, such as subsidies, can directly reduce the BIM application costs, which in turn improve project benefits and attract more project owners to get involved in BIM adoption. On the other hand, a universal BIM standard supported by government will reduce the difficulty to interoperate among different special platforms [11,18,58]. Meanwhile, almost all project owners in our survey believe that government support for policies is very helpful for their BIM adoption.

6.2. Theoretical Implications

This study enriches the theoretical literature in three main areas.

Firstly, based on the theory of the TAM, this study categorizes different dimensions of factors and elements—which impact individuals' perceptions of the usefulness and ease of use of BIM—affecting project owners' BIM adoption behavior in the Chinese AEC sector, with an ultimate aim to explain why project owners might adopt BIM. The factors capable of explaining project owners' BIM adoption behaviors are tested and validated by using a structural equation model, and the intrinsic motivation and action mechanism of project owners' BIM adoption behavior are revealed. The findings provide a deeper understanding for explaining the causal factors leading to BIM adoption behaviors.

Secondly, this study's findings also provide valuable insights into the TOE framework and the theory of the TAM in a specific context. The results show that organizational support has no significant impact on either perceived usefulness or perceived ease of use, which can be explained by Eisenberger's findings "perceived organizational support is the premise of behavior, organizational support without perception cannot work even if the support is already provided [52]", indicating that there seems to be a certain precondition that organizational support could positively affect the behavior. Beyond this, another attracting result is that attitude mediates between perceived usefulness and perceived ease of use, which contracts with the findings of Venkatesh et al. [64], but agrees with the findings of Howard et al. [22].

Thirdly, this study extends the theory of the TAM by integrating the TOE framework, revealing that technical, organizational and environmental variables are significantly related to behavioral intention. These variables are intermediated by two distinct constructs (PU and PEOU) and attitude (AT) in a BIM adoption context. Furthermore, results also demonstrate that most of the proposed hypotheses are well supported and the causal relationships among the postulated constructs in the model are analyzed. As such, the model in our study provides an elaborated explanation of the key factors forming the behavioral intentions of project owners toward BIM adoption. In other words, the model offers important insights into the reasons behind project owners' willingness to adopt BIM. By investigating BIM adoption from project owners' perspectives, this study also responds to and reinforces the concern of Ling et al. [19], i.e., focusing on the adoption behavior of other project participants would add more dimensions and shed more light on construction innovation.

6.3. Practical Implications

BIM is often recognized as a promising platform for project stakeholders (including the project owner) to capture complete information throughout the project lifecycle, and to utilize the available data for sustainable design, sustainability rating analysis and sustainable facilities management. The findings of this study will help project owners to understand the impact and interaction of the external constraints and their own subjective perceptions of BIM adoption, based on which successful BIM adoptions and construction sustainability will be increased by some effective incentives and strategies.

Responding to many previous studies on technology acceptance, attitude, perceived usefulness and perceived ease of use are key determinants of behavioral intention [28,40,43,48], which will lead to the ultimate BIM adoption behavior. Thus, project owners should break the traditional mindset, production-organization mode and work procedures to form a positive attitude to embrace a brand-new or even subversive paradigm based on BIM, leading to long-term sustainable growth not only for the organizations but also for the construction industry.

Among these proposed external antecedents, the technical feature is found to be of the utmost importance to perceived usefulness and perceived ease of use. This finding provides insights revealing that project owners must pay attention to the technical features (such as interoperability and compatibility) of the introduced BIM platform or tools, which would greatly enhance the likelihood of successful BIM adoption and sustainability rating analysis from the point of view of technical feasibility.

Furthermore, social influence has a significant impact on the perceived ease of use of project owners, affecting project owners' BIM adoption through attitudes and behavioral intentions. It is suggested that intensifying the dissemination of BIM's benefits and peer experience exchanges could enhance project owners' acknowledgement of BIM's benefits, thus effectively helping project owners' BIM adoption. Furthermore, as mentioned above, the effect of social influence diminishes as experience is gained. As such, project owners should attach importance to establishing a good corporate environment to embrace BIM adoption.

Lastly, government policies also have a positive impact on the perceived usefulness of BIM to project owners, which in turn indirectly affects their BIM adoption. This finding indicates that external incentives from government will help project owners' BIM adoption. In this regard, launching BIM pilot programs and tax exemption could be effective ways to create a favorable environment for promoting project owners' BIM adoption activities.

7. Conclusions

Based on the theory of the TAM, this study attempts to explain project owners' BIM adoption behaviors by investigating how different dimensions of factors and elements—which impact individuals' perceptions of usefulness and ease of use of BIM—influence project owners' BIM adoption behavior in the Chinese construction sector. The factors affecting project owners' BIM adoption are tested and validated by using a structural equation model, and the intrinsic motivation and action mechanism of project owners' BIM adoption behavior are revealed.

The results indicated that most of the proposed hypotheses are well supported and the causal relationships among the postulated constructs in the model are analyzed. The model in our study provides an elaborated explanation of the key factors influencing the behavioral intentions of project owners toward BIM adoption. Particularly, the results reveal that BIM technical features and government BIM policies have positive effects on perceived usefulness, but social influence and organizational support do not significantly influence perceived usefulness. Furthermore, both social influence and BIM technical features have positive effects on perceived ease of use, while organizational support and government BIM policies do not significantly influence perceived ease of use. Attitude plays a significant intermediary role among perceived usefulness, perceived ease of use and behavior intention. Additionally, attitude significantly affects behavior intention, and behavior intention can also affect BIM adoption behavior. The findings of this study are expected to provide a better understanding of the essential elements of project owners' BIM adoption behaviors and guide industry practitioners in developing proper strategies to achieve more effective BIM implementation.

There are also limitations. Although this study deepens the understanding of project owners' BIM adoption intentions and behavior, a wider range of variables can be considered to enhance the model's robustness to more accurately predict project owners' BIM adoption behaviors. Also, despite some previous studies indicated that project features (such as project size, nature, delivery types, etc.) need to be taken into account when it comes to BIM adopting strategies, the limitation in sample data blocks us to conduct further examinations. Hence, research focusing on the influence of project features (such as project size, nature, delivery types, etc.) on BIM adoption behaviors should be further developed. Besides, extensive studies should be conducted to examine the generality of the proposed model in different countries' practice and background to expand the situations to which it applies.

Author Contributions: Conceptualization, H.Y. and X.X.; methodology, H.Y. and Y.Y.; software, Y.Y.; validation, Y.Y.; formal analysis, Y.Y.; investigation, Y.Y. and H.Y.; resources, H.Y. and X.X.; data collection, Y.Y.; writing—original draft preparation, Y.Y.; writing—review and editing, H.Y.; funding acquisition, H.Y. and X.X.

Funding: This research was funded by the Major Program of the National Social Science Fund of China (Grant number: 18ZDA043), and the National Natural Science Foundation of China (Grant number: 71573216; 71671053).

Conflicts of Interest: There is no conflict of interest.

References

1. HM Government. *Government Construction Strategy*; HMSO: London, UK. Available online: https://www.gov.uk/government/publications/government-construction-strategy/ (accessed on 25 September 2018).
2. MOHURD. The Outline of Construction Informatization Development (2011~2015). MOHURD, PRC. Available online: http://www.gov.cn/gzdt/2011-05/19/content_1866641.htm (accessed on 9 October 2018). (In Chinese)
3. MOHURD. The Outline of Construction Informatization Development (2016~2020). MOHURD, PRC. Available online: http://www.mohurd.gov.cn/wjfb/201609/t20160918_228929.html (accessed on 9 October 2018). (In Chinese)
4. Eastman, C.; Teicholz, P.; Sacks, R.; Liston, K. *BIM Handbook: A Guide to Building Information Modeling for Owners, Managers, Designers, Engineers and Contractors*, 2nd ed.; Wiley: New York, NY, USA, 2011.
5. Adrieli, D.C.; Ariovaldo, G.; Vanessa, D.S. A systematic literature review on integrative lean and sustainability synergies over a building's lifecycle. *Sustainability* **2017**, *9*, 1156.
6. Kibert, C.J. *Sustainable Construction: Green Building Design and Delivery*; Wiley: Hoboken, NJ, USA, 2016.
7. Cao, D.P.; Li, H.; Wang, G.B.; Huang, T. Identifying and contextualizing the motivations for BIM implementation in construction projects: An empirical study in China. *Int. J. Proj. Manag.* **2017**, *35*, 658–669. [CrossRef]
8. Chong, H.Y.; Lopez, R.; Wang, J.; Wang, X.; Zhao, Z. Comparative analysis on the adoption and use of BIM in road infrastructure projects. *J. Manag. Eng.* **2016**, *32*, 05016021. [CrossRef]
9. Succar, B.; Sher, W.; Williams, A. An integrated approach to BIM competency assessment, acquisition and application. *Autom. Constr.* **2013**, *35*, 174–189. [CrossRef]

10. Lee, S.; Yu, J.; Jeong, D. BIM acceptance model in construction organizations. *J. Manag. Eng.* **2015**, *31*, 04014048. [CrossRef]
11. Ozorhon, B.; Karahan, U. Critical Success factors of building information modeling implementation. *J. Manag. Eng.* **2017**, *33*, 04016054. [CrossRef]
12. Song, J.; Migliaccio, G.; Wang, G.; Lu, H. Exploring the influence of system quality, information quality, and external service on BIM user satisfaction. *J. Manag. Eng.* **2017**, *33*, 04017036. [CrossRef]
13. Won, J.; Lee, G.; Dossick, C.; Messner, J. Where to focus for successful adoption of building information modeling within organization. *J. Constr. Eng. Manag.* **2013**, *139*, 04013014. [CrossRef]
14. Chien, K.F.; Wu, Z.H.; Huang, S.C. Identifying and assessing critical risk factors for BIM projects: Empirical study. *Autom. Constr.* **2014**, *45*, 1–15. [CrossRef]
15. Boktor, J.; Hanna, A.; Menassa, C. The state of practice of building information modeling (BIM) in the mechanical construction industry. *J. Manag. Eng.* **2014**, *140*, 05014011.
16. Bryde, D.; Broquetas, M.; Volm, J.M. The project benefits of building information modelling (BIM). *Int. J. Proj. Manag.* **2013**, *31*, 971–980. [CrossRef]
17. Jin, R.; Hancock, C.; Tang, L.; Chen, C.; Wanatowski, D.; Yang, L. Empirical study of BIM implementation–based perceptions among Chinese practitioners. *J. Manag. Eng.* **2017**, *33*, 04017025. [CrossRef]
18. Xu, H.; Feng, J.; Li, S. Users-orientated evaluation of building information model in the Chinese construction industry. *Autom. Constr.* **2014**, *39*, 32–46. [CrossRef]
19. Ling, F.; Hartmann, A.; Kumaraswamy, M.; Dulaimi, M. Influences on innovation benefits during implementation: client's perspective. *J. Constr. Eng. Manag.* **2007**, *133*, 306–315. [CrossRef]
20. Gu, N.; London, K. Understanding and facilitating BIM adoption in the AEC industry. *Autom. Constr.* **2010**, *19*, 988–999. [CrossRef]
21. Son, H.; Lee, S.; Kim, C. What drives the adoption of building information modeling in design organizations? An empirical investigation of the antecedents affecting architect's behavioral intentions. *Autom. Constr.* **2015**, *49*, 92–99. [CrossRef]
22. Howard, R.; Restrepo, L.; Chang, C.Y. Addressing individual perceptions: An application of the unified theory of acceptance and use of technology to building information modelling. *Int. J. Proj. Manag.* **2017**, *35*, 107–120. [CrossRef]
23. Liu, Y.; Nederveen, S.V.; Hertogh, M. Understanding effects of BIM on collaborative design and construction: An empirical study in China. *Int. J. Proj. Manag.* **2016**, *35*, 686–698. [CrossRef]
24. Wang, G.; Song, J. The relation of perceived benefits and organizational supports to user satisfaction with building information model (BIM). *Comput. Hum. Behav.* **2017**, *68*, 493–500. [CrossRef]
25. Cao, D.P.; Li, H.; Wang, G.B. Impacts of isomorphic pressures on BIM adoption in construction projects. *J. Constr. Eng. Manag.* **2014**, *140*, 04014056. [CrossRef]
26. Gledson, B.J.; Greenwood, D. The adoption of 4D BIM in the UK construction industry: An innovation diffusion approach. *Eng. Constr. Archit. Manag.* **2017**, *24*, 950–967. [CrossRef]
27. Davis, F.D. Perceived usefulness, perceived ease of use, and user acceptance of information technology. *Mis Q.* **1989**, *13*, 319–340. [CrossRef]
28. Davis, F.D.; Bagozzi, R.P.; Warshaw, P.R. User acceptance of computer technology: A comparison of two theoretical models. *Manag. Sci.* **1989**, *35*, 982–1003. [CrossRef]
29. Broman, T.M.; Schuitema, G.; Thogersen, J. Responsible technology acceptance: Model development and application to consumer acceptance of smart grid technology. *Appl. Energy* **2014**, *134*, 392–400. [CrossRef]
30. Manis, K.T.; Choi, D. The virtual reality hardware acceptance model (VR-ham): Extending and individuating the technology acceptance model (TAM) for virtual reality hardware. *J. Bus. Res.* **2019**, *100*, 503–513. [CrossRef]
31. Sarah, P.; Oliver, S.; Yasel, C. Acceptance of LNG as an alternative fuel: Determinants and policy implications. *Energy Policy* **2018**, *120*, 259–267.
32. Pratia, G.; Puchades, M.V.; Angelis, M.; Pietrantoni, L.; Fraboni, F.; Decarli, N.; Guerra, A.; Dardari, D. Evaluation of user behavior and acceptance of an on-bike system. *Transp. Res. Part F Traffic Psychol. Behav.* **2018**, *58*, 145–155. [CrossRef]
33. Legris, P.; Ingham, J.; Collerette, P. Why do people use information technology? A critical review of the technology acceptance model. *Inf. Manag.* **2003**, *40*, 191–204. [CrossRef]

34. Kuan, K.K.Y.; Chau, P.Y.K. A perception-based model for EDI adoption in small businesses using a technology–organization–environment framework. *Inf. Manag.* **2001**, *38*, 507–521. [CrossRef]
35. Bosch-Rekveldt, M.; Jongkind, Y.; Mooi, H.; Bakker, H.; Verbraeck, A. Grasping project complexity in large engineering projects: The toe (technical, organizational and environmental) framework. *Int. J. Proj. Manag.* **2011**, *29*, 728–739. [CrossRef]
36. Wei, J.S.; Seedorf, S.; Lowry, P.B. The assimilation of RFID technology by Chinese companies: A technology diffusion perspective. *Inf. Manag.* **2015**, *52*, 628–642. [CrossRef]
37. Wang, Y.S.; Li, H.T.; Li, C.R.; Zhang, D.Z. Factors affecting hotel's adoption of mobile reservation systems: A technology-organization-environment framework. *Tour. Manag.* **2016**, *53*, 163–172. [CrossRef]
38. Yeh, C.C.; Chen, Y.F.; Phillips, F. Critical success factors for adoption of 3d printing. *Technol. Forecast. Soc. Chang.* **2018**, *132*, 209–216. [CrossRef]
39. Cruz-Jesus, F.; Pinheiro, A.; Oliveira, T. Understanding CRM adoption stages: Empirical analysis building on the TOE framework. *Comput. Ind.* **2019**, *109*, 1–13. [CrossRef]
40. Ajzen, I. The theory of planned behavior. *Organ. Behav. Hum. Decis. Process.* **1991**, *50*, 179–211. [CrossRef]
41. Arayici, Y.; Coates, P.; Koskela, L.; Kagioglou, M.; Usher, C.; O'Reilly, K. Technology adoption in the BIM implementation for lean architectural practice. *Autom. Constr.* **2011**, *20*, 189–195. [CrossRef]
42. Etter, W. Attitude theory and decision theory: Where is the common ground? *J. Mark. Res.* **1975**, *12*, 481–483. [CrossRef]
43. Wu, S. The relationship between consumer characteristics and attitude toward online shopping. *Mark. Intell. Plan.* **2003**, *21*, 37–44. [CrossRef]
44. Carolina, L.N.; Molina-Castillo, F.J.; Bouwman, H. An assessment of advanced mobile services acceptance: Contributions from tam and diffusion theory models. *Inf. Manag.* **2008**, *45*, 359–364.
45. Wu, C.S.; Cheng, F.F.; Yen, D.C.; Huang, Y.W. User acceptance of wireless technology in organizations: A comparison of alternative models. *Comput. Stand. Interfaces* **2011**, *33*, 50–58. [CrossRef]
46. Kim, S.; Chin, S.; Han, J.; Choi, C. Measurement of construction BIM value based on a case study of a large-scale building project. *J. Manag. Eng.* **2017**, *33*, 05017005. [CrossRef]
47. Lee, S.; Yu, J. Comparative study of BIM acceptance between Korea and the United States. *J. Constr. Eng. Manag.* **2016**, *142*, 05015016. [CrossRef]
48. Ronny, S.; Fazilat, S.; Jo, T. The technology acceptance model (TAM): A meta-analytic structural equation modeling approach to explaining teachers' adoption of digital technology in education. *Comput. Educ.* **2019**, *128*, 13–35.
49. Deutsch, M. A study of normative and informational social influence on individual judgment. *J. Abnorm. Soc. Psychol.* **1955**, *51*, 629–636. [CrossRef]
50. Bagozzi, R.P.; Lee, K.H. Multiple routes for social influence: The role of compliance, internalization, and social identity. *Soc. Psychol. Q.* **2002**, *65*, 226–247. [CrossRef]
51. Rosenkopf, L.; Abrahamson, E. Modeling reputational and informational influences in threshold models of bandwagon innovation diffusion. *Comput. Math. Organ. Theory* **1999**, *5*, 361–384. [CrossRef]
52. Eisenberger, R.; Fasolo, P.; Davis LaMastro, V. Perceived organizational support and employee diligence, commitment, and innovation. *J. Appl. Psychol.* **1990**, *75*, 51–59. [CrossRef]
53. Shore, L.M.; Wayne, S.J. Commitment and employee behavior: Comparison of affective commitment and continuance commitment with perceived organizational support. *J. Appl. Psychol.* **1993**, *78*, 774–780. [CrossRef]
54. Pearce, C.L.; Herbik, P.A. Citizenship behavior at the team level of analysis: The effects of team leadership, team commitment, perceived team support, and team size. *J. Soc. Psychol.* **2004**, *144*, 293–310. [CrossRef]
55. Eisenberger, R.; Huntington, R.; Hutchison, S.; Sowa, D. Perceived organizational support. *J. Appl. Psychol.* **1986**, *71*, 500–507. [CrossRef]
56. Mcmillan, R. Customer Satisfaction and Organizational Support for Service Providers. Ph.D. Thesis, University of Florida, Gainesville, FL, USA, 1997.
57. Lin, H.-F. An investigation into the effects of is quality and top management support on ERP system usage. *Total Qual. Manag. Bus. Excell.* **2010**, *21*, 335–349. [CrossRef]
58. Dodge Data & Analytics. *Smart Market Report: The Business Value of BIM in China*; Dodge Data & Analytics: Bedford, MA, USA, 2015. (In Chinese)
59. Rogers, E.M. *Diffusion of Innovation*, 1st ed.; The Free Press: New York, NY, USA, 1983.

60. Liu, R.; Issa, R.R.A. Factors influencing the adoption of building information modeling in the AEC Industry. In Proceedings of the International Conference on Computing in Civil and Building Engineering, Nottingham, UK, 30 June–2 July 2010.
61. Succar, B. Building information modelling framework: A research and delivery foundation for industry stakeholders. *Autom. Constr.* **2009**, *18*, 357–375. [CrossRef]
62. Eadie, R.; Browne, M.; Odeyinka, H.; Mckeown, C.; Mcniff, S. BIM implementation throughout the UK construction project lifecycle: An analysis. *Autom. Constr.* **2013**, *36*, 145–151. [CrossRef]
63. Venkatesh, V.; Davis, F.D. A theoretical extension of the technology acceptance model: Four longitudinal field studies. *Manag. Sci.* **2000**, *46*, 186–204. [CrossRef]
64. Venkatesh, V.; Morris, M.G.; Davis, G.B.; Davis, F.D. User acceptance of information technology: Toward a unified view. *MIS Q.* **2003**, *27*, 425–478. [CrossRef]
65. Lee, S.; Yu, J.H. Discriminant model of BIM acceptance readiness in a construction organization. *KSCE J. Civ. Eng.* **2017**, *21*, 555–564. [CrossRef]
66. Nunnally, J.C. *Psychometric Theory*, 2nd ed.; McGraw-Hill: New York, NY, USA, 1978.
67. Hair, J.F.; Anderson, R.E.; Tatham, R.L.; Black, W.C. *Multivariate Data Analysis with Readings*; Prentice-Hall: Upper Saddle River, NJ, USA, 1998.
68. Nunnally, J.C.; Bernstein, I.H. *Psychometric Theory*; McGraw-Hill: New York, NY, USA, 1994.
69. Fornell, C.; Larcker, D.F. Evaluating structural equation models with unobservable variables and measurement error. *J. Mark. Res.* **1981**, *18*, 39–50. [CrossRef]
70. Segars, A.H. Assessing the unidimensionality of measurement: A paradigm and illustration within the context of information systems research. *Omega* **1997**, *25*, 107–121. [CrossRef]
71. Baptista, G.; Oliveira, T. Understanding mobile banking: The unified theory of acceptance and use of technology combined with cultural moderators. *Comput. Hum. Behav.* **2015**, *50*, 418–430. [CrossRef]
72. Yuan, H.P.; Wu, H.Y.; Zuo, J. Understanding factors influencing project managers' behavioral intentions to reduce waste in construction projects. *J. Manag. Eng.* **2018**, *34*, 04018031. [CrossRef]
73. Liu, Y.; Hong, Z.; Zhu, J.; Yan, J.; Qi, J.; Liu, P. Promoting green residential buildings: residents' environmental attitude, subjective knowledge, and social trust matter. *Energy Policy* **2018**, *112*, 152–161. [CrossRef]
74. Myers, G.D. *Social Psychology*, 10th ed.; McGraw-Hill: Holland, MI, USA, 2010.
75. McGraw Hill Construction. *The Business Value of BIM for Construction in Global Markets: How Construction around the World Are Driving Innovation with Building Information Modeling*; McGraw Hill Construction: Bedford, MA, USA, 2014.
76. Hartwick, J.; Barki, H. Explaining the role of user participation in information system use. *Manag. Sci.* **1994**, *40*, 440–465. [CrossRef]
77. Agarwal, R.; Prasad, J. The role of innovation characteristics and perceived voluntariness in the acceptance of information technologies. *Decis. Sci.* **1997**, *28*, 557–582. [CrossRef]
78. Yu, C.; Frenkel, S.J. Explaining task performance and creativity from perceived organizational support theory: Which mechanisms are more important? *J. Organ. Behav.* **2013**, *34*, 1165–1181. [CrossRef]
79. Arayici, Y.; Coates, P. A system engineering perspective to knowledge transfer: A case study approach of BIM adoption. In *Virtual Reality-Human Computer Interaction*; Tan, X.-X., Ed.; InTech: Rijeka, Croatia, 2012; pp. 179–206.

MDPI
St. Alban-Anlage 66
4052 Basel
Switzerland
Tel. +41 61 683 77 34
Fax +41 61 302 89 18
www.mdpi.com

Sustainability Editorial Office
E-mail: sustainability@mdpi.com
www.mdpi.com/journal/sustainability

www.ingramcontent.com/pod-product-compliance
Lightning Source LLC
Chambersburg PA
CBHW061447220926
51850CB00004B/36
9783036526287